THE SPINAL CORD

A CHRISTOPHER AND DANA REEVE FOUNDATION TEXT AND ATLAS

Christopher and
Dana Reeve Foundation

THE SPINAL CORD

A CHRISTOPHER AND DANA REEVE FOUNDATION TEXT AND ATLAS

Editors

Charles Watson
Curtin University of Technology
Perth, Australia

George Paxinos
Prince of Wales Medical Research Institute and The University of New South Wales
Sydney, Australia

Gulgun Kayalioglu
Ege University
Izmir, Turkey

ELSEVIER

AMSTERDAM • BOSTON • HEIDELBERG • LONDON • NEW YORK • OXFORD
PARIS • SAN DIEGO • SAN FRANCISCO • SINGAPORE • SYDNEY • TOKYO

Academic Press is an imprint of Elsevier

THE SPINAL CORD

A CHRISTOPHER AND DANA REEVE FOUNDATION TEXT AND ATLAS

Authors

Colin R Anderson
University of Melbourne, Australia

Ken WS Ashwell
The University of New South Wales, Sydney, Australia

Han Collewijn
Erasmus University Medical Center, Rotterdam, the Netherlands

Amanda Conta
SUNY Upstate Medical University, New York, USA

Alan Harvey
University of Western Australia, Perth, Australia

Claire Heise
Prince of Wales Medical Research Institute, Sydney, Australia

Stuart Hodgetts
University of Western Australia, Perth, Australia

Gert Holstege
University of Groningen, the Netherlands

Gulgun Kayalioglu
Ege University, Izmir, Turkey

Janet R Keast
Kolling Institute University of Sydney at Royal North Shore Hospital, Sydney, Australia

Steve McHanwell
University of Newcastle, UK

Elspeth M McLachlan
Prince of Wales Medical Research Institute and The University of New South Wales, Sydney, Australia

George Paxinos
Prince of Wales Medical Research Institute and The University of New South Wales, Sydney, Australia

Giles Plant
University of Western Australia, Perth, Australia

Oscar Scremin
UCLA, USA

Amandeep Sidhu
Curtin University of Technology, Perth, Australia

Dennis Stelzner
SUNY Upstate Medical University, New York, USA

Charles Watson
Curtin University of Technology, Perth, Australia

Academic Press is an imprint of Elsevier
32 Jamestown Road, London, NW1 7BY, UK
30 Corporate Drive, Suite 400, Burlington, MA 01803, USA
525 B Street, Suite 1900, San Diego, CA 92101-4495, USA

First edition 2009

Cover design by Lewis Tsalis
Book design by Lewis Tsalis

British Library Cataloguing in Publication Data
A catalogue record for this book is available from the British Library

Library of Congress Cataloging-in-Publication Data
A catalog record for this book is available from the Library of Congress

ISBN: 978-0-12-374247-6

We dedicate this book to the Scandinavian pioneers of modern spinal cord anatomy – Alf Brodal, Gunnar Grant, Rolf Nyberg-Hansen, and Bror Rexed

Contents

4 The Spinal Nerves 37

Gulgun Kayalioglu

5 The Spinal Cord Blood Vessels 57

Oscar U Scremin

6 Cytoarchitecture of the Spinal Cord 64

Claire Heise and Gulgun Kayalioglu

7 Localization of Motoneurons in the Spinal Cord 94

Steve McHanwell and Charles Watson

8 Spinal Autonomic Preganglionic Neurons: the visceral efferent system of the spinal cord 115

Colin R Anderson, Janet R Keast and Elspeth M McLachlan

9 Central Nervous System Control of Micturition 130

Gert Holstege and Han Collewijn

10 Projections from the Spinal Cord to the Brain 148

Gulgun Kayalioglu

11 Projections from the Brain to the Spinal Cord 168

Charles Watson and Alan R Harvey

12 The Propriospinal System 180

Amanda C Conta and Dennis J Stelzner

13 Spinal Cord Transmitter Substances 191

Claire Heise and Gulgun Kayalioglu

14 Spinal Cord Injury: experimental animal models and relation to human therapy 209

Stuart I Hodgetts, Giles W Plant and Alan R Harvey

15 Atlas of the Rat Spinal Cord 238

Charles Watson, George Paxinos, Gulgun Kayalioglu and Claire Heise

16 Atlas of the Mouse Spinal Cord 308

Charles Watson, George Paxinos, Gulgun Kayalioglu and Claire Heise

17 Toward a Spinal Cord Ontology 380

Charles Watson and Amandeep Sidhu

Index 384

Preface

Mammalian spinal cord research has been hampered by the lack of a detailed atlas and an authoritative text. We have produced this book in an attempt to fill the gap between elementary accounts of spinal cord anatomy on one hand and the thousands of detailed journal articles on the other.

The project would not have progressed without the very substantial support given by the Christopher and Dana Reeve Foundation and the Office of Science and Medical Research of the Government of NSW. Like all of the atlases we have produced, this part of the book required years of original research, and the generous support from these two bodies made it possible to bring the project to completion.

We have been very fortunate in having been supported by an outstanding group of chapter authors. Their contributions provide excellent coverage of areas in which a good review has previously been difficult to find.

In many of the chapters, the importance of the groundwork done by Scandinavian researchers is very evident. The names of Brodal, Rexed, Nyberg-Hansen, and Gunnar Grant are prominent, even though much of their work is decades old. The reason is that they created good data, which in many cases has not been replaced by modern studies. Brodal's classic textbook, *Neurological Anatomy in Relation to Clinical Medicine*, is a landmark contribution to applied neuroanatomy, and researchers would be well advised to be familiar with this work. Anyone who still uses the unfortunate terms 'extrapyramidal' or 'limbic system' should definitely consult this work.

We had originally planned to include a human spinal cord atlas in this book, but difficulties in obtaining high quality histological material meant that this section did not make the deadline for production of the book. We intend to complete the human atlas in the near future and will publish it either in a journal or the Elsevier Brain Navigator web site.

During the production of the atlas, we became aware that the Allen Institute for Brain Science had begun a project to map gene expression in the spinal cord. We held discussions with Ralph Puchalski of the Allen Institute with a view to sharing the work we had done on the mouse spinal cord. This collaboration has been valuable to both parties, and we have been able to include one of the Allen Institute section series to enhance the mouse atlas. We hope that users of the Allen Institute spinal cord gene data will benefit from the presence of this additional illustrative material.

We were assisted by Professor Elspeth M McLachlan in the delineation of autonomic structures in spinal cord atlas sections. Felicity France assisted us with the delineation of forelimb motoneuron groups.

We thank Hongqin Wang for the fine histological preparations. We thank Peter Zhao and Laksmi Govindasamy for assistance with etching the images and Diana Troiani for help with initial planning of the project. Professor Mete Erturk of Ege University gave us some superb photographs of human spinal dissections, one of which is featured in Chapter 4. We are greatly indebted to Charles Scouten (Myneurolab.com, Leica Biosystems St Louis LLC, 5918 Evergreen Blvd, St. Louis MO 63134) for their generous donation of the CryoJane instrument for minimising distortion during the process of collection of frozen sections of spinal cord on slides.

We are very grateful to Lewis Tsalis for his commitment to excellence in constructing the diagrams and the design and page layout of the book. We were very lucky to have the talented Mari Adams prepare many of the illustrations for this book. Kathryn France, Melissa Read and Paul Watson polished the text with excellent copy editing.

Our work on the spinal cord commenced with a seeding grant from The Prince of Wales Medical Research Institute Spinal Injuries Research Centre. We thank the Office of Science and Medical Research of the Government of NSW for providing a substantial grant to support the construction of the rat spinal cord atlas. The Christopher and Dana Reeve Foundation provided specific support for the construction of the mouse brain atlas, as well as providing more general support for the text chapters.

Finally we wish to acknowledge the major contribution made by our Elsevier senior editor, Johannes Menzel, whose enthusiasm for neuroanatomy knows no bounds.

We welcome feedback on the book, and we will appreciate it if you inform us of any errors you identify.

Reproduction of figures

We are happy for our atlas figures to be reproduced in other publications, but we request that our work be suitably acknowledged. Permission to reproduce may be sought directly from Elsevier Global Rights Department in Oxford, UK: phone: (+44) 1865 843830, fax: (+44) 1865 853333, e-mail: permissions@elsevier.com. You may also complete your request on-line via the Elsevier homepage (http://elsevier.com/locate/permissions). We recommend that you use the nomenclature and abbreviation scheme that we developed for this book. This scheme is part of a comprehensive central nervous system nomenclature system which is widely recognized and which can be used for all mammals (see Paxinos and Watson, 2007).

The editors

1 The Organization of the Spinal Cord

Charles Watson and Gulgun Kayalioglu

Introduction

The spinal cord is the part of the central nervous system that controls the voluntary muscles of the limbs and trunk, and which receives sensory information from these regions. It also controls most of the viscera and blood vessels of the thorax, abdomen and pelvis. The cord is described as being made up of a series of segmental components, but in the adult the cord is actually a continuous cylinder of central nervous tissue. The segments are determined by the pattern of emergence of spinal nerves; the nerve rootlets arising from the cord are bundled together so that one pair of spinal nerves emerges from each segment.

The gross anatomy of the spinal cord

The spinal cord and its meninges (dura mater, arachnoid and pia mater) lie within the vertebral canal. The spinal cord is cylindrical, but slightly flattened dorsoventrally. In humans it occupies the upper two-thirds of the vertebral canal, below which the vertebral canal contains only the spinal nerve roots and meninges. The average length of the spinal cord in humans is 45 cm in males and 42-43 cm in females. During development, the spinal cord extends throughout the vertebral canal to the level of fourth sacral vertebra until the third fetal month, after that the vertebral column grows faster. At the beginning of the sixth month it extends to the base of the sacrum, at birth to the lower border of the second lumbar vertebra and to the adult level about two months after birth (Barson, 1970).

On the ventral surface of the spinal cord, there is a deep longitudinal fissure in the midline, named the ventral (or anterior) median fissure. It extends to nearly one-third of the rostrocaudal length of the spinal cord and contains a prolongation from the pia mater and branches of the anterior spinal artery. In humans the ventral median fissure averages 3 mm in depth and is deepest at C5 (Fountas *et al.*, 1998). Its floor (dorsal limit) is formed by the ventral white commissure. On the dorsal surface of the spinal cord, there is a shallow groove named the dorsal (or posterior) median sulcus. The dorsal septum, composed of pial tissue, extends from the base of this sulcus almost to the commissural gray matter. On the lateral side of the spinal cord there is an indistinct ventrolateral sulcus and a deeper dorsolateral sulcus, these correspond to the line of the origin of ventral and dorsal roots respectively (see Figure 1.1).

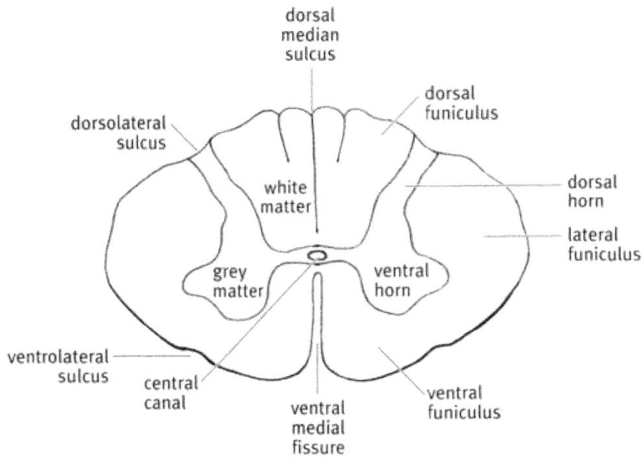

Figure 1.1 Transverse section of human spinal cord

This is a drawing of a transverse section of human spinal cord at the level of C8. On the right hand side the butterfly-shaped gray matter can be seen almost completely surrounded by white matter. On the left hand side the dorsal and ventral horns of gray matter are labeled and the three funiculi of white matter (ventral, lateral, and dorsal) are labeled. In the center of the gray matter is the central canal. The ventral medial fissure, the dorsal median sulcus, the ventrolateral sulcus, and the dorsolateral sulcus are labeled. (Adapted from Cramer and Darby, 2005, p.354)

The widest segments of the spinal cord are found in the cervical region, followed by the lumbar, thoracic and sacral segments (Barson and Sands, 1977). In humans the longest spinal cord segment is at T7, and the segments rostral and caudal to this level become progressively shorter, except for C5 which is slightly longer than other cervical segments (Malinska, 1972). The rostrocaudal length of segments in humans decreases gradually from upper cervical segments towards T1-T2, remains constant in thoracic segments, and gradually increases from T12 to L3 (Ko *et al.*, 2004). The cross-sectional area of the spinal cord in humans increases from C1 to C6 and then decreases rapidly at C8-T2. The white matter occupies most of the cross-sectional area of the cervical enlargement (Kameyama *et al.*, 1996). The cross-sectional area remains constant throughout the middle and lower thoracic segments, but then increases again from T12 to L4, and then begins to decrease rapidly below S1 (Kameyama *et al.*, 1996).

The spinal cord gray matter is greatly enlarged in the regions where the nerves of the limbs (the brachial and lumbosacral plexuses) arise. These 'limb' enlargements are called the cervical enlargement and the lumbar enlargement. Animals with disproportionately large hind limbs, like many dinosaurs, have lumbar enlargements with a diameter rivaling that of the brain. Animals with minimal limb development (like whales) show little tendency to local enlargement in the cervical and lumbar regions (Ranson and Clark, 1953).

The cross-sectional area of the cervical spinal cord in humans is 122 mm^2 at C4, while it is 110 mm2 at C2 and 85 mm^2 at C7

(Sherman *et al.*, 1990). This enlargement is chiefly due to an increase in the transverse diameter (Fountas *et al.*, 1998). Tomographic imaging studies have shown that the cervical enlargement is smaller in life than that observed in postmortem studies, the cross-sectional area being about 78 mm^2 at C5 (Fountas *et al.*, 1998). The sagittal diameter of the cervical enlargement is about 7.5 mm and the transverse diameter is 13 mm in cadavers (Ko *et al.*, 2004) and 7 and 14 mm in computed tomography measurements (Fountas *et al.*, 1998). The cross sectional area of the lumbar enlargement is about 50 mm^2, the sagittal diameter about 7 mm, and the transverse diameter 9 mm in cadavers (Ko *et al.*, 2004).

Spinal cord segments

In the rat the spinal cord is made up of 34 segments: 8 cervical (named C1 to C8), 13 thoracic (T1 to T13), 6 lumbar (L1 to L6), 4 sacral (S1 to S4), and 3 coccygeal (Co1 to Co3). Most mammals have a regional segmental pattern similar to the rat, except those animals with substantial tails have many more coccygeal segments. The human spinal cord differs from that of the rat in having only 12 thoracic segments, 5 lumbar segments, 5 sacral segments and 1 coccygeal segment, making a total of 31 segments. The cervical (brachial) enlargement in most mammals (including rodents and primates) extends from C5 to T1. The lumbosacral enlargement extends from L2 to L6 in rodents and from L2 to S2 in humans. The caudal end of the spinal cord narrows to form the conus medullaris. The caudal continuation of the conus medullaris is a long fibrous strand, called the filum terminale, which stretches from the end of the spinal cord at the level of the L1 vertebra to attach to the coccyx. The upper part of the filum terminale, the filum terminale internum, descends in the dural sac surrounding the cauda equina. It usually starts at the middle L1 level and fuses with the dura mater at the upper S2 level. In humans, the length of the filum terminale internum is about 15 cm, the mean initial diameter 1.5 mm and the midpoint diameter 0.75 mm (Pinto *et al.*, 2002). At the level of S2 vertebra, it perforates the dura and continues as the filum terminale externum that terminates on the dorsal surface of the first coccygeal vertebra together with the coccygeal ligament. The filum terminale is composed of longitudinally arranged collagen bundles and an impressive amount of elastic and elaunin fibers, which give it considerable elasticity (Fontes *et al.*, 2006). In addition to these, there are glial cells, nerve fibers and remnants of an ependyma-lined central canal of the spinal cord (Choi *et al.*, 1992). The central canal continues in the filum terminale for 5-6 mm. In the rat, the spinal cord does not end at the conus medullaris, but its basic components (central canal, gray matter, white matter) continue in the filum terminale (Rethelyi *et al.*, 2004).

Spinal nerves

Pairs of spinal nerves arise from the spinal cord and leave the vertebral column through the intervertebral foramina. Because the vertebral column grows faster than the spinal cord, the spinal cord in the adult human extends down to only the L1 or L2 vertebra. The result is that only the cervical segments of the spinal cord are approximately level with their corresponding vertebrae. Below cervical levels, spinal nerves run increasingly obliquely downwards to their intervertebral foramina. In humans, the C8 segment is level with the C7 vertebra, T12 segment is level with T9-T10, and the L5 segment is level with T11-T12 vertebrae (Ranson and Clark, 1953). The spinal nerves lying beyond the conus medullaris form a bundle of nerves called the cauda equina, because it resembles a horse's tail.

In the rat, the spinal cord ends at the level of L3, and in the cat it ends at the L5 vertebra (Wischnitzer, 1967; Greene, 1968; Gelderd and Chopin, 1977; Padmanabhan and Singh, 1979). In the rat, the filum terminale is traceable into the tail beyond the third caudal nerves. The cauda equina, made up of all the lumbar, sacral, and caudal nerves, conceals the extent of the cord itself in the rat (Greene, 1968; Padmanabhan and Singh, 1979).

In rodents each segment of spinal cord possesses about 15 dorsal rootlets and about 15 ventral rootlets on each side. Humans have only 6-8 dorsal and ventral rootlets. The dorsal rootlets are bundled together to form the dorsal root of a spinal nerve, and the ventral rootlets form the ventral root. There is a fundamental functional difference between the ventral and dorsal roots, which was discovered by Magendie in 1822. Magendie demonstrated that the dorsal roots contained sensory fibers, whereas the ventral roots contained motor fibers. The dorsal roots contain afferent (sensory) fibers from skin, subcutaneous and deep tissues and viscera, along with a small number of efferent fibers, whereas the ventral roots contain somatic efferent (motor) fibers and presynaptic autonomic fibers.

All of the spinal nerves arising from thoracic, lumbar, and sacral segments of spinal cord exit from the vertebral canal caudal to their respective vertebrae. For example, the T5 nerve exits caudal to the T5 vertebra and the L4 nerve exits caudal to the L4 vertebra. The situation is slightly more complicated for the cervical region because there are eight cervical spinal cord segments, but only seven cervical vertebrae. An arbitrary application of nomenclature has resulted in the following plan: the C1 to C7 nerves emerge above their respective vertebrae, and the C8 nerve exits between C7 and T1.

Each dorsal root bears an ovoid swelling named the dorsal root (spinal) ganglion. Dorsal root ganglia consist of pseudo-unipolar neurons. They give rise to a single axon which then bifurcates; one branch connects to the periphery and the other connects with the dorsal horn of the spinal cord.

Spinal cord gray and white matter

The spinal cord is composed of gray matter and white matter. A transverse section of spinal cord shows that the gray matter is arranged in the form of a butterfly or the capital letter H, depending on the level. The cross bar of the H, called the commissural gray matter, encloses the central canal. The dorsally projecting arms of the gray matter are called the dorsal horns, and the ventrally projecting arms are called the ventral horns. The central region, which connects the dorsal and ventral horns, is called the intermediate gray matter. In the thoracic spinal cord and in the upper lumbar segments, there is a small lateral projection of the intermediate gray matter called the intermediolateral horn. The intermediolateral horn contains the cells of origin of the autonomic nervous system. The thoracic intermediolateral horn contains preganglionic sympathetic neurons. In upper sacral segments, the sacral parasympathetic nucleus, which is composed of preganglionic parasympathetic neurons, is present. The dorsal horn of gray matter extends as far as the surface of the spinal cord, and contacts the surface at the point where the dorsal nerve rootlets enter the spinal cord.

White matter of the spinal cord

A layer of white matter surrounds the gray matter except for where the dorsal horn touches the margin of the spinal cord. The white matter consists mostly of longitudinally running axons and also glial cells. A large group of axons which are located in a given area is called a funiculus (e.g. posterior funiculus). Smaller bundles of axons, which share common features within a funiculus are called fasciculus (e.g. fasciculus gracilis). Tracts and pathways are terms applied to nerve fiber bundles that have a functional connotation. A tract is a group of nerve fibers with the same origin, course, termination and function (e.g. spinothalamic tract). A pathway is a group of tracts with a related function (e.g. postsynaptic dorsal column pathway).

The horns of gray matter divide the white matter into three columns (funiculi): dorsal, lateral and ventral. The boundary between the lateral and ventral columns is not distinct. It is generally taken to be in line with the emerging axons of the most lateral motoneurons. The two dorsal columns are located between the two dorsal horns of gray matter and lie side by side with a common medial border. The two ventral columns are separated by a deep fissure, the ventral median fissure, which extends almost as far as the commissural gray matter. Blood vessels travel through the deep fissure as way of reaching the center of the gray matter. At the dorsal limit of the ventral median fissure is a band of white matter (the ventral white commissure) which connects the two ventral columns.

At the region where the dorsal horn reaches the pial surface of the spinal cord, there is a prominent band of fibers called the dorsolateral fasciculus (the tract of Lissauer). This tract contains primary afferent fibers, which either ascend or descend for a few segments before entering the dorsal horn. The dorsal column of white matter is chiefly made up of the central processes of dorsal root ganglion cells. These large myelinated axons form the main pathway conveying skin sensation and position sense (proprioception) from the limbs and trunk to the brain. Because more fibers are added at each segment in a caudorostral order, the dorsal column is very small in sacral segments, and largest in the rostral cervical spinal cord. As the fibers from one segment enter the dorsal column, they move to form a strip as close to the medial edge as possible. Fibers from the next segment move as far medially as they can and form a strip lateral to those of the more caudal segment. When viewed at high cervical levels, the fibers from lower thoracic, lumbar and sacral segments (segments below T6) form a distinct medial strip, the gracile fasciculus. The more lateral group is wedge-shaped and is called the cuneate fasciculus. It primarily contains afferents from the upper thoracic and cervical segments.

In rodents and many non-primate mammals, the ventral-most part of the dorsal column contains a small but very distinct tract made up of small diameter fibers which can be easily distinguished from the cuneate and gracile fasciculi. This group constitutes the dorsal corticospinal tract. Because it contributes fibers to each segment as it descends, this tract diminishes in size from rostral to caudal levels. The dorsal corticospinal tract fibers terminate in the dorsal horn, intermediate gray matter, and, to a lesser extent, on the interneuron pools of the ventral horn. In primates, carnivores, and some other mammals, there is no substantial dorsal corticospinal tract, because almost all of the corticospinal fibers are found in a prominent tract in the dorsal part of the lateral column, a smaller tract in the ventral column. The lateral corticospinal tract of primates is unusual in that some of its fibers make direct synaptic connections with motoneurons in lamina 9. In humans, about 20% of the lateral corticospinal fibers make direct corticomotoneuronal connections – by far the largest percentage in any primate.

The lateral and anterior columns contain a variety of ascending and descending fiber groups. The ascending tracts include the spinothalamic, spinocerebellar and spinotectal tracts and the descending tracts include the corticospinal, vestibulospinal, tectospinal and reticulospinal tracts. As well as these long ascending and descending tracts, there are many fibers in the white columns that connect one spinal cord segment with another. These fibers are called propriospinal because they often lie very close to the gray matter. The largest propriospinal pathways connect the brachial and lumbosacral enlargements to coordinate limb movements.

Gray matter of the spinal cord

The spinal cord gray matter is made up of neuronal cell bodies, dendrites, axons, and glial cells. The neurons are mostly multipolar, but vary greatly in size. The gray can be macroscopically divided into dorsal and ventral horns with and intermediate region between them. Microscopic analysis of the spinal gray matter reveals a complex structure, characterized by successive layers of cells from dorsal to ventral. The landmark description of these layers by Rexed in 1952 and 1954 has formed the basis of the majority of detailed anatomical and physiological studies of the spinal cord in recent times.

The laminae of Rexed

Rexed divided the spinal gray matter into 10 regions on the basis of cytoarchitecture as seen in transverse sections. The first nine laminae are arranged from dorsal to ventral. The tenth is merely the circle of cells surrounding the central canal. Laminae 1-4 are the main cutaneous receptive regions; lamina 5 receives afferents from the viscera, skin and muscles. Lamina 6 receives mostly proprioceptive and some cutaneous afferents Rexed's laminae are described in detail in Chapter 6.

Lamina 1

This very thin layer was previously known as the marginal layer of the dorsal horn (the posteromarginal nucleus). This lamina has a reticular appearance and consists mostly of fusiform neurons.

Lamina 2

This is the substantia gelatinosa of the dorsal horn. The cell density is greater than lamina 1. It can be subdivided into a dorsal outer zone (lamina 2o) and a ventral inner zone (lamina 2i).

Lamina 3

This was previously called the superficial part of the nucleus proprius (proper sensory nucleus) of the dorsal horn. There are many myelinated fibers in this lamina. The neurons are less densely packed and larger than those in lamina 2. Laminae 2 and 3 are functionally related and consist mostly of Golgi type 2 neurons.

Lamina 4

Formerly known as the base of the nucleus proprius, or head of the dorsal horn, the lamina is about twice as thick as lamina 3. Its medial end curves ventrally along the margin of the dorsal horn and makes contact at the midline with its counterpart of the opposite side. In thoracic and upper lumbar levels, its medial extent is interrupted by the dorsal nucleus (of Clarke). Internal basilar nucleus is also present in lamina 4 of C1-C6 segments.

Lamina 5

Formerly called the neck of the dorsal horn, this is the thickest layer of the dorsal horn. The lateral part has a characteristic reticular appearance on account of fiber bundles running through it. From about T1 to L3, the dorsal nucleus (Clarke) is located in the medial part of lamina 5.

The dorsal nucleus receives primary afferents from the dorsal roots of the lower spinal cord and its axons form the dorsal spinocerebellar tract in the lateral funiculus. The axons do not cross the midline. The dorsal nucleus extends from upper thoracic levels to L2, but its cells are concentrated at levels T9 to L2. In some species it may extend rostrally as far as C8 and caudally as far as L4. The dorsal nucleus primarily receives cerebellar afferents from the lower limb and its axons form the dorsal spinocerebellar tract. The equivalent cell groups serving the upper limb are the external cuneate nucleus of the lower medulla.

Lamina 6

This is the most ventral layer of the dorsal horn. It is found in sections through the cervical and lumbosacral enlargements. It does not seem to exist in upper cervical levels or in thoracic levels below T1. It forms a thin layer between the thicker layers 5 and 7 and the boundaries with these layers are often indistinct.

Lamina 7

The dorsal part of this lamina was formerly described as the intermediate zone of spinal gray matter. The ventral part forms the dorsal region of the ventral horn. Lamina 7 cells are less densely stained than those in the adjacent laminae 8 and 9. The cells in the large central part of lamina 7 are the interneurons that connect to motoneuron pools. Lamina 7 neurons are involved in the regulation of posture and movement. Descending motor pathways control motoneurons by means of connection with interneurons in lamina 7, rather than direct connections to motoneurons, which are rare in mammals other than primates. The functional arrangement of lamina 7 interneurons is matched to the topographical arrangement of motoneuron columns – the interneurons supplying trunk musculature are more medially placed, and the interneurons supplying limb musculature are more laterally placed.

In thoracic and upper lumbar segments, lamina 7 contains two notable cell groupings: the intermediolateral nucleus and the intercalated nucleus. The intermediolateral nucleus can be identified at the dorsolateral tip of lamina 7 in levels T2 to L2 in rodents (T2 to L4 in the cat). The cells of the intermediolateral cell column are the preganglionic neurons of the sympathetic nervous system. The axons of these preganglionic neurons leave the spinal cord in the ventral roots and cross to the sympathetic chain via short connecting nerves called gray rami communicantes. Within the sympathetic chain they travel to reach sympathetic ganglion cells on which they synapse. Because the preganglionic sympathetic neurons are restricted to the T1 to L2 region, those serving the rostral and caudal ends of the body must travel some distance in the sympathetic chain before they reach a ganglion close to their target organ or blood vessels. The ganglion cells that receive the preganglionic fibers send their postganglionic axons directly back to the closest spinal nerve for further distribution. The short nerves connecting the ganglion to the spinal nerve are called white rami communicantes. Parasympathetic preganglionic neurons are found in the sacral parasympathetic nucleus in lamina 7 at levels S1 and S2 (and to a lesser extent L6) in the rat. They are found in S2-S4 in humans. At the medial margin of lamina 7 a distinct intermediomedial cell group can also be identified.

In upper cervical levels (C1-C4), the central cervical nucleus is seen in the medial part of lamina 7, immediately ventral to the intermediomedial nucleus. It receives input from neck muscle spindles and projects to the contralateral cerebellum and vestibular nuclei. It also receives a substantial input from the contralateral vestibular nuclei and seems to play an important role in the control of head position (Donevan *et al.*, 1990; Thomson *et al.*, 1996).

Lamina 8

Lamina 8 is found in the ventromedial or ventral region of the ventral horn. The cells of this layer are propriospinal interneurons, heterogeneous in size, and more densely stained than lamina 7. The large cells in this lamina project to motoneurons on the same and opposite side.

Lamina 9

The most prominent feature of lamina 9 is the clusters of large multipolar neurons. These are the alpha motoneurons which supply the extrafusal fibers of the striated muscles of the axial skeleton and the limbs, and gamma motoneurons innervating the intrafusal fibers in muscle spindles. Because of the complexity of limb musculature, the motoneuron clusters are larger and more numerous in the two regions of the cord which supply the limbs. In the thoracic spinal cord, the motoneuron cells are arranged in a medial longitudinal column, which consists of two groups, dorsomedial and ventrolateral. The dorsomedial group supplies spinal muscles, principally the erector spinae group. The ventrolateral group supplies the muscles of the body wall: the intercostal muscles and the abdominal wall muscles.

As well as the large alpha motoneurons referred to above, lamina 9 contains a population of smaller gamma motoneurons, which supply muscle spindles.

Lamina 10 (Area 10)

Lamina 10 is the area surrounding the central canal. It contains the dorsal and ventral commissures, and the substantia gelatinosa centralis. Neurons in this region receive somatic and visceral afferents and are nociceptive. They project to brainstem, midbrain, thalamus and hypothalamus.

Lateral cervical nucleus

This nucleus is lateral to the lateral spinal nucleus in upper cervical levels. It is a sensory nucleus which projects to the cerebellum, midbrain, and thalamus.

Lateral spinal nucleus

This consists of a small group of cells lying ventral to the dorsolateral tip of the dorsal horn. They are thought to receive subcutaneous sensory information and to project to the midbrain, thalamus, and hypothalamus (Kayalioglu *et al.*, 1999).

Onuf's nucleus

A distinct group of motoneurons in the caudal lumbar and rostral sacral spinal cord in mammals, called Onuf's nucleus, is seen in the ventrolateral part of the ventral horn. Onuf's nucleus supplies the perineal muscles and the anal and urethral sphincters. The perineal muscles, bulbocavernous (called bulbospongious in humans) and ischiocavernous, are those involved in penile erection and ejaculation in males. The nucleus is sexually dimorphic, the motoneurons being larger in males. In humans these motoneurons are aggregated into a single column, but in the rat and mouse they are divided into two groups, a ventrolateral group which supplies the external urethral sphincter and the bulbocavernosus muscle, and a dorsomedial linear nucleus, which supplies the external anal sphincter and the ischiocavernosus muscle.

Central canal

The central canal of the spinal cord is a remnant of the embryological ventricular system and is present throughout the whole length of the spinal cord. It is continuous with the fourth ventricle cranially and the terminal ventricle in the conus medullaris caudally. It is lined with columnar ependymal cells. In older adults, the central canal is often blocked with cellular debris.

Spinal cord meninges

The spinal cord is enclosed by a tube of cerebrospinal fluid. The three membranes enclosing this fluid are called the spinal meninges. The meninges are the pia mater, arachnoid mater, and dura mater, separated from each other by the subdural and subarachnoid spaces. The cerebrospinal fluid lies in the subarachnoid space, between arachnoid and pia. Between the dura and the periosteum of the vertebral column is the epidural space filled with fat and lymphatic tissue, small arteries and a venous plexus. On each side of the spinal cord, the pia mater is tethered to the arachnoid and dura between the ventral and dorsal spinal roots at each vertebral level by a saw-toothed fibrous extension of the pia called the denticulate ligament.

Vasculature of the spinal cord

The spinal cord is supplied by a single ventral spinal artery and two dorsal spinal arteries. The ventral spinal artery originates from the vertebral artery and descends within the ventral median fissure of the spinal cord. The dorsal spinal arteries originate either from the vertebral artery or its inferior posterior cerebellar branch, and descend in the dorsolateral sulcus of the spinal cord. Segmental branches from the vertebral, deep cervical, intercostal, and lumbar arteries anastomose with the ventral and dorsal spinal arteries.

The veins of the spinal cord form a surface plexus, which drain rostrally into the cerebellar veins and cranial sinuses and through the intervertebral veins and external venous plexuses to the azygous system.

References

Barson AJ (1970) The vertebral level of termination of the spinal cord during normal and abnormal development. J Anat 106, 489-497.

Barson AJ, Sands J (1977) Regional and segmental characteristics of the human adult spinal cord. J Anat 123, 797-803.

Choi BH, Kim RC, Suzuki M, Choe W (1992) The ventriculus terminalis and filum terminale of the human spinal cord. Hum Pathol 23, 916-920.

Carpenter EM (2002) Hox genes and spinal cord development. Dev Neurosci 24, 24-34.

Donevan AH, Neuber-Hess M, Rose PK (1990) Multiplicity of vestibulospinal projections to the upper cervical spinal cord of the cat: a study with the anterograde tracer Phaseolus vulgaris leucoagglutinin. J Comp Neurol 302, 1-14.

Fontes RB, Saad F, Soares MS, de Oliveira F, Pinto FC, Liberti EA (2006) Ultrastructural study of the filum terminale and its elastic fibers. Neurosurgery 58, 978-984.

Fountas KN, Kapsalaki EZ, Jackson J, Vogel RL, Robinson JS Jr (1998) Cervical spinal cord – smaller than considered? Spine 23, 1513-1516.

Freeman LM, Breedlove SM (1995) Motoneurons innervating guinea pig perineal muscles are sexually dimorphic in size but not number. Brain Res 690, 1-7.

Gelderd JB, Chopin SF (1977) The vertebral level of origin of spinal nerves in the rat. Anat Rec 188, 45-47.

Greene EC (1968) *Anatomy of the Rat.* Hafner, New York.

Kameyama T, Hashizume Y, Sobue G (1996) Morphologic features of the normal human cadaveric spinal cord. Spine 21, 1285-1290.

Kayalioglu G, Robertson B, Kristensson K, Grant G (1999) Nitric oxide synthase and interferon-gamma receptor immunoreactivities in relation to ascending spinal pathways to thalamus, hypothalamus, and the periaqueductal grey in the rat. Somatosens Motor Res 16, 280-290.

Ko HY, Park JH, Shin YB, Baek SY (2004) Gross quantitative measurements of spinal cord segments in human. Spinal Cord 42, 35-40.

Kuzuhara S, Kanazawa I, Nakanishi T (1980) Topographical localisation of the Onuf's nuclear neurons innervating the rectal and vesical striated sphincter muscles: a retrograde fluorescent double labelling in cat and dog. Neurosci Lett 16, 125-130.

Landmesser LT (2001) The acquisition of motoneuron subtype identity and motor circuit formation. Int J Dev Neurosci 19, 175-182.

Malinska J (1972) Volume changes in the human spinal cord during ontogenetic development. Folia Morphol 20, 188-191.

Molander C, Xu Q, Grant G (1984) The cytoarchitectonic organization of the spinal cord in the rat. I. The lower thoracic and lumbosacral cord. J Comp Neurol 230, 133-141.

Nanasaki Y, Sakuma Y (2000) Perineal musculature and its innervation by spinal motoneurons in the male rabbit. Effects of testosterone. J Nippon Med Sch 67, 164-171.

Onuf B (1900) On the arrangement and function of the cell groups of the sacral region of the spinal cord in the man. Arch Neurol Psychopathol 3, 387-412.

Padmanabhan R, Singh S. (1979) Observations on the topographical relations of spinal nerve roots in the rat. Acta Anat 105, 378-380.

Pinto FC, Fontes RB, Leonhardt Mde C, Amodio DT, Porro FF, Machado J (2002) Anatomic study of the filum terminale and its correlations with the tethered cord syndrome. Neurosurgery 51, 725-729.

Ranson SW, Clark SM (1953) *The Anatomy of the Nervous System: Its Development and Function.* 9th edn. WB Saunders, Philadelphia.

Rethelyi M, Lukacsi E, Boros C (2004) The caudal end of the rat spinal cord: transformation to and ultrastructure of the filum terminale. Brain Res 1028, 133-139.

Ribeiro-da-Silva A (2004) Substantia gelatinosa of the spinal cord. In: T*he Rat Nervous System,* 3rd edn. Paxinos G (ed). Elsevier Academic Press, San Diego.

Romanes GJ (1942) The development and significance of the cell columns in the ventral horn of the cervical and upper thoracic spinal cord of the rabbit. J Anat 76, 112-130.

Sherman JL, Nassaux PY, Citrin CM (1990) Measurements of the normal cervical spinal cord on MR imaging. AJNR Am J Neuroradiol 11, 369-372.

Thomson DB, Isu N, Wilson VJ (1996) Responses of neurons of cat central cervical nucleus to natural neck and vestibular stimulation. J Neurophysiol 76, 2786-2789.

Wischnitzer S (1967) *Atlas and Dissection Guide for Comparative Anatomy.* Freeman, San Francisco.

2 Development of the Spinal Cord

Ken WS Ashwell

From neural plate to neural tube

The central nervous system first appears in the embryo as the neural plate, a tadpole-shaped thickening of the ectoderm rostral to the primitive pit (Figure 2.1a). This can be seen at approximately 18 to 19 days pc (days post-conception) in the human (Carnegie stages 6 to 7, see Table 2.1 for comparison with mouse and rat) (Kaufman, 1992).

Induction of the neural plate appears to be due to an inhibition of epidermis formation due to signals released from the primitive node at the cranial end of the primitive streak (Sadler, 2005). In other words, the default option for the ectoderm in this region is to produce epidermis rather than neurectoderm, and the signal for neurulation involves suppression of bone morphogenetic protein (Bmps) and Wnt signaling pathways (Sadler, 2005). In all vertebrates studied, the notochord underlying the future floor plate and the floor plate itself excrete the molecule Sonic hedgehog (Shh), which may be the signal which induces floor plate formation of the neural groove and tube and effectively ventralizes the neural tube (see Lewis and Eisen, 2003 for review).

Within a day of the appearance of the neural plate in the human, the edges of the neural plate elevate to form the neural folds and a neural groove emerges in the midline (Figure 2.1b).

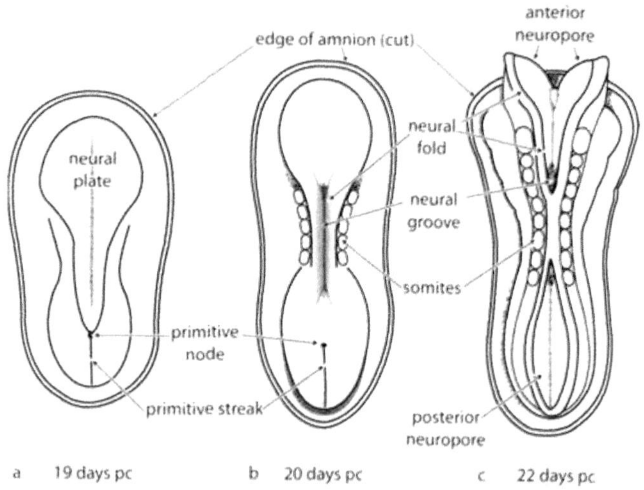

Figure 2.1 Neural plate and neural tube formation

This diagram shows the neural plate and neural tube of human embryos at 19 days pc (a), 20 days pc (b), and 22 days pc (c) showing folding of the neural groove to produce the neural tube. The first point of fusion between the neural folds is at the hindbrain/spinal cord junction.

The initial step in elevation of the neural folds depends on proliferation of the underlying mesoderm and production of hyaluronic acid (Solursh and Morriss, 1977), but later stages involve furrowing and folding at three regions of neurectoderm (one median and two lateral hinge points, see Figure 2.2 and Sadler, 2005 for review).

Shaping of the neural folds through folding requires apical concentrations of microfilaments and lengthening of the cell cycle at the hinge points. The latter ensures that nuclei of dividing cells remain at the base of the neurectoderm for

Table 2.1 Timing of significant events in the development of the spinal cord.

Event	Human Days pc/pn	Mouse Days pc/pn	Rat Days pc/pn
Appearance of neural plate	18-19 pc, C*6-7	7 pc, T†11	7-7.5 pc, WΩ12
First fusion of neural folds	20 pc, C9	8 pc, T12	8 pc, W15
Closure of anterior neuropore	25 pc, C11	8.5 -9 pc, T14	9 pc, W16
Closure of posterior neuropore	27 pc, C12	9.5 to 10 pc, T15	10 pc, W18
Birthdates of motoneurons in brachial (cervical) enlargement	24 to 28 pc?, C11 to C18?	10 to 13 pc, T15 to T21	11 to 14 pc, W20 to W30
Birthdates of motoneurons in lumbosacral enlargement	24 to 28 pc, C11 to C18?	11 to 13 pc, T18 to T21	12 to 14 pc, W22 to W30
Segregation of motoneurons into discrete somatic motor columns	56 to 70 pc	16 to 17 pc, T25	16 to 17 pc, W34
Stretch reflex appears	–	–	19 pc
First appearance of Clarke's column	~ 70 pc	–	16 pc
Growth of corticospinal tract into cervical spinal cord	98 to 112 pc	0 to 2 pn	0 to 1 pn
Elimination of corticospinal tract axons	–	3 to ~ 28 pn	4 to ~ 28 pn
Myelination of corticospinal tract	~ 180 pc to ~ 800 pn	10 to ~ 28 pn	10 to ~ 35 pn

See text for references and comments. * Carnegie stage † Theiler stage Ω Wistchi stage

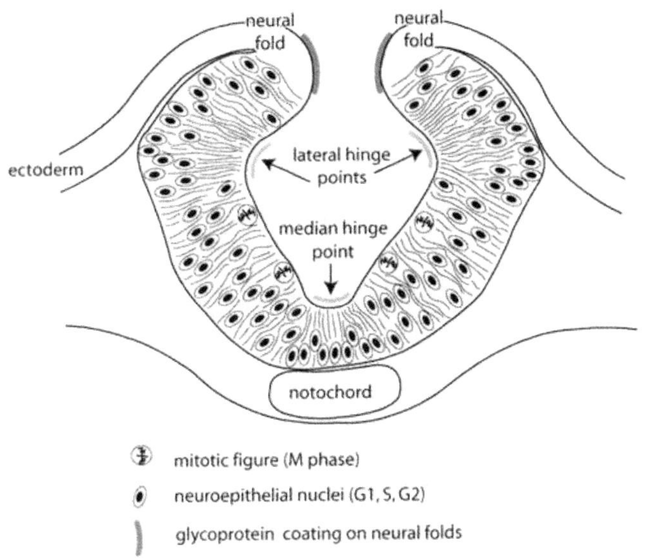

Figure 2.2 Mechanisms involved in the folding of the neural plate to form a neural tube

Most folding occurs at paired lateral and median hinge points where cell division is delayed and nuclei spend more time at the base of the neuroepithelium, thereby narrowing the apical processes of the neuroepithelial cells. Note the aggregation of nuclei at the periphery in these regions and the abundant mitotic figures at non-hinge regions. Glycoprotein on the surface of the adjacent neural folds facilitates adhesion when these points are brought into contact.

longer periods of time, thereby widening the bases and narrowing the apices of neural plate cells at these regions (Figure 2.2, Sadler, 2005).

Fusion of the paired neural folds to form a neural tube first occurs at the junction of the hindbrain and spinal cord (level of the 5th somite) at approximately 20 pc in the human (Carnegie stage 9) and 8 days pc in the mouse and rat (Table 2.1) and depends on glue-like coatings of glycoprotein on the opposing surfaces (Sadler, 1978). Fusion of the neural tube extends rostrally and caudally over the next few days (O'Rahilly and Muller, 2002) to effect complete closure of the neural tube (Figure 2.1c). After initial closure, the remaining open ends of the neural tube are known as the neuropores. In humans, the rostral or anterior neuropore closes at about 25 pc, while the caudal or posterior neuropore seals at 27 to 28 pc. After closure of the neuropores, the neural tube expands rostrally to form the brain vesicles, while the caudal tube begins to differentiate into the primitive spinal cord. The process described above is known as primary neurulation and is responsible for generating the brain and spinal cord as far caudally as S4 or S5. More caudal levels of the spinal cord are generated by a mechanism known as secondary neurulation, whereby mesodermal cells coalesce and epithelialize, form a lumen and become continuous with the remainder of the tube (Sadler, 2005).

Neural crest development

During the elevation of the neural plate, cells appear along the edge (or crest) of the neural folds. These neural crest cells are found along the entire length of the neural tube and initially lie between the neural tube and the overlying ectoderm. Neural crest cells subsequently migrate along two pathways to give rise to a variety of mature cell groups: a dorsolateral pathway to differentiate into pigment cells and a ventrolateral pathway to give rise to neural elements (autonomic ganglia, Schwann cells, adrenal medulla), but their significance in this review is their transformation into sensory (or dorsal root) ganglia. Young neurons of the sensory ganglia develop a central process which invades the dorsal horn (see afferent development below) and a peripheral process, which innervates somatic or visceral structures. The dermamyotome, notochord and ventral spinal cord are all believed to exert chemorepulsive effects on the growing peripheral processes of developing dorsal root ganglia, which direct the initial trajectory of the growing axons (Masuda and Shiga, 2005). In the case of the dermamyotome, the chemorepulsive agent may be semaphorin-3A, while the notochord may exert its effect by means of semaphorin-3A, chondroitin sulphate proteoglycans and an, as yet, unidentified agent. Finally, the factor responsible for the effects exerted by ventral spinal cord remains unknown (Masuda and Shiga, 2005).

Alar and basal plates and their derivatives

Neuropethelial cells provide a thick pseudostratified wall to the early neural tube with abundant junctional complexes between their luminal ends. These complexes are dynamic structures (Bittman *et al.*, 2004), which mediate intercellular communication during the critical early stages of cell type specification and decline in number towards birth in rodents (Bittman *et al.*, 2004). Cell-type-specific coupling (i.e. between cells sharing particular fates) emerges gradually during spinal cord development (Bittman *et al.*, 2004).

The nuclei of the neuroepithelial cells migrate between the neural tube lumen and the outer limiting membrane in a process know as interkinetic nuclear migration. When the nuclei reach the luminal surface of the neural tube they undergo mitotic division, thereby producing either further neuroepithelial cells (during early stages) or primitive nerve cells (the inappropriately named neuroblasts) during later stages. Progressive accumulation of post-mitotic differentiating neuroblasts beneath the external limiting membrane of the neural tube leads to the formation of a mantle layer (future spinal cord gray matter) around the neuroepithelium. The mantle layer on each side of the primitive spinal cord shows dorsal and ventral thickenings, which are known as the alar and basal plates, respectively (Figure 2.3). The paired alar plates will give rise to sensory areas of the spinal cord, while the basal plates contribute to the motor areas of the cord. The neuroepithelium of the early spinal cord also shows a roof plate dorsally and a floor plate ventrally. The ultimate fate of the tissue external to the roof and floor plates is to serve as sites of dorsal and ventral white commissures for crossing axons in the postnatal spinal cord. The region external to the mantle layer is known as the marginal layer and contains nerve fibers emerging from the immature neurons of the mantle layer. The marginal layer will ultimately become the white matter of the fetal and postnatal spinal cord.

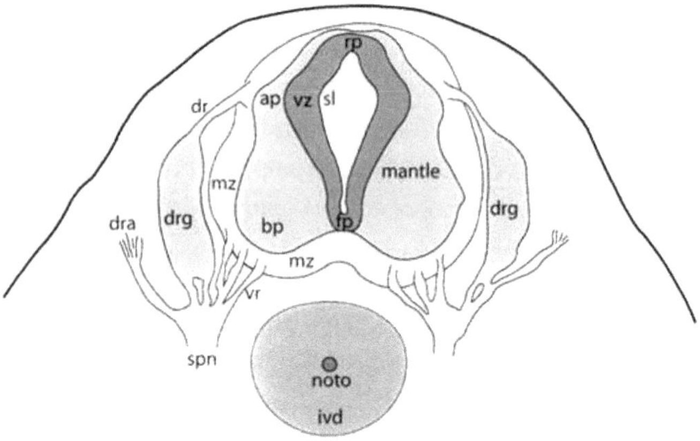

Figure 2.3 Developing human spinal cord at 6 weeks pc

A cross section through the human spinal cord at approximately 6 weeks pc, illustrating alar plates (ap) and basal plates (bp) of the mantle layer (mantle), roof plate (rp) and floor plate (fp) and sulcus limitans (sl). (dr = dorsal roots; dra = dorsal ramus; d = dorsal root ganglion; ivd = intervertebral disc; mz = marginal zone; noto = notochord; spn = spinal nerve; vr = ventral roots).

Naturally, molecular factors must be responsible for controlling this emerging dorsoventral patterning of the cord. The floor plate is induced ventrally by axial mesoderm (see above), whereas the roof plate is thought to be generated dorsally by signals from the overlying ectoderm (Chizhikov and Millen, 2005). The floor plate generates a gradient of Shh that establishes five progenitor domains in the neuroepithelial ventricular zone (Figure 2.4), which in turn give rise to five distinct mature neuronal subtypes (V0-3, MN) in the basal plate mantle zone (Price and Briscoe, 2004; Zhuang and Sockanathan, 2006). Six progenitor domains are present in the dorsal spinal cord neuroepithelium and these give rise to six early-born and two late-born groups of dorsal interneurons (Zhuang and Sockanathan, 2006) (Figure 2.4). At present there are a large number of signaling molecules which have been implicated in dorsal patterning of progenitor domains in the neuroepithelium, including members of the transforming growth factor-β superfamily (Chesnutt *et al.*, 2004). The emergence of alar plate constituents is probably the result of complex interactions between the responsible factors (for review see Zhuang and Sockanathan, 2006).

Segmentation of the developing spinal cord

Rhombomeric and prosomeric organization of the rostral neural tube derivatives is easily recognized even with classical histological techniques, and the molecular factors controlling this segmentation has recently been the subject of intensive research. Segmentation of the developing spinal cord is subtler: most columns of motoneurons, for example, have the superficial appearance of being longitudinally continuous. Nevertheless, discrete rostrocaudally segregated motoneuron pools supply particular muscle groups in the adult and recent studies have shown that families of molecular factors responsible for segmentation in the rostral neural tube are also critical for determining developmental fates of motoneuron populations.

For example, recent studies by Jessell's group in chick embryo spinal cord (Dasen *et al.*, 2005) have shown that two independent sets of Hox regulatory interactions cooperate to determine the fate of motoneurons in the cervical enlargement. One set constrains motoneuron pools to particular rostrocaudal positions (e.g. Hox5 and Hox8 proteins), whereas the other (e.g. Hox4, Hox6, Hox7 and Meis1 proteins) controls diversification of motoneuron pools at a given rostrocaudal level (Dasen *et al.*, 2003, 2005). At present it is not known for certain whether a similar Hox regulatory network operates to specify segmental organization within the dorsal horn, but this would appear likely.

Motoneuron development and cell death

Lumbar motoneurons emerge from the ventricular proliferative zone at about 4 weeks pc in humans and E13 in rodents (Clowry *et al.*, 2005). The most detailed study of the timing of motoneuron generation in the rat was by Altman and Bayer (1984), who found that somatic motoneurons leave the mitotic cycle slightly earlier at cervical compared to lumbosacral levels (cervical levels – 11 to 14 days pc; thoracic levels – 11 to 14 days pc; lumbosacral levels – 12 to 14 days pc) and most sympathetic preganglionic motoneurons are generated on 12 and 13 days pc. Primitive motoneurons migrate into the basal plate trailing a radially oriented and centrally directed process, which transiently extends to the spinal cord lumen as a remnant of their neuroepithelial precursor, but this soon disappears. Motoneurons subsequently develop a primitive axon and dendrites. The axon will break through the marginal zone and emerge from the ventral surface of the cord, collectively forming the ventral roots with other motoneuron axons, while the dendrite will ramify in the emerging neuropil of the ventral horn and intermediate gray matter.

Motoneuron subtype specification and diversification has been the subject of considerable research in recent years. Ventral horn neurons are known to arise from five columnar subtypes: four interneuronal (V0, V1, V2, V3) and one motoneuronal (MN). From the motoneuron group arise three further columns of effector neurons: medial motoneurons throughout the entire cord, lateral motoneurons at the cervical and lumbosacral enlargements and intermediolateral cell column sympathetic neurons (visceral motoneurons) in thoracic and upper lumbar segments. Medial motoneurons, which are the first cells to differentiate, innervate axial musculature, whereas lateral motoneurons innervate limb musculature. Segregation of motoneurons into discrete somatic motor columns occurs at about 8 to 10 weeks pc in humans (Rath *et al.*, 1982) and 16 to 17 days pc in rodents (Clowry *et al.*, 2005). Gene expression studies have shown that divergence of gene expression profiles between motoneuron groups does not strictly correlate with divergence of function as defined by innervation patterns (Cui *et al.*, 2006), suggesting that epigenetic factors may play a role in determining motoneuron functional groups. Some authors have argued that cadherin expression is responsible for segregation of motoneuron pools and pool specific patterns of cadherin expression have been reported (see Guthrie, 2002, for review). Other studies have indicated that differential semaphorin expression may also be a significant factor in sorting motoneuron pools and their connections (Cohen *et al.*, 2005).

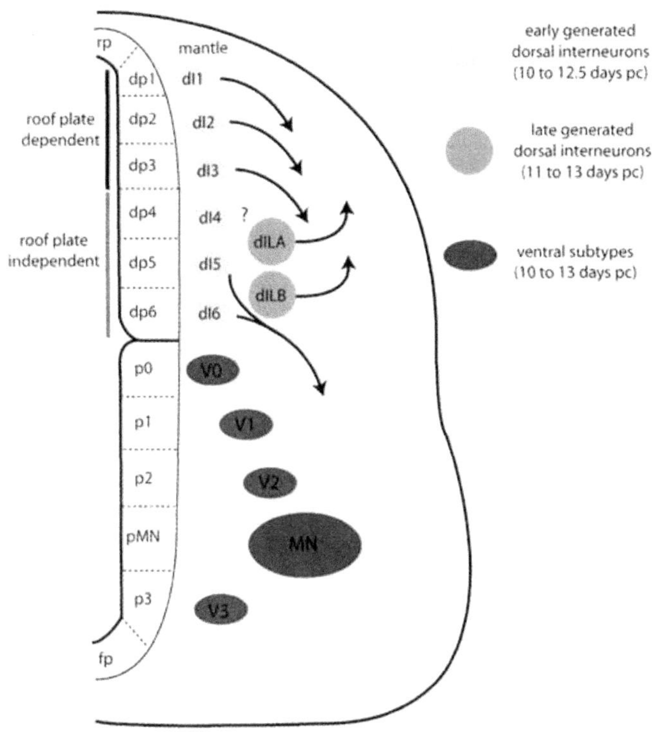

Figure 2.4 Progenitor domains in the developing spinal cord

This diagram shows progenitor domains in the ventricular germinal zone and daughter neuron groups in the alar and basal plates of the mantle layer of a developing rodent spinal cord. Six progenitor domains in the dorsal spinal cord (dp1 to dp6) give rise to 6 early generated (dI1 to dI6) and 2 later generated (dILA, dILB) dorsal horn neuron populations in the alar plate. The 3 most dorsal progenitor domains are dependent on the roof plate while the dp4, dp5 and dp6 are not. Populations dI1, dI2 and dI3 all settle in the deep dorsal horn and give rise to commissural neurons (dI1, dI2), proprioceptor/mechanoreceptor neurons (dI1, dI3), or spinocerebellar neurons (dI1); dI4 may settle in the superficial dorsal horn, whereas dI5 and dI6 appear to be destined for the ventral horn. The fate of dI4, dI5 and dI6 neurons is uncertain at present. Later generated dorsal horn neurons (dILA, dILB) settle in the superficial laminae of the dorsal horn and give rise to GABAergic association neurons (Helms and Johnson, 2003). Ventral progenitor domains p0, p1, p2 and p3 give rise to V0, V1, V2 and V3 ventral horn interneurons, respectively. Progenitor domain pMN gives rise to motoneurons of the ventral horn and visceral motoneurons of the lateral horn.

It is well known that i) more motoneurons are produced by the neuroepithelium than survive to maturity, and ii) developing motoneurons are dependent on trophic support from their target muscles (Oppenheim, 1991). In humans, there is a 35% decline in motoneuron number between 11 and 25 weeks pc (Forger and Breedlove, 1987). Counts of pyknotic cells in developing human spinal cord have indicated that most motoneuronal degeneration occurs between 12 and 16 weeks pc (Forger and Breedlove, 1987). Most motoneuron death appears to be due to competition for trophic support rather than the removal of wiring errors. Molecules identified as having motoneuron survival potential belong to several different gene families including neurotrophins (NT-3, NT-4/5, BDNF), cytokines (cardiotrophin-1, ciliary neurotrophic factor, leucocyte inhibitory factor), TGF-β family members (GDNF, neurturin, persphin), hepatocyte growth factor/scatter factor family members (HGF/SF) and fibroblast growth factors (FGF-1, FGF-2, FGF-5) (Henderson, 1996; Kablar and Belliveau, 2005). Naturally occurring motoneuronal death during development appears to be mediated by oxidative stress and involves reactive oxygen species as signaling molecules for controlling caspase-dependent and caspase-independent mechanisms (Sánchez-Carbente *et al.*, 2005).

The development of some identified motoneuron populations in the spinal cord has been followed in rodents. Phrenic motoneurons in the rat can be identified in the cervical spinal cord ventral horn by 13 days pc, while aggregation of phrenic motoneurons into a column and formation of dendritic bundles become apparent by 16 days pc (Song *et al.*, 2000). The phrenic motoneuron column extends from C2 to C6 at 13 to 14 days pc, but becomes progressively confined to C3 to C5 by birth in the rat.

In human cervical spinal cord, axodendritic synapse formation on motoneurons increases substantially at the end of 8 weeks pc, but axosomatic synapses proliferate rapidly from 10.5 to 13 weeks pc and may continue up to 19 weeks pc (Okado, 1980).

Development of spinal cord afferents and dorsal horn interneurons

The development of dorsal root ganglion cells has been most closely studied in rodents (Altman and Bayer, 1984). The majority of dorsal root ganglion cells in the rat are produced between 12 and 15 days pc with a rostrocaudal gradient of production and larger ganglion cells appear to be produced before smaller ones. By 13 days pc, many dorsal root ganglion cells of the rat adopt a bipolar shape, coinciding with the outgrowth of central processes into the dorsal horn and peripheral processes to somatic targets. Transformation of dorsal root ganglion cells into a pseudo-unipolar morphology occurs on 15 and 16 days pc. Dorsal horn interneurons are generated on 15 and 16 days pc, after the initial ingrowth of dorsal root ganglion cell central processes (Altman and Bayer, 1984), and there appears to be a ventral-to-dorsal gradient of neurogenesis within the dorsal horn interneuron population.

The invasion of the dorsal horn by afferents has also been studied in rodents. The central processes of phrenic nerve afferent fibers invade the dorsal horn at 14 days pc and spindle afferents distribute to the ventral horn and appear to make contact with motoneurons as early as 16 days pc (Song *et al.*, 1999). Some pruning of phrenic nerve afferents may occur during development, in that afferents were seen to cross the midline at birth but these were lost by P4 (Song *et al.*, 1999).

In the developing human cervical spinal cord, central processes of muscle spindle afferents cross the dorsal horn by 7.5 weeks pc and form contacts with motoneurons by 9 weeks pc (Clowry *et al.*, 2005). This coincides with an abrupt increase in the density of axo-dendritic synapses in the ventral horn (Okado, 1980).

Development of glia in the spinal cord

Recent studies in laboratory animals have shown that Olig genes are important in regulating glial differentiation. During late embryonic and early fetal life in rodents, Olig2 expression identifies a domain in developing spinal cord which appears to give rise to a broad range of neural stem and glial progenitor cells (Liu and Rao, 2004). Proliferating stem cells within the neural tube do not express any glial markers until 10.5 days pc. By 11 days pc, glial precursors have begun to differentiate and at least two regions containing glial precursors can be identified in the ventral neural tube. Protoplasmic and fibrous astrocytes develop from radial glia (McDermott *et al.*, 2005) and (as identified by glial fibrillary acidic protein) can first be detected at 16 days pc in rodents (Liu *et al.*, 2002).

Oligodendrocytes are the glial cells responsible for myelination within the central nervous system. In the spinal cord, oligodendrocyte precursors arise from a restricted region in the ventral ventricular zone of both rodents and humans near the floor plate (Noll and Miller, 1993; Hajihosseini *et al.*, 1996), an area which also includes a motoneuron progenitor domain (Richardson *et al.*, 1997). Originally this region was believed not to give rise to astrocytes, but more recent lineage studies have demonstrated that astrocyte and ependymal cells may also be derived from this part of the ventricular zone (Masahira *et al.*, 2006). The emergence of oligodendrocyte precursors is under the influence of inductive signaling by Shh derived from the floor plate (Oh *et al.*, 2005), whereas Wnt proteins have been identified as dorsal factors that directly inhibit oligodendrocyte development (Shimizu *et al.*, 2005). The subsequent dispersal and development of oligodendrocytes appears to be dependent on the guidance molecule netrin-1 (Tsai *et al.*, 2006), which is also secreted from the floor plate region. In the human, oligodendrocyte precursors may be detected in the dorsal spinal cord at 74 days pc and in the ventral roots at 83 pc (Hajihosseini *et al.*, 1996).

Colonization of the developing human spinal cord by microglia appears to coincide with vascularization and neuronal migration, with the invasion of these cells from the meninges following a progression along the vasculature from white to gray matter (Rezaie and Male, 1999). The earliest arrival of microglia is around 9 weeks pc in the human, although the major influx and distribution of microglia occurs from 16 weeks (Rezaie and Male, 1999).

Development of major ascending and descending tracts

In the developing rat spinal cord, the initial step in the development of the dorsal column pathways, i.e. the bifurcation of the central processes of dorsal root ganglion cells, occurs at 14 days pc (Altman and Bayer, 1984). The dorsal columns as a group appear at 17 days and a distinction between the fasciculus cuneatus and gracilis first becomes apparent a day later (Altman and Bayer, 1984).

Immunohistochemical studies in the human spinal cord have shown that non-phosphorylated neurofilament protein appears in the spinocerebellar neurons of Clarke's column as early as 10 weeks pc (Clowry *et al.*, 2005). By 14 weeks pc, dorsal spinocerebellar tract axons can be seen emerging from the nucleus and coursing through the gray matter and by 16 weeks pc these axons can be seen entering the lateral funiculus (Clowry *et al.*, 2005). Spinocerebellar neurons of Clarke's column in the rat leave the mitotic cycle between 13 and 15 days pc, slightly behind the time of generation of spinothalamic neurons in the same segmental level (Beal and Bice, 1994).

Spinothalamic pathways probably develop during the period from 13 to 15 days pc in the rat. The initial outgrowth of these axons is towards the floor plate region to effect decussation and the factors responsible for controlling this initial trajectory have been the subject of considerable recent interest. Netrin-1, a long-range guidance cue expressed by floor plate cells, acts in concert with Shh to attract commissural axons like the spinothalamic tract fibers to the ventral midline (Salinas, 2003). Once these axons have crossed the midline, the pattern of expression of molecules on the growing axons is altered so that the floor plate subsequently exerts a repulsive force for the growth cones (Garbe and Bashaw, 2004).

In the human, the corticospinal tract has been reported to reach the caudal medulla at about 13 weeks pc, with completion of the pyramidal decussation by 15 weeks (for review see ten Donkelaar *et al.*, 2004). Invasion of cervical levels of the cord occurs between 14 and 16 weeks, but caudal spinal cord is not reached until much later (lower thoracic cord – 17 weeks pc; lumbosacral cord – 27 weeks pc). This early contact between the corticospinal tract axons and at least upper spinal cord probably allows activity dependent maturation of spinal motor centers (Eyre *et al.*, 2000), but myelination in the corticospinal tract occurs over a protracted period and is not complete until the age of two to three years. There is also evidence for activity dependent withdrawal of corticospinal projections during human development, much as has been seen in rodents (Eyre *et al.*, 2001).

In contrast to humans, the growth of the corticospinal tract into the rodent spinal cord occurs entirely postnatally. The leading axons of the decussating component of the rat corticospinal tract reach the cervical spinal segments at the time of birth, midthoracic levels at postnatal day 2 (2 pn) and the lumbar enlargement at 5 pn (Gribnau *et al.*, 1986; Joosten *et al.*, 1987; Gorgels, 1990). On the other hand, the murine crossed corticospinal tract does not reach mid-thoracic levels until 4 pn and lumbar levels until the second postnatal week (Gianino *et al.*, 1999; Hsu *et al.*, 2006). The number of viable axons on one side of the murine corticospinal tract peaks at 6 pn and 14 pn at the level of C7 and at 14 pn at the L4 level (Hsu *et al.*, 2006). Axonal degeneration immediately follows the zenith in axon numbers: estimates of degenerating axons show peaks at 6 pn and 14 pn at the C7 level and at about 14 pn at the L4 level. As in other major pathways of the developing central nervous system, exuberant axonal growth followed by substantial axonal loss is evident in the developing corticospinal tract in both rodents and humans.

Myelination of spinal cord pathways

Myelinated fibers can be found in the early fetal human spinal cord (e.g. 10 weeks pc – Okado, 1982; less than 16 weeks pc – Niebroj-Dobosz *et al.*, 1980), but most significant myelination does not occur until the second trimester. In the developing human spinal cord, mRNA for key markers of myelination (i.e. myelin basic protein, proteolipid protein and myelin associated glycoprotein) all undergo rapid rises between 15 and 22 weeks pc (Grever *et al.*, 1997). This corresponds with a transition in the human spinal cord from only sparse myelination to well myelinated tracts, but not all tracts appear to myelinate at the same rate. The descending medial longitudinal fasciculus (medial vestibulospinal tract), for example, myelinates earliest at about 20 weeks pc, whereas the corticospinal tract seems to lag behind other pathways in the extent of myelination and is incompletely myelinated at birth (Tanaka *et al.*, 1995; Weidenheim *et al.*, 1996). There also appear to be anterior-to-posterior and rostral-to-caudal gradients in spinal cord myelination (Weidenheim *et al.*, 1996).

Myelination of the corticospinal tract has been followed in BALB/cByJ mice (Hsu *et al.*, 2006). Pro-myelinated axons (axons surrounded by only one layer of oligodendrocyte process) were first seen at 2 pn and 4 pn at segmental levels C7 and L4, respectively, but a dramatic increase in myelinated axons does not occur until 14 pn at both levels. In the rat, myelination of the corticospinal tract starts around 10 pn and continues into the second postnatal month (Gorgels *et al.*, 1989).

Relative growth of the spinal cord and vertebral column

Up until 14 weeks pc, the human spinal cord extends the entire length of the embryo and spinal nerves exit the vertebral column through intervertebral foramina situated alongside their point of emergence from the spinal cord. With progressive growth during the fetal period, the vertebral column, dura and arachnoid elongate more rapidly than the developing spinal cord so that the caudal end of the spinal cord comes to lie progressively higher up the vertebral column. By the end of the fifth month, the caudal end of the spinal cord is alongside the caudal edge of the S1 vertebra and by birth it lies beside the L3 vertebra. The adult position (alongside the L2 vertebra) is attained by the second year of life. Naturally, this position change necessitates profound lengthening of the dorsal and ventral roots, particularly at the sacral segmental levels.

References

Altman J, Bayer SA (1984) The development of the rat spinal cord. Adv Anat Embryol Cell Biol 85, 1-164.

Beal JA, Bice TN (1994) Neurogenesis of spinothalamic and spinocerebellar tract neurons in the lumbar spinal cord of the rat. Brain Res Dev Brain Res 78, 49-56.

Bittman KS, Panzer JA, Balice-Gordon RJ (2004) Patterns of cell-cell coupling in embryonic spinal cord studied via ballistic delivery of gap-junction-permeable dyes. J Comp Neurol 480, 273-285.

Chesnutt C, Burrus LW, Brown AMC, Niswander L (2004) Coordinate regulation of neural tube patterning and proliferation by TGFb and WNT activity. Dev Biol 274, 334-347.

Chizhikov VV, Millen KJ (2005) Roof-plate dependent patterning of the vertebrate dorsal central nervous system. Dev Biol 277, 287-295.

Clowry GJ, Moss JA, Clough RL (2005) An immunohistochemical study of the development of sensorimotor components of the early fetal human spinal cord. J Anat 207, 313-324.

Cohen S, Funkelstein L, Livet J, Rougon G, Henderson CE, Castellani V, Mann F (2005) A semaphorin code defines subpopulations of spinal motor neurons during mouse development. Eur J Neurosci 21, 1767-1776.

Cui D, Dougherty KJ, Machacek DW, Sawchuk M, Hochman S, Baro DJ (2006) Divergence between motoneurons: gene expression profiling provides a molecular characterization of functionally discrete somatic and autonomic motoneurons. Physiol Genomics 24, 276-289.

Dasen JS, Liu JP, Jessell TM (2003) MN columnar fate imposed by sequential phases of Hox-c activity. Nature 425, 926-933.

Dasen JS, Tice BC, Brenner-Morton S, Jessell TM (2005) A Hox regulatory network establishes motor neuron pool identity and target-muscle connectivity. Cell 123, 477-491.

Eyre JA, Miller S, Clowry GJ, Conway EA, Watts C (2000) Functional corticospinal projections are established prenatally in the human foetus permitting involvement in the development of spinal motor centers. Brain 123, 51-64.

Eyre JA, Taylor JP, Villagra F, Smith M, Miller S (2001) Evidence of activity-dependent withdrawal of corticospinal projections during human development. Neurology 57, 1543-1554.

Forger NG, Breedlove SM (1987) Motoneuronal death during human fetal development. J Comp Neurol 264, 118-122.

Garbe DS, Bashaw GJ (2004) Axon guidance at the midline: From mutants to mechanisms. Crit Rev Biochem Mol Biol 39, 319-341.

Gianino S, Stein SA, Li H, Lu X, Biesiada E, Ulas J, Xu XM (1999) Postnatal growth of corticospinal axons in the spinal cord of developing mice. Dev Brain Res 112, 189-204.

Gorgels TG (1990) A quantitative analysis of axon outgrowth, axon loss, and myelination in the rat pyramidal tract. Dev Brain Res 54 51-61.

Gorgels TG, de Kort EJ, van Aanholt HT, Nieuwenhuys R (1989) A quantitative analysis of the development of the pyramidal tract in the cervical spinal cord in the rat. Anat Embryol 179, 377-385.

Grever WE, Weidenheim KM, Tricoche M, Rashbaum WK, Lyman WD (1997) Oligodendrocyte gene expression in the human fetal spinal cord during the second trimester of gestation. J Neurosci Res 47, 332-340.

Gribnau AAM, de Kort EJM, van Aaholt HTH Nieuwenhuys R (1986) On the development of the pyramidal tract in the rat: II. An anterograde tracer study of the outgrowth of the corticospinal fibers. Anat Embryol 175, 101-110.

Guthrie S (2002) Neuronal development: sorting out motor neurons. Curr Biol 12, R488-R490.

Hajihosseini M, Tham TN, Dubois-Dalcq M (1996) Origin of oligodendrocytes within the human spinal cord. J Neurosci 16, 7981-7994.

Helms AW, Johnson JE (2003) Specification of dorsal spinal cord interneurons. Curr Opin Neurobiol 13, 42-49.

Henderson CE (1996) Role of neurotrophic factors in neuronal development. Curr Opin Neurobiol 6, 64-70

Hsu J-YC, Stein SA, Xu X-M (2006) Development of the corticospinal tract in the mouse spinal cord: A quantitative ultrastructural analysis. Brain Res 1084, 16-27.

Joosten EAJ, Gribnau AAM, Dederen PJWC (1987) An anterograde tracer study of the developing corticospinal tract in the rat: three components. Dev Brain Res 36, 121-130.

Kablar B, Belliveau AC (2005) Presence of neurotrophic factors in skeletal muscle correlates with survival of spinal cord motor neurons. Dev Dyn 234, 659-669.

Kaufman MH (1992) *The Atlas of Mouse Development.* Academic Press: London

Lewis KE, Eisen JS (2003) From cells to circuits: development of the zebrafish spinal cord. Prog Neurobiol 69, 419-449.

Liu Y, Wu Y, Lee JC, Xue H, Pevny LH, Kaprielian Z, Rao MS (2002) Oligodendrocyte and astrocyte development in rodents: an in situ and immunohistological analysis during embryonic development. Glia 40, 25-43.

Liu Y, Rao MS (2004) Olig genes are expressed in a heterogeneous population of precursor cells in the developing spinal cord. Glia 45, 67-74.

Masahira N, Takebayashi H, Ono K, Watanabe K, Ding L, Furusho M, Ogawa Y, Nabeshima Y, Alvarez-Buylla A, Shimizu K, Ikenaka K (2006) Olig2-positive progenitors in the embryonic spinal cord give rise not only to motoneurons and oligodendrocytes, but also to a subset of astrocytes and ependymal cells. Dev Biol 293, 358-369.

Masuda T, Shiga T (2005) Chemorepulsion and cell adhesion molecules in patterning trajectories of sensory axons. Neurosci Res 51, 337-347.

McDermott KW, Barry DS, McMahon SS (2005) Role of radial glia in cytogenesis, patterning and boundary formation in the developing spinal cord. J Anat 207, 241-250.

Niebroj-Dobosz I, Fidzianska A, Rafalowska J, Sawicka E (1980) Correlative biochemical and morphological studies of myelination in human ontogenesis. I. Myelination of the spinal cord. Acta Neuropathologica 49, 145-152.

Noll E, Miller RH (1993) Oligodendrocyte precursors originate at the ventral ventricular zone dorsal to the ventral midline region in the embryonic rat spinal cord. Development 118, 563-573.

Oh S, Huang X, Chiang C (2005) Specific requirements of sonic hedgehog signaling during oligodendrocyte development. Dev Dyn 234, 489-496.

Okado N (1980) Development of the human spinal cord with reference to synapse formation in the motor nucleus. J Comp Neurol 191, 495-513.

Okado N (1982) Early myelin formation and glial cell development in the human spinal cord. Anat Rec 202, 483-490.

Oppenheim RW (1991) Cell death during development of the nervous system. Annu Rev Neurosci 14, 453-501.

O'Rahilly R, Muller F (2002) The two sites of fusion of the neural folds and the two neuropores in the human embryo. Teratology 65, 162-170.

Price SR, Briscoe J (2004) The generation and diversification of spinal motor neurons: signals and responses. Mech Dev 121, 1103-1115.

Rath G, Gopinath G, Bijlani V (1982) Prenatal development of the human spinal cord. I. Ventral motor neurons. J Neurosci Res 7, 437-441.

Rezaie P, Male D (1999) Colonisation of the developing human brain and spinal cord by microglia: A review. Microsc Res Tech 45,359-382.

Richardson WD, Pringle NP, Yu WP, Hall AC (1997) Origins of spinal cord oligodendrocytes: possible developmental and evolutionary relationships with motor neurons. Dev Neurosci 19, 58-68.

Sadler TW (1978) Distribution of surface coat material on fusing neural folds of mouse embryos during neurulation. Anat Rec 191, 345-350.

Sadler TW (2005) Embryology of neural tube development. Am J Med Gen C (Semin Med Genet) 135C, 2-8.

Salinas PC (2003) The morphogen Sonic Hedgehog collaborates with netrin-1 to guide axons in the spinal cord. Trends Neurosci 26, 641-643.

Sánchez-Carbente MR, Castro-Obregón S, Covarrubias L, Narváez V (2005) Motoneuronal death during spinal cord development is mediated by oxidative stress. Cell Death Diff 12, 279-291.

Shimizu T, Kagawa T, Wada T, Muroyama Y, Takada S, Ikenaka K (2005) Wnt signalling controls the timing of oligodendrocyte development in the spinal cord. Dev Biol 282, 397-410.

Solursh M, Morriss GM (1977) Glycosaminoglycan synthesis in rat embryos during formation of the primary mesenchyme and neural folds. Dev Biol 57, 75-86.

Song A, Tracey DJ, Ashwell KWS (1999) Development of the rat phrenic nerve and the terminal distribution of phrenic afferents in the cervical cord. Anat Embryol 200, 625-643.

Song A, Ashwell KWS, Tracey DJ (2000) Development of the rat phrenic nucleus and its connections with brainstem respiratory nuclei. Anat Embryol 202, 159-77

Tanaka S, Mito T, Takashima S (1995) Progress of myelination in the human fetal spinal nerve roots, spinal cord and brainstem with myelin basic protein immunohistochemistry. Early Human Dev 41, 49-59.

ten Donkelaar, HJ, Lammens M, Wesseling P, Hori A, Keyser A, Rotteveel J (2004) Development and malformations of the human pyramidal tract. J Neurol 251, 1429-1442.

Tsai H-H, Macklin WB, Mıller RH (2006) Netrin-1 is required for the normal development of spinal cord oligodendrocytes. J Neurosci 26, 1913-1922.

Weidenheim KM, Bodhireddy SR, Rashbaum WK, Lyman WD (1996) Temporal and spatial expression of major myelin proteins in the human fetal spinal cord during the second trimester. J Neuropathol Exp Neurol 55, 734-745.

Zhuang BQ, Sockanathan S (2006) Dorsal-ventral patterning: a view from the top. Curr Opin Neurobiol 16, 20-24.

3 The Vertebral Column and Spinal Meninges

Gulgun Kayalioglu

The vertebral column

The vertebral column, also known as the spinal column, is the central axis of the skeleton in all vertebrates. The vertebral column provides attachments to muscles, supports the trunk, protects the spinal cord and nerve roots and serves as a site for haemopoiesis. The mammalian vertebral column consists of five morphologically differentiated groups of vertebrae: cervical, thoracic, lumbar, sacral and coccygeal (caudal) (Fig 3.1). In humans, the vertebral column usually consists of 33 vertebrae, placed in series and connected by ligaments and intervertebral discs. However, the number of vertebrae can vary between 32 and 35. Usually there are 7 cervical, 12 thoracic, 5 lumbar, 5 sacral and 4 caudal (coccygeal) vertebrae. In humans, the length of the vertebral column is 71 cm in males and 61 cm in females.

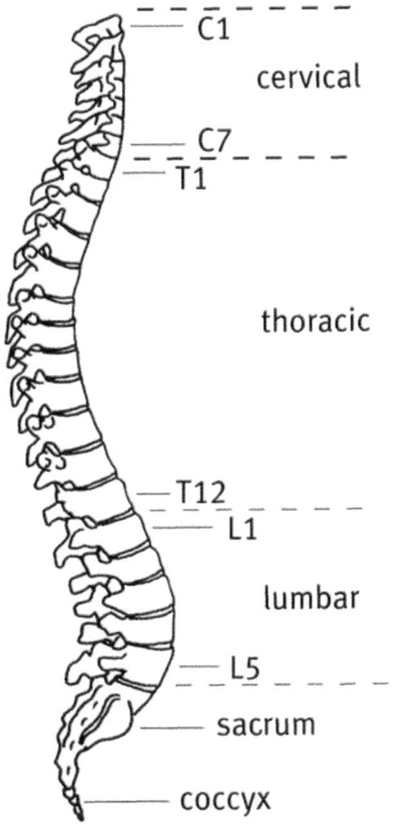

Figure 3.1 Human vertebral column

This diagram shows the curvatures associated with the different regions of the human vertebral column. The primary vertebral curve (concave forwards) is retained in the thoracic and sacral regions. The secondary curvatures in the cervical and lumbar regions are convex anteriorly. (Adapted from Moore and Dalley, 2006, p.479)

General features of the vertebrae in mammals

Vertebrae are classified as irregular bones; they consist largely of spongy bone with a thin outer compact bone covering. Each vertebra consists of two essential parts, a ventral body and a dorsal vertebral arch, which enclose the vertebral foramen.

The vertebral body

The vertebral body (corpus vertebra) supports the weight of the trunk. It is generally cylindrical in shape, but there is a wide range of variation for the shape and size in different regions and in different mammalian species. The most extreme is the first cervical vertebra (atlas) which has no body. In the cervical vertebrae the vertebral body is small and wider from side to side than from front to back. Upper thoracic vertebrae are similar in shape to the cervical, the midthoracic vertebrae are heart-shaped, and the lower thoracic vertebral bodies are kidney-shaped like the lumbar vertebrae. The lumbar vertebral bodies are large, kidney-shaped and relatively wider than the thoracic vertebrae.

The ventral surface of a typical vertebral body is convex from side to side and the dorsal surface is almost flat or slightly concave. The vertebral bodies progressively increase in size from cervical to lumbar segments. In male and female human specimens, the dorsoventral vertebral body length increases gradually from T1 to T12 (Ebraheim *et al.*, 1997b). From this point there is a rapid decrease. Vertebral body height (rostrocaudal length) increases caudally except posteriorly where, after an initial increase, it decreases in the lower lumbar region (Berry *et al.*, 1987). In the lumbar region, the ventral vertebral body height is always bigger than the dorsal height, which emphasizes the lordotic shape of the lumbar region (Nissan and Gilad, 1986). In accordance with the generally larger body size in males, the anterior transverse, the posterior transverse and the anteroposterior diameters of the vertebral body are significantly larger in the male and these features are 90% accurate in predicting sex (MacLaughlin and Oldale, 1992). The bodies of the lumbar vertebrae show a significant trend toward lowering and broadening with age (Ericksen 1978a,b). The vertebral body width in the lumbar vertebrae is higher in males, and vertebral body length higher in females when regressed to body mass or height (Valverde *et al.*, 2005). The vertebral bodies are ventrally wedged in quadriped mammals, but in humans, with an erect posture, the bodies are dorsally wedged at L4-L5. It has been shown that bipedally conditioned rats also show a diminution in ventral wedging at lumbar levels, indicating this wedging is related to the upright posture causing a greater load from body weight (Shi *et al.*, 2007).

The superior and inferior surfaces of a typical vertebral body are nearly flat or slightly concave and are rough suited for the attachment of fibrocartilaginous discs. The margins of the superior and inferior surfaces are formed from a secondary center for ossification, the epiphyseal ring, and are smooth. There are vascular foramina in the vertebral body, some in the ventral surface for the passage of veins, and especially one or more in the dorsal surface serving as exit for basivertebral arteries and veins (Willis, 1949). Nerves also enter the vertebral body via the centrally located posterior vascular foramen and accompany the basivertebral vessels (Antonacci *et al.*, 1998).

The vertebral arch and articular processes

The vertebral arch is composed of a ventral pedicle and a dorsal lamina on each side. From the junction of the pedicle and lamina, the transverse, superior and inferior articular processes project (Fig 3.2). The transverse processes act as levers for the attachments of muscles and also articulate with the ribs in the thoracic region. They become smaller from caudal to cranial cervical segments in humans (Kawashima *et al.*, 2003). In humans, the transverse processes are short and laterally directed in the cervical region. In the thoracic region, the transverse processes are thick and backward directed, along with a surface for articulation with the rib tubercles. The transverse processes of the lumbar vertebrae are thick and spatula-shaped and horizontal, projecting backwards. In the sacrum, the atrophic transverse processes form the lateral sacral crest. The first coccygeal vertebra has a rudimentary transverse process and the remainder have none at all.

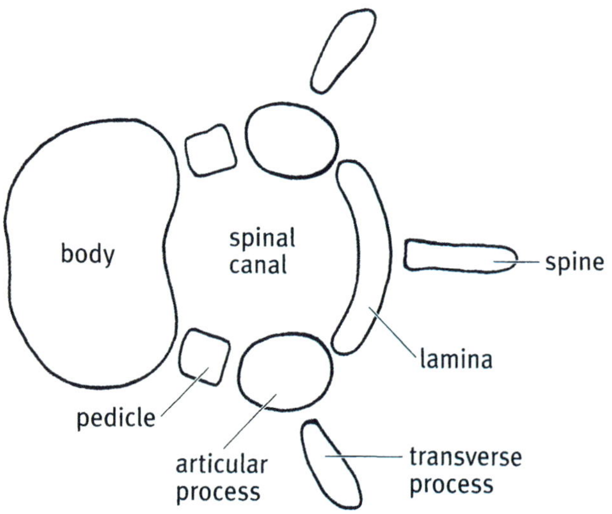

Figure 3.2 Components of a typical vertebra

This exploded view shows the elements that make up a typical vertebra The spinal canal is enclosed by the body, pedicles, articular processes, and laminae These elements protect the spinal cord and the spinal nerve roots. (Adapted from Cramer and Darby, 2005, p.30)

The pedicles are short strong processes that extend posteriorly from the posterolateral surface of the vertebral body. They are continuous with the laminae and articular processes. The concavities of the pedicles above and below make the upper and lower margins of the intervertebral foramen. In humans, the issue of pedicle size has recently become important for the pedicle screw placement technique. In the cervical region, pedicle width and height slightly decrease from C2 to C3 and then gradually increase to C7 (Karaikovic *et al.*, 1997). The greatest pedicle length is at C3 and the greatest pedicle width at C6 (Kayalioglu *et al.*, 2007). The pedicle width is 4.7-5.2 mm, and the pedicle height is 4.7-6.9 mm in C3-C6 (Ebraheim *et al.*, 1997c). In the thoracic region, the pedicle length gradually increases from T1 to T10, then decreases to T12 in the male and the female, except in females there is a decrease at T9 and T11. The smallest pedicle length is found at T3 in the male specimens and at T1 in the female. The pedicle width decreases dramatically from T1 to T4, then increases gradually to T12. The smallest pedicle width is found at T3-T10 for both sexes. The largest and smallest pedicle heights are at T11 and T1, consequently (Ebraheim *et al.*, 1997b). The pedicule height gradually increases from T1 to L5, plateauing between T3 and T9, being widest at the thoracolumbar junction (Chaynes *et al.*, 2001). Pedicle length is at its maximum in the midlumbar region and decreases in the lower lumbar region (Nojiri *et al.*, 2005). The pedicle axis length and pedicle diameter increases with advancing age, but no such increase is observed for pedicle length (Vara and Thompson, 2006).

The articular processes are located at the pediculolaminar junctions and form facet (zygapophyseal) joints between successive vertebrae. The superior articular process projects upwards and has an articular facet on its posterior surface. The inferior articular facet projects downwards and has a facet on its anterior surface. In humans the articular processes are flat in the cervical region. In other regions, the facets have a medial or lateral inclination depending on the vertebral level. In C3-C6, the articular processes are in the form of a bar-like structure and form rod-like columns and the superior and inferior facets lie in the same vertical line (Pal and Routal, 1996). In the thoracic region they are vertical and face ventrally. In the lumbar vertebrae their surfaces lie in the sagittal plane. In the sacrum they form the articular (intermediate sacral) crest, and they are absent in the coccyx.

The vertebral laminae in humans are broad plates directed backwards and medially from the pedicles. They are long and thick in the cervical region, short and strong in the thoracic and lumbar region, fused in the sacrum and absent in the coccyx. The greatest laminar height is at T11 (25.1±2.5 mm), and the least is at C4 (10.4±1.1 mm). The greatest laminar width is at L5 (15.7±2 mm), and the least is at T4 (5.8±0.8

mm). Laminar widths in the cervical region are slightly more than those in the thoracic region. The greatest laminar thickness is at T2 (5.0±0.2 mm), whereas the least is at C5 (1.9±0.6 mm). Laminar thickness tends to increase in the upper thoracic region and to decrease slightly in the lower thoracic region. The mean laminar thickness of the lower cervical region is the least in the whole spine (Xu *et al.*, 1999).

The spinous process or spine arises from the junction of two laminae of the vertebra and projects downwards and backwards. It gives attachments by its surfaces to many muscles and to the supraspinous and interspinous ligaments. The spinous processes vary considerably in shape, size and direction. The spinous processes of C2-C6 are short and bifid. In the thoracic region, the spinous process is long, three-sided, sloping downwards and backwards, in the midthoracic region being almost vertical. In the lumbar region, the spinous process is thick, broad, quadrangular and horizontal. On the sacrum the 3-4 spinous processes become fused to form the median crest. Since no portion of vertebral arch persists in the coccygeal vertebrae, the spinous processes are absent here.

A deep vertebral notch is present on the lower edge of each pedicle, and a shallower notch on the upper edge of each pedicle. The two notches together with the intervening vertebral body and the intervertebral disc form the intervertebral foramen. The intervertebral joint ventrally and the zygapophyseal joint dorsally also contribute to the boundaries of the foramen. The intervertebral foramen transmits neurovascular structures, the ventral and dorsal root (spinal ganglion) of the spinal nerve, spinal branches of the subclavian, intercostal and lumbar arteries and veins, the recurrent meningeal nerves and lymphatics, loose connective and fat tissue and the ligamenta flava (Amonoo-Kuofi *et al.*, 1988; Ruhli *et al.*, 2006). The connective tissue occupies less than 28% and the neural tissue less than 35% of the foramen (Hoyland *et al.*, 1989). Females have larger intervertebral foramina than males (Ruhli *et al.*, 2006). Ligamentous bands named the transforaminal ligaments are present and divide the intervertebral foramen to multiple sub-compartments, possibly serving a protective role in preventing injury to the vasculature passing through them (Amonoo-Kuofi *et al.*, 1988; Gilchrist *et al.*, 2002). The bands are found mostly in the lumbar and lower thoracic regions (Amonoo-Kuofi *et al.*, 1988; Bakkum and Mestan, 1994; Cramer *et al.*, 2002). The transforaminal ligaments are observed in about 80% of the intervertebral foramina, the oblique transforaminal ligament being the most common (Min *et al.*, 2005).

The vertebral foramen is delimited anteriorly by the posterior vertebral body and intervertebral disc, laterally by the pedicles, and posteriorly by the laminae. The vertebral foramen of the T1 vertebra is larger than those of the more caudal thoracic vertebrae. The vertebral foramina successively form the spinal canal. The foramen is small and round in the thoracic, large and triangular in the cervical, lumbar and sacral regions. In the cervical segment, the width of the spinal canal is largest at C6 to accommodate the larger spinal cord. The cross-sectional area ratios of the spinal cord to spinal canal are 1:2.3 at C6, 1:3.7 at C7, 1:4 at T1, and 1:3.7 at T2 (An *et al.*, 1999). The lumbar spinal canal diameters show a gradual decrease from L1 to L5 vertebral levels in both sexes but the relative width of the canal is more in females than in males. The normal value of the lumbar canal to vertebral body ratio varies from 1:2 (wide canal) to 1:5 (stenosis) (Janjua and Muhammad, 1989). The sagittal diameter of the vertebral foramen differs from about 15.3 mm to 16.5 mm in L3-S1, the diameter is larger in the male than the female (Varol *et al.*, 2006).

Interspecific variation in vertebral number

Specification of the vertebral column and axial development is regulated by *Hox* genes along the anterior-posterior axis of the embryo. It has been reported that differences in vertebral specification among animals correlate with different expression patterns of *Hox* genes (Gaunt, 1994, Burke *et al.*, 1995). Birds and reptiles often have variable numbers of cervical vertebrae (Wake, 1979), but in almost all mammals the number is fixed at seven. The numbers of thoracic, lumbar, sacral and caudal vertebrae fluctuate among mammalian orders and families, but almost all have seven cervical vertebrae. In primates there are 7 cervical, 12-13 thoracic, 4-9 lumbar, 1-5 sacral, and 2-27 caudal vertebrae (Narita and Kuratani, 2005). Interestingly, the sum of thoracic and lumbar vertebrae tends to be constant (19) in different families and orders of mammals. By contrast, in reptiles and birds the number of trunk vertebrae, which correspond to mammalian thoracolumbar vertebrae, is highly variable. Another study of the mammalian vertebral formula found that the number of thoracolumbosacral segments is fixed at 22 vertebrae in 80% of species and at 22±1 in the remainder. The shift of a vertebra from one type to another occurs either at the thoracolumbar or at the lumbosacral junction, rarely in the sacrococcygeal junction, but never at the cervicothoracic junction (Abitbol, 1987).

The rodent vertebral column

As detailed above, the rat has 57-60 vertebrae divided into five groups: 7 in the cervical, 13 in the thoracic, 6 in the lumbar, 4 in the sacral, and 8-32 in the coccygeal (tail) region (Narita and Kuratani, 2005). The largest of the cervical vertebrae are the atlas and the axis. The subsequent vertebrae are shorter and wider. As in humans, the typical rat vertebra has a body, a neural arch, and seven processes. These processes are the neural spine, two transverse processes, two anterior and two posterior zygapophyses. In C6 there is a thin plate of bone extending ventrally from the transverse process (carotid tubercle) which is an enlarged costal element. The length of the thoracic vertebrae increases caudally from 2 to 7 mm. The spinous process of T2 is longer than that of any other vertebra and is further extended by a small triangular part which articulates with it. The lumbar vertebrae are well developed and uniform in length (7 mm). The sacral vertebrae are partially fused. The total length of the sacrum is 20-23 mm, with the individual vertebrae being of uniform length. The rat's tail consists of coccygeal vertebrae which makes about 60% of the entire vertebral length (Greene, 1968).

In the cat, the vertebral column consists of 53 vertebrae: 7 cervical, 13 thoracic, 7 lumbar, 3 fused sacral, and 21-23 caudal. The first 8 caudal vertebrae have a distinct vertebral arch, spinous and transverse processes (Wischnitzer, 1967; Wake, 1979). In the atlas of the rat and cat, there is an atlantal foramen on the dorsal part of each facet and wing-like transverse processes (Wischnitzer, 1967; Greene, 1968).

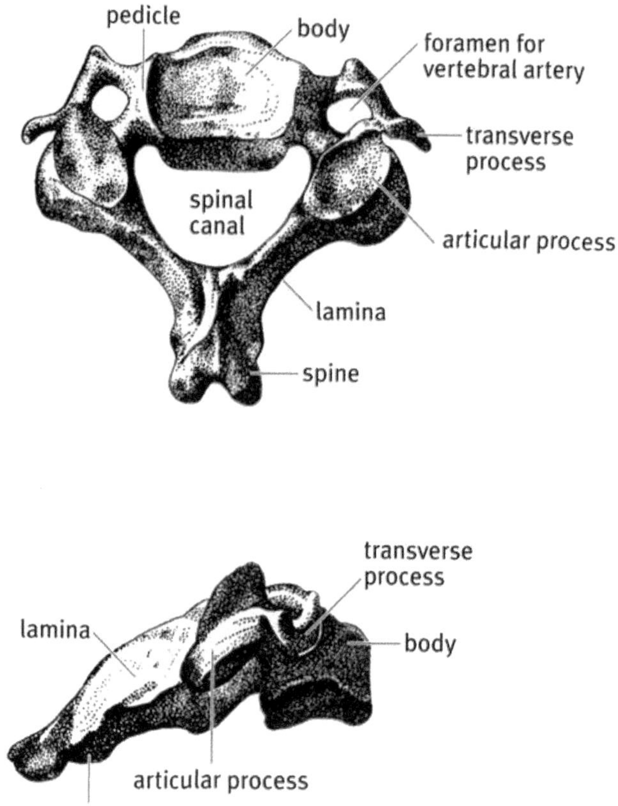

Figure 3.3 Human cervical vertebra

Cervical vertebrae in humans are considerably smaller than lumbar and thoracic vertebrae. The first two cervical vertebrae (the atlas and axis) are specially modified to control rotation of the head, and are shown in Figure 8. The remaining five cervical vertebrae (one of which is shown here) are similar to each other. In this figure the parts of a typical human cervical vertebra are labelled. (Adapted from Bourgery and Jacob, 2005, p.61)

Cervical vertebrae in humans

The seven cervical vertebrae can be distinguished from lumbar and thoracic vertebrae by their relatively small bodies, small and bifid spines in C2-C6, and the arterial foramen in their transverse processes. The atlas (C1) has no vertebral body, and is built around two lateral masses connected by the anterior and posterior arches. The mean width of the lateral mass of the atlas is about 15 mm, and the mean thickness is 17 mm (Dong *et al.*, 2003). (Fig 3.3)

The vertebral canal diameter in the cervical region is 32 mm in the sagittal plane and 29 mm in the transverse plane. The mean thickness of the anterior arch is 6 mm and the posterior ring is 8 mm (Doherty and Heggeness, 1994). There is a groove on the posterolateral margin of the atlas for the passage of the vertebral artery. Superior to this groove, little bridges (ponticuli) or posterolateral tunnels on the superior surface of the atlas, present in other non-primate mammals for the passage of the vertebral artery, have also been reported in the human (Hasan *et al.*, 2001). The atlas articulates above with the occipital bone and below with the odontoid process of the axis (C2). The axis has an odontoid process (dens) projecting superiorly from its body. The vertebra prominens (C7) has a long, non-bifid spinous process which is prominent at the nape of the neck. In C3-C6, and also in T1, there is also an uncinate process located in the superior surface of the vertebral body (Payne and Spillane, 1957). A three-dimensional quantitative study on the middle and lower cervical vertebrae showed that there are two distinct transition regions in the cervical vertebral column; one is toward the upper cervical region with the larger pedicle and spinous process of C2, the other one toward the thoracic spine where the C7 vertebra is wider but the spinal canal is narrower (Panjabi *et al.*, 1991b).

Costal processes in the cervical vertebral column extend laterally from the vertebral body and make the ventral border of the transverse foramen. The costal processes are narrowest (11.3 mm) at C2-C3 and greatest at C6-C7 (24.8 mm) (Pait *et al.*, 1996). The vertebral artery, vein and symphathetic nerves from the inferior cervical ganglion pass through the transverse foramen. The vertebral artery and vein enter together in the transverse canal at the level of C6. The venous organization in the transverse foramen may be in form of a venous plexus, venous sinus, venous plexus, and single or double veins from C1 to C6 (Palombi *et al.*, 2006). The transverse foramen width is 5.5-5.9 mm in C3-C6 (Ebraheim *et al.*, 1997a). In C7, the transverse foramen is relatively small and sometimes absent. It contains an accessory vertebral vein

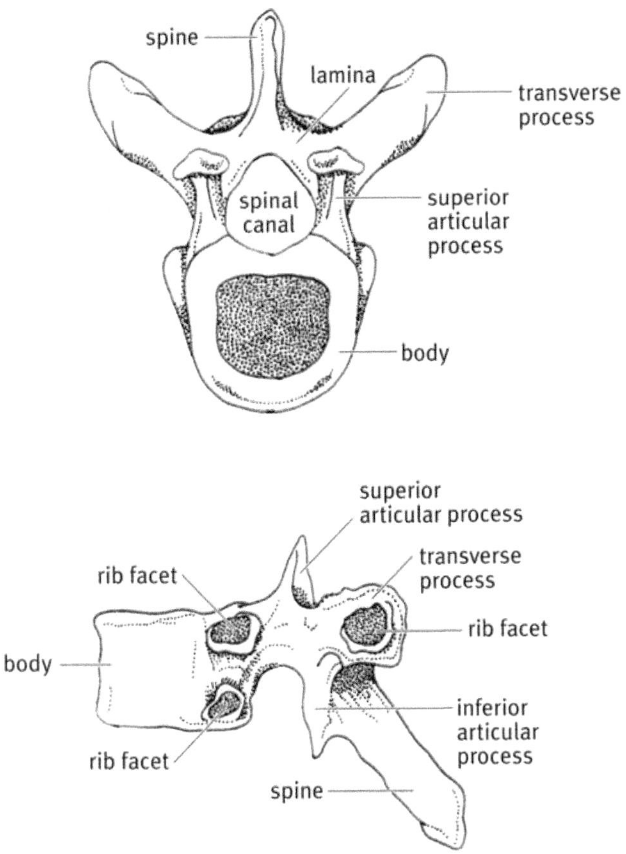

Figure 3.4 Human thoracic vertebra

The bodies and processes of thoracic vertebrae in humans are smaller than those of lumbar vertebrae, but larger than those of cervical vertebrae. The characteristic feature of thoracic vertebrae is the presence of joints that articulate with ribs. A mid-thoracic vertebra (shown here) has two joint facets on the vertebral body for the heads of adjacent ribs, and a third joint facet on the transverse process for the neck of a rib. (Adapted from Cramer and Darby, 2005, p.212)

or veins, is rarely traversed by the vertebral artery, vein and associated nerve fibers (Jovanovic, 1990).

Thoracic vertebrae in humans

The thoracic vertebrae lie in the posterior wall of the thorax with twelve pairs of ribs attached to them. Thoracic vertebrae have bodies of intermediate size. They are distinguished by their long, slender spines and by the presence of facets on the sides of the bodies articulating with the heads of the ribs and by facets on the transverse processes articulating with the tubercles of ribs. (Fig 3.4) The thoracic spine has three distinct regions: upper, middle and lower. Those of the upper region have some features similar to cervical vertebrae and those in the lower region have some features similar to lumbar vertebrae.

Lumbar vertebrae in humans

The lumbar vertebrae lie between the thoracic vertebrae and

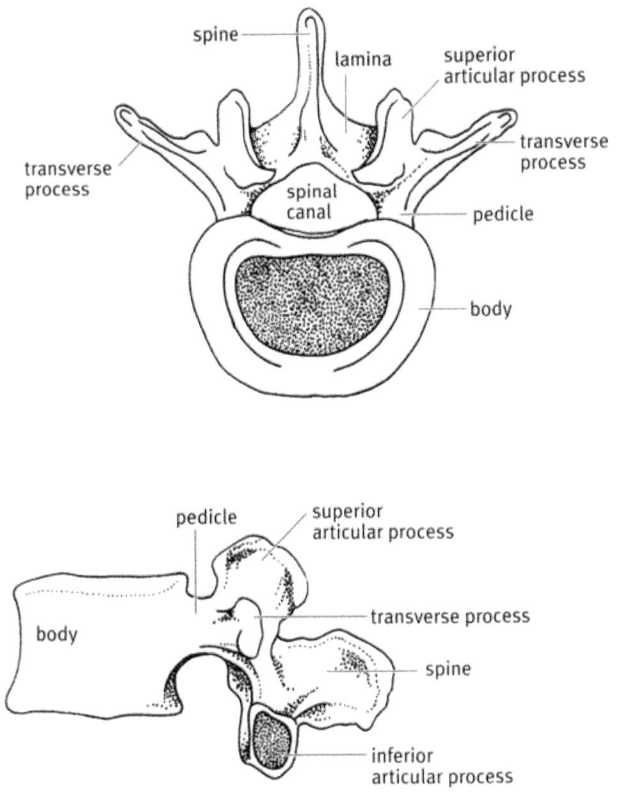

Figure 3.5 Human lumbar vertebra

The bodies and processes of lumbar vertebrae in humans are more massive than those of cervical and thoracic vertebrae. (Adapted from Cramer and Darby, 2005, p.246)

the sacrum. The lumbar vertebrae are distinguished from the cervical and thoracic vertebrae by their large size and the absence of transverse foramina and costal facets. The spinous processes are short and blunt. There is a mamillary process on the dorsal margin of each superior articular process and an accessory process dorsal to the base of each transverse process. (Fig 3.5)

The sacrum in humans

In the adult, the sacrum is a wedge-shaped bone consisting of five vertebrae that have fused, along with the intervening intervertebral discs. The cartilaginous joints between the sacral vertebrae are present in children, but begin fusing at puberty. The sacrum has a characteristically wide body curved upon itself and a triangular foramen. The ventral (pelvic) surface of the sacrum is concave and has four transverse ridges (linea transversaria). On the ventral surface of the sacrum, four pairs of foramina lie between the five sacral vertebrae. These foramina are oriented in a slightly anterolateral direction and give passage to the anterior divisions of respective sacral nerves and the lateral sacral arteries (Cheng and Song, 2003). The corresponding sacral foramina on the dorsal surface transmit the dorsal primary branches of the respective sacral nerve. The

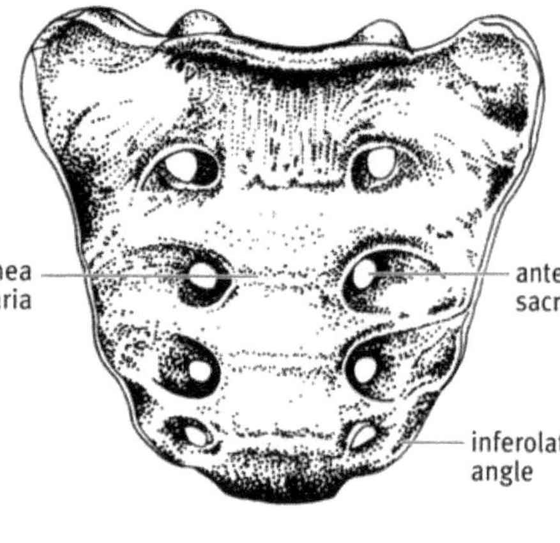

Figure 3.6 Human sacrum – ventral (anterior) view

The sacrum is formed by the fusion of five vertebral elements to form a large bone that contributes to the stability of the bony pelvis. On the dorsal surface of the sacrum is a bony tunnel, which protects the sacral and coccygeal nerve roots. (Adapted from Cramer and Darby, 2005, p.309)

diameters of the upper three sacral vertebral bodies and the sacral canal gradually decrease from S1 to S3 (Lu *et al.*, 2000). The dorsal surface of the sacrum is convex with a median sacral crest in the median plane bearing four spinous tubercles. Lateral to the median crest, the posterior surface is formed by the fused laminae. The lower two laminae are usually unfused in the midline, leaving an opening of variable extent known as the sacral hiatus. The cranial margin of the sacral hiatus is located caudal to the level of S4 and below (Shinohara, 1999; Sekiguchi *et al.*, 2004). The articulate processes of the S5 project downwards as the sacral cornu or horns. The lateral masses of the sacrum are formed by the fusion of transverse processes and costal elements. Auricular surfaces of the lateral masses articulate with the iliac bones on each side. The sacral canal is an opening in the center of the sacrum which is the continuation of the vertebral column through the sacrum. It contains the cauda equina, the filum terminale and the lower portion of the dura and arachnoid. (Fig 3.6)

Except for the sacral ala which is wider in the female, the sacrum is larger in the male. The sacrum articulates with L5 above, the coccyx below and with the iliac bone on either side via the sacroiliac joint. Sacralization of the fifth lumbar vertebra and lumbarization of the first sacral vertebra are common variations in this region.

In comparison to other primates that have only an intermittent upright gait, the human sacrum includes more bone segments and has a wider sacral ala, which offers a firm base of support for the trunk during erect posture (Cheng and Song, 2003). The old world monkeys have seven lumbar vertebrae but humans have only five because the last two 'lumbar' vertebrae in humans have been sacralized (Abitbol, 1987).

The coccyx in humans

The coccyx is a small triangular bone at the bottom of the vertebral column consisting of three to five (usually four) fused rudimentary vertebrae. In other vertebrates, the coccygeal vertebrae form the skeleton of the tail. Human embryos have a tail but this is absorbed in the fetal period. The embryonic tail sometimes persists as a cartilaginous structure containing up to five vertebrae (Bar-Maor *et al.*, 1980). The lower end of the sacrum articulates as a fibrocartilaginous joint or often fuses with the coccyx. The first coccygeal vertebra has short transverse processes that connect with the sacrum and two coccygeal cornua or horns that connect to the sacral cornua. The coccygeal vertebrae lack pedicles and spinous processes, but the first three have a primitive body and transverse processes. The second, third and fourth coccygeal vertebrae diminish successively in size and the last vertebra is a mere small nodule of bone. (Fig 3.7)

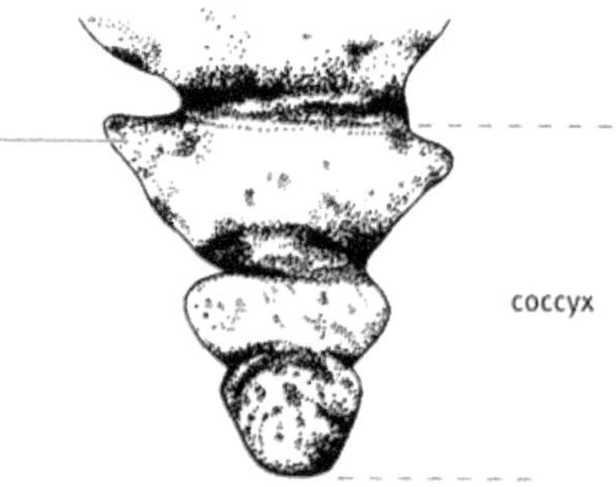

Figure 3.7 The human coccyx

The human coccyx is made up of the fusion of three or more vertebral elements. In mammals with long tails there are large numbers of coccygeal vertebrae (Adapted from Cramer and Darby, 2005, p.322)

Curvatures of the spine

The adult vertebral column in humans presents four curvatures in the sagittal plane, the cervical, thoracic, lumbar and sacral. In the fetus the vertebral column is uniformly concave, with the hollow facing ventrally. The concave thoracic and sacral curvatures are thus said to be primary, because they are present in the fetal period. The primary curvatures are due to the differences in height between the ventral and dorsal parts of the vertebral bodies. The secondary (convex forward) curvatures, cervical and lumbar, are formed as the child begins to support its head and when it begins to walk. (Fig 3.1) The cervical curve, convex forward, begins at the apex of the dens and ends at the middle of T2. The thoracic curve, concave forward, begins at the T2 and ends at T12, most prominent in T7. There is also a slight lateral curvature in the thoracic region, the convexity of which is usually directed toward the right side. The lumbar curve is convex forward and is more prominent in women. It has been shown that the upper lumbar curve, T12-L3, is minimally lordotic (convex forward) whereas the distal lumbar curve, L4-S1, accounts for approximately 65% of the lumbar lordosis (Damasceno *et al.*, 2006). The sacral curve involves the sacral and coccygeal vertebrae and is concave forward.

Joints of the vertebrae

All vertebrae from C2 to S1 are joined by cartilaginous joints (symphyses) between their bodies, synovial joints between their articular processes (zygapophyses), and fibrous joints between their laminae, transverse and spinous processes. There are also synovial joints between the uncinate processes (on the superior border of the vertebral body) of the C3-C7 vertebra and the inferior facet of the superior adjacent vertebrae, known as Luschka joints (Cave *et al.*, 1955; Payne and Spillane, 1957). The joints between the occipital condyles and the atlas, and between the atlas the axis, are highly specialized.

The vertebral joints are held together with spinal ligaments which maintain the stability by mechanical constraint and neuromuscular feedback. Ruffini and Pacinian corpuscles (end organs) and free nerve endings are observed in spinal ligaments (Yahia *et al.*, 1988). The midline spinal ligaments are found in all vertebrates, while lateral spinal ligaments, which are fundamental to the stability of erect posture; are only observed in bipedal (human) and pseudobipedal (avian) species (Jiang *et al.*, 1995a). Human supraspinal and interspinal ligaments are well innervated, innervation being equally distributed along the ligament, symmetrically distributed between left and right sides, and more densely distributed in the periphery. There are Pacinian corpuscles scattered randomly close to blood vessels, and Ruffini corpuscles are located in the periphery, close to the collagen bundles (Jiang *et al.*, 1995b).

Joints between vertebral bodies

The bodies of adjacent vertebrae are joined together with symphyseal joints. The vertebral bodies are united by anterior and posterior longitudinal ligaments and by fibrocartilaginous intervertebral discs between sheets of hyaline cartilage. The anterior longitudinal ligament lies on the ventral surface of the vertebral body through the entire length of the vertebral column; it also sweeps around the lateral aspect of the vertebral bodies (Hayashi *et al.*, 1977). The anterior longitudinal ligament is thicker and narrower in the thoracic than in cervical and lumbar regions. It is also relatively thicker and narrower on the vertebral bodies than in intervertebral symphyses. The posterior longitudinal ligament lies on the posterior surface of the vertebral bodies. It has two layers attached together loosely with fibrous tissue. The deep (anterior) layer sends fibers to the anulus fibrosus, and the superficial (posterior) layer lies directly on the dura mater (Hayashi *et al.*, 1977; Kubo *et al.*, 1994). The posterior longitudinal ligament is attached to the body of axis superiorly and the sacrum inferiorly; above it is continuous with the membrana tectoria. It is broader at the intervertebral discs and narrow at the vertebral bodies.

The intervertebral discs form fibrocartilaginous joints between the vertebral bodies. They give flexibility to the spine, facilitate a range of complex movements, and also serve as shock-absorbers. Each disc is composed of an inner nucleus pulposus surrounded by anulus fibrosus and a cartilaginous end-plate. The anulus fibrosus is a well-hydrated gel having proteoglycan collagen and cartilage cells. The anulus consists of 10-12 concentric rings of mainly type 1 collagen and also elastic fibers, with fibers in the outer lamellae continuing into the longitudinal ligaments and vertebral bodies. This arrangement allows the discs to facilitate movement and flexibility within what would be an otherwise rigid spine (Humzah and Soames, 1988; Roberts *et al.*, 2006). The fibers of the anulus are weakest posterolaterally, herniations of the disc often appear here (Humzah and Soames, 1988). The central gelatinous nucleus pulposus is a remnant of the notochord contained within the anulus fibrosus laterally and the cartilage end plates inferiorly and superiorly. Nucleus pulposus is made up of irregularly oriented type II collagen fibers. With ageing, the water content and the proteoglycan content of nucleus pulposus decreases (Humzah and Soames, 1988). In cervical and lumbar regions the intervertebral discs are thicker anteriorly, contributing to the anterior convexity; in the thoracic region they are nearly uniform, the anterior concavity being largely due to the vertebral bodies. Discs are thinnest in the upper thoracic region and thickest in the lumbar regions. Cartilage end-plate is a disc which covers the superior and inferior aspects of the vertebral body. It is about 1 mm thick at the periphery, and becomes thinner towards the center. The end-plates prevent the highly hydrated nucleus from bulging into the adjacent vertebral bone, while simultaneously absorbing hydrostatic pressure that results from mechanical loading of the spine (Broberg, 1983; Moore, 2000). At birth, the human disc has some vascular supply within both the cartilage end-plates and the anulus fibrosus, but in the adult except for their peripheries, supplied from adjacent blood vessels, discs are avascular and supplied by diffusion through capillaries in the bodies of adjacent vertebrae (Grunhagen *et al.*, 2006; Roberts *et al.*, 2006). Sensory fibers and their endings have been shown in the outer anulus fibrosus of human lumbar intervertebral disc (McCarthy *et al.*, 1991; Palmgren *et al.*, 1999).

Joints between vertebral arches

The vertebral arches are connected by synovial joints between the articular processes, and by a series of ligaments that connect the laminae, the transverse and spinous processes.

The joints between the articular processes are named facet (zygapophyseal) joints. They are planar synovial joints which allow the vertebral bodies to rotate with respect to each other. The outer layer of the fibrous capsule of the facet joint is constituted by dense regular parallel bundles of collagen fibers, and the inner layer by large elastic fibers (Yamashita *et al.*, 1996). The polar recesses, superior and inferior to lumbar facet joints are filled by fat pads from which fat-filled synovial folds project between the articular surfaces (Engel and Bogduk, 1982; Taylor and McCormick, 1991). There are also fibro-adipose meniscoids and connective tissue rims within the zygapopohyseal joints (Engel and Bogduk, 1982). There is an acellular surface lamina (lamina splendens) on the opposing cartilaginous surfaces of the zygapophyseal joints (Giles, 1992).

The supraspinous ligament connects the tips of spinous processes from C7 to the sacrum. Between the spine of C7 and the external occipital protuberance, it is expanded as the nuchal ligament. The supraspinous ligament is thicker and broader at lumbar levels, where it is ultimately blended with neighboring fascia, although only lightly attached to the spines of L3-L5. It terminates between L3 and L5 (Hukins *et al.*, 1990). Interspinous ligaments are thin membranous structures that connect the spinous processes of two adjacent vertebrae. They extend obliquely from the root of one spinous process to the apex of the next. They are narrow and elongated in the thoracic region, broader, thicker and quadrilateral at lumbar region, and poorly developed in the cervical region. In the middle part of the interspinous ligament, there is an S-shaped course of fibrous bundles which represent an available reserve of length in this poorly elastic ligament (Scapinelli *et al.*, 2006). In the dog and cat the interspinous ligaments are rudimentary, with very thin ventral and middle parts (Heylings, 1980). The supraspinous and interspinous ligaments are formed by both muscle tendons of longissimus thoracis and multifidus muscles and aponeuroses (thoracolumbar fascia) along the length of the thoracic and lumbar spine, with regional differences in their connective tissue architecture (Johnson and Zhang, 2002). The dorsal and ventral portions of the nuchal ligament are formed by the aponeurotic fibers of the trapezius, splenius capitis, rhomboideus minor and serratus posterior superior muscles (Johnson *et al.*, 2000). These ligaments limit spinal flexion (Johnson *et al.*, 1975). The intertransverse ligaments run between adjacent transverse processes and limit rotation and lateral flexion (Johnson *et al.*, 1975). These ligaments are weak, relativly strong in the lumbar region.

The ligamentum flavum is a ligament predominantly of elastic fibers which joins the laminae of adjacent vertebrae. The term 'flavum' is used to describe the yellow appearance of this ligament. It consists of a superficial and a deep layer. The superficial layer inserts onto the superior edge and posterosuperior surface of the coccygeal laminae. The deep layer inserts for a variable distance onto the anterosuperior surface of the caudal lamina (Olszewski *et al.*, 1996). The ligamentum flavum is the strongest of all spinal ligaments. It is thinnest in the midline and becomes thicker laterally. The thickest ligament is observed in the lumbar level, progressively weakening in ascending levels. Midline gaps are reported in these ligaments, most frequently in thoracic levels (Lirk *et al.*, 2005).

The craniovertebral joints

The atlas articulates with the occipital condyles above and the axis below. The atlanto-occipital joint on each side is between the occipital condyles and the superior facets on lateral masses of the atlas. (Fig 3.8) This is an ellipsoidal joint supported by the anterior and posterior atlanto-occipital membranes connecting the margins of the foramen magnum to the arches of atlas. The atlanto-axial joint is composed of two lateral joints between the superior facets of the axis and inferior facets of the atlas and a median joint between the dens of the axis and the anterior arch and transverse ligament of atlas. Among the ligaments that connect the atlas to the axis, the anterior and posterior atlanto-axial ligaments are merely the anterior and posterior longitudinal ligaments. The tectorial membrane that covers the capsule of the atlanto-axial joints passes from the back of the lateral mass of the atlas downwards and medially to the back of the body of the axis, and attaches to the margins of the foramen magnum superiorly. This is called the accessory atlanto-axial ligament (Arnold's ligament) (Tubbs *et al.*, 2004). This ligament is observed also in dogs, cats and primates (Hecker, 1922). The median atlanto-axial joint is a pivot joint between the dens and a ring formed by the anterior arch and transverse ligament of atlas. This joint has two synovial cavities which sometimes communicate (Ebraheim *et al.*, 1997d). The transverse ligament of atlas passes behind the dens between the tubercles on the inner side of the lateral masses (Tubbs *et al.*, 2002). It is a thick, strong band variable in length (mean 20.2 mm) (Dvorak *et al.*, 1988). A smaller longitudinal band extends from the dens to the ventral margin of foramen magnum. These transverse and longitudinal bands together constitute the cruciform ligament. The axis is connected to the occipital bone by membrana tectoria, which is the upward continuation of the posterior longitudinal

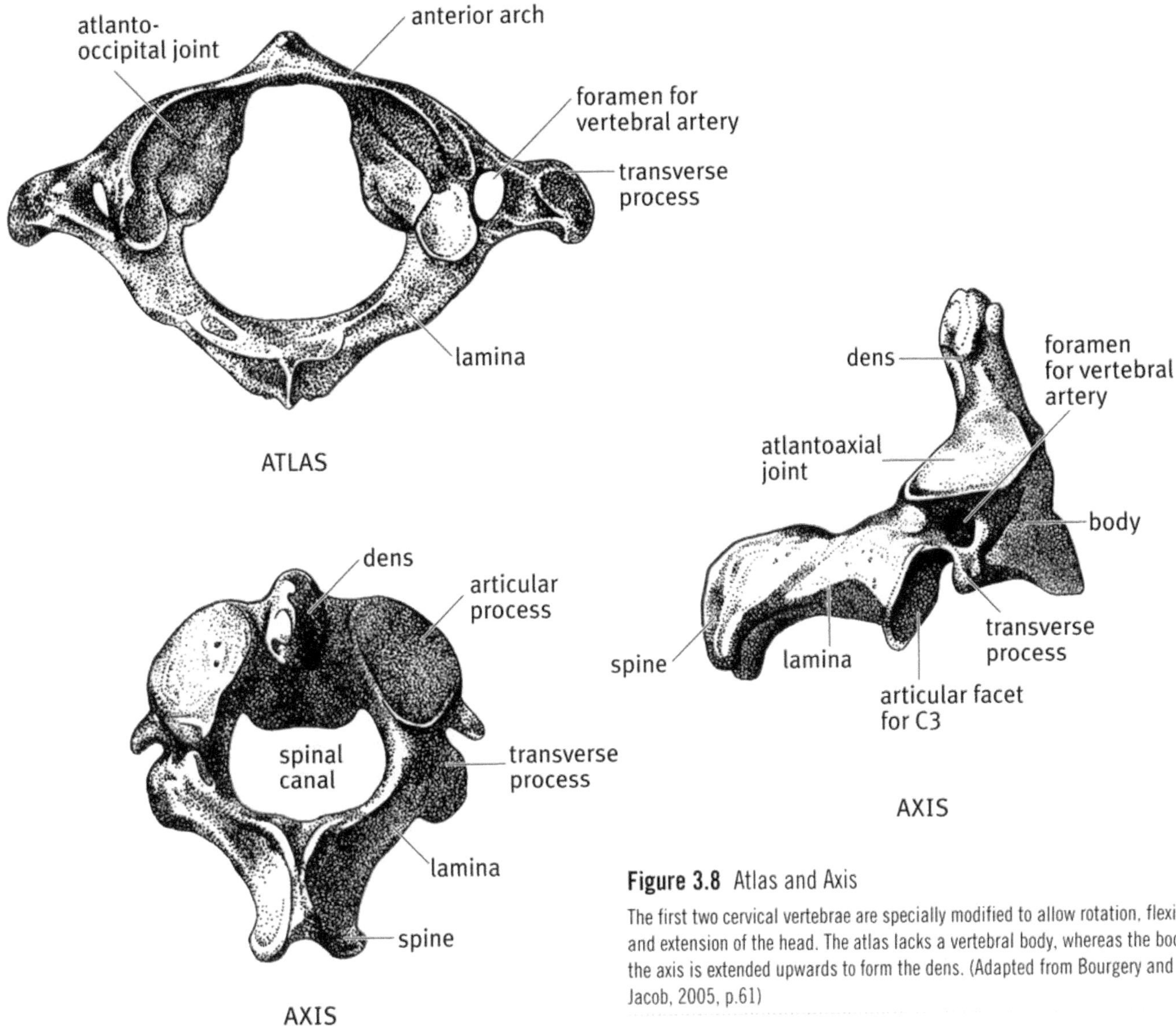

Figure 3.8 Atlas and Axis

The first two cervical vertebrae are specially modified to allow rotation, flexion, and extension of the head. The atlas lacks a vertebral body, whereas the body of the axis is extended upwards to form the dens. (Adapted from Bourgery and Jacob, 2005, p.61)

ligament, the alar ligaments between the lateral aspect of the apex of the dens and medial side of the occipital condyles, the apical ligament of the dens between the apex of the dens and the anterior margin of the foramen magnum, and the ligamentum nuchae. The alar ligaments (Mauchart's ligament) have a length of approximately 10 mm and may occasionally insert to the atlas (Panjabi *et al.*, 1991a). The apical ligament is 23.5 mm, the transverse ligament 21.9 mm, and the alar ligaments are 10.3 mm in length (Panjabi *et al.*, 1991a).

Lumbosacral joints

The lumbosacral joint is between the L5 and the auricular surfaces of sacrum. It has the same characteristics as the lumbar vertebral joints. The vertebral bodies are connected with a large intervertebral disc, and the zygapopohyseal joints have a wider interval than above. The joint is strengthened posteriorly by interosseous and dorsal sacro-iliac ligaments. The weight of the body is transmitted through the sacrum and ilium to the femur during standing and to the ischial tuberosities in sitting. The iliolumbar ligament originates from the transverse process of L5 and inserts to the iliac crest. It has a ventral and a dorsal (sacroiliac) part (Pool-Goudzwaard *et al.*, 2001). This ligament is important for restraining movement at the lumbosacral and the sacroiliac joints.

Sacrococcygeal joint

This joint is between the apex of the sacrum and the base of the coccyx, bound together with the anterior and posterior (superficial and deep), lateral sacrococcygeal ligaments, the intercornual ligaments and the intervertebral disc. The sacrococcygeal joint may be a symphysial, synovial or intermediate type with a disc containing a more or less extensive cleft bordered by annular fibers or synovial cells (Maigne *et al.*, 1992) The sacrococcygeal joint may be obliterated by fusion of the sacrum with the first coccygeal vertebra which is generally regarded to be a characteristic of old age, however Saluja (1988) has shown that this is not related exclusively to age and sex.

Intercoccygeal joints

These joints are symphyses between the coccygeal segments, held together by the anterior and posterior sacrococcygeal ligaments that completely cover the bony nodules. Sacrococcygeal joints are sometimes fused and the joint between the first and second coccygeal vertebrae is occasionally a synovial joint.

Sacro-iliac joints

The sacro-iliac joints are synovial articulations between the auricular surfaces of the sacrum and ilium on each side. The sacrum's rough irregular surface and its wedge shape form an interlocking mechanism with the ilium. Together with its strong ligaments, this makes it a very stable joint transmitting the body weight from the sacrum to the hip bones. The anterior, posterior and interosseous sacro-iliac ligaments firmly attach to the sacrum. Also, the sacrotuberous and sacrospinous ligaments provide flexibility to this joint (Vleeming *et al.*, 1990).

The intrinsic muscles of the vertebral column

The intrinsic muscles of the vertebral column are arranged in a superficial and a deep layer. The superficial layer of back muscles contains the splenius muscles and the erector spinae, and the deep layer is made up of the transversospinal muscle group. The deep muscles of the back have segmental organization and are all innervated by dorsal rami of spinal nerves.

The superficial layer of intrinsic vertebral back muscles in primates

The superficial layer of intrinsic muscles is largely covered by the lattisimus dorsi, trapezius, levator scapulae and rhomboid muscles. The splenius group has two parts; the splenius capitis arises from the spinous processes of C7-T4 and the nuchal ligament and inserts into the mastoid process and lateral third of the superior nuchal line; the splenius cervicis arises from the spinous processes of T3-T6 and inserts into the posterior tubercles of the transverse processes of C1-C3. The splenius muscles wrap around the sides and back of the neck; they extend and laterally bend the neck and head, and rotate head to the same side (Conley *et al.*, 1995). They are innervated by the medial branches of the dorsal rami of spinal nerves C2-C6.

The erector spinae (sacrospinalis) muscle group extends from the sacrum to the skull. It forms a prominent musculotendineous mass on each side of the vertebral column and lies within the thoracolumbar fascia. The erector spinae can be divided into three vertical columns: the laterally placed iliocostalis; the intermediate longissimus; and the medially placed spinalis muscle group. All three originate from the iliac crest, the dorsal aspect of the sacrum, the sacroiliac ligaments, and the sacral and lumbar spinous processes. The erector spinae is largely covered by the erector spinae aponeurosis, which is formed by the tendons of the longissimus thoracis medially and iliocostalis pars thoracis laterally (Bogduk, 1997). The erector spinae contains predominantly type I (slow twitch, fatigue resistant) muscle fibers (Mannion, 1999). The number and density of muscle fibers are highest in the lateral column and lowest in the medial column of the erector spinae. This indicates that the lateral column is predominantly involved in initiating finer movements or maintaining posture, whereas the medial column is involved in initiating gross movement (Amonoo-Kuofi, 1983). The erector spinae receives its blood supply from the segmental arteries – posterior intercostal, subcostal, and lumbar. The lateral and intermediate branches of the dorsal rami of the cervical, thoracic and lumbar spinal nerves innervate the erector spinae.

The iliocostalis muscle column has lumbar, thoracic and cervical parts. The iliocostalis lumborum arises from the sacrum and iliac crests and inserts to the angles of the lower six ribs. The iliocostalis thoracis fibers arise medial to the insertion points of iliocostalis lumborum on the angles of the upper six ribs. The iliocostalis cervicis originates below from the angles of third to sixth ribs medial to the insertion points of iliocostalis thoracis and inserts onto the posterior tubercles of C4 to C6. The muscle bundles of the iliocostalis, inserting into their respective ribs, run parallel to each other along a line which is defined by the caudal tip of the posterior superior iliac spine and the lateral border of the iliocostalis at the twelfth rib (De Foa *et al.*, 1989).

Medial to the iliocostalis lies the largest and longest part of the erector spinae, the longissimus muscle group. The longissimus and the iliocostalis are separated from each other posteriorly but usually not anteriorly in the lumbar region (Jonsson, 1970). The longissimus is divided into three parts according to the regions it traverses. The longissimus thoracis is inserted in to the transverse processes of the upper lumbar and all thoracic vertebrae and into the tubercles of the lower nine or ten ribs. The longissimus cervicis arises medial to the insertion points of longissimus thoracis on the upper six thoracic vertebrae, and in turn terminates on the transverse processes of C2 to C6. The longissimus capitis arises from the transverse processes of the upper four to five thoracic vertebrae and the articular processes of C4 to C7. The muscle separates from the longissimus cervicis as it ascends in the neck and attaches chiefly to the mastoid process of the temporal bone. The longissimus muscle contains a greater percentage of fatigue-resistant type I fibers (Sirca and Kostevc, 1985) and works as a segmental stabilizer (Mayoux-Benhamou *et al.*, 1994). The lumbar part of longissimus thoracis and the lumbar part of iliocostalis lumborum are separated by the lumbar intermuscular aponeurosis (Bogduk, 1980a; Daggfeldt *et al.*, 2000).

The most medial part of the erector spinae is the spinalis muscle group. It is a narrow column, arising from the spinous processes of T11 to L1-L2 and inserting to the T2-T3 to T10. Sometimes a spinalis cervicis muscle is present that arises from the spinous processes of the two upper thoracic and two lower cervical vertebrae and insert onto the spinous processes of the C2-C4. Spinalis capitis arises from the transverse processes of the upper six thoracic vertebrae and from the articular processes of the lower four cervical vertebrae, and inserts on the occipital bone. Spinalis muscles may be blended with the semispinalis muscles and occasionally are completely absent (Greiner *et al.*, 2004).

Acting bilaterally, all three columns of the erector spinae extend the head and part or all of the vertebral column. Acting unilaterally, the erector spinae laterally flexes the head or the vertebral column. In addition, the longissimus capitis rotates the head so that it is turned to the side of the active muscle.

The deep layer of vertebral back muscles in primates – the transversospinal group

The transversospinal muscles lie deep to the erector spinae; its component muscles run obliquely upwards and medially from the transverse processes to spinous processes. The transversospinal group is arranged in three layers: the semispinalis (thoracis, cervicis and capitis), the multifidus group and the rotatatores (with cervical, thoracic and lumbar components). They extend the vertebral column, rotate the trunk to the opposite side, and adjust small movements between individual vertebrae (Donisch and Basmajian, 1972). The transversospinal muscles are innervated by the medial branches of the dorsal rami of spinal nerves.

The most superficial component of the transversospinal muscle group is the semispinalis layer. Each semispinalis bundle traverses several segments. This muscle has three parts named for their insertions: the semispinalis thoracis, cervicis and capitis. The semispinalis thoracis runs from the transverse processes of the sixth to tenth thoracic vertebrae and terminates on the spinous processes of the upper four thoracic and lower two cervical vertebrae. Semispinalis cervicis runs from the transverse processes of the upper six thoracic vertebrae to the spinous processes of the second to fifth cervical vertebrae. The semispinalis capitis is a distinct structure overlying the semispinalis cervicis. It forms the largest muscle mass in the posterior aspect of the neck. It arises from the transverse processes of C4 to T6 vertebrae and inserts to the occipital bone on the medial half of the area between the superior and inferior nuchal lines. The semispinalis capitis, cervicis, and thoracis muscles extend the head and the cervical and thoracic regions of the vertebral column and rotate these regions towards the opposite side (Conley *et al.*, 1995).

The multifidus muscle group lies deep to the semispinalis muscle. It consists of several bundles that cover the vertebral laminae from the fourth sacral to the second cervical vertebra, filling the groove at the sides of the spinous processes from the sacrum to the axis. The muscle arises from the sacrum, the dorsal sacro-iliac ligaments, the mamillary processes of lumbar vertebra, from the transverse processes of the thoracic vertebra, and from the articular processes of the lower four cervical vertebra. Each bundle inserts into the spinous process of vertebra above its origin. Some bundles run as far as three to four vertebral segments, the deeper bundles one to two-three vertebral segments, and the deepest connect the adjacent vertebra. The lumbar part of the multifidus is superficial and thick, and the fibers are vertical. In the thoracic region the multifidus is deeper and thinner, and its fibers are more tendinous and oblique than in the lumbar region (Bojadsen *et al.*, 2000). The fibers of the multifidus run parallel to a line between the posterior superior iliac spine and the L1-L2 interspinous space (De Foa *et al.*, 1989). The multifidus is involved in the posterior sagittal rotation of the lumbar vertebrae and in opposing the flexion effect of abdominal muscles (Macintosh and Bogduk, 1987). It also contributes to extension, lateral flexion and rotation movements of the trunk, and serves to stabilize the vertebral column (Tracy *et al.*, 1989; Solomonow *et al.*, 1998). The multifidus consists of a greater percentage of fatigue-resistant type I fibers (Sirca and Kostevc, 1985), which is consistent with a more significant role in the stabilization of the spine.

The rotatores are the deepest muscles in this group located in the groove between the spinous and transverse processes. They are small and quadrilateral in form. Rotatores are found over the entire length of the vertebral column, arising from the transverse processes of a vertebra and inserting into the lamina of the vertebra directly above. Some fibers extend for two segments (rotatores longi). The rotatores are more developed in the thoracic region where rotation of the vertebral column is greatest. They rotate the transverse process of the vertebra above and also stabilize the joint. The cross-sectional area of the rotatores is small, but the muscle spindle density is five to seven times higher than multifidus and semispinalis which implies that the muscle has an important proprioceptive function (Nitz and Peck, 1986).

The interspinales and intertransversarii are small muscles that unite the spinous and transverse processes of adjacent vertebrae. The interspinales are short muscles and are well developed in the cervical and lumbar regions, but are weak or absent in the thoracic region. They extend the vertebral column when they act bilaterally and rotate towards the opposite side when acting unilaterally. The intertransversarii are slender slips of muscle that extend between the transverse processes. They are best developed in the cervical region. The intertransversarii laterally flex the upper adjacent vertebra, and acting bilaterally help to extend the vertebral column. It has been suggested that like the rotatores, the primary function of the interspinales and the intertransversarii might also be largely proprioceptive, their value not lying in the force they can exert but in the muscle spindles they contain (Bogduk, 1997).

Intrinsic vertebral muscles in non-primate mammals

In most mammals the vertebral column muscles are highly differentiated and are divided into three longitudinal systems, the iliocostalis, the longissimus, and the transversospinalis. A semispinalis muscle group lying medial to the spinalis is also present. These muscles attach to dorsal surfaces of the vertebrae and the ribs and extend in general from the sacrum to the back of the skull continuing into the tail. The short segmental muscles (multifidus, rotatores, intertransversarii) connect various parts of the vertebrae and ribs, mostly traversing only one body segment (Wake, 1979). Caudal muscles are greatly reduced in mammals where the tail is vestigial or absent. In humans, the coccygeal muscles of the pelvic floor are remnants of this group (Quiring, 1950; Wake, 1979).

In the rat, the transversospinalis, the longissimus, and the iliocostalis muscle groups can be identified. The components of the transversospinal muscles are coalesced in the lumbar region, but are distinct and specialized in the thoracic region, and form the tail muscle extensor caudae medialis caudally. The iliocostalis of the lumbar region is fused with one component of the lumbar longissimus to form the lateral longissimus muscle. Anteriorly, the iliocostalis muscle has thoracic and cervical parts. The lumbar longissimus has lateral and medial parts and a short-fiber component. Longissimus dorsi is the anterior continuation of the longissimus segment of the lateral longissimus. The medial longissimus is the continuation of the long, tendinous, tail base muscle. This is called longissimus caudae in the lumbar region. The caudal portion of this muscle is called extensor caudae lateralis. The lumbar transversospinalis muscles are innervated by the medial branches and the lateral longissimus by the lateral branches of the dorsal rami of the spinal nerves (Brink and Pfaff, 1980).

The intrinsic muscles of the vertebral column of the cat are homologous with those in the humans but are relatively thicker. From lateral to medial, the iliocostalis, longissimus and spinalis muscles are present. The spinalis muscle is the most medial of deep muscles and lies along the mid-dorsal line. The longissimus muscle lies lateral to the spinalis dorsi and can be separated into three distinct muscle bundles in the lumbar region. The iliocostalis muscle is located over the most lateral part of the ribs and is covered by the serratus dorsalis muscle. The multifidus is best observed in the sacral region adjacent to the mid-dorsal line. In the thoracic region it is covered by the spinalis dorsi muscles. In the neck, the multifidus continues as the semispinalis cervicis muscle under the splenius (Wischnitzer, 1967). In the cat lumbar region the multifidus, intertransversarii mediales, lumbococcygeus, iliocostalis lumborum and longissimus lumborum muscles are arranged in five parallel polysegmental columns, covered by the dorsal layer of thoracolumbar fascia and the erector spinae aponeurosis. The multifidus muscle connects the mamillary and spinous processes, and the intertransversarii mediales connects the accessory and mamillary processes. The lumbococcygeus muscle arises in the lumbar region from the accessory processes but inserts in the tail. The iliocostalis and longissimus arise from the ilium and from the surfaces of an intermuscular septum (Bogduk, 1980b).

The spinal meninges

The meninges enclosing the spinal cord consist of three layers: the outermost dura mater (pachymenix), the intermediate arachnioid mater and the innermost pia mater. The pia mater and the arachnoid are frequently considered to represent a single membrane, the leptomeninx. An intermediate fenestrated leptomeningeal layer attaching to the inner aspect of the arachnoid has also been described (Nicholas and Weller, 1988). These membranes, and the cerebrospinal fluid in the subarachnoid space, provide protection and nourishment of the spinal cord and spinal nerve roots. In non-mammalian vertebrates, there is an outer dura mater and an inner secondary meninx which correspond to the arachnoid and pia mater of mammals (Wake, 1979). However, the structure of the meninges is basically the same in vertebrates with only minor modifications (Oda and Nakanishi, 1984; Angelov and Vasilev, 1988; Orlin *et al.*, 1991).

There are three spaces within the sheaths of the spinal cord: the epidural, subdural, and subarachnoid spaces. The epidural (extradural) space is located external to the sac of the spinal dura mater and contains loose connective tissue, epidural fat, lymphatics, spinal nerve roots and the internal vertebral venous plexous. This space is bounded by the posterior longitudinal ligament anteriorly, the ligamenta flava and the periosteum of the laminae posteriorly, the pedicles of the vertebrae, and the intervertebral foramina with their contents laterally. The epidural space is closed at the foramen magnum where the spinal dura attaches to the edosteal dura of the cranium and at the sacral hiatus by the sacrcoccygeal ligament. The lumbar epidural space is segmented and discontinuous because the dura is in many areas directly applied to the wall of the vertebral column (Harrison, 1999). The space anterior to the dura is filled with veins and is isolated from the rest of the epidural space by a membranous lateral extension of the posterior longitudinal ligament (Hogan, 1991). Adipose tissue separates the ligamentum flavum and the dura posteriorly (Parkin and Harrison, 1985). In the thoracic and cervical regions, the epidural space contains a large basivertebral vein and a discontinuous ligamentum flavum is present (Hogan, 1996). In classical textbooks of anatomy, the subdural space is described as a potential space between the dura and the arachnoid – the two being apposed due to the pressure of the cerebrospinal fluid. Recently, it has been suggested that the subdural space is not a normal anatomical space but the result of tissue damage where the cell membranes of the dural border cell layer are damaged or the cell junctions are separated (Haines, 1991; Schachenmayr and Friede, 1978; Vandenabeele *et al.*, 1996). It is proposed that, instead of the dural space, there is a compartment between the neurothelial cells named the dural-arachnoid interface (Reina *et al.*, 2002).

The subarachnoid space is a substantial space between the arachnoid and the pia mater, it contains the cerebrospinal fluid. The subarachnoid space also contains extracellular collagen and loosely organized fibroblasts, named the arachnoid trabeculae, that extend from the arachnoid to the pia mater (Haines *et al.*, 1993). Loosely arranged leptomeningeal cells that traverse the subarachnoid space here form the intermediate leptomeningeal layer which resembles the arachnoid trabeculae of the cranial subarachnoid space (Nicholas and Weller, 1988). The subarachnoid space extends caudally to the level of the second sacral vertebra. There is substantial drainage of the cerebrospinal fluid from the subarachnoid space at the spinal level (Bozanovic-Sosic *et al.*, 2001). This drainage may be via the arachnoid proliferations similar to those in the cranium, at the sites of emerging spinal roots by Kido *et al.*, (1976), or via the lymphatic vessels (Boulton *et al.*, 1996).

Dura mater

The dura mater is a tough, dense membrane surrounding the spinal cord. It joins the cranial dura at the foramen magnum and caudally forms the coccygeal ligament that attaches to the coccygeal vertebrae. Laterally, the dura mater extends over the nerve roots and blends with the connective tissue of the epineurium in the dural cuff region. In this region, spinal arteries, veins and lymphatics pierce the dura as they travel toward the spinal cord through the subarachnoid space.

In the past, the periosteum of the vertebral column has been described as representing an outer layer of the dura, but although that there are two layers of cranial dura, only the inner meningeal layer continues beyond the foramen magnum inferiorly (Newell, 1999).

The dura is composed of fibroblasts and dense, interlacing bundles of collagen and elastic fibers. The elastic fibers provide considerable flexibility when subject to streching during movements, and the collagen provides tensile strength and protects the spinal cord. The thin outer layer (outer dural border layer) is composed of extracellular collagen with few elastic fibers. The middle layer is richly vascularized and basically made of extracellular collagen. The inner layer and extracellular spaces, containing an amorphous material (dural border cell layer), is less substantial, being composed of multiple interdigitating fibroblasts, and lacking extracellular collagen. This layer is closely applied to the arachnoid with the dura-arachnoid interface layer (Schachenmayr and Friede, 1978; Haines, 1991; Vandenabeele *et al.*, 1996). The fibers of the dura mater are not arranged in a parallel direction and do not run in a longitudinal direction, but are oriented in concentric laminae around the spinal cord (Patin *et al.*, 1993; Reina *et al.*, 1997). The dura consists of 78-82 laminae of this kind, each layer including 8-12 lamellae (Reina *et al.*, 1997).

Meningovertebral ligaments anchor the dura to the vertebral ligaments. The dura is attached to the posterior longitudinal ligaments by anterior dural (Hofmann) ligaments that are located in segments C7 to L7. The anterior dural ligaments are mostly located in a single segmental level, occasionally crossing several segments (Wadhwani *et al.*, 2004). In the cervical region, the dura is attached to the ligamenta flavum with the posterior cervical epidural ligaments (dorsal meningovertebral ligaments) (Shinomiya *et al.*, 1995).

The dura is separated from the vertebral column by the epidural (extradural) space. There is a soft tissue structure lying between the dura and the vertebral column named the peridural membrane that is located at any point where there is no periosteum, and is a homologue of the periosteum (Wiltse, 2000).

Arachnoid mater

The arachnoid mater, named for its spiderweb-like appearance, is a thin, transparent membrane surrounding the spinal cord like a loosely fitting sac. Continuous with the cerebral arachnoid above, it passes through the foramen magnum and descends caudally to the S2 vertebral level. It is closely applied to the inner surface of the dura mater and is separated from the pia mater by the subarachnoid space containing the cerebrospinal fluid.

The arachnoid is composed of collagen and elastic fibers. It has a variable thickness, in places being formed by several cell layers. Its outer (dural) aspect is smoother than the inner (pial) aspect from which trabeculae emerge to bridge the subarachnoid space (Nicholas and Weller, 1988). The outer layer of the arachnoid is the arachnoid barrier cell layer that consists of closely packed electron-lucent cells. These flat cells are impermeable to the cerebrospinal fluid due to tight intercellular junctions. The collagenous innermost portion of the arachnoid is the arachnoid reticular cell layer that consists of loosely arranged cells anchored by desmosomes to the inner aspect of the arachnoid barrier cell layer. The interwoven arachnoid trabecular cells within this layer possess numerous mitochondria. The arachnoid reticular cell layer is separated from the arachnoid barrier cell layer by a distinct continuous basal lamina. An additional layer of flattened branching cells is present along the inner surface of the arachnoid reticular cell layer named the arachnoid border cell layer (Haines, 1991; Vandenabeele *et al.*, 1996; Weller, 2005).

Arachnoid villi and granulations resembling those in the cranium have been shown at the sites of emerging spinal nerve roots in dogs and sheep (Gomez *et al.*, 1974), monkeys (Welch and Pollay, 1963), and humans (Kido *et al.*, 1976). These may play a role in the drainage of the cerebrospinal fluid.

Intermediate leptomeningeal layer

An intermediate fenestrated leptomeningeal layer, also called the trabecular leptomeninx, is observed attached to the inner aspect of the arachnoid mater. It consists of the loosely arranged leptomeningeal cells that traverse the subarachnoid space and is structurally similar to the trabeculae that cross the cranial subarachnoid space. It is composed of an outer layer adherent to the inner aspect of the arachnoid, and an inner layer lying over the arteries and nerve roots. The intermediate layer is substantial in the dorsal and ventral regions, and forms a series of ligaments connecting the arachnoid membrane to the spinal cord. These ligaments are sparse in the upper cervical region but extensive from the lower cervical to lumbar segments, beyond which they progressively diminish at the filum terminale. Throughout the cauda equina, these fibers connect the roots and support the blood vessels. Laterally, the intermediate layer becomes highly fenestrated as it surrounds the nerves and blood vessels and is not present anteriorly (Nicholas and Weller, 1988; Parkinson, 1991). This layer has also been demonstrated in rats and other laboratory animals (Haller and Low, 1971; Kaar and Fraher, 1986).

Pia mater

The pia mater is the innermost layer of the meninges. The pial cellular layer firmly adheres to the surface of the spinal cord and passes into the anterior median fissure, conus medullaris, filum terminale and nerve roots. This layer contains flat overlapping pial cells with an oval-shaped nucleus and a small nucleolus. Pial cells form a continuous layer joined by desmosomes and gap junctions, making this layer impermeable to the cerebrospinal fluid (Alcolado *et al.*, 1988). Subpial tissue separates the pial cellular layer from neuroglial cells and contains mainly collagen fibers, a few elastic fibers, amorphous intracellular substances, small vessels, fibroblasts, and macrophages (Reina *et al.*, 2004). In the outer layer of the subpial tissue, there is a distinct longitudinal arrangement of collagen fibers carrying the larger branches of spinal vessels and in the inner layer mainly circularly arranged fibers containing capillaries and venules (Fricke *et al.*, 2001). The pial cellular layer is three to six pial cells thick (8-15 μm overall) at the level of the spinal cord and two to five pial cells thick (3-8 μm overall) on the nerve roots. The subpial tissue has a thickness of 130-200 μm at the thoracic and lumbar levels and decreases at other levels to 80-90 μm (conus medullaris), 40-50 μm (initial portion of cauda equina), and 10-12 μm (nerve roots). In humans, there are fenestrations within the pial cellular layer at the thoracic-lumbar junction, conus medullaris, and on spinal nerve roots (Reina *et al.*, 2004). These fenestrations have also been described in the rat (Morse and Low, 1972), cat (Reina *et al.*, 1998) and dog (Cloyd and Low, 1974). Their functional significance has not been determined.

The pia and arachnoid membranes continue with the spinal nerve roots as they leave the spinal cord and exit through the intervertebral foramina, where they blend with the perineurium of the spinal nerves. Midway between the attachments of dorsal and ventral roots, the pia mater gives a series of intermittent lateral extensions known as the denticulate ligaments. These form 21 tooth-like processes on each side that attach the spinal cord to the dura through the arachnoid membrane. The denticulate ligaments are evenly distributed along the length of the spinal cord. The first denticulate ligament attaches to the rim of the foramen magnum between the vertebral artery anteriorly and the hypoglossal nerve posteriorly. The last denticulate ligament ends at the level of the conus medullaris between the roots of the last thoracic and first lumbar nerves (Tubbs *et al.*, 2001). Each denticulate ligament consists of a collagenous core, which is continuous with the subpial connective tissue and is attached at intervals to the dura. The cells of the pia mater and arachnoid cover the surface of the dentate ligaments (Nicholas and Weller, 1988). Denticulate ligaments anchor subpial collagen to the dura mater and stabilize the spinal cord (Weller, 2005).

Innervation of the meninges

Whereas parts of the cranial dura are well innervated, the spinal dura is not richly innervated and thus plays a limited role in the pathogenesis of pain. In the human fetus, it has been shown with acetyl cholinesterase staining that the ventral spinal dura contains a dense, longitudinal nerve plexus which receives its contributions from the sinuvertebral nerves, the nerve plexus of the posterior longitudinal ligament, and the nerve plexus of radicular branches of segmental arteries. The ventral dural nerves may extend up to eight segments, with a great amount of overlap between adjacent nerves. The dorsal dura is poorly innervated, with dural nerves much smaller in number and not forming an evident plexus. The majority of these fibers are autonomic (Groen *et al.*, 1988). Innervation of the dura surrounding spinal nerves is richer in the lumbar and cervical regions, most abundant near the surface of the dura, and on the ventral and lateral aspects of the spinal nerve root dura (Cuatico *et al.*, 1988). In the rat there is a prominent autonomic nerve plexus containing neuropeptide Y. In addition, there is a tyrosine hydroxylase-positive fiber network in the rat spinal dura (Ahmed *et al.*, 1993). Somatosensory (CGRP- and SP-positive) fibers along dural root sleeves in the cervical and lumbar regions have also been demonstrated (Kumar *et al.*, 1996).

There is limited information on the innervation of the spinal leptomeninges. The spinal pia mater receives a rich innervation of small sensory axons via the ventral roots. Afferents containing SP and/or CGRP and sympathetic axons have been shown here, these might mediate pain, and might also have vasomotor, tissue-regulatory and/or mechanoreceptive functions (Hildebrand *et al.*, 1997). Leptomeningeal cells have also been found immunoreactive for trk and trkB genes that encode membrane spanning proteins constituting high affinity binding sites for neurotrophic factors (Risling *et al.*, 1994). The ventral pia mater exhibits a wide nerve plexus of myelinated and unmyelinated fibers in a cranio-caudal direction, some of these fibers are closely related to vessel walls (Clark, 1931). The radicular and the extraparenchymal arteries of spinal cord supplying the leptomeninges are invested with with adrenergic, neuropeptide Y, vasoactive intestinal peptide, substance P, and CGRP containing nerve fibers (Hegde *et al.*, 1988).

References

Abitbol MM (1987) Evolution of the sacrum in hominoids. Am J Phys Anthropol 74, 65-81.

Ahmed M, Bjurholm A, Kreicbergs A, Schultzberg M (1993) Neuropeptide Y, tyrosine hydroxylase, and vasoactive intestinal polypeptide-immunoreactive fibers in the vertebral bodies, discs, dura mater, and spinal ligaments of the rat lumbar spine. Spine 18, 268-273.

Alcolado R, Weller RO, Parrish EP, Garrod D (1988) The cranial arachnoid and pia mater in man: anatomical and ultrastructural observations. Neuropathol Appl Neurobiol 14, 1-17.

Amonoo-Kuofi HS (1983) The density of muscle spindles in the medial, intermediate and lateral columns of human intrinsic postvertebral muscles. J Anat 136, 509-519.

Amonoo-Kuofi HS, el-Badawi MG, Fatani JA (1988) Ligaments associated with lumbar intervertebral foramina. 1. L1 to L4. J Anat 156, 177-183.

An HS, Wise JJ, Xu R (1999) Anatomy of the cervicothoracic junction: a study of cadaveric dissection, cryomicrotomy, and magnetic resonance imaging. J Spinal Disord 12, 519-525.

Angelov DN, Vasilev VA (1988) Electron microscopic study of cat meninges. Anat Anz 166, 1-8.

Antonacci MD, Mody DR, Heggeness MH (1998) Innervation of the human vertebral body: a histologic study. J Spinal Disord 11, 526-531.

Bakkum BW, Mestan M (1994) The effects of transforaminal ligaments on the sizes of T11 to L5 human intervertebral foramina. J Manipulative Physiol Ther 17, 517-522.

Bar-Maor JA, Kesner KM, Kaftori JK (1980) Human tails. J Bone Joint Surg 62-B, 508-510.

Berry JL, Moran JM, Berg WS, Steffee AD (1987) A morphometric study of human lumbar and selected thoracic vertebrae. Spine 12, 362-367.

Bogduk N (1980a) A reappraisal of the anatomy of the human lumbar erector spinae. J Anat 131, 525-540.

Bogduk N (1980b) The dorsal lumbar muscles of the cat. Anat Anz 148, 55-67.

Bogduk N (1997) *Clinical Anatomy of the Lumbar Spine and Sacrum.* 3rd edn. Churchill Livingstone, London.

Bojadsen TW, Silva ES, Rodrigues AJ, Amadio AC (2000) Comparative study of Mm. Multifidi in lumbar and thoracic spine. J Electromyogr Kinesiol 10, 143-149.

Boulton M, Young A, Hay J, Armstrong D, Flessner M, Schwartz M, Johnston M (1996) Drainage of CSF through lymphatic pathways and arachnoid villi in sheep: measurement of 125I-albumin clearance. Neuropathol Appl Neurobiol 22, 325-333.

Bourgery JM and Jacob NH (2005) *Atlas of human anatomy and surgery: The complete coloured plates of 1831-1854.* Le Minor JM and Sick H (eds). Taschen, Koln.

Bozanovic-Sosic R, Mollanji R, Johnston MG (2001) Spinal and cranial contributions to total cerebrospinal fluid transport. Am J Physiol Regul Integr Comp Physiol 281, R909-R916.

Brink EE, Pfaff DW (1980) Vertebral muscles of the back and tail of the albino rat (Rattus norvegicus albinus). Brain Behav Evol 17, 1-47.

Broberg KB (1983) On the mechanical behaviour of the intervertebral disc. Spine 8, 151-165.

Burke AC, Nelson CE, Morgan BA, Tabin C (1995) Hox genes and the evolution of vertebrate axial morphology. Development 121, 333-346.

Cave AJE, Griffiths JD, Whiteley MM (1955) Osteo-arthritis deformans of Luschka joints. Lancet 248, 176-179.

Chaynes P, Sol JC, Vaysse P, Becue J, Lagarrigue J (2001) Vertebral pedicle anatomy in relation to pedicle screw fixation: a cadaver study. Surg Radiol Anat 23, 85-90.

Cheng JS, Song JK (2003) Anatomy of the sacrum. Neurosurg Focus 15, E3.

Clark SM (1931) Innervation of the pia mater of the spinal cord and medulla. J Comp Neurol 53, 129-141.

Cloyd MW, Low FN (1974) Scanning electron microscopy of the subarachnoid space in dog. I. Spinal cord levels. J Comp Neurol 53, 325-368.

Conley MS, Meyer RA, Bloomberg JJ, Feeback DL, Dudley GA (1995) Noninvasive analysis of human neck muscle function. Spine 20, 2505-2512.

Cramer GD and Darby, SA (2005) *Basic and clinical anatomy of the spine, spinal cord, and ANS* (2nd ed.). Elsevier Mosby, St Louis.

Cramer GD, Skogsbergh DR, Bakkum BW, Winterstein JF, Yu S, Tuck NR Jr (2002) Evaluation of transforaminal ligaments by magnetic resonance imaging. J Manipulative Physiol Ther 25, 199-208.

Cuatico W, Parker JC Jr, Pappert E, Pilsl S (1988) An anatomical and clinical investigation of spinal meningeal nerves. Acta Neurochir 90, 139-143.

Daggfeldt K, Huang QM, Thorstensson A (2000) The visible human anatomy of the lumbar erector spinae. Spine 25, 2719-2725.

Damasceno LHF, Catarin SRG, Campos AD, Defino HLA (2006) Lumbar lordosis: a study of angle values and of vertebral bodies and intervertebral discs role. Acta Ortho Bras 14, 193-198.

De Foa JL, Forrest W, Biedermann HJ (1989) Muscle fibre direction of longissimus, iliocostalis and multifidus: landmark-derived reference lines. J Anat 163, 243-247.

Doherty BJ, Heggeness MH (1994) The quantitative anatomy of the atlas. Spine 19, 2497-2500.

Dong Y, Xia Hong M, Jianyi L, Yuan Lin M (2003) Quantitative anatomy of the lateral mass of the atlas. Spine 28, 860-863.

Donisch EW, Basmajian JV (1972) Electromyography of deep back muscles in man. Am J Anat 133, 25-36.

Dvorak J, Schneider E, Saldinger P, Rahn B (1988) Biomechanics of the craniocervical region: the alar and transverse ligaments. J Orthop Res 6, 452-461.

Ebraheim NA, Reader D, Xu R, Yeasting RA (1997a) Location of the vertebral artery foramen on the anterior aspect of the lower cervical spine by computed tomography. J Spinal Disord 10, 304-307.

Ebraheim NA, Xu R, Ahmad M, Yeasting RA (1997b) Projection of the thoracic pedicle and its morphometric analysis. Spine 22, 233-238.

Ebraheim NA, Xu R, Knight T, Yeasting RA (1997c) Morphometric evaluation of lower cervical pedicle and its projection. Spine 22, 1-6.

Ebraheim NA, Yang H, Lu J, Biyani A, Yeasting RA (1997d) Cartilage and synovium of the human atlanto-odontoid joint. An anatomic and histological study. Acta Anat 159, 48-56.

Engel R, Bogduk N (1982) The menisci of the lumbar zygapophysial joints. J Anat 135, 795-809.

Ericksen MF (1978a) Aging in the lumbar spine. II. L1 and L2. Am J Phys Anthropol 48, 241-245.

Ericksen MF (1978b) Aging in the lumber spine. III. L5. Am J Phys Anthropol 48, 247-250.

Fricke B, Andres KH, Von During M (2001) Nerve fibers innervating the cranial and spinal meninges: morphology of nerve fiber terminals and their structural integration. Microsc Res Tech 53, 96-105.

Gaunt SJ (1994) Conservation in the Hox code during morphological evolution. Int J Dev Biol 38, 549-552.

Gilchrist RV, Slipman CW, Bhagia SM (2002) Anatomy of the intervertebral foramen. Pain Physician 5, 372-378.

Giles LG (1992) The surface lamina of the articular cartilage of human zygapophyseal joints. Anat Rec 233, 350-356.

Gomez DG, Chambers AA, Di Benedetto AT, Potts DG (1974) The spinal cerebrospinal fluid absorptive pathways. Neuroradiology 8, 61-66.

Greene EC (1968) *Anatomy of the Rat.* Hafner, New York.

Greiner TM, Bedford ME, Walker RA (2004) Variability in the human m. spinalis capitis and cervicis: frequencies and definitions. Ann Anat 186, 185-191.

Groen GJ, Baljet B, Drukker J (1988) The innervation of the spinal dura mater: anatomy and clinical implications. Acta Neurochir 92, 39-46.

Grunhagen T, Wilde G, Soukane DM, Shirazi-Adl SA, Urban JP (2006) Nutrient supply and intervertebral disc metabolism. J Bone Joint Surg Am 88:S2, 30-35.

Haines DE (1991) On the question of a subdural space. Anat Rec 230, 3-21.

Haines DE, Harkey HL, al-Mefty o (1993) The "subdural" space; a new look at an outdated concept. Neurosurgey 32, 111-120.

Haller FR, Low FN (1971) The fine structure of the peripheral nerve root sheath in the subarachnoid space in the rat and other laboratory animals. Am J Anat 131, 1-19.

Harrison GR (1999) Topographical anatomy of the lumbar epidural region: an in vivo study using computerized axial tomography. Br J Anaesth 83, 229-234.

Hasan M, Shukla S, Siddiqui MS, Singh D (2001) Posterolateral tunnels and ponticuli in human atlas vertebrae. J Anat 199, 339-343.

Hayashi K, Yabuki T, Kurokawa T, Seki H, Hogaki M, Minoura S (1977) The anterior and the posterior longitudinal ligaments of the lower cervical spine. J Anat 124, 633-636.

Hecker P (1922) Appareil ligamenteux occipito-atloïdo-axoïdien: Étude d'anatomie comparée. Arch Anat Histol Embryol 1, 417-433.

Hegde SA, Hara H, Kobayashi S (1988) Innervation of radicular and extraparenchymal arteries of spinal cord. Histochemical and immunohistochemical study in primate. Histochemistry 89, 415-420.

Heylings DJ (1980) Supraspinous and interspinous ligaments in dog, cat and baboon. J Anat 130, 223-228.

Hildebrand C, Karlsson M, Risling M (1997) Ganglionic axons in motor roots and pia mater. Prog Neurobiol 51, 89-128.

Hogan QH (1991) Lumbar epidural anatomy. A new look by cryomicrotome section. Anesthesiology 75, 767-775.

Hogan QH (1996) Epidural anatomy examined by cryomicrotome section. Influence of age, vertebral level, and disease. Reg Anesth 21, 395-406.

Hoyland JA, Freemont AJ, Jayson MI (1989) Intervertebral foramen venous obstruction. A cause of periradicular fibrosis? Spine 14, 558-568.

Hukins DW, Kirby MC, Sikoryn TA, Aspden RM, Cox AJ (1990) Comparison of structure, mechanical properties, and functions of lumbar spinal ligaments. Spine 15, 787-795.

Humzah MD, Soames RW (1988) Human intervertebral disc: structure and function. Anat Rec 220, 337-356.

Janjua MZ, Muhammad F (1989) Measurements of the normal adult lumbar spinal canal. J Pak Med Assoc 39, 264-268.

Jiang H, Moreau M, Raso VJ, Russell G, Bagnall K (1995a) A comparison of spinal ligaments – differences between bipeds and quadrupeds. J Anat 187, 85-91.

Jiang H, Russell G, Raso VJ, Moreau MJ, Hill DL, Bagnall KM. (1995b) The nature and distribution of the innervation of human supraspinal and interspinal ligaments. Spine 20, 869-876.

Johnson GM, Zhang M (2002) Regional differences within the human supraspinous and interspinous ligaments: a sheet plastination study. Eur Spine J 11, 382-388.

Johnson RM, Crelin ES, White AA 3rd, Panjabi MM, Southwick WO (1975) Some new observations on the functional anatomy of the lower cervical spine. Clin Orthop Relat Res 111, 192-200.

Johnson GM, Zhang M, Jones DG (2000) The fine connective tissue architecture of the human ligamentum nuchae. Spine 25, 5-9.

Jonsson B (1970) Topography of the lumbar part of the erector spinae muscle. Anat Embryol 130, 177-191.

Jovanovic MS (1990) A comparative study of the foramen transversarium of the sixth and seventh cervical vertebrae. Surg Radiol Anat 12, 167-172.

Kaar GF, Fraher JP (1986) The sheaths surrounding the attachments of rat lumbar ventral roots to the spinal cord: a light and electron microscopical study. J Anat 148, 137-146.

Karaikovic EE, Daubs MD, Madsen RW, Gaines RW Jr (1997) Morphologic characteristics of human cervical pedicles. Spine 22, 493-500.

Kawashima M, Tanriover N, Rhoton AL Jr, Matsushima T (2003) The transverse process, intertransverse space, and vertebral artery in anterior approaches to the lower cervical spine. J Neurosurg 98, 188-194.

Kayalioglu G, Erturk M, Varol T, Cezayirli E (2007) Morphometry of the cervical vertebral pedicles as a guide for transpedicular screw fixation. Neurol Med Chir 47, 102-107.

Kido DK, Gomez DG, Pavese AM Jr, Potts DG (1976) Human spinal arachnoid villi and granulations. Neuroradiology 11, 221-228.

Kubo Y, Waga S, Kojima T, Matsubara T, Kuga Y, Nakagawa Y (1994) Microsurgical anatomy of the lower cervical spine and cord. Neurosurgery 34, 895-890.

Kumar R, Berger RJ, Dunsker SB, Keller JT (1996) Innervation of the spinal dura. Myth or reality? Spine 21, 18-26.

Lirk P, Colvin J, Steger B, Colvin HP, Keller C, Rieder J, Kolbitsch C, Moriggl B (2005) Incidence of lower thoracic ligamentum flavum midline gaps. Br J Anaesth 94, 852-855.

Lu J, Ebraheim NA, Yang H, Heck BE (2000) Anatomic evaluation of the first three sacral vertebrae and dorsal screw placement. Am J Orthop 29, 376-379.

Macintosh JE, Bogduk N (1987) The morphology of the lumbar erector spinae. Spine 12, 658-668.

MacLaughlin SM, Oldale KN (1992) Vertebral body diameters and sex prediction. Ann Hum Biol 19, 285-292.

Maigne JY, Molinie V, Fautrel B (1992) Anatomie des disques coccygiens. Rev Med Ortho 28, 34-35.

Mannion AF (1999) Fibre type characteristics and function of the human paraspinal muscles: normal values and changes in association with low back pain. J Electromyogr Kinesiol 9, 363-377.

Mayoux-Benhamou MA, Revel M, Vallee C, Roudier R, Barbet JP, Bargy F (1994) Longus colli has a postural function on cervical curvature. Surg Radiol Anat 16, 367-371.

McCarthy PW, Carruthers B, Martin D, Petts P (1991) Immunohistochemical demonstration of sensory nerve fibers and endings in lumbar intervertebral discs of the rat. Spine 16, 653-655.

Min JH, Kang SH, Lee JB, Cho TH, Suh JG (2005) Anatomic analysis of the trans-foraminal ligament in the lumbar intervertebral foramen. Neurosurgery 57, 37-41.

Moore KL and Dalley, AF (2006) *Clinically Oriented Anatomy* (5th ed.). Lippincott Williams and Wilkins, Philadelphia.

Moore RJ (2000) The vertebral end-plate: what do we know? Eur Spine J 9, 92-96.

Morse DE, Low FN (1972) The fine structure of the pia mater of the rat. Am J Anat 133, 349-368.

Narita Y, Kuratani S (2005) Evolution of the vertebral formulae in mammals: a perspective on developmental constraints. J Exp Zool B Mol Dev Evol 304, 91-106.

Newell RL (1999) The spinal epidural space. Clin Anat 12, 375-379.

Nicholas DS, Weller RO (1988) The fine anatomy of the human spinal meninges. A light and scanning electron microscopy study. J Neurosurg 69, 276-282.

Nissan M, Gilad I (1986) Dimensions of human lumbar vertebrae in the sagittal plane. J Biomech 19, 753-758.

Nitz AJ, Peck D (1986) Comparison of muscle spindle concentrations in large and small human epaxial muscles acting in parallel combinations. Am Surg 52, 273-277.

Nojiri K, Matsumoto M, Chiba K, Toyama Y (2005) Morphometric analysis of the thoracic and lumbar spine in Japanese on the use of pedicle screws. Surg Radiol Anat 27, 123-128.

Oda Y, Nakanishi I (1984) Ultrastructure of the mouse leptomeninx. J Comp Neurol 225, 448-457.

Olszewski AD, Yaszemski MJ, White AA 3rd (1996) The anatomy of the human lumbar ligamentum flavum. New observations and their surgical importance. Spine 21, 2307-2312.

Orlin JR, Osen KK, Hovig T (1991) Subdural compartment in pig: a morphologic study with blood and horseradish peroxidase infused subdurally. Anat Rec 230, 22-37.

Pait TG, Killefer JA, James A, Arnautovic K (1996) Surgical anatomy of the anterior cervical spine: The disc space, vertebral artery, and associated bony structures. Neurosurgery 39, 769-776.

Pal GP, Routal RV (1996) The role of the vertebral laminae in the stability of the cervical spine. J Anat 188, 485-489.

Palmgren T, Gronblad M, Virri J, Kaapa E, Karaharju E (1999) An immunohistochemical study of nerve structures in the anulus fibrosus of human normal lumbar intervertebral discs. Spine 24, 2075-2079.

Palombi O, Fuentes S, Chaffanjon P, Passagia JG, Chirossel JP (2006) Cervical venous organization in the transverse foramen. Surg Radiol Anat 28, 66-70.

Panjabi MM, Oxland TR, Parks EH (1991a) Quantitative anatomy of cervical spine ligaments: Part I – Upper cervical spine. J Spinal Disord 4, 270-276.

Panjabi MM, Duranceau J, Goel V, Oxland T, Takata K (1991b) Cervical human vertebrae. Quantitative three-dimensional anatomy of the middle and lower regions. Spine 16, 861-869.

Parkin IG, Harrison GR (1985) The topographical anatomy of the lumbar epidural space. J Anat 141, 211-217.

Parkinson D (1991). Human spinal arachnoid septa, trabeculae, and "rogue strands". Am J Anat 192, 498-509.

Patin DJ, Eckstein EC, Harum K, Pallares VS (1993) Anatomic and biomechanical properties of human lumbar dura mater. Anesth Analg 76, 535-540.

Payne EE, Spillane JD (1957) The cervical spine: An anatomicopathological study of 70 specimens (using a special technique) with particular reference to the problem of cervical spondylosis. Brain 80, 571-597.

Pool-Goudzwaard AL, Kleinrensink GJ, Snijders CJ, Entius C, Stoeckart R (2001) The sacroiliac part of the iliolumbar ligament. J Anat 199, 457-463.

Quiring DP (1950) *Functional Anatomy of the Vertebrates.* McGraw Hill, New York.

Reina MA, Dittmann M, Lopez Garcia A, van Zundert A (1997) New perspectives in the microscopic structure of human dura mater in the dorsolumbar region. Reg Anesth 22, 161-166.

Reina MA, Gorra ME, López A (1998) Presenza di cellule epiteliali nel canale midollare dopo anestesia spinale. Minerva Anestesiol 64, 489-497.

Reina MA, De Leon Casasola O, Lopez A, De Andres JA, Mora M, Fernandez A (2002) The origin of the spinal subdural space: ultrastructure findings. Anesth Analg 94, 991-995.

Reina MA, De Leon Casasola Ode L, Villanueva MC, Lopez A, Maches F, De Andres JA (2004) Ultrastructural findings in human spinal pia mater in relation to subarachnoid anesthesia. Anesth Analg 98, 1479-1485.

Risling M, Dalsgaard CJ, Frisen J, Sjogren AM, Fried K (1994) Substance P-, calcitonin gene-related peptide, growth-associated protein-43, and neurotrophin receptor-like immunoreactivity associated with unmyelinated axons in feline ventral roots and pia mater. J Comp Neurol 339, 365-386.

Roberts S, Evans H, Trivedi J, Menage J (2006) Histology and pathology of the human intervertebral disc. J Bone Joint Surg Am 88;S2,10-14.

Ruhli FJ, Muntener M, Henneberg M (2006) Human osseous intervertebral foramen width. Am J Phys Anthropol 129, 177-188.

Saluja PG (1988) The incidence of ossification of the sacrococcygeal joint. J Anat 156, 11-15.

Scapinelli R, Stecco C, Pozzuoli A, Porzionato A, Macchi V, De Caro R (2006) The lumbar interspinous ligaments in humans: anatomical study and review of the literature. Cells Tissues Organs 183, 1-11.

Schachenmayr W, Friede RL (1978) The origin of subdural neomembranes. I. Fine structure of the dura-arachnoid interface in man. Am J Pathol 92, 53-68.

Sekiguchi M, Yabuki S, Satoh K, Kikuchi S (2004) An anatomic study of the sacral hiatus: a basis for successful caudal epidural block. Clin J Pain 20, 51-54.

Shi C, Matsumura A, Takahashi H, Yamashita M, Kimura T (2007) Effects of erect bipedal standing on the morphology of rat vertebral bodies. Anthropol Sci 115, 9-15.

Shinohara H (1999) The size and position of the sacral hiatus in man. Okajimas Folia Anat Jpn 76, 89-93.

Shinomiya K, Sato T, Spengler DM, Dawson J (1995) Isolated muscle atrophy of the distal upper extremity in cervical spinal cord compressive disorders. J Spinal Disord 8, 311-316.

Sirca A, Kostevc V (1985) The fibre type composition of thoracic and lumbar paravertebral muscles in man. J Anat 141, 131-137.

Solomonow M, Zhou BH, Harris M, Lu Y, Baratta RV (1998) The ligamento-muscular stabilizing system of the spine. Spine 23, 2552-2562.

Taylor JR, McCormick CC (1991) Lumbar facet joint fat pads: their normal anatomy and their appearance when enlarged. Neuroradiology 33, 38-42.

Tracy MF, Gibson MJ, Szypryt EP, Rutherford A, Corlett EN (1989) The geometry of the muscles of the lumbar spine determined by magnetic resonance imaging. Spine 14, 186-193.

Tubbs RS, Salter G, Grabb PA, Oakes WJ (2001) The denticulate ligament: anatomy and functional significance. J Neurosurg 94, 271-275.

Tubbs RS, Salter EG, Oakes WJ (2004) The accessory atlantoaxial ligament. Neurosurgery 55, 400-402.

Tubbs RS, Wellons JC 3rd, Banks J, Blount JP, Oakes WJ (2002) Quantitative anatomy of the transverse ligament tubercles. J Neurosurg 97, 343-345.

Valverde E, Casinos A, Alba-Fernandez C, Del Rio L (2005) Lumbar ontogenetic allometry and dimorphism in humans. A case for comparison between interspecific and intraspecific scaling. Eur J Morphol 42, 185-192.

Vandenabeele F, Creemers J, Lambrichts I (1996) Ultrastructure of the human spinal arachnoid mater and dura mater. J Anat 189, 417-430.

Vara CS, Thompson GH (2006) A cadaveric examination of pediatric cervical pedicle morphology. Spine 31, 1107-1112.

Varol T, Iyem C, Cezayirli E, Erturk M, Kayalioglu G, Hayretdag C (2006) Comparative morphometry of the lower lumbar vertebrae: Osteometry in dry bones and computed tomography images of patients with and without low back pain. J Int Med Res 34, 316-330.

Vleeming A, Stoeckart R, Volkers AC, Snijders CJ (1990) Relation between form and function in the sacroiliac joint. Part I: Clinical anatomical aspects. Spine 15, 130-132.

Wadhwani S, Loughenbury P, Soames R (2004) The anterior dural (Hofmann) ligaments. Spine 29, 623-627.

Wake MH (1979) *Hyman's Comparative Vertebrate Anatomy.* 3rd edn. University of Chicago Press, Chicago.

Welch K, Pollay M (1963) The spinal arachnoid villi of the monkeys Cercopithicus aethiops sabaeus and Macaca irus. Anat Rec 145, 43-48.

Weller RO (2005) Microscopic morphology and histology of the human meninges. Morphologie 89, 22-34.

Willis TA (1949) Nutrient arteries of the vertebral bodies. J Bone Joint Surg 31A, 538-540.

Wiltse LL (2000) Anatomy of the extradural compartments of the lumbar spinal canal. Peridural membrane and circumneural sheath. Radiol Clin North Am 38, 1177-1206.

Wischnitzer S (1967) *Atlas and Dissection Guide for Comparative Anatomy.* Freeman, San Francisco.

Xu R, Burgar A, Ebraheim NA, Yeasting RA (1999) The quantitative anatomy of the laminas of the spine. Spine 24, 107-113.

Yahia LH, Newman N, Rivard CH (1988) Neurohistology of lumbar spine ligaments. Acta Orthop Scand 59, 508-512.

Yamashita T, Minaki Y, Ozaktay AC, Cavanaugh JM, King AI (1996) A morphological study of the fibrous capsule of the human lumbar facet joint. Spine 21, 538-543.

4 The Spinal Nerves

Gulgun Kayalioglu

The anatomy of the dorsal and ventral roots and spinal nerves

The spinal nerves transmit sensory information from the target organs to the central nervous system, and send motor commands from the central nervous system to muscles and target organs. Along the entire length of the spinal cord, spinal nerves arise at regular intervals and leave the vertebral column between the adjacent vertebrae through the intervertebral foramina. In humans there are 31 pairs of spinal nerves: 8 cervical (C1-C8), 12 thoracic (T1-T12), 5 lumbar (L1-L5), 5 sacral (S1-S5) and 1 coccygeal (Co1). There are 34 pairs of spinal nerves in the rat (8 cervical, 13 thoracic, 6 lumbar, 4 sacral, and 3 coccygeal) and 38 pairs (8 cervical, 13 thoracic, 7 lumbar, 3 sacral, and 7 coccygeal) in the cat.

In humans, the first seven cervical spinal nerves emerge from the vertebral canal above their respective vertebrae, the remainder all emerge from below their respective vertebrae. The nerve emerging between C7 and T1 is named the eighth cervical nerve to link the two arrangements. The human cervical nerves increase in size from C1 to C6 and the size stays constant from C7 to T1. Thoracic spinal nerves from T2 to T12 are small and similar in size. Lumbar spinal nerves are large, their size increasing from L1 to L5. The largest spinal nerve is S1, after that the sacral nerves become smaller. The coccygeal nerve is the smallest spinal nerve.

Each spinal nerve is attached to the spinal cord by a ventral (anterior) and a dorsal (posterior) root. Each root is formed by six to eight rootlets that extend throughout the whole length of

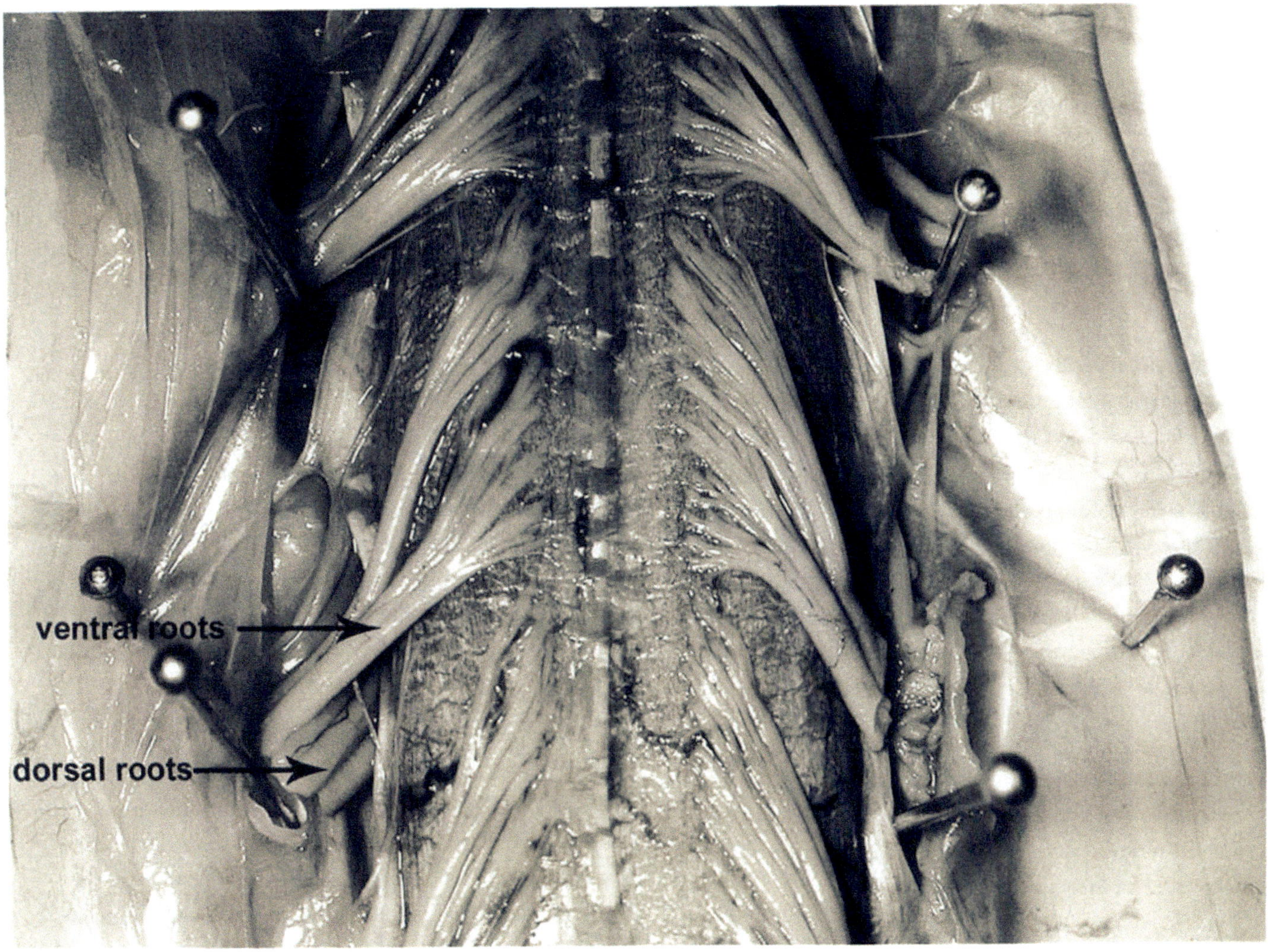

Figure 4.1 Dorsal and ventral roots of spinal nerves

This is a photograph of a dissection showing the ventral surface of the spinal cord and the ventral and dorsal rootlets. Groups of rootlets form the dorsal and ventral roots of each spinal nerve. The dura and arachnoid have been removed to expose the spinal cord. (Photograph courtesy of Professor Mete Erturk, Ege University). *This figure is reproduced in the color insert.*

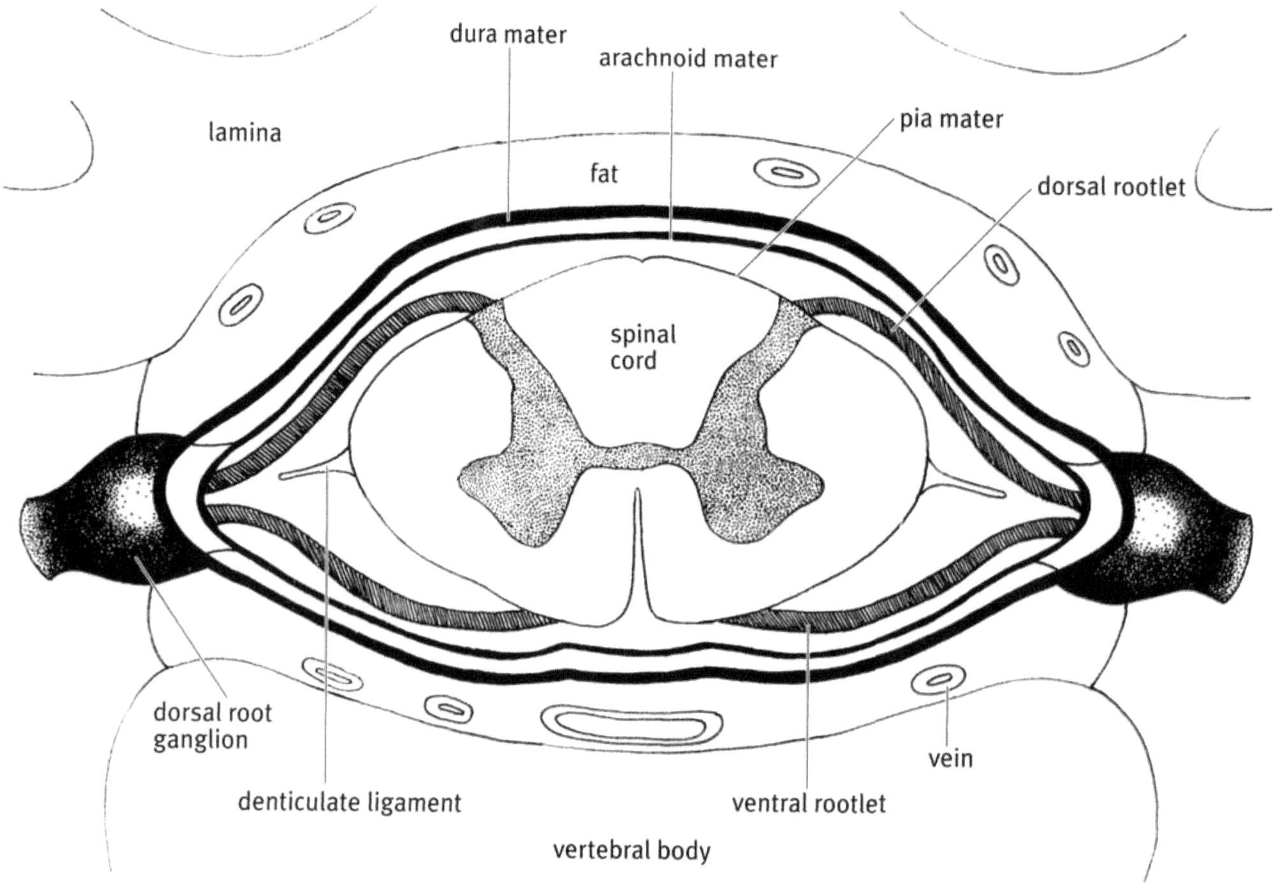

Figure 4.2 Transverse section of cord and meninges

This diagram shows the relationship between the spinal cord and the three meningeal layers – pia mater, arachnoid mater, and dura mater. The dura mater is separated from the surrounding bone by a layer of fat, in which vertebral veins are embedded. The denticulate ligament can be seen projecting from the pia mater. The swelling of the dorsal root ganglion can be seen lying in the intervertebral canal. (Adapted from Moore and Dalley, 2006, p.524)

the corresponding spinal cord segment. The ventral rootlets are made up of axons of motor neurons and dorsal rootlets are made up of the axons of sensory neurons. Each dorsal nerve root bears a dorsal root (spinal) ganglion, the cells of which give rise to peripheral and central nerve fibers. Because of the difference in length between the spinal cord and the vertebral column (the vertebral column is about 20% longer than the spinal cord), the length of the nerve roots increases progressively from rostral to caudal, and the rootlets run more vertically in the caudal segments. The lumbosacral nerve roots form the cauda equina below the termination of the spinal cord.

The pia mater and the arachnoid mater cover the dorsal and ventral roots until they pierce the dura. After that, they receive a dural sheath, which continues with the epineurium of the spinal nerve (see Figures 4.1 and 4.2).

At the attachment point of each rootlet to the spinal cord, there is a transition zone between the central nervous system and the peripheral nervous system; this zone is used extensively to study axon regeneration. Here, the central and peripheral nervous tissues are separated by an irregular but clearly defined interface that consists of astrocytic tissue. Myelin sheaths are formed by oligodendrocytes on the central side of this interface and by Schwann cells on the peripheral side. The central nervous system-peripheral nervous system transitional zone is penetrated only by axons (Fraher, 2000). Thick myelinated fibers in the dorsal rootlet become directed to the center of the rootlet, whereas thinly myelinated and unmyelinated fibers are positioned at the periphery, concentrated on the ventrolateral zone of the rootlet in the transition zone (Carlstedt *et al.*, 2004).

Intradural intersegmental connections may be present between the dorsal or ventral rootlets and the adjacent dorsal root ganglia (Jacob and Weddell, 1975; Parke and Watanabe, 1987). Connections between dorsal rootlets occur most frequently in the cervical region (61%) in humans, compared to 7% in thoracic and 22% in lumbar segments. Similar connections between ventral rootlets have also been noticed but are less frequent (10% in the cervical and 5% in the lumbar segments). No connections have been observed between thoracic ventral rootlets (Moriishi *et al.*, 1989). These connections usually occur by means of a peripheral dorsal or a ventral rootlet converging at the spinal cord surface and forming an irregular group of ectopic rootlets that join the central portion of the next rostral or caudal rootlet (Parke and Watanabe, 1987). Connections between dorsal rootlets may provide a pathway for overlap of sensory dermatomes and motor innervation of the limbs and trunk (Marzo *et al.*, 1987).

Within spinal nerves, there may be somatic sensory fibers from the skin and musculoskeletal system, visceral afferent fibers, motor fibers to voluntary striated muscles, sympathetic fibers to blood vessels and glands, and parasympathetic fibers to pelvic visceral organs. Each of these fiber types has a precise source or destination within the spinal cord.

Ventral roots

Each ventral root (also named the anterior root, radix anterior, radix ventralis, or radix motoria) is attached to the spinal cord by a series of rootlets that emerge from the ventrolateral sulcus of the spinal cord. Unlike the dorsal root fibers that are arranged in a neat line at their emergence from the spinal cord, ventral root fibers form an elliptical area named the anterior root exit zone (AREZ). The ventral roots predominantly consist of efferent somatic motor fibers (thick alpha motor axons and medium-sized gamma motor axons) derived from nerve cells of the ventral column. In thoracic and upper lumbar segments, these are supplemented by thin autonomic preganglionic motor fibers derived from the intermediolateral column. Injury filling of the ventral roots with HRP results in labelling of autonomic preganglionic neurons and their dendritic arbors, indicating there are also afferent fibers terminating on autonomic preganglionic neurons in the ventral roots of cat sacral spinal cord (Mawe *et al.*, 1984; 1986). Unmyelinated sensory afferent fibers innervating the ventral root and its pial sheath have also been shown electrophysiologically in the cat sacral spinal cord (Jänig and Koltzenburg, 1991).

Most of the ventral root fibers are myelinated, but there are also unmyelinated C fibers the majority of which are autonomic. In the rat, cat, mouse and monkey, ventral roots of C1-C8 and L3-S1 segments, which are outside the autonomic (thoracolumbar) outflow, contain C fibers (Clifton *et al.*, 1976; Coggeshall and Ito, 1977; Coggeshall *et al.*, 1977; Biscoe *et al.*, 1982; Chung and Kang, 1987; Karlsson *et al.*, 1991). A significant number (25%) of fibers in human ventral roots are C fibers. It is proposed that a significant fraction of these are sensory afferents (Coggeshall *et al.*, 1975; Hosobuchi, 1980). This has been confirmed by electrophysiological studies on C fibers in the cat ventral root (Clifton *et al.*, 1976). If so, this may explain the failure of dorsal rhizotomy to relieve chronic pain (Coggeshall, 1979). However, the immunohistochemical and electron microscopic studies on this matter are contradictory. Substance P and/or calcitonin gene-related peptide (CGRP) containing axons are found in ventral roots of the mouse (Kimura *et al.*, 1994), rat (Karlsson and Hildebrand, 1996), cat (Risling *et al.*, 1984) and monkey (Hegde *et al.*, 1988), but it has been shown that these do not enter the spinal cord via the ventral roots (Risling *et al.*, 1984, 1994; Kimura *et al.*, 1994; Karlsson and Hildebrand, 1996). According to the available evidence, the presence of unmyelinated fibers in the ventral roots is explained as follows: afferent fibers from the periphery project into the central nervous system using ventral roots as a pathway to the spinal cord; afferent ventral root fibers innervate the ventral spinal pia mater and/or the ventral root itself and enter the cord via dorsal roots; afferent fibers course into the ventral root from the periphery but loop back and enter the spinal cord via the dorsal root; peripheral afferents, which lack a dorsal root projection, course into the ventral root where they diminish gradually; alternatively, these unmyelinated fibers in the ventral root are blindly ending third projections of the peripheral afferent fibers entering the central nervous system via a dorsal root (Hildebrand *et al.*, 1997; Karlsson and Zakrisson, 1998).

The number of ventral root axons increases gradually along its subarachnoid course to the point it is apposed to the dorsal root ganglion. This increase is probably due to axonal sprouting (branching) (Fraher and O'Sullivan, 1989). Axon numbers also change during maturation and throughout maturity. In the rat, the number of axons slightly increases immediately after birth, but after that there is a sharp decline between 4 to 20 days due to a marked loss of unmyelinated axons. A gradual decline in the number of myelinated axons continues up to 500 days (Fraher and O'Sullivan, 1989). An increase in the number of unmyelinated axons in ventral roots has also been shown to occur postnatally in the cat, and it is proposed that this results from axonal sprouting (Risling *et al.*, 1981).

Dorsal roots and primary afferent projections to the spinal cord

Each dorsal root (also named the posterior root, radix posterior, radix dorsalis or radix sensoria) is attached to the dorsolateral sulcus of the spinal cord by a series of rootlets arranged in a line – the dorsal root entry zone (DREZ). In humans, there are 31 pairs of dorsal roots; the dorsal root of C1 spinal nerve is very small, and occasionally absent. Dorsal roots are larger than the ventral roots with thicker and more numerous fibers. In the cervical region of humans, the average number of dorsal rootlets range from 6.7 to 8.7 from C3 to T1, and the length of the DREZ varies from 10.7 mm at C7 to 12.7 mm at C4 (Alleyne *et al.*, 1998).

Each dorsal root fiber is the central process of a dorsal root ganglion cell. The dorsal roots contain sensory fibers from the skin, subcutaneous and deep tissues, and viscera. Primary afferent fibers of the dorsal roots are either myelinated or unmyelinated. Cutaneous, joint and visceral afferents are composed of myelinated Aα/β, Aδ and unmyelinated C fibers. Afferent fibers from muscles are also myelinated (group I, II, III fibers) and unmyelinated (group IV fibers). No efferent fibers have been shown to be present in the dorsal roots. There are also some unmyelinated nerve fibers that do not reach the spinal cord, terminating blindly within the dorsal roots (Carlstedt, 1977).

It is estimated that there are 2-2.5 million afferent fibers in human adult dorsal roots on each side (Schoenen and Grant, 2004). Although former studies suggest the number of dorsal root axons is considerably larger than the number of dorsal root ganglion neurons, possibly due to sprouting of dorsal root axons (Langford and Coggeshall, 1979; Chung and Coggeshall, 1984), a recent stereological study by Tandrup (1995) demonstrated that there is a 1:1 ratio of dorsal root fibers to neurons in the dorsal root ganglion.

In the DREZ, thin unmyelinated and thick myelinated fibers are located in the lateral and medial parts of the dorsal root, respectively (Light and Perl, 1979). Unmyelinated C fibers usually assemble into Remak bundles in the peripheral nervous system, where they are isolated from one another by cytoplasmic processes of Schwann cells. The dorsal roots are an exception to this, here about 7% to 9% of axons are immediately adjacent to other axons (Murinson and Griffin, 2004).

Neuronal tracers, in particular HRP, have been widely used in the rat and cat to study the terminations of primary afferent projections in the spinal cord. Thin unmyelinated fibers reach the spinal cord within the lateral part of the dorsal roots and enter the dorsal horn by way of the dorsolateral fasciculus (Lissauer's tract) to terminate in superficial laminae (laminae 1-2) of the dorsal horn. Thick myelinated fibers enter the spinal cord within the medial part of the dorsal roots and project into the deep layers of dorsal horn and the ventral horn. There are exceptions to this rule, as thin unmyelinated fibers are also observed in laminae 5, 7 and 10 (Schouenborg, 1984; Sugiura *et al.*, 1989; Mizumura *et al.*, 1993). Terminal branches of C-fibers of somatic and visceral origin are distributed mainly in laminae 1-2 in the guinea pig (Sugiura *et al.*, 1993). Cutaneous nociceptive afferents terminate mainly in laminae 1-3 (Maslany *et al.*, 1992), large Aβ fibers from cutaneous mechanoreceptors terminate in laminae 3-5 (Woolf, 1987). Afferents from hair follicles descend through laminae 1-3 to laminae 4-5, then curve back to end in the superficial part of lamina 4 and lamina 3 (Brown *et al.*, 1977). Visceral afferent fibers are composed of mainly unmyelinated C fibers and thin myelinated Aδ fibers, although a population of small rapidly conducting myelinated fibers is also present (Knuepfer and Schramm, 1987; Vera and Nadelhaft, 1990). Their terminations are mainly in laminae 1 and 5 (Cervero and Connell, 1984; Sugiura *et al.*, 1993). The distribution of primary afferents in Rexed's laminae and specific nuclei of the spinal cord are discussed in Chapter 6.

There is also a somatotopic arrangement for the termination of primary afferents on dorsal horn neurons, spinal nerves terminating on characteristic regions in the dorsal horn, distal parts of the limbs represented medially and proximal parts represented laterally (Ygge and Grant, 1983; Molander and Grant, 1985; 1987; Swett and Woolf, 1985; Shortland and Woolf, 1993).

However, there is a similar spinal projection pattern for different muscle nerves, suggesting either a less developed or an essentially different somatotopic organization for muscle afferents compared to cutaneous afferents, as revealed in earlier studies (Molander and Grant, 1987).

Dorsolateral fasciculus (Lissauer's tract)

The dorsolateral fasciculus, first described by Lissauer (1885), is a small bundle of fibers running longitudinally in the white matter between the dorsolateral edge of the dorsal horn and the surface of the spinal cord. Small myelinated Aδ and C primary afferent fibers enter the dorsolateral fasciculus. Axons of the dorsal horn neurons are also observed here (Coggeshall *et al.*, 1981). Most of these fibers enter the substantia gelatinosa at or near the level of their origin, only some ascend or descend for a few segments (Traub *et al.*, 1986).

In the rat, the majority of the axons in the dorsolateral fasciculus at mid-thoracic and lumbosacral levels are primary afferent fibers, a slightly higher percentage of the fibers are unmyelinated, as opposed to the myelinated fibers (Chung *et al.*, 1979). In the monkey, about 40% of the axons in the dorsolateral fasciculus are primary afferent axons from a single segment, and about another 40% are primary afferents from neighbouring cranial and caudal segments, the remaining 20% are propriospinal axons (Coggeshall *et al.*, 1981).

Dorsal root (spinal) ganglia

Each dorsal spinal root bears an ovoid swelling named the dorsal root (spinal) ganglion (DRG). The DRG is located close to the junction of the dorsal and ventral roots. On either side of the DRG, there may be scattered or grouped neurons particularly in the dorsal roots of the lumbar and sacral nerves, named the aberrant spinal ganglion cells. These are DRG neurons displaced during the growth process. The DRG cells are primary sensory neurons. They are pseudo-unipolar neurons with a single axon that after a short course divides into a peripheral and a central branch. The central branch enters the spinal cord via the dorsal roots and carries information from the cell body of DRG neuron to the spinal cord. The peripheral branch conveys sensory information from the body to the DRG cell body. Electron microscopic studies in the rat have shown that these two branches are not formed by the fusion of two opposite processes (as was once thought), but are both derived from the elongated cytoplasm of the cell body (Takahashi and Ninomiya, 1987). The DRG is encapsulated in a condensed connective tissue that is continuous with the sheath of the peripheral nerves.

The DRG neurons start developing as early as the ventral horn neurons, at 7 weeks of gestation (Kato and Takashima, 1994). As their appearance is before the development of osseous spine components, DRGs vary in orientation and location in different segments of the spine (Khorooshi *et al.*, 2001). In the adult human, DRGs lie along the vertebral column within the intervertebral foramina. The exception to this are the DRGs of the first and second cervical nerves, which lie upon the vertebral arches of the atlas and axis, and the DRGs of the second to fifth sacral nerves and that of the coccygeal nerve, which lie within the vertebral canal (see Figure 4.3).

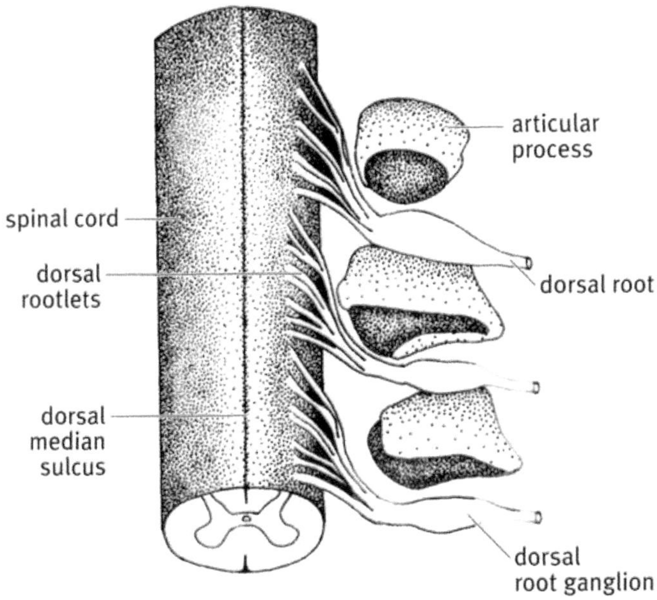

Figure 4.3 Dorsal root ganglia lying in the intervertebral foramen

This diagram shows three dorsal root ganglia lying alongside a section of cervical spinal cord. A number of dorsal rootlets converge on each dorsal root ganglion. Each dorsal root ganglion lies in the intervertebral foramen between the articular processes of adjacent vertebrae. (Adapted from Cramer and Darby, 2005, p.79)

The first DRG, together with the dorsal root, is either reduced in size or absent. The C2 DRG is proximally placed and occupies most of the foramen height, which may render it vulnerable to entrapment (Lu and Ebraheim, 1998). The C7 DRG is more distally situated within the intervertebral foramen C7 when compared to C6 (Yabuki and Kikuchi, 1996). Thoracic DRGs lie within the intervertebral foramina. Most of the lumbar DRGs are located within the intervertebral foramina, with only about 6% in the vertebral canal. The human lumbar DRGs gradually increase in size from L1 to L5, the width increasing from 3.7 mm to 5.9 mm, the length increasing from 4.3 to 9.4 mm, and the height increasing from 4.3 to 8.3 mm (Hasegawa *et al.*, 1996). Double ganglia (biganglia) are frequently observed in L3 and L4 (Shen *et al.*, 2006). The upper sacral DRGs are located within the intervertebral foramina, 55-60% in S1 and 15-50% in S2; the remainder lie within the vertebral canal. All S3 and S4 DRGs are located in the vertebral canal. The S1 DRG is larger than other sacral ganglia (Ebraheim and Lu, 1998). The coccygeal DRGs are located within the vertebral canal, sometimes in the form of scattered nerve cells rather than a swelling.

DRG neurons are divided into two histological types called 'small, dark' and 'large, light' based on their staining properties seen through light microscopy (Lawson, 1979). The large, light DRG cells are rich in neurofilaments, whereas the small, dark neurons are poor in neurofilaments (Lawson *et al.*, 1984; Holford *et al.*, 1994). Electron microscopic studies have classified the large light cells as A cells and the small dark cells as B cells. A detailed electron microscopic study by Rambourg *et al.*, (1983) defined three types of cells (A, B, C) on the basis of their size and distribution of organelles. They were able to further divide into six subtypes on the basis of the arrangement and organization of Nissl bodies and Golgi apparatus.

Large light DRG cells give rise to myelinated $A\alpha$ and $A\beta$ fibers and small-myelinated mechanoreceptive $A\delta$ fibers; the small dark cells give rise to nociceptive $A\delta$ and unmyelinated C fibers (Bergman *et al.*, 1999). The cells of origin of $A\delta$ fibers are larger when compared to those of C fibers (Harper and Lawson, 1985; Tandrup, 1995). However in the cat, unmyelinated afferent fibers do not originate exclusively from the smallest DRG cells but from the whole cell size spectrum (Hoheisel and Mense, 1986).

The small dark cells that give rise to nociceptive fibers contain peptidergic neurotransmitters such as substance P, galanin, leu-enkephalin, CGRP, vasoactive intestinal polypeptide, cholecystokinin and somatostatin (Hayashi *et al.*, 1983; Salt and Hill, 1983; Ju *et al.*, 1987; Garry *et al.*, 1989; Lawson *et al.*, 1993). More neuronal profiles are labeled for substance P in human (44%) than in rat (20%) (Holford *et al.*, 1994). DRG neurons also contain acetylcholine (Sann *et al.*, 1995; Tata *et al.*, 2004) and express both nicotinic and muscarinic receptors (Bernardini *et al.*, 1998, 1999; Genzen *et al.*, 2001). Vesicular acetylcholine transporter (VAChT) has been shown preferentially in medium/large-sized neurons and muscarinic receptors in small to medium-sized neurons and in glial cells (Bernardini *et al.*, 1998, 1999; Tata *et al.*, 2004). The expression of cholinergic markers in DRG neurons starts at early developmental stages, increases gradually and reaches a peak in adulthood (Tata *et al.*, 1994; Biagioni *et al.*, 1999). Some small dark DRG neurons are catecholaminergic, containing tyrosine hydroxylase (Price and Mudge, 1983; Price, 1985). Although some peptides exist alone in a distinct subpopulation of neurons in the DRG, some are co-localized with other transmitter substances. It is suggested that in these neurons one neuropeptide is the main neurotransmitter and the other a neuromodulator regulating the functions of the main neurotransmitter (Ju *et al.*, 1987; Cameron *et al.*, 1988). Nitric oxide synthase immunoreactivity and/or NADPH-diaphorase reactivity has been shown in the thoracolumbar and lumbosacral DRGs of the rat, dog, cat, and guinea pig (Aimi *et al.*, 1991; Vizzard *et al.*, 1994; Zhou *et al.*, 1998; Lukacova *et al.*, 2006;). These neurons contain both substance P and CGRP (Aimi *et al.*, 1991). Nitric oxide possibly has a neuroprotective effect on DRG neurons, and may facilitate regeneration (Thippeswamy and Morris, 2002).

Most lumbar DRG cells are small, dark cells (Perry and Lawson, 1998; Bergman *et al.*, 1999). In humans, the L5 DRG contains 11 000 large, 32 000 intermediate and 25 000 small neurons (Kawamura and Dyck, 1978). In resin-embedded sections in the rat, the number of neurons in C4-C6 DRGs was found to be 6800-7500 (Mille-Hamard *et al.*, 1999). The L4, L5 and L6 DRG contained about 12 000, 15 000, and 14 000 neurons, respectively. In paraffin sections, where the small neurons are difficult to identify, the counts are 23% less (Schmalbruch, 1987). There is controversy on the alteration of the number of DRG neurons with age, probably depending on the species studied and the methodology used. Some studies report an increase in cell number, presumed to be due to neuron proliferation or to late differentiation of neurons that do not present a typical appearance until adulthood (Devor and Govrin-Lipmann, 1991; Popken and Farel, 1997). Others suggest that the total number remains constant during adult life (Aldskogius and Risling, 1989; Pover *et al.*, 1994; Mohammed and Santer, 2001). A small decrease (12%) in the number of DRG neurons has also been reported in aged rats (Bergman and Ulfhake, 1998).

A rostrocaudal organization in the DRG has been shown in the rat in relation to the formation of plexuses (Wessels *et al.*, 1994). A columnar organization has been observed in the cat DRGs; neurons in the lateral border are connected to more rostral body parts, and neurons in the medial border are connected to more caudal body parts (Kausz and Rethelyi, 1985). No somatotopic organization is observed in the thoracic DRGs of the rat, but occasional clustering of labeled cells belonging to the same peripheral nerve was noted (Ygge, 1984). DRG neurons in the rat are arranged in pairs or, less frequently, in groups of three enclosed within a common connective sheath. Sometimes these nerve cell bodies are closely in contact with each other, and sometimes they are separated by neuroglial satellite cells (Pannese *et al.*, 1991). The satellite cells rest on a basal lamina of connective tissue that is continuous with the endoneurium; these cells seem to provide metabolic support to DRG neurons. The number of satellite cells increases along with the volume of ganglion neurons with which they are associated (Ledda *et al.*, 2004).

The sympathetic fibers entering the DRG are vessel-associated fibers located between nerve fiber layers and around perikarya of DRG neurons (McLachlan *et al.*, 1993). Primary sensory afferents have projections to spinal cord neurons that regulate the sympathetic outflow to the corresponding DRG. It has been suggested that imbalance in this circuit resulting from peripheral nerve injury might have a role in neuropathic pain (Hofstetter *et al.*, 2005).

Large DRG neurons relay low-threshold mechanical and proprioceptive stimuli, whereas small DRG neurons respond to high-threshold stimuli (Harper and Lawson, 1985).

Spinal nerves

The ventral and dorsal roots join together and form the spinal nerve at each segmental level. The spinal nerve runs from the lateral aspect of the dural sac to the intervertebral foramen in a tubular canal of variable length; this canal is named the nerve root canal. Extensions of the dural sheath encase all nerve roots here as they depart from the dural sac (Gilchrist *et al.*, 2002). Just lateral to the intervertebral foramen, each spinal nerve gives off a sinuvertebral (recurrent meningeal) branch that re-enters the vertebral canal. The sinuvertebral nerve is formed by a somatic root from a ventral ramus and an autonomic root from a neighbouring gray ramus communicans (Faustmann, 2004). The sinuvertebral nerve supplies the posterior longitudinal ligament, outer 1/3 part of anulus fibrosus, epidural blood vessels, ventral dura mater, dural sleeve, and the dorsal vertebral periosteum. After giving the sinuvertebral nerve, the spinal nerve immediately bifurcates into a small dorsal (posterior) ramus and a much larger ventral (anterior) ramus. The dorsal and ventral rami contain both motor and sensory fibers.

The axons of spinal nerves vary in diameter. Aα fibers are the largest (15-20 μm thick) and most heavily myelinated fibers that conduct action potentials most rapidly. Aβ fibers constitute the medium-sized and well-myelinated axons of Meissner and Pacinian corpuscles and Merkel discs. Thin, poorly myelinated mechanoreceptor fibers with large receptive fields form Aδ fibers. C fibers are very thin, poorly conducting axons that terminate without capsules or other types of end organs in the skin (Zigmond *et al.*, 1999). A cutaneous region innervated by a single spinal nerve is named a dermatome, and muscles innervated by a single spinal nerve as a myotome.

Each spinal nerve contains somatic and visceral fibers that are further classified functionally as somatic afferent, somatic efferent, visceral afferent and visceral efferent fibers. The somatic afferent fibers are peripheral processes of DRG neurons that carry impulses generated by external stimulations and are known as exteroceptive fibers. These fibers carry sensory information from the skin, subcutaneous tissues, skeletal muscles, tendons, and joints. The somatic efferent fibers are axons of ventral horn neurons which carry motor impulses to the skeletal muscles. The visceral afferent fibers are composed of peripheral processes of DRG neurons that convey visceral sensation; these are known as interoceptive fibers. Visceral afferent fibers constitute less than 10% of afferent fibers in dorsal roots. The visceral efferent fibers are autonomic (sympathetic or parasympathetic) nerve fibers that carry motor impulses to the smooth and cardiac muscles and to the glands.

Parasympathetic preganglionic efferent fibers originate from the neurons of the sacral parasympathetic nucleus (or the intermediolateral cell column) (for detailed information, see Chapter 8). Preganglionic parasympathetic fibers arise from the S2-S4 spinal cord segments in the human and cat (Brown and Nolan, 1979), and from the L6-S1 spinal cord segments in the rat (Nadelhaft and Booth, 1984). They leave the spinal cord by way of the ventral roots and pass into the corresponding spinal nerves.

As the spinal nerve branches into dorsal and ventral rami, parasympathetic preganglionic fibers stay in the ventral rami. After they exit through the ventral sacral foramina they diverge from the ventral rami to form the pelvic splanchnic nerves. They pass through the inferior hypogastric (pelvic) plexus to synapse in the terminal ganglia of pelvic parasympathetic ganglia.

Sympathetic preganglionic efferent fibers originate from the intermediolateral cell column of the spinal cord of all thoracic and upper two or three lumbar spinal cord segments, and emerge by way of the corresponding ventral roots. These preganglionic efferent fibers constitute the thoracolumbar (autonomic) outflow. Sympathetic preganglionic axons have a small diameter (1.5-4 μm) and are myelinated, therefore appearing white in fresh specimens. In rats, unlike humans and cats, the great majority of preganglionic fibers are unmyelinated (Hedger and Webber, 1976). These axons soon leave the spinal nerve to form the white rami communicantes and enter the sympathetic chain.

The sympathetic chain (also called the sympathetic trunk, paravertebral chain) is composed of gangliated cords running down along the length of the vertebral column on each side, extending from the skull base to the coccyx. In the human, each sympathetic chain consists of a series of 22 paravertebral ganglia: 3 cervical, 11 thoracic, 4 lumbar, and 4 sacral. The rat sympathetic chain consists of 3 cervical, 10 thoracic, 6 lumbar, 4 sacral and 1 caudal paravertebral ganglia on each side. In the cat, there are 3 cervical and 13 thoracic sympathetic ganglia; the lumbar sympathetic chain consists of 7 ganglia interconnected by nerve fibers (Wischnitzer, 1967). The fibers entering the paravertebral ganglia synapse with ganglionic neurons at the same level, or after ascending or descending for several segments. Neurons that lie rostrally within a spinal cord segment tend to project rostrally, and neurons lying caudally project caudally in the sympathetic chain (Forehand *et al.*, 1994). Some fibers synapse with more than one ganglionic neuron via collateral branches (Brooks-Fournier and Coggeshall, 1981). In addition, more than one preganglionic fiber may synapse with a single ganglion neuron. A large number of sympathetic spinal neurons (1000-2000/ganglion) terminate on the ganglia which control the head or thoracic organs and a relatively small number (100-400/ganglion) terminate on ganglia controlling the gut, kidney, and pelvic organs in the rat (Strack *et al.*, 1988). The great majority of paravertebral ganglion neurons are adrenergic (Jarvi and Pelto-Huikko, 1990; Tajti *et al.*, 1999). They also contain acetylcholine, calcitonin gene-related peptide (CGRP), dopamine-beta-hydroxylase, galanin, neuropeptide Y, nitric oxide, somatostatin, tyrosine hydroxylase, and vasoactive intestinal polypeptide (VIP) (Jarvi *et al.*, 1987; Schmitt *et al.*, 1988, Lindh *et al.*, 1989; Jarvi and Pelto-Huikko, 1990; Baffi *et al.*, 1992; Dun *et al.*, 1992, 1993; Benarroch, 1994; Vizzard *et al.*, 1994; Tajti *et al.*, 1999; Zhou and Ling, 1999). Although the sympathetic ganglia were formerly accepted merely as relay stations for the transfer of information from the central nervous system to the periphery, it is now known that signals traveling along the preganglionic nerve fibers are distributed, integrated, and/or modified here (deGroat and Saum, 1976; Burnstock *et al.*, 1987; Gibbins *et al.*, 2000; Myers, 2001).

Some of the preganglionic symptathetic fibers entering the paravertebral ganglia may pass through without a synapse to form the splanchic nerves that lead to prevertebral ganglia (prevertebral plexuses, collateral ganglia) located in prevertebral plexuses of the thorax, abdomen and pelvis. Structurally, the prevertebral ganglia are very similar to the paravertebral ganglia (Gabella, 1995). However, there are marked differences in nerve growth factor dependence, neuropeptide content, connectivity and electrophysiological properties (Szurszewski, 1981; Hill *et al.*, 1985; Cassell *et al.*, 1986; Dixon and McKinnon, 1994). Neurons of the prevertebral ganglia contain substance P, VIP and enkephalin (Schultzberg, 1983; Herbrecht *et al.*, 1995; Klein and Burden, 1988).

Axons of sympathetic ganglion neurons are unmyelinated and therefore appear gray when demonstrated by dissection. Some of the axons re-enter the spinal nerves by way of the gray rami communicantes which usually join the nerve just proximal to the white ramus. These are segmentally distributed in every spinal nerve through the ventral and dorsal rami, from C1 to the coccygeal nerve, to sweat glands, smooth muscles of blood vessels, or hair (arrector pili) muscles. Each ganglion neuron gives a postganglionic somatic branch to a single spinal nerve, and every spinal nerve receives gray rami communicantes from the sympathetic chain. When formed by the fusion of more than one ganglion, the ganglion may give and receive as many branches as the segments it contains, e.g., the superior cervical ganglion consists of fibers intended for C1-C4 and consequently gives off four branches to these spinal nerves.

Dorsal rami of spinal nerves in humans

Dorsal rami of spinal nerves, with the exception of the first two cervical spinal nerves, are smaller than the ventral rami. They supply the muscles, fascia, ligaments, and skin of the dorsal part of the neck and trunk. Dorsal rami divide into a medial and a lateral branch (except for the first cervical, fourth and fifth sacral and the coccygeal nerves).

Cervical dorsal rami

Dorsal rami of the cervical nerves, with the exception of C1, divide into a medial and a lateral branch, both of which innervate the muscles. The dorsal ramus of the C1 nerve (named the suboccipital nerve) is larger than the ventral ramus and may occasionally be absent. The suboccipital nerve supplies the muscles of the suboccipital triangle and the semispinalis capitis. Sometimes it has a cutaneous branch that joins the greater or lesser occipital nerves. The C2 dorsal ramus gives a large medial branch (the greater occipital nerve) and a small lateral branch. The greater occipital nerve supplies the skin of the back of the scalp as far as the vertex of the skull. The lateral branch supplies adjoining muscles. The C3 dorsal ramus has a superficial medial branch, also known as the third occipital nerve, which innervates the skin of lower occipital region, and a deep lateral branch that enters the multifidus muscle. The suboccipital nerve and the medial branches of the C2-C3 dorsal rami are sometimes joined with nerve fiber loops to form the posterior cervical plexus (Cruveilhier) in the posterior cervical region. The dorsal rami of the lower five cervical nerves also divide into medial and lateral branches. The medial branches of C4-C5 run between the semispinalis capitis and the semispinalis cervicis muscles and finish as cutaneous branches; lateral branches reach the longissimus capitis or splenius capitis muscles. The dorsal rami of C7-C8 (and occasionally C5 and C6) do not have cutaneous branches. Lateral branches of C6-C8 dorsal rami supply the iliocostalis cervicis, longissimus cervicis and longissimus capitis muscles; the medial branches supply the multifidus, semispinalis cervicis, and interspinalis muscles (Becser *et al.*, 1998; Zhang *et al.*, 2003).

Thoracic dorsal rami

The thoracic dorsal rami divide into medial and lateral branches close to the facet joints. The medial branches of the upper thoracic dorsal rami (T1-T6), after traveling through the erector spinae, become superficial and reach the tips of the spinous processes of corresponding vertebrae to innervate the skin. They supply the semispinalis thoracis and multifidus muscles. An anastomosis between the dorsal medial cutaneous branch of T2 with the descending lateral branch of the accessory nerve has been observed (Maigne *et al.*, 1991). The lateral branches of the dorsal rami innervate the lateral paravertebral muscles. From T7 to L1 (or L2), this pattern is reversed: the medial branches of these nerves end up at the tip of the spinous processes, supplying the multifidus and longissimus thoracis muscles, and occasionally innervating the skin near the midline; the lateral branches run through, and supply the longissimus thoracis and the iliocostalis muscles, and also give off cutaneous branches that pierce the serratus posterior inferior and latissimus dorsi muscles. The lateral branch of the dorsal ramus of T12 runs over the iliac crest to the ventral part of the gluteal skin.

Lumbar dorsal rami

The lumbar dorsal rami arise at right angles from the spinal nerves just after their exit from the intervertebral foramina and divide into medial and lateral branches. The medial and lateral branches supply the multifidi and erector spinae muscles, respectively. The medial branches of the lower three lumbar segments give small branches to the skin of the sacral region; the lateral branches of the upper three lumbar dorsal spinal rami give off cutaneous nerves that descend to innervate the skin of the buttock.

According to Bogduk *et al.*, (1982), the L1-L4 dorsal rami form three branches: medial, lateral, and intermediate, which are distributed to the multifidi, iliocostalis and longissimus muscles, respectively. The L4 dorsal ramus regularly forms three branches whereas at L1-L3 the lateral and intermediate branches may alternatively arise from a short common stem. The L5 dorsal ramus is much longer than the other lumbar dorsal rami and forms only a medial and an intermediate branch.

Sacral dorsal rami

The sacral dorsal rami emerge through the dorsal sacral foramina. The upper three dorsal sacral rami divide into medial and lateral branches beneath the multifidi. The medial branches innervate the multifidi. Lateral branches of S1-S3 (S4) and L5 dorsal rami anastomose with each other to form the posterior sacrococcygeal plexus (posterior sacral nerve plexus). Further branches arise from this plexus to form a second series of loops under the gluteus maximus muscle; nerves originating from here pierce this muscle to supply the skin over the lower part of the buttock. The dorsal rami of S4-S5 are small and lie below the multifidi. They unite with each other and with the posterior division of the coccygeal nerve to form loops on the back of the sacrum to supply the skin over the coccyx.

Coccygeal dorsal rami

The dorsal ramus of the coccygeal nerve is small and does not branch. It communicates with a branch from S5 to supply the skin over the dorsal surface of the coccyx.

Ventral rami of spinal nerves in humans

The ventral rami originate from the spinal nerves after their emergence from the intervertebral foramina. They supply the muscles, fascia and ligaments of the ventrolateral aspects of the trunk, limbs and perineum, as well as providing the sympathetic and sensory innervation of the skin. The ventral rami are composed of axons from the ventral and dorsal roots of the spinal cord. The ventral ramus of the spinal nerve contains about twice as many axons and a lower proportion of unmyelinated fibers in comparison with the dorsal ramus (Ygge, 1984). Each ventral ramus receives a gray ramus communicans from the corresponding sympathetic ganglion. The thoracic and upper two or three lumbar nerves also have a white ramus communicans. Branches from ventral rami of the S2-S4 spinal nerves join the pelvic plexus of the autonomic nervous system.

The ventral rami are distributed in a classic segmental pattern in the thoracic region (this pattern is disturbed in the limb enlargement segments). In the thoracic region, each ventral ramus (intercostal nerve) bifurcates into a collateral branch and a lateral cutaneous branch. The collateral branch runs towards the lower rib and ends as the lower anterior cutaneous nerve of the corresponding intercostal space. The lateral cutaneous branch pierces the intercostal muscles and divides into a smaller dorsal and a larger ventral ramus, which supply the skin of the sides and the lateral part of the ventral surface of the body, respectively. The ventral ramus continues ventrally and gives off branches to adjacent muscles and bones to terminate superficially as the anterior cutaneous branch.

In all mammals and in many non-mammalian vertebrates, the ventral rami form plexuses and exchange nerve fibers in the upper cervical, lower cervical (brachial), lumbar, sacral and coccygeal regions.

Cervical plexus

In humans, the cervical plexus is a plexus of the ventral rami of the first four cervical nerves (C1-C4). Each nerve, except for C1, divides into an ascending and a descending branch, which unite with branches of contiguous nerves to form loops. These loops and their branches form the cervical plexus. Each ventral ramus receives one or more gray rami communicantes from the superior cervical ganglion of the sympathetic trunk. The cervical plexus lies ventral to the scalenus medius and levator scapulae muscles, deep to the sternocleidomastoid muscle and the internal jugular vein. From this plexus arise cutaneous and muscular branches to the head, neck, and shoulder, and communicating branches to the vagus, hypoglossal and sympathetic nerves.

The superficial (cutaneous) branches emerge near the middle of the dorsal border of the sternocleidomastoid muscle (the region known as the punctum nervosum). These divide into ascending and descending branches. The ascending branches are the lesser occipital (C2-C3), great auricular (C2-C3) and transverse cervical (C2-C3) nerves. The descending branches are the supraclavicular (C3-C4) nerves. The deep (muscular and communicating) branches are divided into lateral and medial branches. The lateral branches are to the sternocleidomastoid (C2-C4), trapezius (C3-C4), levator scapulae (C3-C4), and scalenus medius muscles (C3-C4), and communicating branches are to the accessory nerve (C2-C4). The medial branches are to prevertebral muscles (C1-C4), to infrahyoid muscles (C1-C3) (by way of ansa cervicalis) and to the diaphragm by way of the phrenic nerve (C3-C5). Communicating branches to the vagus nerve (C1-C2), hypoglossal nerve (C1-C2), and to the superior cervical sympathetic ganglion (by gray rami communicantes, C1-C4) also arise from the medial branches. Although classical textbooks imply that all the motor input to the trapezius muscle is from the spinal accessory nerve, fibers from the cervical plexus (C3-C4 nerves) have been shown to carry an additional motor supply to trapezius, independent of the accessory nerve (Stacey *et al.*, 1995; Kierner *et al.*, 2001).

The ansa cervicalis, formerly called ansa hypoglossi, is formed by the union of the ventral rami of the first three cervical nerves through two main roots, the superior and inferior (Olry and Haines, 2002). The superior root (descendens hypoglossi) contains fibers from C1. It first travels within the sheath of the hypoglossal nerve and than in the carotid sheath to join the inferior root to form the ansa cervicalis. The inferior root, named the descending cervical nerve, contains fibers from C2-C3 spinal nerves. It descends on the dorsolateral side of the internal jugular vein. The ansa cervicalis is located deep to the sternocleidomastoid muscle and overlies or may sometimes lie within the carotid sheath, usually where the superior belly of the omohyoid muscle crosses over the internal jugular vein (Tubbs *et al.*, 2005). Sometimes the ansa is located superior (short ansa) or inferior (long ansa) to the omohyoid muscle (Caliot and Dumont, 1983). The branches of ansa cervicalis innervate the infrahyoid muscles. The ansa cervicalis is usually in a 'concentrated' form, characterized by thick nerve trunks constituting a solitary arch from which one or two branches for the infrahyoid muscles emerge. Rarely, it has a 'diffuse' form characterized by finer nerve trunks in the form of a plexus (Poviraev and Chernikov, 1967).

The phrenic nerve usually arises from C3-C5 nerves approximately at the mid-level of the thyroid cartilage. It gives off sensory fibers to mediastinal pleura and pierces the diaphragm to supply it on the caudal surface. Sometimes, an accessory phrenic nerve is present as a contribution to the phrenic nerve from the ventral ramus of C5, derived as a branch from the nerve to the subclavius muscle. The accessory phrenic nerve may sometimes arise from C4 or C6 segments or from the ansa cervicalis. It travels lateral to the phrenic nerve and joins the phrenic nerve at the root of the neck or in thorax.

Brachial plexus

The ventral rami of the lower four cervical nerves (C5-C8) and the first thoracic nerve (T1) form the brachial plexus in humans. Sometimes a branch of C4 ventral ramus (prefixed brachial plexus) or T2 ventral ramus (postfixed brachial plexus) contributes to the brachial plexus. After the ventral rami leave the intervertebral foramina, they extend ventrolaterally between the anterior and middle scalene muscles. Each ventral ramus of the brachial plexus receives a gray ramus communicans from the middle or inferior cervical or from the first thoracic sympathetic ganglion. From this plexus rise cutaneous branches to the upper limb with the exception of the axillary region supplied by the intercostobrachial nerve, an area on the shoulder supplied by supraclavicular nerves, and the dorsal scapular area supplied by cutaneous branches of dorsal rami. Muscular branches from the brachial plexus innervate all upper limb muscles with the exception of the trapezius muscle (Figure 4.4).

The brachial plexus is said to be composed of roots, trunks, divisions and cords. The ventral rami of spinal nerves C5 to T1 are referred to as the roots of the plexus. The dorsal scapular nerve arises from the C5 root, the long thoracic nerve from C5-C7 roots and nerves to the scalene and longus colli muscles arise from C5-C8 roots. A branch to the phrenic nerve also arises from C5 root.

Shortly after emerging from the intervertebral foramina, the roots of the plexus unite to form three trunks: upper (C5 and C6), middle (C7), and lower (C8 and T1). The nerve to the subclavius muscle and the suprascapular nerve originate from the upper trunk. The trunks of the brachial plexus pass between the anterior and middle scalene muscles. An intraoperative electrophysiological study in human has shown that the roots of the brachial plexus are functionally divided into three groups: C5-C6 roots innervate the shoulder and elbow; C8-T1 roots innervate the hand, and C7 innervates the shoulder, elbow, wrist and hand (Gu, 1996).

At the lateral border of the first rib, inferior to the middle portion of the clavicle, each trunk splits into an anterior and a posterior division. The six divisions later unite in a complex pattern to form the three cords of the brachial plexus. In the developing forelimb, the anterior and posterior divisions supply the anterior compartment (flexor muscles) and posterior compartment (extensor muscles), respectively.

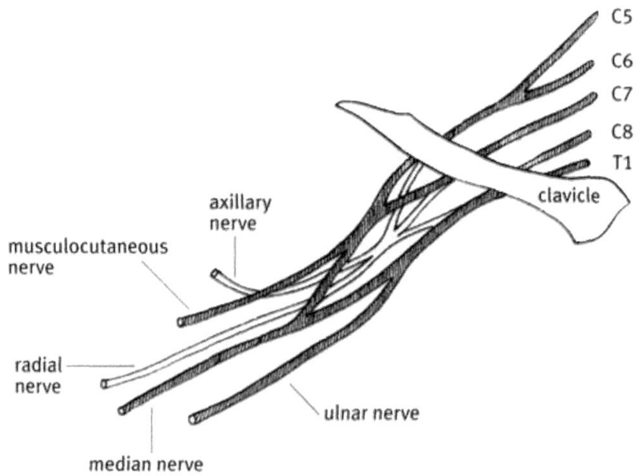

Figure 4.4 Human brachial plexus

The brachial plexus arises from C5, C6, C7, C8, and T1 spinal nerves. The main limb nerves that arise from the plexus are the ulnar, median, radial, musculocutaneous, and axialliary nerves. (Adapted from Moore and Dalley, 2006, p.774)

As the trunks enter the axilla, the posterior divisions of the superior, middle, and inferior trunks unite to form the posterior cord. The anterior divisions of the upper and middle trunks form the lateral cord, and the anterior division of the lower trunk forms the medial cord. The cords pass over the first rib close to the apex of the lung and continue under the clavicle immediately posterior to the subclavian artery. These cords give rise to the major peripheral nerves of the upper limb.

The upper and lower subscapular nerves (C5-C6) and the thoracodorsal nerve (C6-C8) arise from the posterior cord. The posterior cord gives off the axillary nerve (C5-C6) at the lower border of the subscapularis muscle and continues along the inferior and posterior surface of the axillary artery as the radial nerve (C5-T1). The lower trunk continues as the medial cord and gives off the medial pectoral nerve (C8-T1), the medial brachial cutaneous nerve (C8-T1), the medial antebrachial cutaneous nerve (C8-T1), the ulnar nerve (C7-T1), and the medial root of the median nerve (C8-T1). The ulnar nerve innervates numerous flexor muscles of the forearm, wrist and hand. The lateral cord divides into the lateral pectoral nerve (C5-C7), the lateral root of the median nerve (C5-C7) and the musculocutaneous nerve (C5-C7). The medial root of the median nerve originating from the medial cord (C8-T1) and the lateral root originating from the lateral cord (C5-C7) join to form the median nerve. The median nerve continues along the posterior and lateral surface of the axillary artery and innervates several flexor and pronator muscles of the forearm, wrist and hand.

Morphological variations of the brachial plexus are relatively frequent and well documented since the early study by Kerr (1918). The most common variation in brachial plexus is the level of its formation – prefixed or postfixed (Johnson *et al.*, 2006). Branches of brachial plexus may vary considerably in their levels of origin but much less frequently in regard to the fibers they contain.

In the rat, the brachial plexus is formed by the ventral rami of C5-T1 spinal nerves. In most cases there is a small contribution from T2. The plexus is more flattened in the rat and is not divisible into lateral, medial and posterior cords compared to those in human (Greene, 1968). The ventral rami of C6-T1 form the brachial plexus of the cat. C5 takes part in the plexus only through its contribution to the phrenic nerve. The largest number of axons is in C8 ventral root and the radial nerve (Wischnitzer 1967; Aubert *et al.*, 2004).

Lumbosacral plexus

The ventral rami of the lumbar and sacral nerves form the lumbosacral plexus. The human lumbar plexus is formed by the ventral rami of T12-L4. The ventral ramus of the L4, named the furcal nerve, contributes to both lumbar and sacral plexuses. The lumbosacral plexus is termed 'prefixed' when the furcal nerve is located cranial to the L4 nerve root, and 'postfixed' when it is located caudal to L4 nerve root. Lumbar plexus variations are common (20% of cases). The T12-L4 ventral rami enter from the posteromedial aspect of the psoas, continue ventrolaterally and combine in a variable manner to form the lumbar plexus (Samudrala *et al.*, 1999). The lumbar ventral rami are connected to the lumbar sympathetic trunk by rami communicantes within the psoas muscle. The lumbar plexus supplies direct branches to quadratus lumborum (T12-L4) and psoas (L1-L3) muscles. The iliohypogastric (L1), ilioinguinal (L1), and genitofemoral (L1-L2) nerves originate from the cranial portion of the lumbar plexus. The lateral femoral cutaneous (L2-L3), obturator (L2-4) and femoral (L2-L4) nerves originate from the caudal part. Within the psoas, the lateral femoral cutaneous and femoral nerves are usually separated from the obturator nerve by a muscular fold (Sim and Webb, 2004). All cutaneous and muscular branches of the plexus are limited to the thigh except for the saphenous nerve that supplies the medial side of the leg (Figure 4.5).

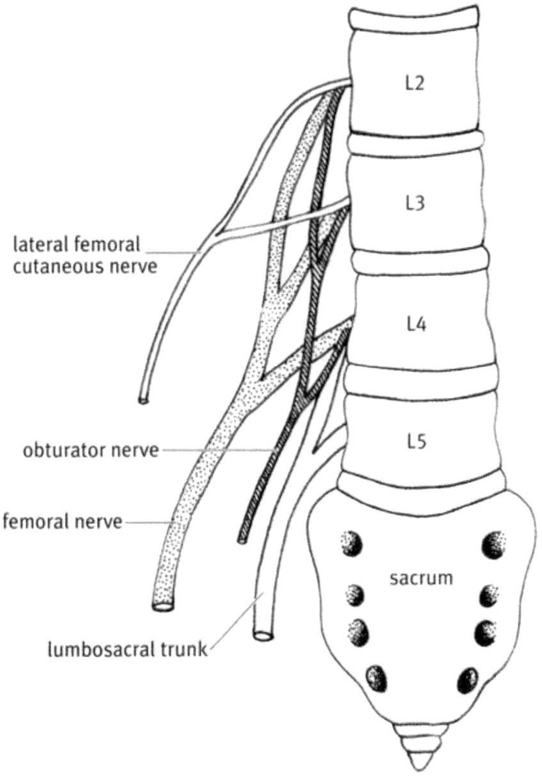

Figure 4.5 Human lumbar plexus

The lumbar plexus in the human arises from T12, L1, L2, L3, and L4 spinal nerves. The main nerves formed by the plexus are the femoral nerve, the obturator nerve, and the lateral femoral cutaneous nerve. Part of the L4 root joins with L5 to form the lumbosacral trunk, which then joins the sacral plexus. (Adapted from Moore and Dalley, 2006, p.366)

The sacral plexus is situated on the dorsal wall of the pelvis between the piriformis muscle and its fascia. The lumbosacral trunk (which comprises the ventral rami of L4-L5), the ventral rami of S1-S3, and a contribution from the ventral ramus of S4, together form the sacral plexus. The ventral rami converge on the lower part of the greater sciatic foramen and unite to form a broad triangular band. The apex of this band is continued through the greater sciatic foramen into the gluteal region as the sciatic nerve. The sciatic nerve (L4-S3) is composed of the tibial nerve and the common peroneal nerve running in a common epineurium. The tibial nerve (L4-S3) innervates the muscles on the posterior compartment of the thigh, leg and foot; the common peroneal nerve (L4-S2) innervates the anterior and lateral compartments of the leg and dorsum of the foot. Other branches of the sacral plexus are the superior gluteal (L4-S1), inferior gluteal (L5-S2), pudendal (S2-S4) and the posterior femoral cutaneous (S2-S3) nerves. The sacral plexus also gives muscular branches to the quadratus femoris and inferior gemellus (L4-S1), obturator internus and superior gemellus (L5-S2), piriformis (S1-S2), levator ani, coccygeus, and sphincter ani externus (S4) muscles, and also branches to pelvic splanchnic nerves (S2-S4) (Figure 4.6).

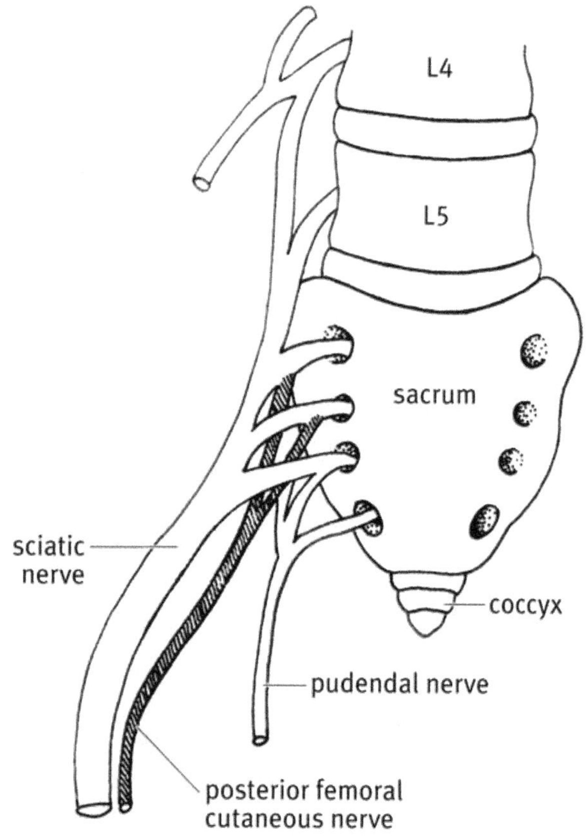

Figure 4.6 Human sacral plexus

The sacral plexus is formed by the lumbosacral trunk (L4 and L5) and sacral nerves S1, S2, and S3. The main nerves arising from the sacral plexus are the sciatic, posterior femoral cutaneous, and pudendal nerves. (Adapted from Moore and Dalley, 2006, p.381)

The lower part of the sacral plexus is sometimes referred to as the pudendal plexus. The pudendal plexus is composed of branches from the anterior divisions of S2-S3, the whole of the anterior divisions of S4-S5 and the coccygeal nerve. The perforating cutaneous (S2-S3), pudendal (S2-S4), and anococcygeal (S4-Co1) nerves are considered to be branches of the pudendal plexus.

In the rat, the sacral plexus is more limited in its extent, formed by the ventral rami of L4-L5 and part of L6 (bigeminal nerve). The ventral rami of the lower part of L6, S1-S4, and the first and second caudal nerves form the pudendal plexus. The bigeminal nerve (L6) represents the boundary between the sacral and pudendal plexuses. The pudendal plexus of the rat supplies the perineum and particularly the tail muscles. It gives rise to parasympathetic branches, the posterior cutaneous nerve of the thigh, the pudendal and perineal nerves, cutaneous and muscular branches, and the inferior caudal trunk which is homologous with the anococcygeal nerve in human (Greene, 1968).

In the cat, there are seven pairs of lumbar nerves and three pairs of sacral nerves. The ventral rami of L1-L3 spinal nerves innervate the muscles and skin of the abdominal wall. The ventral rami of L4-L7 form the lumbar plexus and S1-S3 ventral rami from the sacral plexus form the lumbosacral plexus. Sacral nerves give branches to the gluteal muscles, the rectum, bladder and external genital organs. The caudal nerves innervate the muscles of the tail (Wischnitzer, 1967).

The coccygeal plexus

The coccygeal plexus is formed by the ventral rami of S4-S5 and Co1. It lies on the pelvic surface of the coccygeal muscle and supplies it and part of the levator ani muscle. The anococcygeal nerves arising from this plexus supply the skin over the coccyx.

Dermatomes

A dermatome is the area of skin supplied by a single spinal nerve. Dermatomes have a segmental distribution represented as narrow bands of skin running almost horizontally along the trunk in a regular fashion. In the limbs, due to the growth and rotation of limb buds during development, there is a distortion in the segmental strips, and they are generally oriented longitudinally along the long axis of the limbs (Figure 4.7).

Because a single spinal segmental nerve may be disturbed through a number of different cutaneous nerves, it is generally not possible to identify dermatomes by classical dissection of peripheral nerves. The dermatomes have been mapped by sensory examination after sectioning of dorsal nerve roots,

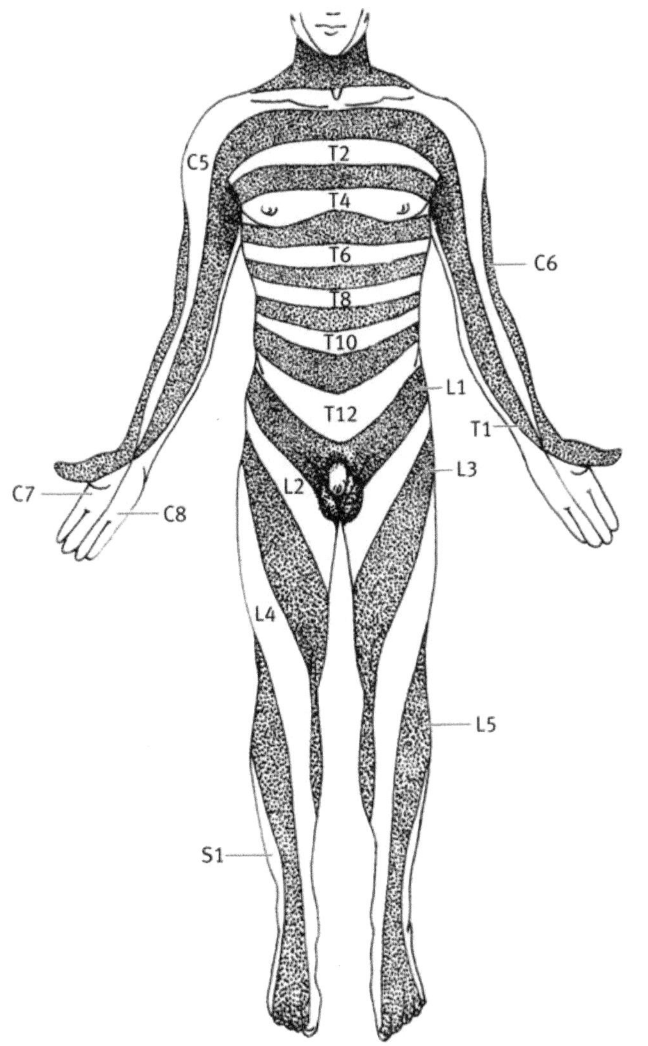

Figure 4.7 Human dermatomes

The area of skin supplied by a single spinal nerve is called a dermatome. In the trunk the dermatomes are roughly horizontal. In the limbs the dermatomes have been stretched out by the advancing limb bud during development. (Adapted from Cramer and Darby, 2005, p.345)

electrical stimulation of dorsal roots, or by recording somatosensory evoked potentials elicited by cutaneous stimuli. They are also revealed by the pattern of lesions in cases of herpes zoster (shingles). The first dermatome maps were made by Charles Sherrington (1893) in cats and monkeys, by Sir Henry Head (1908) and Otfrid Foerster (1933) in humans (see Greenberg, 2003). Dermatomes were described in detail in the earlier works of Bumke and Foerster (1936).

The area of a dermatome innervated by one dorsal root may be twice the size that revealed by its anatomical dissection. Dermatomal sensory loss also depends on the anatomical and physiological properties of multiple adjacent dorsal ganglia and the neighboring spinal cord segments (Denny-Brown *et al.*, 1973).

Dermatome maps are useful in enabling the clinician to determine the level of a lesion of the spinal cord or nerve roots by examining the sensory function on patients with neurological disorders. However, there is considerable individual variation and a variable degree of overlapping between dermatomes of adjacent spinal segments. The segmental overlap is particularly observed in T2-L1 segments. An electrophysiological study of the saphenous and sural nerves in humans has shown that these cutaneous nerves have multiple segmental supply and the sensory area they supply does not lie solely within a single dermatome (Phillips and Park, 1993). The demonstrated dermatome area is more extensive in regard to touch than for pain and temperature (Brodal, 1981). For example, the pain dermatomes of C2 and C3 do not overlap and are smaller than the C2 and C3 tactile dermatomes which do overlap (Poletti, 1991).

Although there are 31 pairs of spinal nerves in humans, there are only 30 dermatomes. The dorsal ramus of the C1 spinal nerve usually has no sensory root; hence the first dermatome corresponds to C2. The sensory innervation of the front of the head comes from the trigeminal nerve that supplies a large area of face and scalp, and is contiguous with the cutaneous area of C2. The dorsal rami of the C5-C8 and T1 nerves do not contribute to the cutaneous innervation of the ventral aspect of the trunk because they are fully committed to innervation of the upper limb. Because of this, the C4 dermatome is contiguous with the T2 dermatome. A similar but less extensive gap is found in the dorsal aspect of the trunk. The cutaneous innervation of the lateral aspect of the upper limb is from C4-C6, and the medial aspect is from C8-T1. The skin of the lower limb is innervated from T12-S3. The lumbar and sacral dorsal rami overlap on the skin of the buttock as a series of clunial nerves. As a memory aid, it is worth noting that the T4 dermatome innervates the skin region around the nipple, T7 the xiphoid process, T10 the umbilicus, L1 the groin, and S2 the perineum.

The pattern of dermatomes in the rat has been investigated by Lombard *et al.*, (1979), Wessels *et al.*, (1994), and mapped in detail by Takahashi *et al.*, (2003). Afferent fibers from L3-L5 innervate superficial and deep tissues of the hindlimb, and fibers rostral to L2 or caudal to L6 innervate the trunk and perineum (Takahashi *et al.*, 1994, 2003; Takahashi and Nakajima, 1996).

Segmental motor distribution

Each spinal nerve originally supplies the muscles originating from the myotome of the same segment; this pattern is called segmental motor (muscular) distribution. The segmental distribution is obscured in the cervical, brachial and lumbosacral regions where axons of the spinal nerves are intermingled and redistributed into a number of peripheral nerves in plexuses. However, the segmental distribution becomes evident again in the distribution of peripheral nerves to the muscles. Most skeletal muscles are innervated by more than one spinal nerve.

Establishment of the adult pattern of connections between motor neurons and their muscles is achieved in stages. An initial guidance mechanism directs the axons with considerable accuracy to their appropriate muscles. This is followed by a stage when many neurons die, even though they are correctly connected. After birth, the final modifications are brought about by elimination of redundant motoneuronal branches (Brown and Booth, 1983).

Pathways taken by motoneuron axons have widely been studied in chick embryos and rats (Lance-Jones and Landmesser, 1981; Whitelaw and Hollyday, 1983; Laskowski and Sanes, 1987). It has been shown that when axons are displaced up to a certain distance, they are still capable of finding their appropriate target (Lance-Jones and Lendmesser, 1981). Axonal ordering, axonal guidance within the plexus and local labels within the limb may all contribute to the specificity of innervation observed (Laskowski and Sanes, 1987). The resulting relationship between motoneuron groups and their target muscles is considered in detail in Chapter 7.

References

Aimi Y, Fujimura M, Vincent SR, Kimura H (1991) Localization of NADPH-diaphorase-containing neurons in sensory ganglia of the rat. J Comp Neurol 306, 382-392.

Aldskogius H, Risling M (1989) Number of dorsal root ganglion neurons and axons in cats of different ages. Exp Neurol 106, 70-73.

Alleyne CH Jr, Cawley CM, Barrow DL, Bonner GD (1998) Microsurgical anatomy of the dorsal cervical nerve roots and the cervical dorsal root ganglion/ventral root complexes. Surg Neurol 50, 213-218.

Aubert L, Carozzo C, Devillaire AC, Crevier-Denoix N, Moissonnier P (2004) Macro- and microanatomical characterization of the cat brachial plexus. Cells Tissues Organs 176, 205-210.

Baffi J, Gorcs T, Slowik F, Horvath M, Lekka N, Pasztor E, Palkovits M (1992) Neuropeptides in the human superior cervical ganglion. Brain Res 570, 272-278.

Becser N, Bovim G, Sjaastad O (1998) Extracranial nerves in the posterior part of the head. Anatomic variations and their possible clinical significance. Spine 23, 1435-1441.

Benarroch EE (1994) Neuropeptides in the sympathetic system: presence, plasticity, modulation, and implications. Ann Neurol 36, 6-13.

Bergman E, Ulfhake B (1998) Loss of primary sensory neurons in the very old rat: neuron number estimates using the disector method and confocal optical sectioning. J Comp Neurol 396, 211-222.

Bergman E, Carlsson K, Liljeborg A, Manders E, Hökfelt T, Ulfhake B (1999) Neuropeptides, nitric oxide synthase and GAP-43 in B4-binding and RT97 immunoreactive primary sensory neurons: normal distribution pattern and changes after peripheral nerve transection and aging. Brain Res 832, 63-83.

Bernardini N, De Stefano ME, Tata AM, Biagioni S, Augusti Tocco G (1998) Neuronal and non-neuronal cell population of the avian dorsal root ganglia express muscarinic acetylcholine receptors. Int J Dev Neurosci 16, 365-367.

Bernardini N, Levey AI, Augusti Tocco G (1999) Rat dorsal root ganglia express m1-m4 muscarinic receptor proteins. J Periph Nerv Syst 4, 222-232.

Biagioni S, Tata AM, Augusti-Tocco G (1999) Expression of cholinergic system components in dorsal root ganglion (DRG) neurons: its possibile dual role in development and nociception. Res Dev Neurochem 2, 443-461.

Biscoe TJ, Nickels SM, Stirling CA (1982) Numbers and sizes of nerve fibres in mouse spinal roots. Q J Exp Physiol 67, 473-494.

Bogduk N, Wilson AS, Tynan W (1982) The human lumbar dorsal rami. J Anat 134, 383-397.

Brodal A (1981) *Neurological Anatomy in Relation to Clinical Medicine.* 3rd edn. Oxford University Press, New York.

Brooks-Fournier R, Coggeshall RE (1981) The ratio of preganglionic axons to postganglionic cells in the sympathetic nervous system of the rat. J Comp Neurol 197, 207-216.

Brown MC, Booth CM (1983) Postnatal development of the adult pattern of motor axon distribution in rat muscle. Nature 304, 741-742.

Brown HK, Nolan MF (1979) Ultrastructure and quantitative synaptology of the sacral parasympathetic nucleus. J Neurocytol 8, 167-179.

Brown AG, Rose PK, Snow PJ (1977) The morphology of hair follicle afferent fibre collaterals in the spinal cord of the cat. J Physiol 272, 779-797.

Bumke O, Foerster O (1936) *Handbuch der Neurologie.* Vol V. Julius Springer, Berlin.

Burnstock G (1987) Autonomic neuroeffector mechanisms: recent developments. Funct Neurol 2, 427-436.

Caliot P, Dumont D (1983) A contribution to the morphological study of the ansa cervicalis. Rev Laryngol Otol Rhinol 104, 441-444.

Cameron AA, Leah JD, Snow PJ (1988) The coexistence of neuropeptides in feline sensory neurons. Neuroscience 27, 969-979.

Carlstedt T (1977) Observations on the morphology at the transition between the peripheral and the central nervous system in the cat. IV. Unmyelinated fibres in S1 dorsal rootlets. Acta Physiol Scand S446, 61-72.

Carlstedt T, Cullheim S, Risling M (2004) Spinal cord in relation to the peripheral nervous system. In: *The Rat Nervous System,* 3rd edn. Paxinos G (ed.) Elsevier Academic Press, San Diego.

Cassell JF, Clark AL, McLachlan EM (1986) Characteristics of phasic and tonic sympathetic ganglion cells of the guinea-pig. J Physiol 372, 457-483.

Cervero F, Connell LA (1984) Distribution of somatic and visceral primary afferent fibres within the thoracic spinal cord of the cat. J Comp Neurol 230, 88-98.

Chung K, Coggeshall RE (1984) The ratio of dorsal root ganglion cells to dorsal root axons in sacral segments of the cat. J Comp Neurol 225, 24-30.

Chung K, Kang HS (1987) Dorsal root ganglion neurons with central processes in both dorsal and ventral roots in rats. Neurosci Lett 80, 202-206.

Chung K, Langford LA, Applebaum AE, Coggeshall RE (1979) Primary afferent fibers in the tract of Lissauer in the rat. J Comp Neurol 184, 587-598.

Clifton GL, Coggeshall RE, Vance WH, Willis WD (1976) Receptive fields of unmyelinated ventral root afferent fibres in the cat. J Physiol 256, 573-600.

Cramer GD and Darby SA (2005) *Basic and clinical anatomy of the spine, spinal cord, and ANS* (2nd ed.). Elsevier Mosby, St Louis.

Coggeshall RE (1979) Afferent fibers in the ventral root. Neurosurgery 4, 443-448.

Coggeshall RE, Ito H (1977) Sensory fibres in ventral roots L7 and SI in the cat. J Physiol 267, 215-235.

Coggeshall RE, Applebaum ML, Fazen M, Stubbs TB 3rd, Sykes MT (1975) Unmyelinated axons in human ventral roots, a possible explanation for the failure of dorsal rhizotomy to relieve pain. Brain 98, 157-166.

Coggeshall RE, Chung K, Chung JM, Langford LA (1981) Primary afferent axons in the tract of Lissauer in the monkey. J Comp Neurol 196, 431-442.

Coggeshall RE, Emery DG, Ito H, Maynard CW (1977) Unmyelinated and small myelinated axons in rat ventral roots. J Comp Neurol 173, 175-184.

deGroat WC, Saum WR (1976) Synaptic transmission in parasympathetic ganglia in the urinary bladder of the cat. J Physiol 256, 137-158.

Denny-Brown D, Kirk EJ, Yanagisawa N. (1973) The tract of Lissauer in relation to sensory transmission in the dorsal horn of spinal cord in the macaque monkey. J Comp Neurol 151, 175-200.

Devor M, Govrin-Lipmann R (1991) Neurogenesis in adult rat dorsal root ganglia: on counting and the count. Somatosens Mot Res 8, 9-12.

Dixon JE, McKinnon D (1994) Expression of the trk gene family of neurotrophin receptors in prevertebral sympathetic ganglia. Brain Res Dev Brain Res 77, 177-182.

Dun NJ, Dun SL, Forstermann U, Tseng LF (1992) Nitric oxide synthase immunoreactivity in rat spinal cord. Neurosci Lett 147, 217-220.

Dun NJ, Dun SL, Wu SY, Forstermann U, Schmidt HH, Tseng LF (1993) Nitric oxide synthase immunoreactivity in the rat, mouse, cat and squirrel monkey spinal cord. Neuroscience 54, 845-857.

Ebraheim NA, Lu J (1998) Morphometric evaluation of the sacral dorsal root ganglia. A cadaveric study. Surg Radiol Anat 20, 105-108.

Faustmann PM (2004) Neuroanatomic basis for discogenic pain. Z Orthop Ihre Grenzgeb 142, 706-708.

Forehand CJ, Ezerman EB, Rubin E, Glover JC (1994) Segmental patterning of rat and chicken sympathetic preganglionic neurons: correlation between soma position and axon projection pathway. J Neurosci 14, 231-241.

Fraher JP (2000) The transitional zone and CNS regeneration. J Anat 196, 137-158.

Fraher JP, O'Sullivan VR (1989) Age changes in axon number along the cervical ventral spinal nerve roots in rats. J Comp Neurol 280, 171-182.

Gabella, G (1995) *The Rat Autonomic Nervous System.* Academic Press, London.

Garry MG, Miller KE, Seybold VS (1989) Lumbar dorsal root ganglia of the cat: a quantitative study of peptide immunoreactivity and cell size. J Comp Neurol 284, 36-47.

Genzen JR, Van Cleve W, McGehee DS (2001) Dorsal root ganglion neurons express multiple nicotinic acetylcholine receptor subtypes. J Neurophysiol 86, 1773-1782.

Gibbins IL, Jobling P, Messenger JP, Teo EH, Morris JL (2000) Neuronal morphology and the synaptic organisation of sympathetic ganglia. J Auton Nerv Syst 81, 104-109.

Gilchrist RV, Slipman CW, Bhagia SM (2002) Anatomy of the intervertebral foramen. Pain Physician 5, 372-378.

Greenberg SA (2003) The history of dermatome mapping. Arch Neurol 60, 126-131.

Greene EC (1968) *Anatomy of the Rat.* Hafner, New York.

Gu Y (1996) Functional motor innervation of brachial plexus roots. An intraoperative electrophysiological study. Chin Med J 109, 749-751.

Harper AA, Lawson SN (1985) Conduction velocity is related to morphological cell type in rat dorsal root ganglion neurones. J Physiol 359, 31-46.

Hasegawa T, Mikawa Y, Watanabe R, An HS (1996) Morphometric analysis of the lumbosacral nerve roots and dorsal root ganglia by magnetic resonance imaging. Spine 21, 1005-1009.

Hayashi M, Edgar D, Thoenen H (1983) The development of substance P, somatostatin and vasoactive intestinal polypeptide in sympathetic and spinal sensory ganglia of the chick embryo. Neuroscience 10, 31-39.

Hedger JH, Webber RH (1976) Anatomical study of the cervical sympathetic trunk and ganglia in the albino rat (Mus norvegicus albinus). Acta Anat (Basel) 96, 206-217.

Hegde SA, Hara H, Kobayashi S (1988) Innervation of radicular and extraparenchymal arteries of spinal cord. Histochemical and immunohistochemical study in primate. Histochemistry 89, 415-420.

Herbrecht F, Bagnol D, Cucumel K, Jule Y, Cupo A (1995) Distribution of enkephalin immunoreactivity in sympathetic prevertebral ganglia and digestive tract of guinea-pigs and rats. Regul Pept 57, 85-95.

Hildebrand C, Karlsson M, Risling M (1997) Ganglionic axons in motor roots and pia mater. Prog Neurobiol 51, 89-128.

Hill CE, Hendry IA, Ngu MC, van Helden DF (1985) Subpopulations of sympathetic neurones differ in their sensitivity to nerve growth factor antiserum. Brain Res 355, 121-130.

Hofstetter CP, Card JP, Olson L (2005) A spinal cord pathway connecting primary afferents to the segmental sympathetic outflow system. Exp Neurol 194, 128-138.

Hoheisel U, Mense S (1986) Non-myelinated afferent fibres do not originate exclusively from the smallest dorsal root ganglion cells in the cat. Neurosci Lett 72, 153-157.

Holford LC, Case P, Lawson SN (1994) Substance P, neurofilament, peripherin and SSEA4 immunocytochemistry of human dorsal root ganglion neurons obtained from post-mortem tissue: a quantitative morphometric analysis. J Neurocytol 23, 577-589.

Hosobuchi Y (1980) The majority of unmyelinated afferent axons in human ventral roots probably conduct pain. Pain 8, 167-180.

Jacob M, Weddell G (1975) Neural intersegmental connection in the spinal root and ganglion region of the rat. J Comp Neurol 161, 115-123.

Jänig W, Koltzenburg M (1991) Receptive properties of pial afferents. Pain 45, 77-85.

Jarvi R, Pelto-Huikko M (1990) Localization of neuropeptide Y in human sympathetic ganglia: correlation with met-enkephalin, tyrosine hydroxylase and acetylcholinesterase. Histochem J 22, 87-94.

Jarvi R, Pelto-Huikko M, Helen P, Hervonen A (1987) Somatostatin-like immunoreactivity in human sympathetic ganglia. Cell Tissue Res 249, 1-5.

Johnson EO, Vekris MD, Zoubos AB, Soucacos PN (2006) Neuroanatomy of the brachial plexus: the missing link in the continuity between the central and peripheral nervous systems. Microsurgery 26, 218-229.

Ju G, Hökfelt T, Brodin E, Fahrenkrug J, Fischer JA, Frey P, Elde RP, Brown JC (1987) Primary sensory neurons of the rat showing calcitonin gene-related peptide immunoreactivity and their relation to substance P-, somatostatin-, galanin-, vasoactive intestinal polypeptide- and cholecystokinin-immunoreactive ganglion cells. Cell Tissue Res 247, 417-431.

Karlsson M, Hildebrand C (1996) Postnatal development and aging of the rat ventral root L5: electron microscopic and immunohistochemical studies. J Comp Neurol 364, 211-218.

Karlsson M, Zakrisson M (1998) Relation between putative afferent axons and the glia limitans in rat motor roots. J Peripher Nerv Syst 3, 47-53.

Karlsson M, Hildebrand C, Warnborg K (1991) Fibre composition of the ventral roots L7 and S1 in the owl monkey (Aotus trivirgatus). Anat Embryol (Berl) 184, 125-132.

Kato M, Takashima S (1994) Immunohistochemical and morphometrical development of the dorsal root ganglion as a neural crest derivative: comparison with the fetal CNS. Early Hum Dev 38, 81-90.

Kausz M, Rethelyi M (1985) Lamellar arrangement of neuronal somata in the dorsal root ganglion of the cat. Somatosens Res 2, 193-204.

Kawamura Y, Dyck PJ (1978) Evidence for three populations by size in L5 spinal ganglion in man. J Neuropathol Exp Neurol 37, 269-272.

Kerr A (1918) The brachial plexus of nerves in man: the variations in its formation and branches. Am J Anat 23, 285-345.

Khorooshi MH, Fischer Hansen B, Keeling J, Nolting D, Kjaer I (2001) Prenatal localization of the dorsal root ganglion in different segments of the normal human vertebral column. Spine 26, 1-5.

Kierner AC, Zelenka I, Burian M (2001) How do the cervical plexus and the spinal accessory nerve contribute to the innervation of the trapezius muscle? As seen from within using Sihler's stain. Arch Otolaryngol Head Neck Surg 127, 1230-1232.

Kimura M, Kishida R, Abe T, Goris RC, Kawai S (1994) Nerve fibers immunoreactive for substance P and calcitonin gene-related peptide in the cervical spinal ventral roots of the mouse. Cell Tissue Res 277, 273-278.

Klein CM, Burden HW (1988) Substance P- and vasoactive intestinal polypeptide (VIP)-immunoreactive nerve fibers in relation to ovarian postganglionic perikarya in para- and prevertebral ganglia: evidence from combined retrograde tracing and immunocytochemistry. Cell Tissue Res 252, 403-410.

Knuepfer MM, Schramm LP (1987) The conduction velocities and spinal projections of single renal afferent fibers in the rat. Brain Res 435, 167-173.

Lance-Jones C, Landmesser L (1981) Pathway selection by embryonic chick motoneurons in an experimentally altered environment. Proc R Soc Lond B Biol Sci 214, 19-52.

Langford LA, Coggeshall RE (1979) Branching of sensory axons in the dorsal root and evidence for the absence of dorsal root efferent fibers. J Comp Neurol 184, 193-204.

Laskowski MB, Sanes JR (1987) Topographic mapping of motor pools onto skeletal muscles. J Neurosci 7, 252-260.

Lawson SN (1979) The postnatal development of large light and small dark neurons in mouse dorsal root ganglia: a statistical analysis of cell numbers and size. J Neurocytol 8, 275-294.

Lawson SN, Harper AA, Harper EI, Garson JA, Anderton BH (1984) A monoclonal antibody against neurofilament protein specifically labels a subpopulation of rat sensory neurones. J Comp Neurol 228, 263-272.

Lawson SN, Perry MJ, Prabhakar E, McCarthy PW (1993) Primary sensory neurones: neurofilament, neuropeptides, and conduction velocity. Brain Res Bull 30, 239-243.

Ledda M, De Palo S, Pannese E (2004) Ratios between number of neuroglial cells and number and volume of nerve cells in the spinal ganglia of two species of reptiles and three species of mammals. Tissue Cell 36, 55-62.

Light AR, Perl ER (1979) Reexamination of the dorsal root projection to the spinal dorsal horn including observations on the differential termination of coarse and fine fibers. J Comp Neurol 186, 117-131.

Lindh B, Lundberg JM, Hökfelt T. (1989) NPY-, galanin-, VIP/PHI-, CGRP- and substance P-immunoreactive neuronal subpopulations in cat autonomic and sensory ganglia and their projections. Cell Tissue Res 256, 259-273.

Lissauer, H (1885) Beitrag zur pathologischen Anatomie der Tabes dorsalis und zum Faserverlauf im menschlichen Ruckenmark. Neur Zentbl 4, 245-246.

Lombard MC, Nashold BSJ, Albe-Fessard D, Salman N, Sakr C (1979) Deafferentation hypersensitivity in the rat after dorsal rhizotomy: a possible animal model of chronic pain. Pain 6, 163-174.

Lu J, Ebraheim NA (1998) Anatomic considerations of C2 nerve root ganglion. Spine 23, 649-652.

Lukacova N, Kolesar D, Marsala M, Marsala J (2006) Immunohistochemical, histochemical and radioassay analysis of nitric oxide synthase immunoreactivity in the lumbar and sacral dorsal root ganglia of the dog. Cell Mol Neurobiol 26, 17-44.

Maigne JY, Maigne R, Guerin-Surville H (1991) Upper thoracic dorsal rami: anatomic study of their medial cutaneous branches. Surg Radiol Anat 13, 109-112.

Marzo JM, Simmons EH, Kallen F (1987) Intradural connections between adjacent cervical spinal roots. Spine 12, 964-968.

Maslany S, Crockett DP, Egger MD (1992) Organization of cutaneous primary afferent fibers projecting to the dorsal horn in the rat: WGA-HRP versus B-HRP. Brain Res 569, 123-135.

Mawe GM, Bresnahan JC, Beattie MS (1984) Primary afferent projections from dorsal and ventral roots to autonomic preganglionic neurons in the cat sacral spinal cord: light and electron microscopic observations. Brain Res 290, 152-157.

Mawe GM, Bresnahan JC, Beattie MS (1986) A light and electron microscopic analysis of the sacral parasympathetic nucleus after labelling primary afferent and efferent elements with HRP. J Comp Neurol 250, 33-57.

McLachlan EM, Janig W, Devor M, Michaelis M (1993) Peripheral nerve injury triggers noradrenergic sprouting within dorsal root ganglia. Nature 363, 543-546.

Mille-Hamard L, Bauchet L, Baillet-Derbin C, Horvat JC (1999) Estimation of the number and size of female adult rat C4, C5 and C6 dorsal root ganglia (DRG) neurons. Somatosens Mot Res 16, 223-228.

Mizumura K, Sugiura Y, Kumazawa T (1993) Spinal termination patterns of canine identified A-delta and C spermatic polymodal receptors traced by intracellular labeling with Phaseolus vulgaris-leucoagglutinin. J Comp Neurol 335, 460-468.

Mohammed HA, Santer RM (2001) Total neuronal numbers of rat lumbosacral primary afferent neurons do not change with age. Neurosci Lett 304, 149-152.

Molander C, Grant G (1985) Cutaneous projections from the rat hindlimb foot to the substantia gelatinosa of the spinal cord studied by transganglionic transport of WGA-HRP conjugate. J Comp Neurol 237, 476-484.

Molander C, Grant G (1987) Spinal cord projections from hindlimb muscle nerves in the rat studied by transganglionic transport of horseradish peroxidase, wheat germ agglutinin conjugated horseradish peroxidase, or horseradish peroxidase with dimethylsulfoxide. J Comp Neurol 260, 246-255.

Moore KL and Dalley AF (2006) *Clinically oriented anatomy* (5th ed.). Lippincott Williams and Wilkins, Philadelphia.

Moriishi J, Otani K, Tanaka K, Inoue S (1989) The intersegmental anastomoses between spinal nerve roots. Anat Rec 224, 110-116.

Murinson BB, Griffin JW (2004) C-fiber structure varies with location in peripheral nerve. J Neuropathol Exp Neurol 63, 246-254.

Myers AC (2001) Transmission in autonomic ganglia. Respir Physiol 125, 99-111.

Nadelhaft I, Booth AM (1984) The location and morphology of preganglionic neurons and the distribution of visceral afferents from the rat pelvic nerve: a horseradish peroxidase study. J Comp Neurol 226, 238-245.

Olry R, Haines DE (2002) Ansa hypoglossi or ansa cervicalis? That is the question. J Hist Neurosci 11, 302-304.

Pannese E, Ledda M, Arcidiacono G, Rigamonti L (1991) Clusters of nerve cell bodies enclosed within a common connective tissue envelope in the spinal ganglia of the lizard and rat. Cell Tissue Res 264, 209-214.

Parke WW, Watanabe R (1987) Lumbosacral intersegmental epispinal axons and ectopic ventral nerve rootlets. J Neurosurg 67, 269-277.

Perry MJ, Lawson SN (1998) Differences in expression of oligosaccharides, neuropeptides, carbonic anhydrase and neurofilament in rat primary afferent neurons retrogradely labelled via skin, muscle or visceral nerves. Neuroscience 85, 293-310.

Phillips LH 2nd, Park TS (1993) Electrophysiological mapping of the segmental innervation of the saphenous and sural nerves. Muscle Nerve 16, 827-831.

Poletti CE (1991) C2 and C3 pain dermatomes in man. Cephalalgia 11, 155-159.

Popken GJ, Farel PB (1997) Sensory neuron number in neonatal and adult rats estimated by means of stereologic and profile-based methods. J Comp Neurol 386, 8-15.

Pover CM, Barnes MC, Coggeshall RE (1994) Do primary afferent cell numbers change in relation to increasing weight and surface area in adult rats? Somatosens Mot Res 11, 163-167.

Poviraev NP, Chernikov YF (1967) Anatomy of the ansa cervicalis. Exerpta Medica 21, 219.

Price J (1985) An immunohistochemical and quantitative examination of dorsal root ganglion neuronal subpopulations. J Neurosci 5, 2051-2059.

Price J, Mudge AW (1983) A subpopulation of rat dorsal root ganglion neurones is catecholaminergic. Nature 301, 241-243.

Rambourg A, Clermont Y, Beaudet A (1983) Ultrastructural features of six types of neurons in rat dorsal root ganglia. J Neurocytol 12, 47-66.

Risling M, Hildebrand C, Aldskogius H (1981) Postnatal increase of unmyelinated axon profiles in the feline ventral root L7. J Comp Neurol 201, 343-351.

Risling M, Dalsgaard CJ, Cukierman A, Cuello AC. (1984) Electron microscopic and immunohistochemical evidence that unmyelinated ventral root axons make u-turns or enter the spinal pia mater. J Comp Neurol 225, 53-63.

Risling M, Dalsgaard CJ, Frisen J, Sjögren AM, Fried K. (1994) Substance P-, calcitonin gene-related peptide, growth-associated protein-43, and neurotrophin receptor-like immunoreactivity associated with unmyelinated axons in feline ventral roots and pia mater. J Comp Neurol 339, 365-386.

Salt TE, Hill RG (1983) Neurotransmitter candidates of somatosensory primary afferent fibres. Neuroscience 10, 1083-1103.

Samudrala S, Khoo LT, Rhim SC, Fessler RG (1999) Complications during anterior surgery of the lumbar spine: an anatomically based study and review. Neurosurg Focus 7, e9.

Sann H, McCarthy PW, Mader M, Schemann M (1995) Choline acetyltransferase-like immunoreactivity in small diameter neurones of the rat dorsal root ganglion. Neurosci Lett 198, 17-20.

Schmalbruch H (1987) The number of neurons in dorsal root ganglia L4-L6 of the rat. Anat Rec 219, 315-322.

Schmitt M, Kummer W, Heym C (1988) Calcitonin gene-related peptide (CGRP)-immunoreactive neurons in the human cervico-thoracic paravertebral ganglia. J Chem Neuroanat 1, 287-292.

Schoenen J, Grant G (2004) Spinal cord: Connections. In *The Human Nervous System.* 2nd edn. (Paxinos G, Mai JK, eds.). Elsevier Academic Press, San Diego.

Schouenborg J (1984) Functional and topographical properties of field potentials evoked in rat dorsal horn by cutaneous C-fibre stimulation. J Physiol 356, 169-192.

Schultzberg M (1983) Bombesin-like immunoreactivity in sympathetic ganglia. Neuroscience 8, 363-374.

Shen J, Wang HY, Chen JY, Liang BL (2006) Morphologic analysis of normal human lumbar dorsal root ganglion by 3D MR imaging. AJNR Am J Neuroradiol 27, 2098-2103.

Shortland P, Woolf CJ (1993) Morphology and somatotopy of the central arborizations of rapidly adapting glabrous skin afferents in the rat lumbar spinal cord. J Comp Neurol 329, 491-511.

Sim IW, Webb T (2004) Anatomy and anaesthesia of the lumbar somatic plexus. Anaesth Intensive Care 32, 178-187.

Stacey RJ, O'Leary ST, Hamlyn PJ (1995) The innervation of the trapezius muscle: a cervical motor supply. J Craniomaxillofac Surg 23, 250-251.

Strack AM, Sawyer WB, Marubio LM, Loewy AD (1988) Spinal origin of sympathetic preganglionic neurons in the rat. Brain Res 455, 187-191.

Sugiura Y, Terui N, Hosoya Y (1989) Difference in distribution of central terminals between visceral and somatic unmyelinated (C) primary afferent fibers. J Neurophysiol 62, 834-840.

Sugiura Y, Terui N, Hosoya Y, Tonosaki Y, Nishiyama K, Honda T (1993) Quantitative analysis of central terminal projections of visceral and somatic unmyelinated (C) primary afferent fibers in the guinea pig. J Comp Neurol 332, 315-325.

Szurszewski JH (1981) Physiology of mammalian prevertebral ganglia. Annu Rev Physiol 43, 53-68.

Tajti J, Moller S, Uddman R, Bodi I, Edvinsson L (1999) The human superior cervical ganglion: neuropeptides and peptide receptors. Neurosci Lett 263, 121-124.

Takahashi Y, Nakajima Y (1996) Dermatomes in the rat limbs as determined by antidromic stimulation of sensory C-fibers in spinal nerves. Pain 67, 197-202.

Takahashi K, Ninomiya T (1987) Morphological changes of dorsal root ganglion cells in the process-forming period. Prog Neurobiol 29, 393-410.

Takahashi Y, Nakajima Y, Sakamoto T (1994) Dermatome mapping in the rat hindlimb by electrical stimulation of the spinal nerves. Neurosci Lett 168, 85-88.

Takahashi Y, Chiba T, Kurokawa M, Aoki Y (2003) Dermatomes and the central organization of dermatomes and body surface regions in the spinal cord dorsal horn in rats. J Comp Neurol 462, 29-41.

Tandrup T (1995) Are the neurons in the dorsal root ganglion pseudounipolar? A comparison of the number of neurons and number of myelinated and unmyelinated fibres in the dorsal root. J Comp Neurol 357, 341-347

Tata AM, Plateroti M, Cibati M, Biagioni S, Augusti-Tocco G (1994) Cholinergic markers are expressed in developing and mature neurons of chick dorsal root ganglia. J Neurosci Res 37, 247-255.

Tata AM, De Stefano ME, Srubek Tomassy G, Vilaro MT, Levey AI, Biagioni S (2004) Subpopulations of rat dorsal root ganglion neurons express active vesicular acetylcholine transporter. J Neurosci Res 75, 194-202.

Thippeswamy T, Morris R (2002) The roles of nitric oxide in dorsal root ganglion neurons. Ann N Y Acad Sci 962, 103-110.

Traub RJ, Sedivec MJ, Mendell LM (1986) The rostral projection of small diameter primary afferents in Lissauer's tract. Brain Res 399, 185-189.

Tubbs RS, Salter EG, Oakes WJ (2005) Anatomic landmarks for nerves of the neck: a vade mecum for neurosurgeons. Neurosurgery 56, 256-260.

Vera PL, Nadelhaft I (1990) Conduction velocity distribution of afferent fibers innervating the rat urinary bladder. Brain Res 520, 83-89.

Vizzard MA, Erdman SL, Erickson VL, Stewart RJ, Roppolo JR, De Groat WC (1994) Localization of NADPH diaphorase in the lumbosacral spinal cord and dorsal root ganglia of the cat. J Comp Neurol 339, 62-75.

Wessels WJ, Feirabend HK, Marani E (1994) The rostrocaudal organization in the dorsal root ganglia of the rat: a consequence of plexus formation? Anat Embryol 190, 1-11.

Whitelaw V, Hollyday M (1983) Neural pathway constraints in the motor innervation of the chick hindlimb following dorsoventral rotations of distal limb segments. J Neurosci 3, 1226-1233.

Wischnitzer S. (1967) *Atlas and Dissection Guide for Comparative Anatomy.* Freeman, San Francisco

Woolf CJ (1987) Central terminations of cutaneous mechanoreceptive afferents in the rat lumbar spinal cord. J Comp Neurol 261, 105-119.

Yabuki S, Kikuchi S (1996) Positions of dorsal root ganglia in the cervical spine. An anatomic and clinical study. Spine 21, 1513-1517.

Ygge J (1984) On the organization of the thoracic spinal ganglion and nerve in the rat. Exp Brain Res 55, 395-401.

Ygge J, Grant G (1983) The organization of the thoracic spinal nerve projection in the rat dorsal horn demonstrated with transganglionic transport of horseradish peroxidase. J Comp Neurol 216, 1-9.

Zhang J, Tsuzuki N, Hirabayashi S, Saiki K, Fujita K (2003) Surgical anatomy of the nerves and muscles in the posterior cervical spine: a guide for avoiding inadvertent nerve injuries during the posterior approach. Spine 28, 1379-1384.

Zhou Y, Ling EA (1999) Neuronal nitric oxide synthase in the neural pathways of the urinary bladder. J Anat 194, 481-496.

Zhou Y, Mack PO, Ling EA (1998) Localization of nicotinamide adenine dinucleotide phosphate-diaphorase reactivity and nitric oxide synthase immunoreactivity in the lumbosacral dorsal root ganglia in guinea pigs. J Hirnforsch 39, 119-127.

Zigmond MJ, Bloom FE, Landis SC, Roberts JL, Squire LR (eds). (1999) *Fundamental Neuroscience.* Academic Press, San Diego.

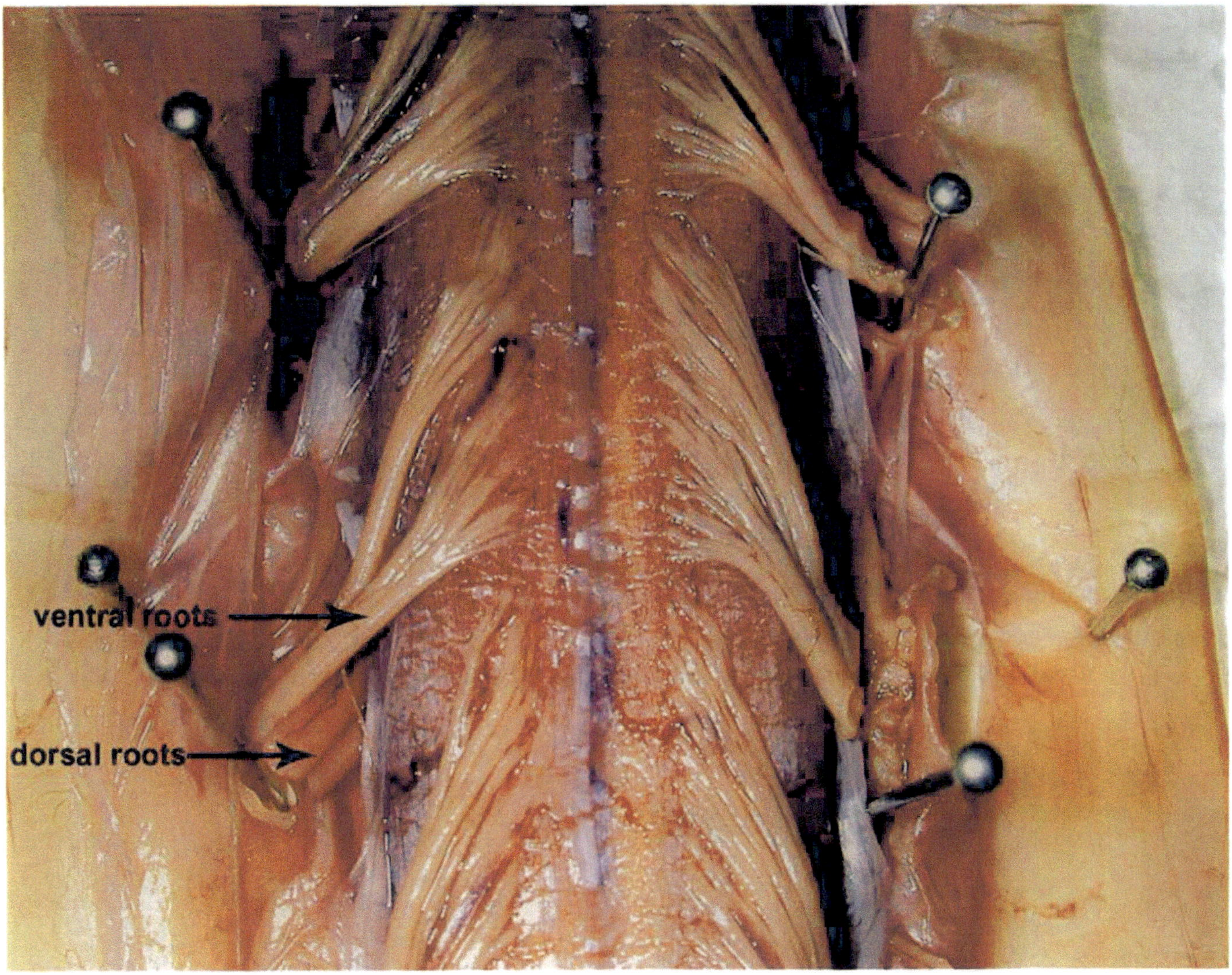

Figure 4.1 Dorsal and ventral roots of spinal nerves

This is a photograph of a dissection showing the ventral surface of the spinal cord and the ventral and dorsal rootlets. Groups of rootlets form the dorsal and ventral roots of each spinal nerve. The dura and arachnoid have been removed to expose the spinal cord. (Photograph courtesy of Professor Mete Erturk, Ege University)

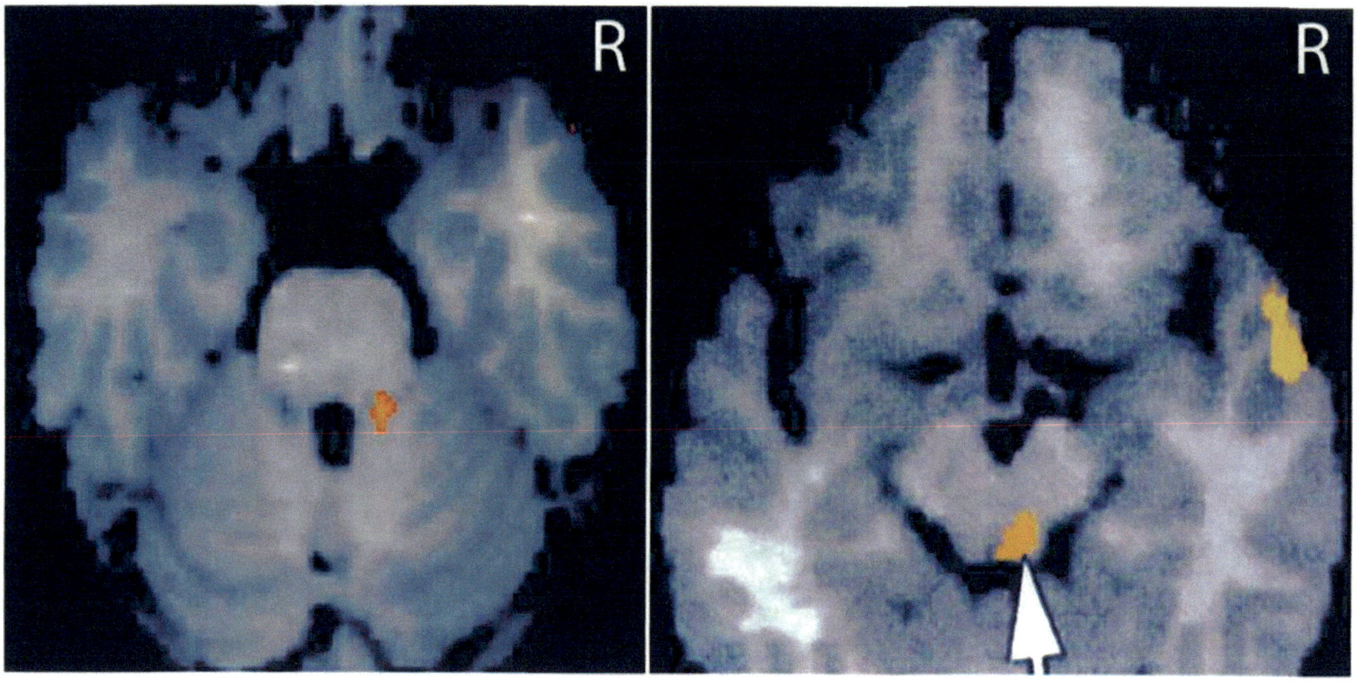

Figure 9.10 Human brain activation during micturition

A PET-study of the brain activation during micturition in humans shows brain activation in the PMC (left) and in the PAG (arrow right) (from Blok et al., 1997b).

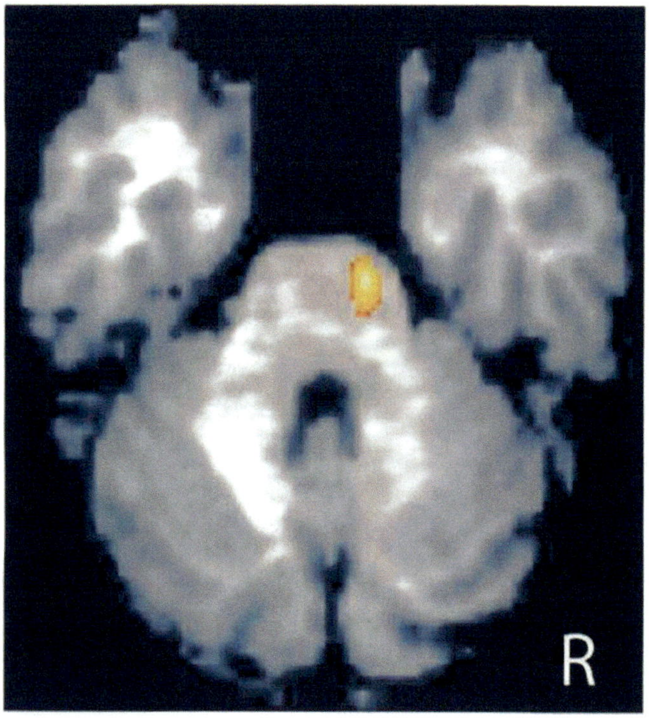

Figure 9.18 PET-scans in humans during failure to micturate

About half of the human volunteers that were asked to micturate while lying with their head in the PET-scanner could, for emotional reasons, not perform. In these cases the PMC was not activated. Instead, a region more ventral in the pons on the right side (artificially indicated with black) was activated, possibly representing the L-region or continence center (from Blok *et al.*, 1997a).

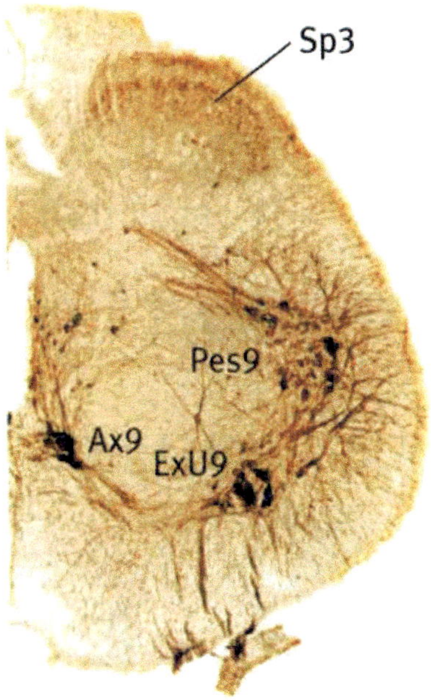

Figure 13.1 ChAT reactivity in lumbar spinal cord

This photomicrograph shows prominent ChAT immuno-staining in the dorsal horn, most prominently in lamina 3 (Sp3). It also shows dense reactivity in three lamina 9 motoneuron groups (Ax9, ExU9, and Pes9).

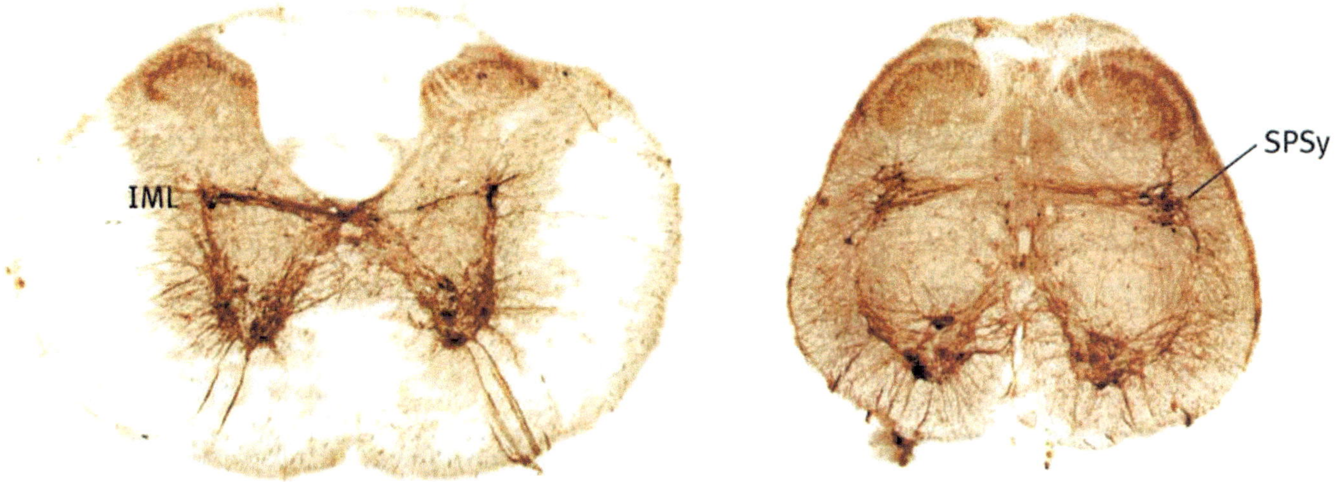

Figure 13.2 ChAT reactivity in sympathetic and parasympathetic preganlionic neurons

This photomicrograph shows prominent ChAT immuno-staining in the thoracic intermediolateral group (IML) and in the sacral parasympathetic group (SPSy).

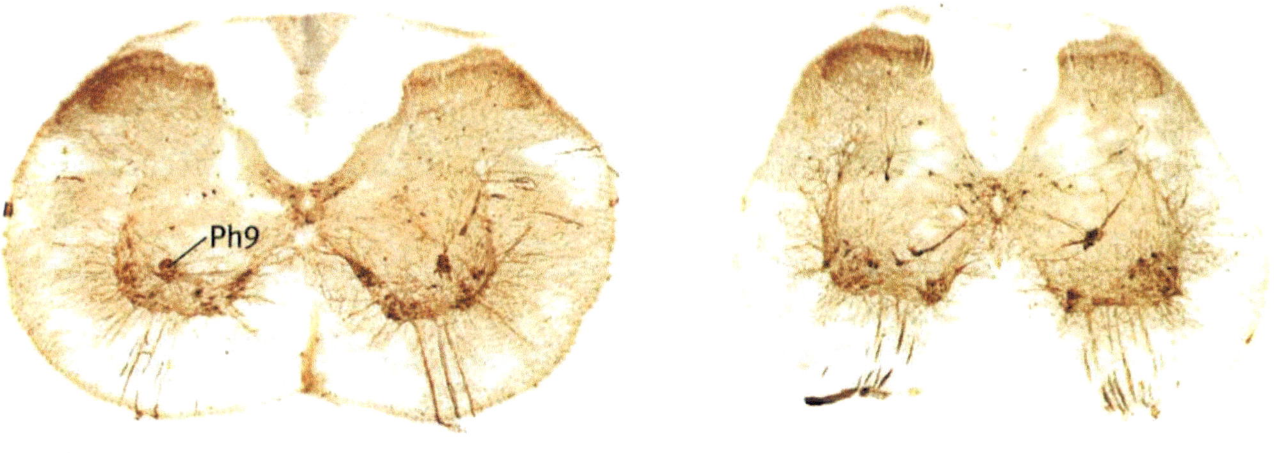

Figure 13.3 ChAT reactivity in cervical and lumbar motoneurons

This photomicrograph shows prominent ChAT immuno-staining in motoneurons at C4 (left) and at L3 (right). Dendritic bundling is a feature of the phrenic nucleus (Ph9).

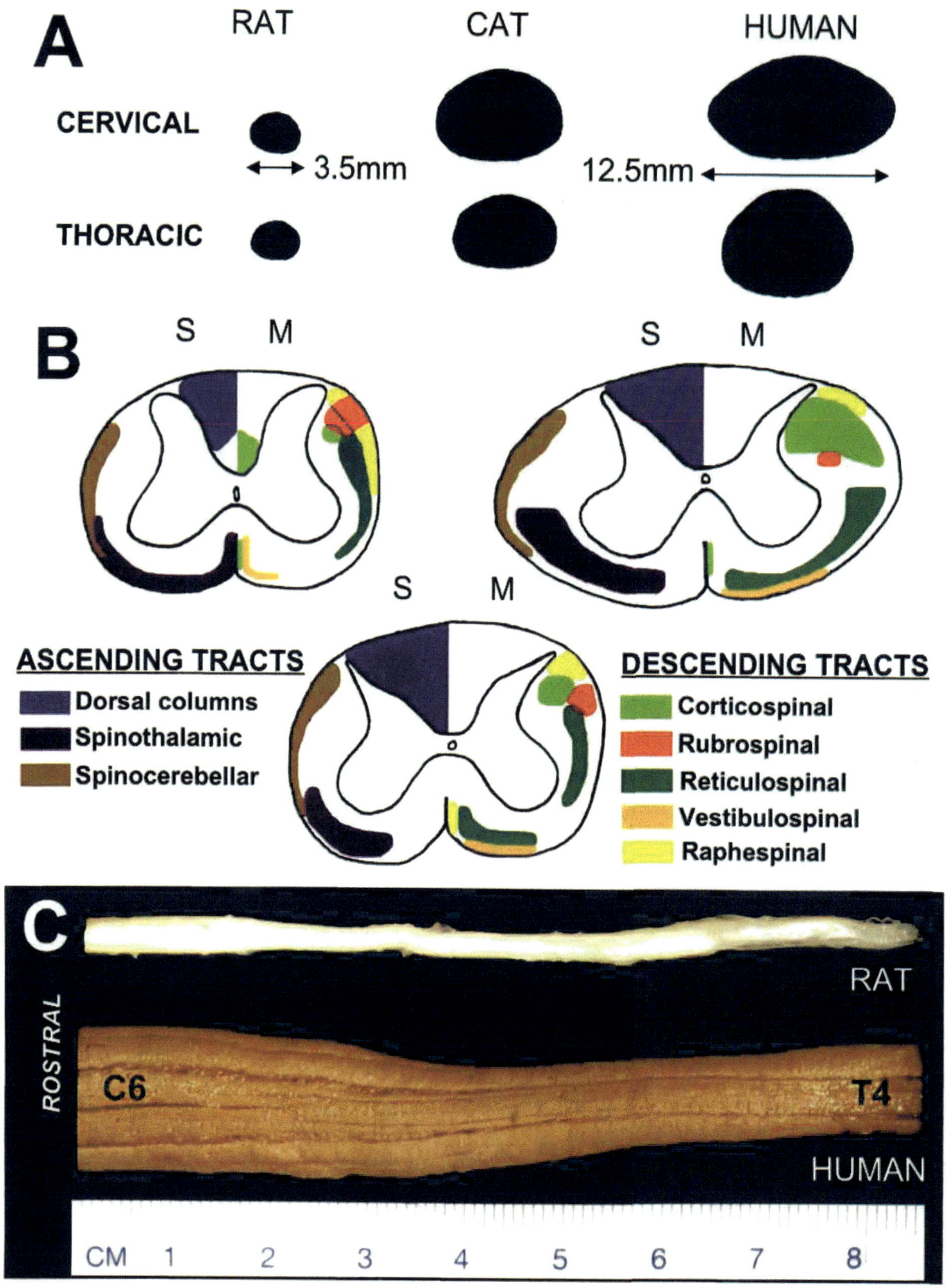

Figure 14.1 Comparative anatomy of the rat, cat and human spinal cord

A: The relative size and shape of cross-sections of rat, cat and human spinal cord at the level of the cervical and lumbar enlargements. B: Approximate location of some ascending and descending tracts in each species. Top row – rat and human; bottom row – cat. Unlike the cross-sections shown in A, these diagrams are not drawn to scale. Dotted lines indicate areas where tracts appear to overlap. Note that, while tracts are drawn as discrete entitities, in some cases (such as the reticulospinal and raphespinal projections), fibres can be quite widely scattered in white matter and the diagrams therefore indicate the position of most axons of a particular type. S = sensory; M = motor. C: Low-power views of the dorsal surface of a complete rat spinal cord and about six segments of human spinal cord. The entire rat spinal cord is about 9 cm in length; in comparison, individual segments of human cervical spinal cord can be 1.5-1.6 cm in length. The longest human spinal cord segment is T6, about 2.2 cm long (Ko *et al.*, 2004) or almost 25% the length of the rat spinal cord.

5 The Spinal Cord Blood Vessels

Oscar U Scremin

Blood flow and spinal cord function

Blood flow is intricately related to function in the spinal cord, as in the rest of the nervous system. Exchange of gases and nutrients is commonly cited as a cardinal function of blood circulation. However, other effects of circulatory convection, are important for the optimal operation of nervous tissue. Metabolic heat in active brain areas need to be cleared by circulatory convection requiring enhancement of blood flow to avoid a local elevation in temperature. Cooling the spinal cord within a narrow range is associated with enhanced excitation, inhibition and post-tetanic potentiation (Koizumi *et al.*, 1954). Thus, local cooling by circulatory convection could result in enhanced synaptic efficacy. In addition, perfused blood vessels provide a 'sink' for the effective dissipation by diffusion of neurotransmitter gradients, thus accelerating the termination of neurotransmitter action. Partial pressure of gases in the neuronal microenvironment is also controlled by the circulation. Given a stable metabolic CO^2 production and arterial PCO^2, local PCO^2 will decrease when local blood flow increases. Decreased PCO^2 is associated with an increase in synaptic efficacy within the spinal cord (Esplin *et al.*, 1973). All of these functions require adaptations of the circulatory system to match the rate of blood flow with the levels of synaptic and metabolic activity. The first of these adaptations occurs during development. The rate of local blood flow is dependent upon, among other factors, hydraulic conductance which in turn correlates directly with the number of blood vessels connected in parallel. It is then not surprising that angiogenesis accompanies neurogenesis. Capillaries are first detected within the neuroepithelium of mice at day 9 (Herken *et al.*, 1989). In rat embryos, vascular development of the spinal cord is well advanced by days 12-13 of development (Simon-Marin *et al.*, 1983).

A capillary network can be found on the surface of the human spinal cord at the 8 mm stage and by the end of the fourth month of pregnancy, capillaries are evenly distributed between gray and white matter. Greater abundance of capillaries in gray matter is later achieved and it appears to reflect the adult pattern at term (Woollam and Millen, 1955).

Angiogenesis is central to the development and differentiation of the nervous system, and a concerted action of a number of angiogenic factors leads to the parallel growth and differentiation of nerve cells and the blood vessel networks

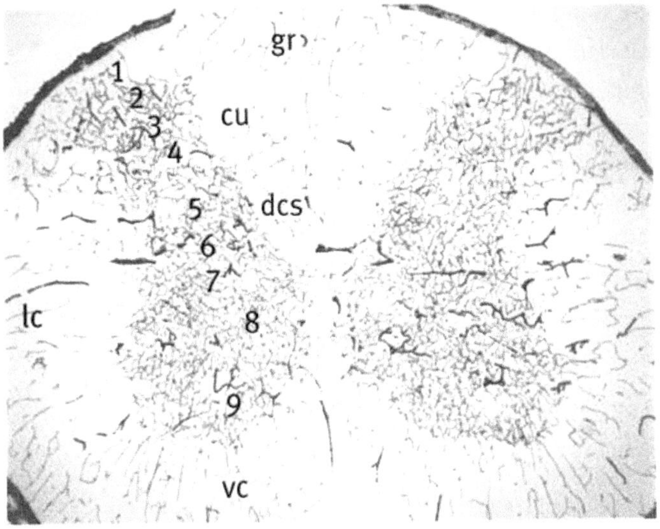

Figure 5.1 Coronal section of spinal cord at C5 level from an animal injected with diluted India ink to fill all vessels

The hemicord on the right shows the contrast between capillary abundance of grey and white matter. Numbers identify Rexed's laminae; gr = gracile fasciculus; cu = cuneate fasciculus; dcs = dorsal corticospinal tract; lc = lateral column; vc = ventral column.

surrounding them (Zagzag, 1995). In fact, vascular endothelial growth factor (VEGF), the main molecule inducing proliferation of endothelial cells has been recently shown to also influence growth and differentiation of nerve cells. The same applies to fibroblast growth factor (FGF). On the other hand, neural growth factors such as nerve growth factor (NGF) and brain-derived neurotrophic factor (BDNF) can also induce angiogenesis (Raab and Plate 2007). Once development is completed, proliferation and differentiation of nerve cells as well as angiogenesis essentially cease and can only occur in restricted regions of vertebrates under physiological conditions (Risau, 1997; Bordey, 2006). In these regions, dividing nerve cells are found associated with dividing endothelial cells (Palmer *et al.*, 2000).

Capillary networks

Early work on vascular microanatomy revealed that the clear anatomical segregation of anterior, lateral and posterior columns, composed mostly of axons and glia and central gray matter, with neuronal bodies and synapses in addition to axons and glia, is paralleled by a spatial variation in blood vessel density (Figure 5.1). A large body of work was devoted in the 1930s and 1940s to detailed descriptions of the microvascular network of the spinal cord in relation to cyto and myeloarchitecture (Fazio, 1938; Zeman and Innes, 1963). Several important features were discovered. The large increase in capillary abundance characteristic of the central gray matter was found associated with the presence of synapses rich in mitochondria more than with nerve cell bodies. This was supported by Scharrer's observations of various central nervous system (CNS) regions in many animal species comparing nerve cell and mitochondria-rich synapses densities (Scharrer, 1945). Thus, the rate of brain tissue energy exchange was early on related to the level of vascularization. Some of the differences in capillary densities within the gray and white matter related to specific functional regions described by earlier workers (Woollam *et al.*, 1958) have been confirmed by functional blood flow imaging as discussed below, such as the greater capillary supply of the pyramidal (cortico-spinal) tract that Craigie (Zeman and Innes, 1963) reported with twice the number of capillaries as the fasciculus cuneatus.

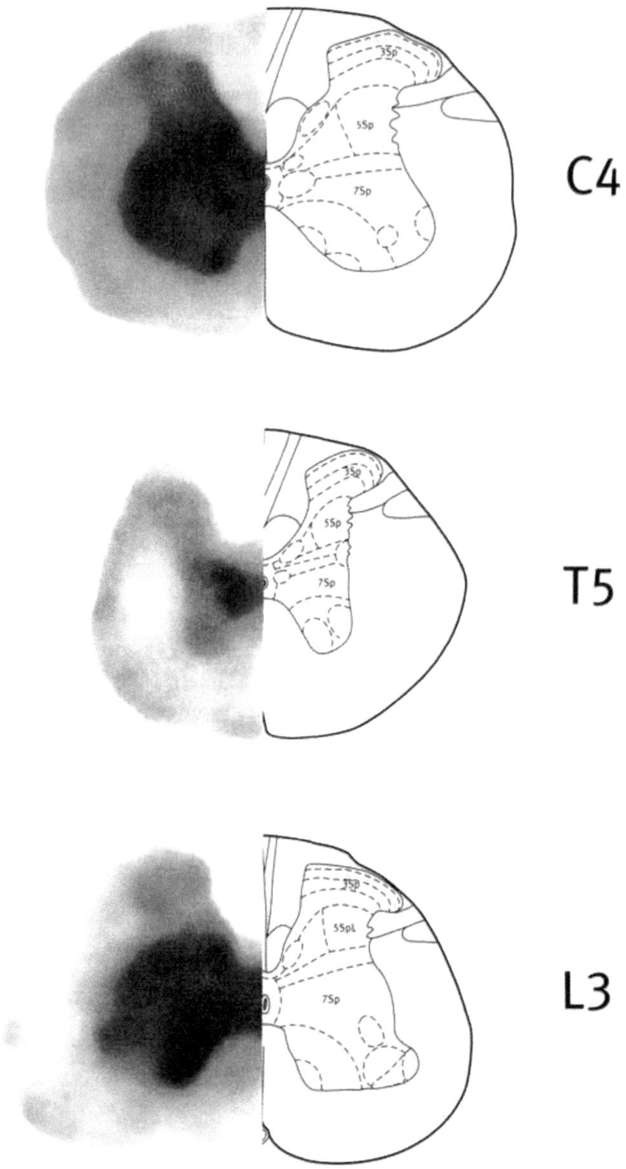

Figure 5.2 Local arterial supply in the spinal cord

Autoradiographs from an animal infused with Iodo-14C-antipyrine to measure blood flow (left side images) paired with anatomical outlines of the corresponding spinal cord level (right side images). Darkness on the autoradiographs correlates with 14C activity. Simultaneous measurement of the time course of arterial 14C activity allows calculation of local blood flow, to be found for different spinal cord regions in Figure 5.3. (Sp 3 = lamina 3 etc.)

Spinal cord blood flow imaging

The assessment of capillary abundance, a rather constant parameter for any given CNS compartment, cannot offer information about moment to moment variations in local blood flow that may reflect neural function. The changing requirements of energy substrates, local thermoregulation and gaseous exchange associated with variations in functional level of the nervous system are met by control of the hydraulic resistance of the blood vessels system mediated by the contractile state of vascular smooth muscle.

The advent the autoradiographic methodology (Kety *et al.*, 1955) allowed visualization of local blood flow with the use of diffusible tracers such as Iodo-antipyrine (Sakurada *et al.*, 1978). This technique provides blood flow 'images', as illustrated in Figure 5.2. These images indicate a variation of blood flow not only between, but also within, the gray and white matter compartments. The lowest levels of blood flow within the white matter are found in the space corresponding to the dorsal columns, with the exception of their deepest portion occupied in the rat by the direct cortico-spinal fibers that show the highest blood flow of all white matter regions (Blisard *et al.*, 1995). Within gray matter, the highest blood flow is found in the ventral horn (Figures 5.2 and 5.3).

Arterial anatomy

The arterial supply to the spinal cord originates from ventral, posterolateral and dorsal systems that extend throughout the entire cord. One ventral spinal artery (vsp) is virtually a continuous channel that extends from the cervical segments to the filum terminale, analogous to the human anterior spinal artery (Figure 5.4). This vessel can be found at the entrance of the ventral median fissure. Two dorsal spinal arteries (dsp) located just ventral to the entrance of the dorsal roots, also tend to form continuous channels along the cervical, thoracic and lumbar cords, but in this case the continuity of these channels is often interrupted and the single vessels are in places replaced by two or more parallel elements. Less consistently, two additional longitudinal arterial channels formed by anastomoses between the lateral spinal arteries (lsp) are found on the lateral surface of the cord. The lateral spinal arteries are found about midway between the attachment of the dorsal and that of the ventral roots. A median dorsal spinal artery (mdosa), situated at or close to the dorsal septum, is frequently found along all segments of the cord (Figure 5.4). The longitudinal channels

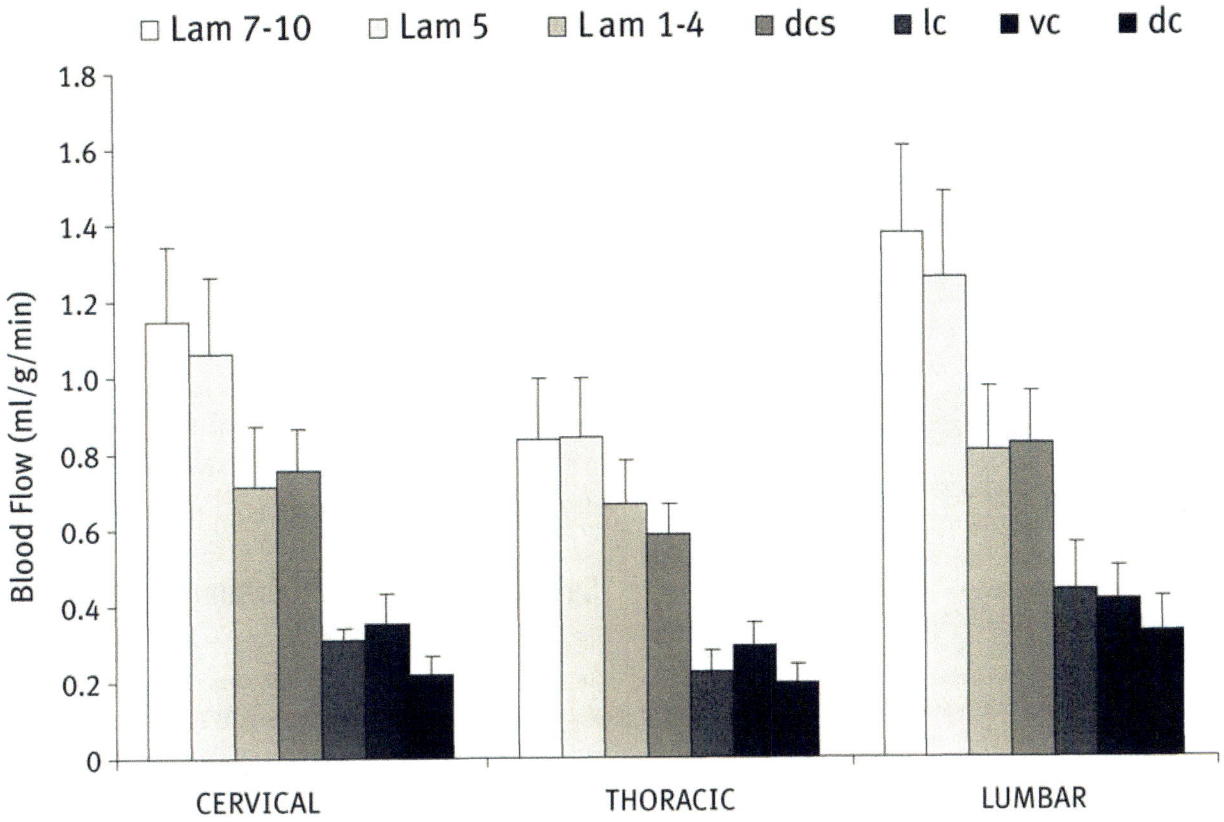

Figure 5.3 Blood flow in different laminae of the spinal cord

Means (bars) and standard errors (brackets) of local spinal cord blood flow (ml/g/min) calculated in eight normal Sprague-Dawley rats with the autoradiographic technique. Tracer activity was acquired from regions of interest in autoradiographs including Rexed's laminae 1-4, 5, 7-10, dorsal column (dc), dorsal corticospinal tract (dcs), lateral column (lc), and ventral column (vc).

described above are joined on the surface of the cord by transverse anastomoses that may take the form of a well-developed artery (arterial transverse anastomotic circle) (Figure 5.4) or an irregular anastomotic network.

The ventral spinal artery originates rostrally from two caudally directed branches stemming off the two vertebral arteries before they join to form the basilar artery. Radicular arteries, branching off deep cervical, intercostal, lumbar and sacral arteries, enter the spinal canal through the intervertebral foraminae along with the spinal nerves and are variable in diameter. They divide, outside the dura mater, into branches that follow the course of ventral and dorsal roots (ventral and dorsal radicular arteries respectively). Some of the central radicular arteries supply blood only to the ventral roots but a number of them are larger and merge into the ventral spinal artery, either directly or after dividing into a rostral and a caudal branch. A similar arrangement is found for the dorsal spinal arteries. The number of the larger ventral and dorsal root arteries that contribute to the ventral and dorsal spinal arteries is quite variable. They are more abundant in the cervical and lumbar enlargements and almost absent in the thoracic cord (Tveten, 1976) found that dorsal radicular arteries were more numerous and evenly distributed, but of a smaller size, than the ventral radicular arteries. The total number of ventral radicular arteries merging into the anterior spinal artery in Tveten's study ranged from 3 to 14, with an average of 7. They were more frequent at C5 and C6, and from T11 to L1. The lowest frequencies of occurrence of these vessels were at C1-C3, T1-T3 and caudal to L2. Dorsal radicular arteries merging into the dorsal spinal arteries ranged from 16 to 35, with an average of 25, and were more abundant at the level of the cervical and lumbar enlargements. Although some authors have reported the consistent presence of a single large ventral radicular artery supplying most of the lumbar enlargement of the rat spinal cord (Tveten 1976), an arrangement similar to that of the human spinal cord (Adamkiewicz, 1881, 1882; Lazorthes *et al.*, 1971). This was contradicted by a later study (Schievink *et al.*, 1988) that identified between three and five such vessels between the T11 and L4 segments in a series of 26 Wistar rats. Our own material from Sprague-Dawley rats confirms the existence of several ventral and dorsal radicular arteries supplying the lumbar cord.

The ventral spinal artery gives off, at regular intervals, the ventromedian or sulcal arteries (Figure 5.4). These vessels

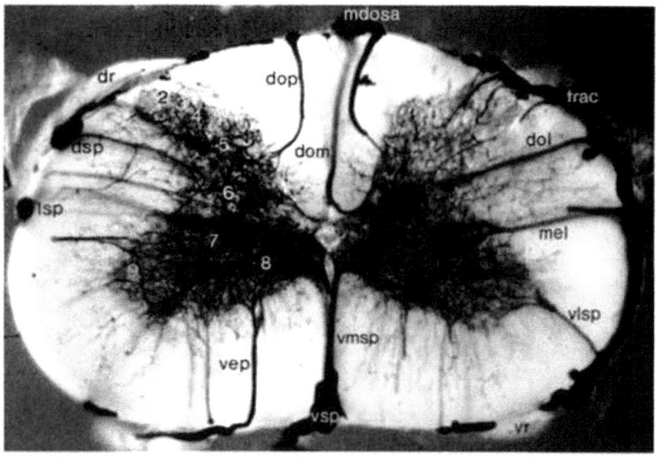

Figure 5.4 Spinal cord arterial system demonstrated with an injected contrast medium

This figure shows a coronal section, 1 mm thick, of the spinal cord (C5) injected intraarterially with neoprene latex-black ink mixture, which does not penetrate capillaries. Ventral (vsp), dorsal (dsp), median dorsal (mdosa), and lateral (lsp) spinal arteries are labelled. Contrast is provided by the presence of injection material in small arteries and precapillary vessels. The remarkable difference in microvascular density of gray and white matter is apparent in this unstained section. Variations in capillary density are shown in Figure 5.1.

ascend in the ventromedian sulcus, usually in pairs destined one to a side, and reach the medial junction between the gray commissure and the ventral horn. They are then distributed widely within the gray matter, giving one or more branches to the ventral horn with collaterals to the commissure, lateral gray column, and base of the dorsal horn. The capillary loops that originate in these distribution vessels travel beyond the boundary between gray and white matter and supply the anterior and ventral portion of the lateral white columns. The rest of the spinal cord is supplied by perforating rami from the pial arterial network that interconnects the dorsal, dorsolateral, and lateral longitudinal arterial channels described above. These perforating arteries are arranged in a radial orientation and travel through the white matter, giving occasional small branches, to a final distribution in the gray matter. They are designated ventral paramedian, ventrolateral, mediolateral, dorsolateral, dorsal paramedian, and dorsomedian arteries (Figure 5.4).

Venous anatomy

An intricate venous plexus is found on the surface of the spinal cord. On the ventral side, a ventral spinal vein (vsv) is found dorsal or adjacent to the ventral spinal artery. Like its arterial counterpart, this vessel spans the entire length of the cord, but with notable variations in calibre. The cervical portion of the vsv parallels in size the vsa, but it is more slender at the thoracic segments. Approximately at the junction between the thoracic and lumbar cords, a large ventral radicular vein divides into a cranial small vessel continuous with the thoracic vsv and a much larger caudal vessel continuous with the similar sized lumbar vsv. This vein ends as a continuous channel that follows a tortuous path along the sacral segments and cauda equina. On the dorsal cord surface, a dorsal spinal vein (dsv) is found ventral or adjacent to the dorso-median spinal artery. The ventral and dorsal spinal veins are connected by a large number of smaller veins that surround the cord, although occasionally, and particularly at the cervical and lumbar enlargements, prominent venous transverse anastomotic circles can be found. Internal spinal veins drain into the ventral or dorsomedian veins and into other vessels of the cord surface plexus.

Spinal cord lymphatic drainage

There is no evidence for lymph vessels inside the spinal cord or any part of the central nervous system. However, it has been known for over a century that large tracer molecules and particles added to the cerebrospinal fluid can be found in the lymphatics of the head, neck and spine at a later time (Brierley and Field 1948). Following the work of Ivanow, apparently the first author to detect migration of India ink from the lumbar subarachnoid space into abdominal and thoracic lymph nodes, Brierley and Field found that carbon particles accumulated in the dural sacs surrounding the nerve roots of the lumbo-sacral and cervical regions. They also observed a leash of fine black lines, possibly corresponding to lymph vessels, passing from the region of the nerve root dural sacs towards the paravertebral lymph nodes. Ink particles were also found in the cervical, posterior thoracic, posterior abdominal and pelvic lymph nodes (Brierley and Field, 1948).

The modern evidence for a route of bulk fluid flow between interstitial and subarachnoid spaces and the lymphatics, as detected by the efflux of large molecules (e.g. radio-iodinated albumin), has been extensively reviewed (Bradbury and Cserr, 1985). The relative magnitude of this efflux mechanism has been variably estimated from one third (Bradbury and Cole, 1980) to one half (Boulton *et al.*, 1996) of that ascribed to transfer of these molecules through blood vessels. The flow through these pathways is slow, with peak concentrations of radio-iodinated albumin in lymph nodes reached several hours after injection into the cerebrospinal fluid.

Bulk flow from cerebrospinal fluid to lymph may be particularly important due to the existence of the blood brain barrier that greatly limits passage of large molecules such as proteins and polar metabolic products through brain capillaries, except for a few areas where the blood brain barrier does not exist. It has been suggested that the lymphatic route described above could represent an important channel of communication between the brain and the immunological system. In addition, as in other territories, the lymphatic system may contribute to regulation of tissue volume.

Experimental spinal cord ischemia

The rat has been extensively used to model the condition of spinal cord ischemia secondary to occlusion of the aorta during reparative surgery of aneurisms of this vessel in humans (LeMay *et al.*, 1987; Follis *et al.*, 1993; Marsala and Yaksh, 1994). In this context, it is important to understand the role of collateral circulation in protecting the spinal cord from ischemia. The longitudinal arterial channels of the rat spinal cord (ventral, dorsal, and median dorsal spinal arteries) receive blood from the vertebral arteries that provide all the radicular arteries of the cervical cord and connect rostrally through the basilar artery and extracranial anastomoses with the internal and external carotid arteries system (Scremin, 2004). The arterial longitudinal channels also receive radicular arteries originating in branches of the aorta at the thoracic, and lumbar levels. Collateral circulation provided by these continuous channels could ameliorate the negative impact of arterial occlusions on spinal cord blood flow. Experimental verification of this possibility was provided by spinal cord blood flow measurements with the autoradiographic technique (Follis *et al.*, 1995). When the subclavian arteries, from which the vertebral arteries originate were occluded, blood flows of gray and white matter of the cervical segment were reduced to about 60% of control. Blood flows of thoracic and lumbar cords on the other hand did not change. Thus it appears that collateral blood flow from the rostral connection of the vertebral arteries to the basilar artery and/or from thoracic radicular arteries via the longitudinal arterial channels were able to partially compensate for elimination of the primary arterial supply of the cervical cord. The phenomenon has functional significance because temporary occlusion of both subclavian arteries does not result in permanent impairment of spinal cord function. When the thoracic aorta was cross-clamped, in addition to subclavian occlusion, blood flow of gray and white matter decreased further to about 40% of control in the cervical cord, and to 15% of control in the thoracic cord, while blood flow was undetectable in the lumbar cord. Again a partial although minimal compensation by collateral circulation from rostral levels was possible for blood flow of the thoracic cord, while no compensation existed for the more distal lumbar cord. The severe ischemia of the thoracic and lumbar cord afforded by subclavian and thoracic aorta cross-clamping, if sustained for a critical occlusion time of 10-11 minutes, results in a permanent spastic paraplegia in most cases (Follis *et al.*, 1993; LeMay *et al.*, 1987). On the other hand, when only the thoracic aorta was occluded by an intravascular balloon without subclavian occlusion, the residual blood flow of the cord was enough to prolong the time of occlusion required to provoke a permanent spastic paraplegia to more than 40 minutes. In order to obtain a similar critical occlusion time as with the combined aortic and subclavian occlusion, it was necessary to reduce collateral blood flow by lowering the arterial blood pressure proximal to the aortic occlusion from 127 to 40 mm Hg (Taira and Marsala, 1996). Thus, the protective effect of collateral circulation is significant even for a complete interruption of the thoracic aorta blood flow in the rat.

Blood flow in spinal cord trauma

Direct experimental trauma to the rat spinal cord induces a decrease in gray and white matter blood flow at the site of injury and variable alterations in segments adjacent to the site of trauma (Rivlin and Tator, 1978; Tei *et al.*, 2005). The severity of blood flow reduction correlates with the functional deficits (Fehlings *et al.*, 1989; Holtz *et al.*, 1990). Time course studies have indicated that blood flow remains at low levels in cases of severe trauma associated with paraplegia or recovers partially over several hours in models of milder injury with functional restoration (Rivlin and Tator, 1978; Holtz *et al.*, 1989). Spinal cord ischemia may play a critical role in the causation of functional deficits associated with trauma and may limit reparative processes.

References

Adamkiewicz A (1881) Die blutgefässe des menschlichen ruckenmarkes. I. Die gefässe der ruckenmarkssubstanz. Akad Wiss (Heidelberg) 84, 469-502.

Adamkiewicz A (1882) Die blutgefässe des menschlichen ruckenmarkes. II. Die gefässe der ruckenmarksoberflache. Akad Wiss (Heidelberg) 85, 101-130.

Blisard KS, Follis F, Wong R, Miller KB, Wernly JA, Scremin OU (1995) Degeneration of axons in the corticospinal tract secondary to spinal cord ischemia in rats. Paraplegia 33, 136-140.

Bordey A (2006) Adult neurogenesis: basic concepts of signaling. Cell Cycle 5, 722-728.

Boulton M, Young A, Hay J, Armstrong D, Flessner M, Schwartz M, Johnston M (1996) Drainage of CSF through lymphatic pathways and arachnoid villi in sheep: measurement of 125I-albumin clearance. Neuropathol Appl Neurobiol 22, 325-333.

Bradbury M, Cserr HF (1985) Drainage of cerebral interstitial fluid and cerebrospinal fluid into lymphatics. In: *Experimental Biology of the Lymphatic Circulation.* Johnston MG (ed). Elsevier, Amsterdam.

Bradbury MWB, Cole DF (1980) The role of the lymphatic system in drainage of cerebrospinal fluid. J Physiol-London 299, 353-365.

Brierley JB, Field EJ (1948) The connexions of the spinal sub-arachnoid space with the lymphatic system. J Anat 82, 153-166.

Esplin DW, Capek R, Esplin B (1973) An intracellular study of the actions of carbon dioxide on the spinal monosynaptic pathway. Can J Physiol Pharmacol 51, 424-436.

Fazio C (1938) L'angioarchitettonica del midollo spinale umano e i suoi rapporti con la cito-mielo-architettonica. Riv Patologia Ner.e Meutali 52, 252-291.

Fehlings MG, Tator CH, Linden RD (1989) The relationships among the severity of spinal cord injury, motor and somatosensory evoked potentials and spinal cord blood flow. Electroencephalogr Clin Neurophysiol 74, 241-259.

Follis F, Scremin OU, Blisard KS, Scremin AME, Pett SB, Scott WJ, Kessler RM, Wernly JA (1993) Selective vulnerability of white matter during spinal cord ischemia. J Cereb Blood Flow Metab 13, 170-178.

Follis F, Miller K, Scremin OU, Pett S, Kessler R, Temes T, Wernly JA (1995) Experimental delayed postischemic spinal cord hypoperfusion after aortic cross-clamping. Can J Neurol Sci 22, 202-207.

Herken R, Gotz W, Wattjes KH (1989) Initial development of capillaries in the neuroepithelium of the mouse. J Anat 164, 85-92.

Holtz A, Nystrom B, Gerdin B (1989) Spinal cord blood flow measured by 14C-iodoantipyrine autoradiography during and after graded spinal cord compression in rats. Surg Neurol 31, 350-360.

Holtz A, Nystrom B, Gerdin B (1990) Relation between spinal cord blood flow and functional recovery after blocking weight-induced spinal cord injury in rats. Neurosurgery 26 (6), 952-957.

Kety SS, Landau WM, Freygang JrWH, Rowland LP, Sokoloff L (1955) Estimation of regional circulation in the brain by the uptake of an inert gas. Fed Proc 14, 85.

Koizumi K, Malcolm JL, Brooks C (1954) Effect of temperature on facilitation and inhibition of reflex activity. Am J Physiol 179, 507-512.

Lazorthes G, Gouaze A, Zadeh JO, Santini JJ, Lazorthes Y, Burdin P (1971) Arterial vascularization of the spinal cord: Recent studies of the anastomotic substitution pathways. J Neurosurg 35, 253-262.

LeMay DR, Neal S, Zelenock GB, D'Alecy LG (1987) Paraplegia in the rat induced by aortic cross-clamping: Model characterization and glucose exacerbation of neurologic deficit. J Vasc Surg 6 (4), 383-390.

Marsala M, Yaksh TL (1994) Transient spinal ischemia in the rat: characterization of behavioral and histopathological consequences as a function of the duration of aortic occlusion. J Cereb Blood Flow Metab 14, 526-535.

Palmer TD, Willhoite AR, Gage FH (2000) Vascular niche for adult hippocampal neurogenesis. J Comp Neurol 425, 479-494.

Raab S, Plate KH (2007) Different networks, common growth factors: shared growth factors and receptors of the vascular and the nervous system. Acta Neuropathol (Berl) 13, 607-626.

Risau W (1997) Mechanisms of angiogenesis. Nature 386, 671-674.

Rivlin AS, Tator CH (1978) Regional spinal cord blood flow in rats after severe cord trauma. J Neurosurg 49, 844-853.

Sakurada O, Kennedy C, Jehle J, Brown JD, Carbin GL, Sokoloff L (1978) Measurement of local cerebral blood flow with iodo[14C]antipyrine. Am J Physiol 234, H59-H66.

Scharrer E (1945) Capillaries and mitochondria in neuropil. J Comp Neurol 83, 237-243.

Schievink WI, Luyendijk W, Los JA (1988) Does the artery of Adamkiewicz exist in the albino rat? J Anat 161, 95-101.

Scremin OU (2004) Cerebral Vascular System. In: Paxinos G (ed) *The Rat Nervous System.* Elsevier, Amsterdam, pp 1165-1202.

Simon-Marin R, Vilanova JR, Aguinagalde A, Barbera-Guillem E (1983) Vascular architecture of the developing spinal cord in the rat: a suggested model. J Embryol Exp Morphol 76, 27-36.

Taira Y, Marsala M (1996) Effect of proximal arterial perfusion pressure on function, spinal cord blood flow, and histopathologic changes after increasing intervals of aortic occlusion in the rat. Stroke 27, 1850-1858.

Tei R, Kaido T, Nakase H, Sakaki T (2005) Secondary spinal cord hypoperfusion of circumscribed areas after injury in rats. Neurol Res 27, 403-408.

Tveten L (1976) Spinal cord vascularity: IV. The spinal cord arteries in the rat. Acta Radiol (Diagn) 17 (4), 385-398.

Woollam DHM, Millen J (1955) The arterial supply of the spinal cord and its significance. J Neu Neurosurg Psychiat 18, 97-102.

Woollam DHM, Millen J, Blackwood W, Pennybacker J (1958) Discussion on vascular disease of the spinal cord. Proc R Soc Med 51, 540-550.

Zagzag D (1995) Angiogenic growth factors in neural embryogenesis and neoplasia. Am J Pathol 146, 293-309.

Zeman W, Innes JRM (1963) *Craigie's Neuroanatomy of the Rat.* Academic Press, New York.

6 Cytoarchitecture of the Spinal Cord

Claire Heise and Gulgun Kayalioglu

The laminae of Rexed

The gray matter of the spinal cord is morphologically heterogeneous. The cells of the spinal cord gray matter differ considerably in terms of size, shape and density. However, the gray matter does show some intrinsic organization, so that zones of regularity or laminae can be identified. These laminae were first described by Rexed (1954) in the cat. Since then, the concept of laminae has been adopted widely for describing cytoarchitectonic boundaries in the spinal cord in many species, though detailed examination and description of these laminae have only been completed in the mouse (Sidman *et al.*, 1971), rat (Molander *et al.*, 1984) and human (Schoenen and Faull, 2004), in addition to Rexed's original work in the cat. In general, the organization of the Rexed's laminae in these mammalian species is similar. Ten laminae are recognized and these are organized in a series of layers from dorsal to ventral, but with lamina 10 (also called area 10) found around the central canal (See Chapter 15). In this chapter, the major cytological characteristics of each of the laminae will be discussed. Many studies have shown that neurons within a lamina often show distinct dendritic architecture, chemoarchitecture and patterns of connections as well as function, and these issues will also be discussed. It is important to note that the borders of the laminae mentioned below generally mark transition zones rather than clear-cut boundaries.

Lamina 1

Cytoarchitecture

Lamina 1 is the most dorsal of the Rexed's laminae. It is a thin layer that covers the dorsal margin of the dorsal horn and bends laterally around its apex to also cover about half the lateral side of the head of the dorsal horn (see Chapters 15 and 16; Rexed, 1954; Molander *et al.*, 1984; Schoenen and Faull, 2004). Lamina 1 is traversed by many myelinated nerve fibers making it appear somewhat reticulated (Rexed, 1954; Molander *et al.*, 1984; Schoenen and Faull, 2004). Because of this, it can be difficult to identify a clear boundary between lamina 1 and the white substance which covers its dorsal aspect (Rexed, 1954; Molander *et al.*, 1984). The cells in lamina 1 are loosely distributed and vary considerably in terms of size and shape (Rexed, 1954; Molander *et al.*, 1984; Schoenen and Faull, 2004). A characteristic feature of lamina 1 is the presence of large elongated cells orientated tangentially to the dorsal curvature of the dorsal horn (Rexed, 1954; Molander *et al.*, 1984; Schoenen and Faull, 2004), however, smaller triangular, fusiform and multipolar shaped cells are more common (Rexed, 1954; Lima and Coimbra, 1986; Molander *et al.*, 1989; Schoenen and Faull, 2004).

Dendritic architecture

The dendritic trees of the cells in lamina 1 differ in the number and orientation of their primary dendrites, the extent and pattern of their branching, and the abundance or scarcity of dendritic spines (Gobel, 1978; Beal *et al.*, 1981; Lima and Coimbra, 1986; Lima *et al.*, 1993; Zhang *et al.*, 1996). Even so, distinctive dendritic structures seem to exist (Gobel, 1978; Lima and Coimbra, 1986; Zhang *et al.*, 1996) and these are generally found associated with certain cell shapes (Gobel, 1978; Lima and Coimbra, 1986; Zhang, 1996). Based on this relationship between cell shape and dendritic structure, Lima and Coimbra (1986), in the rat, were able to classify the neurons of lamina 1 into four morphological groups: fusiform, multipolar, flattened, and pyramidal. These findings were later confirmed (Valtschanoff *et al.*, 1992; Nagy *et al.*, 1994; Lekan and Carlton, 1995) and extended to other species including cat (Zhang *et al.*, 1996; Galhardo and Lima, 1999) and monkey (Zhang and Craig, 1997). The characteristic features of each of the four structural groups will be summarized in the next section.

Fusiform neurons

This cell type is characterized by a spindle shaped cell body oriented longitudinally, generally with two primary dendrites arising from opposite (rostral and caudal) poles (Lima and Coimbra, 1986; Galhardo and Lima, 1999). The dendrites of fusiform neurons branch profusely within lamina 1. In the rat, some distal branches have been reported to reach into outer zone of lamina 2 (Lima and Coimbra, 1986), whereas in cat and monkey, dendrites are confined to the borders of lamina 1 (Gobel, 1978; Beal, 1978; Light *et al.*, 1979; Beal *et al.*, 1981). Spines are common on the dendrites of fusiform neurons and are found predominately on distal dendrites. Cells in the fusiform group can be placed into two subclasses. Most are type A, which have longitudinal oriented dendrites only. The remainder are type B, which also have in addition a distinct ventral dendrite that can reach right down to lamina 3 (Lima

Figure 6.1 Rexed's laminae stained with NADPH-diaphorase

Transverse sections (section thickness 100 micrometer) of NADPH-diaphorase stained rat spinal cord, illustrating the cytological characteristics of Rexed's laminae. (a) lamina 1. (b) laminae 1—2. (c) lamina 2. (d) lamina 3. (e) lamina 4. (f) lamina 5 medial. (g) lamina 5 lateral. (h) lamina 7. (i) lamina 8. (j) lamina 10.

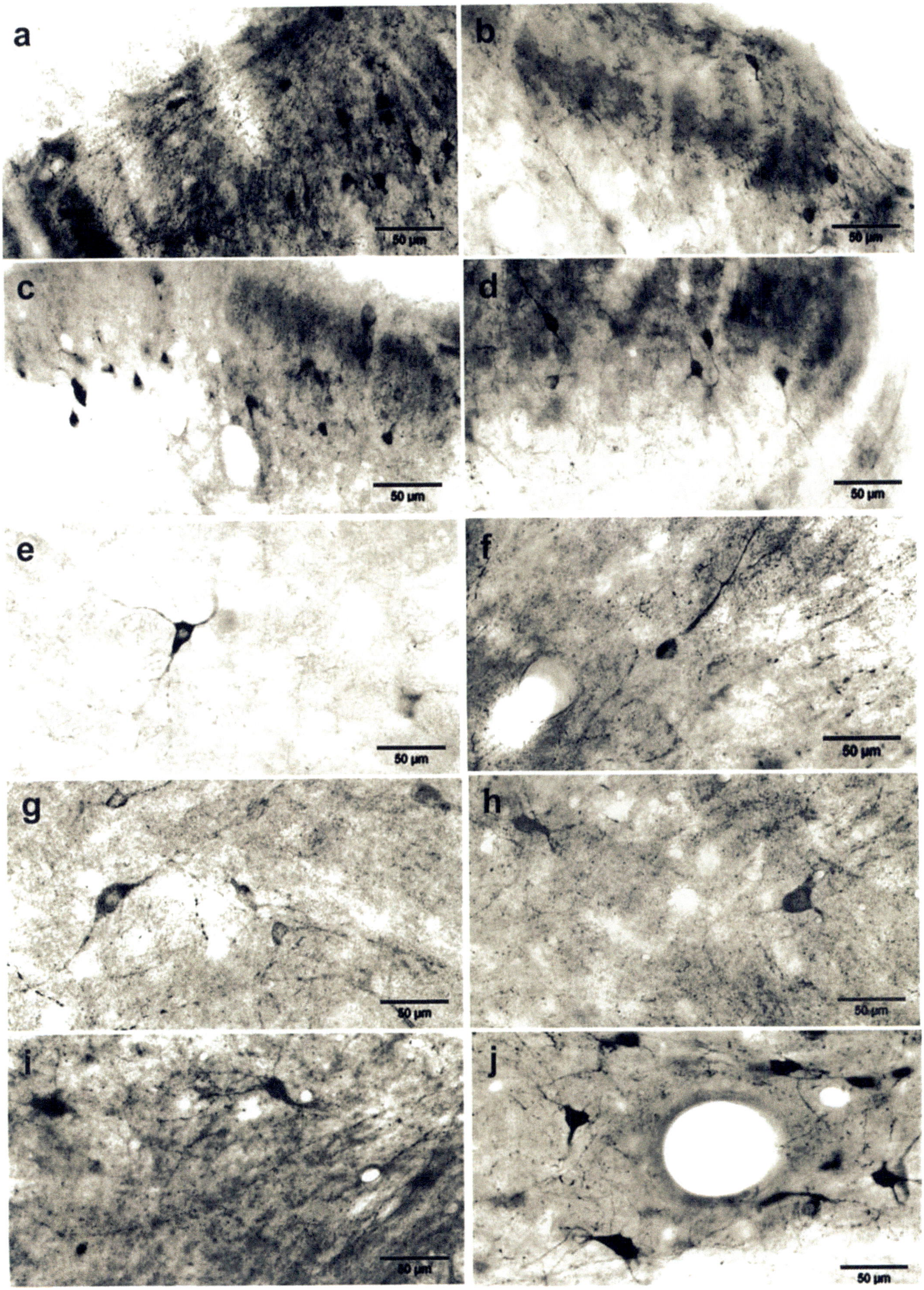
a
50 µm
b
50 µm
c
50 µm
d
50 µm
e
50 µm
f
50 µm
g
50 µm
h
50 µm
i
50 µm
j
50 µm

and Coimbra, 1986; Galhardo and Lima, 1999). Neurons with prominent ventral extending dendrites have also been found in the human lamina 1 (Schoenen and Faull, 2004). Fusiform cells are the most numerous cell types in lamina 1 and are located predominately in the lateral third of the lamina (Lima and Coimbra, 1986; Galhardo and Lima, 1999).

Multipolar neurons

A typical multipolar cell has an ovoid shaped cell body from which several primary dendrites arise. The dendrites spread out in all directions with characteristic tortuous branches (Lima and Coimbra, 1986; Galhardo and Lima, 1999). This cell type has also been divided into two subgroups (Lima and Coimbra, 1986; Galhardo and Lima, 1999). Type A multipolar cells are characterized by the presence of spines all over their dendritic trees, which extend mostly in the rostrocaudal plane. In general, multipolar cells in the type A class have densely arborized dendritic trees. In contrast, type B cells have loose dendritic arborizations and very few spines. Their cells bodies are also generally larger than type A. Multipolar cells are usually confined to the medial portion of lamina 1 (Lima and Coimbra, 1986; Galhardo and Lima, 1999).

Pyramidal neurons

The cell bodies of pyramidal neurons are triangular in shape and their dendrites typically arise from each corner of the cell body. Three to five dendritic trunks are generally found and these extend predominately in the rostrocaudal plane (Lima and Coimbra, 1986; Galhardo and Lima, 1999). A significant mediolateral extension is also a common feature of pyramidal neurons. These can be seen to expand out into the dorsal column medially or the dorsolateral funiculus laterally. Pyramidal cells are located along the dorsal border of lamina 1 across the entire mediolateral extent of the lamina (Lima and Coimbra, 1986; Galhardo and Lima, 1999).

Flattened neurons

Flattened neurons have discoid cell bodies and are characterized by their small, sparsely branched dendritic trees (Lima and Coimbra, 1986; Galhardo and Lima, 1999). In general, two to five primary trunks are seen which expand mediolaterally parallel to the surface of the dorsal horn. Flattened neurons rarely display dendritic spines and are predominately found in middle third of lamina 1 (Lima and Coimbra, 1986; Galhardo and Lima, 1999). This cell type is less frequently identified than the other three and could not be identified in the monkey (Zhang *et al.*, 1996; Zhang and Craig, 1997).

Chemoarchitecture

Many different neurochemical substances have been identified in lamina 1 (Hunt *et al.*, 1981; Nagy *et al.*, 1981; Barber *et al.*, 1982; Rökaeus *et al.*, 1984; Cruz and Basbaum, 1985; Ribeiro-da-Silva and Cuello, 1990; Todd and Sullivan, 1990). These include γ-aminobutyric acid (GABA) (Hunt *et al.*, 1981; Barber *et al.*, 1982; Todd and Sullivan, 1990; Lima *et al.*, 1993;), avian pancreatic neuropeptide (Hunt *et al.*, 1981), calbindin (Ren and Ruda 1994), calcitonin-gene related peptide (CGRP) (Conrath *et al.*, 1989), calretinin (Anelli and Heckman, 2005), cholecystokinin (Conrath-Verrier *et al.*, 1984; Leah *et al.*, 1988; Cho and Basbaum, 1989); dopamine (Holstege *et al.*, 1996), dynorphin (Cruz and Basbaum, 1985; Miller and Seybold, 1987; Nahin, 1987; Leah *et al.*, 1988; Lima *et al.*, 1993), enkephalin (Aronin *et al.*, 1981; Glazer and Basbaum, 1981; Hunt *et al.*, 1981; de Lanerolle and LaMotte, 1982; Cruz and Basbaum, 1985; Standaert *et al.*, 1986; Ribeiro-da-Silva *et al.*, 1991; Lima *et al.*, 1993), galanin (Rokaeus *et al.*, 1984; Melander *et al.*, 1986), glutamate (Antal *et al.*, 1991), glycine (Campistron *et al.*, 1986; Ottersen and Storm-Mathisen, 1987; Van den Pol and Gorcs, 1988; Todd and Sullivan, 1990; Spike *et al.*, 1997), neuropeptide Y (Hunt *et al.*, 1981), neurotensin (Hunt *et al.*, 1981), nitric oxide (Dun *et al.*, 1993; Saito *et al.*, 1994; Maiskii *et al.*, 1998; Reuss and Reuss, 2001; Dagci *et al.*, 2008), serotonin (Ruda *et al.*, 1982), somatostatin (Riberio-da-Silva and Cuello, 1990), substance P (Hunt *et al.*, 1981; Nagy *et al.*, 1981; de Lanerolle and LaMotte, 1982; Yashpal *et al.*, 1990; Ribeiro-da-Silva *et al.*, 1991; Lima *et al.*, 1993), and vasoactive intestinal polypeptide (LaMotte and de Lanerolle, 1986).

Furthermore, it appears that each of the morphologically distinct cell types in lamina 1 may have a unique neurochemical identity (Lima *et al.*, 1993). In the rat, fusiform cells are found to contain GABA and dynorphin, multipolar cells contain GABA, flattened cells contain substance P or dynorphin, and pyramidal cells contain enkephalin and, in smaller amounts, dynorphin (Lima *et al.*, 1993). Such an organization suggests the likelihood of functional segregation between the morphologically different cell types; different cell types may exert differential effects on target cells (either locally or supraspinally) through differing chemical systems.

Connections

The neurons in lamina 1 receive input from small-diameter fibers, principally A-delta and C-fibers innervating the skin, viscera, muscles and joints throughout the body (Christensen and Perl, 1970; Cervero and Tattersall, 1986; Hoheisel *et al.*, 1989). These fibers travel through the dorsal roots to terminate in lamina 1 and carry information about pain and temperature

(Perl, 1984; Willis, 1985; Light 1992; Craig, 1994, 2000). Lamina 1 neurons also respond to innocuous mechanical stimuli conveyed by A-β fibers (Willis and Coggeshall, 1991). Lamina 1 receives input from supraspinal sources, in particular from the hypothalamus and brainstem (Dawson *et al.*, 1981; Dostrovsky *et al.*, 1983; Holstege, 1987; Mokha *et al.*, 1987); projections that are involved in modulating ascending pain information (Perl, 1984; Willis, 1985; Light 1992; Craig and Serrano, 1994). It is suggested that lamina 1 cells might activate brainstem circuits, which in turn influence deep dorsal horn neurons (McMahon and Wall, 1988).

The cells in lamina 1 are the major source of supraspinal projections from the dorsal horn of the spinal cord. The primary target of lamina 1 is the thalamus (Carstens and Trevino, 1978; Granum, 1986; Lima and Coimbra, 1988; Kayalioglu *et al.*, 1996, 1999; Maiskii *et al.*, 1998; Mouton and Holstege, 1998) and axons from lamina 1 make up about half of the spinothalamic tract (Carstens and Trevino, 1978; Willis *et al.*, 1979; Apkarian and Hodge, 1989; Craig *et al.*, 1989). This projection is important for discriminative aspects of pain and temperature sensation.

Lamina 1 has also been reported to project rostrally to the lateral parabrachial area (Cecchetto *et al.*, 1985; Todd *et al.*, 2000), nucleus accumbens and septal nuclei (Kayalioglu *et al.*, 1996), the nucleus of the solitary tract (Menétrey and Basbaum, 1987), hypothalamus (Burstein *et al.*, 1987, 1990a; Kayalioglu *et al.*, 1999), the periaqueductal gray matter (Liu, 1983; Keay and Bandler, 1993; Kayalioglu *et al.*, 1996, 1999; Maiskii *et al.*, 1998), the lateral reticular nucleus (Todd *et al.*, 2000), the ventrolateral medulla and the cuneiform nucleus (Wiberg and Blomqvist, 1984; Cechetto *et al.*, 1985; Craig, 1995; Vanderhorst *et al.*, 1996; Andrew *et al.*, 2003). Further, intraspinal projecting neurons are found in lamina 1. These target the thoracolumbar sympathetic nuclei including the intermediolateral nucleus, the intermediomedial nucleus and the intercalated nucleus (Foreman *et al.*, 1988). Studies in the rat suggest the different morphological cell types in lamina 1 may also have distinctive patterns of projections (Lima and Coimbra, 1988, 1989, 1990; Lima *et al.*, 1991; Esteves *et al.*, 1993). However, other studies in cat and monkey do not support this (Hylden *et al.*, 1986; Light *et al.*, 1993; Zhang *et al.*, 1996).

Function

The morphological diversity of neurons in lamina 1 is suggestive of neurons with differing functional roles. Indeed, the cells of lamina 1 have been classified functionally into three major classes of modality-selective neurons in the cat and monkey, including nociceptive-specific neurons (NS), polymodal nociceptive (HPC; heat, pinch and cold) and thermoreceptive-specific (COLD) neurons (Christensen and Perl, 1970; Craig and Hunsley, 1991; Craig and Serrano, 1994; Dotrovsky and Craig, 1996). In monkeys, wide dynamic range neurons (WDR), excited by nociceptors and innocuous mechanoreceptors have also been identified in lamina 1 (Feerington *et al.*, 1987). Moreover, Han *et al.*, (1998), in the cat, showed that these functionally segregated types correlate with the morphological cell classes of lamina 1. NS neurons were identified as fusiform cells, COLD neurons as pyramidal cells and HPC neurons as multipolar cells.

Physiological evidence also suggests functional segregation between the lamina 1 morphological cell types. Prescott and De Koninck (2002) classified four types of lamina 1 cells based on how they respond to sensory stimulation. They identified: (1) tonic cells, that fire comparatively slowly but continuously throughout stimulation; (2) phasic cells, that fire a high frequency burst of variable duration; (3) delayed onset cells, that fire irregularly and with a marked delay to the first spike; and (4) single spike cells that typically fire only one action potential even when strongly depolarized. They also found a significant correlation between the morphological cell types and these physiological types; tonic cells were mostly fusiform, phasic cells were pyramidal, and delayed onset and single spike cells were multipolar.

In summary, the heterogenity of cells in lamina 1, in terms of morphology, neurochemistry, connectivity, and physiology, seems to underly a functional system that is capable of identifying and transmitting modality specific information important for maintaining homeostatic integrity.

Lamina 2

Cytoarchitecture

Lamina 2, commonly known as the substantia gelatinosa (Rolando, 1824), lies ventral to lamina 1. High neuronal density gives lamina 2 a dark appearance in Nissl-stained sections. Lamina 2 follows the shape of lamina 1 in transverse sections, bending around the apex so that lamina 1 forms its dorsal and lateral borders (Rexed, 1954). At the medial side, it is bordered by the white substance of the dorsal funiculus (Rexed, 1954). Lamina 2 is thicker than lamina 1 and does not have a uniform thickness; it is thickest laterally in humans and medially in rodents (Molander *et al.*, 1984, 1989; Schoenen and Faull, 2004). At the levels of cervical and lumbar enlargements, lamina 2 is forced ventrally away from lamina 1, presumably due to an increase in the number of afferent inputs from distal skin (Woodbury *et al.*, 2000). In most species, lamina 2 consists of a dorsal outer zone (lamina 2o) and a ventral inner zone (lamina 2i) (Rexed, 1954; Molander *et al.*, 1984, 1989). The outer zone makes up about one quarter of lamina 2 and contains slightly smaller and more densely packed cells than the inner zone (Rexed, 1954; Molander *et al.*, 1984). The outer and inner zones occupy discrete, non-overlapping regions of neuropil (Woodbury *et al.*, 2000). In the human, however, there is no obvious separation between an outer zone and an inner zone; instead the neuron population of lamina 2 appears homogeneous (Schoenen and Faull, 2004). Cells in lamina 2 are generally small, rounded or slightly elongated (Rexed, 1954; Molander *et al.*, 1984; Schoenen and Faull, 2004).

Dendritic architecture

Two main cell types are commonly recognized in lamina 2: the stalked and islet cells (Gobel; 1975, 1978). The islet cells have very long dendritic trees extending exclusively along the rostrocaudal axis of the cord (Todd and Lewis, 1986). These dendrites usually arise from only a limited area of the soma, either grouping together and extending out from one point or sometimes from opposite poles of the cell (Schoenen, 1982a). Recurrent branches are also a typical feature of islet cells (Todd and Lewis, 1986). Islet cells are found mainly in the central part of lamina 2 in the human and throughout this lamina in the rat. They generally have only a few dendritic spines. Todd and Lewis (1986) identified three types of islet cells in the rat: those in which the soma lie dorsal to the major part of the dendritic tree, usually situated in the dorsal part of the lamina; those in which both the soma and the dendritic tree lie in the same plane, and those with a soma ventral to most of the dendrites. The last two types are more common in the ventral part of lamina 2. The axons of islet cells arborize close to the cell body and dendritic tree. The dendrites and axons of islet cells generally do not extend out of lamina 2 (Schoenen, 1982a).

The second type of cell found in lamina 2, the stalked cells, are most numerous in the outer layer of lamina 2 (Schoenen, 1982a). A typical stalked cell has a round or oval shaped soma with ventrally directed dendrites that stream out at oblique angles to form a cone-like shape (Schoenen, 1982a). The characteristic feature of stalked cells is dendrites covered with spines and stalk-like branches (Todd and Lewis, 1986). The dendrites of stalked cells can pass ventrally toward deeper spinal laminae and their axons may extend dorsally into lamina 1 (Schoenen, 1982a; Todd and McKenzie, 1989). Stalked cells have been identified in the rat, cat and monkey. In humans, Schoenen (1982a) described a group of curly cells that resemble stalked cells in morphology.

The stalked and islet cells are widely recognized as the main cell types in lamina 2, but many cells identified in lamina 2 do not fit into either of these classes, and some have been shown to display features of both classes (Todd and McKenzie, 1989). Some of these cells in the rat have axons that either remain in lamina 2 or pass to lamina 3. It is suggested that these may correspond to the stellate cells in the human spinal cord or the 2-3 borders in cat medulla. These neurons possess presynaptic dendrites and function as inhibitory neurons in lamina 2 (Todd and Lewis, 1986). Indeed, in the human, Schoenen (1982a) identified two more classes of neurons in lamina 2, filamentous and stellate cells. Filamentous cells, located in both lamina 2o and 2i, have fine expansions giving a filamentous appearance to their dendritic trees. Stellate cells are mostly located in lamina 2i and have numerous dendritic spines that spread in all directions (Willis and Coggeshall, 1991). Thus lamina 2, like lamina 1, appears to contain a morphologically heterogeneous population of cells.

Chemoarchitecture

Approximately one third (31%) of all lamina 2 neurons are GABAergic (Todd and Sullivan, 1990) and these are generally reported to be islet cells (Todd and McKenzie, 1989; Heinke *et al.*, 2004). Further, islet cells that are GABAergic are for the most part large, while islet cells that are not GABAergic are usually small (Todd and McKenzie, 1989; Spike and Todd, 1992; Todd and Spike, 1993; Heinke *et al.*, 2004). Glycine, neuropeptide Y and parvalbumin have all been found to be co-localized in GABAergic, presumably large islet, neurons in lamina 2 (Antal *et al.*, 1991; Todd *et al.*, 1992; Rowan *et al.*, 1993). Conversely, glutamate is found in stalked and small islet cells only (Light and Kavookjian, 1988; Spike and Todd, 1992). Somatostatin, which is found in both islet and stalked cells (Mather and Ho, 1992), is restricted to those that do not contain GABA (Proudlock *et al.*, 1993). Neurotensin, present in islet cells (Seybold and Elde, 1982), is also found in cells that do not contain GABA (Todd *et al.*, 1992). Taken together, these findings suggest that there are two neurochemically distinct populations of islet cells, large and small.

Many other neurochemicals have been identified in cells in lamina 2, including calbindin and calretinin (Ren and Ruda, 1994; Anelli and Heckman, 2005), calcitonin gene-related peptide (Conrath *et al.*, 1989), cholecystokinin (Gibson *et al.*, 1981; Schroder, 1983; Conrath-Verrier *et al.*, 1984; Fuji *et al.*, 1985; Nahin, 1987; Hökfelt *et al.*, 1988; Leah *et al.*, 1988), dynorphin (Cruz and Basbaum, 1985; Miller and Seybold, 1987; Nahin, 1987; Leah *et al.*, 1988; Cho and Basbaum, 1989), enkephalin (Aronin *et al.*, 1981; Finely *et al.*, 1981; Glazer and Basbaum, 1981; Hunt *et al.*, 1981; de Lanerolle and LaMotte, 1982; Leah *et al.*, 1988; Senba *et al.*, 1988; Ribeiro-da-Silva *et al.*, 1991; Todd and Spike 1992), galanin (Rokaeus *et al.*, 1984; Melander *et al.*, 1986), neurotensin (Gibsin *et al.*, 1981), nitric oxide (Saito *et al.*, 1994; Maiskii *et al.*, 1998; Reuss and Reuss, 2001;), parvalbumin (Ren and Ruda, 1994), thyrotropin-releasing hormone (Coffield *et al.*, 1986; Harkness and Brownfield, 1986; Ulfhake *et al.*, 1987), somatostatin (Krukoff *et al.*, 1986), substance P (Gibson *et al.*, 1981; Ribeiro-da-Silva *et al.*, 1991; Todd *et al.*, 2000) and vasoactive intestinal polypeptide (Gibson *et al.*, 1981).

Connections

Lamina 2, like lamina 1, is a principal region of termination of A-δ and/or C-fiber dorsal root afferents (Bernardi *et al.*, 1995; Heinke *et al.*, 2004). In lamina 2 these projections show some topography so that the outer zone of lamina 2 receives predominately terminations from C-fibers (Light and Perl, 1979a; Bernardi *et al.*, 1995; Woodbury *et al.*, 2000; Hantman *et al.*, 2004), whilst the inner part receives endings from both C- and A-δ fibers (Light and Perl, 1979a; Sugiura *et al.*, 1986; Alvarez *et al.*, 1993; Bernardi *et al.*, 1995; Woodbury *et al.*, 2000; Woodbury and Koerber, 2003). Also, there is a somatotopical organization in lamina 2 (Swett and Woolf, 1985). Cutaneous afferents from the ventral median line of the trunk and extremities project to the medial margin and cutaneous afferents from the dorsal median line project to the lateral margin of the dorsal horn (Takahashi *et al.*, 2002).

Cells in lamina 2 form mostly local connections. Light and Kavookjian (1988) showed, in the cat and monkey, nociceptive neurons in lamina 2 have axons that distribute terminal collaterals to deeper laminae of the spinal cord, including laminae 3, 4 and 5. As such, lamina 2 may relay nociceptive information to deeper laminae of the spinal dorsal horn. Supra-spinal projections from cells in lamina 2 to the thalamus and brainstem reticular formation have also been described (Lima and Coimbra, 1991).

Function

Lamina 2 neurons primarily function as interneurons that modulate nociceptive transmission in the dorsal horn (Fields *et al.*, 1995). Lamina 2 neurons have been implicated in intergrating primary afferent fiber input and modulating the output of ascending projection neurons in the surrounding laminae (Réthelyi and Szentágothai, 1969; Kumazawa and Perl, 1978; Light *et al.*, 1979c; Millan, 1999). In this scheme, it has been proposed that islet cells function as inhibitory interneurons while stalked cells are excitatory (Gobel, 1978; Todd and McKenzie, 1989). Electrophysiological studies have shown that lamina 2o neurons are nociceptive, responding to high intensity nociceptive stimuli, while lamina 2i neurons are innocuous mechanoreceptive cells responding to brush stimuli (Light and Wilcockson, 1999). However, as mentioned above, not all lamina 2 neurons are locally projecting interneurons, and ascending projection neurons have also been described. It is therefore likely that the morphological heterogeneity of lamina 2 reflects a system involved in several spinal functions rather than a single one.

Lamina 3

Cytoarchitecture

Lamina 3 runs straight across the dorsal horn roughly parallel with laminae 1 and 2. Its border bends around the apex of the dorsal horn, though not as sharply as lamina 1 and 2 (Rexed, 1954; Molander *et al.*, 1984, 1989). In the rat, lamina 3 also has a medial deflection (Molander *et al.*, 1984, 1989). It is thicker than the first two layers and is bordered by lamina 2 laterally and the white column medially (Rexed, 1954; Molander *et al.*, 1984, 1989). The presence of myelinated fibers distinguishes lamina 3 from lamina 2o which is myelin-free (Molander *et al.*, 1984). The neuropil of lamina 3 is heterogeneous when compared to the homogeneous texture of lamina 2o (Molander *et al.*, 1989). In the cat and human, the cells in lamina 3 vary more in size and are less densely packed than the cells in lamina 2 (Beal and Cooper, 1978; Rexed, 1954; Schoenen and Faull, 2004). This is not the case in the rat where cells are of similar size and density to the overlying lamina 2 make the border between the two laminae difficult to demarcate (Molander *et al.*, 1984, 1989). In general, rounded, slightly elongated or spindle shaped cells are found in lamina 3 (Rexed, 1954).

Dendritic architecture

Studies across several species have revealed lamina 3 to contain a morphologically heterogeneous population of cells (Réthelyi and Szentágothai, 1969; Mannen and Sugiura, 1976; Beal and Cooper, 1978; Schoenen, 1982a; Maxwell *et al.*, 1983; Maxwell, 1985; Réthelyi *et al.*, 1989). Nevertheless, in terms of dendritic structure they can be placed roughly into two broad classes: neurons with dendritic trees extending along the rostrocaudal axis that are confined within lamina 3 (Maxwell, 1985) and neurons with dendrites oriented dorsoventrally, where the dorsal extension is generally longest (Schoenen and Faull, 2004). Dendrites of the latter type often extend out of lamina 3 into lamina 2 (Schoenen 1982a; Maxwell *et al.*, 1983). Considering that lamina 3 consists of a population of neurons with dendrites that remain within the lamina itself and/or extend into surrounding laminae, it is in a position to receive as well as integrate information from multiple sources.

Chemoarchitecture

A large proportion (46%) of cells in lamina 3 have been reported to be GABAergic (Magoul *et al.*, 1987; Todd and McKenzie, 1989; Todd and Sullivan, 1990; Lima *et al.*, 1993). Most of these GABAergic cells have dendrites that extend rostrocaudally. Cells with dorsally or ventrally directed dendrites are rarely GABAergic. This suggests these two morphological types may subserve different functional roles (Powell and Todd, 1992). In addition, many neurochemical substances have been found to be co-localized in GABAergic neurons in lamina 3. These include calbindin (Antal *et al.*, 1991), gylcine (Campistron *et al.*, 1986; Ottersen and Storm-Mathisen, 1987; van den Pol and Gorcs, 1988; Todd and Sullivan, 1990), neuropeptide Y (Todd *et al.*, 1992; Rowan *et al.*, 1993) and parvalbumin (Antal *et al.*, 1991). Acetylcholine is also reported to co-localize with GABA in cells in lamina 3, but is not present in those GABAergic cells that contain glycine (Todd, 1991). In contrast, some neurochemicals, such as neurotensin, exist only in those cells that do not contain GABA (Todd *et al.*, 1992).

Several other neurochemicals have been identified in neurons in lamina 3. These include calcitonin gene-related peptide (Conrath *et al.*, 1989), calretinin (Anelli and Heckman, 2005), cholecystokinin (Gibson *et al.*, 1981; Conrath-Verrier *et al.*, 1984; Fuji *et al.*, 1985; Hökfelt *et al.*, 1988), dopamine (Holstege *et al.*, 1996), dynorphin (Cruz and Basbaum, 1985; Miller and Seybold, 1987; Nahin, 1987; Leah *et al.*, 1988; Cho and Basbaum, 1989), enkephalin (de Lanerolle and LaMotte, 1982; Ribeiro-da-Silva *et al.*, 1991; Lima *et al.*, 1993), neurotensin (Gibson *et al.*, 1981), nitric oxide (Saito *et al.*, 1994; Maiskii *et al.*, 1998; Reuss and Reuss, 2001), somatostatin (Krukoff *et al.*, 1986; Ribeiro-da-Silva and Cuello, 1990), thyrotropin-releasing hormone (Coffield *et al.*, 1986; Harkness and Brownfield, 1986; Ulfhake *et al.*, 1987) and substance P (Gibson *et al.*, 1981).

Connections

Lamina 3 receives large diameter cutaneous afferent fibers of A-β type (Brown, 1981) and smaller A-δ fibers which innervate down hairs (Light and Perl, 1979b). Further, while neurons with rostrocaudal dendritic trees limited to lamina 3 are only capable of receiving direct input from A-beta and A-delta cutaneous fibers, those with dorsoventral dendrites that extend into lamina 2 may also receive a direct input from C-fibers terminating in lamina 2 (Réthelyi, 1977; Perl, 1980; Maxwell, 1985). Indeed, cells in lamina 3 have been shown to respond to C-fiber input (Szentágothai, 1964; Surmeier *et al.*, 1988; Todd, 1989, De Koninck *et al.*, 1992; Ma *et al.*, 1996). Neurons in lamina 3 that possess the NK1 receptor (substance P receptor) have dendrites that extend dorsally into lamina 2, where they are contacted by substance P-immunoreactive boutons (Naim *et al.*, 1997).

Projection neurons within lamina 3 relay the incoming sensory information either in the spinocervical tract (Brown *et al.*, 1977) or in the post-synaptic dorsal column pathway (Brown *et al.*, 1977; Brown and Fyffe, 1981). In the rat, some spinothalamic, spinohypothalamic, spinoparabrachial and spinoreticular neurons have also been identified (Giesler *et al.*, 1979a; Kevetter and Willis, 1983; Kayalioglu *et al.*, 1999; Todd *et al.*, 2000). In addition, lamina 3 contains a population of smaller neurons, most likely propriospinal and/or interneurons, whose axons do not appear to project far rostrally (Maxwell *et al.*, 1983).

Function

Physiological data, albeit limited, suggest there are two functional types in lamina 3. Some neurons in lamina 3 respond to light tactile stimulation (Bennett *et al.*, 1984; Réthelyi *et al.*, 1989), while others are excited by strong pressure (Maxwell *et al.*, 1983). There is no evidence to suggest that these functional classes can be correlated with the morphological classes mentioned above, rather it seems morphologically heterogeneous cells are found in both classes.

Lamina 4

Cytoarchitecture

Lamina 4 runs straight across the base of the head of the dorsal horn and does not bend laterally in conformity with the above three layers. It does however, like lamina 3, have a medial bend in the rat (Molander *et al.*, 1984). Lamina 4 is thicker than laminae 1-3 and is bordered by white substance medially and by laminae 1-3 laterally (Rexed, 1954). The dorsal and ventral borders of lamina 4 are not distinct in most species, however the lower density of cells and the presence of large cells distinguishes lamina 4 from the above three laminae (Molander *et al.*, 1984, Schoenen and Faull, 2004). Lamina 4 is interrupted by the dorsal column (Clarke) in T1-L3 segments. Internal basilar nucleus is located within lamina 4 of C1-C6 segments.

Dendritic architecture

Lamina 4 contains neurons of varying size and shape, but all are characterized by the dendritic trees which look like antennae (Szentágothai, 1964; Szentágothai and Réthelyi, 1973; Schoenen, 1982a; Schoenen and Faull, 2004). These dendritic trees are oriented dorsally, with many spreading into overlying lamina 3 and lamina 2, and some even reaching lamina 1. The mediolateral spread of the dendritic trees is generally greater in neurons located more laterally (Schoenen, 1982a). Cells in lamina 4 are richly covered with dendritic spines, which are most numerous on dorsal dendrites (Schoenen 1982a).

Chemoarchitecture

GABAergic cells have been identified in lamina 4 and these are situated mostly in the lateral one third of the lamina (Barber *et al.*, 1982). Calbindin and calretinin (Anelli and Heckman, 2005), dopamine (Holstege *et al.*, 1996), dynorphin (Lima *et al.*, 1993), glycine (Hossaini *et al.*, 2007), enkephalin (de Lanerolle and LaMotte, 1982; Lima *et al.*, 1993), nitric oxide (Saito *et al.*, 1994), somatostatin (Krukoff *et al.*, 1986), substance P (de Lanerolle and LaMotte, 1982; Todd, 2002) and cholecystokinin (Gibson *et al.*, 1981) have also been identified in cell bodies in lamina 4.

Connections

The afferent and efferent projections of lamina 4 are similar to lamina 3. Lamina 4 receives mostly non-nociceptive information and projects mainly in the spinocervical (Willis and Coggeshall, 1991), spinocerebellar (Edgley and Gallimore, 1988; Rivero-Melián and Grant, 1990), spinohypothalamic (Giesler *et al.*, 1994; Kayalioglu *et al.*, 1999), spinomesencephalic (Kayalioglu *et al.*, 1996, 1999), spinothalamic tracts (Kayalioglu *et al.*, 1996, 1999; Schoenen and Faull, 2004), and the postsynaptic dorsal column pathway (Bennett *et al.*, 1983). Neurons projecting to nucleus accumbens and septal nuclei (Kayalioglu *et al.*, 1996), the lateral parabrachial area and reticular nuclei (Todd *et al.*, 2000) were also observed in lamina 4.

Function

Lamina 4 neurons respond to light mechanical stimuli (Cervero *et al.*, 1988). Wide dynamic range neurons responding to noxious stimuli are also observed here (De Koninck *et al.*, 1992). Additionally as the majority of the dendrites of lamina 4 neurons reach more superficial laminae, they are probably contacted by nociceptive afferents as well as being modulated by lamina 2 interneurons.

The dorsal nucleus

The dorsal nucleus (D), or Clarke's column, is found in lamina 5 dorsolaterally to lamina 10. It is present in segments T1-L3 in most species. The location of the dorsal nucleus in transverse sections varies somewhat between species, being found at the medial border of laminae 5, 6 and 7 in the human and cat and laminae 4 and 5 in the rat. It consists of large, medium, and small-sized neurons. In the rat and mouse, the neurons are smaller and more fusiform than the cat and monkey, and the nucleus is more dorsoventrally flattened (Snyder *et al.*, 1978).

The dendritic architecture of cells in the dorsal nucleus has been studied in the cat. Loewy (1970) identified three neuronal classes: A, B and C. Class A is characterized by small cells with variable dendritic branching patterns. Class B contains multipolar and fusiform cells. The majority of the multipolar cells have dendrites that project in a radial fashion, while the fusiform neurons generally have their long axis oriented perpendicularly. Class C is made up of large cells that have dendrites extending in the rostrocaudal plane. Dendrites of all three cell types cross the boundaries of the dorsal nucleus and enter laminae 5, 6, 7 and 10. The dorsal nucleus primarily receives afferents from the lower limb and its axons form the dorsal spinocerebellar tract. The equivalent cell group serving the upper limb is the external cuneate nucleus of the lower medulla. Neurons of the dorsal nucleus project ipsilaterally to the cerebellum (Snyder *et al.*, 1978). There is a somatotopic organization in the dorsal nucleus, hindlimb nerves innervating distal muscles project medially and nerves innervating proximal muscles project laterally in the dorsal nucleus (Rivero-Melián, 1996).

Internal basilar nucleus

The internal basilar nucleus (IB) is located in the ventromedial dorsal horn, within the limits of lamina 4 of C1-C6 segments. It was first described by Ramón y Cajal (1909) and Torvik (1956). The IB is continuous rostrally with the cuneate nucleus (Cliffer and Giesler, 1989). There are primary afferent fibers from the dorsal root ganglia to the IB (Rivero-Melián and Arvidsson, 1992). Central projections of the median and ulnar nerves to IB have also been shown in C1-C6 segments (LaMotte *et al.*, 1991). A large number of IB neurons project to the thalamus (Granum, 1986; Kemplay and Webster, 1986). Also, postsynaptic afferent fibers from the cervical enlargement project to the IB (Cliffer and Giesler, 1989). Descending projections are present from the sensorimotor cortex to the IB (Antal, 1984; Valtschanoff *et al.*, 1993). Calmodulin, parvalbumin, calretinin (Ren and Ruda, 1994), glutamate (Valtschanoff *et al.*, 1993; Persson *et al.*, 2006) immunoreactivities have been shown in IB neurons.

Lateral spinal nucleus

The lateral spinal nucleus (LSp) consists of neurons in the white matter ventrolateral to the lateral edge of the dorsal horn in rodents and is present at all levels of the spinal cord (Molander and Grant, 2004). It was first described as 'a nucleus in the dorsolateral funiculus of the rat' by Gwyn and Waldron (1968). LSp neurons are mainly multipolar and occasionally spindle-shaped, measuring 20-35 μm in their longest diameter (Réthelyi, 2003). A Golgi study by Réthelyi (2003) showed that the dendrites of LSp neurons have three patterns: several of them turning laterally and approaching the spinal cord, a second group running longitudinally within the column and a third group coursing medially to the reticulated portion of the gray matter. The dendritic arborization of the LSp neurons overlap that of the 'neck cells' which are neurons displaced into the dorsolateral fasciculus from the spinal gray matter.

The LSp neurons have been shown to project bilaterally to midbrain (Menétrey *et al.*, 1982; Kayalioglu *et al.*, 1999), hypothalamus (Burstein *et al.*, 1987, Kayalioglu *et al.*, 1999), thalamus (Burstein *et al.*, 1990b; Kayalioglu *et al.*, 1999), amygdala (Burstein and Potrebic, 1993), nucleus accumbens (Burstein and Giesler, 1989; Kayalioglu *et al.*, 1996), parabrachial nucleus (Ding *et al.*, 1995), the tractus solitarius nucleus (Esteves *et al.*, 1993), and to the nucleus of the solitary tract and medullary reticular formation (Leah *et al.*, 1988). LSp neurons from thoracic segments have been shown to project to LSp of cervical segments (Verburgh *et al.*, 1990), and reciprocally from cervical to thoracic segments (Jansen and Loewy, 1997). Also, LSp neurons project to laminae 1, 2, 5, and 7 neurons and to the sympathetic preganglionic neurons of the spinal cord (Jansen and Loewy, 1997). LSp neurons receive descending projections from the midbrain, medullary and pontine reticular formation and dorsal column nuclei (Carlton *et al.*, 1985) and peptidergic input from local spinal cord neurons (Cliffer *et al.*, 1988).

LSp neurons contain a variety of neurotransmitters including bombesin (Leah *et al.*, 1988), calcitonin gene-related peptide (Conrath *et al.*, 1989), calbindin and calretinin (Ren and Ruda, 1994), dynorphin (Nahin, 1987), enkephalin (Cliffer *et al.*, 1988), glycine (Hossaini *et al.*, 2007), glutamate (Alvarez *et al.*, 2000), neuropeptide Y (Zhang *et al.*, 1995), nitric oxide (Herdegen *et al.*, 1994; Maiskii *et al.*, 1998; Kayalioglu *et al.*, 1999; Dagci *et al.*, 2008), somatostatin (Cliffer *et al.*, 1988), substance P (Battaglia and Rustioni, 1992), and vasoactive intestinal polypeptide (Nahin, 1987; Leah *et al.*, 1988). An electrophysiological analysis of the LSp neurons disclosed several similar features between LSp and the deep dorsal horn neurons (Jiang *et al.*, 1999).

LSp neurons are known to be responsible for mediating afferent input, including nociceptive information. Menétrey *et al.* (1980) first suggested a possible role of this nucleus in transmission of nociceptive information, showing that cells in this nucleus generally respond to stimulation of structures that are subcutaneous, deep, or both. They have also shown c-fos expression in the LSp neurons following noxious somatic, articular, and also visceral stimulation (Menétrey *et al.*, 1989).

Lateral cervical nucleus

The lateral cervical nucleus (LatC) is found within the dorsal part of the lateral funiculus, lateral to the lateral spinal nucleus of the upper cervical segments (C1-C3) (see Chapters 15 and 16). LatC has been identified in the spinal cords of several species including the rat, cat, dog, and monkey, and although not consistently, human (Mizuno *et al.*, 1967; Truex *et al.*, 1970). LatC neurons have rounded shapes and their dendrites longitudinally extend within the spinal cord, differing from the multipolar LSp neurons with dendrites extending to all directions (Giesler *et al.*, 1979b). LatC and LSp neurons differ also with regard to their peptidergic contents (Giesler and Elde, 1985). The LatC of the rat consists of 300-500 neurons (Baker and Giesler, 1984).

The spinocervical pathway, originating predominately from lamina 4, but also observed in laminae 1-3 and 5, ascends within the dorsolateral fasciculus and projects to the LatC at C1-C3 segments. LatC projects to the periaqueductal gray (Berkley *et al.*, 1980; Mouton *et al.*, 2004) and to the contralateral thalamus (Craig and Burton, 1979; Granum, 1986). The majority (85%) of LatC neurons projecting to the thalamus are located in C2 (Baker and Giesler, 1984). LatC neurons respond to hair movement, noxious cutaneous, mechanical and thermal stimulation (Cervero *et al.*, 1977; Craig and Tapper, 1978; Giesler *et al.*, 1979b; Kajander and Giesler, 1987).

LatC is somatotopically organized with the hindlimb located dorsolaterally, forelimb ventromedially and the face units most medially. The most medial part of the LatC is histologically and functionally different from the ventromedial and dorsolateral parts. The lateral part consists of dense clusters of thick fibers bearing large numbers of boutons, whereas the terminal axons in the medial part display a reticulated network of longitudinally oriented, fine fibers with well-spaced varicosities (Craig *et al.*, 1992).

LatC neurons contain calbindin, calretinin (Ren and Ruda, 1994) and glycine (Hossaini *et al.*, 2007). The medial part contains substance P and GABA immunoreactivity (Giesler and Elde, 1985; Broman and Westman, 1989). It is hypothesized that this medial part is composed of local inhibitory interneurons (Craig *et al.*, 1992). The LatC receives serotonergic innervation, suggesting that the transmission of somatosensory information through the LatC is modulated by a descending pathway (Broman and Blomqvist, 1990).

Lamina 5

Cytoarchitecture

Lamina 5 is located at the neck of the dorsal horn (Rexed, 1954; Molander *et al.*, 1984, 1989; Schoenen and Faull, 2004). It is characterized by the presence of thick nerve fibers running through it, particularly in its lateral half, where the lamina appears reticulated in transverse sections. The reticulation of lamina 5 divides the lamina into a medial (5M) and a lateral (5L) zone (Rexed, 1954). The size of these zones varies between species with the lateral zone occupying about one third of lamina 5 in the cat (Rexed, 1954) and approximately two thirds of the lamina in the rat (Molander *et al.*, 1984, 1989). The medial zone tends to be more homogenous and denser than the lateral zone (Molander *et al.*, 1984, 1989). Triangular, star shaped, and spindle shaped cells have been identified in lamina 5, and these cells vary greatly in size. In general, large cells tend to be located in the lateral zone, while small and medium sized cells are most commonly found in the medial zone (Rexed, 1954). In the human, lamina 5 cannot be clearly distinguished from lamina 6, instead these laminae are cytologically homogenous (Schoenen and Faull, 2004)

Dendritic architecture

In humans, laminae 5 and 6 share a similar dendritic architecture (Schoenen and Faull, 2004). Their dendrites extend predominately dorsoventrally and the dorsal and ventral extensions tend to be symmetrical (Schoenen and Faull, 2004). Dorsal dendrites can extend into laminae 3 and 2 and ventral dendrites often reach down into lamina 7 (Schoenen and Faull, 2004). In cats, dendrites in lamina 5 extend mostly in the mediolateral plane (Réthelyi, 1976). Further, Ritz and Greenspan (1985) showed that some neurons in lamina 5 of the cat have a longitudinal dendritic spread, whereas in the human longitudinal branching dendrites are not common here.

Chemoarchitecture

GABA is found in cells evenly distributed throughout lamina 5. It has been suggested that GABAergic cells play a role in switching activity in lamina 5 between cutaneous and visceral afferent systems (Barber *et al.*, 1982). In addition, calretinin (Ren and Ruda, 1994; Anelli and Heckman, 2005), cholecystokinin (Gibson *et al.*, 1981), dopamine (Holstege *et al.*, 1996), dynorphin (Cruz and Basbaum, 1986), enkephalin (Gibson *et al.*, 1981; Glazer and Basbaum, 1981; de Lanerolle and LaMotte, 1982), glycine (Cabot *et al.*, 1992; Hossaini *et al.*, 2007), nitric oxide (Saito *et al.*, 1994; Maiskii *et al.*, 1998), parvalbumin (Anelli and Heckman, 2005) somatostatin (Krukoff *et al.*, 1986; Leah *et al.*, 1988) and substance P (Gibson *et al.*, 1981,de Lanerolle and LaMotte, 1982) have also been identified in cell bodies in lamina 5.

Projections

Lamina 5 receives direct input from large A-β and small A-δ myelinated fibers and polysynaptic inputs from C-fibers from distal dendrites that extend into superficial laminae (Willis and Coggeshall, 1991). Lamina 5 also receives input from the red nucleus (Waldron and Gwyn, 1969; Brown, 1974). These rubrospinal fibers have been shown to contact excitatory and inhibitory interneurons within lamina 5.

The majority of projection neurons in lamina 5 are located in the medial zone (Molenaar and Kuypers, 1978). They project to several supraspinal targets including the cerebellum (Edgley and Gallimore, 1988; Rivero-Melián and Grant, 1990, Matsushita, 1999), dorsal column nuclei, rostromedial medulla and adjacent reticular formation (Abols and Basbaum, 1981; Basbaum and Fields, 1984; Mason *et al.*, 1985; Holstege, 1988), the thalamus (Burstein *et al.*, 1990b; Kayalioglu *et al.*, 1996, 1999), the parabrachial nucleus (Kitamura *et al.*, 1993), the cerebellum (Matsushita and Ikeda, 1970), the periaqueductal gray (Liu, 1983; Keay and Bandler, 1993), the pontomedulllary reticular formation (Chaouch *et al.*, 1983), the lateral cervical nucleus (Baker and Giesler, 1984; Willis and Coggeshall, 1991), the nucleus of the solitary tract (Menétrey and Basbaum, 1987) and the superior colliculus (Morrell and Pfaff, 1983).

In contrast, cells in the lateral reticulated zone of lamina 5 project locally, particularly to the preganglionic spinal neurons (Cabot *et al.*, 1994) where they overlap with visceral primary afferents (Kuo *et al.*, 1983; Cervero and Connell, 1984; Morgan *et al.*, 1986).

Functions

Lamina 5 is characterized by multireceptive wide dynamic range neurons (Ritz and Greenspan, 1985; Willis and Coggeshall, 1991). In addition, nociceptive and mechanoreceptive neurons have been identified in lamina 5. These three cell types appear to be morphologically distinct. Multireceptive cells appear to have large cell bodies and extensive dendritic spreads in all directions and axons that ascend in the controlateral ventral white matter. Nociceptive cells have smaller cell bodies but a similar dendritic spread. Mechanoreceptive cells have less extensive dendritic trees than the other two types (Ritz and Greenspan, 1985).

Lamina 6

Cytoarchitecture

Lamina 6, the deepest layer of the dorsal horn, is prominent at the limb enlargements, but absent in other regions of the cord (see Chapters 15 and 16; Rexed, 1954; Molander *et al.*, 1984, 1989). It is thickest medially and has a slightly curved ventral surface (Rexed, 1954). Lamina 6 can be distinguished from laminae 5 and 7 by its more compact appearance (Molander *et al.*, 1989). Lamina 6, like lamina 5, can be divided into a medial and a lateral zone in many species (Rexed, 1954; Schoenen and Faull, 2004), but in the rat medial and lateral zones can only be identified at lumbar segments 4-6 (Molander *et al.*, 1984). The medial zone covers about one third of the lamina and the cells are found to be small and tightly packed. In the lateral zone, the cells are larger and less numerous (Rexed, 1954). Moreover, the medial and lateral zones are often separated by a cell free zone (Molander *et al.*, 1984). Small and medium sized cells in lamina 6 are usually elongated, spindle shaped, or long-triangular in shape, whereas large cells are typically star shaped (Rexed, 1954).

Dendritic architecture

See details for lamina 5.

Chemoarchitecture

A few neurochemicals have been identified in the cell bodies of lamina 6, these include calbindin and calretinin (Ren and Ruda, 1994; Anelli and Heckman, 2005), enkephalin (Gibson *et al.*, 1981; Glazer and Basbaum, 1981), GABA (Barber *et al.*, 1982), glycine (Hossaini *et al.*, 2007), nitric oxide (Saito *et al.*, 1994; Maiskii *et al.*, 1998) and parvalbumin (Anelli and Heckman, 2005).

Projections

Collateral axons of group Ia muscle spindle afferent fibers form profuse terminal arbors in lamina 6 (Brown and Fyffe, 1978). Further, lamina 6 receives cutaneous and nociceptive input from cells in lamina 6 with dorsally projecting dendrites that reach the superficial laminae. Lamina 6 also receives descending input from neurons in the red nucleus (Waldron and Gwyn, 1969; Brown, 1974).

Cells in lamina 6 have been shown to project to the cerebellum (Rivero-Melián and Grant, 1990) and the dorsal reticular nucleus of the medulla (Villanueva *et al.*, 1991). However, the majority of lamina 6 cells are either propriospinal neurons or interneurons with short locally projecting axons, with many that target the ventral horn motoneurons.

Function

The major function of cells in lamina 6 appears to be as interneurons involved in reflex pathways.

Lamina 7

Cytoarchitecture

Lamina 7 corresponds to the large area in the center of the gray matter commonly known as the intermediate gray (Rexed, 1954; Molander *et al.*, 1984, 1989). It extends medially to border lamina 10 and laterally to the white matter of the lateral funiculus (see Chapters 15 and 16; Rexed, 1954). At the limb enlargements, the size of lamina 7 increases and reaches ventrally to the ventral border of the gray matter between laminae 8 and 9 (see Chapters 15 and 16; Molander *et al.*, 1984, 1989; Schoenen and Faull, 2004). Cells are mostly medium-sized and evenly distributed throughout lamina 7, giving the lamina a homogenous appearance (Schoenen and Faull, 2004). However, a few large star shaped cells resembling large motoneurons, though slightly more rounded, are often seen (Rexed, 1954; Molander *et al.*, 1984, 1989). Triangular, multipolar, and fusiform shaped cells have also been identified in lamina 7 (Rexed, 1954; Schoenen and Faull, 2004). The central cervical nucleus is located in lamina 7, lateral to lamina 10 in C1-C6 segments.

Dendritic architecture

In the human, lamina 7 is characterized by cells with dendrites that extend mainly in the mediolateral plane. These dendrites can extend over the entire width of gray matter (Schoenen and Faull, 2004). In the rat, small neurons in lamina 7 have been described as having short, sparsely branched dendritic trees (Ritz and Greenspan, 1985).

Chemoarchitecture

A few neurochemical substances have been identified in cell bodies in lamina 7, including calbindin and calretinin (Anelli and Heckman, 2005), enkephalin (Gibson *et al.*, 1981; Glazer and Basbaum, 1981), GABA (Barber *et al.*, 1982), glycine (Cabot *et al.*, 1992), nitric oxide (Dun *et al.*, 1993; Saito *et al.*, 1994; Maiskii *et al.*, 1998) and parvalbumin (Anelli and Heckman, 2005). Large GABAergic neurons in lamina 7 have been found to lie in close association with the soma of motoneurons lying at the medial border of lamina 9 and probably function as premotor interneurons (see below) (Barber *et al.*, 1982).

Connections

Lamina 7 is targeted by regions in the brain involved in motor functions including the motor cortex, the red nucleus and the periaqueductal gray. These projections terminate medially in lamina 7, probably on interneurons (Waldron and Gwyn, 1969; Brown, 1974). Lamina 7 also receives a small contingent of primary afferents, originating chiefly from muscles and joints (Eccles *et al.*, 1954; Menétrey *et al.*, 1977). These afferents are also concentrated in medial portion where they have been shown to contact inhibitory interneurons. In contrast, visceral afferent inputs in the thoracic cord terminate laterally in lamina 7 (Kuo *et al.*, 1983; Cervero and Connell, 1984; Morgan *et al.*, 1986).

A majority of cells in lamina 7 are premotor interneurons projecting to the motor cells in lamina 9. Lamina 7 has also been shown to have supraspinal projections, including projections to the amygdala and hypothalamus (Menétrey and De Pommery, 1991), cerebellum (Rivero-Melián and Grant, 1990; Matsushita, 1999;), the contralateral superior colliculus (Morrell and Pfaff, 1983), parabrachial nucleus (Kitamura *et al.*, 1993), periaqueductal gray (Liu, 1983; Keay and Bandler, 1993), and the pontomedulllary reticular formation (Chaouch *et al.*, 1983; Krutki *et al.*, 1999). Large lamina 7 cells in the lumbar spinal cord, have been shown to be cells projecting into the ventral spinocerebellar tract (Yamada *et al.*, 1991).

Function

The cells in the large central part of lamina 7 are the interneurons that connect to motoneuron pools. In almost all mammals, descending motor pathways control motoneurons by means of connection with interneurons in lamina 7, rather than direct connections to motoneurons. The advantage of using the interneuron system is that a single lamina 7 interneuron or small group of interneurons may be able to execute an entire movement involving excitation or inhibition of a number of different muscles. For example, kicking something with one foot primarily involves stimulation of hip flexors and inhibition of hip extensors, but also requires extension of the knee of the same side. It also requires the hip extensors and abductors of the opposite side to be stimulated to maintain balance. Many of the interneurons of lamina 7 are pre-wired to execute all components of a desired movement. This modular movement control system provides a very efficient way for brain centers to execute desired movements. The functional arrangement of lamina 7 interneurons is matched to the topographical arrangement of motoneuron columns; the interneurons supplying trunk musculature are more medially placed, and the interneurons supplying limb musculature are more laterally placed.

Central cervical nucleus

The central cervical nucleus (CeCv) is located in lamina 7 just lateral to lamina 10 at segmental levels C1-C4 (see Chapters 15 and 16). Central cervical nucleus (CeCV) is mainly composed of large multipolar neurons with dendrites radiating dorsolaterally, laterally and ventrally, and occasionally small triangular and spindle shaped neurons with dendrites following the rounded countour of the CeCv (Wiksten, 1979).

CeCv receives primary afferents from upper cervical muscles (Amman *et al.*, 1983; Neuhuber and Zenker, 1989), cervical spine joints (Hirai *et al.*, 1978), also from lower cervical dorsal roots, and semicircular canals (Thomson *et al.*, 1996). The afferent input relayed here is mediated directly to the contralateral vestibular nuclei. Vestibulospinal neurons project to the contralateral CeCv (Carleton and Carpenter, 1983; Donevan *et al.*, 1990), this connection serves for the tonic neck (cervicovestibulospinal) reflex (Matsushita *et al.*, 1995). CeCv neurons have been shown to project to the cerebellum, mostly contralaterally (Matsushita and Tanami, 1987; Matsushita, 1991). Projections from upper cervical segments to lateral, medial and descending vestibular nuclei also originate from the CeCv (Carleton and Carpenter, 1983; Bankoul and Neuhuber 1992; Matsushita *et al.*, 1995; Xiong and Matsushita, 2001). CeCv neurons respond to cutaneous, proprioceptive and vestibular stimuli (Popova *et al.*, 1995). Calretinin and calmodulin (Ren and Ruda, 1994), enkephalin (Ikeda *et al.*, 1995), GABA and glycine (Ragnarson *et al.*, 2002), glutamate (Ornung *et al.*, 1995; Ragnarson *et al.*, 2002; Persson *et al.*, 2006) and parvalbumin (Ren and Ruda, 1994) are found within neurons of the CeCv.

Autonomic cell columns

The preganglionic cell columns of the autonomic nervous system are found in lamina 7 (see Chapters 15 and 16). These include the intermediolateral nucleus (IML) and the intercalated nucleus (ICl) (Poljak, 1924; Cummings, 1969; Petras and Cummings, 1972; Chung *et al.*, 1975; Barber *et al.*, 1982).

The IML contains the preganglionic sympathetic neurons of the spinal cord. It consists of a principal intermediolateral (IMLp) and a funicular intermediolateral (IMLf) part (see Chapters 15 and 16). The IMLp is located at very lateral edge of lamina 7 forming the intermediolateral horn in segments T1 to L1 in the human, and as far down to L3 in the rat (Molander *et al.*, 1984) and L4 in the cat (Rexed, 1954). It is arranged as rounded clusters (up to 30 neurons) following each other in the rostrocaudal direction. Their dendrites extend longitudinally within the column (Light and Metz, 1978),

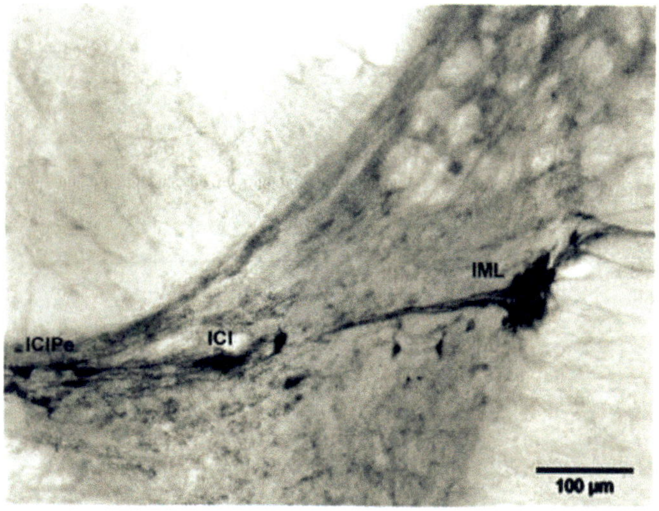

Figure 6.2 The intermediolateral nucleus and related cell groups

Transverse section of NADPH-diaphorase stained rat spinal cord showing the intermediolateral nucleus (IML), intercalated nucleus (ICl), and paraependymal part of the intercalated nucleus (IC1Pe).

dendritic projections are also present to the intermediate area and into the lateral funiculus (Deuschl and Illert, 1981). Within the IMLp, neurons form subgroups that also give rise to prominent, medially projecting dendritic bundles, as well as much smaller laterally projecting bundles that can extend right to the lateral edge of the spinal cord (Light and Metz, 1978). The IMLf runs along the side of the IMLp, embedded in the lateral funiculus. The dendrites of neurons in the IMLf extend into the IMLp medially and into the white matter laterally (Light and Metz, 1978).

ICl forms a series of bridges in the intermediate zone, connecting the IML with the central autonomic area (CA, named also the paraependymal part of the intercalated nucleus, ICLPe) which is located in lamina 10, dorsolateral to the central canal (see Chapters 15 and 16).

Sympathetic preganglionic neurons contain acetylcholine (ACh), cholecystokinin, enkephalin, neurotensin, neurophysin, nitric oxide, oxytocin, substance P, serotonin, somatostatin, and vasoactive intestinal polypeptide (VIP) (Swanson and McKellar, 1979; Holets and Elde, 1982; Krukoff *et al.*, 1985; Schoenen *et al.*, 1985; LaMotte and de Lanerolle, 1986; Romagnano *et al.*, 1987; Vizzard *et al.*, 1993, 1994; Tang *et al.*, 1998).

The intermediomedial nucleus

The intermediomedial nucleus (IMM) is located immediately lateral and partly embedded to lamina 10. In C1-C4 segments it is located more caudally, dorsal to the central cervical nucleus and in C5-C8 more ventrally in lamina 6, and in lamina 7 in T1 and below. The neurons of IMM are small to medium-sized and triangular in shape. Afferent autonomic neurons are considered to relay visceral information to the IMM at all spinal levels. Bulbospinal fibers of the baroreceptor reflex arc are shown to terminate on and excite interneurons in the vicinity of IMM. It is speculated that these neurons directly inhibit components of sympathoexcitatory pathways located in the IML (McCall *et al.*, 1977). IMM neurons contain acetylcholine (Schäfer *et al.*, 1998) and vasoactive intestinal polypeptide (LaMotte and de Lanerolle, 1986).

Lumbar dorsal commissural nucleus

The lumbar dorsal commissural nucleus (LDCom) is located in lamina 7 of the upper lumbar spinal cord (L1-L2) along the midline, in the dorsal gray commissure. It was first identified as a midline preganglionic autonomic cell column by Hancock and Peveto (1979). LDCom neurons are round, oval, or spindle in shape. Dendrites of these neurons form a loose, longitudinal plexus throughout the nucleus (Hosoya *et al.*, 1994). The axons of this nucleus project to the hypogastric nerve and pelvic ganglia (Nadelhaft and McKenna, 1987). This nucleus shows sexual dimorphism, being considerably larger in the male (Nadelhaft and McKenna, 1987). LDCom neurons contain choline acetyltransferase (Barber *et al.*, 1984), enkephalin (Romagnano *et al.*, 1987), neurophysin, and oxytocin (Swanson and McKellar, 1979).

Sacral dorsal commissural nucleus

The sacral dorsal commissural nucleus (SDCom) is located in the medial parts of laminae 2i and 3-7 in spinal cord segments L6-S4. It is involved in mediating pelvic visceral information,

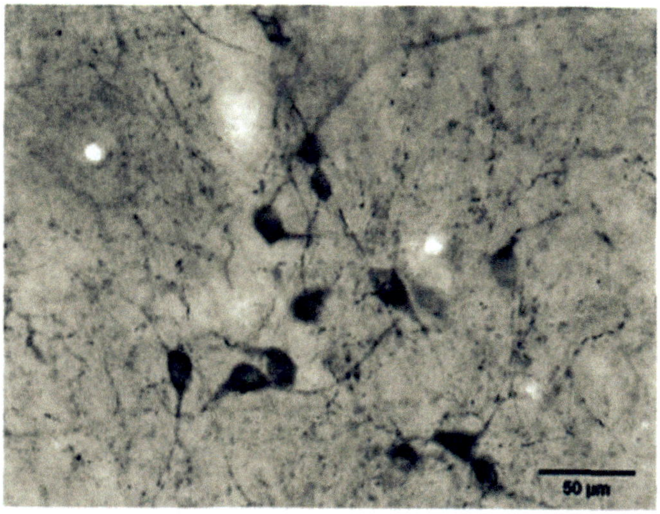

Figure 6.3 Sacral dorsal commissural nucleus

Transverse section of NADPH-diaphorase stained rat spinal cord showing the neurons of the sacral dorsal commissural nucleus.

including nociception (Wang *et al.*, 1999). SDCom neurons also receive somatic afferents (Honda, 1985; Al Chaer *et al.*, 1996). Electrical stimulation of SDCom neurons has been shown to evoke relaxation of the external urethral sphincter in the cat (Blok *et al.*, 1998). SDCom neurons contain enkephalin (Romagnano *et al.*, 1987), GABA and glycine (Feng *et al.*, 2005), glutamate (Xu *et al.*, 1999), nitric oxide (Tan *et al.*, 2006), and substance P (Lü *et al.*, 1995).

The sacral precerebellar nucleus (Stilling's sacral nucleus)

The sacral precerebellar nucleus (SPrCb) is located in lamina 7of S1-S4 segments. SPrCb is present in the rat and squirrel monkey, but not in the cat (Snyder *et al.*, 1978). Molander *et al.* (1984) identified this nucleus as the medial group of large neurons in the intermediate zone in sacral segments of the rat spinal cord. SPrCb consists of large multipolar neurons and is a round nucleus in transverse sections. The dendrites of this nucleus are longitudinally oriented. SPrCb projects to the cerebellum. In many ways, the nucleus resembles the dorsal nucleus (Clarke) which is located in a similar location in the gray matter of upper segments, and it was initially thought that SPrCb was a continuation of the dorsal nucleus. However, the SPrCb neurons project to the contralateral cerebellum in contrast to the ipsilaterally projecting dorsal nucleus neurons (Snyder *et al.*, 1978). SPrCb neurons, like those of the dorsal nucleus, are activated by group I muscle afferents (Edgley and Grant, 1991). Electrophysiological properties of SprCb neurons also resemble that of central cervical nucleus neurons (Edgley and Grant, 1991).

Some SPrCb neurons also project to the thalamus, but these neurons are not influenced by cutaneous afferents, and have the same electrophysiological properties of spinocerebellar neurons. It is suggested that these neurons are involved in spatial orientation of the tail (Milne *et al.*, 1982).

In upper lumbar segments, a lumbar precerebellar nucleus (LPrCb) is also identified, presumably similar to SPrCb in structure and function. LPrCb has been described in L1-L3 segments of the rat as 'the medial group of large neurons in the intermediate zone' by Molander *et al.* (1984). The location of the neurons in the LPrCb corresponds well with contralateral projecting spinocerebellar neurons described by Matsushita and Hosoya (1979).

Sacral parasympathetic nucleus

The sacral parasympathetic nucleus (SPSy) is located in the lateral part of lamina 7 in S1-S2 segments in rodents. It consists 7 of parasympathetic preganglionic neurons that innervate the large intestine, urethral and anal sphincters, urinary bladder, and reproductory system organs (De Groat *et al.*, 1981, 1982). It receives projections from pelvic visceral primary afferents (Mawe *et al.*, 1984). Burstein *et al.*, (1990a) have shown that SPSy also contains a distinct population of spinohypothalamic neurons. SPSy neurons contain substance P, enkephalin, cholecystokinin, somatostatin (Schoenen *et al.*, 1985) and vasoactive intestinal polypeptide (LaMotte and de Lanerolle, 1986). The Barrington nucleus of the hindbrain, which initiates micturition, projects to the SPSy (see Chapter 9).

Lamina 8

Cytoarchitecture

The size and shape of lamina 8 varies at different levels of the spinal cord. At the cervical and lumbar enlargements, it is limited to the medial base of the ventral horn (see Chapters 15 and 16; Rexed, 1954; Molander *et al.*, 1984, 1989). In all other regions of the spinal cord, lamina 8 extends across the whole base of the ventral horn (see Chapters 15 and 16; Molander *et al.*, 1984, 1989). Cells of all sizes and shapes, including very large cells almost identical to motoneurons, are found in lamina 8 (Schoenen and Faull, 2004). Spindle shaped cells are also seen, most commonly, along the medial border (Rexed, 1954). The heterogeneity of lamina 8 distinguishes it from the more cytologically homogenous lamina 7 (Molander *et al.*, 1984, 1989; Schoenen and Faull, 2004).

Dendritic architecture

While cells in lamina 8 vary considerably in size and shape, they appear, in the human at least, to have a consistent organization of their dendritic trees. The majority of cells in lamina 8 have dendritic trees that are oriented dorsoventrally. Dorsal dendrites are directed towards lamina 7 and/or the ventral gray commisure, but do not cross the midline. Ventral branches run towards the ventromedial tip of the ventral horn. The dorsal and ventral arborizations are generally symmetrical with only a very small mediolateral spread (Schoenen and Faull, 2004).

Chemoarchitecture

There is limited data on the neurochemical identity of cells in lamina 8 neurons, though the majority are likely to be GABAergic (Barber *et al.*, 1981); some may also contain nitric oxide (Saito *et al.*, 1994), calbindin and parvalbumin (Anelli and Heckman, 2005).

Connections

The majority of neurons in lamina 8 do not project supraspinally. Some neurons in lamina 8, known as the spinal commissural neurons, have axons projecting to the contralateral ventral horn, to laminae 7, 8, and 9 (Ramón y Cajal, 1909; Scheibel and Scheilbel, 1969; Matsushita, 1970; Harrison *et al.*, 1986; Stokke *et al.*, 2002; Nissen *et al.*, 2005), while many others have axons that ascend or descend in the white matter immediately adjacent to the gray matter (the so-called long propriospinal fibers). In general, short propriospinal fibers are derived from neurons in the lateral halves of laminae 5, 6, and 7, and generally connect nearby segments, while long propriospinal fibers that connect the lumbar and cervical segments, as above, are derived in the medial part of laminae 7 and 8.

While the majority of cells in lamina 8 are interneurons or propriospinal neurons, lamina 8 does contain some projection neurons. Cells in lamina 8 have been reported to project to the contralateral cerebellum (Matsushita *et al.*, 1979), to the dorsal reticular nucleus of medulla (Villanueva *et al.*, 1991), periaqueductal gray (Kayalioglu *et al.*, 1996, 1999), the pontomedullary reticular formation (Chaouch *et al.*, 1983) and the thalamus (Craig *et al.*, 1989). A projection to the controlateral superior colliculus has also been reported (Morrell and Pfaff, 1983) and this appears to be reciprocal (Waldron and Gwyn, 1969; Zemlan *et al.*, 1979; Leong *et al.*, 1984). Also, there are ascending propriospinal connections from the lamina 8 neurons of the lumbar and sacral segments to mostly contralateral upper cervical segments (Dutton *et al.*, 2006).

Function

Neurons in lamina 8 seem to play a pivotal role in the coordination of motor activity. The spinal commissural neurons lie predominately in lamina 8 (Ramon y Cajál, 1909; Scheibel and Scheibel, 1969; Matsushita, 1970; Harrison *et al.*, 1986; Stokke *et al.*, 2002; Nissen *et al.*, 2005). These are thought to be critically important for coordinating locomotor activity on the two sides of the body (Ramón y Cajal, 1909; Scheibel and Scheibel, 1969; Matsushita and Ikeda, 1970; Harrison *et al.*, 1986; Stokke *et al.*, 2002; Nissen *et al.*, 2005), probably by mediating crossed reflex actions and/or crossed descending synaptic actions upon motoneurons (Perl, 1958, 1959; Nyberg-Hansen and Mascitti, 1964; Nyberg-Hansen, 1965; Grillner and Hongo, 1972; Skinner and Remmel, 1978). Further, the majority of neurons in long propriospinal pathways connecting the cervical and lumbar enlargements are found here and function in the coordination of front and hindlimb movements, for example, reciprocal arm and leg swing.

Lamina 9

Cytoarchitecture

Lamina 9, located at the base of the ventral horn, is the site of the motoneurons of the spinal cord. The α-motoneurons, whose axons innervate striated muscles, are the largest of all cells in the spinal cord and are usually star-shaped. Amongst these large α-motoneurons, small cells are also found. In the cat, small cells account for approximately 30% of the total number of neurons in lamina 9. These small cells correspond to γ-motoneurons which innervate contractile elements of the muscle spindles (Bryan *et al.*, 1972; Burke *et al.*, 1977). The axons of both α and γ-motoneurons exit through the ventral root.

The cells in lamina 9 are organized into a series of longitudinal motor columns, the number of which varies at different segments of the spinal cord (see Chapters 15 and 16). Two groups of motoneurons, the medial and lateral cell columns, are found throughout the length of the spinal cord (Molander *et al.*, 1984). In thoracic regions, the medial and lateral cell columns are hard to distinguish and may appear as a single cell column that occupies the most ventral part of the ventral horn (Rexed 1954; Barber *et al.*, 1984; Molander *et al.*, 1984; Routal and Pal, 1999). In the brachial and lumbosacral enlargements the number of motoneuron columns is much larger as to accommodate the supply of limb muscles. The medial and lateral groups are still present and easily distinguished at these levels, but additional groups are also found (see Chapters 15 and 16). These additional groups are generally found in lateral parts of the ventral horn with the number of groups identified differing somewhat between studies (Rexed, 1954; Schoenen 1982b; Nicolopoulos-Stournaras and Iles, 1983; Barber *et al.*, 1984; Molander *et al.*, 1984; Rivero-Melián, 1996; Routal and Pal, 1999; Schoenen and Faull, 2004).

Dendritic architecture

Many studies have identified longitudinal branching dendrites on motoneurons and these often run together with those of nearby motoneurons to form longitudinal bundles (Ramón y Cajal, 1909; Laruelle, 1937, 1948; Scheibel and Scheibel, 1970; Matthews *et al.*, 1971; Schoenen, 1982b). In some species longitudinal dendritic bundling is found in all of the motor columns (Ramón y Cajal, 1911; Laruelle 1937, 1948; Scheibel and Scheibel 1970; Matthews *et al.*, 1971; Schoenen 1982b). Dendrites of motoneurons are extensive in the rostrocaudal, mediolateral and dorsoventral planes and may extend dorsally into laminae 3-4 in the rat (Cook and Woolf, 1985). In the human, however, the dendrites of cells located dorsolaterally have very little dendritic bundling. In the human, the longitudinal dendritic bundles are exclusively formed by motoneurons of the same column in the human, while in the cat, branches of longitudinal bundles can arise from different motoneuronal columns as well as with dendrites of lamina 8. The functional significance of motoneuron dendritic bundling is largely speculative. It has been suggested that they may act to synchronize the action of a specific group of muscles (Scheibel and Scheibel, 1970; Schoenen, 1982b).

Further, in most species, motoneurons also display some transverse dendritic bundling. These dendritic bundles have been reported to extend into the dorsal horn as well as deep into the ventral, lateral, and dorsolateral gray matter. Some transversely oriented dendritic bundles are formed between several different motoneuron groups. In addition, in the cat and rat, but not in human, some transverse dendritic bundles from medially located motoneurons cross the midline in the ventral commissure and mix together with dendrites of corresponding contralateral cells (Light and Metz, 1978). The function of the contralateral dendrites has been suggested to relate to reciprocal motoneuronal interactions such as those involved in walking (Scheibel and Scheibel, 1973).

Chemoarchitecture

Motoneurons in the ventral horn are cholinergic (see Chapters 15 and 16; Satoh *et al.*, 1983; Barber *et al.*, 1984). Other substances that have been identified in motoneurons include calmodulin (Ren and Ruda, 1994), calcitonin gene-related peptide (Gibson *et al.*, 1984) and dopamine (Holstege *et al.*, 1996).

Projections

Lateral group of lamina 9 neurons are known to project to the cerebellum (Rivero-Melián and Grant, 1990; Matsushita, 1999).

Function

Functions of lamina 9 neurons are discussed in Chapter 7.

Lamina 10

Cytoarchitecture

Lamina 10, also known as area 10 or the central gray of the spinal cord, corresponds to the area around the central canal (see Chapters 15 and 16; Molander *et al.*, 1984, 1989; Schoenen and Faull, 2004). It is bordered dorsally by lamina 4 and in sacral and coccygeal segments with lamina 5 which cross the midline, and ventrally by the ventral white commissure. The intermediomedial nucleus (IMM) is adjacent to or in some segments partly embedded in lamina 10. Lamina 10 consists of triangular or star shaped cells (Rexed, 1954). Spindle shaped cells are also seen in the dorsal part of lamina 10. The cells in lamina 10 are more densely packed and small to medium-sized neurons, generally smaller than those in the adjacent lamina 7 (Molander *et al.*, 1989; Schoenen and Faull, 2004). The central autonomic area (CA) (also known as rhe paraependymal part of the intercalated nucleus, IClPe) and cholinergic central canal cluster neurons are also located in lamina 10 (Petras and Cummings, 1972; Phelps *et al.*, 1988).

Dendritic architecture

Two types of cells have been found in lamina 10 in the human. In the middle and rostral portions of lamina 10, cells with a bipolar dendritic tree extending in the dorsoventral plane are common. The branches of these can extend out into laminae 6 and 8. In the ventral portion of lamina 10, dendritic trees are found to extend mostly in the rostrocaudal plane. In the cat, neurons in lamina 10 appear to be more heterogeneous than in the human and these cannot be grouped to any obvious morphological classes (Honda and Perl, 1985).

Chemoarchitecture

GABAergic cells are found in lamina 10 and these are organized into two longitudinal columns, one located dorsolateral and one ventrolateral to the central canal (Barber *et al.*, 1982). Further, neurons immunoreactive for calbindin and calretinin (Anelli and Heckman, 2005), calcitonin gene-related peptide (Ròssler *et al.*, 1993), choline acetyltransferase (Barber *et al.*, 1984), cholecystokinin (Gibson *et al.*, 1981), dopamine (Holstege *et al.*, 1996), enkephalin (Glazer and Basbaum, 1981), GABA (Barber *et al.*, 1982), glycine (Spike *et al.*, 1993) neuropeptide Y (Weihe, 1992), neurotensin (Gibson *et al.*, 1981), nitric oxide (Dun *et al.*, 1993; Saito *et al.*, 1994; Maiskii *et al.*, 1998; Reuss and Reuss, 2001), parvalbumin (Anelli and Heckman, 2005), somatostatin (LaMotte, 1988), substance P (Gibson *et al.*, 1981, LaMotte and de Lanerolle, 1986), serotonin (LaMotte, 1988; Weihe, 1992) and vasoactive intestinal polypeptide (Gibson *et al.*, 1981; LaMotte, 1988) have been identified in lamina 10.

Projections

Lamina 10 receives input from C-fibers and A-δ fibers and projects to the lateral parabrachial nucleus, the nucleus of the solitary tract and amygdala (Menétrey and Basbaum, 1987; Menétrey and De Pommery, 1991), medullary and pontine reticular formation (Nahin *et al.*, 1983; Villanueva *et al.*, 1991), nucleus of the solitary tract (Menétrey and Basbaum, 1987), periaqueductal gray (Liu, 1983; Keay and Bandler, 1993; Kayalioglu *et al.*, 1996, 1999; Mouton and Holstege, 2000), thalamus (Menétrey *et al.*, 1989; Burstein *et al.*, 1990b; Kayalioglu *et al.*, 1996, 1999), and hypothalamus (Burstein *et al.*, 1987; Menétrey and De Pommery, 1991; Kayalioglu *et al.*, 1996, 1999). They have also been shown to project to the contralateral lamina 10 (Nahin and Miceyvich, 1986). Ascending propriospinal axons have also been shown to project to lamina 10 at all spinal levels (Matsushita, 1998). Lamina 10 receives a direct projection from nuclei in the brainstem known to be involved in descending inhibition of nociceptive transmission in the spinal cord, such as the raphe magnus nucleus (Holstege, 1987).

Function

Lamina 10 receives somatic (Neuhuber *et al.*, 1986) and visceral afferents (Ciriello and Calaresu, 1983; Nadelhaft and Booth, 1984). Anatomical and physiological studies in the cat and monkey (Light and Perl, 1979a,b; Honda 1985; Honda and Perl, 1985) suggest that lamina 10 plays a role in nociceptive transmission and visceroception including visceral nociception (Matsushita 1998; Wang *et al.*, 1999). Honda and Perl (1985) showed that there are at least three functional categories of neurons in lamina 10 – multireceptive, nociceptive, and mechanoreceptive. In light of the diversity of lamina 10 in terms of morphology, connections and the response characteristic of its cells, they suggested that the functional organization of lamina 10 may be similar to the central gray of the brainstem.

References

Abols IA, Basbaum AI (1981) Afferent connections of the rostral medulla of the cat, A neural substrate for midbrain-medullary interactions in the modulation of pain. J Comp Neurol 201, 285-297.

Al-Chaer ED, Lawand NB, Westlund KN, Willis WD (1996) Pelvic visceral input into the nucleus gracilis is largely mediated by the postsynaptic dorsal column pathway. J Neurophysiol 76, 2675-2690.

Alvarez FJ, Kavookjian AM, Light AR (1993) Ultrastructural morphology, synaptic relationships, and CGRP immunoreactivity of physiologically identified C-fiber terminals in the monkey spinal cord. J Comp Neurol 329, 472-490.

Alvarez FJ, Villalba RM, Carr PA, Grandes P, Somohano PM (2000) Differential distribution of metabotropic glutamate receptors 1a, 1b, and 5 in the rat spinal cord. J Comp Neurol 422, 464-487.

Ammann B, Gottschall J, Zenker W (1983) Afferent projections from the rat longus capitis muscle studied by transganglionic transport of HRP. Anat Embryol (Berl) 166, 275-289.

Andrew D, Krout KE, Craig AD (2003) Differentiation of lamina I spinomedullary and spinothalamic neurons in the cat. J Comp Neurol 458, 257-271.

Anelli R, Heckman CJ (2005) The calcium binding proteins calbindin, parvalbumin, and calretinin have specific patterns of expression in the gray matter of cat spinal cord. J Neurocytol 34, 369-385.

Antal M (1984) Termination areas of corticobulbar and corticospinal fibres in the rat. J Hirnforsch 25, 647-659.

Antal M, Polgár E, Chalmers J, Minson JB, Llewellyn-Smith I, Heizmann CW, Somogyi P (1991) Different populations of parvalbumin- and calbindin-D28k-immunoreactive neurons contain GABA and accumulate 3H-D-aspartate in the dorsal horn of the rat spinal cord. J Comp Neurol 314, 114-124.

Apkarian AV, Hodge CJ (1989) A dorsolateral spinothalamic tract in macaque monkey. Pain 37, 323-333.

Aronin N, DiFiglia M, Liotta AS, Martin JB (1981) Ultrastructural localization and biochemical features of immunoreactive LEU-enkephalin in monkey dorsal horn. J Neurosci 1, 561-577.

Baker ML, Giesler GJ Jr (1984) Anatomical studies of the spinocervical tract of the rat. Somatosens Res 2, 1-18.

Bankoul S, Neuhuber WL (1992) A direct projection from the medial vestibular nucleus to the cervical spinal dorsal horn of the rat, as demonstrated by anterograde and retrograde tracing. Anat Embryol (Berl) 185, 77-85.

Barber RP, Vaughn JE, Roberts E (1982) The cytoarchitecture of GABAergic neurons in rat spinal cord. Brain Res 238, 305-328.

Barber RP, Phelps EP, Houser CR, Crawford DG, Salvaterra MP, Vaughn JE (1984) The morphology and distribution of neurons containing choline acetyltransferase in the adult rat spinal cord: an immunocytochemical study. J Comp Neurol 229, 329-346.

Basbaum AI, Fields HL (1984) Endogenous pain control systems: brainstem spinal pathways and endorphin circuitry. Annu Rev Neurosci 7, 309-338.

Battaglia G, Rustioni A (1992) Substance P innervation of the rat and cat thalamus. II. Cells of origin in the spinal cord. J Comp Neurol 315, 473-486.

Beal JA, Cooper MH (1978) The neurons in the gelatinosal complex (Laminae II and III) of the monkey (Macaca mulatta): a golgi study. J Comp Neurol 179, 89-121.

Beal JA, Penny JE, Bicknell HR (1981) Structural diversity of marginal (lamina I) neurons in the adult monkey (Macaca mulatta) lumbosacral spinal cord: a golgi study. J Comp Neurol 202, 237-254.

Bennett GJ, Seltzer Z, Lu GW, Nishikawa N, Dubner R (1983) The cells of origin of the dorsal column postsynaptic projection in the lumbosacral enlargements of cats and monkeys. Somatosens Res 1, 131-149.

Bennett GJ, Nishikawa N, Lu GW, Hoffert MJ, Dubner R (1984) The morphology of dorsal column postsynaptic spinomedullary neurons in the cat. J Comp Neurol 224, 568-578.

Berkley KJ, Blomqvist A, Pelt A, Flink R (1980) Differences in the collateralization of neuronal projections from the dorsal column nuclei and lateral cervical nucleus to the thalamus and tectum in the cat: an anatomical study using two different double-labeling techniques. Brain Res 202, 273-290.

Bernardi PS, Valtschanoff JG, Weinberg RJ, Schmidt HH, Rustioni A (1995) Synaptic interactions between primary afferent terminals and GABA and nitric oxide-synthesizing neurons in superficial laminae of the rat spinal cord. J Neurosci 15, 1363-1371.

Blok BF, van Maarseveen JT, Holstege G (1998) Electrical stimulation of the sacral dorsal gray commissure evokes relaxation of the external urethral sphincter in the cat. Neurosci Lett 249, 68-70.

Broman J, Blomqvist A (1990) Serotonergic innervation of the lateral cervical nucleus: an immunohistochemical study in cats and monkeys (Aotus trivirgatus). Synapse 6, 55-62.

Broman J, Westman J (1988) GABA-immunoreactive neurons and terminals in the lateral cervical nucleus of the cat. J Comp Neurol 274, 467-482.

Brown AG (1981) *Organization in the Spinal Cord.* New York, Verlag Springer.

Brown AG, Fyffe RE (1978) Synaptic contacts made by identified Ia afferent fibres upon motoneurones. J Physiol 284, 43P-44P.

Brown AG, Fyffe RE (1981) Direct observations on the contacts made between Ia afferent fibres and alpha-motoneurones in the cat's lumbosacral spinal cord. J Physiol 313, 121-140.

Brown AG, Rose PK, Snow PJ (1977) The morphology of hair follicle afferent fibre collaterals in the spinal cord of the cat. J Physiol 272, 779-797.

Brown LT (1974) Rubrospinal projections in the rat. J Comp Neurol 154,169-187.

Bryan RN, Trevino DL, Willis WD (1972) Evidence for a common location of alpha and gamma motoneurons. Brain Res 38, 193-196.

Burke RE, Strick PL, Kanda K, Kim CC, and Walmsley B (1977) Anatomy of medial gastrocnemius and soleus motor nuclei in cat spinal cord. J Neurophysiol 40, 667-680.

Burstein R, Giesler GJ Jr (1989) Retrograde labeling of neurons in spinal cord that project directly to nucleus accumbens or the septal nuclei in the rat. Brain Res 497, 149-154.

Burstein R, Potrebic S (1993) Retrograde labeling of neurons in the spinal cord that project directly to the amygdala or the orbital cortex in the rat. J Comp Neurol 335, 469-485.

Burstein R, Cliffer KD, Giesler GJ Jr (1987) Direct somatosensory projections from the spinal cord to the hypothalamus and telencephalon. J Neurosci 7, 4159-4164.

Burstein R, Cliffer KD, Giesler GJ Jr (1990a) Cells of origin of the spinohypothalamic tract in the rat. J Comp Neurol 291, 329-344.

Burstein R, Dado RJ, Giesler GJ Jr (1990b) The cells of origin of the spinothalamic tract of the rat: a quantitative reexamination. Brain Res 511, 329-337.

Burstein R, Wang JL, Elde RP, Giesler GJ Jr (1990c) Neurons in the sacral parasympathetic nucleus that project to the hypothalamus do not also project through the pelvic nerve – a double labeling study combining Fluoro-gold and cholera toxin B in the rat. Brain Res 506, 159-165.

Cabot JB, Alessi V, Bushnell A (1992) Glycine-like immunoreactive input to sympathetic preganglionic neurons. Brain Res 571, 1-18.

Cabot JB, Alessi V, Bushnell A (1994) A quantitative ultrastructural immunogold labeling study of the relationships between glutamate (GLU) and GABA containing boutons synapsing on sympathetic preganglionic neurons. Soc Neurosci Abstr 20, 843.

Campistron G, Buijs RM, Geffard M (1986) Glycine neurons in the brain and spinal cord. Antibody production and immunocytochemical localization. Brain Res 376, 400-405.

Carleton SC, Carpenter MB (1983) Afferent and efferent connections of the medial, inferior and lateral vestibular nuclei in the cat and monkey. Brain Res 278, 29-51.

Carlton SM, Chung JM, Leonard RB, Willis WD (1985) Funicular trajectories of brainstem neurons projecting to the lumbar spinal cord in the monkey (Macaca fascicularis): a retrograde labeling study. J Comp Neurol 241, 382-404.

Carstens E, Trevino DL (1978) Laminar origins of spinothalamic projections in the cat as determined by the retrograde transport of horseradish peroxidase. J Comp Neurol 182, 161-165.

Cechetto DF, Standaert DG, Saper CB (1985) Spinal and trigeminal dorsal horn projections to the parabrachial nucleus in the rat. J Comp Neurol 240, 153-160.

Cervero F, Connell LA (1984) Distribution of somatic and visceral primary afferent fibres within the thoracic spinal cord of the cat. J Comp Neurol 230, 88-98.

Cervero F, Iggo A, Molony V (1977) Responses of spinocervical tract neurones to noxious stimulation of the skin. J Physiol 267, 537-558.

Cervero F, Tattersall JEH (1986) Somatic and visceral sensory integration in the thoracic spinal cord. Prog Brain Res 67, 189-205.

Cervero F, Handwerker HO, Laird JM (1988) Prolonged noxious mechanical stimulation of the rat's tail: responses and encoding properties of dorsal horn neurones. J Physiol 404, 419-436.

Chaouch A, Menétrey D, Binder D, Besson JM (1983) Neurons at the origin of the medial component of the bulbopontine spinoreticular tract in the rat: an anatomical study using horseradish peroxidase retrograde transport. J Comp Neurol 214, 309-320.

Cho HJ, Basbaum AI (1989) Ultrastructural analysis of dynorphin B-immunoreactive cells and terminals in the superficial dorsal horn of the deafferented spinal cord of the rat. J Comp Neurol 281, 193-205.

Christensen BN, Perl ER (1970) Spinal neurons specifically excited by noxious or thermal stimuli: marginal zone of the dorsal horn. J Neurophysiol 33, 293-307.

Chung JM, Chung K, Wurster R (1975) Sympathetic preganglionic neurons of the cat spinal cord: horseradish peroxidase study. Brain Res 91, 126-131.

Ciriello J, Calaresu FR (1983) Central projections of afferent renal fibers in the rat: an anterograde transport study of horseradish peroxidase. J Auton Nerv Syst 8, 273-285.

Cliffer KD, Giesler GJ Jr (1989) Postsynaptic dorsal column pathway of the rat. III. Distribution of ascending afferent fibers. J Neurosci 9, 3146-3168.

Cliffer KD, Urca G, Elde RP, Giesler GJ Jr (1988) Studies of peptidergic input to the lateral spinal nucleus. Brain Res 460, 356-360.

Coffield JA, Miletic V, Zimmermann E, Hoffert MJ, Brooks BR (1986) Demonstration of thyrotropin-releasing hormone immunoreactivity in neurons of the mouse spinal dorsal horn. J Neurosci 6, 1194-1197.

Conrath M, Taquet H, Pohl M, Carayon A (1989) Immunocytochemical evidence of calcitonin gene-related peptide-like neurons in the dorsal horn and lateral spinal nucleus of the rat cervical spinal cord. J Chem Neuroanat 2, 335-347.

Conrath-Verrier M, Dietl M, Tramu G (1984) Cholecystokinin-like immunoreactivity in the dorsal horn of the spinal cord of the rat: a light and electron microscopic study. Neuroscience, 13, 871-885.

Cook AJ, Woolf CJ (1985) Cutaneous receptive field and morphological properties of hamstring flexor alpha-motoneurones in the rat. J Physiol. 364, 249-263.

Craig AD (1994) Spinal and supraspinal processing of specific pain and temperature. In: T*ouch, temperature, and pain in health and disease: mechanisms and assessments. Progress in pain research and management*, Vol 3. Boivie J, Hansson P, Lindblom U (Eds.) IASP Press, Seattle.

Craig AD (1995) Distribution of brainstem projections from spinal lamina I neurons in the cat and the monkey. J Comp Neurol 361, 225-248.

Craig AD (2000) Spinal location of ascending lamina I axons in the macaque monkey. J Pain 1, 33-45.

Craig AD Jr, Burton H (1979) The lateral cervical nucleus in the cat: anatomic organization of cervicothalamic neurons. J Comp Neurol. 185, 329-346.

Craig AD, Hunsley SJ (1991) Morphine enhances the activity of thermoreceptive cold-specific lamina I spinothalamic neurons in the cat. Brain Res 558, 93-97.

Craig AD, Serrano LP (1994) Effects of systemic morphine on lamina I spinothalamic tract neurons in the cat. Brain Res 636, 233-244.

Craig AD, Linington AJ, Kniffki KD (1989) Significant differences in the retrograde labeling of spinothalamic tract cells by horseradish peroxidase and the fluorescent tracers fast blue and diamidino yellow. Exp Brain Res 74, 431-436.

Craig AD, Broman J, Blomqvist A (1992) Lamina I spinocervical tract terminations in the medial part of the lateral cervical nucleus in the cat. J Comp Neurol 322, 99-110.

Craig AD Jr, Tapper DN (1978) Lateral cervical nucleus in the cat, functional organization and characteristics. J Neurophysiol 41, 1511-1534.

Craig AD Jr, Linington AJ, Kniffki KD (1989) Cells of origin of spinothalamic tract projections to the medial and lateral thalamus in the cat. J Comp Neurol 289, 568-85.

Cruz L, Basbaum AI (1985) Multiple opioid peptides and the modulation of pain: immunohistochemical analysis of dynorphin and enkephalin in the trigeminal nucleus caudalis and spinal cord of the cat. J Comp Neurol 240, 331-348.

Cummings JF (1969) Thoracolumbar preganglionic neurons and adrenal innervation in the dog. Acta Anat 73, 27-37.

Dagci T, Okur BE, Kayalioglu G (2008) Lateral spinal nucleus of the rat, NADPH-diaphorase activity and Fos expression after noxious peripheral stimulation. Neurophysiology 40, 34-38.

Dawson NJ, Dickenson AH, Hellon RF, Woolf CJ (1981) Inhibitory controls on thermal neurones in the spinal trigeminal nucleus of cats and rats. Brain Res 209, 440-445.

De Groat WC, Nadelhaft I, Milne RJ, Booth AM, Morgan C, Thor K (1981) Organization of the sacral parasympathetic reflex pathways to the urinary bladder and large intestine. J Auton Nerv Syst 3, 135-160.

De Groat WC, Booth AM, Milne RJ, Roppolo JR (1982) Parasympathetic preganglionic neurons in the sacral spinal cord. J Auton Nerv Syst 5, 23-43.

De Koninck Y, Ribeiro-da-Silva A, Henry JL, Cuello AC (1992) Spinal neurons exhibiting a specific nociceptive response receive abundant substance P-containing synaptic contacts. Proc Natl Acad Sci USA 89, 5073-5077.

Deuschl G, Illert M (1981) Cytoarchitectonic organization of lumbar preganglionic sympathetic neurons in the cat. J Auton Nerv Syst 3, 193-213.

Ding YQ, Takada M, Shigemoto R, Mizuno N (1995) Spinoparabrachial tract neurons showing substance P receptor-like immunoreactivity in the lumbar spinal cord of the rat. Brain Res 674, 336-340.

Donevan AH, Neuber-Hess M, Rose PK (1990) Multiplicity of vestibulospinal projections to the upper cervical spinal cord of the cat: a study with the anterograde tracer Phaseolus vulgaris leucoagglutinin. J Comp Neurol 302, 1-14.

Dostrovsky JO, Craig AD (1996) Cooling-specific spinothalamic neurons in the monkey. J Neurophysiol 76, 3656-3665.

Dostrovsky JO, Shah Y, Gray BG (1983) Descending inhibitory influences from periaqueductal gray, nucleus raphe magnus, and adjacent reticular formation, II. Effects on medullary dorsal horn nociceptive and nonnociceptive neurons. J Neurophysiol 49, 948-960.

Dun NJ, Dun SL, Wu SY, Forstermann U, Schmidt HH, Tseng LF (1993) Nitric oxide synthase-immunoreactivity in the rat, mouse, cat and squirrel monkey spinal cord. Neuroscience 54, 845-857.

Dutton RC, Carstens MI, Antognini JF, Carstens E (2006) Long ascending propriospinal projections from lumbosacral to upper cervical spinal cord in the rat. Brain Res 1119, 76-85.

Eccles JC, Fatt P, Koketsu K (1954) Cholinergic and inhibitory synapses in a pathway from motor-axon collaterals to motoneurones. J Physiol 126, 524-562.

Edgley SA, Gallimore CM (1988) The morphology and projections of dorsal horn spinocerebellar tract neurones in the cat. J Physiol 397, 99-111.

Edgley SA, Grant GM (1991) Inputs to spinocerebellar tract neurones located in stilling's nucleus in the sacral segments of the rat spinal cord. J Comp Neurol 305, 130-138.

Esteves F, Lima D, Coimbra A (1993) Structural types of spinal cord marginal (lamina I) neurons projecting to the nucleus of the tractus soiltarius in the rat. Somatosens Mot Res 10, 203-216.

Feng YP, Li YQ, Wang W, Wu SX, Chen T, Shigemoto R, Mizuno N (2005) Morphological evidence for GABA/glycine-cocontaining terminals in synaptic contact with neurokinin-1 receptor-expressing neurons in the sacral dorsal commissural nucleus of the rat. Neurosci Lett 388, 144-148.

Ferrington DG, Sorkin LS, Willis WD Jr (1987) Responses of spinothalamic tract cells in the superficial dorsal horn of the primate lumbar spinal cord. J Physiol 388, 681-703.

Fields HL, Malick A, Burstein R (1995) Dorsal horn projection targets of ON and OFF cells in the rostral ventromedial medulla. J Neurophysiol 74, 1742-1759.

Finley JC, Maderdrut JL, Petrusz P (1981) The immunocytochemical localization of enkephalin in the central nervous system of the rat. J Comp Neurol 198, 542-565.

Foreman RD, Hobbs SF, Oh UT, Chandler MJ (1988) Differential modulation of thoracic and lumbar spinothalamic tract cell activity during stimulation of cardiopulmonary sympathetic afferent fibers in the primate. A new concept for visceral pain? In: *Proceedings of the 5th World Congress on Pain.* Dubner R, Gebhart GF, Bond MR (Eds.). Elsevier, Amsterdam.

Fuji K, Senba E, Fujii S, Nomura I, Wu JY, Ueda Y, Tohyama M (1985) Distribution, ontogeny and projections of cholecystokinin-8, vasoactive intestinal polypeptide and gamma-aminobutyrate-containing neuron systems in the rat spinal cord: an immunohistochemical analysis. Neuroscience 4, 881-894.

Galhardo V, Lima D (1999) Structural characterization of marginal (lamina I) spinal cord neurons in the cat: a Golgi study. J Comp Neurol, 315-333.

Gibson SJ, Polak JM, Bloom SR, Wall PD (1981) The distribution of nine peptides in rat spinal cord with special emphasis on the substantia gelatinosa and on the area around the central canal (lamina X). J Comp Neurol 201, 65-79.

Giesler GJ Jr, Elde RP (1985) Immunocytochemical studies of the peptidergic content of fibers and terminals within the lateral spinal and lateral cervical nuclei. J Neurosci 5, 1833-1841.

Giesler GJ Jr, Menétrey D, Basbaum AI (1979a) Differential origins of spinothalamic tract projections to medial and lateral thalamus in the rat. J Comp Neurol 184, 107-126.

Giesler GJ Jr, Urca G, Cannon JT, Liebeskind JC (1979b) Response properties of neurons of the lateral cervical nucleus in the rat. J Comp Neurol 186, 65-77.

Giesler GJ Jr, Katter JT, Dado RJ (1994) Direct spinal pathways to the limbic system for nociceptive information. Trends Neurosci 17, 244-250.

Glazer EJ, Basbaum AI (1981) Immunohistochemical localization of leucine-enkephalin in the spinal cord of the cat: enkephalin-containing marginal neurons and pain modulation. J Comp Neurol 196, 377-389.

Gobel S (1975) Golgi studies in the substantia gelatinosa neurons in the spinal trigeminal nucleus. J Comp Neurol 162, 397-415.

Gobel S (1978) Golgi studies of the neurons in layer II of the dorsal horn of the medulla (trigeminal nucleus caudalis). J Comp Neurol 180, 395-413.

Granum SL (1986) The spinothalamic system of the rat. I. Locations of cells of origin. J Comp Neurol 247, 159-180.

Grillner S, Hongo T (1972) Vestibulospinal effects on motoneurones and interneurones in the lumbosacral cord. Prog Brain Res 37, 243-262.

Gwyn DG, Waldron HA (1968) A nucleus in the dorsolateral funiculus of the spinal cord of the rat. Brain Res 10, 342-351.

Han ZS, Zhang ET, Craig AD (1998) Nociceptive and thermoreceptive lamina I neurons are anatomically distinct. Nat Neurosci 1, 218-225.

Hancock MB, Peveto CA (1979) A preganglionic autonomic nucleus in the dorsal gray commissure of the lumbar spinal cord of the rat. J Comp Neurol 183, 65-72.

Hantman AW, van den Pol AN, Perl ER (2004) Morphological and physiological features of a set of spinal substantia gelatinosa neurons defined by green fluorescent protein expression. J Neurosci 24, 836-842.

Harkness DH, Brownfield MS (1986) A thyrotropin-releasing hormone-containing system in the rat dorsal horn separate from serotonin. Brain Res 384, 323-333.

Harrison PJ, Jankowska E, Zytnicki D (1986) Lamina VIII interneurones interposed in crossed reflex pathways in the cat. J Physiol 371, 147-166.

Heinke B, Ruscheweyh R, Forsthuber L, Wunderbaldinger G, Sandkuhler J (2004) Physiological, neurochemical and morphological properties of a subgroup of GABAergic spinal lamina II neurons identified by expression of green fluorescent protein in mice. J Physiol 560, 249-266.

Herdegen T, Rudiger S, Mayer B, Bravo R, Zimmermann M (1994) Expression of nitric oxide synthase and colocalisation with Jun, Fos and Krox transcription factors in spinal cord neurons following noxious stimulation of the rat hindpaw. Brain Res Mol Brain Res 22, 245-258.

Hirai N, Hongo T, Sasaki S (1978) Cerebellar projection and input organizations of the spinocerebellar tract arising from the central cervical nucleus in the cat. Brain Res 157, 341-345.

Hoheisel U, Lehmann-Willenbrock E, Mense S (1989) Termination patterns of identified group II and III afferent fibres from deep tissues in the spinal cord of the cat. Neuroscience 28, 495-507.

Holets V, Elde R (1982) The differential distribution and relationship of serotoninergic and peptidergic fibers to sympathoadrenal neurons in the intermediolateral cell column of the rat, a combined retrograde axonal transport and immunofluorescence study. Neuroscience 7, 1155-1174.

Holstege G (1987) Some anatomical observations on the projections from the hypothalamus to brainstem and spinal cord: an HRP and autoradiographic tracing study in the cat. J Comp Neurol 260, 98-126.

Holstege G (1988) Brainstem-spinal cord projections in the cat, related to control of head and axial movements. Rev Oculomot Res 2, 431-470.

Holstege JC, Van Dijken H, Buijs RM, Goedknegt H, Gosens T, Bongers CM (1996) Distribution of dopamine immunoreactivity in the rat, cat and monkey spinal cord. J Comp Neurol 376, 631-652.

Honda CN (1985) Visceral and somatic afferent convergence onto neurons near the central canal in the sacral spinal cord of the cat. J Neurophysiol 53, 1059-1078.

Honda CN, Perl ER (1985) Functional and morophological features of neurons in the midline region of the caudal spinal cord of the cat. Brain Res 340, 285-295.

Hosoya Y, Nadelhaft I, Wang D, Kohno K (1994) Thoracolumbar sympathetic preganglionic neurons in the dorsal commissural nucleus of the male rat: an immunohistochemical study using retrograde labeling of cholera toxin subunit B. Exp Brain Res 98, 21-30.

Hossaini M, French PJ, Holstege JC (2007) Distribution of glycinergic neuronal somata in the rat spinal cord. Brain Res 1142, 61-69.

Hökfelt T, Herrera-Marschitz M, Seroogy K, Ju G, Staines WA, Holets V, Schalling M, Ungerstedt U, Post C, Rehfeld JF, *et al.*, (1988) Immunohistochemical studies on cholecystokinin (CCK)-immunoreactive neurons in the rat using sequence specific antisera and with special reference to the caudate nucleus and primary sensory neurons. J Chem Neuroanat 1, 11-51.

Hunt SP, Kelly JS, Emson PC, Kimmel JR, Miller RJ, Wu JY (1981) An immunohistochemical study of neuronal populations containing neuropeptides or gamma-aminobutyrate within the superficial layers of the rat dorsal horn. Neuroscience 6, 1883-1898.

Hylden JL, Hayashi H, Dubner R, Bennett GJ (1986) Physiology and morphology of the lamina I spinomesencephalic projection. J Comp Neurol 247, 505-515.

Ikeda M, Houtani T, Nakagawa H, Baba K, Kondoh A, Ueyama T, Yamamoto T, Gemba H, Sugimoto T (1995) Enkephalin-immunoreactive fastigial neurons in the rat cerebellum project to upper cervical cord segments. Brain Res 690, 225-230.

Jansen AS, Loewy AD (1997) Neurons lying in the white matter of the upper cervical spinal cord project to the intermediolateral cell column. Neuroscience 77, 889-898.

Jiang MC, Liu L, Gebhart GF (1999) Cellular properties of lateral spinal nucleus neurons in the rat L6-S1 spinal cord. J Neurophysiol 81, 3078-3086.

Kajander KC, Giesler GJ Jr (1987) Effects of repeated noxious thermal stimuli on the responses of neurons in the lateral cervical nucleus of cats: evidence for an input from A-nociceptors to the spinocervicothalamic pathway. Brain Res 436, 390-395.

Kayalioglu G, Hariri NI, Govsa F, Erdem B, Peker G, Maiskii VA (1996) Laminar distribution of the cells of origin of the spinocerebral pathways involved in nociceptive transmission and pain modulation in the rat. Neurophysiology 28, 111-118.

Kayalioglu G, Robertson B, Kristensson K, Grant G (1999) Nitric oxide synthase and interferon-gamma receptor immunoreactivities in relation to ascending spinal pathways to thalamus, hypothalamus, and the periaqueductal gray in the rat. Somatosens Mot Res 16, 280-290.

Keay KA, Bandler R (1993) Deep and superficial noxious stimulation increases Fos-like immunoreactivity in different regions of the midbrain periaqueductal gray of the rat. Neurosci Lett 154, 23-26.

Kemplay SK, Webster KE (1986) A qualitative and quantitative analysis of the distributions of cells in the spinal cord and spinomedullary junction projecting to the thalamus of the rat. Neuroscience 17, 769-789.

Kevetter GA, Willis WD (1983) Collaterals of spinothalamic cells in the rat. J Comp Neurol 215, 453-464.

Kitamura T, Yamada J, Sato H, Yamashita K (1993) Cells of origin of the spinoparabrachial fibers in the rat: a study with fast blue and WGA-HRP. J Comp Neurol 328, 449-461.

Krukoff TL, Ciriello J, Calaresu FR (1985) Segmental distribution of peptide-like immunoreactivity in cell bodies of the thoracolumbar sympathetic nuclei of the cat. J Comp Neurol 240, 90-102.

Krukoff TL, Ciriello J, Calaresu FR (1986) Somatostatin-like immunoreactivity in neurons, nerve terminals, and fibers of the cat spinal cord. J Comp Neurol 243, 13-22.

Krutki P, Grottel K, Mrówczyński W (1999) Divergence of lamina VII and VIII neurones of S1 and S2 segments of the cat's spinal cord to the cerebellum and the reticular formation. Acta Neurobiol Exp 59, 81-88.

Kumazawa T, Perl ER (1978) Excitation of marginal and substantia gelatinosa neurons in the primate spinal cord: indications of their place in dorsal horn functional organization. J Comp Neurol 177, 417-434.

Kuo DC, Nadelhaft I, Hisamitsu T, de Groat WC (1983) Segmental distribution and central projections of renal afferent fibers in the cat studied by transganglionic transport of horseradish peroxidase. J Comp Neurol 216, 162-174.

LaMotte CC (1988) Lamina X of primate spinal cord: distribution of five neuropeptides and serotonin. Neuroscience 25, 639-658.

LaMotte CC, de Lanerolle NC (1986) VIP terminals, axons, and neurons: distribution throughout the length of monkey and cat spinal cord. J Comp Neurol 249, 133-145.

LaMotte CC, Kapadia SE, Shapiro CM (1991) Central projections of the sciatic, saphenous, median, and ulnar nerves of the rat demonstrated by transganglionic transport of choleragenoid-HRP (B-HRP) and wheat germ agglutinin-HRP (WGA-HRP). J Comp Neurol 311, 546-562.

de Lanerolle NC, LaMotte CC (1982) The human spinal cord: substance P and methionine-enkephalin immunoreactivity. J Neurosci 2, 1369-1386.

Laruelle L (1937) La structure de la moelle epiniere en coupes longitudinales. Rev Neurol 44, 695-725.

Laruelle L (1948) Etude d'anatomie microscopique du nevraxe sur coupes longitudinales. Acta Neurol Psychiatr Belg 48, 138-280.

Leah J, Menétrey D, de Pommery J (1988) Neuropeptides in long ascending spinal tract cells in the rat: evidence for parallel processing of ascending information. Neuroscience 24, 195-207.

Lekan HA, Carlton SM (1995) Glutamatergic and GABAergic input to rat spinothalamic tract cells in the superficial dorsal horn. J Comp Neurol 361, 417-428.

Leong SK, Shieh JY, Wong WC (1984) Localizing spinal-cord-projecting neurons in adult albino rats. J Comp Neurol 228, 1-17.

Light AR (1992) The initial processing of pain and its descending control: spinal and trigeminal systems. In: *Pain and Headache.* Gildenberg PL (Ed). Karger, Basel.

Light AR, Kavookjian AM (1988) Morphology and ultrastructure of physiologically identified substantia gelatinosa (lamina II) neurons with axons that terminate in deeper dorsal horn laminae (III-V) J Comp Neurol 267, 172-189.

Light AR, Metz CB (1978) The morphology of the spinal cord efferent and afferent neurons contributing to the ventral roots of the cat. J Comp Neurol 179, 501-515.

Light AR, Perl ER (1979a) Reexamination of the dorsal root projection to the spinal dorsal horn including observations on the differential termination of coarse and fine fibers. J Comp Neurol 186, 117-131.

Light AR, Perl ER (1979b) Spinal termination of functionally identified primary afferent neurons with slowly conducting myelinated fibers. J Comp Neurol 186, 133-150.

Light AR, Willcockson HH (1999) Spinal laminae I-II neurons in rat recorded in vivo in whole cell, tight seal configuration: properties and opioid responses. J Neurophysiol 82, 3316-3326.

Light AR, Trevino DL, Perl ER (1979) Morphological features of functionally defined neurons in the marginal zone and substantia gelatinosa of the spinal dorsal horn. J Comp Neurol 186, 151-171.

Light AR, Sedivec MJ, Casale EJ, Jones SL (1993) Physiological and morphological characteristics of spinal neurons projecting to the parabrachial region of the cat. Somatosens Mot Res 10, 309-325.

Lima D, Coimbra A (1986) A Golgi study of the neuronal population of the marginal zone (lamina I) of the rat spinal cord. J Comp Neurol 244, 53-71.

Lima D, Coimbra A (1988) The spinothalamic system of the rat: structural types of retrogradely labelled neurons in the marginal zone (lamina I). Neuroscience 27, 215-230.

Lima D, Coimbra A (1989) Morphological types of spinomesencephalic neurons in the marginal zone (lamina I) of the rat spinal cord, as shown after retrograde labelling with cholera toxin subunit B. J Comp Neurol 279, 327-339.

Lima D, Coimbra A (1990) Structural types of marginal zone (lamina I) neurons projecting to the dorsal reticular nucleus of the medulla oblongata. Neuroscience 34, 591-606.

Lima D, Coimbra A (1991) Neurons in the substantia gelatinosa rolandi (lamina II) project to the caudal ventrolateral reticular formation of the medulla oblongata in the rat. Neurosci Lett 132, 16-18.

Lima D, Mendes-Ribeiro JA, Coimbra A (1991) The spino-latero-reticular system in the rat: projections from the superficial dorsal horn and structural characterization of marginal neurons involved. Neuroscience 45, 137-152.

Lima D, Avelino A, Coimbra A (1993) Morphological characterization of marginal (lamina I) neurons immunoreactive for substance P, enkephalin, dynorphin and gamma-aminobutyric acid in the rat spinal cord. J Chem Neuroanat 6, 43-52.

Liu RP (1983) Laminar origins of spinal projection neurons to the periaqueductal gray of the rat. Brain Res 264, 118-122.

Loewy AD (1970) A study of neuronal types in Clarke's column in the adult cat. J Comp Neurol 139, 53-79.

Lü Y, Jin SX, Xu TL, Qin BZ, Li JS, Ding YQ, Shigemoto R, Mizuno N (1995) Expression of c-fos protein in substance P receptor-like immunoreactive neurons in response to noxious stimuli on the urinary bladder: an observation in the lumbosacral cord segments of the rat. Neurosci Lett 198, 139-142.

Ma W, Ribeiro-da-Silva A, De Koninck Y, Radhakrishnan V, Henry JL, Cuello AC (1996) Quantitative analysis of substance P-immunoreactive boutons on physiologically characterized dorsal horn neurons in the cat lumbar spinal cord. J Comp Neurol 376, 45-64.

Magoul R, Onteniente B, Geffard M, Calas A (1987) Anatomical distribution and ultrastructural organization of the GABAergic system in the rat spinal cord. An immunocytochemical study using anti-GABA antibodies. Neuroscience 20, 1001-1009.

Maiskii VA, Kayalioglu G, Govsa F, Erdem B, Hariri NI (1998) Nitric oxide contributes to the spinal nociceptive processing. Neurophysiology 30, 370-376.

Mannen H, Sugiura Y (1976) Reconstruction of neurons of dorsal horn proper using Golgi-stained serial sections. J Comp Neurol 168, 303-312.

Mason P, Strassman A, Maciewicz R (1985) Pontomedullary raphe neurons: monosynaptic excitation from midbrain sites that suppress the jaw opening reflex. Brain Res 329, 384-389.

Mather CS, Ho RH (1992) Golgi impregnated somatostatin immunoreactive neurons in lamina II of the rat spinal cord. Brain Res Bull 28, 305-309.

Matsushita M (1970) The axonal pathways of spinal neurones in the cat. J Comp Neurol 138, 391-417.

Matsushita M (1991) Cerebellar projections of the central cervical nucleus in the rat: an anterograde tracing study. Neurosci Res 12, 201-216.

Matsushita M (1998) Ascending propriospinal afferents to area X (substantia grisea centralis) of the spinal cord in the rat. Exp Brain Res 119, 356-366.

Matsushita M (1999) Projections from the upper lumbar cord to the cerebellar nuclei in the rat, studied by anterograde axonal tracing. J Comp Neurol 412, 633-648.

Matsushita M, Hosoya Y (1979) Cells of origin of the spinocerebellar tract in the rat, studied with the method of retrograde transport of horseradish peroxidase. Brain Res 173, 185-200.

Matsushita M, Ikeda M (1970) Spinal projections to the cerebellar nuclei in the cat. Exp Brain Res 10, 501-511.

Matsushita M, Tanami T (1987) Spinocerebellar projections from the central cervical nucleus in the cat, as studied by anterograde transport of wheat germ agglutinin-horseradish peroxidase. J Comp Neurol 266, 376-397.

Matsushita M, Hosoya Y, Ikeda M (1979) Anatomical organization of the spinocerebellar system in the cat, as studied by retrograde transport of horseradish peroxidase. J Comp Neurol 184, 81-106.

Matsushita M, Gao X, Yaginuma H (1995) Spinovestibular projections in the rat, with particular reference to projections from the central cervical nucleus to the lateral vestibular nucleus. J Comp Neurol 361, 334-344.

Matthews MA, Willis WD, Williams V (1971) Dendrite bundles in lamina IX of cat spinal cord: a possible source for electrical interaction between motoneurons? Anat Rec 171, 313-327.

Maxwell DJ (1985) Combined light and electron microscopy of Golgi-labelled neurons in lamina III of the feline spinal cord. J Anat 141, 155-169.

Maxwell DJ, Fyffe RE, Réthelyi M (1983) Morphological properties of physiologically characterized lamina III neurones in the cat spinal cord. Neuroscience 10, 1-22.

Mawe GM, Bresnahan JC, Beattie MS (1984) Primary afferent projections from dorsal and ventral roots to autonomic preganglionic neurons in the cat sacral spinal cord: light and electron microscopic observations. Brain Res 290, 152-157.

McCall RB, Gebber GL, Barman SM (1977) Spinal interneurons in the baroreceptor reflex arc. Am J Physiol 232, H657-665.

McMahon SB, Wall PD (1988) Descending excitation and inhibition of spinal cord lamina I projection neurons. J Neurophysiol 59, 1204-1219.

Melander T, Hökfelt T, Rökaeus Å (1986) Distribution of galaninlike immunoreactivity in the rat central nervous system. J Comp Neurol 248, 475-517.

Menétrey D, Basbaum AI (1987) Spinal and trigeminal projections to the nucleus of the solitary tract: a possible substrate for somatovisceral and viscerovisceral reflex activation. J Comp Neurol 255, 439-450.

Menétrey D, De Pommery J (1991) Origins of spinal ascending pathways that reach central areas involved in visceroception and visceronociception in the rat. Eur J Neurosci 3, 249-259.

Menétrey D, Giesler GJ Jr, Besson JM (1977) An analysis of response properties of spinal cord dorsal horn neurones to nonnoxious and noxious stimuli in the spinal rat. Exp Brain Res 27, 15-33.

Menétrey D, Chaouch A, Besson JM (1980) Location and properties of dorsal horn neurons at origin of spinoreticular tract in lumbar enlargement of the rat. J Neurophysiol 44, 862-877.

Menétrey D, Chaouch A, Binder D, Besson JM (1982) The origin of the spinomesencephalic tract in the rat: an anatomical study using the retrograde transport of horseradish peroxidase. J Comp Neurol 206, 193-207.

Menétrey D, Gannon A, Levine JD, Basbaum AI (1989) Expression of c-fos protein in interneurons and projection neurons of the rat spinal cord in response to noxious somatic, articular, and visceral stimulation. J Comp Neurol 285, 177-195.

Millan MJ (1999) The induction of pain: an integrative review. Prog Neurobiol 57, 1-164.

Miller KE, Seybold VS (1987) Comparison of met-enkephalin-, dynorphin A-, and neurotensin-immunoreactive neurons in the cat and rat spinal cords: I. Lumbar cord. J Comp Neurol 255, 293-304.

Milne RJ, Foreman RD, Willis WD (1982) Responses of primate spinothalamic neurons located in the sacral intermediomedial gray (Stilling's nucleus) to proprioceptive input from the tail. Brain Res 234, 227-236.

Mizuno N, Nakano K, Imaizumi M, Okamoto M (1967) The lateral cervical nucleus of the Japanese monkey (Macaca fuscata). J Comp Neurol 129, 375-384.

Mokha SS, Goldsmith GE, Hellon RF, Puri R (1987) Hypothalamic control of nocireceptive and other neurons in the marginal layer of the dorsal horn of the medulla (trigeminal nucleus caudalis) in the rat. Exp Brain Res 65, 427-436.

Molander C, Grant G (2004) Spinal cord cytoarchitecture. In: *The Rat Nervous System.* 3rd edn. Paxinos, G (Ed). Elsevier Academic Press, San Diego.

Molander C, Xu Q, Grant G (1984) The cytoarchitectonic organization of the spinal cord in the rat. I. The lower thoracic and lumbosacral cord. J Comp Neurol 230, 133-141.

Molander C, Xu Q, Rivero-Melian C, Grant G (1989) Cytoarchitectonic organization of the spinal cord in the rat: II. The cervical and upper thoracic cord. J Comp Neurol 289, 375-385.

Molenaar I, Kuypers HG (1978) Cells of origin of propriospinal fibers and of fibers ascending to supraspinal levels. A HRP study in cat and rhesus monkey. Brain Res 152, 429-450.

Morgan C, deGroat WC, Nadelhaft I (1986) The spinal distribution of sympathetic preganglionic and visceral primary afferent neurons that send axons into the hypogastric nerves of the cat. J Comp Neurol 243, 23-40.

Morrell JI, Pfaff DW (1983) Retrograde HRP identification of neurons in the rhombencephalon and spinal cord of the rat that project to the dorsal mesencephalon. Am J Anat 167, 229-240.

Mouton LJ, Holstege G (1998) Three times as many lamina I neurons project to the periaqueductal gray than to the thalamus: a retrograde tracing study in the cat. Neurosci Lett 255, 107-110.

Mouton LJ, Holstege G (2000) Segmental and laminar organization of the spinal neurons projecting to the periaqueductal gray (PAG) in the cat suggests the existence of at least five separate clusters of spino-PAG neurons. J Comp Neurol 428, 389-410.

Mouton LJ, Klop EM, Broman J, Zhang M, Holstege G (2004) Lateral cervical nucleus projections to periaqueductal gray matter in cat. J Comp Neurol 471, 434-445.

Nadelhaft I, Booth AM (1984) The location and morphology of preganglionic neurons and the distribution of visceral afferents from the rat pelvic nerve: a horseradish peroxidase study. J Comp Neurol. 226, 238-245.

Nadelhaft I, McKenna KE (1987) Sexual dimorphism in sympathetic preganglionic neurons of the rat hypogastric nerve. J Comp Neurol 256, 308-315.

Nagy I, Woolf CJ, Dray A, Urbán L (1994) Cobalt accumulation in neurons expressing ionotropic excitatory amino acid receptors in young rat spinal cord: morphology and distribution. J Comp Neurol 344, 321-335.

Nagy JI, Hunt SP, Iversen LL, Emson PC (1981) Biochemical and anatomical observations on the degeneration of peptide-containing primary afferent neurons after neonatal capsaicin. Neuroscience 6, 1923-1934.

Nahin RL (1987) Immunocytochemical identification of long ascending peptidergic neurons contributing to the spinoreticular tract in the rat. Neuroscience 23, 859-869.

Nahin RL, Micevych PE (1986) A long ascending pathway of enkephalin-like immunoreactive spinoreticular neurons in the rat. Neurosci Lett 65, 271-276.

Nahin RL, Madsen AM, Giesler GJ Jr (1983) Anatomical and physiological studies of the gray matter surrounding the spinal cord central canal. J Comp Neurol 220, 321-335.

Naim M, Spike RC, Watt C, Shehab SA, Todd AJ (1997) Cells in laminae III and IV of the rat spinal cord that possess the neurokinin-1 receptor and have dorsally directed dendrites receive a major synaptic input from tachykinin-containing primary afferents. J Neurosci 17, 5536-5548.

Neuhuber WL, Zenker W (1989) Central distribution of cervical primary afferents in the rat, with emphasis on proprioceptive projections to vestibular, perihypoglossal, and upper thoracic spinal nuclei. J Comp Neurol 280, 231-253.

Neuhuber WL, Sandoz PA, Fryscak T (1986) The central projections of primary afferent neurons of greater splanchnic and intercostal nerves in the rat. A horseradish peroxidase study. Anat Embryol (Berl) 174, 123-144.

Nicolopoulos-Stournaras S, Iles JF (1983) Motor neuron columns in the lumbar spinal cord of the rat. J Comp Neurol 217, 75-85.

Nissen UV, Mochida H, Glover JC (2005) Development of projection-specific interneurons and projection neurons in the embryonic mouse and rat spinal cord. J Comp Neurol 483, 30-47.

Nyberg-Hansen R (1965) Sites and mode of termination of reticulo-spinal fibers in the cat. An experimental study with silver impregnation methods. J Comp Neurol 124, 71-99.

Nyberg-Hansen R, Mascitti TA (1964) Sites and mode of termination of fibers of the vestibulospinal tract in the cat. An experimental study with silver impregnation methods. J Comp Neurol 122, 369-383.

Ornung G, Ragnarson B, Grant G, Ottersen OP, Storm-Mathisen J, Ulfhake B (1995) Ia boutons to CCN neurones and motoneurones are enriched with glutamate-like immunoreactivity. Neuroreport 6, 1975-1980.

Ottersen OP, Storm-Mathisen J (1987) Distribution of inhibitory amino acid neurons in the cerebellum with some observations on the spinal cord: An immunocytochemical study with antisera against fixed GABA, glycine, taurine, and alanine. J Mind Behav 8, 503-518.

Perl ER (1959) Effects of muscle stretch on excitability of contralateral motoneurones. J Physiol 145, 193-203.

Perl ER (1980) Afferent basis of nociception and pain: evidence from the characteristics of sensory receptors and their projections to the spinal dorsal horn. Res Publ Assoc Res Nerv Ment Dis 58, 19-45.

Perl ER (1984) Pain and nociception. In: *Handbook of Physiology. The Nervous System. Sensory Processes.* Section 1, volume 3. Darian-Smith I (Ed). Am Physiol Soc, Bethesda.

Persson S, Boulland JL, Aspling M, Larsson M, Fremeau RT Jr, Edwards RH, Storm-Mathisen J, Chaudhry FA, Broman J (2006) Distribution of vesicular glutamate transporters 1 and 2 in the rat spinal cord, with a note on the spinocervical tract. J Comp Neurol 497, 683-701.

Petras JM, Cummings JF (1972) Autonomic neurons in the spinal cord of the Rhesus monkey: a correlation of the findings of cytoarchitectonics and sympathectomy with fiber degeneration following dorsal rhizotomy. J Comp Neurol 146, 189-218.

Phelps PE, Barber RP, Vaughn JE (1988) Generation patterns of four groups of cholinergic neurons in rat cervical spinal cord: a combined tritiated thymidine autoradiographic and choline acetyltransferase immunocytochemical study. J Comp Neurol 273, 459-472.

Poljak, SI (1924) Die Struktureigentümlichkeiten des Rückenmarkes bei den Chiropteren. Z Ges Anat Abt 1, Z Anat Entwicklungsgesch 74, 509-576.

Popova LB, Ragnarson B, Orlovsky GN, Grant G (1995) Responses of neurons in the central cervical nucleus of the rat to proprioceptive and vestibular inputs. Arch Ital Biol 133, 31-45.

Powell JJ, Todd AJ (1992) Light and electron microscope study of GABA-immunoreactive neurons in lamina III of rat spinal cord. J Comp Neurol 315, 125-136.

Prescott SA, De Koninck Y (2002) Four cell types with distinctive membrane properties and morphologies in lamina I of the spinal dorsal horn of the adult rat. J Physiol 539, 817-836.

Proudlock F, Spike RC, Todd AJ (1993) Immunocytochemical study of somatostatin, neurotensin, GABA and glycine in rat spinal dorsal horn. J Comp Neurol 327, 289-297.

Ragnarson B, Yi SJ, Ulfhake B, Grant G (2002) GABA-, glycine-, and glutamate-immunoreactive bouton profiles in apposition to neurons of the central cervical nucleus in the rat. Anat Rec 266, 226-233.

Ramón y Cajal S (1909) *Histologie du Système Nerveux de l'Homme et des Vertébrés.* Maloine, Paris.

Ranson SW, Davenport HK, Doles EA (1932) Intramedullary course of the dorsal root fibers of the first three cervical nerves. J Comp Neurol 54, 1-12.

Ren K, Ruda MA (1994) A comparative study of the calcium-binding proteins calbindin-D28K, calretinin, calmodulin and parvalbumin in the rat spinal cord. Brain Res Brain Res Rev 19, 163-179.

Réthelyi M (1976) Central core in the spinal gray matter. Acta Morphol Acad Sci Hung 24, 63-70.

Réthelyi M (1977) Preterminal and terminal axon arborizations in the substantia gelatinosa of cat's spinal cord. J Comp Neurol 172, 511-521.

Réthelyi M (2003) Neurons of the lateral spinal nucleus in the rat spinal cord – A Golgi study. Eur J Anat 7, 1-8.

Réthelyi M, Light AR, Perl ER (1989) Synaptic ultrastructure of functionally and morphologically characterized neurons of the superficial spinal dorsal horn of cat. J Neurosci 9, 1846-1863.

Reuss MH, Reuss S (2001) Nitric oxide synthase neurons in the rodent spinal cord: distribution, relation to Substance P fibers, and effects of dorsal rhizotomy. J Chem Neuroanat 21, 181-196.

Réthelyi M, Szentágothai J (1969) The large synaptic complexes of the substantia gelatinosa. Exp Brain Res 7, 258-274.

Rexed, B (1954) A cytoarchitectonic atlas of the spinal cord in the cat. J Comp Neurol 100, 297-379.

Ribeiro-da-Silva A, Cuello AC (1990) Choline acetyltransferase-immunoreactive profiles are presynaptic to primary sensory fibers in the rat superficial dorsal horn. J Comp Neurol 295, 370-384.

Ribeiro-da-Silva A, Pioro EP, Cuello AC (1991) Substance P- and enkephalin-like immunoreactivities are colocalized in certain neurons of the substantia gelatinosa of the rat spinal cord: an ultrastructural double-labeling study. J Neurosci 11, 1068-1080.

Ritz LA, Greenspan JD (1985) Morphological features of lamina V neurons receiving nociceptive input in cat sacrocaudal spinal cord. J Comp Neurol 238, 440-452.

Rivero-Melián C (1996) Organization of hindlimb nerve projections to the rat spinal cord: a choleragenoid horseradish peroxidase study. J Comp Neurol 364, 651-663.

Rivero-Melián C, Arvidsson J (1992) Brain stem projections of rat lumbar dorsal root ganglia studied with choleragenoid conjugated horseradish peroxidase. Exp Brain Res 91, 12-20.

Rivero-Melián C, Grant G (1990) Distribution of lumbar dorsal root fibers in the lower thoracic and lumbosacral spinal cord of the rat studied with choleragenoid horseradish peroxidase conjugate. J Comp Neurol 299, 470-481.

Rolando L (1824) *Ricerche Anatomiche sulla Struttura del Midollo Spinale.* Staperia Reale, Torino.

Romagnano MA, Braiman J, Loomis M, Hamill RW (1987) Enkephalin fibers in autonomic nuclear regions: intraspinal vs. supraspinal origin. J Comp Neurol 266, 319-331.

Routal RV, Pal GP (1999) A study of motoneuron groups and motor columns in the human spinal cord. J Anat 195, 211-224.

Rowan S, Todd AJ, Spike RC (1993) Evidence that neuropeptide Y is present in GABAergic neurons in the superficial dorsal horn of the rat spinal cord. Neuroscience. 53, 537-545.

Rökaeus A, Melander T, Hökfelt T, Lundberg JM, Tatemoto K, Carlquist M, Mutt V (1984) A galanin-like peptide in the central nervous system and intestine of the rat. Neurosci Lett 47, 161-166.

Rössler W, Gerstberger R, Sann H, Pierau FK (1993) Distribution and binding sites of substance P and calcitonin gene-related peptide and their capsaicin-sensitivity in the spinal cord of rats and chicken: a comparative study. Neuropeptides 25, 241-253.

Ruda MA, Coffield J, Steinbusch HW (1982) Immunocytochemical analysis of serotonergic axons in laminae I and II of the lumbar spinal cord of the cat. J Neurosci 2, 1660-1671.

Saito S, Kidd GJ, Trapp BD, Dawson TM, Bredt DS, Wilson DA, Trytsman RJ, Snyder SH, Hanley DF (1994) Rat spinal cord neurons nittain nitric oxide synthase. Neuroscience 59, 447-456.

Satoh K, Armstrong DM, Fibiger HC (1983) A comparison of the distribution of central cholinergic neurons as demonstrated by acetylcholinesterase pharmacohistochemistry and choline acetyltransferase immunohistochemistry. Brain Res Bull 11, 693-720.

Schäfer MK, Eiden LE, Weihe E (1998) Cholinergic neurons and terminal fields revealed by immunohistochemistry for the vesicular acetylcholine transporter. I. Central nervous system. Neuroscience 84, 331-359.

Scheibel ME, Scheibel AB (1969) Terminal patterns in cat spinal cord. 3. Primary afferent collaterals. Brain Res 13, 417-443.

Scheibel ME, Scheibel AB (1970) Developmental relationship between spinal motoneuron dendrite bundles and patterned activity in the hind limb of cats. Exp Neurol 29, 328-335.

Scheibel ME, Scheibel AB (1973) Dendrite bundles as sites for central programs: an hypothesis. Int J Neurosci 6, 195-202.

Schoenen J (1982a) The dendritic organization of the human spinal cord: the dorsal horn. Neuroscience 7, 2057-2087.

Schoenen J (1982b) Dendritic organization of the human spinal cord: the motoneurons. J Comp Neurol 211, 226-247.

Schoenen J, Faull RLM (2004) Spinal cord: Cyto- and Chemoarchitecture. In: Paxinos G, Mai JK (Eds), *The Human Nervous System.* 2nd ed, Elsevier Academic Press, San Diego, pp 190-232.

Schoenen J, Lotstra F, Vierendeels G, Reznik M, Vanderhaeghen JJ (1985) Substance P, enkephalins, somatostatin, cholecystokinin, oxytocin, and vasopressin in human spinal cord. Neurology 35, 881-890.

Senba E, Yanaihara C, Yanaihara N, Tohyama M (1988) Co-localization of substance P and Met-enkephalin-Arg6-Gly7-Leu8 in the intraspinal neurons of the rat, with special reference to the neurons in the substantia gelatinosa. Brain Res 453, 110-116.

Seybold VS, Elde RP (1982) Neurotensin immunoreactivity in the superficial laminae of the dorsal horn of the rat. I. Light microscopic studies of cell bodies and proximal dendrites. J Comp Neurol 205, 89-100.

Sidman RL, Angevine Jr JB, Pierce ET (1971) *Atlas of the Mouse Brain and Spinal Cord.* Harvard University Press, Cambridge.

Skinner RD, Remmel RS (1978) Monosynaptic inputs to lumbar interneurons from the lateral vestibulospinal tract and the medial longitudinal fasciculus. Neurosci Lett 10, 259-264.

Snyder RL, Faull RL, Mehler WR (1978) A comparative study of the neurons of origin of the spinocerebellar afferents in the rat, cat and squirrel monkey based on the retrograde transport of horseradish peroxidase. J Comp Neurol 181, 833-852.

Spike RC, Todd AJ (1992) Ultrastructural and immunocytochemical study of lamina II islet cells in rat spinal dorsal horn. J Comp Neurol 323, 359-369.

Spike RC, Todd AJ, Johnston HM (1993) Coexistence of NADPH diaphorase with GABA, glycine, and acetylcholine in rat spinal cord. J Comp Neurol 335, 320-333.

Spike RC, Watt C, Zafra F, Todd AJ (1997) An ultrastructural study of the glycine transporter GLYT2 and its association with glycine in the superficial laminae of the rat spinal dorsal horn. Neuroscience 77, 543-551.

Standaert DG, Saper CB, Rye DB, Wainer BH (1986) Colocalization of atriopeptin-like immunoreactivity with choline acetyltransferase- and substance P-like immunoreactivity in the pedunculopontine and laterodorsal tegmental nuclei in the rat. Brain Res 382, 163-168.

Stokke MF, Nissen UV, Glover JC, Kiehn O (2002) Projection patterns of commissural interneurons in the lumbar spinal cord of the neonatal rat. J Comp Neurol 446, 349-359.

Sugiura Y, Lee CL, Perl ER (1986) Central projections of identified, unmyelinated (C) afferent fibers innervating mammalian skin. Science 234, 358-361.

Surmeier DJ, Honda CN, Willis WD (1988) Natural groupings of primate spinothalamic neurons based on cutaneous stimulation. Physiological and anatomical features. J Neurophysiol 59, 833-860.

Swanson LW, McKellar S (1979) The distribution of oxytocin- and neurophysin-stained fibers in the spinal cord of the rat and monkey. J Comp Neurol 188, 87-106.

Swett JE, Woolf CJ (1985) The somatotopic organization of primary afferent terminals in the superficial laminae of the dorsal horn of the rat spinal cord. J Comp Neurol 231, 66-77.

Szentágothai J (1964) Neuronal and synaptic arrangement in the substantia gelatinosa Rolandi. J Comp Neurol 122, 219-239.

Szentágothai J, Réthelyi M (1973) Cyto- and neuropil architecture of the spinal cord. In: *New developments in electromyography and clinical neurophysiology.* Vol. 1. Desmedt JE (Ed). Karger, Basel.

Takahashi Y, Chiba T, Sameda H, Ohtori S, Kurokawa M, Moriya H (2002) Organization of cutaneous ventrodorsal and rostrocaudal axial lines in the rat hindlimb and trunk in the dorsal horn of the spinal cord. J Comp Neurol 445, 133-144.

Tan H, He J, Wang S, Hirata K, Yang Z, Kuraoka A, Kawabuchi M (2006) Age-related NADPH-diaphorase positive bodies in the lumbosacral spinal cord of aged rats. Arch Histol Cytol 69, 297-310.

Tang H, Hammond P, Brimijoin S (1998) Acetylcholinesterase immunolesioning: regional vulnerability of preganglionic sympathetic neurons in rat spinal cord. Exp Neurol 152, 167-176.

Thomson DB, Isu N, Wilson VJ (1996) Responses of neurons of the cat central cervical nucleus to natural neck and vestibular stimulation. J Neurophysiol 76, 2786-2789.

Todd AJ (1989) Cells in lamina III and IV of rat spinal dorsal horn receive monosynaptic primary afferent input in lamina II. J Comp Neurol 289, 676-686.

Todd AJ (1991) Immunohistochemical evidence that acetylcholine and glycine exist in different populations of GABAergic neurons in lamina III of rat spinal dorsal horn. Neuroscience 44, 741-746.

Todd AJ, Lewis SG (1986) The morphology of Golgi-stained neurons in lamina II of the rat spinal cord. J Anat 149, 113-119.

Todd AJ, McKenzie J (1989) GABA-immunoreactive neurons in the dorsal horn of the rat spinal cord. Neuroscience 31, 799-806.

Todd AJ, Spike RC (1992) Co-localization of Met-enkephalin and somatostatin in the spinal cord of the rat. Neurosci Lett 145, 71-74.

Todd AJ, Spike RC (1993) The localization of classical transmitters and neuropeptides within neurons in laminae I-III of the mammalian spinal dorsal horn. Prog Neurobiol 41, 609-645.

Todd AJ, Sullivan AC (1990) Light microscope study of the coexistence of GABA-like and glycine-like immunoreactivities in the spinal cord of the rat. J Comp Neurol 296, 496-550.

Todd AJ, Russell G, Spike RC (1992) Immunocytochemical evidence that GABA and neurotensin exist in different neurons in laminae II and III of rat spinal dorsal horn. Neuroscience 47, 685-691.

Todd AJ, McGill MM, Shehab SA (2000) Neurokinin 1 receptor expression by neurons in laminae I, III and IV of the rat spinal dorsal horn that project to the brainstem. Eur J Neurosci 12, 689-700.

Torvik A (1956) Afferent connections to the sensory trigeminal nuclei, the nucleus of the solitary tract and adjacent structures. An experimental study in the rat. J Comp Neurol 106, 51-142.

Truex RC, Taylor MJ, Smythe MQ, Gildenberg PL (1970) The lateral cervical nucleus of cat, dog and man. J Comp Neurol 139, 93-104.

Ulfhake B, Arvidsson U, Cullheim S, Hokfelt T, Brodin E, Verhofstad A, Visser T (1987) An ultrastructural study of 5-hydroxytryptamine-, thyrotropin-releasing hormone-and substance P-immunoreactive axonal boutons in the motor nucleus of spinal cord segments L7-S1 in the adult cat. Neuroscience 23, 917-929.

Valtschanoff JG, Weinberg RJ, Rustioni A (1993) Amino acid immunoreactivity in corticospinal terminals. Exp Brain Res 93, 95-103.

Valtschanoff JG, Weinberg RJ, Rustioni A, Schmidt HH (1992) Nitric oxide synthase and GABA colocalize in lamina II of rat spinal cord. Neurosci Lett 148, 6-10.

Van den Pol AN, Gorcs T (1988) Glycine and glycine receptor immunoreactivity in brain and spinal cord. J Neurosci 8, 472-492.

Vanderhorst VG, Mouton LJ, Blok BF, Holstege G (1996) Distinct cell groups in the lumbosacral cord of the cat project to different areas in the periaqueductal gray. J Comp Neurol 376, 361-385.

Verburgh CA, Voogd J, Kuypers HG, Stevens HP (1990) Propriospinal neurons with ascending collaterals to the dorsal medulla, the thalamus and the tectum: a retrograde fluorescent double-labeling study of the cervical cord of the rat. Exp Brain Res 80, 577-590.

Villanueva L, de Pommery J, Menétrey D, Le Bars D (1991) Spinal afferent projections to subnucleus reticularis dorsalis in the rat. Neurosci Lett 134, 98-102.

Vizzard MA, Erdman SL, de Groat WC (1993) Localization of NADPH-diaphorase in pelvic afferent and efferent pathways of the rat. Neurosci Lett 152, 72-76.

Vizzard MA, Erdman SL, Erickson VL, Stewart RJ, Roppolo JR, De Groat WC (1994) Localization of NADPH diaphorase in the lumbosacral spinal cord and dorsal root ganglia of the cat. J Comp Neurol 339, 62-75.

Waldron HA, Gwyn DG (1969) Descending nerve tracts in the spinal cord of the rat. I. Fibers from the midbrain. J Comp Neurol 137, 143-153.

Wang CC, Willis WD, Westlund KN (1999) Ascending projections from the area around the spinal cord central canal: A Phaseolus vulgaris leucoagglutinin study in rats. J Comp Neurol 415, 341-367.

Weihe E (1992) Neurochemical anatomy of the mammalian spinal cord: functional implications. Ann Anat 174, 89-118.

Westman J (1989) Light and electron microscopical studies of the substance P innervation of the dorsal column nuclei and the lateral cervical nucleus in the primate. Ups J Med Sci 94, 123-128.

Wiberg M, Blomqvist A (1984) The spinomesencephalic tract in the cat: its cells of origin and termination pattern as demonstrated by the intraaxonal transport method. Brain Res 291, 1-18.

Wiksten B (1979a) The central cervical nucleus in the cat. I. A Golgi study. Exp Brain Res 36, 143-154.

Wiksten B (1979b) The central cervical nucleus in the cat. III. The cerebellar connections studied with anterograde transport of 3H-leucine. Exp Brain Res 36, 175-189.

Willis WD (1985) *The Pain System. The Neural Basis of Nociceptive Transmission in the Mammalian Nervous System.* Karger, Basel.

Willis WD, Coggeshall RE (1991) *Sensory Mechanisms of the Spinal Cord.* 2nd edn. Plenum Press, New York.

Willis WD, Trevino DL, Coulter JD, Maunz RA (1974) Responses of primate spinothalamic tract neurons to natural stimulation of hindlimb. J Neurophysiol 37, 358-372.

Willis WD, Kenshalo DR Jr, Leonard RB (1979) The cells of origin of the primate spinothalamic tract. J Comp Neurol 188, 543-573.

Woodbury CJ, Koerber HR (2003) Widespread projections from myelinated nociceptors throughout the substantia gelatinosa provide novel insights into neonatal hypersensitivity. J Neurosci 23, 601-610.

Woodbury CJ, Ritter AM, Koerber HR (2000) On the problem of lamination in the superficial dorsal horn of mammals: a reappraisal of the substantia gelatinosa in postnatal life. J Comp Neurol 417, 88-102.

Woolf CJ, Fitzgerald M (1983) The properties of neurones recorded in the superficial dorsal horn of the rat spinal cord. J Comp Neurol 221, 313-328.

Xiong G, Matsushita M (2001) Ipsilateral and contralateral projections from upper cervical segments to the vestibular nuclei in the rat. Exp Brain Res 141, 204-217.

Xu TL, Li JS, Akaike N (1999) Functional properties of ionotropic glutamate receptor channels in rat sacral dorsal commissural neurons. Neuropharmacology 38, 659-670.

Yamada T, Placzek M, Tanaka H, Dodd J, Jessell TM (1991) Control of cell pattern in the developing nervous system: polarizing activity of the floor plate and notochord. Cell 64, 635-647.

Yashpal K, Dam TV, Quirion R (1990) Quantitative autoradiographic distribution of multiple neurokinin binding sites in rat spinal cord. Brain Res 506, 259-266.

Zemlan FP, Kow LM, Morrell JI, Pfaff DW (1979) Descending tracts of the lateral columns of the rat spinal cord: a study using the horseradish peroxidase and silver impregnation techniques. J Anat 128, 489-512.

Zhang ET, Craig AD (1997) Morphology and distribution of spinothalamic lamina I neurons in the monkey. J Neurosci 17, 3274-3284.

Zhang ET, Han ZS, Craig AD (1996) Morphological classes of spinothalamic lamina I neurons in the cat. J Comp Neurol 367, 537-549.

Zhang X, Ji RR, Nilsson S, Villar M, Ubink R, Ju G, Wiesenfeld-Hallin Z, Hökfelt T (1995) Neuropeptide Y and galanin binding sites in rat and monkey lumbar dorsal root ganglia and spinal cord and effect of peripheral axotomy. Eur J Neurosci 7, 367-380.

7 Localization of Motoneurons in the Spinal Cord

Steve McHanwell and Charles Watson

Introduction – motoneuron types

The motoneurons in the spinal cord belong to two functional groups: somatic and visceral. Somatic motoneurons innervate skeletal or voluntary muscle while visceral (autonomic) motoneurons innervate smooth muscle and glands.

Somatic motoneurons are located in the ventral horn of the gray of the spinal cord and innervate the striated muscles of the axial skeleton (neck and trunk) and the muscles of the upper and lower limbs. The majority of these motoneurons are alpha motoneurons innervating the extrafusal muscle fibers within skeletal muscle. A smaller proportion of motoneurons, the gamma motoneurons, innervate the intrafusal muscle fibers within muscle spindles. A third class of somatic motoneuron has been described, the beta motoneuron, which sends axon branches within the muscle to both extrafusal and intrafusal muscle fibers. These beta motoneurons are less numerous than the other two types and are found predominantly innervating distal muscles of the limbs. Two groups of somatic motoneurons need to be singled out because they exhibit certain properties distinct from other somatic motoneurons. These are the motoneurons that innervate the muscles of the perineum (including the external anal sphincter and external urethral sphincter) and the cremaster muscle. One important characteriztic of these motoneurons is that they are relatively spared in motoneuron disease in man.

Autonomic motoneurons are located in the intermediate horn of the spinal cord between the dorsal and ventral horns. These motoneurons are further subdivided into two functional groups, sympathetic and parasympathetic. Sympathetic preganglionic motoneurons are located in the thoracic and upper lumbar parts of the spinal cord; they provide a vasodilator innervation to smooth muscle of the blood vessels both within viscera and to the tissues, innervation to the sweat glands, to the arrector pilii muscles and to the adrenal medulla. The parasympathetic preganglionic motoneurons are found within the sacral spinal cord; they provide a parasympathetic innervation to the pelvic viscera and the hindgut as far as the splenic flexure of the colon. Both these groups of motoneurons synapse not directly with their target organs, but via a synapse with a postganglionic neuron located in an autonomic ganglion. More detail on the localization of autonomic motoneurons in the spinal cord can be found in Chapter 8.

Cellular organization of neurons within the ventral and intermediate horns

Within the ventral horn is a population of large neurons which are the alpha motoneurons. These multipolar neurons are amongst the largest in the nervous system; in man they can reach up to 120 μm in diameter. In light microscope sections, they are characterized by large and prominent Nissl bodies. Their axons leave in the ventral roots and in some species, though not all, undergo chromatolysis following axonal section or crush. Scattered amongst these larger neurons are populations of small and intermediate sized neurons. The majority of these smaller neurons are interneurons, but a proportion of them will be gamma motoneurons and an even smaller proportion beta motoneurons. Some will be Renshaw cells. Autonomic motoneurons belong to the intermediate and smaller size ranges of neuron within the spinal cord.

The large motoneurons are organized in discrete, longitudinal columns throughout the length of the spinal cord. Although the existence of these columns has been known for over 100 years (Sprague, 1948), there has been remarkably little subsequent agreement on their detailed nomenclature. This is because these columns are not composed of uniform aggregations of cells, but instead tend to be clumped. This, together with the fact that some of the columns can be quite close together, can make the columns difficult to trace when reconstructing them from transverse sections. It might be thought that this problem could be addressed through the use of longitudinal or horizontal sections, but the columns curve laterally, especially at the edges of the ventral horn and also change their mediolateral and dorsoventral positions as well. Consequently, reconstruction from transverse sections has been the method most commonly employed in describing the columnar organization of the ventral horn. However, in smaller mammals, such as mice and rats, the lower resolution with which columns can be reconstructed means that it is difficult to identify the existence of the smaller subdivision of columns that have been described in larger mammals such as cat or man.

At the simplest level, three columnar groups may be recognized (Gray, 1980): medial, central and lateral, with each being further divided into a dorsal and a ventral group. The medial group is present in all segments of spinal cord above L5. From T1 to L4 both the dorsal and ventral divisions of the medial group are present, while above T1, in the cervical spinal cord, only the ventromedial division of the medial group can be recognized. The central column is far less extensive than the other two, being chiefly restricted to the cervical and lower lumbosacral segments of the spinal cord. The lateral columns are confined to the cervical and lumbosacral enlargements of the spinal cord and innervate the upper and lower limbs respectively. In addition to a dorsal and ventral columns, a retrodorsal column has also been described.

However, this very simple account of the columnar structure of the large motoneurons has been the subject of considerable elaboration, especially in descriptions of the human spinal cord. For example, Routal and Pal (1999) described a total of 11 columns throughout the human spinal cord, characterizing their appearance and disappearance throughout. These authors highlighted the considerable discrepancies between previous studies and their own.

In the lateral motor columns of the cervical and lumbosacral enlargements, the general columnar structure has been considerably subdivided on the basis of evidence, deriving both from reconstructing columns from Nissl-stained transverse sections and from experiments localizing motoneurons innervating muscles using experimental methods. The cat lumbosacral enlargement has been intensively investigated; the original detailed description of Romanes (1951), which has been confirmed in its broad outlines by many subsequent investigators, has recently been further refined by Vanderhorst and Holstege (1997) in a detailed and thorough study using retrograde tracing.

The functional significance of these motoneuron cell columns has become apparent from a variety of experimental studies carried out in the last 60 years. These studies have shown that the motoneurons innervating a particular muscle or muscle group are located in a particular region of a longitudinal column in a more or less constant relationship to the motoneurons innervating other muscles in patterns that are similar between individuals of a species. Furthermore, these patterns appear to be highly conserved as they exhibit broad similarities across vertebrate species, once uncertainties in determining muscle homologies are taken into account. These studies have further shown that these patterns might be even more fine-grained in that the motoneurons of a particular region of a muscle or a particular functional type may occupy a defined position within a motor pool for an individual muscle (Burke *et al.*, 1977; English *et al.*, 1993). However, where such compartmentalization of a muscle occurs, leading to a partitioning of a motor pool, then the localizations show a degree of overlap and are seldom absolutely discrete. It is these patterns that will form the subject of the remainder of this short review.

Whichever method is used to characterize motor columns in the spinal cord, there are always fewer columns than there are muscles in the body. Consequently, except in a few special cases, the motoneurons innervating different muscles must overlap one another in the spinal cord. This fact, combined with the very real difficulties in identifying motoneuron columns, means that it is not possible to ascribe to any motoneuron observed in a histological section a particular peripheral innervation unless a parallel retrograde double-labeling has been carried out. Only if this is done can the identity of a motoneuron as innervating a particular muscle be made with any certainty. However, familiarity with the columnar structure of the spinal cord can allow a useful degree of approximation as to the muscle group innervated by a motoneuron to be made. This level of approximation may be sufficient in some functional or pathological studies.

Experimental approaches to motoneuron localization

The classical (pre-1970) approach to localizing motoneurons innervating specific muscles has been to section or crush a nerve and then observe the subsequent degenerative, chromatolytic changes. As Romanes (1964) has observed, the difficulty with this approach is the variability in both degree of reaction and time course of the changes between species; with some species, such as rats and mice, showing only very slight changes in response to such injury. These difficulties are likely to lead to an underestimation of the number, and possibly also the topographic extent, of motoneurons within a motor pool.

The advent of retrograde tracing techniques in the early 1970s has revolutionized the identification of neuronal connections and has led to a huge upsurge in studies localizing motoneurons innervating a wide range of muscles. Many studies have employed horseradish peroxidase, a technique still favored by many because of its proven reliability. However, the availability of a range of fluorescent markers has allowed double- and multiple-labeling studies to be carried out, permitting more refined investigations into the relationships between different motoneuron pools. The key problem with all retrograde techniques is to avoid false-positives through the spread of the tracer both peripherally from the original injection site and centrally from the retrogradely-labelled motoneurons themselves. Possible methods to avoid peripheral spread have been discussed by McHanwell and Biscoe (1981) and Vanderhorst and Holstege (1997) among others. Aschoff and Holländer (1982) and Illert *et al.*, (1982) have discussed the problems of interpretation of retrograde labeling experiments that are caused by anterograde labeling and diffusion when certain fluorescent tracers are employed. When suitable precautions are taken, retrograde techniques can be both qualitatively and quantitatively reliable, though it would be fair to say that not all studies have adopted a rigorous methodological approach in this regard.

A further characteriztic which distinguishes different retrograde labeling techniques is their ability to label all or most of the motoneurons supplying a given muscle. Vanderhorst and Holstege (1997) found that HRP reliably filled over 70% of the motoneurons supplying a particular muscle that was injected, whereas smaller yields were gained from fluorescent tracers. They speculated that HRP might diffuse widely throughout a muscle, whereas the other tracers do not.

Whatever method is employed, one remaining problem is the aligning of results of motoneuron pool localization between experiments. This difficulty arises because of the variability in spinal cord segment length, and even the number of segments in a given spinal cord region in some cases. Multiple-labeling studies can provide some information about interrelationships between motor pools, but this information is strictly limited. One solution to this has been to align motor pools from different animals to a readily identifiable external landmark (Vanderhorst and Holstege, 1997). A comparable approach was adopted by McHanwell and Biscoe (1981).

The relationship between motoneuron pools on the one hand and reliable landmarks on the other is not a trivial point. Although there have been numerous studies upon the localization of motoneurons, these have tended to concentrate on a few muscles, sometimes comparing the innervation of these few muscles across a range of species. There have been very few attempts at comprehensive description of the innervation of all the muscles from a limb. A notable exception in this regard is the study of Vanderhorst and Holstege (1997) which demonstrates the potential power of a comprehensive approach. However, such studies are rare because of the need to use a very large number of experimental animals. Consequently, it can be difficult to compare the results of different studies. This review will attempt to survey a wide range of studies in order to provide as full a description of motoneuron localization as is possible within a single chapter. Some muscle motoneurons still remain to be localized, either because of technical constraints or simply because they have not been a major focus of interest. However, it is possible to provide a summary of the localization of motoneurons for the great majority of muscle groups by combining the results obtained from a variety of species.

Topography of motoneuron pools in the upper cervical spinal cord

Several groups have examined the localization of motoneurons innervating the muscles of the neck in mammals. Investigating the localization of motoneurons innervating these muscles can present significant technical challenges, which explains why studies on these muscle groups are relatively few in number.

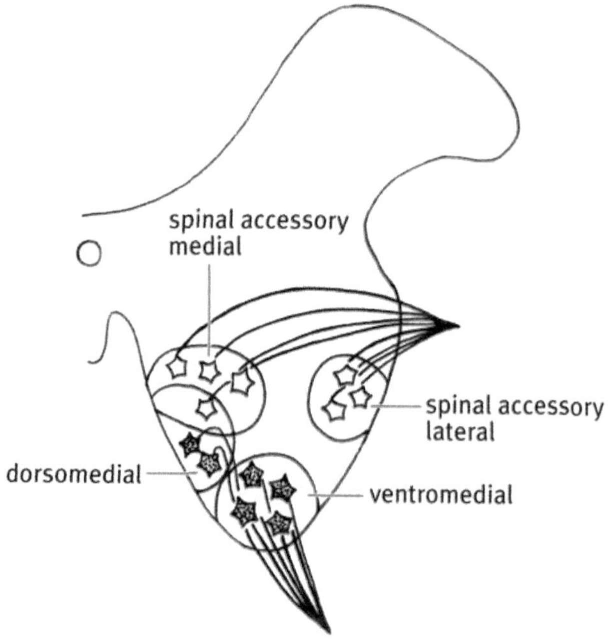

Figure 7.1 Motoneuron pools in the upper cervical spinal cord

This diagram shows the motoneuron groups which supply the trapezius and sternocleidomastoid (spinal accessory medial and spinal accessory lateral) and the vertebral muscles (dorsomedial and ventromedial). (Adapted from Callister *et al.*, 1987).

The motoneurons in this region of the spinal cord can be divided into five main groups (Kitamura *et al.*, 1986; Callister *et al.*, 1987) (see Figure 7.1). A ventromedial column is located in the lower half of the ventral horn of the spinal cord and dorsomedial column can be identified along the medial margin of the ventral horn close to its white matter border. Motoneurons in both these columns send their axons into the ipsilateral ventral roots. A medial column of commissural neurons can be identified further dorsally and extending almost as far as the midline. These motoneurons send their axons to the contralateral ventral roots. Finally, two group of motoneurons have been shown to be associated with the spinal accessory nerve. The spinal accessory group is divided by Callister *et al.*, (1987) into a medial group located in C1 and C2 and a lateral group located in C2 to C6. Other studies have suggested that these two latter groups are continuous with each other.

Studies of the cervical spinal muscles in the cat

The most comprehensive studies on the muscles that move the head have been carried out in the cat. A series of studies (Richmond *et al.*, 1978; Abrahams and Keane, 1984; Bakker *et al.*, 1984; Gordon and Richmond, 1991; Kitamura and Richmond, 1994; Liinamaa *et al.*, 1997) has been carried out to determine the localization of the motoneuron pools innervating muscles that run between the skull and the cervical vertebrae (the ventral, lateral and dorsal suboccipital muscles, the intervertebral muscles and the long dorsal extensor muscles) and the muscles which run between attached between the shoulder girdle and the head and cervical vertebrae (rhomboideus, levator scapulae, trapezius, sternomostoideus, cleidomastoideus).

Within the first cervical segment, motoneurons supplying ventral and lateral suboccipital muscles were located dorsomedially. Thus, rectus capitis lateralis motoneurons were located in the dorsomedial motoneuron column where they were confined to C1. Rectus capitis anterior motoneurons were also located dorsomedially where they showed substantial overlap with those of rectus capitis lateralis, but then they extended more caudally to C4. Motoneurons innervating the dorsal suboccipital muscles had a more ventral location. The motoneurons of rectus capitis posterior were located ventromedially and were exclusively found within C1. Within this ventromedial nucleus the motoneurons of obliquus capitis inferior were also found overlapping those of rectus capitis posterior. The motoneurons supplying obliquus capitis inferior muscle were not confined to C1, but extended caudally to C3. Between these two groups of motoneurons, and also in a ventromedial location, the motoneurons of complexus were situated. This motonucleus started at C1 and extended as far caudally as C4. Within C1, complexus motoneurons overlapped with those of the medial head of obliquus capitis superior. A small proportion of motoneurons supplying rectus capitis superior and complexus were reported as being located in the commissural nucleus and thus the motoneurons of these muscles have a dual location.

The motoneuron nuclei of the intervertebral muscles spinalis dorsi and semispinalis cervicis were located within C3 in a ventromedial region of the ventral horn similar to the location of the dorsal suboccipital motoneurons in more rostral levels. The motoneurons of the long dorsal extensors, biventer cervicis and splenius, were located ventromedially predominantly within C2 to C4. As with complexus, some motoneurons innervating these muscles were found in locations outside the ventral nucleus, mainly the commissural nucleus and the spinal accessory complex (Abrahams and Keane, 1984).

Kitamura and Richmond (1994) have suggested that the observed distributions can be explained according to a musculotopic pattern in which motoneurons innervating the dorsal muscles have the most ventral locations, whereas those innervating more lateral and ventral muscles are situated more dorsomedially within the ventral horn. This is consistent with the observations of Gutman *et al.*, (1993), which are summarized below in the section on thoracic and upper lumbar spinal cord.

Motoneurons innervating the muscles with scapular attachments involved in head movements in the cat were located more caudally (Liinamaa *et al.*, 1997). Rhomboideus capitis and rhoboideus minor motoneurons were located in a single column within C4-C6 more ventral than those of the other neck muscles described above, while levator scapulae motoneurons were located in a column lateral to those of rhomboideus. The location of motoneurons in the spinal accessory complex are described separately below. Figure 7.3 from the paper of Liinamaa *et al.*, (1997) shows the interrelationships between the motoneurons innervating the neck muscles, spinal accessory complex and those attached to scapula with those more rostral motonuclei innervating muscles of the upper limb.

Studies of the cervical spinal muscles in the rat

In a similar but less extensive study to those described above, Callister *et al.*, (1987) detailed the locations of the dorsal neck muscles splenius, biventer cervicis and complexus in the rat. The motoneurons innervating these muscles were located in the ventral half of the ventral horn from the lowest part of the medulla rostrally to C4 caudally. The splenius motoneurons were located more dorsally, while the biventer cervicis motoneurons were located at the ventral tip of the ventral horn overlapping with those of complexus. This result is essentially similar to those described in cats above except that in the rat there is no evidence for any dual innervation of these muscles – a result which is in agreement with that reported by Gutman *et al.*, (1993) in the chick. However, when Callister *et al.*, (1987) repeated the experiments of Richmond *et al.*, (1978), using a similar injection method, they obtained similar results, including clear evidence of contralateral motoneurons innervating these muscles, suggesting that there is a species difference in this regard between rat and cat.

Infrahyoid muscles

The infrahyoid muscles of mammals are a group of strap-like muscles in the ventral aspect of the neck with a rather complex pattern of innervation. Three of these muscles (sternohyoid, sternothyroid and omohyoid) are innervated by the ansa cervicalis – a nerve loop formed of axons exiting the spinal cord within the ventral rami of C1 to C3 and then joining the cervical plexus. One other infrahyoid muscle receives an innervation from C1 axons that travel with the hypoglossal nerve; these hypoglossal axons also innervate one suprahyoid muscle, the geniohyoid. The remaining suprahyoid muscles are innervated by cranial nerves. Kitamura *et al.*, (1986) investigated the localization of the motoneurons innervating these muscles in the rat.

They showed that the motoneurons of sternohyoid and sternothyroid are located in the dorsomedial cell columns within the ventral horn in C1 and C2 spinal segments. Rostrally the motoneurons innervating the omohyoid muscle are located in a similar position but more caudally this motonucleus is located more ventrolaterally. Motoneurons innervating geniohyoid are located in the ventrolateral subnucleus of the hypoglossal nucleus in the rat. The hypoglossal nucleus extends to the spinal cord medullary junction where it occupies a similar location to the dorsomedial cell columns with the ventral horn of C1. Thyrohyoid motoneurons are distributed between the caudal end of the hypoglossal motonucleus and the more dorsal and medial part of the ventral horn of C1. A similar location for these motoneurons has been reported by Yoshida *et al.*, (1983) in the cat.

Spinal accessory complex

The story of the innvervation of the sternocleidomastoid and trapezius muscles is an intruiging one. In a comprehensive review of the anatomy of the spinal accessory nerve, Straus and Howell (1936) noted that many researchers had concluded on the basis of studies of non-experimental material that the motoneurons supplying trapezius muscle (which in some species is represented by two separate muscles, the acromiotrapezius and the cleidotrapezius) and sternocleidomastoid muscle (represented by cleidomastoid and sternomastoid parts in some species) were located in the dorsolateral part of the ventral horn in the rostral cervical spinal cord. Straus and Howell (1936) confirmed this themselves by cutting the nerve to trapezius muscle in a cat and observing 'degenerative changes' in motoneurons of the dorsolateral motoneuron group in 'the middle cervical region'. Since the advent of modern retrograde tracing methods, a large number of papers on the musculotopic pattern of spinal accessory motoneurons in a range of different mammals have been published (Liinamaa *et al.*, 1997 – cat; Baba *et al.*, 1997 – mouse; Satomi *et al.*, 1985 – cat; Ueyama *et al.*, 1990 – macaque monkey; Brichta *et al.*, 1987 – rat; Kitamura and Sakai, 1982 – rat; Ullah *et al.*, 2007 – rat; Rapoport, 1978 – cat; Matesz and Szekely, 1983 – rat; Clavenzani *et al.*, 1994 – sheep; Augustine and White, 1986 – baboon; Routal and Pal, 2000 – man). While there are differences in the patterns reported, the recent report by Ullah *et al.*, (2007) in the rat can be taken as typical the findings of previous studies. Ullah *et al.*, (2007) found that accessory nerve motoneurons extended from the level of the pyramidal decussation to the upper half of C6. They found that sternocleidomastoid muscle motoneurons were located from the caudal medulla to rostral C3, and that trapezius muscle motoneurons were located from caudal C2 to rostral C6. From C2 to C6, both trapezius muscle and sternocleidomastoid muscle motoneurons were mostly located in a well-defined cell cluster the dorsolateral part of the ventral horn, but sternocleidomastoid muscle motoneurons were also found in the dorsomedial and central groups in C1 and C2.

The fibers that arise in the cervical cord that are destined to supply trapezius muscle and sternocleidomastoid muscle take an unusual course. Instead of emerging from the ventral aspect of the cord, they emerge from the lateral aspect immediately dorsal to the ligamentum denticulatum and travel rostrally through the foramen magnum. They then join with the cranial root of the accessory nerve for a short distance, after which they break away and travel through the jugular foramen to reach the neck. Some fibers from the upper cervical cord reach the trapezius muscle and sternocleidomastoid muscle by the 'standard' direct route, traveling in the spinal nerves C2, C3, and C4. It is believed that these 'direct' spinal connections are mostly afferent in nature, although there may be some motor fibers amongst them. Straus and Howell (1936) point out that in the giraffe, all the innervation of the sternocleidomastoid and trapezius travels direct from the spinal segments to the muscles, and do not travel via the foramen magnum and the jugular foramen.

The unusual course of the spinal accessory nerve suggests a branchiomeric origin and this has been a subject of debate for over a century. Straus and Howell (1936) reviewed the comparative anatomical evidence for a branchial arch versus somite origin and summarized the arguments put forward by proponents of the two points of view. They concluded that the weight of evidence was on the side of a branchial origin, citing the developmental origin of the motoneurons from the vagal complex, the fact that afferents from the trapezius muscle and sternocleidomastoid muscle end in the nucleus of the solitary tract, and the emergence of the motor rootlets of the spinal accessory from the lateral border of the cervical spinal cord.

Recent data from gene expression supports a branchial (non-somitic) origin for the trapezius muscle and sternocleidomastoid muscle and the spinal accessory nerve. The homeobox gene Nkx2.9 is expressed in spinal accessory motoneurons and deletion of this gene results in hyomorphism of the spinal accessory nerve (Pabst *et al.*, 2003). Schubert and Kaprellian (2001) reported that a monoclonal antibody SAC1 selectively labeled Isl1 positive motoneurons in cervical spinal cord were identified on the basis of axon trajectory as spinal accessory motoneurons. Lieberam et *et al.*, (2005) showed that the chemokine receptor Cxcr4 is expressed in somatomotor spinal neurons but not in brachial motoneurons, including spinal accessory motoneurons. Finally, observations by Watson and Tvrdik (unpublished) show that the Phox2b gene, a known marker for branchial and visceral motoneurons, is expressed in the trapezius muscle and sternocleidomastoid muscle motoneurons.

The phrenic nucleus

The phrenic motonucleus has been located in the rat (Goshgarian and Rafols, 1981), guinea pig (Johnson and Getting, 1988), rabbit (Rikard-Bell and Bystrzycka, 1980), and cat (Rikard-Bell and Bystrzycka, 1980; Berger *et al.*, 1984; Takahashi and Ninomiya, 1985). These studies are all in broad agreement in showing that, while the precise segmental level will vary between species, the majority of motoneurons are located in C5 and C6 spinal segments. The nucleus is located centrodorsally within the ventral horn between the medial and lateral motor columns. Devries and Goshgarian, (1989) have shown that axons leaving the cervical spinal cord in the accessory phrenic nerve are located in a caudal extension of the main motonucleus in C5 spinal segment and the rostral part (the main phrenic nucleus of the rat is located in C3-C5). This contribution to the phrenic motonucleus was quite significant in the rat, representing some 11% of the total motoneuron population at those levels. The topography of the phrenic nucleus has been investigated by a number of investigators who have shown that the motoneurons within it are clustered (Rikard-Bell and Bystrzycka, 1980; Berger *et al.*, 1984), with some internal topographical partitioning of the nucleus (Gordon *et al.*, 2004).

Topography of forelimb motoneuron pools in the cervical enlargement

The motoneuron pools of the forelimb have not been the subject of such comprehensive investigation as those of the hindlimb. Although there have been a large number of studies on individual muscles or groups, there has been no comprehensive study of the motoneuron localization within entire cervical lateral motor column in a single species comparable to those of Romanes (1951) or Vanderhorst and Holstege (1997) in the lower limb of the cat, or McHanwell and Biscoe (1981) in the lower limb of the mouse. Most of the forelimb muscle studies describe the topographical localization of selected muscles or muscle groups in one or more species. Thus, a picture of the entire forelimb motoneuron pool has to be constructed by combining data from different studies. The reasons for the difference between forelimb and hindlimb muscle studies are largely technical. The forelimb muscles are much smaller and more difficult to isolate, especially in rodents. This makes retrograde tracing studies problematic due to the difficulties in being certain that tracer spread has been limited. Despite this, two major conclusions can be drawn in relation to forelimb muscle innervation. First, as in the hindlimb, there is very good agreement between studies in different species on the relative positions of motoneurons innervating the different forelimb muscles. Second, there are striking similarities in the organization of the motoneuron pools between the forelimb and the hindlimb. As in many of the examples discussed in this review, the majority of studies have been carried out in the cat and the rat. Consequently, the available data for these two species will be discussed first before brief mention, by way of comparison, is made of the some of the studies carried in other species.

Forelimb motoneuron groups in the cat

The organization of forelimb motoneuron pools in the cat has been studied by Fritz *et al.*, (1986a,b), Horner and Kummel (1993) and Streling and Kuypers (1996).

Horner and Kummel (1993) investigated the location of shoulder muscle motonuclei in the forelimb of the cat by injection of their peripheral nerves. Using multiple pressure injection of up to four fluorescent tracers into the nerves of 12 muscles they were able to study the somatotopic relationships

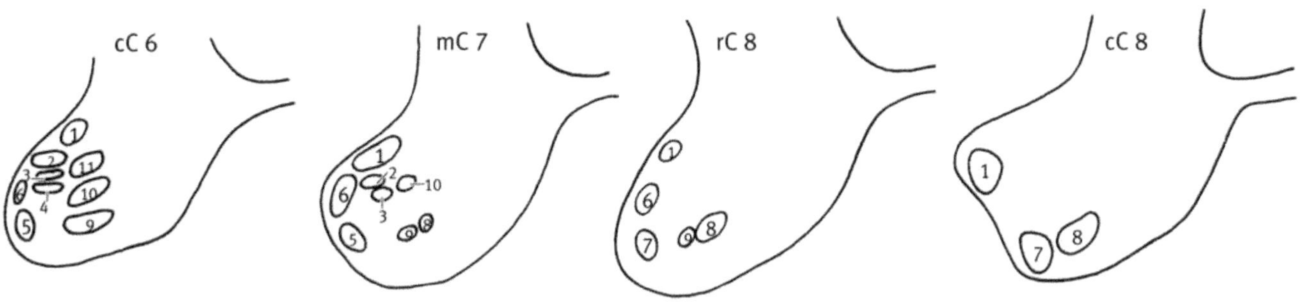

Figure 7.2 Motoneurons supplying shoulder muscles in the cat

This diagram shows motoneuron groups supplying shoulder muscles of the cat. (1 = radial profundus; 2 = deltoid; 3 = teres minor; 4 = cromiodeltoid; 5 = subscapularis; 6 = teres major; 7 = latissimus dorsi; 8 = pectoralis minor; 9 = pectoralis major; 10 = supraspinatus; cC6 = caudal C6 segment; mC7 = middle C7 segment; rC8 = rostral C8 segment; cC8 = caudal C8 segment). (Adapted from Horner and Kummel, 1993).

of muscles that act on the scapulo-humeral joint. Motoneurons innervating these muscles were located in spinal segments C5-T2 in discrete motonuclei that overlapped little. The distributions of spinodeltoideus, teres minor, acromiodeltoideus, subscapularis, teres major, latissimus dorsi, pectoralis major and minor, pectoantebrachialis, supraspinatus and infraspinatus were investigated (see Figure 7.2).

The motonuclei for subscapularis, latissimus dorsi, and the pectoral muscles were located ventrally in the ventral horn. The latissimus dorsi group was found to be a large cluster located on the ventrolateral margin of the ventral horn, and ventral to all other motor groups identified in the cat. It is situated caudally in the brachial enlargement, extending from caudal C7 to rostral T1. Lateral to it were located the motoneurons innervating subscapularis. Medial to the latissimus dorsi motonucleus were the motonuclei for the pectoral muscles. The pectoralis major group is a very large cluster which extends from caudal C5 to rostral C8. The pectoralis minor motonucleus was located between that of latissimus dorsi and pectoralis major. The location reported by Horner and Kummel (1993) for latissimus dorsi muscle is similar to that described by Holstege *et al.*, (1987) (however see discussion below). The location of the pectoralis major group reported by Horner and Kummel (1993) is similar to that described by Charlton *et al.*, (1988).

Dorsal to these groups were located the motonuclei for teres major caudally from C6 to C8, acromiodeltoideus, rostrally from C5 to C6, spinodeltoideus from C6 to C7, and teres minor from C6 to C7. The supraspinatus and infraspinatus groups were located medial to these groups within the ventral horn in a rostral group. Together with the pectoralis major group, these motoneuron groups are the rostral of all major forelimb motoneuron groups in the cat. The supraspinatus motoneurons extend from caudal C5 to mid C7. The supraspinatus motor group is located medially in the ventral horn, ventral to the infraspinatus and dorsal to the pectoralis major motor group. The infraspinatus motor group is located from caudal C5 to mid C7. The infraspinatus is the most dorsomedial of the shoulder muscle groups in the cat. More caudally the infraspinatus group occupies a more ventral position in the ventral horn.

Fritz *et al.*, (1986) studied the motoneuron groups associated with the deep radial, median, and ulnar nuclei. These three nerves supply the distal forelimb muscles of the cat. They showed that the deep radial nerve motoneuron pools were located from the caudal end of C5 to the rostral part of T1. In more rostral segments this pool is located dorsolaterally, but more caudally (in C8 and T1 spinal cord segments) it is located more ventrally, although it never extends as far as the ventral tip of the ventral horn. The median nerve motoneuron pool was located within C7 to T1 spinal cord segments medial to that of the deep radial group. In C7 there is a single motoneuron pool associated with the median nerve, but more caudally (in C8 and T1) there are two distinct motoneuron pools, one placed dorsally in the lateral motor column and a second located ventral to it. The ulnar motoneuron group is located within C8 and T1 spinal segments. It is located in the dorsal part of the ventral horn and becomes confined to a more dorsolateral location more caudally.

The projections of some individual muscle motonuclei were also investigated by Fritz *et al.*, (1986). Within the deep radial motoneuron pool, the motoneurons supplying three muscles, brachioradialis, extensor carpi radialis, and supinator occupy the rostral part of the motoneuron pool in C6 and C7. The motoneurons innervating the remaining muscles supplied by the radial nerve (forelimb extensors) were located within C8 and T1. There seems to be some overlap among these latter groups even though they were described in this study as having centers clearly distinct from one another.

In the case of the median nerve motoneuron pool, the motoneurons of the dorsal region projected solely to the intrinsic muscles of the hand, while the motoneurons supplying forelimb flexors are located more ventrally. Pronator teres and flexor digitorum longus motoneurons were located solely in C7 with an almost complete overlap in their distribution. Plantaris, pronator quadratus and flexor digitorum profundus 3, 4, and 5 motoneurons are located ventrally in C7 and C8, with a considerable amount of overlap between the motonuclei.

In the case of the ulnar motoneuron pool, the motoneurons innervating the intrinsic muscles of the hand are located dorsally in C8 to rostral T2 spinal segments (similar to the pattern of intrinsic hand muscles supplied by the median nerve). The motoneurons innervating flexor carpi ulnaris and flexor digitorum profundus 1 and 2 were located ventromedially, deeper in the ventral horn. These results were in good agreement with other available data on the cat and with the observations in the monkey by Jenny and Inukai (1983).

Forelimb motoneuron groups in the rat

There have been a number of studies investigating the localization of motoneurons innervating the forelimb in the rat (Curfs *et al.*, 1993; Choi and Hoover, 1996; McKenna *et al.*, 2000).

McKenna *et al.*, (2000) used injections of the carbocyanine dye tracers (DiI and DiD) to map the localization of motoneurons innervating 14 forelimb muscles. Their results were essentially similar to those of the cat. Motoneurons innervating acromiotrapezius and levator claviculae were located in C2 and C3 with lower levels of labeling in C4. The motoneurons innervating levator claviculae being located relatively more ventral and medial to those of acromiotrapezius.

The biceps brachii motoneurons formed a dense cluster which extended from rostral C4 to caudal C5. It was located rostral to that of the triceps, extensor carpi ulnaris and flexor carpi ulnaris motor groups within the ventral horn but lay ventral and slightly medial to those of extensor carpi radialis. The spinodeltoid motoneursons were located in the spinal cord from rostral C4 to caudal C5 and overlapped those of biceps brachii. Extensor carpi radialis longus and brevis motoneurons were located at the same rostrocaudal level as that of biceps brachii and were situated slightly dorsal and lateral to it with some overlap. The location of the spinodeltoid nucleus in this study was broadly similar to that of Choi and Hoover (1996), although the latter study suggested a slightly more caudal and more extensive location for the spinodeltoid motoneurons.

The triceps motoneuron column was located caudal to the biceps, extending from mid C6 to the caudal C8 segments with the largest part lying between rostral C7 and mid-C8. The extensor carpi ulnaris motor column extended from rostral C7 to rostral T1. The flexor carpi ulnaris column is situated between caudal C6 and rostral T1. Both these motor groups are situated right on the lateral margin of the ventral horn and showed extensive overlap with each other, with the triceps motonucleus, and with other forelimb wrist and finger extensors. Similar results were reported by Curfs *et al.*, (1993).

The results confirmed three well-recognized general patterns of organization within the forelimb motor columns (see below). However there was evidence of much more overlap between these motonuclei in the rat in comparison to that described by Jenny and Inukai (1983) in the monkey (see below). Whether this represents real overlap because motonuclei may be closer together in rat compared to monkey, or whether it is due to small leakages of tracer is not clear (see Figure 7.3).

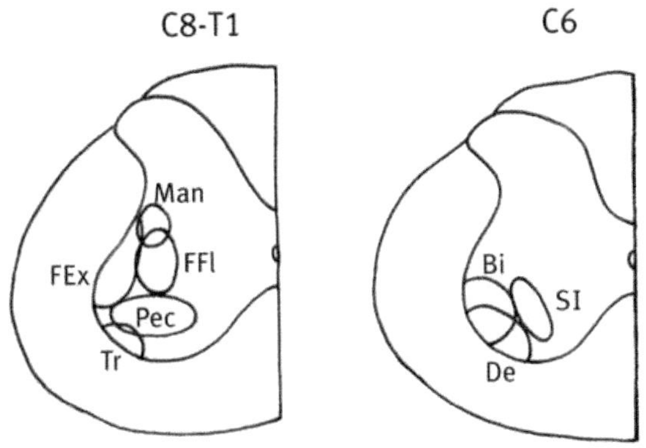

Figure 7.3 Forelimb enlargement of the rat

This diagram shows the motoneuron groups supplying hte main muscle groups in the forelimb of the rat. Bi = biceps; De = deltoid; FEx = forelimb extensors; FFl = forelimb flexors; Man = manus; Pec = pectoralis; SI = supraspinatus and infraspinatus; Tr = tricepts. (Adapted from McKenna and Nadelhaft, 1986).

Forelimb motoneuron groups in the mouse

Ryan *et al.*, (1998) studied forelimb motoneuron pool organization in the mouse using the retrograde neural tracer WGA-HRP injected into seven forelimb muscles – the spinodeltoideus, biceps brachii, triceps brachii, pectoralis major, supaspinatus, and infraspinatus.

The biceps motoneuron group was located on the lateral margin of the ventral horn, dorsal to all other motoneuron groups in the mouse. This column extended from mid-C4 to caudal C5. The deltoid group was relatively small, located on the ventrolateral margin of the ventral horn just ventral of the biceps and dorsal to that of the triceps. The deltoid motor column extended from caudal C4 to rostral C6. The infraspinatus and supraspinatus groups were located medial to that of the biceps and deltoid motor groups and dorsolateral to those of the pectoralis. The infraspinatus column extended from caudal C4 to rostral C6 and the supraspinatus column extended from mid-C4 to caudal C5.

The triceps and pectoralis groups lay caudal to those of the biceps, deltoid, supraspinatus and infraspinatus. The pectoralis group was the largest of the forelimb motoneuron groups that were studied by Ryan *et al.*, (1998), and was located ventromedially to all other groups in the ventral horn. The pectoralis motor group extended from mid-C5 to caudal C7. The triceps motoneuron group was located on the ventrolateral margin of the ventral horn and extended from mid-C6 to rostral C8.

Forelimb motoneuron groups in the monkey

Jenny and Inukai (1983) studied the musculotopic pattern in the *Macaca mulatta and Macaca fascicularis* monkeys. They identified the motoneuron groups supplying 18 selected arm and forearm muscles using HRP, a retrograde tracer technique. Motoneurons innervating biceps brachii were located in the lateral margin of the ventral horn from caudal C5 to C7 and dorsal to all other motoneuron groups at these levels. Caudally, at levels C6 to C7, its position shifts to lie more medially in the anterior horn. The triceps motonucleus was found to be located in segments C7, C8, and T1, with small numbers of motoneurons extending into C5 and C6 rostrally and T2 caudally. In segments C6 and C7, the triceps motoneuron group was found at the lateral margin of the ventral horn, but more caudally the triceps column was more medially located.

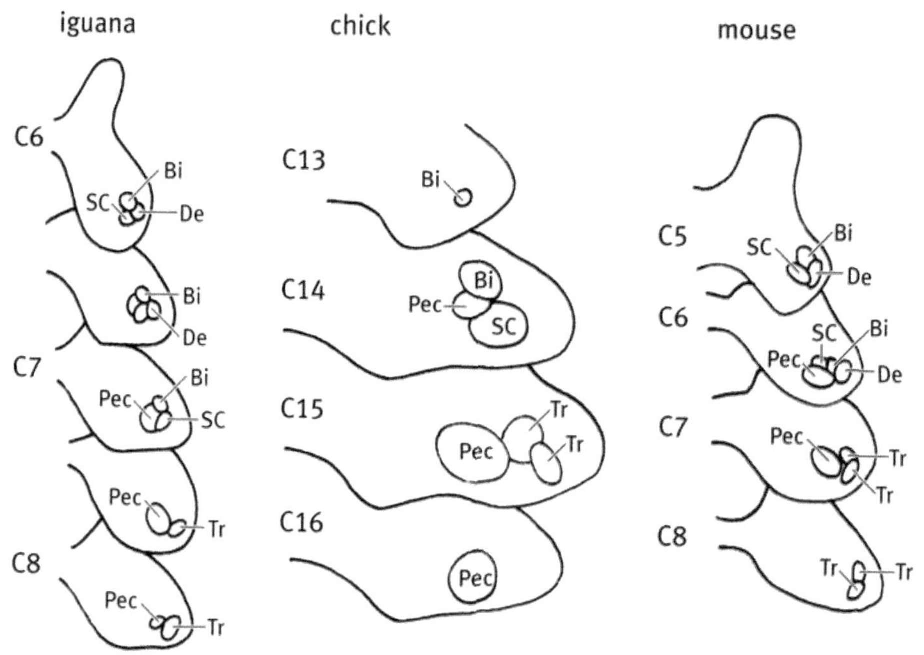

Figure 7.4 Topographic position of forelimb motoneuron pools in the reptile, bird and mammal

This diagram shows the position of motoneuron groups supplying the biceps (Bi) supracoracoideus (SC), deltoid (De), pectoralis (Pec), and triceps (Tr) in a reptile (iguana), a bird (chick), and a mammal (mouse). (Adapted from Ryan *et al.*, 1998).

Jenny and Inukai (1983) studied four forearm muscles that act on the wrist: flexor carpi ulnaris, flexor carpi radialis, extensor carpi ulnaris and extensor carpi radialis. Extensor carpi radialis motoneurons were located in C5, C6, and C7, and extensor carpi ulnaris motoneurons were located in C8 and T1. The extensor carpi radialis group is positioned along the dorsolateral margin at its rostral end. In more caudal levels the extensor carpi radialis column assumes a more medial and ventral position in the ventral horn. The extensor carpi ulnaris is a small group of motoneurons positioned on the lateral margin of the ventral horn. In C7, the extensor carpi ulnaris group is dorsally placed, but in more caudal sections the column became progressively more ventrally placed.

The flexor carpi radialis motoneurons were found in C7 and rostral C8, and the flexor carpi ulnaris motoneurons were in C8 and rostral T1. In C7 and rostral C8, the flexor columns overlap each other. The flexor carpi ulnaris column is medial to the columns supplying the other forearm muscles. It begins in C7 in the most dorsal part of the ventral horn but moves ventrally in succeeding levels. The flexor carpi radialis has a dorsolateral position in C7 and C8.

The finger flexor and extensor muscle motoneuron pools located in C7-T1 had a more complex relationship to another. Rostrally the motonuclei of the finger extensors (extensor digitorum longus and abductor pollicis longus) were lateral to the motonuclei of the finger flexors (flexor digitorum superficialis and profundus). However, in more caudal regions of the spinal cord the flexor pools expanded and came to lie dorsally to those of the finger extensors.

The motoneurons innervating the intrinsic muscles of the hand were located dorsally and caudally with respect to the other motor pools innervating the forelimb muscles predominantly within C8 and T1 spinal segments.

Forelimb motoneuron groups in other species

As part of the study referred to above in the mouse, Ryan *et al.*, (1998) also studied the forelimb motoneuron pool organization in the bat and iguana. They showed that forelimb motoneuron pool organization is a conserved feature in amniote evolution. Studying the location of the motoneuron pools of seven homologous muscles they concluded that the relative positions of these motoneuron pools were similar in the three species studied (see Figure 7.4)

Straznicky and Tay (1982) and Hollyday and Jacobson (1990) examined the location of motoneurons innervating the wing muscles of the chick. Their results were broadly similar. As with the studies of mouse, bat, and iguana, Ryan *et al.*, (1998) found that homologous pools occupy similar relative locations in the cervical lateral motor column. This provides further support to the proposition that motoneuron pool organization in a highly conserved feature of vertebrate evolution, at least in amniotes. The two studies on the chick also highlight the fact that in the forelimb pools there is a clear relationship between the position of a motonucleus within the lateral motor column and the embryonic origin of the muscle it innervates. The muscles that originate from the ventral muscle mass are innervated by motoneurons in medially-located motor pools whilst those originating from the dorsal muscle mass are innervated by motoneurons in laterally-located motor pools.

Summary of forelimb motoneuron pool organization

Forelimb motoneuron groups in mammals can be divided into a rostral group that includes deltoid, supraspinatus, infraspinatus, and biceps brachii, and a caudal group that includes the forearm flexors and extensors, triceps, pectoralis, and the intrinsic muscles of the hand. The rostral group is located within spinal segments C4 to C6, and the caudal group is located within spinal segments C7 to T1. Three further general principles of organization can be identified: first, that extensor motoneurons are located more laterally than flexor motoneurons; second, that rostral motoneurons innervate more proximal muscles than caudal motoneurons; and third, that more dorsally located motoneurons innervate more distal muscles. The diversity of studies in the forelimb make detailed comparisons very difficult, but in these studies, the motoneurons supplying four muscles have been most frequently studied and it is worth summarizing the locations of these motoneurons:

- The deltoid motoneuron nucleus is placed laterally and dorsally in the ventral horn, between segments C4 to C6.
- The biceps brachii motoneuron group occupies a dorsal position at the lateral margin of the ventral horn. The biceps group extends from C5 to C7 in both the monkey and the bat, but in the mouse the biceps column extends from C4 to caudal C5.
- The pectoralis motoneuron nucleus is by far the largest group in the brachial enlargement of the iguana, chick, bat, cat and mouse. In mammals the pectoralis motor group usually stretches from between mid-C5 to C8. The pectoralis group is similarly postioned in those vertebrates that have been studies. In both the iguana and mouse the pectoralis group occupies a ventro-medial position within the ventral horn. In the chick, bat and cat the pectoralis group occupies a medial position.
- The triceps motoneuron group lies in the ventrolateral quadrant of the ventral horn in the caudal segments of the brachial enlargement. In rodents it forms a dense cluster at the ventrolateral tip in transverse sections of the ventral horn in segment C6 to C8.

Finally, as in the hindlimb, there is a clear relationship between the embryonic origin of a muscle and the position of the motoneurons that innervate it within the lateral motor column.

Topography of motoneuron pools in the thoracic and upper lumbar spinal cord

With the exception of the first thoracic spinal segment (which supplies upper limb muscles – see above), motoneurons in the thoracic and upper lumbar spinal cord innervate predominantly the muscles of the thorax and abdomen. A small number of studies, confined mainly to the cat, that have investigated the precise topography of these motoneuron pools. These studies have confined their attention to the main muscle groups and there remain a number of smaller muscles whose precise motoneuron localizations remain unknown except in general terms. It is also the case that many studies have investigated the localization of one or two muscle groups in isolation, so that the relationships between those motoneurons and motoneurons in the relevant segment or segments have to be inferred.

Motoneurons of the thoracic and upper lumbar segments in the cat

Tani *et al.*, (1994) investigated the localization of the motoneurons innervating both thoracic and abdominal muscles. Using retrograde transport of cholera toxin B bound to latex beads, they studied the localization of motoneurons innervating the intercostal muscles, the serratus dorsalis muscles, and muscles of the anterior abdominal wall. As would be expected from the segmental location of these muscles, the motoneurons innervating them were located between T2 and L2 spinal segments. The motoneurons supplying the muscles of the anterior abdominal wall, as a group, tend to lie more caudally in this spinal cord region.

The external intercostal muscle motoneurons were located between T1 and T10 segments in the ventromedial motoneuron column of the thoracic spinal cord. The serratus dorsalis cranialis motoneurons occupied a similar position in the thoracic cord but were restricted to segments T5-T9. The motoneurons innervating the internal intercostal muscles were located lateral to those of the external intercostal muscles in the ventrolateral column on the lateral margin of the ventral horn and within spinal segments T2-T11. Also located in this column, but confined to its caudal portion, were the motoneurons innervating the serratus dorsalis caudalis muscle, which were restricted to segments T10-T12. Broadly similar results, with respect to the intercostal muscles in the cat, were reported by Larnicol *et al.*, (1982) and Lipski and Martin-Body (1987). They described the external intercostal muscle motoneuron pool as being located ventromedially. They found that the internal intercostal muscle motoneuron pools were located laterally and were more widely distributed, with some overlap medially with the external intercostal muscle motoneuron pool. They also observed, as might be predicted, that the motoneurons were almost exclusively confined to the spinal segment from which the intercostal nerve innervating the relevant intercostal space had originated. Larnicol *et al.*,

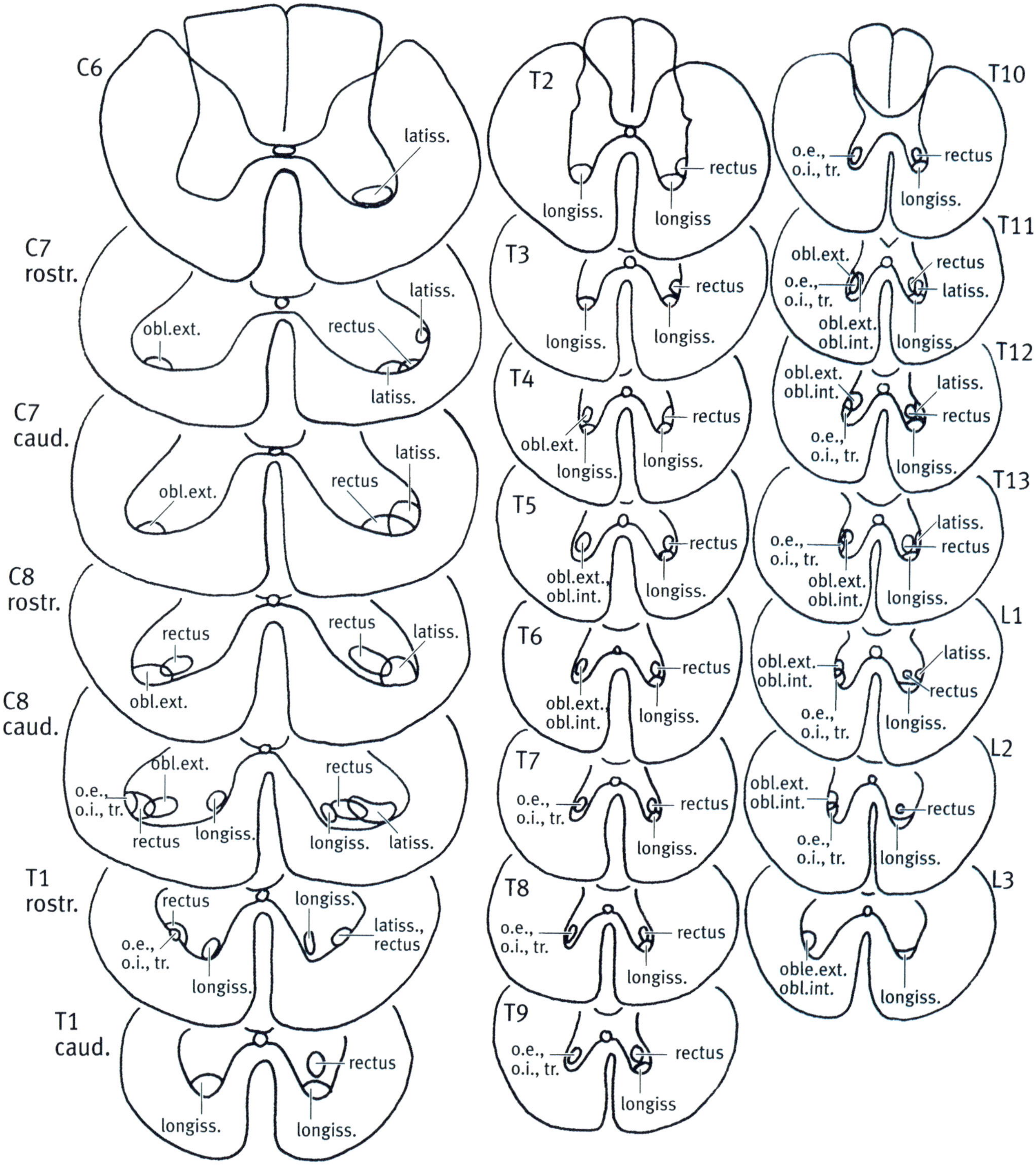

Figure 7.5 Motoneuron groups of the thoracic and upper lumbar segments in the cat

This diagram shows the motoneuron groups supplying the muscles of the thorax and abdomen in the cat. (latiss = latissimus dorsi, longiss = longissimus; rectus = rectus abdominus, o e and obl ext = external abdominal oblique, o i. and obl.int. = internal abdominal oblique; tr = transverse abdominal oblique) (Adapted from Holstege *et al*, 1987)

(1982) also studied the distribution of motoneurons innervating the triangularis sterni muscle and the intercartilagenous muscle and the motoneurons. Both these muscle groups were reported as lying in the ventrolateral column of the thoracic spinal cord. It is not easy to reconcile the differences in results from these two studies; part of the explanation may lie in the slightly different muscle groups labeled in the earlier study and the possibility of spread of the injected tracer.

In the case of the muscles of the anterior abdominal wall, Tani *et al.*, (1994) reported that the external and internal oblique muscle motoneuron pools were located ventrolaterally in extensively overlapping pools from T6-L2 and T11-L2 respectively. Tani *et al.*, (1994) also report that the location of these abdominal motoneuron pools differ markedly from those of the intercostal muscles. While there would appear to be little or no overlap with the external intercostal muscle motoneurons lying ventromedially, it is difficult, from the data they present, to exclude the possibility of some overlap with the internal intercostal muscle motoneuron pool at their mediolateral border. Tani *et al.*, (1994) also studied the location of rectus abdominis muscle motoneurons, showing that they were located in a ventromedial column distinct from those of the other muscles studied and extending from T4-L2.

Holstege and his colleagues (1987) also described the location of motoneurons innervating the anterior abdominal wall of the cat, and studied the relationships of these motoneuron pools to those containing motoneurons innervating two muscles of the back, latissimus dorsi and longissimus dorsi (see Figure 7.5). The locations they described for the external oblique, internal oblique, and rectus abdominis muscle motoneuron pools were similar to those of Tani *et al.*, (1994) and in the case of external oblique, similar to those of Charlton *et al.*, (1988). Holstege *et al.*, (1987) also studied the location of transversus motoneurons, finding these to be co-located with the other motoneurons innervating the anterior abdominal wall muscles. Latissimus dorsi motoneurons were found by Holstege *et al.*, (1987) in a ventrolateral column in C6-C8 segments, as would be expected from its segmental innervation, but also in segments T9-L3, overlapping with the motoneurons pools innervating the muscles of the anterior abdominal wall. Longissimus muscle motoneurons were found in the most ventral parts of the ventral horn from T1-L4. Holstege *et al.*, (1987) also studied the location of the motoneurons innervating the cat cutaneous maximus muscle. These motoneurons were situated in a ventrolateral nucleus at the edge of the ventral horn within C8 and the rostral part of C1 segments. A similar localization has been reported for the rat cutaneous trunci motonucleus by Theriault and Diamond in 1988. However, Holstege *et al.*, (1987) also observed that labeled motoneurons were found in the same position as the cutaneous maximus motonucleus following injections of the anterior abdominal wall muscles and in caudal parts of the latissimus dorsi muscle. They were not able to resolve whether this anomalous location of these motoneurons and the anomalous location of latissimus dorsi motoneurons in lower thoracic spinal segments were due to leakage of the injected tracer or were due to the particular functional organization of the motor pools innervating these muscle groups.

Motoneurons of the thoracic and upper lumbar segments in the rat

There have been relatively few studies of the motoneuron localization in the rat thoracic spinal cord. Brink *et al.*, (1979) studied the localization of motoneurons in the rat innervating the lumbar epaxial muscles lateral and medial longissimus and transverso-spinalis muscles, and reported them to be in similar locations to that described for the cat longissimus muscle by Holstege *et al.*, (1987).

Smith and Hollyday (1983) investigated the distribution of motoneurons innervating thoracic muscles in the rat. Their results were broadly similar to those in the cat. They showed that motoneurons whose axons left the spinal cord in the ventral primary rami were located in the lateral regions of the ventral horn, while motoneurons whose axons left the ventral horn in the medial part of the spinal cord were located in the medial region of the ventral horn. There was, however, considerable overlap between these two motoneuron populations in the intermediate parts of the ventral horn. There also appeared to be a considerable amount of overlap between the motonuclei innervating individual muscles. Smith and Hollyday (1983) suggested that the superposition of maps from different regions of the thoracic spinal cord might have exaggerated the degree of overlap, although it is worth pointing out that similar degrees of overlap have been reported in studies on the cat. Direct comparisons of the work of Smith and Hollyday (1983) in cases where similar muscle motonucleus distributions have been investigated reveal strong similarities between these and the other studies in the cat described above. Thus, the external intercostal muscle motoneurons were located dorsally in the ventral horn, medial to the motoneurons innervating the internal intercostal motoneurons. In the rat there appears to be no overlap between these two motonuclei, as there seems to be in the cat. In the rat, longissimus muscle motoneurons were located in an extreme ventral location in the ventral horn while the motoneurons innervating the external oblique muscle were located ventromedially.

As with other motor neuron groups in the spinal cord, the location of these motoneuron pools in the thoracic spinal cord can best be explained on the basis of their origin in the embryo (Smith and Hollyday, 1982). Axial muscles can be divided in to two groups, epaxial and hypaxial, based on their relationship to the transverse processes of the vertebra in land vertebrates. Epaxial muscles lie dorsally in the body, situated between the spinous processes and transverse processes of the vertebrae, while hypaxial muscles lie ventral to the transverse processes. Gutman *et al.*, (1993) have shown that in the chick there is a clear relationship between the position of motoneuron pools innervating hypaxial and epaxial muscles with hypaxial muscle motoneurons located more medially in the ventral horn. A similar relationship can be seen in other species including cat, rat and monkey. This somatotopic relationship of motoneurons in the ventral horn correlates with the embryonic origin of these two groups of muscles. Epaxial muscles are derived solely from somites while hypaxial muscles have a dual embryonic origin with the muscle fibers derived from somites and the connective tissues derived from the lateral plate. Thus, the relationship between the position of motoneuron pools in the ventral horn and the embryonic origin of the muscles they innervates first described in the lumbosacral lateral motor column may represent a consistent feature of the organization of motoneurons in the spinal cord.

Topography of hindlimb motoneuron pools in the lumbosacral spinal cord

Motoneuron groups supplying the lower limb muscles in the cat

The lower limb enlargement is the region of the spinal cord for which the most comprehensive studies of motoneuron localization are available. There have been extensive studies of motoneuron localization in cat (Romanes, 1951; Vanderhorst and Holstege, 1997), mouse (McHanwell and Biscoe, 1981a, b), rat (Nicolopoulos-Stournas and Iles, 1983), and man (Sharrard, 1955). In addition, there are many other studies of individual motoneuron pools in a variety of species. Of these the work of Burke *et al.*, (1977) on the gastrocnemius and soleus muscles, Gordon *et al.*, (2004) on sartorius and tensor fasciae latae, and Horscholle-Bossavit *et al.*, (2004) on the peroneal (fibular) muscles are notable for the evidence they provide on the detailed organization of individual motoneuron pools and the subdivision of those pools in the case of individual muscles.

The most detailed work has been done on the cat and this work will be reviewed first and the results from other species discussed in the light of these comprehensive findings.

Romanes (1951), in his classic study using chromatolysis, was the first worker to undertake a comprehensive investigation of the topography of the motoneuron pools in any single region of the spinal cord. In the cephalic part of the lumbosacral enlargement of the cat spinal cord, Romanes (1951) described three columns: a ventrolateral group consisting of motoneurons innervating the rectus femoris and sartorius muscles; a ventromedial group consisting of motoneurons innervating the vastus lateralis and vastus medialis muscles; and a central group consisting of motoneurons innervating adductor magnus and gracilis. In the more caudal parts of the enlargement these groups were described as continuing in roughly similar positions, although both tended to split into medial and lateral parts. In the caudal part of the enlargement the ventrolateral group innervates the gluteal muscles, while the central group innervates the hamstring muscles. The dorsolateral group was described as turning dorsally in more caudal regions of the enlargement, with the motoneurons within it innervating the muscles of the anterior compartment of the leg (including tibialis anterior and extensor digitorum longus). In the caudal part of the enlargement three further columns were identified by Romanes (1951). A dorsal column (dorsal to the partially-divided central group and medial to the dorsolateral group) consists of motoneurons that innervate muscles posterior compartment of the leg, including the gastrocnemeii and soleus. A more ventral group consists of motoneurons that innervate the long flexors of the foot and plantaris. A third group dorsal to all the others contains motoneurons innervating the intrinsic muscles of the foot.

The results of the comprehensive study of Vanderhorst and Holstege (1997) on the hindlimb motoneuron pools of the cat are broadly similar to those of Romanes (1951), although there are some differences of detail (Figure 7.6). The most likely explanation for these differences is that the retrograde tracer techniques employed in the later study enabled a more complete labeling of the motor pools in question. This later study also studied a greater number of muscles than Romanes (1951). Another possibility is that in some instances there had been some spread of injected tracer, though the descriptions of the procedures employed by the authors suggest this is unlikely. There is in addition other evidence (see below) indicating the likelihood that the distributions observed by Vanderhorst and Holstege (1997) are a more complete representation of the actual patterns. The key differences between the two studies can be summarized as follows.

In the rostral part of the enlargement sartorius motoneurons were thought by Romanes (1951) to be co-located with those of psoas, but Vanderhorst and Holstege (1997) consistently found them to be more medially placed. The apparent continuity between the dorsolateral column innervating vastus medialis and lateralis and the muscles of the anterior compartment of the leg could not be confirmed in the study by Vanderhorst and Holstege (1997), who suggested the two columns were separate. The positions of the motoneuron pools for the adductor muscles and the hamstrings were similar in both studies but the detailed relationships between the motoneurons innervating individual hamstring muscles were found to be different in the latter study, and this confirms the results of other recent studies carried out by a variety of authors. Similarly, the motoneuron pools innervating the muscles of the posterior compartment of the leg were located medially and ventrally in both studies but there were differences in the detailed relationships of motoneurons innervating the individual muscles. Thus, plantaris muscle motoneurons were found to be localized in this column rostral, rather than dorsal, to lateral gastrocnemius muscle motoneurons (as was found by Romanes (1951). The groups of motoneurons supplying plantaris and lateral gastrocnemius were rostral to those of medial gastrocnemius and soleus, whose motoneurons also lie more laterally. Finally, Vanderhorst and Holstege (1997) found that a single dorsal column innervates the muscles controlling the foot, with the long flexor muscle motoneurons lying in the rostral one third of this group and the motoneurons innervating the intrinsic foot muscles comprising the caudal two thirds.

Motoneuron groups supplying the lower limb muscles in the mouse and rat

A similar study was carried out in the mouse by McHanwell and Biscoe (1981) using horseradish peroxidase as a retrograde tracer (Figure 7.7). The smaller size of this animal meant that it was not technically feasible to inject many individual muscles as in the cat because of the limitations posed by spread of the tracer from the injection site. Instead spread of the tracer was controlled by the denervation of surrounding muscles. Taking into account these limitations, the results showed that the topographic localization of motoneurons supplying the mouse hindlimb to be broadly similar to that of the cat. The detailed results in relation to columnar organization tend to show greater similarities to the work of Vanderhorst and Holstege (1997), suggesting that some of Romanes' (1951) descriptions may have underestimated the extent of the motoneuron pools, and so misinterpreted their relationships because of the limitations inherent in the chromatolysis method.

The next section will summarize the data relating to the motoneuron groups supplying each of the main muscle groups in the lower limb in rodents. The point should be made at this juncture that the spinal segments contributing to the nerves of the hindlimb in rodents differ from that in the cat and man and this needs to be remembered when results are compared from the different species.

In the rostral part of the mouse, lumbosacral enlargement motoneurons innervating the quadriceps muscle were located laterally in the ventral horn confined mainly to segment L2 but extending into the caudal part of L1 and rostral part of L3. The quadriceps motoneuron pool occupies a ventrolateral position throughout all segments, but as it extends caudally it expands both dorsally and laterally. Motoneurons innervating the adductor muscles were located medially to the quadriceps motoneurons and over a similar segmental extent occupying a more ventral and medial location more caudally. The gracilis motoneuron pool overlaps significantly with that of the

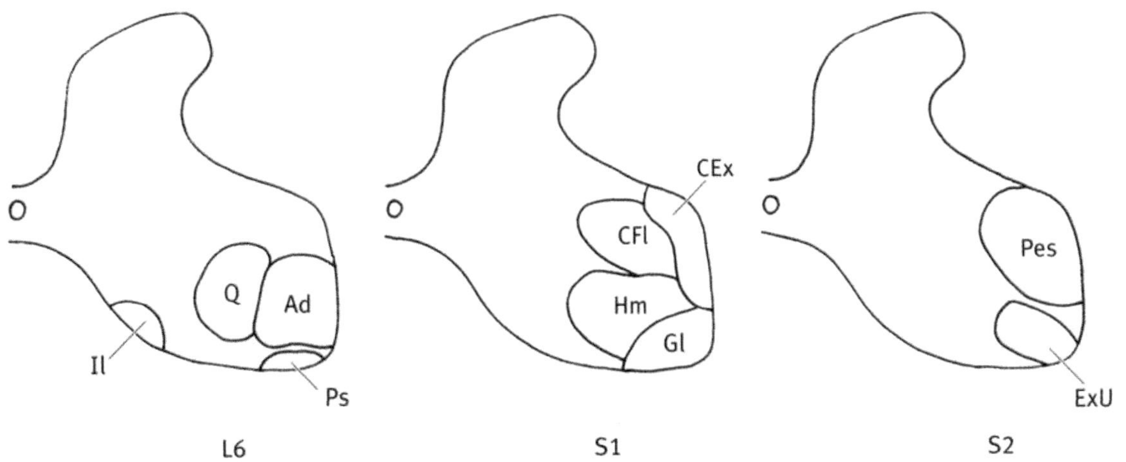

Figure 7.6 Motoneurons supplying hindlimb muscles in the cat

This diagram shows the motoneuron groups at upper (L6), middle (S1), and lower (S2) levels of the lower limb enlargement in the cat. (Ad= adductors; CEx = crural extensors; CFl = crural flexors; Gl = gluteal muscles; Hm = hamstrings; Il = iliacus; Q= quadriceps femoris; Pes = foot muscles; Ps = Psoas; ExU = external urethral sphincter). (Adapted from Vanderhorst and Holstege, 1997).

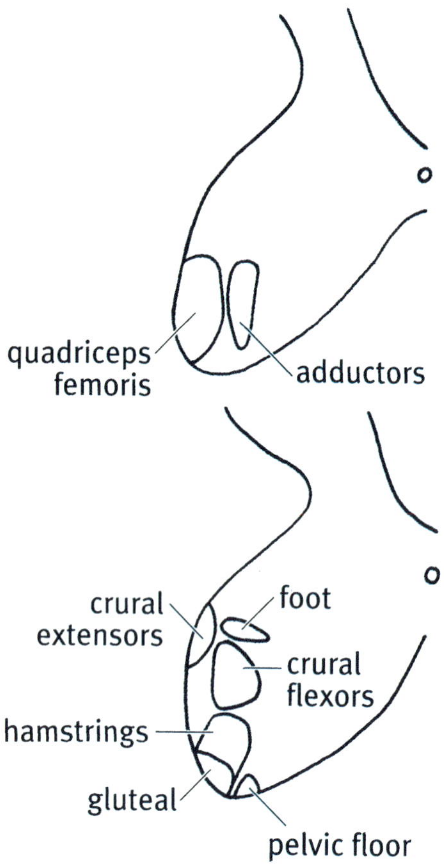

Figure 7.7 Motoneurons supplying hindlimb muscles in the mouse

This diagram shows the topographic position of motoneuron groups supplying the crural extensors, crural flexors, foot, hamstring, gluteal, quadriceps femoris, adductors, and pelvic floor muscles in the mouse (Adapted from McHanwell and Biscoe, 1981)

adductor muscles, although more caudally gracilis motoneurons occupy a more dorsal position in the pool. This distribution is similar to the results obtained in the cat. However, because it was not possible to fractionate the quadriceps motoneuron pool by injecting the individual muscles of this muscle group, one could not show by retrograde tracing that both a ventrolateral and dorsolateral column were present at this level in mice. However, reconstruction of the columnar organization of motoneurons in Nissl-stained sections showed that there were indeed two columns, ventrolateral and dorsolateral present in the rostral parts of the lateral motor column in mice as in the cat.

Within the more caudal part of the lateral motor column in mice, the ventrolateral column (that contains motoneurons innervating quadriceps motoneurons in the rostral lumbar enlargment) continues in roughly the same position but now consists of motoneurons innervating the gluteal muscles. This motoneuron pool is located in L3 and L4 and as it extends into the caudal part of L3 and rostral L4 it occupies a more medial and dorsal position. The medial column that contains adductor motoneurons rostrally also continues in approximately the same position caudally but now contains motoneurons innervating the hamstring muscles. The hamstring muscle group includes the biceps femoris, the principal and accessory heads of semitendinosus, caudofemoralis and semimembranosus, and is the largest muscle group in the mouse lower limb. Its motoneurons are located in segments L3 and L4. At its rostral end, this motoneuron group is positioned dorsomedially, but caudally it expands laterally and ventrally. In caudal L4 there may be some overlap with this motoneuron pool and that of the gluteal motoneuron pool. It is likely that the rostral part of motonucleus in L3 supplies the semimembranosus while the caudal motonucleus in L4 supplied the caudofemoralis. The biceps femoralis motonucleus was located in caudal L3 and in L4. The biceps femoralis group is a small cluster located in a far ventral and lateral position within the ventral horn. The semitendinosus motonucleus was located caudally within the main hamstring motoneuron group and was medial in position. There was some overlap between the semitendinosus and biceps femoris groups at their mediolateral border. In Nissl-stained sections reconstructions show that the hamstring motoneuron pool is represented by two motor columns rostrally in L4 though this is not readily apparent in retrograde labeling experiments.

In the caudal part of the lateral motor column of the mouse three further motor columns appear in a position dorsal to those already described above. A dorsolateral column appears at the rostral border of L3 and contains motoneurons innervating muscles in the anterior and lateral compartment of the leg (tibialis anterior, extensor digitorum longus, extensor hallucis longus and the peroneii muscles). The motoneurons innervating these muscles were found in the L3 and L4 segments, with the majority being located in L3. The rostral portion of this motor nucleus expands ventrally in the lateral column, but at caudal levels the group is located at the extreme dorsal margin of the lateral column. At all levels motoneurons supplying the anterior and lateral crural muscles are situated lateral to those posterior crural muscles. Motoneurons whose axons travel in the deep peroneal nerve are located predominantly within L3, whereas those motoneurons whose axons travel in the superficial peroneal nerve are located more caudal and dorsolaterally. There is some overlap between the motoneuron groups supplying the two groups of muscles. There appeared to be no continuity between this motor column and that of the dorsolateral column found in the rostral part of the lateral motor column. In this respect these results are similar to those of Vanderhorst and Holstege (1997) and differ from those of Romanes (1951).

Medial to this motoneuron pool is a motoneuron column supplying muscles within the posterior compartment of the leg (lateral and medial gastrocnemius, soleus, plantaris, flexor digitorum longus, flexor hallucis longus and tibialis posterior). The motoneuron pool supplying these muscles extends from L2 to L5, with the majority being located in L3 and L4. The rostral part of this main motor nucleus is located in the dorsomedial region of the lateral column and it expands dorsally and ventrally as it extends through L3. The motoneurons supplying the deep posterior muscles of the crus were found from rostral L3 to caudal L4. They were dorsally positioned within the main crural motoneuron group. The motoneuron group supplying superficial crural flexors was located between caudal L3 and rostral L5. This group was located ventrally within the main crural motoneuron group. There was no attempt to fractionate this motoneuron pool further although it should be noted that a separate study describing the position of the soleus motonucleus has been published (Parry *et al.*, 1982).

The most dorsal of the motor columns appears in the lower part of L4 and supplies the intrinsic muscles of the foot (pes). These include the dorsal and the plantar muscles. The motoneurons supplying the intrinsic muscles start in rostral L4 and extend as far as the caudal limit of L5. The rostral part of this motoneuron group occupies a dorsal and medial position in the lateral column, but at more caudal levels it occupies a more lateral and ventral position. The motoneurons of the dorsal muscles of the pes are located in caudal L3, and are positioned in the dorsolateral extreme of the lateral ventral horn. The motoneurons of the plantar muscles are located in the rostral two-thirds of the motoneuron group that supplies the pes. It also needs to be noted that in rodents a significant fraction of motoneuron axons reach the muscles of the foot via the sural nerve. Thus, the sural nerve is not a purely cutaneous nerve as it is in many other mammalian species. The motor axons traveling to the foot in this nerve tend to occupy a more caudal position in the motor pool. These studies in the mouse did not examine the distribution of the motoneurons innervating the long flexors of the foot separately from those of other motonuclei. Thus, it is not possible to determine whether these motoneurons are located in the same column as the intrinsic muscles of the foot as the studies of Vanderhorst and Holstege (1997) in the cat indicate or whether they occupy a distinct column as suggested by Romanes (1951). However, it may be noted that the distribution of motoneurons innervating the deep posterior muscles of the leg overlaps caudally with those of the motoneurons innervating the foot, suggesting similarities with the findings of the later study of Vanderhorst and Holstege (1997).

Nicolopoulos-Stournas and Iles (1983) examined the distribution of motoneurons suppling the hindlimb muscles in the rat. Their results are very similar to those obtained in the mouse two years earlier by McHanwell and Biscoe (1981).

Motoneuron groups supplying the lower limb muscles in man

The other comprehensive study of the distribution of motoneurons innervating hindlimb muscles that should be mentioned is that of Sharrard (1955) in man. Using the technique of comparing the distribution of permanent paralysis in the lower limb muscles in seven patients with the residual cells observed in the lateral motor column post-mortem, Sharrard (1955) was able to build up a map of the topographic organization of the motoneuron pools in man. The patterns he described show strong similarities with the patterns described in experimental animals. The difficulties with this approach are considerable as typically in each case several muscles or muscle groups were affected and the degree of paralysis was not uniform. It is therefore a tribute to Sharrard's (1955) careful analysis that it was possible for him to produce such a convincing set of results from less than ideal material.

Summary of hindlimb motoneuron pool organization

It is possible to draw general conclusions about the topographic organization of motoneuron pools innervating hindlimb muscles from this diverse set of data. The first and most striking fact is the strong similarities between the various sets of results in different mammalian species, described above, demonstrate how strongly this pattern has been conserved in evolution. This conclusion is strengthened when similar studies in other species such as chick (Landmesser, 1975; Hollyday, 1980) and Xenopus (Lamb, 1977) are examined, even when the inevitable uncertainties about muscle homologies are accepted. More recent studies have provided further evidence for the conservation of these patterns in forelimb (Ryan *et al.*, 1998). Many authors have tried to draw out conclusions about the functional significance of the pattern. Thus, Romanes (1951, 1964) has observed how there is a tendency for proximal muscles to be innervated by more rostral muscles, though this is far from absolute. He has also noted that flexors and extensors acting on the same joint tend to be in the same transverse plane. However, the key principle is that the relationship between motoneuron position within the lateral motor column and muscles they innervate is determined by the position that muscle occupied in the embryonic hindlimb, as has been shown to be the case for the forelimb motor pools.

Topography of remaining motoneuron pools in the lumbosacral spinal cord

Caudal to the motoneuron pools of the lumbosacral enlargement that innervate the hindlimb are motor neuronal groups innervating the muscles of the pelvic floor, perineum, muscles of the lower back, and the muscles that move the tail. With the exception of the motoneurons of the muscles of the perineum, these motoneuron nuclei have been less well-characterized than most others in the spinal cord.

Muscles of the perineum

The muscles of the perineum lie below the pelvic diaphragm and regulate the opening of the anus and urethra, and cause penile and clitoral erection. A review of the location of the motoneurons has been written by Holstege (2004).

Vanderhorst and Holstege (1997), in their study of the entire lumbosacral spinal cord have located the motoneurons of these muscles within a column of motoneurons in the sacral spinal cord lying ventral and slightly medial to the motoneurons innervating the intrinsic musculature of the foot. This is the nucleus Y of Romanes (1951), but is more commonly known as Onuf's nucleus, first described by Onufrowicz (1900). Similar results have been described in other species including man (Schrøder 1981), monkey (Roppollo *et al.*, 1985), and dog (Kuzuhara *et al.*, 1980). Vanderhorst and Holstege (1997) have shown that Onuf's nucleus in the cat is divided into a dorsomedial part containing motoneurons innervating the muscles of the external anal sphincter, and a ventrolateral part containing motoneurons innervating the external urethral sphincter and ischiocavernosus. Similar results have been obtained in man, monkey and dog. In the rat, a similar spatial relationship is seen between the motoneurons of the external anal and urethral sphincters but the two nuclei are entirely separated within the rat L6 and S1 segments (Schrøder, 1980; McKenna and Nadelhaft, 1986). In the rat Schrøder (1980) demonstrated that the ischicavernosus, bulbospongiosus, sphincter ani and sphincter urethrae were each innervated by two motoneuron populations in the spinal cord with different locations and different histochemical properties. Schrøder (1980) suggested that this difference in rat this was simply because these nuclei had become widely separated in this species while in other species such as cat the nuclei were located closer to one another. This apparent duality of location of motoneurons innervating one muscle or muscle group in one species but not others is similar to that observed in the neck muscles of the cat. In this species, such as the pig (Blok *et al.*, 1996) and the Mongolian gerbil (Ulibarri *et al.*, 1995), these motoneurons are located in a different part of the spinal cord gray matter close to the central canal.

There is some debate as to how these motoneurons should be classified because they exhibit some properties distinct from those of other somatic motoneurons. Thus, though they innervate striated muscles of the perineum they are spared in human motoneuron disease (Mannen, *et al.*, 1977) but degenerate in Shy-Dräger syndrome (Mannen *et al.*, 1982) and show distinct patterns of afferent input (Holstege, 2004) more characteristic of autonomic motoneurons and distinct neurochemical properties (Xu *et al.*, 2007), suggesting a class of motoneurons intermediate between the two (Holstege *et al.*, 2004) (Figure 7.8).

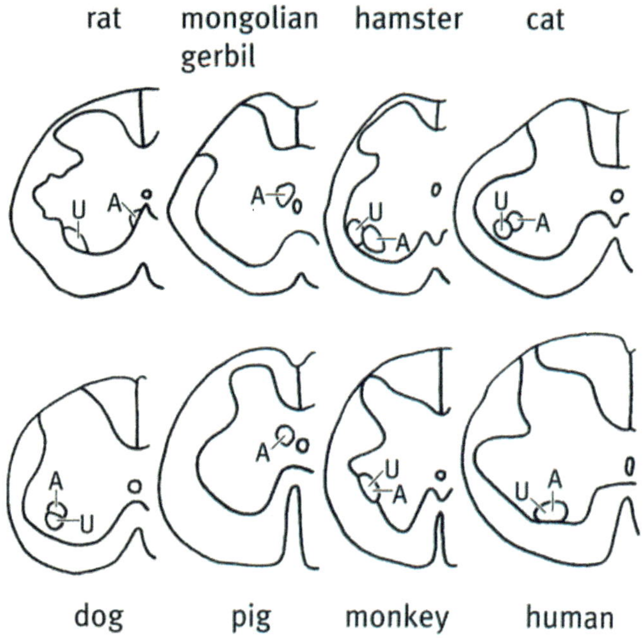

Figure 7.8 Onuf's nucleus in a range of different mammals

This diagram shows the motoneuron groups supplying the external urethral sphincter (U) and external anal sphincter (A) in a range of different mammals. (Adapted from Holstege, 2004).

Muscles of the pelvic diaphragm

The muscles of the pelvic floor are the levator ani and, when present, coccygeus. Vanderhorst and Holstege (1997) have identified the motoneurons innervating this muscle as distinct group more caudal than Onuf's nucleus, lying initially ventromedially in the S2 segment of the spinal cord, but the motoneurons are more scattered in more caudal levels. Similar results were obtained by Sato *et al.*, (1978) and Wada *et al.*, (1990).

Deep muscles of the back and tail

There have been relatively few studies of deep muscles of the lower back and tail. Vanderhorst and Holstege (1997), in their study of the cat, observed motoneurons innervating multifidi and extensor caudalis medialis in a dorsomedial location within the spinal cord; those of longissimus medialis and extensor caudalis lateralis were in a ventrolateral location. These authors cautioned that they had only injected parts of some muscles in this study.

A more comprehensive study of the location of muscles innervating the tail of the cat has been published by Wada *et al.*, (1990) (Figure 7.9). These authors show that motoneurons innervating these muscles are in the sacral and coccygeal spinal cord from S2 to Co3. The motoneurons innervating the extensor caudae medialis and lateralis muscles, responsible for tail raising or extension, are located in the dorsomedial and ventromedial columns of the cat spinal cord. In more caudal regions of the spinal cord these columns shift laterally in position. Motoneurons innervating the abductor caudae externalis and abductor caudae internalis muscles, responsible for bending the tail sideways or abducting it, are located centrally in the sacral and coccygeal spinal cord and overlap in their position. The motoneurons innervating flexor caudalis longus and flexor caudalis brevis are located respectively ventrolaterally and dorsomedially in the ventral horn in this region of the cat spinal cord.

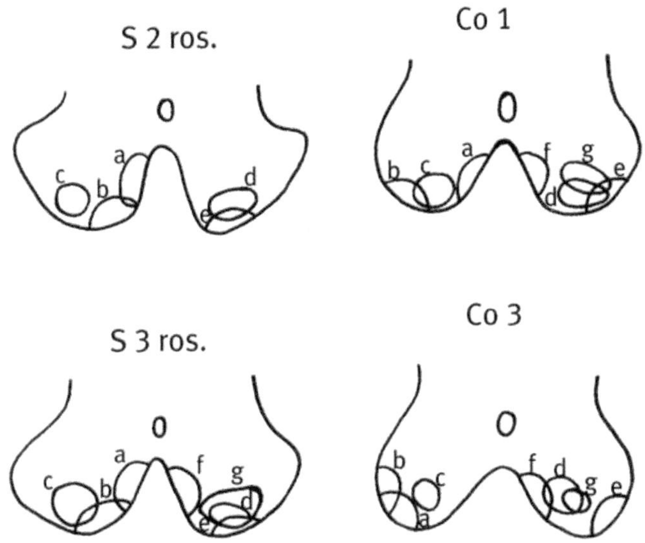

Figure 7.9 Motoneurons supplying the muscles of the tail

This diagram shows the motoneuron groups that supply the tail muscles in the cat. The motoneurons innervating the extensor caudae medialis and lateralis muscles are located in the motoneuron columns labelled (a) and (b) respectively. Columns (a) and (b) are medial at S2 but shift laterally in more caudal sections. Motoneurons innervating the abductor caudae externalis and abductor caudae internalis muscles are located in groups (c) and (d) respectively. The motoneurons innervating flexor caudalis longus and flexor caudalis brevis are located in (e) and (f) respectively. The motoneurons innervating the iliocaudalis and levator ani are located in group (g). (Adapted from Wada et al., 1990). (ros = rostral; S = sacral; Co = coccygeal).

Conclusions

Since the work of Romanes (1951) there have been many attempts to study the localization within the ventral horn of motoneurons innervating many of the muscles. Much or this work has focused on the key mammalian species of interest to laboratory investigators but data have been accumulated from other species, although this is often more fragmentary in nature. The results of these studies have shown that motoneuron localization projection patterns onto muscles have been highly conserved in evolution and that these projection patterns within the ventral horn of the spinal cord can be related to the positions of muscles in the embryo rather than to their positions in the adult body.

References

Abrahams VC, Keane J (1984) Contralateral, midline and commissural motoneurons of neck muscle: a retrograde study in the cat. J Comp Neurol 223, 448-456.

Aschoff A, and Holländer H (1982) Fluorescent compounds as retrograde tracers compared with horseradish peroxidise (HRP) I. A parametric study in the central visual system of the albino rat. J Neurosci Meth 6 179-197.

Augustine JR and White JF (1986) The accessory nerve nucleus in the baboon. Anat Rec 214, 321-320.

Baba H, Maezawa Y, Uchida K, Imura S, Kawahara N, Tomita K, Kudo M (1997) Three-dimensional topographic analysis of spinal accessory motoneurons under chronic mechanical compression: an experimental study in the mouse. J Neurol 244, 222-229.

Bakker DA, Richmond FJR, Abrahams VC (1984) Central projections from cat suboccipital muscles: a study using transganglionic transport of horseradish peroxidise. J Comp. Neurol 228, 409-421.

Berger AJ, Cameron WE, Averill DB, Kramis, RC, Binder MD (1984) Spatial distribution of phrenic and medial gastrocnemius motoneurons in the cat spinal cord. Exp Neurol 86, 559-575.

Blok BFM, Roukema G, Geerdes B, Holstege G (1996) Location of external anal sphincter motoneurons in the sacral cord of the female domestic pig. Neurosci Lett 216, 203-206.

Brichta AM, Callister RJ, Peterson EH (1987) Quantitative analysis of cervical musculature in rats: histochemical composition and motor pool organization. I. muscles of the spinal accessory complex. J Comp Neurol 255, 351-368.

Brink EE, Morrell JI, Pfaff DW (1979) Localization of lumbar epaxial motoneurons in the rat. Brain Res 170, 23-41.

Burke RE, Strick PL, Kanda K, Kim CC, Walmsley B (1977) Anatomy of medial gastrocnemius and soleus motor nuclei in the cat spinal cord. J Neurophys 40, 667-680.

Callister RJ, Brichta AM, Peterson EH (1987). Quantitative analysis of cervical musculature in rats: histochemical composition and motor pool organization. II. Deep dorsal muscles. J Comp Neurol 255, 369-385.

Chandrasekhar A. (2004) Turning heads: development of vertebrate branchiomotor neurons. Dev Dyn 229, 143-161.

Charlton CG, Crowell Jr B, Benson R (1988) Identification of motor neurons for accessory muscles of inspiration and expiration, pectoralis, trapezius and external oblique: comparison with non-respiratory skeletal muscle. Synapse 2, 219-224.

Choi JY, Hoover JE (1996) The organization of the acromiodeltoid and spinodeltoid motor nuclei in rat spinal cord. Brain Res 738, 146-149.

Clavenzani P, Scapolo PA, Callegari E, Barazzoni AM, Petrosino G, Lucchi ML, Bortolami R (1994) Motoneuron organization of the muscles spinal accessory complex of the sheep investigated with the fluorescent retrograde tracer technique. J Anat 184, 381-385.

Curfs MH, Gribnau AA, Dederen PJ (1993) Postnatal maturation of the dendritic fields of motoneuron pools supplying flexor and extensor muscles of the distal forelimb in the rat. Development 117, 535-541.

Devried KL, Goshgarian HG (1989) Spinal cord localization and characterization of the neurons which give rise to the accessory phrenic nerve in the adult rat. Exp Neurol 104, 88-90.

Fritz N, Illert M, Saggau P (1986a) Location of motoneurones projecting to the cat distal forelimb. I. Deep radial motornuclei. J Comp Neurol 244, 286-301.

Fritz N, Illert M, Reeh P (1986b) Location of motoneurones projecting to the cat distal forelimb. II. Median and ulnar motornuclei. J Comp Neurol 244, 302-312.

English AW, Wolf SL, Segal RL (1993) Compartmentalization of muscles and their motor nuclei: the partitioning hypothesis. Phys Ther 73, 857-867.

Gordon DC, Richmond FJR (1990) Topography in the phrenic motoneuron nucleus demonstrated by retrograde mulitiple-labeling techniques. J Comp Neurol 292, 424-434.

Gordon DC, Richmond FJR (1991) Distribution of motoneurons supplying dorsal suboccipital and intervertebral muscles in the cat neck. J Comp Neurol 304, 343-356.

Gordon DC, Loeb GE, Richmond FJR (2004) Distribution of motoneurons supplying cat sartorius and tensor fasciae latae, demonstrated by retrograde multiple-labeling methods. J Comp Neurol 304, 357-372.

Goshgarian HG, Rafols JA, (1981) The phrenic nucleus of the albino rat: a correlative HRP and Golgi study. J Comp Neurol 201, 441-456.

Gray H (1980) *Gray's Anatomy.* 36th ed. Williams PL, Warwick R (Eds in chief). Churchill Livingstone, Edinburgh.

Gutman CR, Ajmere MK, Hollyday M (1993) Organization of motor pools supplying axial muscles in the chicken. Brain Res 609, 129-136.

Hollyday M (1980) Organization of motor pools in the chick lumbar lateral motor column. J Comp Neurol 194, 143-170.

Hollyday M, Jacobson RD (1990) Location of motor pools innervating the chick wing. J Comp Neurol 302, 575-588.

Holstege G (2004) Central nervous system control of micturition. In: *The Rat Nervous System.* 3rd edn. Paxinos G (ed). Elsevier, Amsterdam.

Holstege G, van Neervan J, Evertse F (1987) Spinal cord location of the motoneurons innervating the abdominal, cutaneous maximus, latissimus dorsi and longissimus dorsi muscles in the cat. Exp Brain Res 67, 179-194.

Horner M, Kummel H (1993) Topographical representation of shoulder motor nuclei in the cat spinal cord as revealed by retrograde fluorochrome tracers. J Comp Neurol 335, 309-319.

Horscholle-Bossavit G, Jami L, Thiesson D, Zytnicki D (2004) Motor nuclei of peroneal muscles in the cat spinal cord. J Comp Neurol 277, 430-440.

Illert M, Fritz N, Aschoff A, Hollander H (1982) Fluorescent compounds as retrograde tracers compared with horseradish peroxidise. II. A parametric study in the peripheral motor system of the cat. J Neurosci Meth 6, 199-218.

Jenny AB, Inukai J (1983) Principles of organization of the monkey spinal cord. J Neurosci 3, 567-575.

Johnson SM, Getting PS (1988) Phrenic motor nucleus of the guinea pig: dendrites are bundled without clustering of cell somas. Exp Neurol 101, 208-220.

Kitamura S, Richmond FJR (1994) Distribution of motoneurons supplying dorsal and ventral suboccipital mucles in the feline neck. J Comp Neurol 347, 25-35.

Kitamura S, Sakai A (1982) A study on the localization of the sternocleidomastoid and trapezius motoneurons in the rat by means of the HRP method. Anat Rec 202, 527-536.

Kitamura S, Nishiguchi T, Okubo J, Chen KL, Sakai A (1986) An HRP study of the motoneurons supplying the rat hypobranchial muscles: central location, peripheral axon course and soma size. Anat Rec 216, 73-81.

Kuzuhara M, Kanazawa I, Nakanishi T (1980) Topographical localization of the Onuf's nuclear neurons innervating the rectal and vesical striated sphincter muscles: a retrograde fluorescent double labeling in cat and dog. Neurosci Lett 16, 125-130.

Lamb AH (1977) Retrograde axonal transport of horseradish peroxidase for determining motor projection patterns to the developing limb in Xenopus. Brain Res 134, 197-212.

Landmesser L (1978) The distribution of motoneurones in chick hindlimb. J Physiol 284, 391-414.

Larnicol N, Rose D, Marlot D, Duron B (1982) Spinal localization of the intercostal motoneurons innervating the upper thoracic spaces. Neurosci Lett 31, 13-18.

Liebram I, Agalliu D, Nagasawa T, Ericson J, Jessell T (2005) A Cxcl12-Cxcr4 chemokine signalling pathway defines the initial trajectory of mammalian motor axons. Neuron 47, 667-679.

Liinamaa TL, Keane J, Richmond FJR (1997) Distribution of motoneurons supplying feline neck muscles taking origin from the shoulder girdle. J Comp Neurol 377, 298-312.

Lipiski J, Martin-Body RL (1987) Morphological properties of respiratory intercostal motoneurons in cats as revealed by intracellular injection of horseradish peroxidise. J Comp Neurol 260, 423-434.

Mannen T, Iwata M, Toyokura Y, Nagashima K (1977) Preservation of a certain motoneurone group of the sacral cord in amyotrophic lateral sclerosis and its clinical significance. J Neurol Neurosurg Pyschiat 40, 464-469.

Mannen T, Iwata M, Toyokura Y, Nagashima K (1982) The Onuf's nucleus and the external anal sphincter muscles in amyotrophic lateral sclerosis and Shy-Drager syndrome. Acta Neuropathol 58, 255-260.

Matesz C and Szekely G (1983) The motor nuclei of the glossopharyngeal-vagal and the accessorius nerves in the rat. Acta Biol Hung 34, 215-229.

McHanwell S, Biscoe TJ (1981a) The localization of the motoneurons supplying the hindlimb muscles of the mouse. Phil Trans R Soc B 293, 477-508.

McHanwell S, Biscoe TJ (1981b) The sizes of motoneurons supplying the hindlimb muscles in the mouse. Proc R Soc Lond B 213, 201-216.

McKenna KE, Nadelhaft I (1986) The organization of the pudendal nerve in the male and female rat. J Comp Neurol 248, 532-549.

McKenna KE, Prusky GT, Whishaw IQ (2000) Cervical motoneuron topography reflects proximodistal organization of muscles and movements of the rat forelimb: a retrograde carbocyanine dye analysis. J Comp Neurol 419, 286-296.

Nicolopoulos-Stournas S, Iles JF (1993) Motor neuron columns in the lumbar spinal cord of the rat. J Comp Neurol 217, 27-85.

Onufrowicz B (1900) On the arrangement and function of the cell groups of the sacral spinal cord in man. Arch Neurol Psychopathol 3, 387-411.

Pabst O, Rummelies J, Winter B, Arnold H-H (2003) Targeted disruption of the homeobox gene Nkx2.9 reveals a role in development of the spinal accessory nerve. Development 130, 1193-1202.

Parry DJ, McHanwell S, Haas N (1982) The numbers and sizes of motoneurones in the soleus motornucleus of the normal and dystrophic (C57BLdy2J/dy2J) mouse. Exp Neurol 75, 743-754.

Rapoport S. (1978) Location of sternocleidomastoid and trapezius motoneurons in the cat. Brain Res 156, 339-344.

Richmond FJR, Scott DA, Abrahams VC (1978) Distribution of motoneurones to biventer cervicis, splenius and complexus in the cat. J Comp Neurol 181, 451-464.

Rikard-Bell GC, Bystrzycka EK (1980) Localization of phrenic motor nucleus in the cat and rabbit studied with horseradish peroxidase. Brain Res 194, 479-483.

Romanes GJ (1951) The motor cell columns of the lumbosacral spinal cord of the cat. J Comp Neurol 94, 313-363.

Romanes GJ (1964) The motor pools of the spinal cord. In: *Progress in Brain Research. Vol 11, Organization of the Spinal Cord.* Eccles JC, Schadé JP (eds). Elsevier, New York.

Roppollo JR, Nadelhaft I, deGroat WC (1985) The organization of pudendal motoneurons and primary afferent projections in the spinal cord of the rhesus monkey revealed by horseradish peroxidise. J Comp Neurol 234, 475-488.

Routal RV, Pal GP (1999) A study of motoneuron groups and motor columns of the human spinal cord. J Anat 195, 211-224.

Routal, RV, Pal GP (2000) Location of the spinal nucleus of the accessory nerve in the human spinal cord. J Anat 196, 263-268.

Ryan JM, Cushman J, Jordan B, Samuels A, Frazer H, Baier C (1998) Topographic position of forelimb motoneuron pools is conserved in vertebrate evolution. Brain Behav Evol 51, 90-99.

Sato M, Mizuno N, Konishi A (1978) Localization of motoneurons innervating perineal muscles: a HRP study. Brain Res 140, 149-154.

Satomi H, Takahashi K, Aoki M, Kasaba T, Kurosawa Y, Otsuka K (1985) Localization of the spinal accessory motoneurons in the cervical cord in connection with the phrenic nucleus: an HRP study in cats. Brain Res 344, 227-230.

Schrøder HD (1980) Organization of the motoneurons innervating the pelvic muscles of the male rat. J Comp Neurol 192, 567-587.

Schrøder HD (1981) Onuf's nucleus X: a morphological study of a human spinal nucleus. Anat Embryol 162, 443-453.

Schubert W, Kaprielian Z (2001) Identification and characterization of a cell surface marker for embryonic rat spinal accessory motor neurons. J Comp Neurol 439, 368-383.

Sharrard WJW (1955) The distribution of the permanent paralysis in the lower limb in poliomyelitis. A clinical and pathological study. J Bone Joint Surg 37B, 540-558.

Smith CL, Hollyday M (1983) The development and organization of motor nuclei in the rat spinal cord. J Comp Neurol 220, 16-28.

Sprague JM (1948) A study of motor cell localization in the spinal cord of the rhesus monkey. Am J Anat 82, 1-25.

Sterling P, Kuypers HGJM (1966) Anatomical organization of the brachial spinal cord of the cat. II. the motoneuron plexus. Brain Research 4, 16-32.

Stracznicky C, Tay D (1982) Location of motoneuron pools innervating wing muscles in the chick. Anat Embryol 166, 209-218.

Strauss WL, Howell AB (1936) The spinal accessory nerve and its musculature. Quart Rev Biol 11, 387

Takahashi K, Ninomiya T (1985) Observations on the fine structure of the phrenic nucleus in the cervical spinal cord of the cat with special reference to its dendritic bundles. J Anat 140, 491-498.

Tani M, Kida MY, Akita K (1994) Relationship between the arrangement of motoneuron pools in the ventral horn and ramification pattern of the spinal innervating trunk muscles in the cat *(Felis domestica)*. Exp Neurol 128, 290-300.

Theriault E, Diamond J (1988) Intrinsic organization of the rat cutaneous trunci motor neucleus. J Neurophysiol 60, 463-477.

Ulibarri C, Popper P, Micevych PE (1995) Motoneurons dorsolateral to the central canal innervate perineal muscles in the Mongolian gerbil. J Comp Neurol 222, 225-237.

Ullah M, Mansor O, Ismail ZIM, Kapitonova MY, Sirajudeen KNS (2007) Localization of the spinal nucleus of the accessory nerve in rat: a horseradish peroxidase study. J Anat 210, 428-438.

Uevama T, Satoda T, Tashiro T, Sugimoto T, Matsushima R, Mizuno N (1990) Infrahyoid and accessory motoneurons in the Japanese monkey (Macaca fuscata). J Comp Neurol 291, 373-382.

Vanderhorst VGJM, Holstege G (1997) Organization of lumbosacral motoneuronal cell groups innervating hindlimb, pelvic floor, and axial muscles in the cat. J Comp Neurol 382, 46-76.

Wada N, Sugita S, Kolblinger G (1990) Spinal cord location of the motoneurons innervating the tail muscles of the cat. J Anat 173, 101-107.

Watanabe T, Ohmori Y. (1988) Location of motoneurons supplying upper neck muscles in the chicken studied by means of horseradish peroxidase. J Comp Neurol 270, 271-278.

Watson C, Tvrdik P (unpublished) Spinal accessory motoneurons in the mouse express Phox2b. Unpublished data. The University of Utah.

Xu C, Giuliano F, Sun XQ, Brisorgueil M-J, Lecclerc P, Vergé D, Conrath M (2007) Serotonin 5-HT2A and 5-HT5A receptors are expressed by different motoneuron populations in rat's Onuf's nucleus. J Comp Neurol 502, 620-634.

Yoshida T, Miyazaki T, Hirano M, Shin A, Kanaseki T (1983) Topographic arrangement of motoneurons innervating the suprahyoid and infrahyoid muscles. Ann Otol Rhinol Laryngol 92, 259-266.

8 Spinal Autonomic Preganglionic Neurons: the visceral efferent system of the spinal cord

Colin R Anderson, Janet R Keast and Elspeth M McLachlan

Visceral efferent pathways

The spinal cord contains the output neurons that carry signals from the nervous system to the peripheral targets – skeletal muscle fibers are innervated by motoneurons lying in the ventral horns (Chapter 7) whereas all other effector tissues (cardiac and smooth muscles, exocrine and some endocrine glands, fat, immune cells, etc.) receive input from autonomic pathways. In the spinal cord, the pathways of the sympathetic nervous system arise in the thoracic and upper lumbar segments while the parasympathetic pathways controlling the pelvic organs arise in the lowest lumbar and sacral segments. There are no autonomic neurons in the cervical and lumbar enlargements where the many motoneurons supplying the limbs are located. The segmental levels of the spinal autonomic outflow vary a little between species (see Table 8.1).

The autonomic neurons within the cord are preganglionic neurons while neurons in peripheral ganglia are postganglionic neurons. Like skeletal motorneurons, the axons of preganglionic neurons (at least in mammals) leave the cord via the ventral roots. Both types of efferent neurone are cholinergic and at their first synapses activate nicotinic acetylcholine receptors. However the autonomic pathways differ from skeletomotor pathways because each axon leaving the spinal cord does not directly contact cells of the effector tissue but forms synapses with multiple peripheral neurons that are clustered together in autonomic ganglia. The number of peripheral neurons that are contacted varies from a few in some parasympathetic ganglia to hundreds in the sympathetic ganglia of larger mammals including humans. This allows the central command signals to be distributed via suprathreshold connections to multiple functionally appropriate postganglionic neurons innervating the effector tissues.

Table 8.1 Levels of autonomic outflow from spinal cord in different species.

Species	Sympathetic	Parasympathetic
Mouse	T1-L2	L6-S1
Rat	T1-L3	L6-S1
Guinea Pig	T1-L4	S1-3
Rabbit	T1-L5	S2-4
Cat	T1-L5	S2-3
Dog	T1-L5	S1-3
Monkey	C8-L4	S1-3
Human	T1-L3	S2-4

The caudal limit of the sympathetic outflow varies over a segment's length and the segmental level shown is the most caudal segment in which sympathetic preganglionic neurones may be found, usually in the most rostral part. Most of these mammalian species have 13 thoracic segments but primates have only 12.

Spatial distribution of preganglionic neurons

The cell bodies of autonomic preganglionic neurons lie in the intermediate zone of the spinal cord. They are mainly clustered together in clumps at the edge of the intermediate gray matter, forming two intermediolateral columns (IML). During development, the neurons migrate dorsolaterally from the central canal; a few end up in the dorsolateral funiculus (the 'lateral horn') whereas others trail behind to remain in the intercalated nucleus (IC) or central autonomic nucleus (CA, see Figure 8.3). The clumps of neurons link to the central autonomic nucleus to form horizontal 'ladders' made up primarily of preganglionic dendrites and the terminals of inputs to them (see below). The average interval between 'ladder rungs' is ~250 µm in the rat (Anderson *et al.*, 1989) and ~300 µm in the cat (Krukoff *et al.*, 1985a). These ladders are prominent in the thoracolumbar cord and much less obvious in the sacral cord.

The distribution of preganglionic neurons within the cord can be revealed by immunohistochemical staining for choline acetyltransferase, which also demonstrates skeletal motorneurons and cholinergic interneurons, or for the neuronal isoform of nitric oxide synthase (nNOS) which is present in 65-80% of sympathetic preganglionic neurons in the rat (Anderson, 1992), depending on the segmental level. Many rat parasympathetic preganglionic neurons also contain nNOS (Burnett *et al.*, 1995), but the proportion is unknown.

The location of preganglionic somata can also be identified by retrograde labeling with tracers via their axons in the periphery. Neurons with their terminals outside the CNS (i.e. skeletal motoneurons and preganglionic neurons) can be labeled after injection of the fluorescent tracer, FluoroGold, into the systemic circulation (Anderson and Edwards, 1994). Injection of tracers into specific peripheral ganglia or applied to transected white rami communicantes, parts of the paravertebral chain or splanchnic nerves has revealed the location of the cell bodies of subpopulations of preganglionic neurons, particularly in the thoracic (Oldfield and McLachlan, 1981; Anderson *et al.*, 1989; Sah and McLachlan, 1995), lumbar (Jänig and McLachlan, 1986a; Strack *et al.*, 1988; Anderson *et al.*, 1989) and sacral (Nadelhaft *et al.*, 1980; Nadelhaft and Booth, 1984) segments of

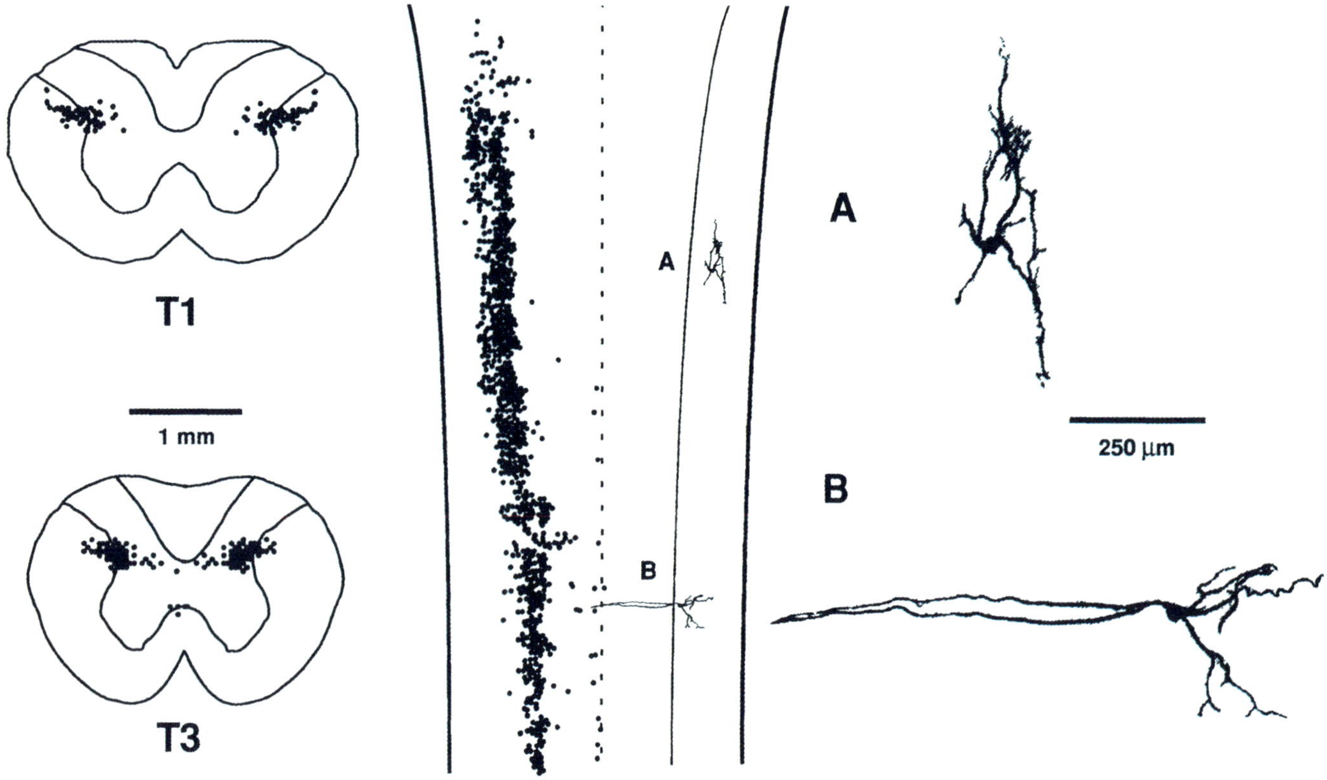

Figure 8.1 Sympathetic preganglionic neurons in the upper thoracic spinal cord of the rat

Plots to the left and in the middle show reconstructions of the location of preganglionic somata in T1 to T3 spinal cord in transverse and horizontal section, respectively On the right are two preganglionic neurones filled with biocytin during intracellular recording in horizontal slices *in vitro.* The location of these cells is illustrated at A (in the lateral funiculus) and B (in the IML oriented medially) in the middle diagram Adapted from from Sah and McLachlan (1995)

the cat and rat (Figure 8.1). The preganglionic cell bodies projecting to different targets are topographically organized and form distinct columns that resemble those of the motorneurons supplying different skeletal muscles (Appel and Elde, 1988; Pyner and Coote, 1994). Reconstructions of the longitudinal distribution of preganglionic neurons projecting in different pathways have been made from horizontal sections in several species (cat (Baron *et al.*, 1985; Jánig and McLachlan, 1986a, b); rabbit, (Anderson *et al.*, 1987); rat, (Anderson *et al.*, 1989; Sah and McLachlan, 1995) (Figure 8.1). The major sites where a general functional distinction exists are in the lumbar cord where the IML and lateral funiculus (LF) contain neurons projecting in the paravertebral chain and supplying hindlimb blood vessels (also sweat glands and piloerector muscles) whereas the central autonomic area (CA) and intercalated nucleus (IC) project in the lumbar splanchnic nerves to the pelvic viscera. This has been demonstrated in rat, guinea pig and cat (Jänig and McLachlan, 1986a; Jänig and McLachlan, 1987; Baron and Jänig, 1991).

Preganglionic neurons receive segmental inputs via interneurons in the dorsal horn and intermediate zone, as well as direct projections from the medulla, pons and hypothalamus and other spinal segments (propriospinal inputs). Many descending inputs make connections not only ipsilaterally but also contralaterally after crossing the cord along the "ladder rungs". Thus preganglionic neurons integrate information of primary afferent, propriospinal and supraspinal origin to determine the pattern of activity leaving the cord in each functional pathway. The segmental outflow to different target organs is summarized in Figure 8.2.

Morphology of preganglionic neurons and arrangement of their dendrites

The morphology of individual preganglionic neurons has been revealed by intracellular staining using dyes such as horseradish peroxidase (Dembowsky *et al.*, 1985; Inokuchi *et al.*, 1993), biocytin (Sah and McLachlan, 1995) or neurobiotin (Morgan *et al.*, 1993), or by retrograde labeling with cholera toxin subunit B which also demonstrates the dendritic processes (Hosoya *et al.*, 1991). The cell bodies are fusiform, ovoid or triangular (20 to 30 µm long) with 2-8 primary dendrites that branch only a few times and extend up to >1 mm (Light and Metz, 1978; Dembowsky *et al.*, 1985; Sah and McLachlan, 1995). Some examples in the upper thoracic spinal cord of the rat are shown in Figure 8.1. The somata of parasympathetic neurons in the sacral segments are smaller than those of sympathetic neurons

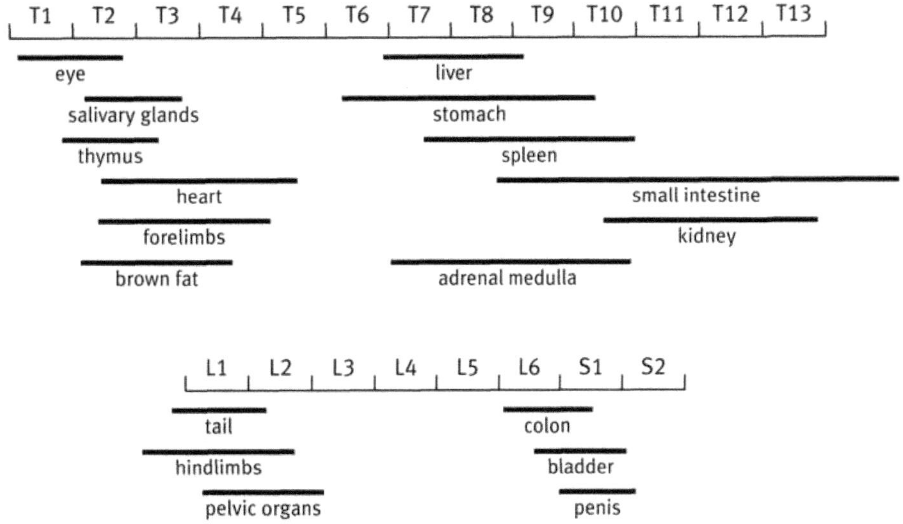

Figure 8.2 Segmental outflow from the spinal cord involved in functional control of some major target organs in the periphery

The bars indicate levels of the cord which either generate the greatest functional effects when their ventral roots are stimulated or contain the largest proportion of the preganglionic cell bodies identified by retrograde labelling (adrenal medulla) or transneuronal labelling with viral tracers injected into the target organ.

in the more rostral segments. The dendrites of preganglionic neurons lying laterally in the IML extend mainly rostrocaudally along the IML or (less frequently) dorsolaterally into the lateral funiculi, while some that lie medially in the IML extend their processes medially. The dendrites of parasympathetic neurons in the IML of sacral segments extend around the lateral edge of the dorsal horn where they intermingle with the terminations of primary afferent axons (Light and Metz, 1978; Nadelhaft and Booth, 1984). The axon usually arises from one of the dendritic processes and passes around the lateral edge of the ventral horn to join the ventral root. Axon collaterals can form synapses in many laminae, particularly in the sacral spinal cord (Morgan *et al.*, 1991).

Definitive evidence is not available at this stage as to whether or not preganglionic neurons receive direct monosynaptic inputs of primary afferent origin as suggested by the close proximity of some primary afferent terminals to their cell bodies and dendrites (Light and Metz, 1978). The general consensus is that there are not. This view comes primarily from electrophysiological evidence based on assumed values for conduction velocity within the cord, synaptic delay, etc. Direct monosynaptic contacts have not been demonstrated morphologically.

Sympathetic preganglionic neurons

A summary of the arrangement of preganglionic neurons along the thoracolumbar cord is shown in Figure 8.3. The cell bodies of sympathetic preganglionic neurons in the thoracolumbar cord lie in distinct subnuclei, described originally by Petras and Cummings in Rhesus monkeys (Petras and Cummings, 1972). The majority of neurons lie in the IML or nucleus thoracolumbalis intermediolateralis pars principalis, a compact, longitudinally-oriented nucleus at the boundary of the gray matter and the lateral funiculus on each side. About 95% of preganglionic neurons lie in the IML. The IML extends into the lateral horn, becoming the nucleus intermediolateralis thoracolumbalis pars funicularis, formed by flattened neurons lying in the lateral funiculus (LF) with their dendrites oriented transversely. Dendrites of preganglionic neurons can be found touching the pia. This nucleus is most prominent in the rostral thoracic and, to a lesser extent, in the upper lumbar segments. However, even in these segments, only about 2% of neurons lie in the LF. Between these regions, the IML is triangular in transverse section in the middle of the clumps of neurons, with the base of the triangle on the border with the lateral funiculus and the apex facing the central canal. However, in the regions between

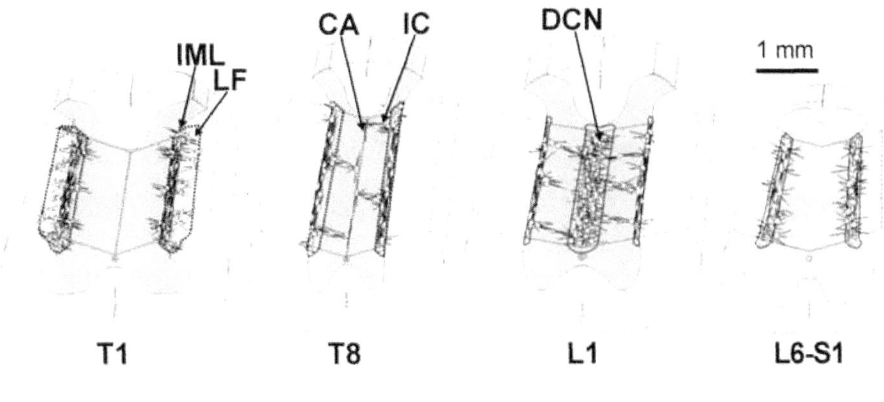

Figure 8.3 Distribution of preganglionic neuron somata in the rat spinal cord

This figure shows the distribution of preganglionic neurons at three sympathetic (T1, T8 and L1) levels and in the sacral parasympathetic nucleus (L6-S1). Cell bodies with dendrites lie in the subnuclei (IML = intermediolateral column; LF = lateral funiculus; CA = central autonomic area; IC = intercalated nucleus; DCN = dorsal commissural nucleus). Note the characteristic clusters at intervals along the IML separated by medially projecting "ladder like" structures. The DCN is only present in the upper lumbar segments.

Table 8.2 Substances present in sympathetic preganglionic somata and/or their terminals in autonomic ganglia.

Substance	Species	Reference
Atrial natriuretic factor	Rat	Debinski *et al.*, 1988
Bombesin/Gastrin releasing peptide	Rat	Helen *et al.*, 1984
Calbindin	Rat	Grkovic and Anderson, 1997
Calcitonin gene-related peptide CGRP	Rat	Senba and Tohyama, 1988
	Guinea pig	Gibbins, 1992
	Cat	Samano *et al.*, 2006
Calretinin	Rat	Grkovic and Anderson, 1995
	Cat	Edwards *et al.*, 1995
Cholecystokinin	Rat	Schröder, 1983
Choline acetyltransferase	Rat	Barber *et al.*, 1984
	Guinea pig	Lindh *et al.*, 1986
	Cat	Samano *et al.*, 2006
Cocaine and amphetamine regulated transcript	Rat	Burman *et al.*, 2004; Fenwick *et al.*, 2006
	Guinea pig	Locknar *et al.*, 2002
Corticotropin-releasing factor CRF	Rat	Mercanthaler *et al.*, 1983
	Cat	Krukoff, 1986 ; Shafton *et al.*, 1992
Cyclic guanosine monophosphate	Rat	Powers-Martin *et al.*, 2006
Growth associated protein-43 GAP-43	Rat	Michael and Priestley, 1995
	Cat	Shafton *et al.*, 1992
Glutaminase	Rat	Senba *et al.*, 1991
Opioid peptides	Rat	Sasek *et al.*, 1984; Kondo *et al.*, 1985; Häppöla *et al.*, 1987; Folan and Heym, 1989; Forehand, 1989; Kanagawa-Terayama *et al.*, 1989; Forehand, 1991; Soinila *et al.*, 1991; Chanthaphavong *et al.*, 2003
	Guinea pig	Lindh *et al.*, 1988 ; Gibbins, 1992; Colombo-Benkmann *et al.*, 1995
	Cat	Krukoff *et al.*, 1985b; Morales *et al.*, 1995; Samano *et al.*, 2006
Neuronal nitric oxide synthase nNOS	Rat	Anderson, 1992
	Mouse	Brüning, 1992
	Guinea pig	Anderson *et al.*, 1995
	Cat	Vizzard *et al.*, 1994
Neurotensin	Cat	Krukoff *et al.*, 1985b
Neurokinin 1 receptor	Rat	Grkovic and Anderson, 1996
Parvalbumin	Rat	Endo and Onaya, 1988
Pituitary adenylate cyclase activating polypeptide PACAP	Rat	Klimaschewski *et al.*, 1996a
	Guinea pig	Locknar *et al.*, 2002
Secretoneurin	Rat	Klimaschewski *et al.*, 1996b
Somatostatin	Cat	Krukoff *et al.*, 1985b
Substance P	Rat	Heym *et al.*, 1993
	Guinea pig	Gibbins, 1992
	Cat	Krukoff *et al.*, 1985b
Vasoactive intestinal peptide VIP/peptide histidine isoleucine PHI	Rat	Baldwin *et al.*, 1991
	Guinea pig	Gibbins, 1992
	Cat	Krukoff, 1986

the clumps, transverse sections may contain few or no preganglionic neurons.

The intercalated nucleus (IC or nucleus intercalatus spinalis) extends along each 'ladder rung' from the medial margin of the IML dorsal to the central canal. Dendrites from the IML and IC form an open plexus above the central canal together with dendrites from the contralateral preganglionic neurons. A few preganglionic neurons are also located immediately above and beside the central canal in an area (in Lamina X) known as the central autonomic area, (CA-Cummings, 1969) or nucleus intercalatus pars paraependymalis, (ICpe-Petras and Cummings, 1972). Less than 4% of preganglionic neurons lie in the IC and central autonomic area.

In rodents, the central autonomic area contains relatively few neurons from T1 to T12 (Anderson *et al.*, 1989) and should be distinguished from a densely packed dorsal commissural nucleus (DCN) of preganglionic neurons in L1-2 (Hancock and Peveto, 1979). Neurons in this nucleus are preferentially labeled from the hypogastric nerve or, in males, the main pelvic ganglion (Hancock and Peveto, 1979; Strack *et al.*, 1988). The DCN is also present in the upper lumbar cord of hamsters (Hancock, 1982), guinea pigs (McLachlan *et al.*, 1985) and rabbits (Beyer and Gonzalez-Mariscal, 1994) but not cats (Baron *et al.*, 1985). In the rat, 20-30% of preganglionic neurons in L1/2 lie in the DCN (Strack *et al.*, 1988; Grkovic and Anderson, 1996).

Thus sympathetic preganglionic neurons are distributed across the intermediate zone of the thoracolumbar spinal cord, from the white matter on one side, through the gray matter above the central canal, to the contralateral white matter. The great majority lie in the IML in the thoracic segments in most species.

Parasympathetic preganglionic neurons

The cell bodies of parasympathetic preganglionic neurons in the sacral cord are also distributed across the intermediate zone of 2-3 segments, forming a compact grouping. In rats, the majority that are involved in regulating the pelvic viscera are located in the IMLs (Nadelhaft and Booth, 1984). In cats, the neurons on each side are aligned to form a V-shaped longitudinal column intermingled with ascending projection neurons (Lu *et al.*, 1993). In horizontal sections from the more rostral part of this column, the 'dorsal band' lies medially (ventrolateral to the dorsal commissural nucleus, see below) and is separated from the 'lateral band' that extends around the dorsolateral edge of the ventral horn (Nadelhaft *et al.*, 1980). However the IC and CA subnuclei are much less prominent in the sacral segments and "ladder rungs" cannot readily be distinguished.

There is some functional topography in the sacral cord. Bladder neurons in the rat lie slightly more caudally than penis-regulating preganglionic neurons (Banrezes *et al.*, 2002). In the cat, the colonic neurons lie in the 'dorsal band', a group of preganglionic neurons that lies medially from the bladder projecting neurons in a 'lateral band'. The lateral band contains 65% of all sacral preganglionic neurons (Nadelhaft *et al.*, 1980). The functional arrangement of these pathways is described in several reviews (e.g. de Groat *et al.*, 1981; Jänig and McLachlan, 1987).

Table 8.3 Putative markers of preganglionic terminals within the pelvic ganglion.

Substance	Species	Reference
Atrial natriuretic factor	Rat	Papka *et al.*, 1987
Cocaine- and amphetamine-regulated transcript (CART)	Rat	Fenwick *et al.*, 2006
Cholecystokinin (CCK)	Rat	Papka *et al.*, 1987; Keast, 1991; Keast, 1994 *
Calcitonin gene-related peptide (CGRP)	Rat	Papka and McNeill, 1993
Galanin	Rat	Keast, 1991; Papka and McNeill, 1993; Keast, 1994 *
Neuronal nitric oxide synthase (nNOS)	Rat	Anderson *et al.*, 1993
Neuropeptide Y (NPY)	Guinea pig	Morris and Gibbins, 1987
Opioid peptides	Rat	Dail *et al.*, 1986; Wang *et al.*, 1990; Keast, 1991; Keast, 1994 *
	Mouse	Rogers and Henderson, 1990
	Guinea pig	Morris and Gibbins, 1987
	Cat	Kawatani *et al.*, 1983; Kawatani *et al.*, 1989
Secretoneurin	Rat	Li *et al.*, 1998
Somatostatin	Rat	Keast, 1991; Keast, 1994
Substance P	Rat	Dail and Dziurzynski, 1985; Papka and McNeill, 1993
	Guinea pig	Morris and Gibbins, 1987

It should be noted that a prominent cylindrical dorsal commissural nucleus (also known as the dorsal gray commissure) is located in the sacral cord between the central canal and the dorsal corticospinal tract, making the sacral segments easy to distinguish from the lumbar ones in which there is very little gray matter dorsal to the central canal. This nucleus consists mainly of second order afferent neurons involved in pelvic organ reflexes, as revealed by c-fos activation and viral tracing studies (Birder and de Groat, 1993; Marson *et al.*, 1993; Marson, 1997; Vizzard, 2000; Lu *et al.*, 2001; Marson *et al.*, 2003). It is located dorsal to many dendritic extensions of the parasympathetic preganglionic neurons that traverse the cord.

Chemistry of preganglionic neurons

As mentioned above, all preganglionic neurons are cholinergic and many express nNOS. Thus they can be demonstrated with immunohistochemical staining for nNOS or choline acetyltransferase (Figure 8.4). In addition, they can express a diverse array of neuropeptides, amines and other substances. Peptides are often not readily detected by immunohistochemistry in the somata of central neurons that contain their mRNA, presumably due to low levels of storage, but can be demonstrated when axonal transport is blocked after colchicine application. Neuropeptides can readily be demonstrated in preganglionic terminals in peripheral ganglia (Gibbins, 1995). Tables 8.2 and 8.3 provide sources of data about neuropeptides and other substances reported to be present in preganglionic neurons of several species. These data have been obtained from animals treated to block retrograde transport or the substances have been localized to varicose terminals surrounding neurons in autonomic ganglia, visualised using immunohistochemistry. In the pelvic ganglion, for most of these substances, it is not known if these chemical classes of terminals arise from preganglionic neurons in lumbar or sacral segments, or both (asterisks indicate where lesion studies on hypogastric and pelvic nerves have shown that these terminals originate exclusively from sacral levels). Some of these substances may be located in collaterals of sensory fibers arising from lumbosacral dorsal root ganglia (e.g. substance P, calcitonin gene-related peptide (CGRP) or in non-varicose postganglionic axons, but the appropriate nerve lesion studies have not yet been performed to ascertain their origin.

Chemistry of synaptic inputs to preganglionic neurons

Preganglionic neurons receive and integrate information that arises in primary afferent axons, segmental interneurons, other propriospinal neurons (e.g. lumbosacral and sacrolumbar reflex connections involved in the regulation of micturition and defecation) and supraspinal neurons, particularly in the particular nuclei of the brainstem and hypothalamus. Many of the terminals of these inputs contain monoamines or neuropeptides that are readily identified by

Table 8.4 Substances likely to be in bulbospinal terminals associated with preganglionic neurones.

Substance	Species	Reference
5-hydroxytryptamine serotonin	Rat	Anderson *et al.*, 1989; Tang *et al.*, 1998
	Cat	Krukoff *et al.*, 1985a
Adrenaline	Rat	Bernstein-Goral and Bohn, 1988; Yaici *et al.*, 2002
	Guinea pig, rabbit	McLachlan *et al.*, 1989
Angiotensin	Rat, guinea pig	Galabov *et al.*, 1990
Calcitonin gene-related peptide (CGRP)	Rat, mouse, guinea pig, cat, monkey, human, etc.	Gibson *et al.*, 1984a
Corticotropin-releasing factor (CRF)	Rat	Puder and Papka, 2001
Enkephalins	Rat	Romagnano *et al.*, 1987; Llewellyn-Smith *et al.*, 2005
Galanin	Rat, pig	Ch'ng *et al.*, 1985
Neuropeptide Y	Rat, guinea pig, cat	Gibson *et al.*, 1984c
Noradrenaline	Rat	Westlund *et al.*, 1983; Yaici *et al.*, 2002
	Cat	McLachlan and Oldfield, 1981
Oxytocin	Rat	Hosoya *et al.*, 1995; Tang *et al.*, 1998
Substance P	Rat	Newton *et al.*, 1988
	Rabbit, cat monkey	Oldfield *et al.*, 1985
Vasoactive intestinal polypeptide (VIP)	Rat, mouse, guinea pig, cat, monkey, etc.	Gibson *et al.*, 1984b

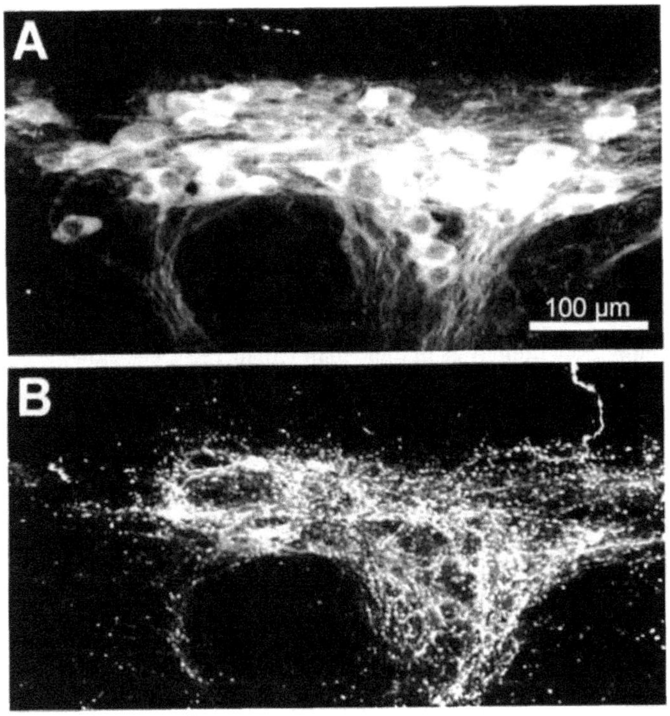

Figure 8.4 Chemistry of preganglionic neurones and their synaptic inputs

A: Somata of preganglionic neurones demonstrated in a horizontal section of T5 of the rat using immunohistochemistry for choline acetyltransferase (ChAT).
B: In the same section, terminals immunopositive for cocaine- and amphetamine-regulated transcript (CART) surround most of the neurons in this cluster. Horizontal section oriented with the midline below. These terminals originate in the rostroventrolateral medulla (Burman *et al.*, 2004). The calibration bar applies to both images.

immunohistochemistry. Where rostrocaudal comparisons have been made, peptide levels are much higher in sacral cord than in the thoracolumbar cord. In sections of spinal cord, the arrangement of the dendrites of the preganglionic neurons is mirrored by the pattern displayed by the terminals of their inputs (Figure 8.4) (Anderson *et al.*, 1989; Hosoya *et al.*, 1991). Table 8.4 provides sources of data about the diversity of terminals detected in the autonomic regions of the cord.

Spinal cord transection has helped to identify the supraspinal contribution to these terminals (Giuliano and Rampin, 2004; Llewellyn-Smith *et al.*, 2006). A great many of these inputs disappear, revealing the intraspinal and primary afferent contributions. For example, many terminals immunoreactive for substance P, neuropeptide Y, enkephalin or vasoactive intestinal polypeptide persist after more rostral cord transection and could arise from either more local sources (Gibson *et al.*, 1984b; Romagnano *et al.*, 1987; Cassam *et al.*, 1997; Llewellyn-Smith *et al.*, 2006). Further, the remaining pathways can become remodelled, expressing different substances and forming novel connections under the influence of neurotrophins released from reactive glia (Cassam *et al.*, 1997, 1999; Krenz and Weaver, 1998). This means that, more than a few days after lesions, the distribution of terminals of a particular chemical class may not be representative of the arrangement in the intact cord and will need to be interpreted carefully.

The effects of dorsal rhizotomy and capsaicin treatment have indicated that many of the peptidergic terminals in the vicinity of sacral preganglionic neurons have a primary afferent origin. These include calcitonin gene-related peptide (Gibson *et al.*, 1984a), vasoactive intestinal polypeptide (Gibson *et al.*, 1984b), galanin (Ch'ng *et al.*, 1985) and neuropeptide Y (Gibson *et al.*, 1984c). There is a small loss of axons containing these substances after cord lesion, indicating that some originate from supraspinal levels. This is supported by evidence showing expression of these substances in spinally-projecting brainstem neurons (Sasek *et al.*, 1990; Minson *et al.*, 1994; Stornetta *et al.*, 1999, 2001).

Descending catecholamine and serotonin projections arise exclusively from supraspinal sources (McLachlan and Oldfield, 1981; Anderson *et al.*, 1989), as they disappear following rostral spinal cord transection. Catecholaminergic terminals are relatively sparse in the sacral cord but form functional synapses on preganglionic neurons whereas serotonergic terminals may have a more indirect action (Ranson *et al.*, 2006). Oxytocin terminals associated with sacral preganglionic neurons are also of supraspinal origin (Tang *et al.*, 1998). After spinal transection, some monoamine enzymes and peptides reappear below the lesion too early to have regenerated, so it is concluded that these substances are expressed *de novo* by intrinsic spinal neurons (Cassam *et al.*, 1997).

It is not known which of these various populations of nerve terminals closely associated with preganglionic neurons form functional synapses. Physiological studies have demonstrated excitatory glutamate and inhibitory gamma-amino-butyric acid (GABA) and glycine inputs to preganglionic neurons (Inokuchi *et al.*, 1992a, b) and all synapses have been shown ultrastructurally to contain at least one amino acid transmitter (Llewellyn-Smith *et al.*, 1998). Slow post-synaptic potentials have also been recorded, some of which appear to be catecholaminergic (Yoshimura *et al.*, 1987a, b). Many synapses on preganglionic neurons contain both an amino acid and a monoamine or neuropeptide.

Autonomic spinal cord circuitry includes interneurons presynaptic to preganglionic neurons (Schramm, 2006). Interneurons can be visualized by neurotrophic virus injections into autonomic ganglia (Strack *et al.*, 1989; Joshi *et al.*, 1995; Clarke *et al.*, 1998; Tang *et al.*, 2004; Deuchars *et al.*, 2005) or by filling during intracellular recording (Deuchars *et al.*, 2001). Autonomic interneurons are small neurons that can contain GABA (Deuchars *et al.*, 2005) or glycine (Spanswick *et al.*,

1994) and are found around and within the IML, IC and CA. Evidence suggests some interneurons contain enkephalin (Llewellyn-Smith *et al.*, 2005) or galanin (Ohmachi *et al.*, 1996).

Regulation of pelvic organ function

Continence and voiding reflexes: bladder and bowel

There are three main groups of spinal motor pathways involved in regulation of the lower urinary tract and lower bowel. Sympathetic preganglionic neurons (mainly in L1-L2, although some are found as far rostral as T12) which are active during continence (storage), parasympathetic preganglionic neurons (mainly in L6-S1) which are active during voiding (micturition and defecation), and somatic motorneurons in L5-L6 which innervate the external urethral and anal sphincters and are active during continence. Function-specific spinal neurons have been identified by fos activation (e.g. after bladder distension) or viral tracing (injection of virus into organ of interest). Together these studies have demonstrated involvement of preganglionic neurons in the central autonomic area of L1 and L2, and intermediolateral column of L1, L2, L6 and S1. Small numbers of labeled neurons are also found in these areas in adjacent segments. Whether these are interneurons or neurons involved with sensory processing or ascending projection neurons is not known. Somatic motoneurons innervating pelvic floor muscles and the external urethral and anal sphincters have been described in most detail for cats (Holstege and Tan, 1987; Sasaki, 1994) but the pattern in rats is fundamentally similar (Marson, 1997), with most neurons located in the dorsolateral nucleus of L5 and L6 ventral horn which is sometimes known as Onuf's nucleus (McKenna and Nadelhaft, 1986; Collins *et al.*, 1991). These motoneurons project in the pudendal nerve and are sexually dimorphic (see below).

Penile erection

Penile erection involves coordination between sympathetic, parasympathetic and somatic systems (Giuliano and Rampin, 2004). Erection requires activation of pelvic parasympathetic neurons by sacral preganglionic neurons, located as described above for other pelvic visceral functions. Detumescence is promoted by sympathetic activity, arising in preganglionic neurons located in L1 and L2. Somatic motoneurons active during copulation are located in the ventral horn of L5 and L6 and supply ischiocavernosus, bulbospongiosus and ischiourethralis striated muscles as well as the external urethral and anal sphincters (McKenna and Nadelhaft, 1986; Collins *et al.*, 1991). Contraction of perineal striated muscles does not cause erection but increases penile rigidity and intracavernous pressure, while rhythmic activity of these muscles is involved in saccadic expulsion of semen during ejaculation. A spinal center controlling and coordinating smooth and skeletal muscles involved in ejaculation (the 'spinal ejaculation generator') is located in lamina 7 and 10 of L3-4 (Truitt and Coolen, 2002; Coolen *et al.*, 2004). These neurons receive afferent inputs from somatic and visceral targets and project to the thalamus and other supraspinal centres. Some of the neurons in this region are sexually dimorphic, showing differences in size and being more numerous in male than female rats (see below).

Sexual dimorphism in spinal cord circuits

There is extensive sexual dimorphism in the lumbosacral spinal cord, particularly in neurons involved with reproductive activities. For example, there are at least twice as many preganglionic neurons in the upper lumbar cord that project in the hypogastric nerve of male guinea pigs as there are in female (McLachlan, 1985). This reflects the dense sympathetic innervation of the male internal reproductive organs.

Motoneurons located in the dorsolateral and dorsomedial nuclei of the ventral horn in L5-6 are referred to as 'Onuf's nucleus' (from the original Onufrowicz'). In humans, cats and dogs these neurons are located in the sacral cord. This term is not used in rodents, where the dorsomedial nucleus is more commonly referred to as the spinal nucleus of the bulbocavernosus/ bulbospongiosus (SNB). These neurons innervate the perineal striated muscles as well as the external urethral and anal sphincters. They are often referred to as 'pudendal motoneurons' as they project in the pudendal nerve. This nucleus is sexually dimorphic in many features but most notably in that the total number and the soma size of neurons are both greater in males than females. This dimorphism is established during development and refined during puberty (Cooke *et al.*, 1998) under the influence of androgens which affect soma size, dendritic complexity and synaptic inputs (Matsumoto *et al.*, 1988). There may also be a genetic contribution (Arnold and Lubischer, 1995). Androgen receptors (Jordan *et al.*, 1991, 1997) and estrogen receptors (Williams and Papka, 1996; Papka *et al.*, 2001) are prominent in the adult lumbosacral spinal cord, including in many autonomic neurons. These receptors continue to mediate structural and other changes in spinal circuits during adulthood (Matsumoto *et al.*, 1988).

Summary

Autonomic preganglionic neurons are concentrated in two intermediolateral columns running down the thoracic cord into the first few lumbar segments (sympathetic) and in the segments immediately below the lumbar enlargement (parasympathetic). However, they lie in clumps from which smaller numbers are distributed across the intermediate zone at intervals with their dendrites crossing dorsal to the central canal, forming a ladder-like structure. The terminals of descending inputs from the dorsolateral columns follow the same ladder arrangement. Longitudinal columns of preganglionic neurons project to particular target ganglia but it has not been easy to relate these directly to particular functions. The most prevalent functional type is vasoconstrictor and these lie laterally in the IML and the lateral funiculi, whereas neurons projecting to visceral targets are more often located medially.

References

Anderson CR (1992) NADPH diaphorase-positive neurons in the rat spinal cord include a sub-population of autonomic preganglionic neurons. Neurosci Lett 139, 280-284.

Anderson CR, Edwards SL (1994) Intraperitoneal injections of fluorogold reliably label all sympathetic preganglionic neurons in the rat. J Neurosci Methods 53, 137-141.

Anderson CR, McLachlan EM, Srb-Christie O (1987) The relationship of terminals containing neuropeptide Y-like immunoreactivity to lumbar sympathetic vasoconstrictor neurons in the rabbit. Neurosci Lett 80, 33-38.

Anderson CR, McLachlan EM, Srb-Christie O (1989) The distribution of sympathetic preganglionic neurons and monoaminergic nerve terminals in the spinal cord of the rat. J Comp Neurol 283, 269-284.

Anderson CR, Edwards SL, Furness JB, Bredt DS, Snyder SH (1993) The distribution of nitric oxide synthase-containing autonomic preganglionic terminals in the rat. Brain Res 614, 78-85.

Anderson CR, Furness JB, Woodman HL, Edwards SL, Crack PJ, Smith AI (1995) Characterization of neurons with nitric-oxide synthase immunoreactivity that project to prevertebral ganglia. J Auton Nerv Syst 52, 107-116.

Appel NM, Elde RP (1988) The intermediolateral cell column of the thoracic spinal cord is comprised of target-specific subnuclei: evidence from retrograde transport studies and immunohistochemistry. J Neurosci 8, 1767-1775.

Arnold AP, Lubischer JL (1995) Evidence for target regulation of the development of androgen sensitivity in rat spinal motoneurons. Dev Neurosci 17, 106-117.

Baldwin C, Sasek CA, Zigmond RE (1991) Evidence that some preganglionic sympathetic neurons in the rat contain vasoactive intestinal peptide-like or peptide histidine isoleucine amide-like immunoreactivities. Neuroscience 40, 175-184.

Banrezes B, Andrey P, Maschino E, Schirar A, Peytevin J, Rampin O, Maurin Y (2002) Spatial segregation within the sacral parasympathetic nucleus of neurons innervating the bladder or the penis of the rat as revealed by three-dimensional reconstruction. Neuroscience 115, 97-109.

Barber RP, Phelps PE, Houser CR, Salvaterra PM, Crawford GD, Vaughn JE (1984) The morphology and distribution of neurons containing choline acetyltransferase in the adult rat spinal cord: an immunocytochemical study. J Comp Neurol 229, 329-346.

Baron R, Janig W (1991) Afferent and sympathetic neurons projecting into the lumbar visceral nerves of the male rat. J Comp Neurol 314, 429-436.

Baron R, Jänig W, McLachlan EM (1985) The afferent and sympathetic components of the lumbar spinal outflow to the colon and pelvic organs in the cat: III. The colonic nerves, incorporating an analysis of all components of the lumbar prevertebral outflow. J Comp Neurol 238, 158-168.

Bernstein-Goral H, Bohn MC (1988) Ontogeny of adrenergic fibers in rat spinal cord in relationship to adrenal preganglionic neurons. J Neurosci Res 21, 333-351.

Beyer C, Gonzalez-Mariscal G (1994) Effects of sex steroids on sensory and motor spinal mechanisms. Psychoneuroendocrinology 19, 517-527.

Birder LA, de Groat WC (1993) Induction of c-fos expression in spinal neurons by nociceptive and nonnociceptive stimulation of LUT. Am J Physiol 265, R326-R333.

Bruning G (1992) Localisation of NADPH diaphorase, a histochemical marker for nitric oxide synthase in the mouse spinal cord. Acta Histochem 93, 397-401.

Burman KJ, Sartor DM, Verberne AJ, Llewellyn-Smith IJ (2004) Cocaine- and amphetamine-regulated transcript in catecholamine and noncatecholamine presympathetic vasomotor neurons of rat rostral ventrolateral medulla. J Comp Neurol 476, 19-31.

Burnett AL, Saito S, Maguire MP, Yamaguchi H, Chang TSK, Hanley DF (1995) Localisation of nitric oxide synthase in spinal nuclei innervating pelvic ganglia. J Urol 153, 212-217.

Cassam AK, Llewellyn-Smith IJ, Weaver LC (1997) Catecholamine enzymes and neuropeptides are expressed in fibers and somata in the intermediate gray matter in chronic spinal rats. Neuroscience 78, 829-841.

Cassam AK, Rogers KA, Weaver LC (1999) Co-localization of substance P and dopamine beta-hydroxylase with growth-associated protein-43 is lost caudal to a spinal cord transection. Neuroscience 88, 1275-1288.

Ch'ng JLC, Christofides ND, Anand P, Gibson SJ, Allen YS, Su HC, Tatemoto K, Morrison JFB, Polak JM, Bloom SR (1985) Distribution of galanin immunoreactivity in the central nervous system and the responses of galanin-containing neuronal pathways to injury. Neuroscience 16, 343-354.

Chanthaphavong RS, Murphy SM, Anderson CR (2003) Chemical coding or sympathetic neurons controlling the tarsal muscle of the rat. Auton Neurosci 105, 77-89.

Clarke HA, Dekaban GA, Weaver LC (1998) Identification of lamina V and VII interneurons presynaptic to adrenal sympathetic preganglionic neurons in rats using a recombinant herpes simplex virus type 1. Neuroscience 85, 863-872.

Collins WF, Erichsen JT, Rose RD (1991) Pudendal motor and premotor neurons in the male rat: a WGA transneuronal study. J Comp Neurol 308, 28-41.

Colombo-Benkmann M, Heym C, Schemann M (1995) Preganglionic sympathetic neurons, innervating the guinea pig adrenal medulla, immunohistochemically contain choline acetyltransferase and also leu-enkephalin. Neurosci Lett 190, 155-158.

Cooke B, Hegstrom CD, Villeneuve LS, Breedlove SM (1998) Sexual differentiation of the vertebrate brain: principles and mechanisms. Front Neuroendocrinol 19, 323-362.

Coolen LM, Allard J, Truitt WA, McKenna KE (2004) Central regulation of ejaculation. Physiol Behav 83, 203-215.

Cummings JF (1969) Thoracolumbar preganglionic neurons and adrenal innervation in the dog. Acta Anat (Basel) 73, 27-37.

Dail WG, Dziurzynski R (1985) Substance P immunoreactivity in the major pelvic ganglion of the rat. Anat Rec 212, 103-109.

Dail WG, Minorsky N, Moll MA, Manzanares K (1986) The hypogastric nerve pathway to penile erectile tissue: histochemical evidence supporting a vasodilator role. J Auton Nerv Syst 15, 341-349.

de Groat WC, Nadelhaft I, Milne RJ, Booth AM, Morgan C, Thor K (1981) Organization of the sacral parasympathetic reflex pathways to the urinary bladder and large intestine. J Auton Nerv Syst 3, 135-160.

Debinski W, Kuchel O, Nemer M, Buu NT, Genest J (1988) Atrial natriuretic factor in sympathetic ganglia of the rat: dependence on cholinergic innervation. Neuroscience 27, 965-968.

Dembowsky K, Czachurski J, Seller H (1985) Morphology of sympathetic preganglionic neurons in the thoracic spinal cord of the cat: an intracellular horseradish peroxidase study. J Comp Neurol 238, 453-465.

Deuchars SA, Brooke RE, Frater B, Deuchars J (2001) Properties of interneurons in the intermediolateral cell column of the rat spinal cord: role of the potassium channel subunit Kv3.1. Neuroscience 106, 433-446.

Deuchars SA, Milligan CJ, Stornetta RL, Deuchars J (2005) GABAergic neurons in the central region of the spinal cord: a novel substrate for sympathetic inhibition. J Neurosci 25, 1063-1070.

Edwards SL, Anderson SR, Southwell BR, McAllen RM (1995) Distinct preganglionic neurons innervate noradrenaline and adrenaline cells in the cat adrenal medulla. Neuroscience 70, 825-832.

Endo T, Onaya T (1988) Immunohistochemical localization of parvalbumin in rat and monkey autonomic ganglia. J Neurocytol 17, 73-77.

Fenwick NM, Martin CL, Llewellyn-Smith IJ (2006) Immunoreactivity for cocaine- and amphetamine-regulated transcript in rat sympathetic preganglionic neurons projecting to sympathetic ganglia and the adrenal medulla. J Comp Neurol 495, 422-433.

Folan JC, Heym C (1989) Immunohistochemical evidence for different opioid systems in the rat superior cervical ganglion as revealed by imipramine treatment and receptor blockade. J Chem Neuroanat 2, 107-118.

Forehand CJ (1989) Peptidergic specificity in the innervation of rat sympathetic ganglion cells. Anat Rec 223, 394.

Forehand CJ (1991) Development of specific, leu-enkephalinergic input to rat stellate ganglion cells that project to sweat glands. J Auton Nerv Syst 33, 217-218.

Galabov PG, Davidoff MS, Dimitrov T (1990) Angiotensin II immunoreactivity in the vegetative regions of the thoracic and sacral spinal cord of guinea pig and rat. J Hirnforsch 31, 723-730.

Gibbins IL (1992) Vasoconstrictor, vasodilator and pilomotor pathways in sympathetic ganglia of guinea pigs. Neuroscience 47, 657-672.

Gibbins IL (1995) Chemical neuroanatomy of sympathetic ganglia. In: *Autonomic Ganglia.* EM McLachlan (ed). Harwood Academic Publishers Luxembourg.

Gibson SJ, Polak JM, Bloom SR, Sabate IM, Mulderry PM, Ghatei MA, McGregor GP, Morrison JFB, Kelly JS, Evans RM, Rosenfeld MG (1984a) Calcitonin gene-related peptide immunoreactivity in the spinal cord of man and of eight other species. J Neurosci 4, 3101-3111.

Gibson SJ, Polak JM, Anand P, Blank MA, Morrison JFB, Kelly JS, Bloom SR (1984b) The distribution and origin of VIP in the spinal cord of six mammalian species. Peptides 5, 201-207.

Gibson SJ, Polak JM, Allen JM, Adrian TE, Kelly JS, Bloom SR (1984c) The distribution and origin of a novel brain peptide, neuropeptide Y, in the spinal cord of several mammals. J Comp Neurol 227, 78-91.

Giuliano F, Rampin O (2004) Neural control of erection. Physiol Behav 83, 189-201.

Grkovic I, Anderson C (1996) Distributions of immunoreactivity for the NK1-receptor on different subpopulations of sympathetic preganglionic neurons in the rat. J Comp Neurol 374, 376-386.

Grkovic I, Anderson CR (1995) Calretinin-containing preganglionic terminals in the rat superior cervical ganglion surround neurons projecting to the submandibular gland. Brain Res 684, 127-136.

Grkovic I, Anderson CR (1997) Calbindin D28K-immunoreactivity identifies distinct subpopulations of sympathetic pre- and postganglionic neurons in the rat. J Comp Neurol 386, 245-259.

Hancock MB (1982) Separate populations of lumbar preganglionic neurons identified with retrograde transport of horseradish peroxidase (HRP) and 4,6-diamidino-2-phenylindole (DAPI). J Auton Nerv Syst 5, 135-143.

Hancock MB, Peveto CA (1979) Preganglionic neurons in the sacral spinal cord of the rat: an HRP study. Neurosci Lett 11, 1-5.

Häppölä O, Soinila S, Päivärinta H, Panula P (1987) [Met^5]-enkephalin-Arg^6-Phe^7- and [Met^5]-enkephalin-Arg^6-Gly^7-Leu^8-immunoreactive nerve fibers and neurons in the superior cervical ganglion of the rat. Neuroscience 21, 283-295.

Helen P, Panula P, Yang HY, Rapoport SI (1984) Bombesin/gastrin-releasing peptide (GRP)- and Met5-enkephalin-Arg6-Gly7-Leu8-like immunoreactivities in small intensely fluorescent (SIF) cells and nerve fibers of rat sympathetic ganglia. J Histochem Cytochem 32, 1131-1138.

Heym C, Common B, Klimaschewski L, Preissler U, Kummer W (1993) Immunohistochemical evidence from co-localization and denervation studies for four types of substance P-containing nervous structures in the rat superior cervical ganglion. Anat Embryol (Berl) 187, 485-492.

Holstege G, Tan J (1987) Supraspinal control of motorneurons innervating the striated muscles of the pelvic floor including urethral and anal sphincters in the cat. Brain 110, 1323-1344.

Hosoya Y, Okado N, Sugiura Y, Kohno K (1991) Coincidence of ladder-like patterns in distributions of monoaminergic terminals and sympathetic preganglionic neurons in the rat spinal cord. Exp Brain Res 86, 224-228.

Hosoya Y, Matsukawa M, Okado N, Sugiura Y, Kohno K (1995) Oxytocinergic innervation to the upper thoracic sympathetic preganglionic neurons in the rat. A light and electron microscopical study using a combined retrograde transport and immunocytochemical technique. Exp Brain Res 107, 9-16.

Inokuchi H, Masuko S, Chiba T, Yoshimura M, Polosa C, Nishi S (1993) Membrane properties and dendritic arborization of the intermediolateral nucleus neurons in the guinea-pig thoracic spinal cord in vitro. J Auton Nerv Syst. 43, 97-106.

Inokuchi, H., M. Yoshimura, A. Trzebski, C. Polosa and S. Nishi (1992a) Fast inhibitory postsynaptic potentials and responses to inhibitory amino acids of sympathetic preganglionic neurons in the adult cat. J Auton Nerv Syst 41, 53-59.

Inokuchi H, Yoshimura M, Yamada S, Polosa C, Nishi S (1992b) Fast excitatory postsynaptic potentials and the responses to excitant amino acids of sympathetic preganglionic neurons in the slice of the cat spinal cord. Neuroscience 46, 657-667.

Jänig W, McLachlan EM (1986a) Identification of distinct topographical distributions of lumbar sympathetic and sensory neurons projecting to end organs with different functions in the cat. J Comp Neurol 246, 104-112.

Jänig W, McLachlan EM (1986b) The sympathetic and sensory components of the caudal lumbar sympathetic trunk in the cat. J Comp Neurol 245, 62-73.

Jänig W, McLachlan EM (1987) Organization of lumbar spinal outflow to distal colon and pelvic organs. Physiol Rev 67, 1332-1404.

Jordan CL, Breedlove SM, Arnold AP (1991) Ontogeny of steroid accumulation in spinal lumbar motoneurons of the rat: implications of androgen's site of action during synapse elimination. J Comp Neurol 313, 441-448.

Jordan CL, Padgett B, Hershey J, Prins G, Arnold A (1997) Ontogeny of androgen receptor immunoreactivity in lumbar motorneurons and in the sexually dimorphic levator ani muscle of male rats. J Comp Neurol 379, 88-98.

Joshi S, Levatte MA, Dekaban GA, Weaver LC (1995) Identification of spinal interneurons antecedent to adrenal sympathetic preganglionic neurons using trans-synaptic transport of herpes simplex virus type 1. Neuroscience 65, 893-903.

Kanagawa-Terayama Y, Wanaka A, Yamasaki H, Matsuyama T, Matsumoto M, Kamada T, Yanaihara C, Tohyama M (1989) Immunocytochemical analysis of [Met5]enkephalin-Arg6-Gly7-Leu8 immunoreactive structures in the rat superior cervical ganglion. Brain Res 494, 75-84.

Kawatani M, Lowe I, Booth AM, de Groat WC (1983) The presence of leucine-enkephalin in the sacral preganglionic pathways to the urinary bladder of the cat. Neurosci Lett 39, 143-148.

Kawatani M, Shioda S, Nakai Y, Takashige C, de Groat WC (1989) Ultrastructural analysis of enkephalinergic terminals in parasympathetic ganglia innervating the urinary bladder of the cat. J Comp Neurol 288, 81-91.

Keast JR (1991) Patterns of co-existence of peptides and differences of nerve fibre types associated with noradrenergic and non-noradrenergic (putative cholinergic) neurons in the major pelvic ganglion of the male rat. Cell Tissue Res 266, 405-415.

Keast JR (1994) Neuropeptide-containing axon terminals in the male rat major pelvic ganglion are primarily of sacral origin. J Auton Nerv Syst 47, 151-158.

Klimaschewski L, Hauser C, Heym C (1996a) PACAP immunoreactivity in the rat superior cervical ganglion in comparison to VIP. Neuroreport 7, 2797-2801.

Klimaschewski L, Kroesen S, Eder U, Leitner B, Fischer Colbrie R (1996b) Localization and axotomy-induced regulation of the peptide secretoneurin in the rat superior cervical ganglion. Eur J Neurosci 8, 1953-1964.

Kondo H, Kuramato H, Wainer BH, Yanaihara N (1985) Evidence for the coexistence of acetylcholine and enkephalin in the sympathetic preganglionic neurons of rats. Brain Res 335, 309-314.

Krenz NR, Weaver LC (1998) Sprouting of primary afferent fibers after spinal cord transection in the rat. Neuroscience 85, 443-458.

Krukoff TL (1986) Segmental distribution of corticotropin-releasing factor-like and vasoactive intestinal peptide-like immunoreactivities in presumptive sympathetic preganglionic neurons in the cat. Brain Res 382, 153-157.

Krukoff TL, Ciriello J, Calaresu FR (1985a) Segmental distribution of peptide- and 5HT-like immunoreactivity in nerve terminals and fibers of the thoracolumbar sympathetic nuclei of the cat. J Comp Neurol 240, 103-116.

Krukoff TL, Ciriello J, Calaresu FR (1985b) Segmental distribution of peptide-like immunoreactivity in cell bodies of the thoracolumbar sympathetic nuclei of the cat. J Comp Neurol 240, 90-102.

Li J.-Y, Leitner B, Winkler H, Dahlstrom A (1998) Distribution of chromogranins A and B and secretogranin II (secretoneurin) in rat pelvic neurons. Neuroscience 84, 281-294.

Light AR, Metz CB (1978) The morphology of the spinal cord efferent and afferent neurons contributing to the ventral roots of the cat. J Comp Neurol 179, 501-515.

Lindh B, Staines W, Hökfelt T, Terenius L, Salvaterra PM (1986) Immunohistochemical demonstration of choline acetyltransferase-immunoreactive preganglionic nerve fibers in guinea pig autonomic ganglia. Proc Natl Acad Sci USA 83, 5316-5320.

Lindh B, Hökfelt T, Elfvin L-G (1988) Distribution and origin of peptide containing nerve fibers in the celiac superior mesenteric ganglion of the guinea-pig. Neuroscience 26, 1037-1071.

Llewellyn-Smith IJ, Arnolda LF, Pilowsky PM, Chalmers JP, Minson JB (1998) GABA- and glutamate-immunoreactive synapses on sympathetic preganglionic neurons projecting to the superior cervical ganglion. J Auton Nerv Syst 71, 96-110.

Llewellyn-Smith IJ, DiCarlo SE, Collins HL, Keast JR (2005) Enkephalin-immunoreactive interneurons extensively innervate sympathetic preganglionic neurons regulating the pelvic viscera. J Comp Neurol 488, 278-289.

Llewellyn-Smith IJ, Weaver LC, Keast JR (2006) Effects of spinal cord injury on synaptic inputs to sympathetic preganglionic neurons. Prog Brain Res 152, 11-26.

Locknar SA, Calupca MA, Parsons RL (2002) Innervation of guinea-pig stellate ganglia by nitric oxide synthase, cocaine- and amphetamine-regulated transcript protein- and pituitary adenylate cyclase activating polypeptide-immunoreactive fibers. Neuroscience 112, 525-533.

Lu Y, Du YJ, Qin BZ, Li JS (1993) The subdivisions of the intermediolateral nucleus in the sacral spinal cord of the cat. Brain Res 632, 351-355.

Lu Y, Inokuchi H, McLachlan EM, Li J-S, Higashi H (2001) Correlation between electrophysiology and morphology of three groups of neurons in the dorsal commissural nucleus of lumbosacral spinal cord of mature rats studied in vitro. J Comp Neurol 437, 156-169.

Marson L (1997) Identification of central nervous system neurons that innervate the bladder body, bladder base, or external urethral sphincter of female rats: a transneuronal tracing study using pseudorabies virus. J Comp Neurol 389, 584-602.

Marson L, Platt KD, McKenna KE (1993) Central nervous system innervation of the penis as revealed by the transmeuronal transport of pseudorabies virus. Neuroscience 55, 263-280.

Marson L, Cai R, Makhanova N (2003) Identification of spinal neurons involved in the urethrogenital reflex in the female rat. J Comp Neurol 462, 355-370.

Matsumoto A, Arnold AP, Zampighi GA, Micevych PE (1988) Androgenic regulation of gap junctions between motoneurons in the rat spinal cord. J Neurosci 8, 4177-4183.

McKenna KE, Nadelhaft I (1986) The organization of the pudendal nerve in the male and female rat. J Comp Neurol 248, 532-549.

McLachlan EM (1985) The components of the hypogastric nerve in male and female guinea pigs. J Auton Nerv Syst 13, 327-342.

McLachlan EM, Oldfield BJ (1981) Some observations on the catecholaminergic innervation of the intermediate zone of the thoracolumbar spinal cord of the cat. J Comp Neurol 200, 529-544.

McLachlan EM, Oldfield BJ, T. Sittiracha T (1985) Localization of hindlimb vasomotor neurons in the lumbar spinal cord of the guinea pig. Neurosci Lett 54, 269-275.

McLachlan EM, Anderson CR, Sinclair AD (1989) Are there bulbospinal catecholaminergic neurons in the guinea pig equivalent to the C1 cell group in the rat and rabbit? Brain Res 481, 274-285.

Mercanthaler I, Hynes MA, Vigh S, Shally AV, Petrusz P (1983) Immunocytochemical localisation of corticotrophin releasing factor (CRF) in the rat spinal cord. Brain Res 275, 373-377.

Michael GJ, Priestley JV (1995) Expression of GAP-43 mRNA in preganglionic sympathetic neurons of the adult rat spinal cord. Neuroreport 7, 338-42.

Minson JB, Llewellyn-Smith IJ, Pilowsky PM, Chalmers JP (1994) Bulbospinal neuropeptide Y-immunoreactive neurons in the rat: comparison with adrenaline-synthesising neurons. J Auton Nerv Syst 47, 233-243.

Morales MA, Holmberg K, Xu ZQ, Cozzari C, Hartman BK, Emson P, Goldstein M, Elfvin LG, Hökfelt T (1995) Localisation of choline acyltransferase in rat peripheral sympathetic neurons and its coexistence with nitric oxide synthase and neuropeptides. Proc Natl Acad Sci USA 92, 11819-11823.

Morgan C, de Groat WC, Felkins LA (1991) Axon collaterals indicate broad intraspinal role for sacral preganglionic neurons. Proc Natl Acad Sci USA 88, 6888-6892.

Morgan CW, de Groat WC, Felkins LA, Zhang SJ (1993) Intracellular injection of neurobiotin or horseradish peroxidase reveals separate types of preganglionic neurons in the sacral parasympathetic nucleus of the cat. J Comp Neurol 331, 161-182.

Morris JL, Gibbins IL (1987) Neuronal colocalization of peptides, catecholamines, and catecholamine-synthesizing enzymes in guinea pig paracervical ganglia. J Neurosci 7, 3117-3130.

Nadelhaft I, Booth AM (1984) The location and morphology of preganglionic neurons and the distribution of visceral afferents from the rat pelvic nerve: a horseradish peroxidase study. J Comp Neurol 226, 238-245.

Nadelhaft I, de Groat WC, Morgan C (1980) Location and morphology of parasympathetic preganglionic neurons in the sacral spinal cord of the cat revealed by retrograde axonal transport of horseradish peroxidase. J Comp Neurol 193, 265-281.

Newton BW, Burkhart AB, Romagnano MA, Hammill RW (1988) The postnatal ontogeny of substance P-immunoreactive fibers in the sympathetic preganglionic nuclei of the rat. Devel Brain Res 44, 161-180.

Ohmachi T, Nakamura T, Zhang FZ, Tani I, Takagi H (1996) Morpholological analyses of galaninergic inputs to the rat spinal parasympathetic nucleus. Exp Brain Res 109, 399-406.

Oldfield BJ, McLachlan EM (1981) An analysis of the sympathetic preganglionic neurons projecting from the upper thoracic spinal roots of the cat. J Comp Neurol 196, 329-345.

Oldfield BJ, Sheppard A, Nilaver G (1985) A study of the substance P innervation of the intermediate zone of the thoracolumbar spinal cord. J Comp Neurol 236, 127-140.

Papka RE, McNeill DL (1993) Light- and electron-microscopic study of synaptic connections in the paracervical ganglion of the female rat: special reference to calcitonin gene-related peptide-, galanin- and tachykinin (substance P and neurokinin A)-immunoeractive nerve fibers and terminals. Cell Tissue Res 271, 417-428.

Papka RE, Traurig HH, Klenn P (1987) Paracervical ganglia of the female rat: histochemistry and immunohistochemistry of neurons, SIF cells, and nerve terminals. Am J Anat 179, 243-252.

Papka RE, Storey-Workley M, Shughrue PJ, Merchenthaler I, Collins JJ, Usip S, Saunders PT, Shupnik M (2001) Estrogen receptor-alpha and beta- immunoreactivity and mRNA in neurons of sensory and autonomic ganglia and spinal cord. Cell Tissue Res 304, 193-214.

Petras JM, Cummings JF (1972) Autonomic neurons in the spinal cord of the Rhesus monkey: a correlation of the findings of cytoarchitectonics and sympathectomy with fibre degeneration following dorsal rhizotomy. J Comp Neurol 146, 189-218.

Powers-Martin K, McKitrick DJ, Arnolda LF, Phillips JK (2006) Distinct subpopulations of cyclic guanosine monophosphate (cGMP) and neuronal nitric oxide synthase (nNOS) containing sympathetic preganglionic neurons in spontaneously hypertensive and Wistar-Kyoto rats. J Comp Neurol 497, 566-574.

Puder BA, Papka RE (2001) Distribution and origin of corticotropin-releasing factor-immunoreactive axons in the female rat lumbosacral spinal cord. J Neurosci Res 66, 1217-1225.

Pyner S, Coote JH (1994) Evidence that sympathetic preganglionic neurons are arranged in target-specific columns in the thoracic spinal cord of the rat. J Comp Neurol 342, 15-22.

Ranson RN, Santer RM, Watson AHD (2006) The relationship between serotonin, dopamine beta hydroxylase and GABA immunoreactive inputs and spinal preganglionic neurons projecting to the major pelvic ganglion of Wistar rats. Neuroscience 141, 1935-1940.

Rogers H, Henderson G (1990) Activation of μ- and ∂-opioid receptors present on the same nerve terminals depresses transmitter release in the mouse hypogastric ganglion. Br J Pharmacol 101, 505-512.

Romagnano MA, Braiman J, Loomis M, Hamill RW (1987) Enkephalin fibers in autonomic nuclear regions: intrapsinal vs. supraspinal origin. J Comp Neurol 266, 319-331.

Sah P, McLachlan EM (1995) Membrane properties and synaptic potentials in rat sympathetic preganglionic neurons studied in horizontal spinal cord slices in vitro. J Auton Nerv Syst 53, 1-15.

Samano C, M.E. Zetina ME, Marin MA, Cifuentes F, Morales MA (2006) Choline acetyl transferase and neuropeptide immunoreactivities are colocalized in somata, but preferentially localized in distinct axon fibers and boutons of cat sympathetic preganglionic neurons. Synapse 60, 295-306.

Sasaki M (1994) Morphological analysis of external urethral and external anal sphincter motoneurons of cat. J Comp Neurol 349, 269-287.

Sasek CA, Seybold VS, Elde RP (1984) The immunohistochemical localisation of nine peptides in the sacral parasympathetic nucleus and the dorsal gray commissure in the rat spinal cord. Neuroscience 12, 855-873.

Sasek CA, Wessendorf MW, Helke CJ (1990) Evidence for co-existence of thyrotropin-releasing hormone, substance P and serotonin in ventral medullary neurons that project to the intermediolateral cell column in the rat. Neuroscience 35, 105-119.

Schramm LP (2006) Spinal sympathetic interneurons: their identification and roles after spinal cord injury. Prog Brain Res 152, 27-37.

Schroder HA (1983) Localisation of cholecystokinin-like immunoreactivity in the rat spinal cord, with particular reference to the autonomic innervation of the pelvic organs. J Comp Neurol 217, 176-186.

Senba E, Tohyama M (1988) Calcitonin gene-related peptide containing autonomic efferent pathways to the pelvic ganglia of rat. Brain Res 449, 386-390.

Senba E, Kaneko T, Mizuno N, Tohyama M (1991) Somato-, branchio- and viscero-motoneurons contain glutaminase-like immunoreactivity. Brain Res Bull 26, 85-97.

Shafton AD, Oldfield BJ, McAllen RM (1992) CRF-like immunoreactivity selectively labels preganglionic sudomotor neurons in cat. Brain Res 599, 253-260.

Soinila J, Happola O, Yanaihara N, Soinila S (1991) Immunohistochemical localisation of [met5]enkephalin and [met5]enkephalin-arg6-gly7-leu8 in sympathetic and parasympathetic neurons and nerve fibers projecting to the rat submandibular gland. Neuroscience 40, 545-554.

Spanswick D, Pickering AE, Gibson IC, Logan SD (1994) Inhibition of sympathetic preganglionic neurons by spinal glycinergic interneurons. Neuroscience 62, 205-216.

Stornetta RL, Akey PJ, Guyenet PG (1999) Location and electrophysiological characterization of rostral medullary adrenergic neurons that contain neuropeptide Y mRNA in rat medulla. J Comp Neurol 415, 482-500.

Stornetta RL, Schreihofer AM, Pelaez NM, Sevigny CP, Guyenet PG (2001) Preproenkephalin mRNA is expressed by C1 and non-C1 barosensitive bulbospinal neurons in the rostral ventrolateral medulla of the rat. J Comp Neurol 435, 111-126.

Strack AM, Sawyer WB, Marubio LM, Loewy AD (1988) Spinal origin of sympathetic preganglionic neurons in the rat. Brain Res 455, 187-191.

Strack AW, Sawyer WB, Hughes JH, Platt KB, Loewy AD (1989) A general pattern of CNS innervation of the sympathetic outflow demonstrated by transneuronal pseudorabies viral infection. Brain Res 491, 156-162.

Tang X, Neckel ND, Schramm LP (2004) Spinal interneurons infected by renal injection of pseudorabies virus in the rat. Brain Res 1004, 1-7.

Tang Y, Rampin O, Calas A, Facchinetti P, Giuliano F (1998) Oxytocinergic and serotonergic innervation of identified lumbosacral nuclei controlling penile erection in the male rat. Neuroscience 82, 241-254.

Truitt WA, Coolen LM (2002) Identification of a potential ejaculation generator in the spinal cord. Science 297, 1566-1569.

Vizzard MA (2000) Increased expression of spinal cord Fos protein induced by bladder stimulation after spinal cord injury. Am J Physiol 279, R295-305.

Vizzard MA, Erdman SL, Erickson VL, Stewart RJ, Roppolo JR, de Groat WC (1994) Localisation of NADPH diaphorase in the lumbosacral spinal cord and dorsal root ganglia of the cat. J Comp Neurol 339, 62-75.

Wang BR, Senba E, Tohyama M (1990) Met5-enkephalin-Arg6-Gly7-Leu8- like immunoreactivity in the pelvic ganglion of the male rat: a light and electron microscopic study. J Comp Neurol 293, 26-38.

Westlund KN, Bowker RM, Ziegler MG, Coulter JD (1983) Noradrenergic projections to the spinal cord of the rat. Brain Res 263, 15-31.

Williams SJ, Papka RE (1996) Estrogen receptor-immunoreactive neurons are present in the female rat lumbosacral spinal cord. J Neurosci Res 46, 492-501.

Yaici ED, Rampin O, Tang Y, Calas A, Jestin A, Leclerc P, Benoit G, Giuliano F (2002) Catecholaminergic projections onto spinal neurons destined to the pelvis including the penis in the rat. Internat J Impot Res 14, 151-166.

Yoshimura M, Polosa C, Nishi S (1987a) Slow EPSP and the depolarizing action of noradrenaline on sympathetic preganglionic neurons. Brain Res 414, 138-142.

Yoshimura M, Polosa C, S. Nishi S (1987b) Slow IPSP and the noradrenaline-induced inhibition of the cat sympathetic preganglionic neuron in vitro. Brain Res 419, 383-386.

9 Central Nervous System Control of Micturition

Gert Holstege and Han Collewijn

Introduction

The goal of the central nervous system is survival of the individual and survival of the species. The kidneys also play an important role because they filter hazardous substances out of the bloodstream into urine, a fluid the body needs to get rid of. In order not to lose urine incessantly, it is collected in the bladder. The bladder is kept closed at all times by means of the external urethral sphincter. Only when it is appropriate for emptying the bladder, the bladder contracts and the sphincter relaxes. Urine is not only used for the disposal of hazardous substances, but also, because of its olfactory characteristics, for territorial demarcation and for a female signaling to males that she is in estrus (survival of the species).

In most mammals micturition is such a complex activity that it cannot be combined with other activities, i.e. during urination animals cannot escape from danger. Thus, micturition requires a 'safe' place, which explains the requisite strong brain control. In humans, with the exception of infants, it is not considered socially acceptable to have no control over time and place of micturition. Urinary incontinence is a disease in which the individual has no control over when and where to micturate. Suffering from urinary incontinence is more embarrassing than suffering from most other diseases. Pet animals are deemed pets because of their ability to control their micturition, similar to humans. However, very old cats, similar to human elderly, may lose this ability, usually resulting in the owners getting rid of them as soon as possible. Thus, the mammalian brain needs to have total control over when and where micturition takes place. This chapter is an attempt to explain how this control is organized. First the spinal-brainstem-spinal micturition will be explained, second the brain structures that have access to this circuit and, third, why certain lesions can cause incontinence or the reverse, retention of urine.

Afferent fibers from bladder to sacral spinal cord

In order to properly control micturition, the central nervous system has to be informed how much urine the bladder contains, not only when the bladder is full, but at all times. Therefore, precise bladder filling information is conveyed from the bladder to the sacral cord. Incoming information to the central nervous system is always relayed by neurons in ganglia outside the central nervous system itself, in this case ganglion cells of the upper sacral cord. These sacral ganglion cells have peripheral and central fibers. The peripheral axons of the dorsal root pass through the pelvic and hypogastric nerves enter the bladder, where it contacts elements in the bladder wall that measure bladder filling.

Sensory endings in the bladder wall

The organization of the sensory endings of the afferent fibers from the bladder is complex and only partly understood. Morphologically, afferent fibers have been found in a suburothelial plexus, with the highest density in the bladder neck and the trigone, as well as in the detrusor muscle itself, extending throughout the bladder (Gabella and Davis, 1998). The terminal axons in the suburothelial plexus include chains of varicosities that contain small agranular (clear) and larger granulated (dense cored) vesicles (Dixon and Gilpin, 1987). These findings are corroborated by ultrastructural analysis of human biopsy material by Wiseman *et al.*, (2002), who, in addition, found in the lamina propria contacts of such (unmyelinated) varicosities with a particular cell type, called myofibroblasts. Wiseman *et al.*, (2003) proposed that these myofibroblasts, in combination with their associated nerve endings, function as stretch receptors. These myofibroblasts seem to be related to the regulatory interstitial cells of Cajal in the gastrointestinal tract. This finding led to the hypothesis that the myofibroblasts serve as a mediator in purinergic modulation of afferent activity after release of ATP by the urothelium (Andersson, 2002; Wiseman *et al.*, 2003). The contribution of myelinated versus unmyelinated fibers to the afferent bladder innervation is difficult to assess from these studies, although it appears that naked axons (about 1.5 μm diameter) predominate in the basal urothelium, while myelinated fibers (3-4 μm diameter) are more often seen close to the detrusor muscle layer (Fowler, 2002). On the other hand, a recent study (Zagorodnyuk *et al.*, 2006) of low-threshold stretch-sensitive mechanoreceptors of the bladder, presumably located in the muscular layers, found no evidence for ATP as a mediator in the mechanotransduction.

Taken together, the findings suggest that low threshold mechanoreceptors, sensitive to normal bladder filling, are mainly located in the muscular layers, and are directly sensitive to mechanical stretch and connected to myelinated (A-δ) fibers. On the other hand, the unmyelinated C-fibers are mainly associated with urothelial structures and with purinergic, peptidergic and other chemical neuromediators, that are activated by high filling pressures, changes in urine composition, inflammations and other non-physiological conditions (see also Wyndaele and De Wachter, 2003 for review).

In simple terms, the A-δ fibers relay information concerning bladder filling to the sacral cord, while the C-fibers convey nociceptive information, i.e. information that informs the sacral cord that something is wrong with the bladder. Some examples of this are overfilled bladder and bladder inflammation.

Sensory endings in the urethra

Sacral cord ganglion cells also send peripheral axons through the pudendal nerve that enter the urethra

These relatively fast conducting A-delta fibers are activated by mild mechanical stimuli and urine flow through the urethra (Bahns *et al.*, 1986; Buss and Shefchyk, 1999; Gustafson *et al.*, 2004; Kenton *et al.*, 2007). In all likelihood, they are not activated during filling of the bladder, except at high pressures but only when urine enters the urethra. They might play an important role in causing a strong urge to start micturition, and are also thought to reinforce micturition once urine passes through the urethra.

Sacral cord

The sacral cord plays a central role in bladder and bladder sphincter control. Not only does it receive afferent information from bladder and bladder sphincter and relays it to supraspinal levels, it also receives information from supraspinal levels, because it contains the motoneurons innervating bladder and bladder sphincter.

Bladder C-fibers

There are two classes of fibers entering the sacral cord: thin myelinated A-δ fibers and unmyelinated C-fibers. The C-fibers behave as all other C-fibers, entering the spinal cord and caudal medulla, they specifically relay nociceptive information and terminate on neurons in Rexed's (1954) laminae 1, outer 2 and 5-8. These capsaicin sensitive C-fibers contain a large number of neuromodulatory neuropeptides as substance P, neurokinin A, calcitonin-gene-related peptide (CGRP), vaso-active peptide (VIP), pituitary adenylate cyclase-activating peptide, enkephalins and glial-derived neurotrophic factors (deGroat and Yoshimura, 2001; Forrest and Keast, 2008).

The cells in laminae 1, outer 2 and 5-8 relay the information received from the C-fibers to the brainstem, i.e. to the ventromedial tegmentum of caudal pons and medulla, to the lateral tegmentum of the pons, the dorsal parabrachial nucleus and to the dorsomedial, lateral and especially the ventrolateral parts of the periaqueductal gray (PAG). They also send fibers to the medial hypothalamus, medial preoptic area and thalamus. Rodella *et al.*, (1998), after injecting 5% formalin in the bladder, i.e. after noxious bladder stimulation, observed Fos-positive neurons in these regions in brainstem and diencephalon. Importantly, they found a virtually identical distribution of Fos-positive neurons after noxious stimulation of neck muscles, indicating that the main task of C-fibers is to inform the central nervous system that 'something is wrong'. Cells in the sacral cord, but not GN (see below), also project to the most lateral part of the ventral posterior lateral thalamic nucleus (VPL), which notifies the brain that the problem is in the bladder or urethra, because the most lateral VPL, in turn, projects to the "bladder and urethral" part of the somatosensory cortex. Thus, C-fibers from the bladder inform the central nervous system that "something" is wrong in the bladder, but not about the amount of bladder filling. According to Morgan *et al.*, (1999) in the sacral cord there are also many VIP-containing C-fibers that terminate on the dendrites of parasympathetic preganglionic motoneurons in the area of the lateral collateral pathway (see below).

Bladder A-delta fibers

Information about bladder filling is relayed by A-δ fibers, using glutamic acid as their main neurotransmitter. After entering the sacral cord, A-delta fibers do not terminate in the same laminae as the C-fibers, but in a separate cell group, which in rats Morgan *et al.*, (1981) called the lateral collateral pathway of Lissauer's tract (LCP). At thoracic and lumbar levels the LCP consists of more or less well-defined visceral afferent axon bundles emanating from the lateral Lissauer's tract and curving around the lateral edge of the dorsal horn and through the dorsolateral funiculus (Neuhuber *et al.*, 1986). However, at sacral levels the LCP does not only contain fibers, but also neurons. Holstege and Kuypers (1982) demonstrated in the cat that a cell group in the most lateral part of the upper sacral dorsal horn receives afferent projections from the pontine micturition center (see below), but not from the lateral pontine tegmentum (Figure 9.1, right). The absence of afferents from this latter region is important, because the lateral pontine tegmentum projects to laminae I, outer II, and V throughout the length of the spinal cord, including the sacral cord, but the projection did not involve the LCP (figure 1 left). VanderHorst *et al.*, (1996) showed that this "lateral collateral pathway" contains neurons projecting to the central parts of the periaqueductal gray (PAG; Figure 9.2). Part of these cells are estrogen receptor-α immunoreactive (VanderHorst *et al.*, 2001). It means that the 'lateral collateral pathway' is not a pathway but a cell group (Figure 9.3). For this reason Pfizer Urology research group in Sandwich (UK) called it Gert's nucleus (GN). Thus, GN receives projections from the pontine

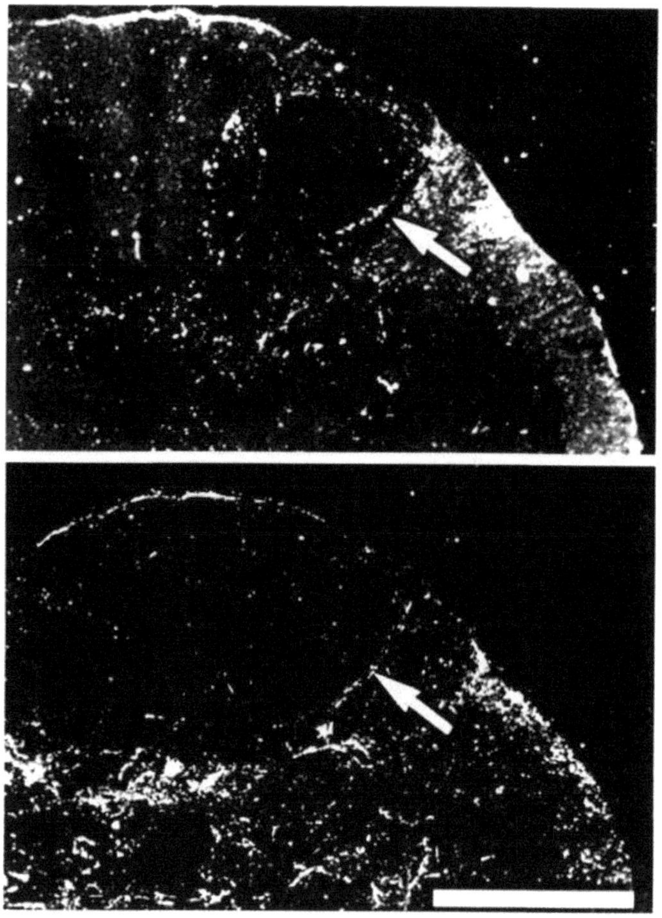

Figure 9.1 Descending fibers from the brainstem target different parts of the sacral dorsal horn

Upper figure: the lateral pontine tegmentum projects to lamina I throughout the length of the spinal cord, including the sacral cord but avoids Gert's nucleus (GN; arrow), which lies lateral to the dorsal horn. Lower figure: the pontine micturition center (PMC) projects to GN (arrow), and to the parasympathetic motoneurons, which are located caudal to this level (see Fig. 9.14), but avoids lamina I. (from Holstege and Kuypers, 1982).

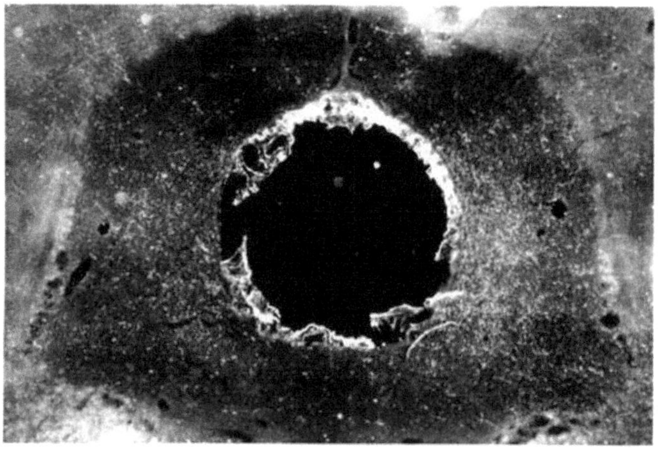

Figure 9.2 Sacral cord projection to periaqueductal gray

The sacral cord projects to the same regions of the periaqueductal gray (PAG) as the other parts of the spinal cord, i.e. to its dorsomedial, lateral and ventrolateral parts. However, only the sacral cord also projects very strongly to the central parts of the PAG. These fibers originate in GN (see next figure; from VanderHorst *et al.*, 1996).

micturition center, as well as from the bladder via the A-delta fibers (Nadelhaft and Booth, 1984; Hwang *et al.*, 2005), while it also contains axon collaterals (Morgan, 2001), dendrites (Morgan and Ohara, 2001), but no cell bodies of the parasympathetic preganglionic motoneurons innervating the detrusor muscle of the bladder.

Ascending projections

Both the neurons in GN and the cells in laminae I, II, and V project to the PAG, but to different parts of it. GN projects specifically to the central PAG (Figure 9.2), while the laminae I, II and V neurons project to the dorsomedial, lateral and ventrolateral PAG. In contrast to the cells in laminae I, II and V, GN does not project to the thalamus (Klop *et al.*, 2005). One might interpret this finding as that the individual does not have to be aware of the precise filling status of the bladder, as long as it is not completely full. Only an overloaded bladder represents 'danger', in which case C-fibers become active to alert the individual via sacral cord and thalamus that it might be time for 'action'. The PAG, on the other hand, is continuously fully informed by the GN-neurons about the exact amount of urine in the bladder. In conclusion, a separate cell group in the sacral cord (GN), lateral to the dorsal horn, is specifically involved in the relay of bladder filling information. It cannot be excluded that cells in GN are also involved in relaying information concerning defecation or sexual activity to the PAG, since pudendal afferents from pelvic viscera as colon descendens, rectum, vagina, vaginal cervix and uterine cervix also terminate in GN (Morgan *et al.*, 1981, Kawatani *et al.*, 1990; Figure 9.4).

Motor innervation of bladder and bladder sphincter

Autonomic innervation

Preganglionic parasympathetic motoneurons in the sacral parasympathetic nucleus (SPN) innervate, via ganglion cells in the bladder wall, the detrusor smooth muscle fibers that are active during bladder contraction. The myocytes in the detrusor muscle are connected by means of so-called gap junctions, which allow sustaining and transferring excitation between them. In cat SPN motoneurons are located in the second and third sacral spinal segments, and have myelinated axons with conduction velocities between 3.3 and 13 m/s. They are located in the lateral band of the SPN, (deGroat *et al.*, 1982). Other parasympathetic motoneurons have unmyelinated axons with conduction velocities of 0.5-1.4 m/s. They are located in the dorsal band of the nucleus and innervate the colon descendens and rectum (Leedy *et al.*, 1988; Figure 9.5).

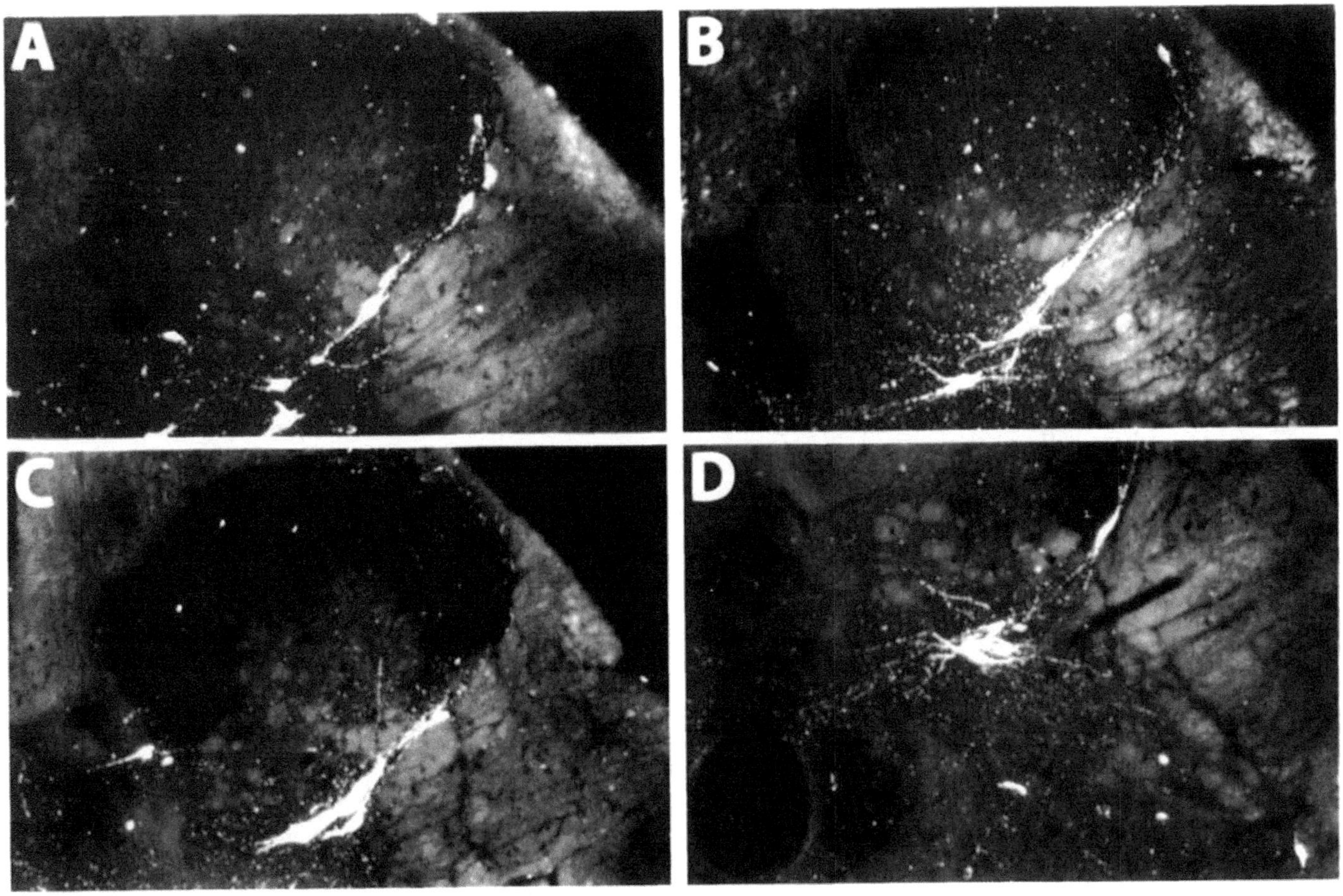

Figure 9.3 Periaqueductal gray projection to sacral cord

After a WGA-HRP retrograde tracer injection in the cat, also involving the central PAG, many retrogradely labeled neurons were found in GN from rostral S1 (A) to caudal S2 (D) (from VanderHorst *et al.*, 1996).

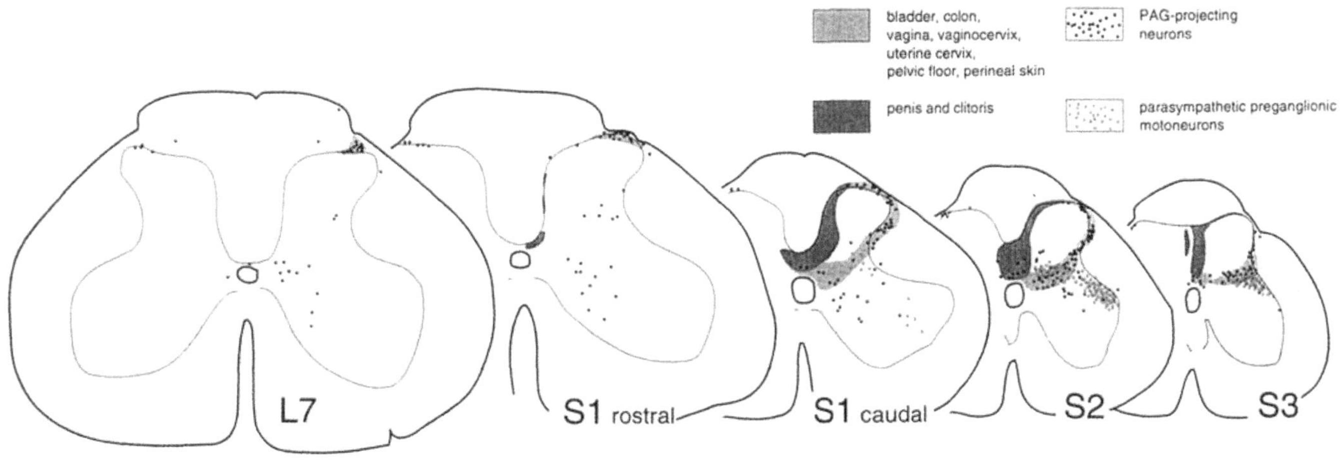

Figure 9.4 Sacral cord projection to the periaqueductal gray

Summary diagram of the incoming information from the sacral cord. The neurons in GN project to the PAG (from VanderHorst *et al.*, 1996).

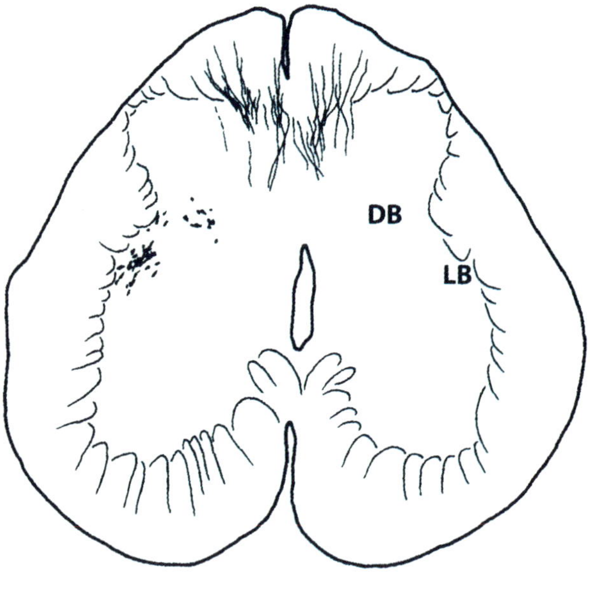

Figure 9.5 Sacral parasympathetic motoneurons

The parasympathetic motoneurons in the cat can be divided in a dorsal band (DB) and in a lateral band (LB) (from Leedy et al , 1988) The motoneurons in the dorsal band innervate the distal colon and rectum, those in the lateral band the bladder

The bladder is also innervated by sympathetic autonomic motoneurons. They are located in the upper lumbar spinal cord, and innervate the bladder through ganglion cells in the sympathetic chain and the hypogastric nerve. It has been reported that one of the major functions of the sympathetic bladder innervation is to decrease bladder pressure during the filling phase in order to limit the number of micturition episodes (Morrison, 1999). In how far this sympathetic innervation also decreased the afferent information from the bladder to the sacral cord is not known. Recently, it has been suggested that the sympathetic innervation might also play a role in bladder hyperreflexia induced by bladder inflammation (Dmitrieva and Berkley, 2006).

Somatomotor innervation of the external bladder sphincter

In contrast to the bladder itself, the external sphincter of the bladder, as part of the pelvic floor, consists of striated musculature. The motoneurons innervating the sphincter are also located in the sacral cord, in a distinct cell group ventromedial and slightly rostral to the parasympathetic bladder motoneurons. This cell group has first been described in 1899 by Onufrowitz, known as Onuf. Onuf suggested that the cell group (ON) was involved in sexual behavior (Onuf, 1899). In all likelihood, this is indeed the case, since the external urethral sphincter is not exclusively involved in micturition. Romanes (1951) called ON group Y, and only in 1978 in the cat Sato *et al.*, identified ON as containing pelvic floor muscle motoneurons. In 1981 Schrøder demonstrated that this was also the case in humans. Already eight years previously Dekker *et al.*, (1973) had characterized ON as having a great many rostrocaudally oriented dendrites with many crest-like synapses. In the rat ON is split up into two separate nuclei. In both nuclei many gap junctions were found, which number is influenced by testosterone levels, (Matsumoto *et al.*, 1989; Coleman and Sengelaub, 2002). Although gap junctions have not yet been found in ON of cat (Takahashi and Yamamoto, 1979) or humans, in all likelihood they are present in these species also. The reason is that all Onuf motoneurons act together, i.e. continuously contract the pelvic floor muscles in order to close both bladder and the anal sphincters. Such a function corresponds with an involvement in long-term enhancement in synaptic transmission, as suggested by the presence of B-50 (GAP-43) in this nucleus (Nacimiento *et al.*, 1993). Only when there is strong reason not to, i.e. during micturition or defecation, and, at least in males, during sexual activity, the pelvic floor sphincters have to relax, i.e. only during these activities ON motoneurons have to be inhibited. Motoneurons in the ventrolateral ON innervate the bladder sphincter muscles, those in the dorsomedial ON the anal sphincter (Kuzuhara *et al.*, 1980; Iwata *et al.*, 1993; Kuipers *et al.*, 2004). In the pig, however, these anal sphincter motoneurons are located in a different location, dorsolateral to the central canal (Blok *et al.*, 1996; Holstege and Sie, 2002; Figure 9.6). ON also contains motoneurons innervating other pelvic floor muscles as ischiocavernosus and bulbospongiosus. ON motoneurons represent a distinct class of motoneurons, because on the one hand they are somatic motoneurons innervating striated muscles, and are under voluntary control, but on the other hand they behave as autonomic motoneurons. For example, similar to all other autonomic motoneurons including the parasympathetic bladder motoneurons, they receive direct afferents from the paraventricular nucleus of the hypothalamus (Holstege, 1987; Figure 9.7) and in humans, like autonomic motoneurons, they are not or to only a limited extent, affected by amyotrophic lateral sclerosis (ALS) (Mannen *et al.*, 1977; Carvalho *et al.*, 1995). Moreover, also comparable with other autonomic motoneurons, ON motoneurons are more affected in Shy-Drager syndrome than somatic motoneurons (Chalmers and Swash; 1987). ON motoneurons, in contrast to all other somatic motoneurons, do not receive afferent projections from premotor interneurons in the intermediate zone (Figure 9.8), but only from the intermediomedial cell group (Blok *et al.*, 1998b; Sie *et al.*, 2001). Thus, ON motoneurons occupy a separate position between somatic and autonomic motoneurons. Similar motoneurons with such distinct properties are not known in brainstem and spinal cord.

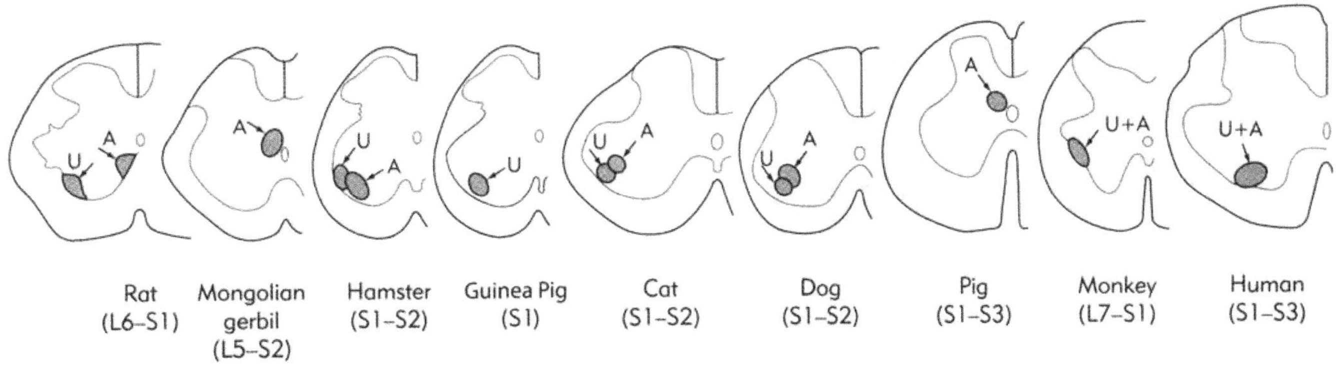

Figure 9.6 Onuf's nucleus

The motoneurons innervating the external urethral (U) and the external anal (A) sphincter are adjacent in most mammals, but in the rat they are widely separated. Those innervating the urethral sphincter are usually located ventrolateral to the anal sphincter ones. However, in Mongolian gerbil and pig, the anal sphincter motoneurons are located dorsolateral to the central canal (from Holstege and Sie, 2002 and Kuipers *et al.*, 2004).

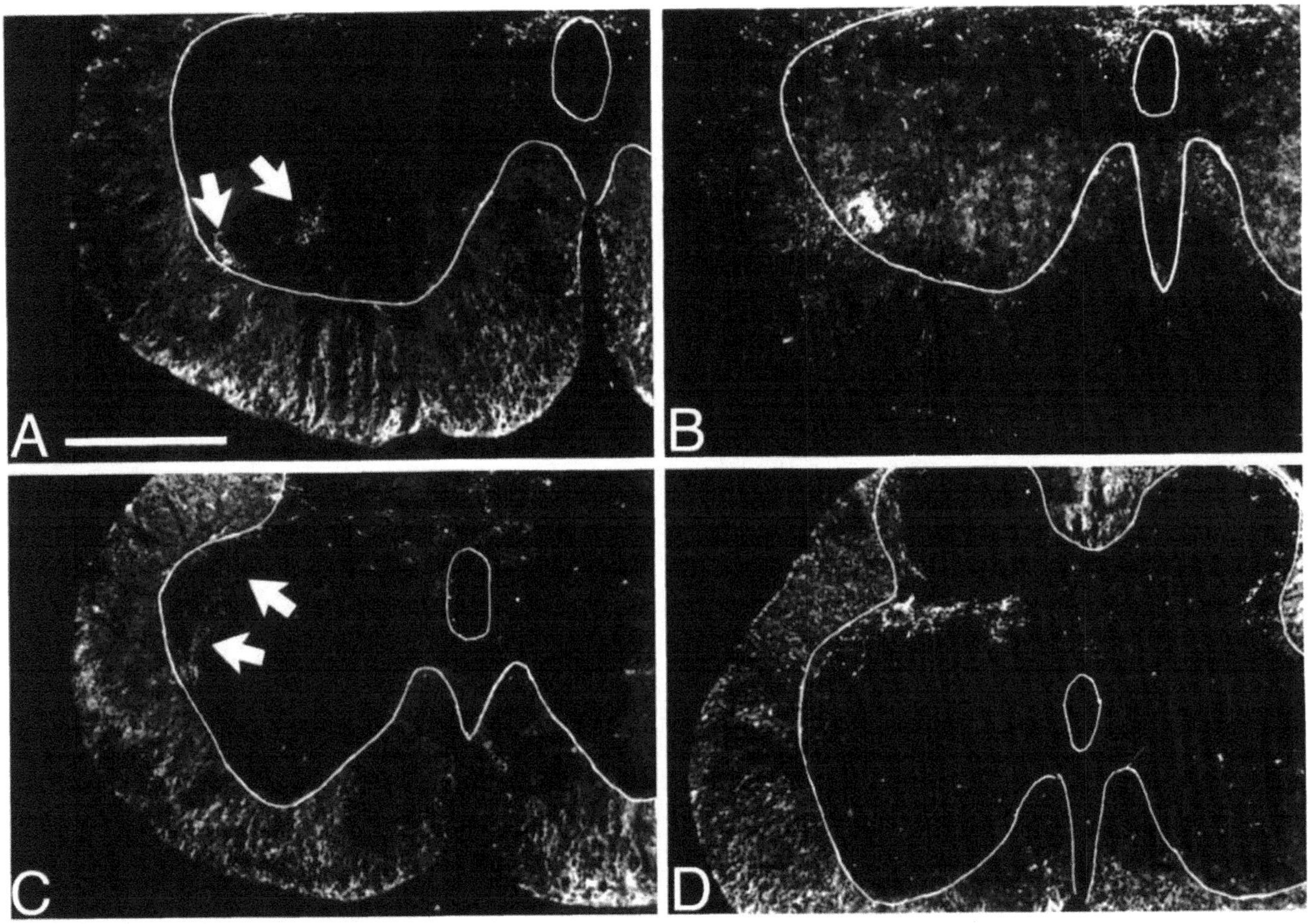

Figure 9.7 Hypothalamic projections to sacral cord

This figure shows the projections from the paraventricular hypothalamic nucleus to the sacral cord, as revealed by autoradiographic tracing. Note the projections to the ON, rostrally split into two parts (A), but further caudally forming one nucleus (B). Between ON and the parasympathetic cells there are projections to the dendrites of ON (C), and at S2 (D) there are projections to the parasympathetic motoneurons (dorsal and lateral band) (from Holstege, 1987).

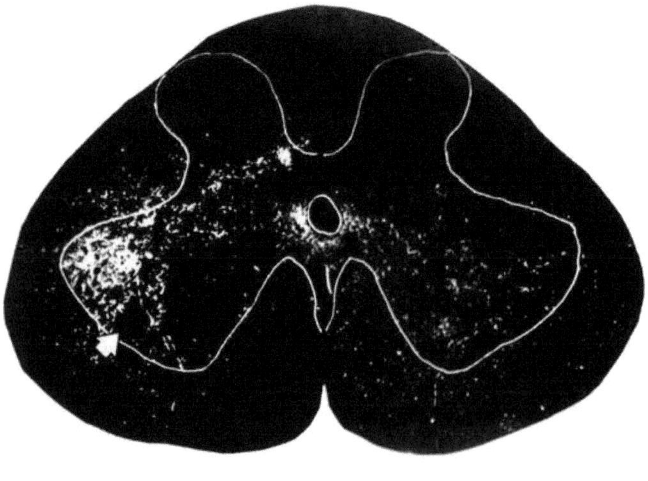

Figure 9.8 Interneurons at L7 project to hindlimb motoneurons but not Onuf's nucleus at S1

In the cat the motoneurons at the level of the first sacral segment that innervate the hindlimb receive very strong projections from the interneurons located at the L7-level. However, no such projections go to ON, which only receives projections from the sacral intermediomedial cell group and from brainstem structures.

Sacral micturition reflexes

The first evidence for the existence of sacral micturition reflexes was given by De Groat (1975) and De Groat *et al.*, (1981), who observed that micturition as well as defecation are elicited in neonatal kittens when the mother licks the perineal region. This stimulation is crucial, because separation of the kittens from the mother results in urinary retention. The reason why the mother cat licks the perineum of the kittens is an interesting question that still remains to be answered. This perineal-to-bladder reflex is quite prominent during the first four postnatal weeks, after which it becomes less effective and usually disappears at 7-8 weeks postnatally, the approximate age of weaning. Although during this early postnatal period there is a weak supraspinal bladder reflex present (Kruse and De Groat, 1990), the perineal-to-bladder reflex is so prominent that thoracic cord transections of the spinal cord did not abolish it. Thus, the perineal-to-bladder reflex is a sacral cord reflex. After 7-8 weeks the supraspinal bladder reflexes have replaced the perineal-to-bladder reflex. Transection of the spinal cord in older kittens or adult cats causes reemergence of perineally induced micturition within 1-2 weeks. In humans this spinal cord reflex system is functionally non-existent, except in patients with spinal cord transection rostral to the sacral cord. Thus, although in adult animals and humans pathways exist within the sacral cord that can produce bladder and sphincter contractions, they are usually not well coordinated and often dyssynergic. Bladder dyssynergia means that when the bladder contracts, the sphincter contracts also, preventing micturition.

Periaqueductal gray

The PAG is one of the phylogenetically oldest structures in the brain that coordinates motor output in the context of survival of the individual and the species. It maintains basic projection systems that innervate the lower brainstem systems, involved in the control of respiration, vocalization, blood pressure, nociception, and even locomotion and receptive behavior (Figure 9.9). In the context of these basic mechanisms it also plays a crucial role in micturition control, because the PAG decides when micturition takes place, on the basis of information from more rostral brain structures about the safety situation of the individual. There exist relatively few studies about the position of the PAG in the micturition circuit. The best evidence for its importance in the micturition circuit is that a lesion in the PAG of a human patient resulted in a complete retention for urine for several weeks (Yaguchi *et al.*, 2004). This result corresponds with Matsuura *et al.*, (1998) who found in rats that blockade of synaptic neurotransmission in the ventrolateral PAG by means of injecting cobalt chloride interrupted the micturition reflex. Another patient, diagnosed with Wernicke's encephalopathy with a lesion in the dorsal PAG, suffered from urge-incontinence (Sakibara *et al.*, 1997). Furthermore, scientists studying cats, looking at the role of the PAG in vocalization and respiratory control noticed, as an irritating side effect, that stimulation in the PAG not only caused changes in respiration, but also produced micturition (H.H. Subramanian personal communication). Interestingly, according to Liu *et al.*, (2004) high frequency stimulation in the PAG, but also in the tegmental area ventrolateral to it produced an inhibition of the micturition reflex. On the other hand, the same authors, using single-unit recording in the PAG not only found neurons that were active during storage of urine, but also many that were specifically active during micturition. Thus, it is clear that the PAG plays a critical role in micturition control, also because in humans neuro-imaging studies of Blok *et al.*, (1997a, 1998a; Figure 9.10, right) have shown that the PAG is also activated during micturition. On the other hand, how this PAG control of micturition is precisely organized remains to be explained.

Spinal projections to the PAG

VanderHorst and Holstege (1992) were the first to draw attention to PAG playing an important role in micturition control, because it receives a specific projection from GN in the sacral cord, that receives information regarding bladder filling (see above). Later studies in the cat (VanderHorst *et al.*, 1996) and in the rat (Keay *et al.*, 1997) showed that this projection originated in GN, while other neurons in the sacral and other parts of the spinal cord projected to the dorsomedial, lateral and ventrolateral parts of the PAG.

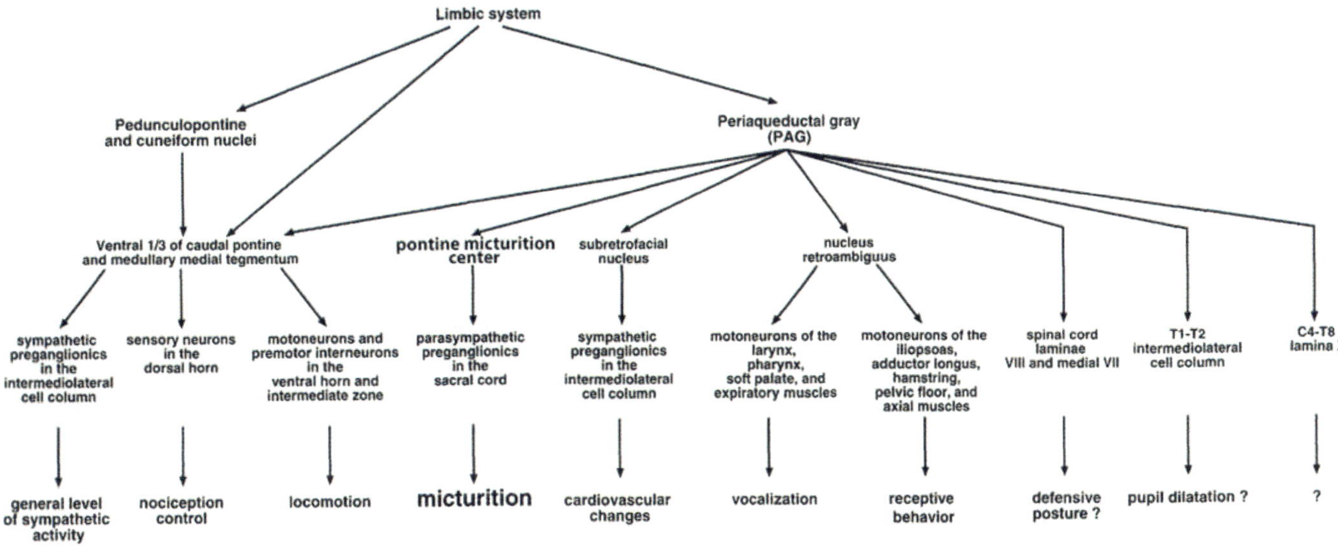

Figure 9.9 Projections from the periaqueductal gray

Summary diagram of the various projections from the periaqueductal gray (PAG). The projections involving micturition control only form a small part of this highly organized survival system.

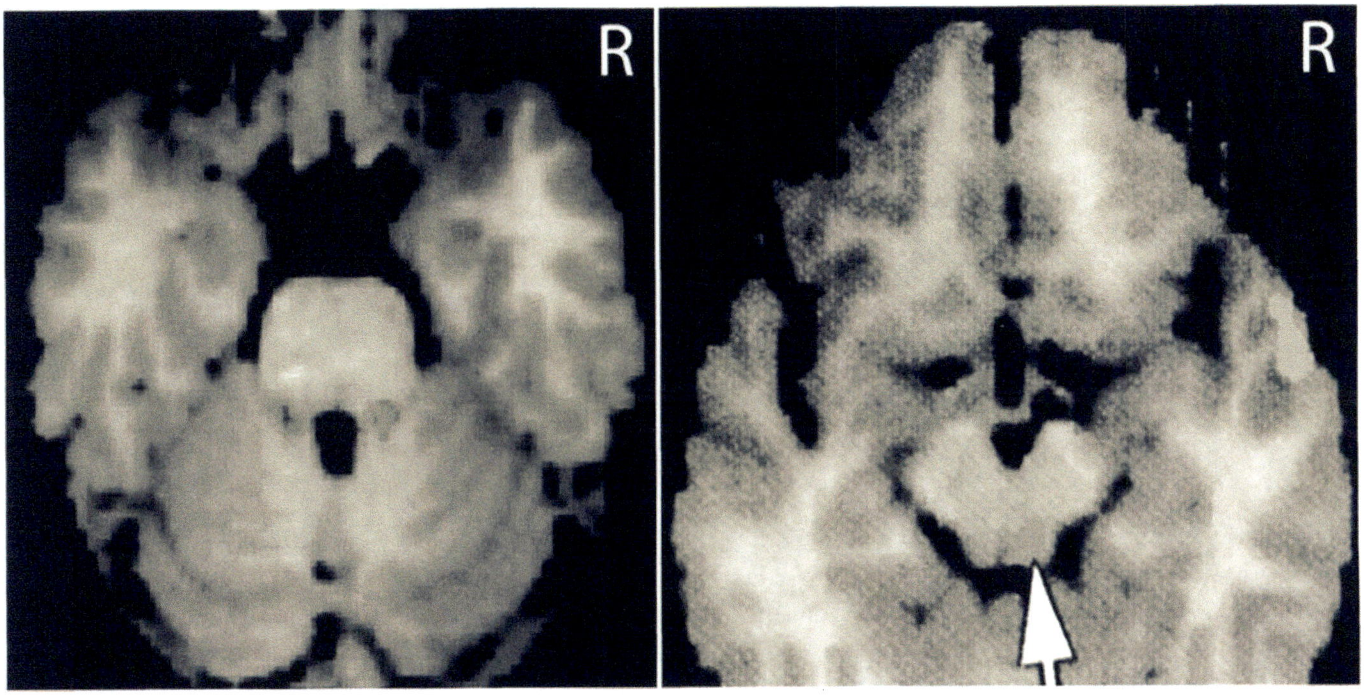

Figure 9.10 Human brain activation during micturition

A PET-study of the brain activation during micturition in humans shows brain activation in the PMC (left) and in the PAG (arrow right) (from Blok *et al.*, 1997b). *This figure is reproduced in the color insert.*

Descending PAG projections in relation to micturition

The strong impact of the PAG on micturition is based on its projections to the pontine micturition center (PMC) (Blok and Holstege, 1994; Kuipers *et al.*, 2006a; Figure 9.11). The PAG has no direct control of the sacral cord motoneurons (Mouton and Holstege, 1994). The PAG projections to the PMC originate partly from the neurons in the central PAG, that receive afferents from GN in the sacral cord. In addition there are also neurons in the dorsomedial, lateral and ventrolateral parts of the PAG, and, importantly, many in the mesencephalic tegmentum lateral and ventrolateral to the PAG that project to the PMC (Blok and Holstege, 1994; Figure 9.12). All these PMC projecting neurons also receive strong afferents connections from forebrain regions. The moment that the central PAG neurons are informed by GN that the bladder is filled to an extent that micturition is appropriate, they relay this

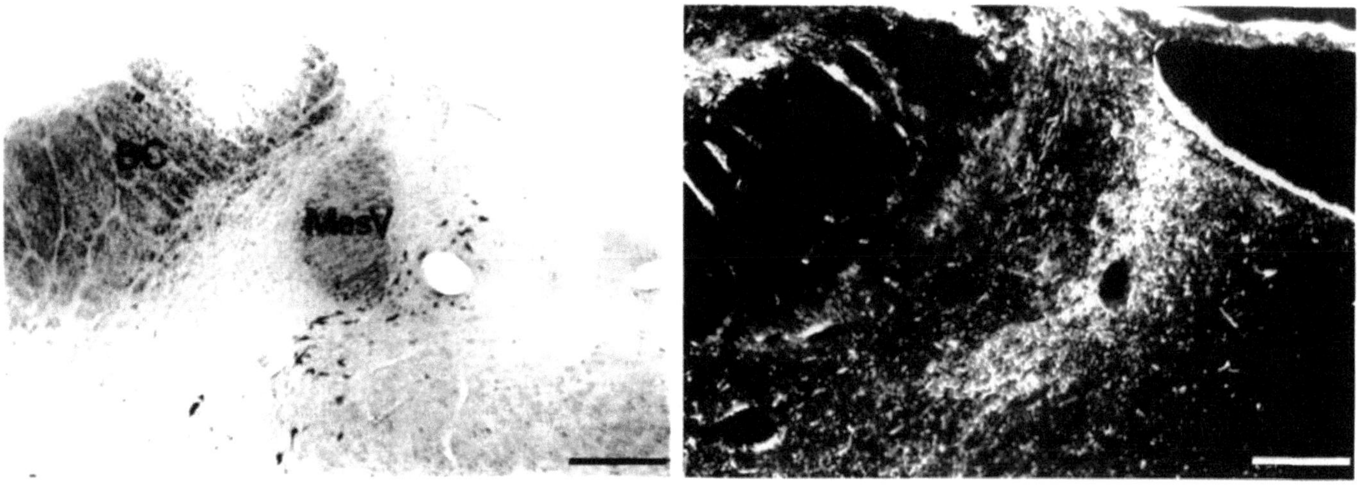

Figure 9.11 PMC projections to sacral cord

The image on the left shows retrogradely labeled neurons in the PMC after sacral cord injection of a retrograde tracer. The image on the right is a darkfield photograph showing the projection to the PMC after injection of an anterograde tracer in the PAG (from Blok *et al.*, 1994). BC = Brachium conjunctivum; MesV = Mesencephalic trigeminal tract.

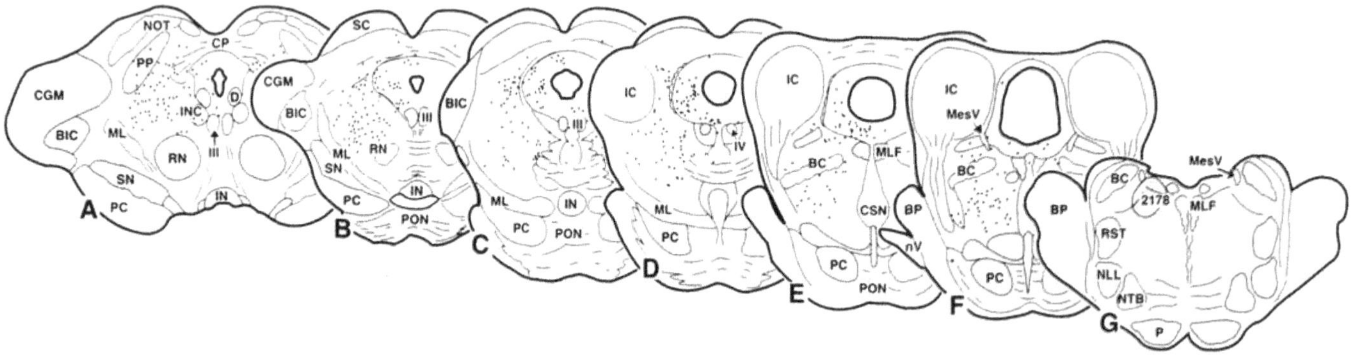

Figure 9.12 PAG projections to PMC

After a retrograde tracer injection in the area of the PMC, retrogradely labeled cells were found in the central PAG, and also in more lateral parts of the PAG. Cells were also found in the mesencephalic tegmentum lateral and ventrolateral to the PAG (from Blok and Holstege, 1994).

information to the other PMC projecting cells in PAG and adjoining areas. Since these same cells also receive information from various forebrain structures about whether the situation in which the individual finds itself is appropriate for micturition, they decide whether or not to excite the PMC cells. Usually, this is not the case and micturition will not take place. Thus, neurons in PAG and adjacent regions decide, based on information from GN and forebrain, whether or not to excite the PMC, i.e. whether or not to start micturition. How these PAG neurons excite the PMC, remains to be investigated.

Pontine micturition center (PMC)

In 1925, Barrington demonstrated that in the dorsolateral pontine tegmentum, a crucial structure for micturition is located, because bilateral lesions in this area in cat resulted in urinary retention. Sixty years later Holstege and co-workers showed that there are two different systems, which they called M- (=medial) and L- (=lateral) region (Holstege *et al.*, 1986). The M-region is also known as Barrington's nucleus, but the most commonly used name nowadays is pontine micturition center (PMC). It is a group of neurons in the medial part of the dorsolateral pons, located just ventral to the mesencephalic trigeminal nucleus and locus coeruleus. PMC neurons display corticotropin-releasing factor (CRF)-like immunoreactivity (Vincent and Satoh, 1984). The L-region, on the other hand, is a group of cells in the lateral part of the dorsolateral pons, also known as pontine continence center (PCC) (Griffiths *et al.*, 1990).

Pontine micturition center in animals

The PMC has complete control of micturition, because it is the only group of neurons that excites the detrusor muscle of the bladder and simultaneously inhibits its sphincter muscle, leading to complete so-called synergic micturition. Holstege *et al.*, (1986) showed in the cat that electrical stimulation of the PMC leads to complete micturition, which entails stimulation of the detrusor muscle of the bladder and relaxation of the external urethral sphincter (Figure 9.13). Neurons in the PMC send fibers throughout the length of the spinal cord that terminate directly on the parasympathetic preganglionic motoneurons in the sacral cord (Holstege *et al.*, 1979, 1986; Figure 9.14, left). An ultrastructural study (Blok and Holstege, 1997) showed that the PMC projections to the bladder motoneurons are monosynaptic and excitatory. However, electrical and chemical stimulation in the PMC not only elicits a steep rise in the intravesical pressure, but also an immediate and sharp decrease of the urethral pressure and pelvic floor electromyogram (Holstege *et al.*, 1986 in the cat, Mallory *et al.*, 1989 in the rat; Figure 9.13). This decrease is the result of the PMC projection to GABA- and glycinergic interneurons in the sacral intermediomedial cell group (Blok *et al.*, 1997b; Sie *et al.*, 2001). Retrograde tracing studies with the pseudorabies virus have shown that the sacral intermediomedial cell group indeed contains interneurons projecting to the nucleus of Onuf (Nadelhaft and Vera, 1996), and electrical stimulation in this region results in an inhibition of the external urethral sphincter via GABA ergic interneurons in the intermediomedial cell group (Blok *et al.*, 1998b; Figure 9.15).

Bilateral lesions in the PMC result in total urinary retention leading to depressed detrusor activity and increased bladder capacity (Griffiths *et al.*, 1990). In rats, unilateral chemical lesions of this area already attenuate the bladder response (Mallory *et al.*, 1989).

Pontine micturition center in humans

The importance of the PMC in humans is illustrated by the finding that patients with a transection of the spinal cord almost always suffer from great problems with micturition, often leading to so-called dyssynergic micturition. It means that contraction of the bladder is accompanied by simultaneous contraction of the bladder sphincter. Consequently, in order to expel urine through the tonically closed urethral sphincter, the bladder has to produce an extremely high intravesical pressure. The result is a thick bladder wall and a small bladder capacity, the so-called overflow bladder. Patients suffering from dyssynergic micturition often require manual catherization of the bladder. Brain lesions rostral to the pons never result in bladder-sphincter dyssynergia, but in urge incontinence (Blaivas, 1982).

Holstege and co-workers (Blok *et al.*, 1997a in men; 1998a in women), have demonstrated that micturition in humans seems to be similarly organized as in cats. In these PET-scan studies,

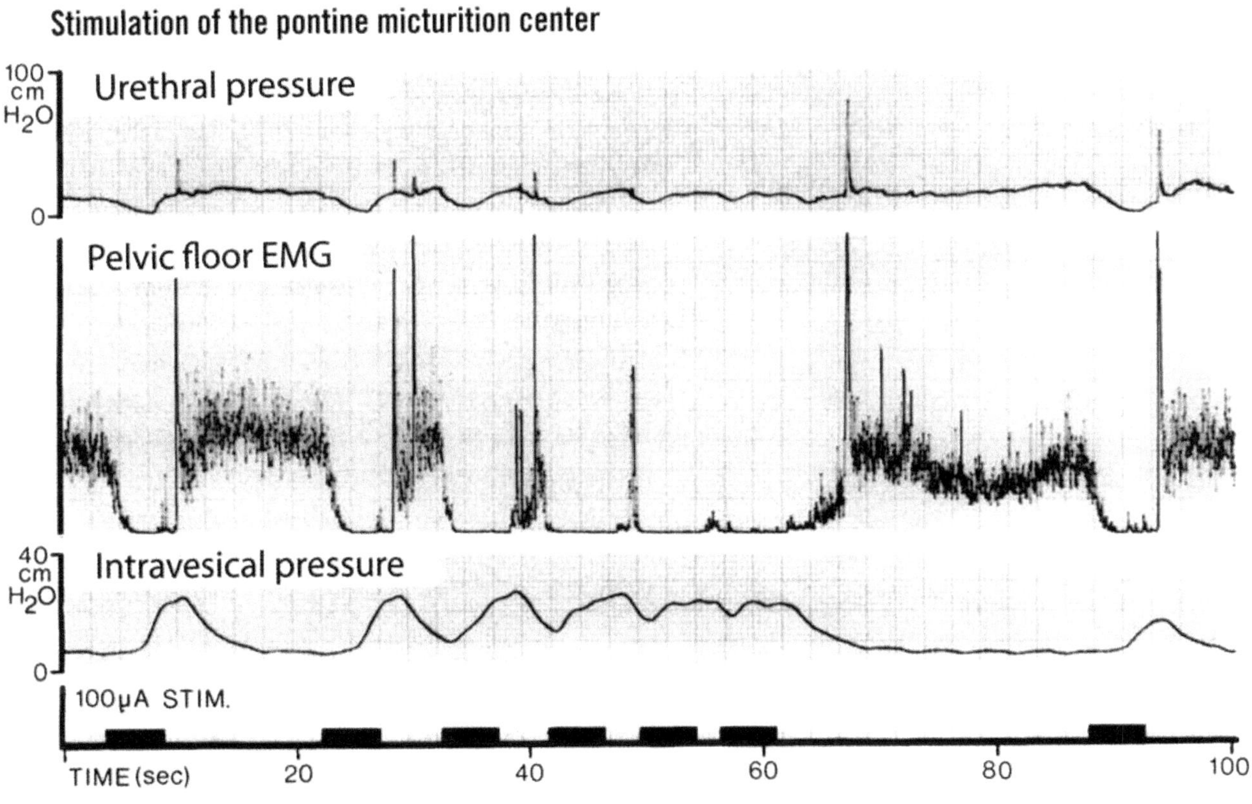

Figure 9.13 Electrical stimulation of PMC

Electrical stimulation in the PMC in a cat results in an increase of the intravesical pressure, representing contraction of the bladder. The same stimulation also generates an immediate decrease of the urethral pressure and pelvic floor EMG, representing relaxation of the external urethral sphincter (from Holstege *et al.*, 1986).

volunteers were asked to urinate, while lying with their heads in the scanner. In the group that could satisfactorily perform the task, increased regional cerebral blood flow (rCBF) was observed in an area in the dorsal pontine tegmentum, which was interpreted as being the PMC (Figure 9.10, left). Similar to animals, also in humans the PMC contains corticotrophin-releasing factor (CRF) neurons (Benarroch and Schmeichel, 2001). These authors also showed that in patients with multiple system atrophy the number of these CRF cells was strongly decreased, which probably explains why these patients suffer from severe bladder dysfunction.

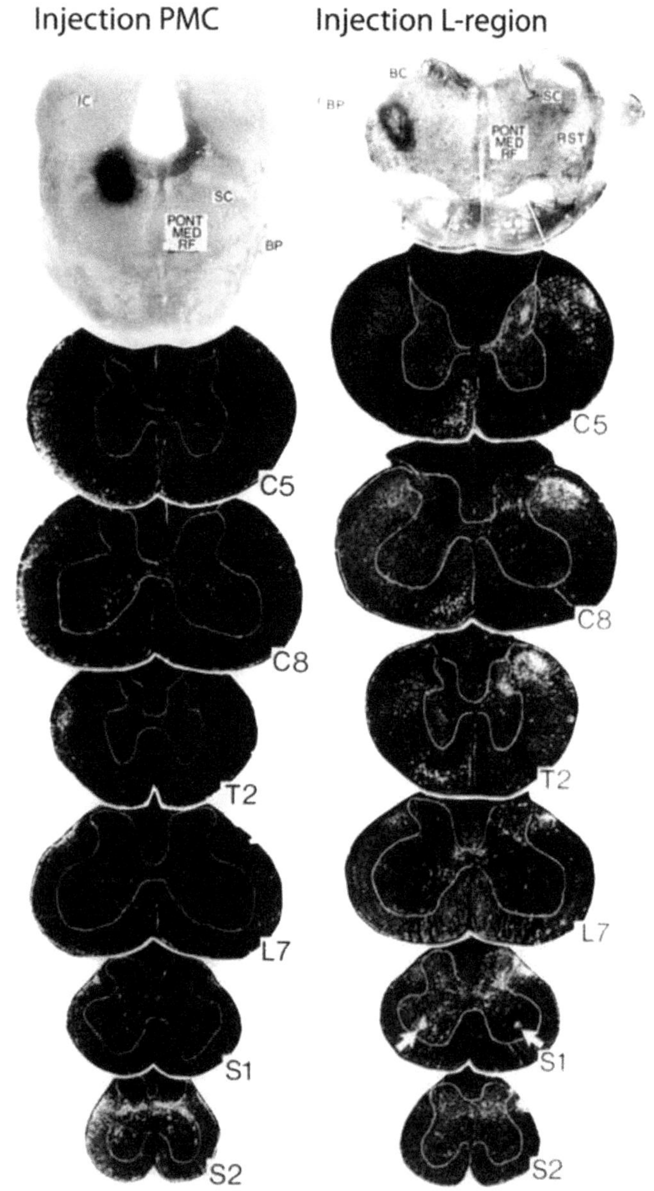

Figure 9.14 Spinal projections from the PMC and the L-region

Autoradiographic tracing in the cat of the projections of the pontine micturition center (PMC) on the left and of the continence center (L-region) on the right. Note on the left that the PMC fibers first descend through the ipsilateral ventrolateral funiculus (C5), but slowly shift to a dorsolateral position (L7 and sacral), where they terminate bilaterally on the parasympathetic motoneurons and in the intermediomedial cell groups. The cells of the L-region do not project to the sacral parasympathetic motoneurons, but to the ON motoneurons in S1 (see arrows). BC = Brachium coniunctivum; BP = Brachium pontis; IC = Inferior colliculus; Pont Med RF = Pontine medial reticular formation; RST = Rubrospinal tract; SC = Nucleus subcoeruleus.

PMC-afferents

As indicated above, the PMC is under enormous influence of the nearby PAG and PAG stimulation leads to micturition. However, there are two other afferent pathways to the PMC. The first originates in the preoptic region (Figure 9.16A,B) and the second in the caudal hypothalamus (Figure 9.16C). Both areas also project heavily to the PAG. The function of these pathways to the PMC is unknown, but, because the preoptic region plays an important part in sexual behavior, one might speculate that the first projection is involved in inhibition of micturition during sex (Kuipers *et al.*, 2006a). In rats a strong projection from the magnocellular bed nucleus of the stria terminalis to the PMC has been described (Dong and Swanson, 2006), but such projections have not been found in the cat (Holstege *et al.*, 1985).

Continence center or L-region

The L- (=lateral) region was first demonstrated by Holstege *et al.*, (1986), when they showed a direct projection from a region in the lateral pons to Onuf's nucleus, but avoiding the sacral parasympathetic bladder motoneurons (Figure 9.14, right). Stimulation in the L-region results in strong excitation of the pelvic floor musculature and an increase in the urethral pressure (Holstege *et al.*, 1986; Figure 9.17). Bilateral lesions in the L-region give rise to an inability to store urine; bladder capacity is reduced and urine is expelled prematurely by excessive detrusor activity, accompanied by urethral relaxation (Griffiths *et al.*, 1990). Outside the episodes of detrusor activity, the urethral pressure is not depressed below normal values (Mallory *et al.*, 1989). These observations suggest that during the filling phase the L-region has a continuous excitatory effect on the nucleus of Onuf, resulting in inhibition of urethral relaxation coupled with detrusor contraction. In the PET-study of Holstege and co-workers (Blok *et al.*, 1997a, 1998a), about half of the volunteers were willing to micturate, but, for emotional reasons, could not perform and tightly contracted their pelvic floor. The PET-scan results in these volunteers revealed increased rCBF in an area in the ventrolateral pontine tegmentum, which might represent the L-region (Figure 9.18). Perhaps the L-region should be considered as a "continence" center, especially since the PET-scan results suggest the existence of such a center in humans (Blok *et al.*, 1997a, 1998a).

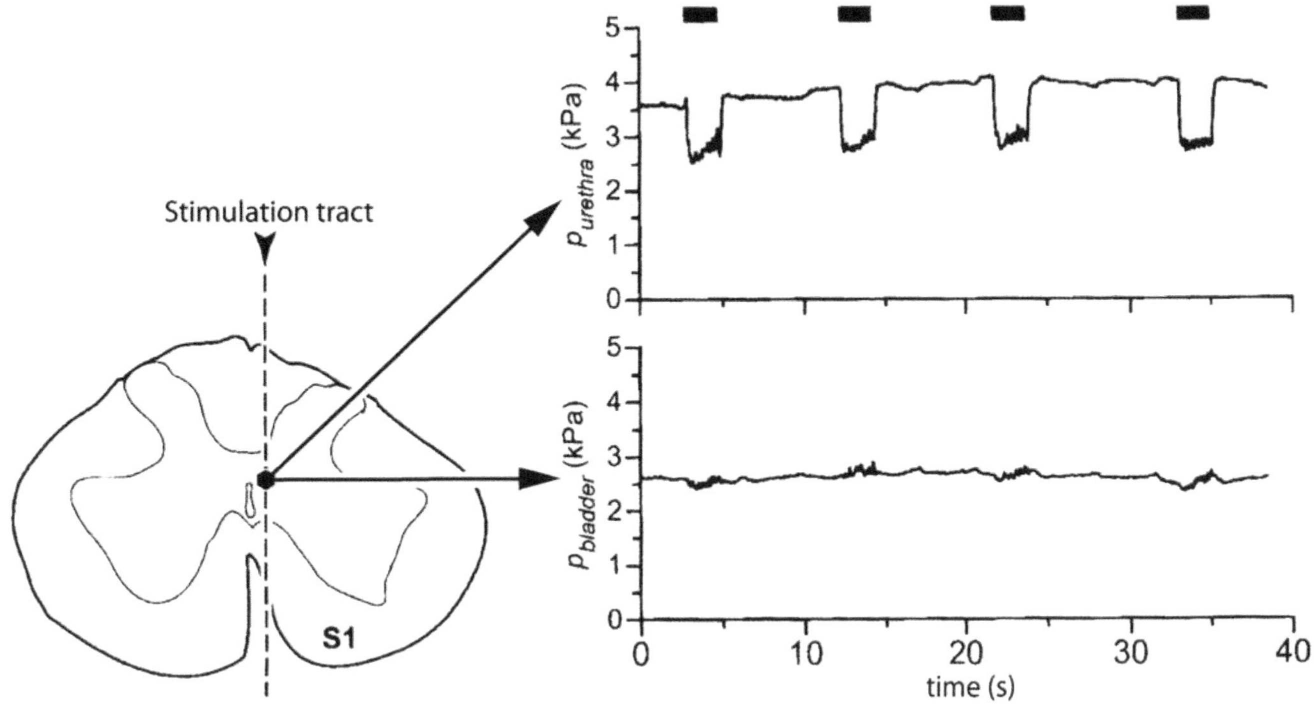

Figure 9.15 Electrical stimulation of the sacral intermediomedial cell group

Electrical stimulation of the sacral intermediomedial cell group results in a clearcut decrease of the urethral pressure, but without any substantial effect on bladder pressure (from Blok *et al.*, 1998b).

Other brainstem-spinal pathways possibly involved in bladder and sphincter motoneuronal control

Diffuse descending systems

SPN and ON not only receive projections from the M- or L-region, but also from other cell groups such as the level-setting mechanisms in the ventromedial caudal pons and medulla (Holstege and Kuypers, 1982; see Holstege, 1991 for review). The neurons in this region send fibers to all parts of the gray matter of the spinal cord, thus also to SPN and ON motoneurons. These diffuse serotonergic and other projections are sometimes mistaken for specific micturition control systems. Also the nor-adrenergic projections from the dorsolateral pontine tegmentum (locus coeruleus and nucleus subcoeruleus) and the dopamine projections from the A11 cell group in the rostral mesencephalon (Holstege, JC *et al.*, 1996) take part in these diffuse descending systems but don't seem to play a specific role in micturition control.

Abdominal pressure control systems

The pelvic floor forms the bottom of the abdominal cavity.

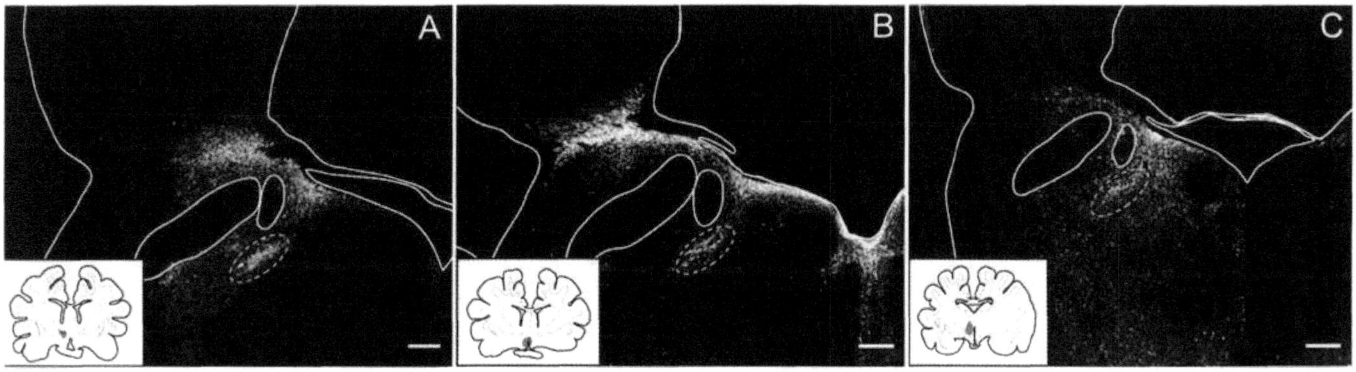

Figure 9.16 Preoptic and hypothalamic projections to PMC

After injecting an anterograde tracer in the medial preoptic area (A,B), or in the caudal hypothalamus (C), labeled fibers were present in the PMC (from Kuipers *et al.*, (2006a).

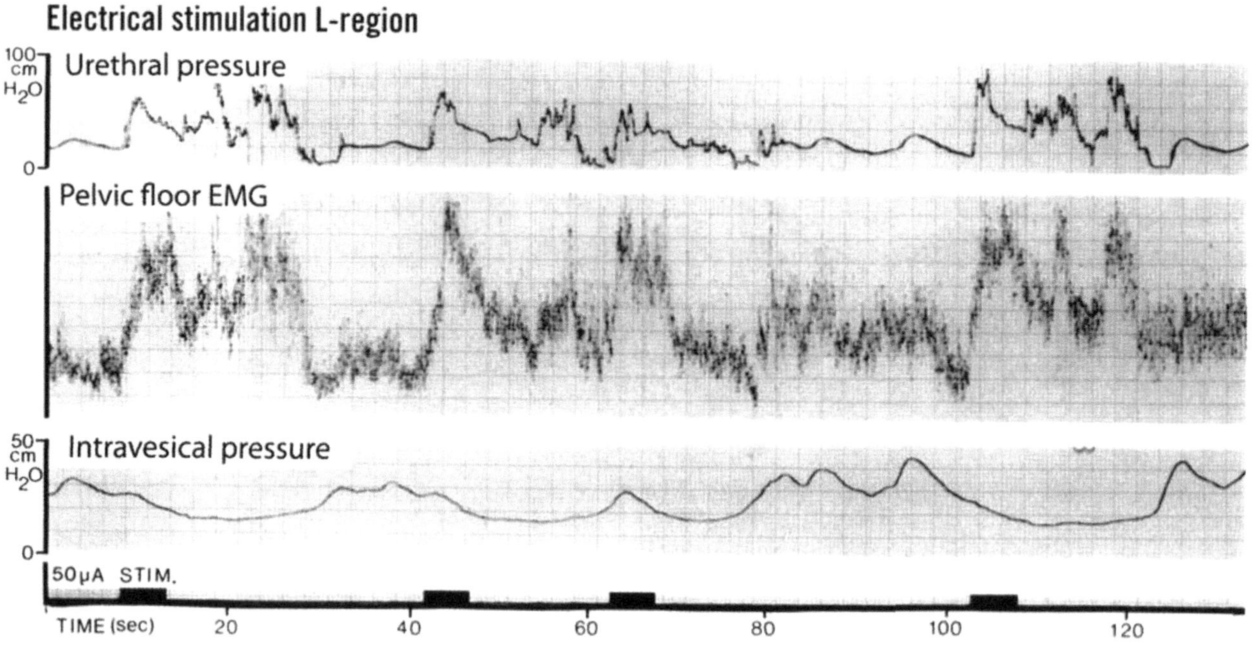

Figure 9.17 Electrical stimulation of the L-region

Electrical stimulation in the L-region or continence center in the cat produces an immediate increase of the urethral pressure and pelvic floor EMG during the time of stimulation. There is no obvious effect on the bladder pressure (from Holstege *et al.*, 1986).

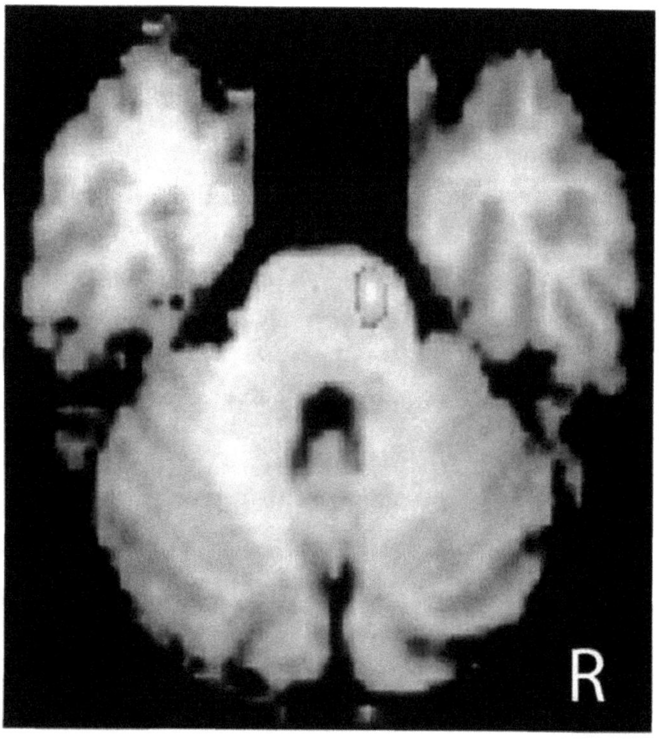

Figure 9.18 PET-scans in humans during failure to micturate

About half of the human volunteers that were asked to micturate while lying with their head in the PET-scanner could, for emotional reasons, not perform. In these cases the PMC was not activated. Instead, a region more ventral in the pons on the right side (artificially indicated with black) was activated, possibly representing the L-region or continence center (from Blok *et al.*, 1997a).
This figure is reproduced in the color insert.

Increased abdominal pressure is needed for strong expiration, vocalization, vomiting, and parturition. The pelvic floor plays an important role in these motor activities. It is not surprising, therefore, that the nucleus retroambiguus, which controls abdominal pressure by innervating abdominal wall muscle motoneurons, also innervates ON motoneurons (Holstege and Tan, 1987; Figure 9.19). It has also been demonstrated that the nucleus retroambiguus is involved in mating behavior in cat, because it projects to the lower limb and other motoneuronal cell groups concerned with the posture of the female to enable the male to copulate. These projections grow and become almost ten times as strong when the female cat gets in estrus (VanderHorst and Holstege, 1997). It is not yet known whether this process of growth involves the retroambiguus projections to ON as well.

Forebrain micturition control

Forebrain projections to the PAG

In general, the PAG receives input from many limbic structures as the preoptic region (Holstege, 1987), central nucleus of the amygdala (Hopkins and Holstege, 1978), lateral bed nucleus of the stria terminalis (Holstege *et al.*, 1985; Figure 9.20) and lateral hypothalamus (Holstege, 1987) and from the tegmental field ventrolaterally adjoining to the PAG. Cortical projections originate from the medial prefrontal cortex and the insula (Paredes *et al.*, 2000; Dujardin and Jurgens, 2005). Especially as

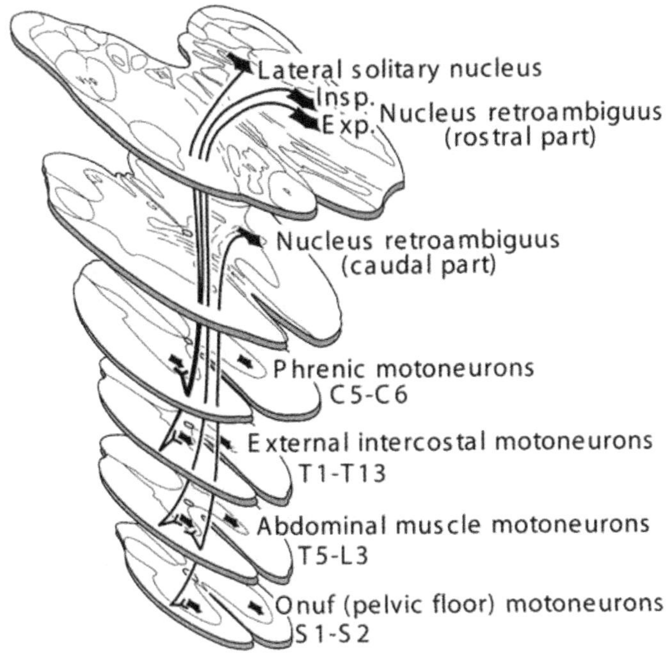

Figure 9.19 Projections from solitary and retroambiguus nuclei to spinal cord

This diagram shows the projections from the lateral solitary nucleus and from the rostral and caudal parts of the nucleus retroambiguus (NRA) to the spinal cord. Note that the caudal NRA projects to ON as well as the abdominal muscle motoneurons.

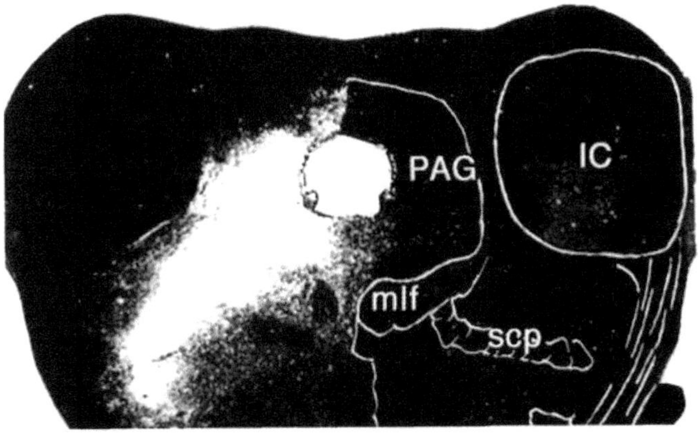

Figure 9.20 Forebrain projections from PAG

An anterograde tracer injection in the bed nucleus of the stria terminalis in the cat reveals strong projections to the dorsomedial, lateral and ventrolateral PAG (from Holstege *et al.*, 1985). Similar projections were found after injections in the central nucleus of the amygdala and lateral hypothalamus. (IC = Inferior colliculus; mlf = medial longitudinal fasciculus; PAG = Periaqueductal gray; scp = Superior cerebellar peduncle).

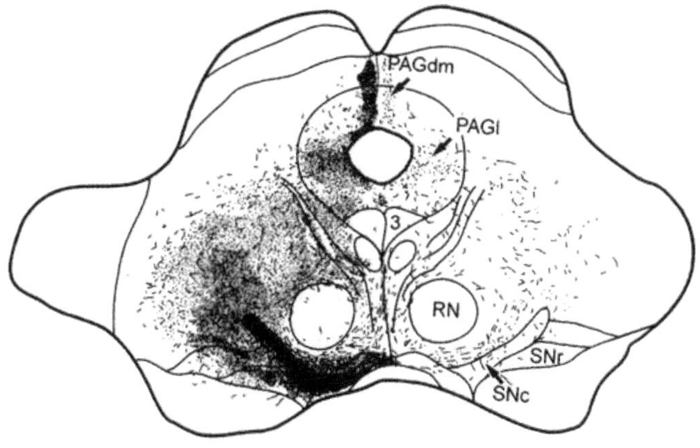

Figure 9.21 Projection from the infralimbic cortex to PAG

After an anterograde tracer injection in the infralimbic cortex in the cat a strong projection to the more central parts of the PAG is observed (from Kuipers *et al.*, 2006). (PAGdm = dorsomedial PAG; PAGl = lateral PAG; RN = Red nucleus; SNc = Substantia nigra pars compacta; SNr = Substantia nigra pars reticulata).

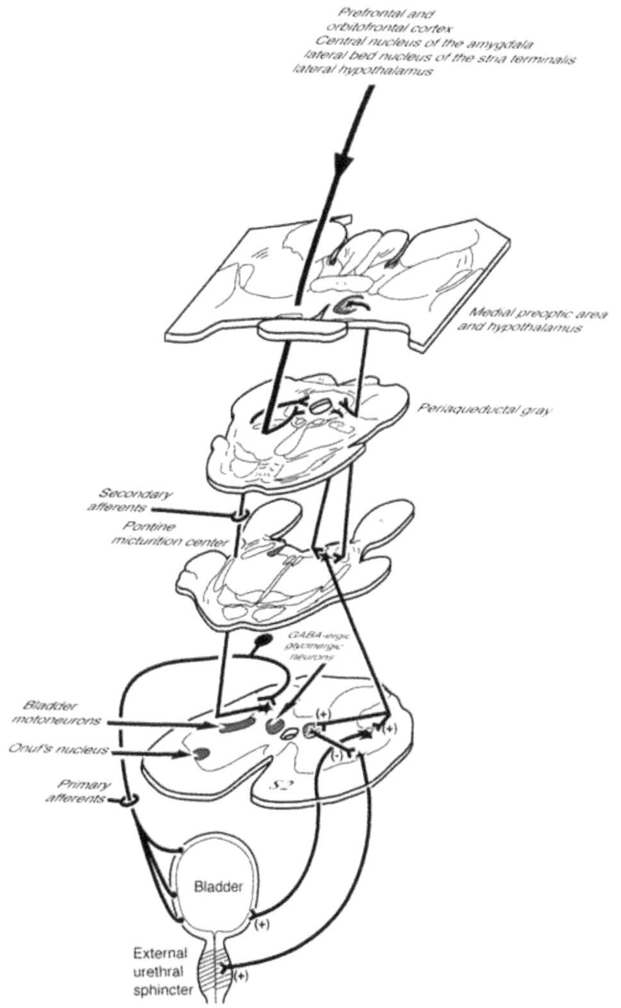

Figure 9.22 Centers involved in the control of micturition in humans

This diagram summarizes the most important elements involved in the central nervous system control of micturition in humans. BC = brachium coniunctivum; CA = anterior commissure; IC = Inferior colliculus; PON = Pontine nuclei; SC = superior colliculus.

the ventromedial prefrontal cortex has a strong relationship with the PAG. For example, in humans, in the case of increasing threat, brain activity shifts from the ventromedial prefrontal cortex to the PAG (Mobbs *et al.*, 2007). The infralimbic cortex (Kuipers *et al.*, 2006b; Figure 9.21) and the paraventricular hypothalamic nucleus (Holstege, 1987) project specifically to the central PAG. As mentioned before, PVN also projects to all autonomic (sympathetic and parasympathetic) motoneurons in brainstem and spinal cord as well as to ON (Figure 9.7).

In conclusion, the cells in the central PAG, receiving bladder filling information from GN as well as the cells in the PAG and in adjoining regions, projecting to the PMC are under strong control of the limbic system. Furthermore, one might expect that central PAG neurons are also influenced by transmitter substances in the aqueductal cerebrospinal fluid, released by more rostrally located paraventricular cell groups into the lateral or third ventricles (McMillan *et al.*, 1998).

Conclusion

The micturition circuit and the forebrain of this circuit, as described above, allows the individual, animals as well as humans, to decide where and when to micturate. When this control is failing great problems arise. If the lesion is caudal to the PAG, retention of urine may develop; when the lesion is rostral to the PAG, so-called overactive bladder (OAB) and urge-incontinence may occur. OAB and urge-incontinence are extremely frequent diseases in the elderly, and a thorough knowledge of this micturition control system is crucial in order to fight these awful diseases. We are convinced, however, that further research on this control system will lead to much better therapies than are available at present.

References

Andersson KE (2002) Bladder activation: afferent mechanisms. Urology 59 (Suppl 5A), 43-50.

Bahns E, Ernsberger U, Jánig W, Nelke A (1986) Functional characteristics of lumbar visceral afferent fibres from the urinary bladder and the urethra in the cat. Pflúgers Arch 407, 510-518.

Barrington FJ.F (1925) The effect of lesions of the hind- and mid-brain on micturition in the cat, Quart J Exp Physiol Cogn Med 15, 81-102.

Benarroch EE, Schmeichel AM (2001) Depletion of corticotrophin-releasing factor neurons in the pontine micturition area in multiple system atrophy. Ann Neurol 50, 640-645.

Blaivas JG (1982) The neurophysiology of micturition: a clinical study of 550 patients. J Urol 127, 958-63.

Blok BFM, Holstege G (1994) Direct projections from the periaqueductal gray to the pontine micturition center (M-region). An anterograde and retrograde tracing study in the cat. Neurosc Lett 166, 93-96

Blok BFM, Holstege G (1997) Ultrastructural evidence for a direct pathway from the pontine micturition center to the parasympathetic preganglionic motoneurons of the bladder of the cat. Neurosci Lett 222, 195-198

Blok BFM, Roukema G, Geerdes B, Holstege G (1996) Location of external anal sphincter motoneurons in the sacral cord of the female domestic pig. Neurosci Lett 216, 203-206.

Blok BFM, Willemsen AT, Holstege G (1997a) A PET study on brain control of micturition in humans. Brain 120, 111-121.

Blok BFM, deWeerd H, Holstege G (1997b) The pontine micturition center projects to sacral cord GABA immunoreactive neurons in the cat. Neurosci Lett 233, 109-112

Blok BFM, Sturms LM, Holstege G (1998a) Brain activation during micturition in women. Brain 121, 2033-2042.

Blok BFM, Maarseveen JTPW, Holstege G (1998b) Electrical stimulation of the sacral dorsal gray commissure evokes relaxation of the external urethral sphincter in the cat. Neurosc Lett 249, 68-70.

Buss RR, Shefchyk SJ (1999) Excitability changes in sacral afferents innervating the urethra, perineum and hindlimb skin of the cat during micturition. J Physiol 514, 593-607.

Carvalho M, Schwartz MS, Swash M (1995) Involvement of the external anal sphincter in amyotrophic lateral sclerosis. Muscle Nerve 18, 848-53.

Chalmers D, Swash M (1987) Selective vulnerability of urinary Onuf motoneurons in Shy-Drager syndrome. J Neurol 234, 259-260.

Coleman AM, Sengelaub DR (2002) Patterns of dye coupling in lumbar motor nuclei of the rat. J Comp Neurol 454, 34-41

De Groat WC (1975) Nervous control of the urinary bladder of the cat. Brain Res 87, 201-11.

De Groat WC, Yoshimura N (2001) Pharmacology of the lower urinary tract. Annu Rev Pharmacol Toxicol 41, 691-721.

De Groat WC, Nadelhaft I, Milne RJ, Booth AM, Morgan C, Thor K (1981) Organization of the sacral parasympathetic reflex pathways to the urinary bladder and large intestine. J Auton Nerv Syst 3, 135-160.

De Groat WC, Booth AM, Milne RJ, Roppolo JR (1982) Parasympathetic preganglionic neurons in the sacral spinal cord. J Auton Nerv Syst 23-43.

Dekker JJ, Lawrence DG, Kuypers HGJM (1973) The location of longitudinally running dendrites in the ventral horn of the cat spinal cord. Brain Res 51, 319-325.

Dixon JS, Gilpin CJ (1987) Presumptive sensory axons of the human urinary bladder: a fine structural study. J Anat 151, 199-207.

Dmitrieva N, Berkley KJ (2005) Influence of estradiol on micturition thresholds in the rat: involvement of the hypogastric nerve. Am J Physiol Regul Integr Comp Physiol 289, R1724-1728.

Dong HW, Swanson LW (2006) Projections from bed nuclei of the stria terminalis, magnocellular nucleus: implications for cerebral hemisphere regulation of micturition, defecation, and penile erection. J Comp Neurol 494, 108-41.

Dujardin E, Jurgens U (2005) Afferents of vocalization-controlling periaqueductal regions in the squirrel monkey. Brain Res 1034, 114-131.

Forrest SL, Keast JR (2008) Expression of receptors for glial cell line-derived neurotrophic factor family ligands in sacral spinal cord reveals separate targets of pelvic afferent fibers. J Comp Neurol 506, 989-1002.

Fowler CJ (2002) Bladder afferents and their role in the overactive bladder. Urology 59 (Suppl 5A), 37-42.

Gabella C, Davis S (1998) Distribution of afferent axons in the bladder of rats. J Neurocytol 27, 141-155.

Griffiths D, Holstege G, Dalm E, De Wall H (1990) Control and coordination of bladder and urethral function in the brainstem of the cat. Neurourology and Urodynamics 9, 63-82

Gustafson KJ, Creasey GH, Grill WM (2004) A urethral afferent mediated excitatory bladder reflex exists in humans. Neurosci Lett 360, 9-12.

Holstege G (1987) Some anatomical observations on the projections from the hypothalamus to brainstem and spinal cord: an HRP and autoradiographic tracing study in the cat. J Comp Neurol 260, 98-126.

Holstege G (1991) Descending motor pathways and the spinal motor system. Limbic and non-limbic components. In: *Role of the forebrain in sensation and behavior.* G Holstege (ed). Elsevier, Amsterdam.

Holstege G, Kuypers HGJM (1982) The anatomy of brain stem pathways to the spinal cord in cat. A labeled amino acid tracing study. In: *Descending pathways to the spinal cord.* HGJM Kuypers and GF Martin (eds). Elsevier, Amsterdam.

Holstege G, Sie JAML (2002) The central control of the pelvic floor. In: *The pelvic floor, its functions and Disorders.* J Pemberton, M Schwash and MM Henry (eds) WB Saunders, London.

Holstege G, Tan J (1987) Supraspinal control of motoneurons innervating the striated muscles of the pelvic floor including urethral and anal sphincters in the cat. Brain 110, 1323-1344.

Holstege G, Kuypers HGJM, Boer RC (1979) Anatomical evidence for direct brain stem projections to the somatic motoneuronal cell groups and autonomic preganglionic cell groups in cat spinal cord. Brain Res 171, 329-333.

Holstege G, Meiners L, Tan K (1985) Projections of the bed nucleus of the stria terminalis to the mesencephalon, pons, and medulla oblongata in the cat. Exp Brain Res 58, 379-391.

Holstege G, Griffiths D, De Wall H, Dalm E (1986) Anatomical and physiological observations on supraspinal control of bladder and urethral sphincter muscles in the cat. J Comp Neurol 250, 449-461.

Holstege JC, Van Dijken H, Buijs RM, Goedknegt H, Gosens T, Bongers CM (1996) Distribution of dopamine immunoreactivity in the rat, cat and monkey spinal cord. J Comp Neurol 376, 631-652.

Hopkins DA, Holstege G (1978) Amygdaloid projections to the mesencephalon, pons and medulla oblongata in the cat. Exp Brain Res 32, 529-547.

Hwang SJ, Oh JM, Valtschanoff JG (2005) The majority of bladder sensory afferents to the rat lumbosacral spinal cord are both IB4- and CGRP-positive. Brain Res 1062, 86-91

Iwata M, Inoue K, Mannen T (1993) Functional localization in the Onufrowicz nucleus in man. Clin Neuropathol 12, 112-116

Kawatani M, Takeshige C, De Groat WC (1990) Central distribution of afferent pathways from the uterus of the cat. J Comp Neurol 302, 294-304.

Keay KA, Feil K, Gordon BD, Herbert H, Bandler R (1997) Spinal afferents to functionally distinct periaqueductal gray columns in the rat: an anterograde and retrograde tracing study. J Comp Neurol 385, 207-229.

Kenton K, Simmons J, FitzGerald MP, Lowenstein L, Brubaker L (2007) Urethral and bladder current perception thresholds: normative data in women. J Urol 178, 189-192.

Klop EM, Mouton LJ, Kuipers R, Holstege G (2005) Neurons in the lateral sacral cord of the cat project to periaqueductal grey, but not to thalamus. Eur J Neurosci 21, 2159-2166.

Kruse MN, De Groat WC (1990) Micturition reflexes in decerebrate and spinalized neonatal rats. Am J Physiol 258, 1508-1511.

Kuipers R, Izhar Z, Gerrits PO, Miner W, Holstege G (2004) Location of bladder and urethral sphincter motoneurons in the male guinea pig (Cavia porcellus). Neurosci Lett 362, 57-60.

Kuipers R, Mouton LJ, Holstege G (2006a) Afferent projections to the pontine micturition center or Barringtons nucleus in the cat. J Comp Neurol 494, 36-53

Kuipers R, Mensinga GM, Boers J, Klop EM, Holstege G (2006b) Infralimbic cortex projects to all parts of the pontine and medullary lateral tegmental field in cat. Eur J Neurosci 23, 3014-24.

Kuzuhara S, Kanazawa I, Nakanishi T (1980) Topographical localization of the Onuf's nuclear neurons innervating the rectal and vesical striated sphincter muscles: a retrograde fluorescent double labeling in cat and dog. Neurosci Lett 16, 125-130.

Leedy MG, Bresnahan JC, Mawe GM, Beattie MS (1988) Differences in synaptic inputs to preganglionic neurons in the dorsal and lateral band subdivisions of the cat sacral parasympathetic nucleus. J Comp Neurol 268, 84-90.

Liu Z, Sakakibara R, Nakazawa K, Uchiyama T, Yamamoto T, Ito T, Hattori T (2004) Micturition-related neuronal firing in the periaqueductal gray area in cats. Neuroscience 126, 1075-1082.

MacMillan SJ, Mark MA, Duggan AW (1998) The release of beta-endorphin and the neuropeptide-receptor mismatch in the brain. Brain Res 794, 127-36.

Mallory DS, Bona-Gallo A, Gallo RV (1989) Neurotransmitter involvement in naloxone-induced stimulation of pulsatile LH release on day 8 of pregnancy in the rat. Brain Res Bull 22 1015-1021

Mannen T, Iwata M, Toyokura Y, Nagashima K (1977) Preservation of a certain motoneurone group of the sacral cord in amyotrophic lateral sclerosis: its clinical significance. J Neurol Neurosurg Psychiatry 40, 464-469.

Matsumoto A, Arnold AP, Micevych PE (1989) Gap junctions between lateral spinal motoneurons in the rat. Brain Res 495, 362-366.

Matsuura S, Allen GV, Downie JW (1998) Volume-evoked micturition reflex is mediated by the ventrolateral periaqueductal gray in anesthetized rats. Am J Physiol 275, R2049-2055

Mobbs D, Petrovic P, Marchant JL, Hassabis D, Weiskopf N, Seymour B, Dolan RJ, Frith CD (2007) When fear is near: threat imminence elicits prefrontal-periaqueductal gray shifts in humans. Science 317, 1079-1083.

Morgan CW (2001) Axons of sacral preganglionic neurons in the cat: II. Axon collaterals. J Neurocytol 30, 767-87.

Morgan CW, Ohara PT (2001) Quantitative analysis of the dendrites of sacral preganglionic neurons in the cat. J Comp Neurol 437, 56-69.

Morgan CW, Nadelhaft I, De Groat WC (1981) The distribution of visceral primary afferents from the pelvic nerve to Lissauer's tract and the spinal gray matter and its relationship to the parasympathetic nucleus. J Comp Neurol 201, 415-440.

Morgan CW, Ohara PT, Scott DE (1999) Vasoactive intestinal polypeptide in sacral primary sensory pathways in the cat. J Comp Neurol 407, 381-94.

Morrison J (1999) The activation of bladder wall afferent nerves. Exp Physiol 84, 131-136.

Mouton LJ, Holstege G (1994) The periaqueductal gray in the cat projects to lamina VIII and the medial part of lamina VII throughout the length of the spinal cord. Exp Brain Res 101, 253-264.

Nacimiento W, Töpper R, Fischer A, Mobius E, Oestreicher AB, Gispen WH, Nacimiento AC, Noth J, Kreutzberg GW (1993) B-50 (GAP-43) in Onuf's nucleus of the adult cat. Brain Res 613, 80-87.

Nadelhaft I, Booth AM (1984) The location and morphology of preganglionic neurons and the distribution of visceral afferents from the rat pelvic nerve: a horseradish peroxidase study. J Comp Neurol 226, 238-45.

Nadelhaft I, Vera PL (1996) Neurons in the rat brain and spinal cord labeled after pseudorabies virus injected into the external urethral sphincter. J Comp Neurol 375, 502-517.

Neuhuber WL, Sandoz PA, Fryscak T (1986) The central projections of primary afferent neurons of greater splanchnic and intercostal nerves in the rat. A horseradish peroxidase study. Anat Embryol (Berl) 174, 123-144.

Onuf (Onufrowicz) B (1899) Notes on the arrangement and function of the cell groups in the sacral region of the spinal cord. J Nerv Ment Disease 26, 363-369.

Paredes J, Winters RW, Schneiderman N, McCabe PM (2000) Afferents to the central nucleus of the amygdala and functional subdivisions of the periaqueductal gray: neuroanatomical substrates for affective behavior. Brain Res 887, 157-173.

Rexed B (1954) A cytoarchitectonic atlas of the spinal cord in the cat. J Comp Neurol 100, 297-379.

Rodella L, Rezzani R, Gioia M, Tredici G, Bianchi R (1998) Expression of Fos immunoreactivity in the rat supraspinal regions following noxious visceral stimulation. Brain Res Bull 47, 357-366.

Romanes GJ (1951) The motor cell columns of the lumbo-sacral spinal cord of the cat. J Comp Neurol 94, 313-363.

Sakakibara R, Nakazawa K, Shiba K, Nakajima Y, Uchiyama T, Yoshiyama M, Yamanishi T, Hattori T (2002) Firing patterns of micturition-related neurons in the pontine storage centre in cats. Auton Neurosci 99, 24-30.

Sato M, Mizuno N, Konishi A (1978) Localization of motoneurons innervating perineal muscles: a HRP study in cat. Brain Res 149-54.

Schrøder HD (1981) Onuf's nucleus X: a morphological study of a human spinal nucleus. Anat Embryol (Berl) 162, 443-53.

Sie JA, Blok BFM, de Weerd H, Holstege G (2001) Ultrastructural evidence for direct projections from the pontine micturition center to glycine-immunoreactive neurons in the sacral dorsal gray commissure in the cat. J Comp Neurol 429, 631-637.

Takahashi K, Yamamoto T (1979) Ultrastructure of the cell group X of Onuf in the cat sacral spinal cord. Z Mikrosk Anat Forsch 93, 244-256.

VanderHorst VGJM, Holstege G (1992) Afferent projections from the spinal cord to the periaqueductal gray in the cat. Soc Neurosci Abstr 18, 311.

VanderHorst VGJM, and Holstege G (1997) Estrogen induces axonal outgrowth in the nucleus retroambiguus-lumbosacral motoneuronal pathway in the adult female cat. J Neuroscience 17, 1122-1136.

VanderHorst VGJM, Mouton LJ, Blok BFM, Holstege G (1996) Somatotopical organization of input from the lumbosacral cord to the periaqueductal gray in the cat; possible implications for aggressive and defensive behavior, micturition, and lordosis. J Comp Neurol 376, 361-385.

VanderHorst VGJM, Meijer E, Holstege G (2001) Estrogen receptor-alpha immunoreactivity in parasympathetic preganglionic neurons innervating the bladder in the adult ovariectomized female cat. Neurosci Lett 298, 147-150.

Vincent SR, Satoh K (1984) Corticotropin-releasing factor (CRF) immunoreactivity in the dorsolateral pontine tegmentum: further studies on the micturition reflex system. Brain Res 308, 387-391.

Wiseman OJ, Brady CM, Hussain IF, Dasgupta P, Watt H, Fowler CJ, Landon DN (2002) The ultrastructure of bladder lamina propria nerves in healthy subjects and patients with detrusor hyperreflexia. J Urology 168, 2040-2045.

Wiseman OJ, Fowler CJ, Landon DN (2003) The role of the human bladder lamina propria myofibroblast. BJU Int 91, 89-93.

Wyndaele JJ, De Wachter S (2003) The basics behind bladder pain: a review of data on lower urinary tract sensations. Int J Urol 10, S49-S55.

Yaguchi H, Soma H, Miyazaki Y, Tashiro J, Yabe I, Kikuchi S, Sasaki H, Kakizaki H, Moriwaka F, Tashiro K (2004) A case of acute urinary retention caused by periaqueductal grey lesion. J Neurol Neurosurg Psychiatry 75, 1202-1203.

Zagorodnyuk VP, Costa M, Brookes SJH (2006) Major classes of sensory neurons to the urinary bladder. Auton Neurosci Bas Clin 126/127, 390-397.

10 Projections from the Spinal Cord to the Brain

Gulgun Kayalioglu

Introduction

The ascending spinal projections connect the spinal cord to supraspinal levels and transmit sensory information such as pain, temperature, position sense and touch from somatic structures and pressure, pain and visceral information from internal organs. Spinal cord neurons project to the brainstem, cerebellum, midbrain, diencephalon and telencephalon. Some of the projections are directly to supraspinal structures, such as the spinothalamic, spinomesencephalic and spinohypothalamic tracts. Some projections, such as the postsynaptic dorsal column pathway and the spinocervical pathway, synapse with second-order neurons which in turn project to higher centers. Ascending spinal projections are composed of mostly myelinated dorsal root fibers from the dorsal root ganglia and axons of spinal neurons. Anatomically, the ascending spinal projections are located in the ventral, lateral and dorsal funiculi on each side of the spinal cord. In experimental animals, electrophysiological studies and anatomical tract tracing methods using tracers and Wallerian degeneration technique have collectively provided extensive and accurate information on ascending spinal projections. In humans, on the contrary, information is limited to data obtained from patients with localized traumatic and inflammatory spinal lesions, or surgical interventions.

Ascending spinal projections in the ventrolateral funiculus

The ascending spinal projections in the ventrolateral funiculus transmit nociceptive, thermal, non-discriminative touch and pressure information to supraspinal levels. The fibers located in the ventrolateral spinal cord maintain their position throughout the spinal cord and in the brainstem (see Figure 10.1). The spinothalamic tract is a major projection with axons terminating in several thalamic nuclei including the ventral posterolateral nucleus, the intralaminar nuclei and the posterior thalamic nucleus. The ventrolateral ascending projections also include the spinoreticular, spinomesencephalic, spinotectal and spinohypothalamic fibers.

The spinothalamic tract

The spinothalamic tract conveys nociception, temperature, non-discriminative (crude) touch and pressure information to the somatosensory region of the thalamus. It is composed of a ventral (anterior, paleospinothalamic) and a lateral (neospinothalamic) pathway. A dorsolateral spinothalamic tract is also described in the rat, cat and macaque monkey, and clinical evidence suggests it is also present in humans. The dorsolateral spinothalamic tract may contain about one fourth of the total spinothalamic population in primates (Apkarian and Hodge, 1989a).

The ventral spinothalamic tract, located in the anterior funiculus, transmits crude touch and pressure sensations. The lateral spinothalamic tract lies in the ventral part of the lateral funiculus and transmits pain and temperature. Clinical evidence suggests the pathway for pain and temperature conduction is organized as two distinct components (Friehs *et al.*, 1995). The dorsolateral spinothalamic tract lies in the dorsolateral funiculus and is a major nociceptive-specific ascending spinal pathway (Martin *et al.*, 1990). The ventral and lateral spinothalamic tracts ascend separately in the spinal cord accompanied with the spinomesencephalic, spinoreticular and spinohypothalamic tracts. In the medulla they merge to form the spinal lemniscus.

The spinothalamic tract axons migrate ventrally as they ascend the length of the spinal cord. In segments rostral to the cervical enlargement, axons do not continue to migrate further ventrally but continue a position ventral to that in which they ascend through thoracic segments (Zhang *et al.*, 2000). The ventral spinothalamic tract joins the medial lemniscus in the medulla and pons, while the lateral spinothalamic tract continues as the spinal lemniscus. There is somatotopic organization of axons within the spinothalamic tract; fibers entering from rostral and caudal segments are located in the medial and lateral parts of the tract, respectively. In the cat, the ventral-to-dorsal distribution of spinothalamic tract axons is bimodal, located as the ventrolateral and dorsolateral groups. In monkeys, the distribution is unimodal, extending from the ventral surface of the spinal white matter to the ventralmost part of the dorsolateral funiculus. In both animals, ventrally located spinothalamic tract axons are large, coarse, and primarily located peripherally, whereas dorsal spinothalamic tract axons are of fine caliber and are equally distributed in the medial and lateral white matter (Stevens *et al.*, 1991).

The spinothalamic tract terminates mainly in the ventroposterolateral nucleus, ventroposteromedial nucleus, the intralaminar nuclei, mainly the central lateral nucleus, and the posterior complex. Spinothalamic tract projections to the central lateral nucleus of the thalamus play a part in motivational-affective responses to pain, and the projection to lateral thalamus (the ventrobasal complex) is involved in sensory-discriminative aspects of pain (Albe-Fessard *et al.*,

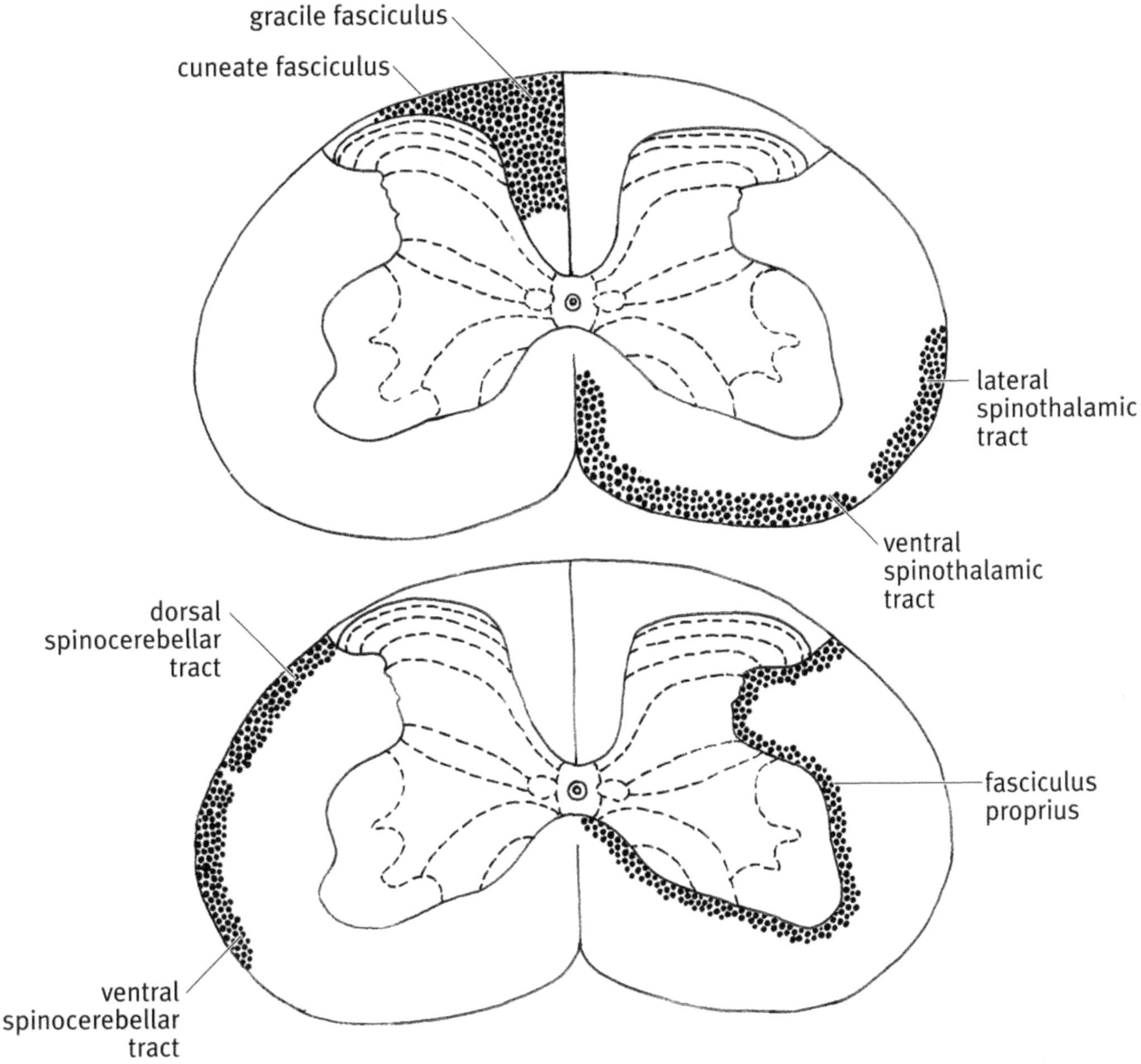

Figure 10.1 Ascending tracts in the spinal cord

A diagram of a transverse section of cervical spinal cord showing the position of the major ascending pathways in a mammal. (adapted from Cramer and Darby, 2005, p.357)

1985). Spinothalamic tract neurons send collateral branches to the medullary reticular formation (Kevetter and Willis, 1983), the parabrachial area (Hylden *et al.*, 1989), the periaqueductal gray (Harmann *et al.*, 1988), and the nucleus accumbens (Kayalioglu *et al.*, 1996). These projections distribute information to multiple brainstem sites, which might in turn activate autonomic or affective responses or descending pain modulatory mechanisms (Hylden *et al.*, 1989).

The total number of spinothalamic tract neurons is estimated to be approximately 6000 in cats (Klop *et al.*, 2005), 9500 in rats (Burstein *et al.*, 1990), and 18000 in monkeys (Apkarian and Hodge, 1989b). The largest concentration of spinothalamic tract neurons is in the upper cervical spinal cord. In rats, the number of neurons projecting from C1-C3 is about 30% of the total number of the spinothalamic tract neurons, and about 24% are located in C4-C8, 26% in T1-T13, 13% in L1-L5, and 7% in L6-Co3 segments (Burstein *et al.*, 1990b). The distribution is similar in cats (Klop *et al.*, 2005) and primates (Apkarian and Hodge, 1989b). The vast majority of spinothalamic tract neurons are located in the upper cervical segments (C1-C3), 30% in rats (Burstein *et al.*, 1990b), 45% in cats (Klop *et al.*, 2005) and 35% in monkeys (Apkarian and Hodge, 1989b).

Neurons that give rise to the spinothalamic tract are localized in different layers and are found in all segments of the spinal cord. Axons of most spinothalamic tract neurons decussate in the ventral white commissure. In the rat, the majority of spinothalamic tract neurons are located contralaterally, mainly in laminae 1, 3-7, 10, and in the lateral spinal nucleus. In lumbar and sacral segments, the neurons are very prominent in the medial part of lamina 5, in contrast to the cervical and thoracic segments where the localization is mostly in the lateral part of laminae 4-5. A few spinothalamic tract neurons are also present in the ventral horn (Kayalioglu *et al.*, 1996, 1999). In cats, spinothalamic tract neurons are located primarily contralaterally in medial laminae 7 and 8, while in monkeys they are found in the contralateral laminae 1 and 4-5, with the rest in laminae 6-8 (Trevino and Carstens, 1975). In cats, ventral spinothalamic tract neurons are found in laminae 4-5, 7-9 and 10, and the dorsolateral spinothalamic tract neurons in lamina 1 (Jones *et al.*, 1987). The spinothalamic tract neurons were observed predominantly in contralateral laminae 1 and 4 in the human following cordotomy (Kuru, 1949).

In the rat, the majority of spinothalamic tract neurons (67-81%) have been found to be located contralaterally (Burstein *et al.*, 1990b) and less than 2% to project bilaterally (Kevetter and Willis, 1983). At C1-C4 levels, ipsilateral cells are prominent in the dorsal portion of the ventral horn (lamina 8) in rats (Granum, 1986), cats (Klop *et al.*, 2005) and monkeys (Apkarian and Hodge, 1989b).

Lamina 1 neurons have the smallest receptive fields of all spinothalamic tract neurons and respond maximally to noxious peripheral stimulation. Lamina 4 neurons have larger receptive fields and respond most commonly to both innocuous and noxious stimuli. Spinothalamic tract neurons in laminae 7-10 have large, frequently bilateral receptive fields and respond to deep somatic and innocuous or noxious cutaneous stimuli (Hodge and Apkarian, 1990).

The spinothalamic tract neurons located in laminae 1-6 project primarily to the lateral thalamus and deeper spinothalamic tract neurons to the intralaminar and medial nuclei of the thalamus. It is hypothesized that the deep neurons are related to aversive behaviors in response to pain, while the more superficial layer neurons are related to the sensory-discriminative aspects of pain (Hodge and Apkarian, 1990). Spinothalamic tract neurons also respond to noxious visceral stimulation. Electrophysiological studies have shown that spinothalamic tract neurons respond to stimulation of visceral organs as in coronary artery occlusion (Blair *et al.*, 1984), testicle compression and urinary bladder, renal pelvic, gall bladder and colorectal distension (Milne *et al.*, 1981; Ammons *et al.*, 1984, Ammons, 1989; Al-Chaer *et al.*, 1999). Responsive neurons are primarily located in laminae 5 and 7 (Ammons *et al.*, 1984). Stimulation of greater splanchnic and cardiopulmonary sympathetic fibers increases (Hobbs *et al.*, 1992), while vagal stimulation (Ammons *et al.*, 1983) decreases activity in spinothalamic tract neurons.

GABA, glycine, serotonin, norepinephrine, dopamine and acetylcholine have an inhibitory effect on spinothalamic tract neurons, whereas glutamate has an excitatory role (Willcockson *et al.*, 1984). It has been suggested that co-release of excitatory amino acids (e.g. glutamate and NMDA) and neuropeptides may contribute to hyperalgesia in sensory transmission in the spinothalamic tract (Dougherty and Willis, 1991). Galanin and cholecystokinine are also found in spinothalamic tract neurons (Ju *et al.*, 1987). Dorsal horn spinothalamic tract neurons show neurokinin-1 (substance P) immunoreactivity, with the vast majority in lamina 1 (Marshall *et al.*, 1996). Although previous studies suggest an involvement of nitric oxide in nociceptive transmission (Maiskii *et al.*, 1998; Lin *et al.*, 1999), recent studies failed to nitric oxide immunoreactivity in major ascending pathways including the spinothalamic tract (Kayalioglu *et al.*, 1999; Usunoff *et al.*, 1999).

The spinoreticular tract

The spinoreticular tract (SRT) ascends in the ventrolateral funiculus and terminates in several nuclei of the reticular formation of the brainstem, including the lateral, dorsal and gigantocellular reticular nuclei, the oral and caudal pontine reticular nuclei, the dorsal and lateral paragigantocellularis nuclei, the raphe magnus nucleus, and the central reticular nucleus (Mehler *et al.*, 1960; Hanckok and Fougerousse, 1976; Chaouch *et al.*, 1983; Menetrey *et al.*, 1983; Lima, 1990; Willis and Westlund, 1997). Some spinoreticular tract axons are collateral branches of the spinothalamic tract neurons (Kevetter and Willis, 1983). There is no clear somatotopic organization of the spinoreticular tract axons and most of the axons are myelinated.

The spinoreticular tract is involved in the control of descending modulation, motivational-affective aspects of pain, and also motor and neurovegetative responses to pain (Haber *et al.*, 1982; Mense, 1983; Chapman *et al.*, 1985; Zhang *et al.*, 1990; Millan, 1999).

The cells of origin of the spinoreticular tract are located throughout the length of the spinal cord, mostly in the cervical and lumbar segments. Neurons are located mainly in contralateral laminae 7 and 8, also in the lateral reticulated part of lamina 5. There are also some neurons in laminae 1 and 10 and the lateral spinal nucleus (Kevetter *et al.*, 1982; Chaouch *et al.*, 1983; Menetrey *et al.*, 1983; Peschanski and Besson, 1984). The spinoreticular tract neurons in the lumbosacral spinal cord mainly project to the contralateral brainstem, but a bilateral distribution is observed from the cervical segments (Kevetter *et al.*, 1982; Chaouch *et al.*, 1983).

Electrophysiological studies have shown that the spinoreticular projection has considerable functional heterogeneity (Sahara *et al.*, 1990). Although some spinoreticular tract neurons are not activated by noxious and innocuous peripheral stimuli, some are activated by high and low threshold cutaneous stimulation and high threshold stimulation of muscle afferents (Haber *et al.*, 1982; Sahara *et al.*, 1990). The largest population of spinoreticular tract neurons are high threshold, requiring noxious stimulation for their activation (Haber *et al.*, 1982; Ness *et al.*, 1998). The spinoreticular tract neurons are also under the influence of descending inhibitory control originating from the nucleus raphe magnus and bulbar reticular formation (Menetrey *et al.*, 1980). This suggests that the brainstem and the spinoreticular tract play a role in diffuse noxious inhibitory control (De Broucker *et al.*, 1990). Spinoreticular neurons especially in lamina 10 have been shown to contain various immunoreactive peptides including substance P, vasoactive intestinal polypeptide, bombesin, dynorphin, enkephalin, and cholecystokinin (Nahin and Micevych, 1986; Nahin, 1987; Leah *et al.*, 1988).

The spinomesencephalic tract

The spinomesencephalic tract projects to different areas in the midbrain including the periaqueductal gray, the cuneiform nucleus, the intercollicular nucleus, the deep layers of the superior colliculus, nucleus of Darkschewitsch, the anterior and posterior pretectal nuclei, the red nucleus, Edinger-Westphal nucleus and the interstitial nucleus of Cajal (Mehler *et al.*, 1960; Kerr, 1975; Wiberg *et al.*, 1987; Yezierski, 1988; Kayalioglu et al,. 1996, 1999). Cells of origin are located throughout the whole length of the spinal cord, with the largest population in C1-C4. In the rat and the cat, the spinomesencephalic tract originates from laminae 1, 4-6, 10 and the lateral spinal nucleus neurons. A small number of neurons are also observed in the ventral horn (Mantyh, 1982; Menetrey *et al.*, 1982; Kayalioglu *et al.*, 1996, 1999; Wiberg and Blomqvist, 1984; Wiberg *et al.*, 1987; Yezierski and Mendez, 1991).

Most axons forming the spinomesencephalic tract cross the midline and ascend in the ventrolateral funiculus together with the spinothalamic and spinoreticular tracts, but axons of lamina 1 spinomesencephalic tract neurons ascend bilaterally in the dorsolateral funiculus (Hylden *et al.*, 1986). Some spinomesencephalic tract axons are collaterals of spinothalamic tract neurons (Harmann *et al.*, 1988). There is a rough somatotopic organization in the spinomesencephalic tract; axons from the upper spinal cord segments terminate more rostrally in the midbrain than axons from the lower segments (Wiberg and Blomqvist, 1984). The periaqueductal gray has a columnar organization; afferents from deep somatic and visceral structures terminate in the ventrolateral column of the periaqueductal gray, whereas afferents from the skin terminate in the lateral column (Keay *et al.*, 1997; Clement *et al.*, 2000). In the upper cervical segments, there is a rough topographical separation of neurons projecting to the ventrolateral and lateral periaqueductal gray columns, whereas in lower segments the neurons are similarly distributed (Keay *et al.*, 1997).

Different components of the spinomesencephalic tract have different functions. Projections to the periaqueductal gray are responsible for motivational-affective responses to pain and for descending control of nociception (Sewards and Sewards, 2002).

Projections to the superior colliculus, the intercollicular nucleus and the pretectal nuclei constitute the spinotectal tract, which ascends in the lateral funiculus ventral to the spinothalamic tract (Antonetty and Webster, 1975; Zemlan *et al.*, 1978; Yezierski, 1988). The cells of origin of the spinotectal tract are located in contralateral laminae 1, 3-5, 7-8 and the lateral spinal, lateral cervical and internal basilar nuclei (Rhoades, 1981; Menetrey *et al.*, 1982; Morrell and Pfaff, 1983). Antonetty and Webster (1971) reported two overlapping pathways in the lateral funiculus projecting to the contralateral superior colliculus. One of them is more ventrally placed and crosses immediately in the spinal cord and remains crossed. The other pathway, lying more dorsally, ascends ipsilaterally but crosses in the brainstem (especially the intertectal) commissures to reach the contralateral colliculus. The projections from the cervical segments are located rostrally and projections from the sacrococcygeal segments caudally in the lateral funiculus (Antonetty and Webster, 1975). This tract provides afferent information for spinovisual reflexes.

The spinomesencephalic projections to the interstitial nucleus of Cajal, nucleus of Darkschewitsch and Edinger-Westphal nucleus are all connected to the oculomotor nucleus. The anterior pretectal nuclei projections have also a role in inhibition of nociception (Villarreal *et al.*, 2004). Spinomesencephalic tract projections to the cuneiform nucleus and the red nucleus play a functional role in motor control (Yezierski, 1988; Vinay and Padel, 1990).

The spinomesencephalic tract neurons have large and complex receptive fields (Yezierski *et al.*, 1987). In the upper cervical spinal cord, the neurons of the spinomesencephalic tract have simple (e.g. ipsilateral forelimb or face) to complex (e.g. excitatory and/or inhibitory responses from large portions of the body) peripheral receptive fields on widely separated areas of the body. These neurons are wide dynamic range, high or low-threshold (Yezierski and Broton, 1991). Electrophysiological studies have shown that most of the spinomesencephalic tract neurons are nociceptive (Hylden *et al.*, 1986; Yezierski and Schwartz, 1986).

The spinomesencephalic tract neurons contain a variety of neuropeptides including the vasoactive intestinal polypeptide, bombesin, substance P, dynorphin, enkephalin, cholecystokinin and somatostatin (Leah *et al.*, 1988).

The spinoparabrachial tract

Although often described as part of the spinomesencephalic or spinoreticular tracts by some authors, the spinoparabrachial tract is a distinct nociceptive pathway. This pathway originates predominantly from laminae 1, 2, 5, 7 and 10, and the lateral spinal nucleus of the spinal cord, and terminates in the parabrachial nuclei in the pontomesencephalic junction (Cechetto *et al.*, 1985; Hylden *et al.*, 1986; Kıtamura *et al.*, 1993; Feil and Herbert, 1995; Wang *et al.*, 1999). Laminae 1-2 neurons of the upper cervical segments project specifically to the ventral part of the external lateral parabrachial subnucleus. Neurons in the superficial dorsal horn of thoracic and lumbar spinal segments project mainly to the dorsal lateral and central lateral parabrachial subnuclei. The projections of spinoparabrachial tract neurons in the lateral reticulated area of lamina 5 and the lateral spinal nucleus of all spinal segments are almost exclusively to the internal parabrachial subnucleus. In addition, the corresponding neurons of upper cervical segments project to another subnucleus, the Kölliker-Fuse nucleus (Feil and Herbert, 1995). This nucleus is involved in respiratory and cardiac regulation (Cechetto *et al.*, 1985).

Most of the spinoparabrachial tract projections are contralateral, except for a small number of neurons located ipsilaterally in laminae 2-4 in the upper cervical segments (Hylden *et al.*, 1989; Menetrey and De Pommery, 1991; Kitamura *et al.*, 1993). Lamina 1 spinoparabrachial tract neurons ascend in the dorsal part of the lateral funiculus and send collaterals to the thalamus (Hylden *et al.*, 1989).

Spinoparabrachial tract neurons show substance P, vasoactive intestinal polypeptide, bombesin, dynorphin, and enkephalin immunoreactivity (Leah *et al.*, 1988; Blomqvist and Mackerlova, 1995).

Electrophysiological studies in the rat and cat have shown that the majority of spinoparabrachial tract neurons, predominantly those in lamina 1, respond to somatic and visceral noxious stimuli (Hylden *et al.*, 1985; Bernard *et al.*, 1994; Bester *et al.*, 2000). The parabrachial nuclei project to the thalamus (Kitamura *et al.*, 1993), hypothalamus (spinoparabrachiohypothalamic pathway) (Bester *et al.*, 1995), amygdala (spinopontoamygdaloid pathway) (Bernard and Besson, 1990), periaqueductal gray, and the ventrolateral medulla (Gauriau and Bernard, 2002). The connections of the spinoparabrachial tract to these regions suggest its involvement in the motivational-affective, autonomic and endocrine responses to pain.

The spinohypothalamic tract

The spinohypothalamic tract ascends in the lateral funiculus and terminates in several hypothalamic areas, including the lateral and dorsal hypothalamic areas, dorsomedial nucleus, suprachiasmatic, paraventricular, and supraoptic nuclei (Cliffer *et al.*, 1991). Although earlier studies report ascending degeneration in the hypothalamus after spinal cord lesions (Kerr, 1975), the first electrophysiological study describing the spinohypothalamic tract was done by Burstein *et al.*, (1987) in rats, followed by the studies of Katter et al. (1991) in cats and Zhang *et al.*, (1999) in monkeys.

The spinohypothalamic tract neurons are located throughout the length of the spinal cord, predominantly in lamina 1, the lateral part of laminae 3-4, 10 and the lateral spinal nucleus. A small number of neurons are also observed in the intermediate zone and the ventral horn (Kayalioglu *et al.*, 1999).

Burstein *et al.*, (1990a) estimated a total number of 9000 spinohypothalamic tract neurons throughout the length of the spinal cord in the rat, 4700 neurons projecting both to the medial and lateral hypothalamus, 3000 neurons to the medial and 3200 to the lateral hypothalamus. The pattern of distribution is the same for neurons projecting to the medial and lateral hypothalamus. More than 70% of the spinohypothalamic tract neurons are located in the marginal zone, the lateral reticulated area including laminae 5 and 10, 20% in the lateral spinal nucleus, and some in the intermediate zone and ventral horn. Neurons in the deep dorsal horn are most numerous in the upper cervical segments (Burstein *et al.*, 1990a). In cats, the total number of spinohypothalamic tract neurons is lower, but the distribution is similar to that in rats (Katter *et al.*, 1991).

About 60% of the spinohypothalamic tract projections are contralateral in the rat and 70% in the cat (Burstein *et al.*, 1987; 1990a; Katter *et al.*, 1991). Collateral projections from the spinohypothalamic tract to the thalamus, medulla, pons and midbrain have also been described (Burstein *et al.*, 1996; Dado *et al.*, 1994; Li *et al.*, 1997). Axons of many spinohypothalamic tract neurons decussate in the hypothalamus, and then descend into the ipsilateral posterior thalamus, midbrain, pons, and rostral medulla (Zhang *et al.*, 1995).

Electrophysiological studies have shown that

spinohypothalamic tract neurons respond either preferentially or specifically to noxious mechanical stimuli (Burstein *et al.*, 1991; Kostarczyk *et al.*, 1997; Zhang *et al.*, 1999). About half of the spinohypothalamic tract neurons are wide dynamic range and 40% are high threshold. Therefore about 90% of the spinohypothalamic tract neurons respond preferentially or exclusively to noxious mechanical stimulation. About 9% respond exclusively to innocuous manipulation of joints and muscles, and 4% only to innocuous tactile stimuli. Spinohypothalamic tract neurons that respond to stimulation of muscle, tendon, or joints are located deep in the gray matter (Burstein *et al.*, 1991). Spinohypothalamic tract neurons have also been shown to respond to visceral stimuli (Katter *et al.*, 1996).

The distribution and electrophysiological properties of spinohypothalamic tract neurons suggest the involvement of this pathway in autonomic, endocrine, and motivational-affective responses to somatic and visceral stimulation, including noxious stimuli. Some spinohypothalamic tract neurons have been found to be substance P immunoreactive, supporting the role of spinohypothalamic tract in nociception (Li *et al.*, 1997).

The spinocervical tract

The lateral cervical nucleus (LatC) has been identified in the spinal cords of several species including the rat, cat, dog and monkey, and although not consistently human (Mizuno *et al.*, 1967; Truex *et al.*, 1970). The spinocervical tract ascends in the dorsalateral part of the ipsilateral funiculus and synapses in the lateral cervical nucleus in the upper cervical segments (C1-C4). The spinocervical tract neurons are localized mainly in lamina 4, but also observed in laminae 1-3 and 5 at all levels of the spinal cord, predominantly in the cervical enlargement (Bryan *et al.*, 1974; Craig, 1978; Brown *et al.*, 1980; Baker and Giesler, 1984). The spinocervical tract neurons and the neurons in the lateral cervical nucleus are fewer in number in the rat than in the cat (Baker and Giesler, 1984, Giesler *et al.*, 1988). There is a somatotopic organization of the lateral cervical nucleus in the cat, with rostral parts of the body represented medially and caudal part laterally (Craig and Burton, 1979); no such organization was found in the rat (Giesler *et al.*, 1988). There is evidence for a functional link between the lateral cervical nucleus and dorsal column nuclei. Spinocervical collaterals terminate in dorsal column nuclei and there are also projections from dorsal column nuclei to the lateral cervical nucleus (Craig, 1978).

Axons of the lateral cervical nucleus neurons decussate in the ventral white commissure of upper cervical spinal cord, ascend to reach the contralateral thalamus by way of the medial lemniscus. This projection, named the cervicothalamic tract, terminates in the contralateral ventroposterolateral nucleus and the medial part of the posterior thalamic nucleus (Berkley *et al.*, 1980; Boivie, 1980). There are also collateral projections from these neurons to the midbrain (Djouhri *et al.*, 1997). The major projections are to the thalamus in the cat, and to he midbrain in the monkey (Smith and Apkarian, 1991). In the rat, projections from the lateral cervical nucleus to the midbrain are less prominent than those in the cat, and projections to the thalamus are comparatively small (Giesler *et al.*, 1988). Spinocervical tract neurons show substance P immunoreactivity (Craig *et al.*, 1992). Electrophysiological studies have shown that the spinothalamic, spinomesencephalic, spinoreticular and postsynaptic dorsal column neurons receive excitation by way of spinocervical tract collaterals (Cao *et al.*, 1993; Djouhri and Jankowska, 1998). The spinocervical tract neurons can also be inhibited by electrical stimulation of some brainstem regions including the periaqueductal gray, the raphe magnus and the cuneiform nucleus (Dostrovsky, 1984).

The majority of the spinocervical tract neurons are low-threshold or wide dynamic range, with receptive fields on hairy skin (Downie *et al.*, 1988). The spinocervical tract neurons respond to a variety of sensory stimuli, primarily from cutaneous receptors. They are excited by hair movement, noxious mechanical and thermal stimulation (Cervero *et al.*, 1977; Brown *et al.*, 1989). Many spinocervical tract neurons also respond to noxious muscle stimulation (Hamann *et al.*, 1978). Thus, it is possible that the spinocervical tract may serve as a potential pathway for nociceptive transmission.

The spinovestibular tract

The spinovestibular tract originates mainly from the central cervical nucleus neurons in C1-C4 spinal cord segments (Matsushita *et al.*, 1995; Xiong and Matsushita, 2001). There are also projections from laminae 4-8 spinal neurons (McKelvey-Briggs *et al.*, 1989). The axons ascend in the ventral funiculus and project mainly to the lateral vestibular nucleus, and also to the spinal vestibular nucleus, the parvocellular, magnocellular and caudal parts of the medial vestibular nuclei (Matsushita *et al.*, 1995; Xiong and Matsushita, 2001). There are also projections to the spinal vestibular nucleus from all spinal levels to the descending vestibular nucleus (McKelvey-Briggs *et al.*, 1989). The spinovestibular projection is bilateral (Xiong and Matsushita, 2001). Ipsilateral cervical projections to the vestibular nuclei originate from neurons in the medial part of the dorsal horn, whereas contralateral projections are from the central cervical nucleus and neurons of lamina 8 (Matsushita *et al.*, 1995; McKelvey-Briggs *et al.*, 1989; Xiong and Matsushita, 2001). Central cervical nucleus receives primary afferent fibers from neck muscles, joints and ligaments, and from the semicircular canals (Hirai *et al.*, 1978; Ammann *et al.*, 1983; Thomson *et al.*, 1996). The spinovestibular pathway has a possible role in the tonic neck reflex or cervicovestibulospinal reflex (by connecting the upper cervical segments to the lateral vestibular nucleus) and also a role in postural reflexes (Xiong and Matsushita, 2001).

There are also projections from the spinal cord to nucleus Z, which like nucleus X and Y, is described as an accessory vestibular nucleus. This nucleus is a somatosensory relay to the contralateral ventrobasal thalamus and cortex. The majority of the axons projecting to nucleus Z (up to 92% in the rat) are collaterals of the dorsal spinocerebellar tract axons (Low *et al.*, 1986). Collaterals of primary afferents from the hindlimb also terminate in nucleus Z (Leong and Tan, 1987). Electrophysiological studies have shown that nucleus Z receives proprioceptive afferents from hind limb muscles (Landgren and Silfvenius, 1971).

The spinoolivary tract

The spinoolivary tract (Helweg's tract) synapses with neurons in the primary olivary and the medial and dorsal accessory olivary nuclei of the inferior olivary complex. Climbing fibers from the inferior olivary complex terminate in the cerebellum (Richmond *et al.*, 1982; Azizi and Woodward, 1987).

The spinoolivary tract is present throughout the spinal cord. The cells of origin are located in the medial part of the nucleus proprius and in the central cervical nucleus within a few segments of the dorsal root entrance. Axons of these neurons decussate, ascend in the contralateral ventral funiculus (Oscarsson and Sjolund, 1977; Swenson and Castro, 1983) to reach the accessory olivary nucleus (Oscarsson and Sjolund, 1977). There is also a dorsal spinoolivary tract ascending in the dorsal funiculus and projecting to the contralateral inferior olivary nucleus (Molinari *et al.*, 1996). The inferior olivary nucleus is a source of climbing fibers to Purkinje cells in the cerebellar cortex (Matsushita and Ikeda, 1970). The axons enter the cerebellum by way of the inferior cerebellar peduncle (Iwata and Hirano, 1978).

Degeneration in the inferior olivary nucleus following infarction in the contralateral cerebellum implies the presence of the spinoolivary tract in the human (Iwata and Hirano, 1978), but its function has not been precisely described. The inferior olivary nucleus is a source of climbing fibers to Purkinje cells in the cerebellar cortex cortex (Matsushita and Ikeda, 1970). Thus, the spinoolivary tract may be important in the control of movements of the body and limbs. In the cat, information is carried by five different paths in the spinoolivary tract. The presence of the distinct paths implies this pathway may be involved with segmental motor control (Oscarsson and Sjolund, 1977).

Other ascending projections in the ventrolateral funiculus

There are also afferent projections from the spinal cord to the solitary nucleus and the sensory trigeminal complex. The solitary nucleus receives sensory inputs from visceral organs of cardiovascular, respiratory, genital and digestive systems (Hubscher and Berkley, 1994, 1995). Information to this nucleus is carried by spinal neurons as well as by the glossopharyngeal and vagus nerves (Hubscher and Berkley, 1995). The cells of origin in the spinal cord are located in laminae 1, 5, 10 and the lateral spinal nucleus (Menetrey and Basbaum, 1987; Wang *et al.*, 1999). There is evidence that lamina 1 spinal neurons projecting to the solitary nucleus are involved in the modulation of somatic and/or visceral nociceptive transmission (Traub *et al.*, 1996; Guan *et al.*, 1998).

A direct ipsilateral projection from the spinal cord neurons to the sensory and motor nuclei of the spinal trigeminal complex has also been shown in the rat (Phelan and Falls, 1991; Xiong and Matsushita, 2000) and the dog (Marsala *et al.*, 1989). The spinal afferent fibers which terminate in the dorsolateral parts of the spinal trigeminal complex ascend in the dorsal funiculus, while those terminating in its ventral parts ascend in both the dorsal and lateral funiculi (Phelan and Falls, 1991). Since these regions are concerned with the processing of sensory information from the lateral and posterior parts of the face, it is proposed that this tract is primarily involved with the integration of head and neck functions (Phelan and Falls, 1991).

Direct projections were found from the lumbosacral parasympathetic nuclei and dorsal commissural nuclei to Barrington's nucleus of pons (Ding *et al.*, 1997), from lamina 10 to locus coeruleus and subcoerular region (Wang *et al.*, 1999) and from lamina 6-7 of the lumbar spinal cord to basilar pontine nuclei (Mihailoff *et al.*, 1989).

There are also direct connections from the spinal cord to several telencephalic regions such as the medial and lateral septal nuclei, ventral pallidum, globus pallidus, nucleus accumbens, amygdala, and the infralimbic and medial orbital cortex (Burstein and Giesler, 1989; Cliffer *et al.*, 1991; Kayalioglu *et al.*, 1996). The projection neurons are located predominantly in the deep dorsal horn, the lateral reticulated area of lamina 5, the lateral spinal nucleus and lamina 10 at all segmental levels of the spinal cord. The projections are mainly contralateral. These regions probably play a role in motivational-affective responses to pain.

The spinal cord connections to the medial thalamus, hypothalamus and other limbic system structures are by way of the spinoreticulothalamic, the spinoamygdalar and the spinohypothalamic tracts. These tracts are sometimes referred to collectively as the 'spino-limbic tract'. It is assumed that the various projections of this functional pathway are involved in the endocrine, autonomic and motivational-affective aspects of pain.

Projections from the spinal cord to the cerebellum

The spinocerebellar tracts occupy the periphery of the lateral funiculus and carry proprioceptive and cutaneous information from Golgi tendon organs and muscle spindles to the cerebellum for the coordination of movements. There are two principal spinocerebellar tracts which carry information from the lower extremities, the dorsal (posterior) spinocerebellar and the ventral (anterior) spinocerebellar tracts (See Figure 10.1). The cuneocerebellar and rostral spinocerebellar tracts are the upper extremity homologues of the dorsal and the ventral spinocerebellar tracts, respectively. There are also projections from the central cervical nucleus to the cerebellum in the upper cervical segments.

Spinocerebellar axons terminate mainly in lobules 1-5 of the anterior lobe and lobule 8 of the posterior lobe (Grant, 1962; Wiksten and Grant, 1986; Matsushita and Tanami, 1987; Berretta *et al.*, 1991; Xu and Grant, 1994, 2005). The terminal fields of spinocerebellar projections from each spinal cord level have different distribution patterns. Mossy fiber terminals that originate from the cervical enlargement are observed mainly in the vermal area of the anterior lobe (including the most anterior part of lobule 6), to lobules 7b and 8 and to the ipsilateral paramedian lobule. Terminals from the thoracic spinal cord are observed mainly in lobules 2b-5b, and from the lumbar and sacrococygeal spinal cord in lobules 1-5 (Wiksten and Grant, 1986; Matsushita and Ikeda, 1987; Okado *et al.*, 1987; Yaginuma and Matsushita, 1987; Xu and Grant, 1990). Some neurons projecting to the anterior lobe have divergent axon collaterals to the posterior vermis and the paramedian lobule (Xu and Grant, 1988).

The spinocerebellar tracts are laminated, with the fibers from the lower segments located superficially. These tracts are mostly composed of large-diameter myelinated fibers, but fine-caliber fibers are also present in the ventral spinocerebellar tract.

The dorsal spinocerebellar tract

The dorsal spinocerebellar tract (Flechsig's tract) is located at the dorsal aspect of the dorsolateral funiculus, adjacent to the lateral corticospinal tract. In the cat and rat, it originates from the large neurons of the dorsal nucleus (column of Clarke) and laminae 4-6, starting from L3-L4 to upper thoracic spinal cord segments (Matsushita and Ikeda, 1980; Grant *et al.*, 1982; Edgley and Gallimore, 1988; Rivero-Melián and Grant, 1990). In the human, the dorsal spinocerebellar tract arises from the dorsal nucleus (Smith, 1976). The dorsal nucleus is present from the T1 through the L3 spinal segments, but is largest in the lower thoracic and upper lumbar segments (Matsushita and Hosoya, 1979). Axons entering the spinal cord from the lower segments ascend in the dorsal funiculus to reach the dorsal nucleus in L3-L4 segments. In the sacral and coccygeal segments of the rat, squirrel, but not of the cat; spinocerebellar afferents are found in Stilling's nucleus (Snyder *et al.*, 1978; Grant *et al.*, 1982). This nucleus is similar in position and orientation to the dorsal nucleus. The axons of these neurons travel and project to the cerebellum in the controlateral lateral funiculus. Electrophysiological studies have shown that neurons of Stilling's nucleus, resembling the neurons of the dorsal and the central cervical nuclei, are excited by group I muscle afferents (Snyder *et al.*, 1978; Edgley and Grant, 1991; Matsushita, 1999a). These neurons are strongly activated by passive tail movement, indicating that proprioceptors from the tail provide a powerful input to these neurons (Edgley and Grant, 1991).

The dorsal spinocerebellar tract projection is predominantly ipsilateral, but a contralateral projection is also present (Xu and Grant, 1994; Matsushita and Gao, 1997; Matsushita, 1999b). The dorsal tract shifts dorsally as it ascends and enters the cerebellum by way of the inferior cerebellar peduncle (restiform body). A somatotopic arrangement is present in the dorsal spinocerebellar tract (Xu and Grant, 1994). There are also collateral projections from this tract to nucleus Z (Low *et al.*, 1986).

Neurons of the dorsal nucleus are excited monosynaptically by group Ia afferents from muscle spindles, group Ib afferents from tendon organs (Aoyama *et al.*, 1988), and also group II muscle and cutaneous touch and pressure afferents (Edgley and Jankowska, 1988). Inhibition of the dorsal spinocerebellar tract neurons by group Ia and group Ib afferents by way of interneurons has also been shown (Hongo et al, 1983). The dorsal spinocerebellar tract transmits narrow, low range of stimuli for fine coordination of individual hindlimb muscles (Kim *et al.*, 1986), and signals information about the position and movement of the hindlimb (Edgley and Jankowska, 1988; Bosco and Poppele, 2000).

The ventral spinocerebellar tract

The ventral spinocerebellar tract (Gower's tract) arises mostly contralaterally from the lower thoracic, lumbar and more caudal segments of the spinal cord. The tract is located in the ventral funiculus at sacral and lower lumbar levels and peripherally in the ventrolateral funiculus at more rostral levels. There is a somatopic arrangement in the ventral spinocerebellar tract (Xu and Grant, 1994, 2005). The axons of the ventral spinocerebellar neurons cross the midline, enter the cerebellum by way of the superior cerebellar peduncle, and mostly recross to terminate in the ipsilateral cerebellum; however a minor portion of the ventral spinocerebellar tract, originating from the sacrococcygeal region, enters the cerebellum by way of the inferior cerebellar peduncle (Grant and Xu, 1988; Xu and Grant, 1994).

The ventral spinocerebellar tract originates mainly from the medial part of lamina 7 in the lumbosacral segments and from the dorsolateral nucleus of lamina 9 at L3-L6, and also from the neurons of the ventrolateral nucleus of lamina 9 and the lateral part of lamina 7 at L4-L5 segments (Xu and Grant, 2005). The group of large cells in laminae 7, 8 and the dorsolateral motoneuronal column of the lumbar spinal cord were first described by Cooper and Sherrington (1940) as the spinal border cells. Spinal border cells project mainly to the ipsilateral anterior lobe of the cerebellum (Matsushita and Yagunima, 1989). Axons originating from the medial part of lamina 7 have both ipsilateral and contralateral projections (Grant *et al.*, 1982).

The ventral spinocerebellar tract neurons are activated by group Ia and Ib afferents (Bras *et al.*, 1988). They transmit information for coordinated movement and posture of the entire lower limb. In contrast to the dorsal spinocerebellar tract, they lack subdivisions for different modalities and transmit information from large receptive fields that include different spinal cord segments, e.g. from the entire extremity (Kim *et al.*, 1986).

The cuneocerebellar tract

The cuneocerebellar (spinocuneocerebellar) tract is the forelimb homolog of the dorsal spinocerebellar tract. It carries proprioceptive and exteroceptive information from the forelimb, and associated neck and upper trunk regions to the cerebellum. The cells of origin are located in lamina 1, the deep dorsal horn on the lamina 5/7 border, and the lateral and medial part of lamina 6 of the spinal cord from above C8 segment (Abrahams and Swett, 1986; Nyberg and Blomqvist, 1984). The axons of the neurons then enter the cuneate fasciculus and project to the external and rostral cuneate nuclei. Axons arising from these nuclei project to the cerebellum via the inferior cerebellar peduncle to form the cuneocerebellar (spinocuneocerebellar) tract (Grant, 1962; Tolbert and Gutting, 1997). The cerebellar projections of the external cuneate nucleus are mainly to lobules 4 and 6, lobules 1-2 and lobule 9 of the cerebellum (Grant, 1962; Somana and Walberg, 1980). The cuneocerebellar tract is predominantly ipsilateral (Grant, 1962; Somana and Walberg, 1980).

The external cuneate nucleus receives its main afferents from brachial and upper cervical dorsal root ganglia supplying muscle afferent input (Abrahams and Swett, 1986; Abrahams *et al.*, 1988; Murakami and Kato, 1983; Nyberg and Blomqvist, 1984), and also non-primary afferents from the dorsal columns from levels above T6. Some non-primary spinal afferents traveling in the dorsolateral funiculus to the external cuneate nucleus have also been found. These are derived from levels extending caudally to between T6 and T10-T11 (Gordon and Grant, 1982). Electrophysiological studies have shown that the majority of cells in the cuneate nucleus respond to stimulation of joints, the remaining neurons receive convergent input from joint, muscle and cutaneous receptors (Tracey, 1980; Cerminara *et al.*, 2003)

The rostral spinocerebellar tract

The rostral spinocerebellar tract appears to be the upper extremity homolog of the ventral spinocerebellar tract. The cells of origin of this tract are located in laminae 5-7 at C5-C8 (Wiksten, 1985; Matsushita and Ikeda, 1987; Matsushita and Xiong, 1997; Xu and Grant, 2005). The projection is predominantly ipsilateral, but there is also a minor bilateral projection (Matsushita and Xiong, 1997). The axons of the rostral spinocerebellar tract neurons terminate mainly in cerebellar lobules 4-5 of the anterior lobe, also some laminae 5-6 neurons terminate in the ipsilateral paramedian lobule (Wiksten, 1985).

Projections from the central cervical nucleus to the cerebellum

The central cervical nucleus is located just lateral to the central canal in C1-C4 spinal cord segments. The axons of the central cervical nucleus neurons cross the midline at the same segmental level, ascend initially in the ventral funiculus, next in the lateral funiculus at C1 and in the lateral border of the medulla (Hirai *et al.*, 1984). Axons enter the cerebellum through mainly the superior and a few axons through the inferior cerebellar peduncles (Wiksten, 1979, 1987; Matsushita and Yaginuma, 1995). The projection to the cerebellar cortex is mainly to lobules 1-4, 7b, 8 and 9 (Wiksten, 1979, 1987; Matsushita and Tanami, 1987; Matsushita and Yaginuma, 1995). The central cervical nucleus receives primary afferent input from the labyrinth and muscle spindle afferents of deep dorsal neck muscles. There are also projections from the central cervical nucleus to the vestibular nuclei (Matsushita and Yaginuma, 1995; Thomson *et al.*, 1996). The primary neck afferent input relayed at the central cervical nucleus is transmitted directly to the contralateral vestibular nuclei and this connection serves as an important linkage from the upper cervical segments to the lateral vestibulospinal tract, serving the tonic neck reflex (Matsushita *et al.*, 1995).

Dorsal column ascending pathways

The of the spinal cord, also known as the dorsal column, consists of fibers from dorsal root afferents and from second-order neurons of the spinal cord. In the dorsal column, there are also axons that form the postsynaptic dorsal column pathway (Al-Chaer *et al.*, 1996), several descending tracts in the dorsal column, and descending fibers from the dorsal column nuclei (Burton and Loewy, 1977). In the rat, the dorsal corticospinal tract also descends in the dorsal column (Antal, 1984). The axons from the dorsal root ganglia entering the spinal cord pass directly to the dorsal column of the same side and divide into ascending and descending branches. The descending fibers are short and less well organized, with only 3% descending up to two spinal cord segments (Smith and Bennett, 1987). The long ascending fibers terminate on the dorsal column nuclei (the gracile and cuneate nuclei) and make the direct dorsal column pathway.

Most of the ascending fibers are short and leave the dorsal columns within two or three segments of their site of entry (Horch *et al.*, 1976; Davidoff, 1989). These short projections are mostly small myelinated and unmyelinated fibers (Burgess and Horch, 1978). Some of the short ascending fibers project to the neurons of the dorsal nucleus (column of Clarke) which gives rise to the dorsal spinocerebellar tract (e.g. Rivero-Melián and Grant, 1990). Others project to the dorsal horn, including neurons whose axons ascend in the dorsal columns to form the postsynaptic dorsal column pathway. Intermediate projections ascend for 4-12 segments, these include mostly large and some small myelinated fibers (Burgess and Horch, 1978). Only a small percentage of axons of dorsal root ganglion axons reach the dorsal column nuclei and these belong exclusively to the large cells of dorsal root ganglia (Giuffrida and Rustioni, 1992). These are mostly large myelinated fibers (Horch *et al.*, 1976), whilst a small number of unmyelinated fibers are also present (Tamatani *et al.*, 1989).

A propriospinal dorsal intersegmental pathway (interfascicular tract; comma tract of Schultz) consists of axons arising from dorsal horn neurons that divide into short ascending and descending branches. This pathway, located between the gracile and cuneate fasciculi, occupies the deepest part of the dorsal funiculus and has the shape of a comma in transverse sections. Dorsal intersegmental pathway provides intersegmental communication and is involved in intersegmental reflexes.

The direct dorsal column pathway (The gracile and cuneate fasciculi)

The direct dorsal column pathway includes two large ascending pathways, the gracile and cuneate fasciculi. (see Figure 10.1). The gracile fasciculus (tract of Goll) is present throughout the length of the spinal cord, and contains afferents from the lower trunk and extremities below the T6 spinal cord segment. The cuneate fasciculus (tract of Burdach), located laterally in the upper thoracic and cervical (C1 to T6) spinal cord segments, contains afferents from the upper trunk and extremities. In animals with a prominent tail, such as rats, alligators and some monkeys, there is an additional dorsal column nucleus, the nucleus of Bishoff. In these animals, the dorsal column contains afferent projections from the tail.

The long dorsal column fibers are gradually displaced medially while ascending. There is no, or minimal, overlapping of fibers of the gracile and cuneate fasciculi (Smith and Deacon, 1984). The direct dorsal column pathways ascend ipsilaterally and terminate by synapsing on the second-order neurons in the gracile and cuneate nuclei. There is a high degree of somatotopic organization in the dorsal column pathway, such as the fibers entering at successive rostral levels are located lateral to those from the lower segments. This somatotopic organization also continues in the dorsal column nuclei (e.g. Nyberg and Blomqvist, 1982). Whitsel *et al.*, (1970) have shown, in squirrel monkeys, that the organization of the fibers within the upper part of the gracile tract is somatotopic, whereas in the caudal part the organization is dermatomal. This is explained as fiber resorting as the axons in the dorsal column ascend toward the medulla (Willis and Coggeshall, 1991).

The axons of the neurons of the dorsal column nuclei, also named the internal arcuate fibers, cross the midline to form the medial lemniscus and terminate in the ventroposterolateral nucleus of the thalamus. Thus, the dorsal columns of the spinal cord and the medial lemniscus in the brainstem form a dorsal column-medial lemniscus pathway. The dorsal column-medial lemniscus pathway is involved in transmitting sensations of discriminatory touch, deep pressure, proprioception, sense of position of joints, stereognosis and vibration. At upper cervical levels, the gracile fasciculus contains a larger proportion of afferents from cutaneous receptors than from deep proprioceptors. These leave the dorsal column at lower segments to synapse on the neurons of the dorsal nucleus (Clarke) (Whitsel *et al.*, 1970). A safe method for intrathecal recording from the cervical and lumbosacral spinal cord and recorded evoked potentials from the gracile and cuneate fasciculi was introduced by Ertekin (1973, 1976a, b).

The postsynaptic dorsal column pathway

Some of the primary afferents entering the spinal cord synapse on dorsal horn neurons. The axons of these neurons constitute the postsynaptic dorsal column pathway (see Figure 10.1). The postsynaptic dorsal column pathway neurons are located mainly in laminae 3-4 and 10, and also in laminae 5-7 (Rustioni *et al.*, 1979; Bennett *et al.*, 1983, 1984; De Pommery *et al.*, 1984; Wang *et al.*, 1999). In rats, the neurons are located more superficially in rats than those in cats (Bennett *et al.*, 1984; Giesler *et al.*, 1984). The number of postsynaptic dorsal column neurons was estimated as 1700-2000 in the cervical and 800-1000 in the lumbosacral enlargements of cats (Enevoldson and Gordon, 1989), and 750-1000 in the cervical and 500-700 in the lumbar enlargements of rats (Giesler *et al.*, 1984). Bennett *et al.*, (1983) estimated approximately 800-1100 postsynaptic dorsal column neurons in the cervical and lumbar enlargements of both cats and monkeys.

The neurons of the postsynaptic dorsal column pathway project to the ipsilateral gracile and cuneate nuclei. These constitute over 30% of neurons projecting to the gracile and 38% of neurons projecting to the cuneate nuclei in rats (Giesler *et al.*, 1984). Cliffer and Giesler (1989) have shown that postsynaptic dorsal column neurons project to the cuneate nucleus from the cervical enlargement, to the medial cuneate and lateral gracile nuclei from the thoracic spinal cord, to the gracile nucleus from the lumbar enlargement, and to the medial gracile nucleus from the sacral spinal cord in rats. In monkeys, projections from the lumbar segments terminate mainly in the rostral part of gracile nucleus, and projections from the cervical enlargement in the cuneate and external cuneate nuclei (Rustioni *et al.*, 1979).

The axons of the dorsal column neurons constitute the medial lemniscus and relay information to the contralateral thalamus. As in the direct dorsal column pathway, there is a somatotopic organization in the postsynaptic dorsal column pathway (Giesler *et al.*, 1984). The axons of neurons relaying information from pelvic visceral organs travel in the dorsal column near the midline, whereas axons from thoracic and abdominal organs travel between the gracile and cuneate fasciculi (Willis, 2007).

Postsynaptic dorsal column pathway neurons have been shown to respond to innocuous mechanical (Giesler and Cliffer, 1985) and noxious peripheral stimuli (Bennett *et al.*, 1984). About half of the postsynaptic dorsal column neurons respond to noxious stimuli and all to low-threshold mechanical stimuli (Bennett *et al.*, 1984). The presence of projections from the dorsal column nuclei to the ventroposterolateral nucleus and posterior thalamic nuclei suggests the postsynaptic dorsal column pathway may also play a role both in the sensory-discriminative and motivational-affective components of pain (Millan, 1999).

Although it is known from earlier reports that the spinothalamic tract is involved in the transmission of visceral nociceptive stimuli, in contrast to earlier reports (Giesler and Cliffer, 1985), the postsynaptic dorsal column pathway is now accepted as the major afferent pathway for visceral nociception (Berkley and Hubscher, 1995; Al-Chaer *et al.*, 1996, 1999; Willis *et al.*, 1999). Clinical studies in the human have shown that visceral pain of thoracic or pelvic origin can be relieved by surgical lesioning of the fibers of the lateral edge or medial edge of the gracile fasciculus, respectively (Hirshberg *et al.*, 1996; Nauta *et al.*, 1997, 2000; Westlund, 2000; Palecek, 2004). Dorsal column lesions in rats also reduce nociceptive behavioral responses to visceral stimulation, i.e. significantly reverse the reduction of rearing behavior in pancreatitis (Houghton *et al.*, 1997) or decrease exploratory activity induced by noxious visceral stimuli in colorectal distension (Palecek *et al.*, 2002). Mainly lamina 10 neurons are involved in visceral nociception (Al-Chaer *et al.*, 1996). Electrophysiological studies have shown that dorsal column lesions significantly reduce the responses of the neurons of the gracile nucleus and thalamus (Al-Chaer *et al.*, 1996, 1998), suppress inhibition of exploratory activity induced by visceral noxious stimulation and prevent potentiation of visceromotor reflex (Palecek and Willis, 2003).

Postsynaptic dorsal column neurons contain glycine and GABA (Maxwell *et al.*, 1995). The majority of postsynaptic dorsal column neurons are apposed by serotonin-immunoreactive varicosities in the spinal cord (Wu and Wessendorf, 1992).

References

Abrahams VC, Swett JE (1986) The pattern of spinal and medullary projections from a cutaneous nerve and a muscle nerve of the forelimb of the cat: a study using the transganglionic transport of HRP. J Comp Neurol 246, 70-84.

Abrahams VC, Downey ED, Hammond CG (1988) Organization of segmental input from neck muscles to the external cuneate nucleus of the cat. Exp Brain Res 71, 557-562.

Al-Chaer ED, Lawand NB, Westlund KN, Willis WD (1996) Pelvic visceral input into the nucleus gracilis is largely mediated by the postsynaptic dorsal column pathway. J Neurophysiol 76, 2675-2690.

Al-Chaer ED, Feng Y, Willis WD (1998) A role for the dorsal column in nociceptive visceral input into the thalamus of primates. J Neurophysiol 79, 3143-3150.

Al-Chaer ED, Feng Y, Willis WD (1999) Comparative study of viscerosomatic input onto postsynaptic dorsal column and spinothalamic tract neurons in the primate. J Neurophysiol 82, 1876-1882.

Albe-Fessard D, Berkley KJ, Kruger L, Ralston HJ III, Willis WD Jr (1985) Diencephalic mechanisms of pain sensation. Brain Res 356, 217-296.

Ammann B, Gottschall J, Zenker W (1983) Afferent projections from the rat longus capitis muscle studied by transganglionic transport of HRP. Anat Embryol 166, 275-289.

Ammons WS (1989) Responses of primate spinothalamic tract neurons to renal pelvic distension. J Neurophysiol 62, 778-788.

Ammons WS, Blair RW, Foreman RD (1983) Vagal afferent inhibition of primate thoracic spinothalamic neurons. J Neurophysiol 50, 926-940.

Ammons WS, Blair RW, Foreman RD (1984) Responses of primate T1-T5 spinothalamic neurons to gallbladder distension. Am J Physiol 247, 995-1002.

Antal M (1984) Termination areas of corticobulbar and corticospinal fibres in the rat. J Hirnforsch 25, 647-659.

Antonetty CM, Webster KE (1975) The organisation of the spinotectal projection. An experimental study in the rat. J Comp Neurol 163, 449-465.

Aoyama M, Hongo T, Kudo N (1988) Sensory input to cells of origin of uncrossed spinocerebellar tract located below Clarke's column in the cat. J Physiol 398, 233-257.

Apkarian AV, Hodge CJ (1989a) Primate spinothalamic pathways: II. The cells of origin of the dorsolateral and ventral spinothalamic pathways. J Comp Neurol 288, 474-492.

Apkarian AV, Hodge CJ (1989b) Primate spinothalamic pathways: I. A quantitative study of the cells of origin of the spinothalamic pathway. J Comp Neurol 288, 447-473.

Azizi SA, Woodward DJ (1987) Inferior olivary nuclear complex of the rat: morphology and comments on the principles of organization within the olivocerebellar system. J Comp Neurol 263, 467-484.

Baker ML, Giesler GJ Jr (1984) Anatomical studies of the spinocervical tract of the rat. Somatosens Res 2, 1-18.

Bennett GJ, Seltzer Z, Lu GW, Nishikawa N, Dubner R (1983) The cells of origin of the dorsal column postsynaptic projection in the lumbosacral enlargements of cats and monkeys. Somatosens Res 1, 131-149.

Bennett GJ, Nishikawa N, Lu GW, Hoffert MJ, Dubner R (1984) The morphology of dorsal column postsynaptic spinomedullary neurons in the cat. J Comp Neurol 224, 568-578.

Berkley KJ, Hubscher CH (1995) Are there separate central nervous system pathways for touch and pain? Nat Med 1, 766-773.

Bernard JF, Besson JM (1990) The spino (trigemino) pontoamygdaloid pathway: electrophysiological evidence for an involvement in pain processes. J Neurophysiol 63, 473-490.

Bernard JF, Huang GF, Besson JM (1994) The parabrachial area: electrophysiological evidence for an involvement in visceral nociceptive processes. J Neurophysiol 71, 1646-1660.

Berkley KJ, Blomqvist A, Pelt A, Flink R (1980) Differences in the collateralization of neuronal projections from the dorsal column nuclei and lateral cervical nucleus to the thalamus and tectum in the cat: an anatomical study using two different double-labeling techniques. Brain Res 202, 273-290.

Berretta S, Perciavalle V, Poppele RE (1991) Origin of spinal projections to the anterior and posterior lobes of the rat cerebellum. J Comp Neurol 305, 273-281.

Bester H, Menendez L, Besson JM, Bernard JF (1995) Spino (trigemino) parabrachiohypothalamic pathway: electrophysiological evidence for an involvement in pain processes. J Neurophysiol 73, 568-585.

Bester H, Chapman V, Besson JM, Bernard JF (2000) Physiological properties of lamina I spinoparabrachial neurons in the rat. J Neurophysiol 83, 2239-2259.

Blair RW, Ammons WS, Foreman RD (1984) Responses of thoracic spinothalamic and spinoreticular cells to coronary artery occlusion. J Neurophysiol 51, 636-648.

Blomqvist A, Mackerlova L (1995) Spinal projections to the parabrachial nucleus are substance P-immunoreactive. Neuroreport 6, 605-608.

Boivie J (1980) Thalamic projections from lateral cervical nucleus in monkey. A degeneration study. Brain Res 198, 13-26.

Bosco G, Poppele RE (2000) Reference frames for spinal proprioception: kinematics based or kinetics based? J Neurophysiol 83, 2946-2955.

Bras H, Cavallari P, Jankowska E (1988) Demonstration of initial axon collaterals of cells of origin of the ventral spinocerebellar tract in the cat. J Comp Neurol 273, 584-592.

Brown AG, Fyffe REW, Noble R, Rose PK, Snow PJ (1980) The density, distribution and topographical organization of spinocervical tract neurones in the cat. J Physiol 300, 409-428.

Brown AG, Maxwell DJ, Short AD (1989) Receptive fields and in-field afferent inhibition of neurones in the cat's lateral cervical nucleus. J Physiol 413, 119-137.

Bryan RN, Coulter JD, Willis WD (1974) Cells of origin of the spinocervical tract in the monkey. Exp Neurol 42, 574-586.

Burgess PR, Horch KW (1978) The distinction between the short and intermediate ascending pathways in the fasciculus gracilis of the cat. Brain Res 151, 579-580.

Burstein R, Cliffer KD, Giesler GJ Jr (1987) Direct somatosensory projections from the spinal cord to the hypothalamus and telencephalon. J Neurosci 7, 4159-4164.

Burstein R, Giesler GJ Jr (1989) Retrograde labeling of neurons in spinal cord that project directly to nucleus accumbens or the septal nuclei in the rat. Brain Res 497, 149-154.

Burstein R, Cliffer KD, Giesler GJ Jr (1990) Cells of origin of the spinohypothalamic tract in the rat. J Comp Neurol 291, 329-344.

Burstein R, Dado RJ, Giesler GJ Jr (1990) The cells of origin of the spinothalamic tract of the rat: a quantitative reexamination. Brain Res 511, 329-337.

Burstein R, Dado RJ, Cliffer KD, Giesler GJ Jr (1991) Physiological characterization of spinohypothalamic tract neurons in the lumbar enlargement of rats. J Neurophysiol 66, 261-284.

Burstein R, Falkowsky O, Borsook D, Strassman A (1996) Distinct lateral and medial projections of the spinohypothalamic tract of the rat. J Comp Neurol 373, 549-574.

Burton H, Loewy AD (1977) Projections to the spinal cord from medullary somatosensory relay nuclei. J Comp Neurol 173, 773-792.

Cao CQ, Djouhri L, Brown AG (1993) Lumbosacral spinal neurons in the cat that are candidates for being activated by collaterals from the spinocervical tract. Neuroscience 57, 153-165.

Cechetto DF, Standaert DG, Saper CB (1985) Spinal and trigeminal dorsal horn projections to the parabrachial nucleus in the rat. J Comp Neurol 240, 153-160.

Cerminara NL, Makarabhirom K, Rawson JA (2003) Somatosensory properties of cuneocerebellar neurones in the main cuneate nucleus of the rat. Cerebellum 2, 131-145.

Cervero F, Iggo A, Molony V (1977) Responses of spinocervical tract neurones to noxious stimulation of the skin. J Physiol 267, 537-558.

Chaouch A, Menetrey D, Binder D, Besson JM (1983) Neurons at the origin of the medial component of the bulbopontine spinoreticular tract in the rat: An anatomical study using horse-radish peroxidase retrograde transport. J Comp Neurol 214, 309-320.

Chapman CD, Ammons WS, Foreman RD (1985) Raphe magnus inhibition of feline T1-T4 spinoreticular tract cell responses to visceral and somatic inputs. J Neurophysiol 53, 773-785.

Cliffer KD, Giesler GJ Jr (1989) Postsynaptic dorsal column pathway of the rat. III. Distribution of ascending afferent fibers. J Neurosci 9, 3146-3168.

Cliffer KD, Burstein R, Giesler GJ Jr (1991) Distribution of spinothalamic, spinohypothalamic, and spinotelencephalic fibers revealed by anterograde transport of PHA-L in rats. J Neurosci 11, 852-868.

Cooper JA, Sherringon CS (1940) Gower's tract and spinal border cells. Brain Res 63, 123-134.

Craig AD Jr (1978) Spinal and medullary input to the lateral cervical nucleus. J Comp Neurol 181, 729-744.

Craig AD Jr, Burton H (1979) The lateral cervical nucleus in the cat: anatomic organization of cervicothalamic neurons. J Comp Neurol 185, 329-346.

Craig AD, Broman J, Blomqvist A (1992) Lamina I spinocervical tract terminations in the medial part of the lateral cervical nucleus in the cat. J Comp Neurol 322, 99-110.

Dado RJ, Katter JT, Giesler GJ Jr (1994) Spinothalamic and spinohypothalamic tract neurons in the cervical enlargement of rats. I. Locations of antidromically identified axons in the thalamus and hypothalamus. J Neurophysiol 71, 959-980.

Davidoff RA (1989) The dorsal columns. Neurology 39, 1377-1385.

De Broucker T, Cesaro P, Willer JC, Le Bars D (1990) Diffuse noxious inhibitory controls in man: Involvement of the spinoreticular tract. Brain 113, 1223-1234.

De Pommery J, Roudier F, Menetrey D (1984) Postsynaptic fibers reaching the dorsal column nuclei in the rat. Neurosci Lett 50, 319-323.

Ding YQ, Zheng HX, Gong LW, Lu Y, Zhao H, Qin BZ (1997) Direct projections from the lumbosacral spinal cord to Barrington's nucleus in the rat: a special reference to micturition reflex. J Comp Neurol 389, 149-160.

Djouhri L, Meng Z, Brown AG, Short AD (1997) Electrophysiological evidence that spinomesencephalic neurons in the cat may be excited via spinocervical tract collaterals. Exp Brain Res 116, 477-484.

Djouhri L, Jankowska E (1998) Indications for coupling between feline spinocervical tract neurones and midlumbar interneurones. Exp Brain Res 119, 39-46.

Dostrovsky JO (1984) Brainstem influences on transmission of somatosensory information in the spinocervicothalamic pathway. Brain Res 292, 229-238.

Dougherty PM, Willis WD (1991) Modification of the responses of primate spinothalamic neurons to mechanical stimulation by excitatory amino acids and an N-methyl-D-aspartate antagonist. Brain Res 542, 15-22.

Downie JW, Ferrington DG, Sorkin LS, Willis WD Jr (1988) The primate spinocervicothalamic pathway: responses of cells of the lateral cervical nucleus and spinocervical tract to innocuous and noxious stimuli. J Neurophysiol 59, 861-865.

Edgley SA, Jankowska E (1988) Information processed by dorsal horn spinocerebellar tract neurones in the cat. J Physiol 397, 81-97.

Edgley SA, Gallimore CM (1988) The morphology and projections of dorsal horn spinocerebellar tract neurones in the cat. J Physiol 397, 99-111.

Edgley SA, Grant GM (1991) Inputs to spinocerebellar tract neurones located in Stilling's nucleus in the sacral segments of the rat spinal cord. J Comp Neurol 305, 130-138.

Enevoldson TP, Gordon G (1989) Postsynaptic dorsal column neurons in the cat: a study with retrograde transport of horseradish peroxidase. Exp Brain Res 75, 611-620.

Ertekin C (1973) Human evoked electrospinogram. In: *Developments in Electromyography and Clinical Neurophysiology.* Desmedt JE (ed) Vol 2, Karger, Basel.

Ertekin C (1976a) Studies on the human evoked electrospinogram. Part I. The origin of the segmental evoked potentials. Acta Neurol Scand 53, 3-20.

Ertekin C (1976b) Studies on the human evoked electrospinogram. Part II. The conduction velocity along the dorsal funiculus. Acta Neurol Scand 53, 21-38.

Feil K, Herbert H (1995) Topographic organization of spinal and trigeminal somatosensory pathways to the rat parabrachial and Kolliker-Fuse nuclei. J Comp Neurol 353, 506-528.

Friehs GM, Schrottner O, Pendl G (1995) Evidence for segregated pain and temperature conduction within the spinothalamic tract. J Neurosurg 83, 8-12.

Gauriau C, Bernard JF (2002) Pain pathways and parabrachial circuits in the rat. Exp Physiol 87, 251-258

Giesler GJ, Nahin RL, Madsen AM (1984) Postsynaptic dorsal column pathway of the rat. I. Anatomical studies. J Neurophysiol 51, 260-275.

Giesler GJ, Cliffer KD (1985) Postsynaptic dorsal column pathway of the rat. II. Evidence against an important role in nociception. Brain Res 326, 347-356.

Giesler GJ Jr, Bjorkeland M, Xu Q, Grant G (1988) Organization of the spinocervicothalamic pathway in the rat. J Comp Neurol 268, 223-233.

Giuffrida R, Rustioni A (1992) Dorsal root ganglion neurons projecting to the dorsal column nuclei of rats. J Comp Neurol 316, 206-220.

Gordon G, Grant G (1982) Dorsolateral spinal afferents to some medullary sensory nuclei. An anatomical study in the cat. Exp Brain Res 46, 12-23.

Grant G (1962) Projection of the external cuneate nucleus onto the cerebellum in the cat: an experimental study using silver methods. Exp Neurol 5, 179-195.

Grant G, Wiksten B, Berkley KJ, Aldskogius H (1982) The location of cerebellar-projecting neurons within the lumbosacral spinal cord in the cat. An anatomical study with HRP and retrograde chromatolysis. J Comp Neurol 204, 336-348.

Grant G, Xu Q (1988) Routes of entry into the cerebellum of spinocerebellar axons from the lower part of the spinal cord. An experimental anatomical study in the cat. Exp Brain Res 72, 543-561.

Granum SL (1986) The spinothalamic system of the rat. I. Locations of cells of origin. J Comp Neurol 247, 159-180.

Guan ZL, Ding YQ, Li JL, Lu BZ (1998) Substance P receptor-expressing neurons in the medullary and spinal dorsal horns projecting to the nucleus of the solitary tract in the rat. Neurosci Res 30, 213-218.

Haber LH, Moore BD, Willis WD (1982) Electrophysiological response properties of spinoreticular neurons in the monkey. J Comp Neurol 207, 75-84.

Hamann WC, Hong SK, Hong SK, Kniffki KD, Schmidt RF (1978) Projections of primary afferent fibres from muscle to neurones of the spinocervical tract of the cat. J Physiol 283, 369-378.

Hancock MB, Fougerousse CL (1976) Spinal projections from the nucleus locus coeruleus and nucleus subcoeruleus in the cat and monkey as demonstrated by the retrograde transport of horseradish peroxidase. Brain Res Bull 1, 229-234.

Harmann PA, Carlton SM, Willis WD (1988) Collaterals of spinothalamic tract cells to the periaqueductal gray: a fluorescent double-labeling study in the rat. Brain Res 441, 87-97.

Hirai N, Hongo T, Sasaki S (1978) Cerebellar projection and input organizations of the spinocerebellar tract arising from the central cervical nucleus in the cat. Brain Res 157, 341-345.

Hirai N, Hongo T, Sasaki S (1984) A physiological study of identification, axonal course and cerebellar projection of spinocerebellar tract cells in the central cervical nucleus of the cat. Exp Brain Res 55, 272-285.

Hirshberg RM, Al-Chaer ED, Lawand NB, Westlund KN, Willis WD (1996) Is there a pathway in the posterior funiculus that signals visceral pain? Pain 67, 291-305.

Horch KW, Burgess PR, Whitehorn D (1976) Ascending collaterals of cutaneous neurons in the fasciculus gracilis of the cat. Brain Res 117, 1-17.

Hubscher CH, Berkley KJ (1994) Responses of neurons in caudal solitary nucleus of female rats to stimulation of vagina, cervix, uterine horn and colon. Brain Res 664, 1-8.

Hubscher CH, Berkley KJ (1995) Spinal and vagal influences on the responses of rat solitary nucleus neurons to stimulation of uterus, cervix and vagina. Brain Res 702, 251-254.

Hobbs SF, Chandler MJ, Bolster DC, Foreman RD (1992) Segmental organization of visceral and somatic input onto C3-T6 spinothalamic tract cells of the monkey. J Neurophysiol 68, 1575-1588.

Hodge CJ Jr, Apkarian AV (1990) The spinothalamic tract. Crit Rev Neurobiol 5, 363-397.

Hongo T, Jankowska E, Ohno T, Sasaki S, Yamashita M, Yoshida K (1983) Inhibition of dorsal spinocerebellar tract cells by interneurones in upper and lower lumbar segments in the cat. J Physiol 342, 145-159.

Horch KW, Burgess PR, Whitehorn D (1976) Ascending collaterals of cutaneous neurons in the fasciculus gracilis of the cat. Brain Res 117, 1-17.

Houghton AK, Kadura S, Westlund KN (1997) Dorsal column lesions reverse the reduction of homecage activity in rats with pancreatitis. Neuroreport 8, 3795-3800.

Hylden JL, Hayashi H, Bennett GJ, Dubner R (1985) Spinal lamina I neurons projecting to the parabrachial area of the cat midbrain. Brain Res 336, 195-198.

Hylden JL, Hayashi H, Bennett GJ (1986) Lamina I spinomesencephalic neurons in the cat ascend via the dorsolateral funiculi. Somatosens Res 4, 31-41.

Hylden JL, Anton F, Nahin RL (1989) Spinal lamina I projection neurons in the rat: collateral innervation of parabrachial area and thalamus. Neuroscience 28, 27-37.

Hylden JL, Anton F, Nahin RL (1989) Spinal lamina I projection neurons in the rat: collateral innervation of parabrachial area and thalamus. Neuroscience 28, 27-37.

Iwata M, Hirano A (1978) Localization of olivo-cerebellar fibers in inferior cerebellar peduncle in man. J Neurol Sci 38, 327-335.

Jones MW, Apkarian AV, Stevens RT, Hoge CJ Jr (1987) The spinothalamic tract: an examination of the cells of origin of the dorsolateral and ventral spinothalamic pathways in cats. J Comp Neurol 260, 349-361.

Ju G, Melander T, Ceccatelli S, Hökfelt T, Frey P (1987) Immunohistochemical evidence for a spinothalamic pathway co-containing cholecystokinin- and galanin-like immunoreactivities in the rat. Neuroscience 20, 439-456.

Katter JT, Burstein R, Giesler GJ Jr (1991) The cells of origin of the spinohypothalamic tract in cats. J Comp Neurol 303, 101-112.

Katter JT, Dado RJ, Kostarczyk E, Giesler GJ Jr (1996) Spinothalamic and spinohypothalamic tract neurons in the sacral spinal cord of rats. II. Responses to cutaneous and visceral stimuli. J Neurophysiol 75, 2606-2628.

Kayalioglu G, Hariri NI, Govsa F, Erdem B, Peker G, Maiskii VA (1996) Laminar distribution of the cells of origin of the spinocerebral pathways involved in nociceptive transmission and pain modulation in the rat. Neurophysiology 28, 111-122.

Kayalioglu G, Robertson B, Kristensson K, Grant G (1999) Nitric oxide synthase and interferon-gamma receptor immunoreactivities in relation to ascending spinal pathways to thalamus, hypothalamus, and the periaqueductal gray in the rat. Somatosens Motor Res 16, 280-290.

Keay KA, Feil K, Gordon BD, Herbert H, Bandler R (1997) Spinal afferents to functionally distinct periaqueductal gray columns in the rat: an anterograde and retrograde tracing study. J Comp Neurol 385, 207-229.

Kerr FWL (1975) The ventral spinothalamic tract and other ascending systems of the ventral funiculus of the spinal cord. J Comp Neurol 159, 335-356.

Kevetter GA, Haber LH, Yezierski RP, Chung JM, Martin RF, Willis WD (1982) Cells of origin of the spinoreticular tract in the monkey. J Comp Neurol 207, 61-74.

Kevetter GA, Willis D (1983) Collaterals of spinothalamic cells in the rat. J Comp Neurol 215, 453-464.

Kim JH, Ebner TJ, Bloedel JR (1986) Comparison of response properties of dorsal and ventral spinocerebellar tract neurons to a physiological stimulus. Brain Res 369, 125-135.

Kitamura T, Yamada J, Sato H, Yamashita K (1993) Cells of origin of the spinoparabrachial fibers in the rat: a study with fast blue and WGA-HRP. J Comp Neurol 328, 449-461.

Klop EM, Mouton LJ, Holstege G (2005) Segmental and laminar organization of the spinothalamic neurons in cat: evidence for at least five separate clusters. J Comp Neurol 493, 580-595.

Kostarczyk E, Zhang X, Giesler GJ Jr (1997) Spinohypothalamic tract neurons in the cervical enlargement of rats: locations of antidromically identified ascending axons and their collateral branches in the contralateral brain. J Neurophysiol 77, 435-451.

Kuru M (1949) *Sensory Pathways in the Spinal Cord and Brain Stem of Man.* Sogensya, Tokyo.

Landgren S, Silfvenius H (1971) Nucleus Z, the medullary relay in the projection path to the cerebral cortex of group I muscle afferents from the cat's hind limb. J Physiol 218, 551-557.

Leah J (1988) Neuropeptides in long ascending spinal tract cells in the rat: evidence for paralel processing of ascending information. Neuroscience 24, 195-207.

Leong SK, Tan CK (1987) Central projection of rat sciatic nerve fibres as revealed by Ricinus communis agglutinin and horseradish peroxidase tracers. J Anat 154, 15-26.

Li JL, Kaneko T, Shigemoto R, Mizuno N (1997) Distribution of trigeminohypothalamic tract neurons and spinohypothalamic tract neurons displaying substance P receptor-like immunoreactivity in the rat. J Comp Neurol 378, 508-521.

Lima D (1990) A spinomedullary projection terminating in the dorsal reticular nucleus of the rat. Neuroscience 34, 577-589.

Lin Q, Palecek J, Paleckova V, Peng YB, Wu J, Cui M, Willis WD (1999) Nitric oxide mediates the central sensitization of primate spinothalamic tract neurons. J Neurophysiol 81, 1075-1085.

Low JS, Mantle-St John LA, Tracey DJ (1986) Nucleus Z in the rat: spinal afferents from collaterals of dorsal spinocerebellar tract neurons. J Comp Neurol 243, 510-526.

Maiskii VA, Kayalioglu G, Govsa F, Erdem B, Hariri NI (1998) Nitric oxide contributes to the spinal nociceptive processing. Neurophysiology 30, 370-376.

Mantyh PW (1982) The ascending input to the midbrain periaqueductal gray of the primate. J Comp Neurol 211, 50-64.

Marsala J, Mechirova E, Sulla I, Santa M, Jalc P (1989) Spinal ascending projections to the nucleus tractus spinalis nervi trigemini of the dog. Folia Morphol 37, 64-70.

Marshall GE, Shehab SA, Spike RC, Todd AJ (1996) Neurokinin-1 receptors on lumbar spinothalamic neurons in the rat. Neuroscience 72, 255-263.

Martin RJ, Apkarian AV, Hodge CJ Jr (1990) Ventrolateral and dorsolateral ascending spinal cord pathway influence on thalamic nociception in cat. J Neurophysiol 64, 1400-1412.

Matsushita M (1999a) Projections from the lowest lumbar and sacral-caudal segments to the cerebellar nuclei in the rat, studied by anterograde axonal tracing. J Comp Neurol 404, 21-32.

Matsushita M (1999b) Projections from the upper lumbar cord to the cerebellar nuclei in the rat, studied by anterograde axonal tracing. J Comp Neurol 412, 633-648.

Matsushita M, Ikeda M (1970) Olivary projections to the cerebellar nuclei in the cat. Exp Brain Res 10, 488-500.

Matsushita M, Hosoya Y (1979) Cells of origin of the spinocerebellar tract in the rat, studied with the method of retrograde transport of horseradish peroxidase. Brain Res 173, 185-200.

Matsushita M, Ikeda M (1980) Spinocerebellar projections to the vermis of the posterior lobe and the paramedian lobule in the cat, as studied by retrograde transport of horseradish peroxidase. J Comp Neurol 192, 143-162.

Matsushita M, Ikeda M (1987) Spinocerebellar projections from the cervical enlargement in the cat, as studied by anterograde transport of wheat germ agglutinin-horseradish peroxidase. J Comp Neurol 263, 223-240.

Matsushita M, Tanami T (1987) Spinocerebellar projections from the central cervical nucleus in the cat, as studied by anterograde transport of wheat germ agglutinin-horseradish peroxidase. J Comp Neurol 266, 376-397.

Matsushita M, Yagınuma H (1989) Spinocerebellar projections from spinal border cells in the cat as studied by anterograde transport of wheat germ agglutinin-horseradish peroxidase. J Comp Neurol 288, 19-38.

Matsushita M, Yaginuma H (1995) Projections from the central cervical nucleus to the cerebellar nuclei in the rat, studied by anterograde axonal tracing. J Comp Neurol 353, 234-246.

Matsushita M, Gao X, Yaginuma H (1995) Spinovestibular projections in the rat, with particular reference to projections from the central cervical nucleus to the lateral vestibular nucleus. J Comp Neurol 361, 334-344.

Matsushita M, Gao X (1997) Projections from the thoracic cord to the cerebellar nuclei in the rat, studied by anterograde axonal tracing. J Comp Neurol 386, 409-421.

Matsushita M, Xiong G (1997) Projections from the cervical enlargement to the cerebellar nuclei in the rat, studied by anterograde axonal tracing. J Comp Neurol 377, 251-261.

Maxwell DJ, Todd AJ, Kerr R (1995) Colocalization of glycine and GABA in synapses on spinomedullary neurons. Brain Res 690, 127-132.

McKelvey-Briggs DK, Saint-Cyr JA, Spence SJ, Partlow GD (1989) A reinvestigation of the spinovestibular projection in the cat using axonal transport techniques. Anat Embryol 180, 281-291.

Mehler WR, Feferman ME, Nauta WJ (1960) Ascending axon degeneration following anterolateral cordotomy. An experimental study in the monkey. Brain 83, 718-750.

Menetrey D, Chaouch A, Besson JM (1980) Location and properties of dorsal horn neurons at origin of spinoreticular tract in lumbar enlargement of the rat. J Neurophysiol 44, 862-877.

Menetrey D, Chaouch A, Binder D, Besson JM (1982) The origin of the spinomesencephalic tract in the rat: an anatomical study using the rertograde transport of horseadish peroxidase. J Comp Neurol 206, 193-207.

Menetrey D, Roudier F, Besson JM (1983) Spinal neurons reaching the lateral reticular nucleus as studied in the rat by retrograde transport of horseradish peroxidase. J Comp Neurol 220, 439-452.

Menetrey D, Basbaum AI (1987) Spinal and trigeminal projections to the nucleus of the solitary tract: a possible substrate for somatovisceral and viscerovisceral reflex activation. J Comp Neurol 255, 439-450.

Menetrey D, De Pommery J (1991) Origins of spinal ascending pathways that reach central areas involved in visceroception and visceronociception in the rat. Eur J Neurosci 3, 249-259.

Mense S (1983) Basic neurobiologic mechanisms of pain and analgesia. Am J Med 75, 4-14.

Mihailoff GA, Kosinski RJ, Azizi SA, Border BG (1989) Survey of noncortical afferent projections to the basilar pontine nuclei: a retrograde tracing study in the rat. J Comp Neurol 282, 617-643.

Millan MJ (1999) The induction of pain: an integrative review. Prog Neurobiol 57, 1-164.

Milne RJ, Foreman RD, Giesler GJ Jr, Willis WD (1981) Convergence of cutaneous and pelvic visceral nociceptive inputs onto primate spinothalamic neurons. Pain 11, 163-183.

Mizuno N, Nakano K, Imaizumi M, Okamoto M (1967) The lateral cervical nucleus of the Japanese monkey (Macaca fuscata). J Comp Neurol 129, 375-384.

Molinari HH, Schultze KE, Strominger NL (1996) Gracile, cuneate, and spinal trigeminal projections to inferior olive in rat and monkey. J Comp Neurol 375, 467-480.

Morrell JI, Pfaff DW (1983) Retrograde HRP identification of neurons in the rhombencephalon and spinal cord of the rat that project to the dorsal mesencephalon. Am J Anat 167, 229-240.

Murakami S, Kato M (1983) Central projection of nuchal group I muscle afferent fibers of the cat. Exp Neurol 79, 472-487.

Nahin RL (1987) Immunocytochemical identification of long ascending peptidergic neurons contributing to the spinoreticular tract in the rat. Neuroscience 23, 859-869.

Nahin RL, Micevych PE (1986) A long ascending pathway of enkephalin-like immunoreactive spinoreticular neurons in the rat. Neurosci Lett 65, 271-276.

Nauta HJ, Hewitt E, Westlund KN, Willis WD Jr (1997) Surgical interruption of a midline dorsal column visceral pain pathway. Case report and review of the literature.J Neurosurg 86, 538-542.

Nauta HJ, Soukup VM, Fabian RH, Lin JT, Grady JJ, Williams CG, Campbell GA, Westlund KN, Willis WD Jr (2000) Punctate midline myelotomy for the relief of visceral cancer pain. J Neurosurg 92, 125-130.

Ness TJ, Follett KA, Piper J, Dirks BA (1998) Characterization of neurons in the area of the medullary lateral reticular nucleus responsive to noxious visceral and cutaneous stimuli. Brain Res 802, 163-174.

Nyberg G, Blomqvist A. (1982) The termination of forelimb nerves in the feline cuneate nucleus demonstrated by the transganglionic transport method. Brain Res 248, 209-222.

Nyberg G, Blomqvist A (1984) The central projection of muscle afferent fibres to the lower medulla and upper spinal cord: an anatomical study in the cat with the transganglionic transport method. J Comp Neurol 230, 99-109.

Okado N, Ito R, Homma S (1987) The terminal distribution pattern of spinocerebellar fibers. An anterograde labelling study in the posthatching chick. Anat Embryol (Berl) 176, 175-182.

Oscarsson O, Sjolund B (1977) The ventral spino-olivocerebellar system in the cat. III. Functional characteristics of the five paths. Exp Brain Res 28, 505-520.

Palecek J (2004) The role of dorsal columns pathway in visceral pain. Physiol Res 53, S125-130.

Palecek J, Paleckova V, Willis WD (2002) The roles of pathways in the spinal cord lateral and dorsal funiculi in signaling nociceptive somatic and visceral stimuli in rats. Pain 96, 297-307.

Palecek J, Willis WD (2003) The dorsal column pathway facilitates visceromotor responses to colorectal distention after colon inflammation in rats. Pain 104, 501-507.

Peschanski M, Besson JM (1984) A spino-reticulo-thalamic pathway in the rat: an anatomical study with reference to pain transmission. Neuroscience 12, 165-178.

Phelan KD, Falls WM (1991) The spinotrigeminal pathway and its spatial relationship to the origin of trigeminospinal projections in the rat. Neuroscience 40, 477-496.

Richmond FJ, Courville J, Saint-Cyr JA (1982) Spino-olivary projections from the upper cervical spinal cord: an experimental study using autoradiography and horseradish peroxidase. Exp Brain Res 47, 239-251.

Rivero-Melián C, Grant G (1990) Lumbar dorsal root projections to spinocerebellar cell groups in the rat spinal cord: a double labeling study. Exp Brain Res 81, 85-94.

Rhoades RW (1981) Cortical and spinal somatosensory input to the superior colliculus in the golden hamster: an anatomical and electrophysiological study. J Comp Neurol 195, 415-432.

Rustioni A, Hayes NL, O'Neill S (1979) Dorsal column nuclei and ascending spinal afferents in macaques. Brain 102, 95-125.

Sahara Y, Xie YK, Bennett GJ (1990) Intracellular records of the effects of primary afferent input in lumbar spinoreticular tract neurons in the cat. J Neurophysiol 64, 1791-1800.

Sewards TV, Sewards MA (2002) The median pain system: neural representations of the motivational aspect of pain. Brain Res Bull 59, 160-180.

Smith MC (1976) Retrograde cell changes in human spinal cord after anterolateral cordotomies. Location and identification after different periods of survival. Adv Pain Res Ther 1, 91-98.

Smith MC, Deacon P (1984) Topographical anatomy of the posterior columns of the spinal cord in man. The long ascending fibres. Brain 107, 671-698.

Smith KJ, Bennett BJ (1987) Topographic and quantitative description of rat dorsal column fibres arising from the lumbar dorsal roots. J Anat 153, 203-215.

Smith MV, Apkarian AV (1991) Thalamically projecting cells of the lateral cervical nucleus in monkey. Brain Res 555, 10-18.

Snyder RL, Faull RL, Mehler WR (1978) A comparative study of the neurons of origin of the spinocerebellar afferents in the rat, cat and squirrel monkey based on the retrograde transport of horseradish peroxidase. J Comp Neurol 181, 833-852.

Somana R, Walberg F (1980) A re-examination of the cerebellar projections from the gracile, main and external cuneate nuclei in the cat. Brain Res 186, 33-42.

Stevens RT, Apkarian AV, Hodge CJ Jr (1991) The location of spinothalamic axons within the spinal cord white matter in cat and squirrel monkey. Somatosens Mot Res 8, 97-102.

Swenson RS, Castro AJ (1983) The afferent connections of the inferior olivary complex in rats: a study using the retrograde transport of horseradish peroxidase. Am J Anat 166, 329-341.

Tamatani M, Senba E, Tohyama M (1989) Calcitonin gene-related peptide- and substance P-containing primary afferent fibers in the dorsal column of the rat. Brain Res 495, 122-130.

Thomson DB, Isu N, Wilson VJ (1996) Responses of neurons of the cat central cervical nucleus to natural neck and vestibular stimulation. J Neurophysiol 76, 2786-2789.

Tolbert DL, Gutting JC (1997) Quantitative analysis of cuneocerebellar projections ın rats: differential topography in the anterior and posterior lobes. Neuroscience 80, 359-371.

Tracey DJ (1980) The projection of joint receptors to the cuneate nucleus in the cat. J Physiol 305, 433-449.

Traub RJ, Sengupta JN, Gebhart GF (1996) Differential c-fos expression in the nucleus of the solitary tract and spinal cord following noxious gastric distention in the rat. Neuroscience 74, 873-884.

Trevino DL, Carstens E (1975) Confirmation of the location of spinothalamic neurons in the cat and monkey by the retrograde transport of horseradish peroxidase. Brain Res 98, 177-182.

Truex RC, Taylor MJ, Smythe MQ, Gildenberg PL (1970) The lateral cervical nucleus of cat, dog and man. J Comp Neurol 139, 93-104.

Usunoff K, Kharazia VN, Valtschanoff JG, Schmidt HH, Weinberg RJ (1999) Nitric oxide synthase-containing projections to the ventrobasal thalamus in the rat. Anat Embryol 200, 265-281.

Villareal CF, Kina VA, Prado WA (2004) Antinociception induced by stimulating the anterior pretectal nucleus in two models of pain in rats. Clin Exp Pharmacol Physiol 31, 608-613.

Vinay L, Padel Y (1990) Spatio-temporal organization of the somaesthetic projections in the red nucleus transmitted through the spino-rubral pathway in the cat. Exp Brain Res 79, 412-426.

Wang CC, Willis WD, Westlund KN (1999) Ascending projections from the area around the spinal cord central canal: A Phaseolus vulgaris leucoagglutinin study in rats. J Comp Neurol 415, 341-367.

Westlund KN (2000) Visceral nociception. Curr Rev Pain 4, 478-487.

Whitsel BL, Petrucelli LM, Sapiro G, Ha H (1970) Fiber sorting in the fasciculus gracilis of squirrel monkeys. Exp Neurol 29, 227-242.

Wiberg M, Blomqvist A (1984) The spinomesencephalic tract in the cat: its cells of origin and termination pattern as demonstrated by the intraaxonal transport method. Brain Res 291, 1-18.

Wiberg M, Westman J, Blomqvist A (1987) Somatosensory projection to the mesencephalon: an anatomical study in the monkey. J Comp Neurol 264, 92-117.

Wiksten B (1979) The central cervical nucleus in the cat. III. The cerebellar connections studied with anterograde transport of 3H-leucine. Exp Brain Res 36, 175-189.

Wiksten B (1985) Retrograde HRP study of neurons in the cervical enlargement projecting to the cerebellum in the cat. Exp Brain Res 58, 95-101.

Wiksten B (1987) Further studies on the fiber connections of the central cervical nucleus in the cat. Exp Brain Res 67, 284-290.

Wiksten B, Grant G (1986) Cerebellar projections from the cervical enlargement: an experimental study with silver impregnation and autoradiographic techniques in the cat. Exp Brain Res 61, 513-518.

Willcockson WS, Chung JM, Hori Y, Lee KH, Willis WD (1984) Effects of iontophoretically released amino acids and amines on primate spinothalamic tract cells. J Neurosci 4, 732-740.

Willis WD Jr (2007) The somatosensory system, with emphasis on structures important for pain. Brain Res Rev 55, 297-313.

Willis WD, Coggeshall RE (1991) *Sensory Mechanisms of the Spinal Cord.* 2nd edn, Plenum, New York.

Willis WD, Westlund KN (1997) Neuroanatomy of thr pain system that modulate pain. J Clin Neurophysiol 14, 2-31.

Willis WD, Al-Chaer ED, Quast MJ, Westlund KN (1999) A visceral pain pathway in the dorsal column of the spinal cord. Proc Natl Acad Sci USA 96, 7675-7679.

Wu W, Wessendorf MW (1992) Organization of the serotonergic innervation of spinal neurons in rats – I. Neuropeptide coexistence in varicosities innervating some spinothalamic tract neurons but not in those innervating postsynaptic dorsal column neurons. Neuroscience 50, 885-898.

Xiong G, Matsushita M (2000) Upper cervical afferents to the motor trigeminal nucleus and the subnucleus oralis of the spinal trigeminal nucleus in the rat: an anterograde and retrograde tracing study. Neurosci Lett 286, 127-130.

Xiong G, Matsushita M (2001) Ipsilateral and contralateral projections from upper cervical segments to the vestibular nuclei in the rat. Exp Brain Res 141, 204-217.

Xu Q, Grant G (1988) Collateral projections of neurons from the lower part of the spinal cord to anterior and posterior cerebellar termination areas. A retrograde fluorescent double labeling study in the cat. Exp Brain Res 72, 562-576.

Xu Q, Grant G (1990) The projection of spinocerebellar neurons from the sacrococcygeal region of the spinal cord in the cat. An experimental study using anterograde transport of WGA-HRP and degeneration. Arch Ital Biol 128, 209-228.

Xu Q, Grant G (1994) Course of spinocerebellar axons in the ventral and lateral funiculi of the spinal cord with projections to the anterior lobe: an experimental anatomical study in the cat with retrograde tracing techniques. J Comp Neurol 345, 288-302.

Xu Q, Grant G (2005) Course of spinocerebellar axons in the ventral and lateral funiculi of the spinal cord with projections to the posterior cerebellar termination area: an experimental anatomical study in the cat, using a retrograde tracing technique. Exp Brain Res 162, 250-256.

Yaginuma H, Matsushita M (1987) Spinocerebellar projections from the thoracic cord in the cat, as studied by anterograde transport of wheat germ agglutinin-horseradish peroxidase. J Comp Neurol 258, 1-27.

Yezierski RP (1988) Spinomesencephalic tract: projections from the lumbosacral spinal cord of the rat, cat and monkey. J Comp Neurol 267, 131-146.

Yezierski RP, Schwartz RH (1986) Response and receptive-field properties of spinomesencephalic tract cells in the cat. J Neurophysiol 55, 76-96.

Yezierski RP, Sorkin LS, Willis WD (1987) Response properties of spinal neurons projecting to midbrain or midbrain-thalamus in the monkey. Brain Res 437, 165-170.

Yezierski RP, Broton JG (1991) Functional properties of spinomesencephalic tract (SMT) cells in the upper cervical spinal cord of the cat. Pain 45, 187-196.

Yezierski RP, Mendez CM (1991) Spinal distribution and collateral projections of rat spinomesencephalic tract cells. Neuroscience 44, 113-130.

Zemlan FP, Leonard CM, Kow LM, Pfaff DW (1978) Ascending tracts of the lateral columns of the rat spinal cord: a study using the silver impregnation and horseradish peroxidase techniques. Exp Neurol 62, 298-334.

Zhang DX, Carlton SM, Sorkin LS, Willis WD (1990) Collaterals of primate spinothalamic neurons to the periaqueductal gray. J Comp Neurol 296, 277-290.

Zhang X, Kostarczyk E, Giesler GJ Jr (1995) Spinohypothalamic tract neurons in the cervical enlargement of rats: descending axons in the ipsilateral brain. J Neurosci 15, 8393-8407.

Zhang X, Wenk HN, Gokin AP, Honda CN, Giesler GJ Jr (1999) Physiological studies of spinohypothalamic tract neurons in the lumbar enlargement of monkeys. J Neurophysiol 82, 1054-1058.

Zhang X, Honda CN, Giesler GJ Jr (2000) Position of spinothalamic tract axons in upper cervical spinal cord of monkeys. J Neurophysiol 84, 1180-1185.

11 Projections from the Brain to the Spinal Cord

Charles Watson and Alan R Harvey

Introduction

Nudo and Masterton (1988) identified 27 brain centers which send descending tracts to the spinal cord in mammals. The best known of these descending tracts is the corticospinal tract, which is a huge fiber bundle in humans and the other anthropoid primates. Apart from the corticospinal tract, which originates from a number of cortical areas, the most prominent descending tracts are those that arise from the red nucleus (rubrospinal), the vestibular nuclei (lateral and medial vestibulospinal), and the hindbrain reticular formation (lateral and medial reticulospinal) (see Figure 11.1). The remaining sites of origin of descending tracts, which are smaller in size, include the hypothalamus (paraventricular nucleus), prethalamus (zona incerta), pretectum (nucleus of Darkschewitsch), midbrain (superior colliculus, periaqueductal gray, supraoculomotor nucleus, interstitial nucleus of Cajal, cuneiform nucleus), and rhombencephalon (parabigeminal nucleus, medial cerebellar nucleus, locus coeruleus, subcoeruleus nuclei, nucleus gigantocellularis, raphe, and the nucleus of the solitary tract). The origin and topography of supraspinal projections in humans is beginning to be elucidated using modern diffusion tensor imaging and tractography (e.g. Aravamuthan *et al.*, 2007).

The corticospinal tract

All mammals possess a corticospinal tract, although there is considerable variation in the position of the main corticospinal tract in the spinal cord (Verhaart, 1962). The major corticospinal bundle is found in the dorsal column in rodents (Brown, 1971), tree shrews (Jane *et al.*, 1969), and marsupials (Watson, 1971; Martin *et al.*, 1972), in the lateral column in primates and carnivores (Petras, 1969), and in the ventral column in insectivores (Linowiecki, 1914; Michaloudi *et al.*, 1988), certain edentates (Strominger, 1969), and in an elephant (Verhaart, 1963).

On the basis of the consistent anatomy of the corticospinal tract in marsupials (Martin and Fisher, 1968; Watson, 1971; Martin *et al.*, 1972; Watson and Freeman, 1977; Frost *et al.*, 2000), tree shrews (Jane *et al.*, 1969) and rodents (Brown, 1971), it seems likely that the prototypical features of the mammalian corticospinal tract are dorsal funicular position, uniformly small diameter of the fibers (Verhaart, 1962), and termination primarily in the medial part of laminae 3 to 6 of the spinal cord. A further 'primitive' feature found in marsupials, tree shrews and edentates (but not in rodents) is a failure to reach the lumbosacral enlargement of the spinal cord. Even in rodents (rat) the cortical projection to this hindlimb-related region appears to derive mainly from sensory cortical areas rather than 'pure' motor cortex (Li *et al.*, 1990; Tracey 2004; Plant, Clarkson and Harvey, unpublished observations).

There has been an evolutionary expansion of a large fiber, direct corticomotoneuronal component in anthropoid primates that is correlated with the development of 'skilled' motor capacities, such as an increased dexterity of distal musculature, particularly in humans (Heffner and Masterton, 1975, 1983; Phillips and Porter, 1977; Bortoff and Strick, 1993; Nudo and Frost, 2007). The importance of corticospinal connections in motor control in humans has led many neuroscientists to assume that the tract is important to motor control (locomotion, manipulation) in non-primate mammals. However, the evidence for a major role of the corticospinal tract in non-primate mammals is generally poor, especially for projections below the cervical cord.

A projection from the pallium to the spinal cord has been described in passerine birds (Wild and Williams, 2000), but such a projection was not found in a previous experimental study on parrots (Webster *et al.*, 1990).

Pyramidal tract or corticospinal tract?

Because the corticospinal fibers form the medullary pyramids in mammals, it has traditionally been called the 'pyramidal tract'. The use of this term has the potential to cause some confusion because, by coincidence, the cells of origin are pyramidal neurons in the cerebral cortex. Moreover, it seems that the corticospinal fibers in monotreme mammals do not assemble in a paramedian pyramid as they do in other mammals (Goldby, 1939). A further reason to avoid the term 'pyramidal tract' is the unfortunate and now discredited 'pyramidal/extrapyramidal' dichotomy that was created in an attempt to explain the organization, and presumed voluntary versus involuntary components, of the different parts of the motor systems (see discussion in Brodal, 1981). For all of these reasons, we prefer to use the term 'corticospinal tract'.

Development

During early development, cortical efferents from many cortical areas reach the spinal cord, but many of these are eliminated as development proceeds. For example, there is a transient corticospinal projection from the occipital cortex in rodents, but in normal animals this gradually disappears

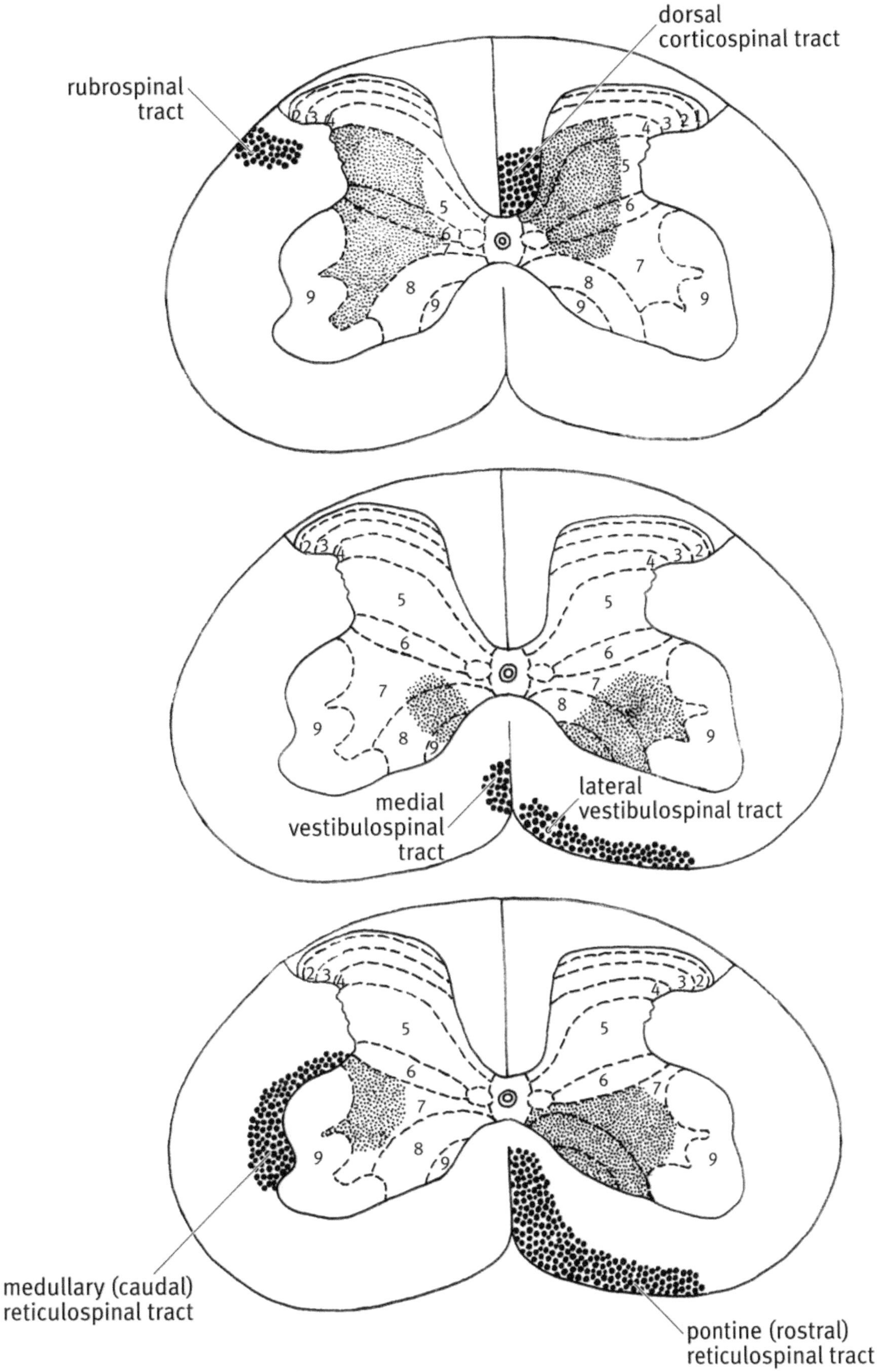

Figure 11.1 Spinal cord descending tracts

Three diagrams showing the funicular position of the major descending tracts in the rat. The termination of each of the tracts in the laminae of Rexed is shown in each case.

during the second postnatal week (O'Leary and Stanfield, 1986; Joosten *et al.*, 1987). However, visuocortical projections to the brainstem are retained. A study of the loss of corticospinal fibers in the three weeks after birth in rats showed that most of the deleted fibers arose from lateral parietal, lateral frontal, and cingulate cortex (Schreyer and Jones, 1988; Oudega *et al.*, 1994).

Cortical origin of corticospinal fibers

In mammals there are important inter-species variations in the cortical regions that contribute to the corticospinal tract (Nudo and Masterton, 1990a,b), with increasing emphasis on projections from motor cortex in primate evolution. More 'primitive' mammals have been said to possess a 'sensorimotor amalgam' (Lende, 1963) which is another way of saying that they lack a typical motor cortex. The number of corticospinal neurons is also not very great in these species. A motor cortex rostral to somatosensory cortex evolved later in association with increasing need for fine control of distal musculature, especially in the forelimbs (Frost *et al.*, 2000).

The fibers that give rise to the corticospinal tract in adult mammals arise chiefly from the primary motor, premotor and somatosensory cortical areas. These corticospinal fibers derive from medium-to-large pyramidal neurons in layer 5. In monkeys, corticospinal fibers from the frontal lobe arise from the primary motor cortex (40%) and the six major premotor areas (60%). Fibers also originate in the limb and trunk areas of somatosensory cortex. However, in some studies, small numbers of labeled pyramidal cells have also been identified in parietal association areas, secondary visual areas, insular, and prefrontal cortex (Miller, 1987).

The medullary pyramid and pyramidal decussation

The corticospinal fibers sequentially pass through the internal capsule, cerebral peduncle, longitudinal fibers of the pons, and medullary pyramid to reach the caudal end of the brain stem, where most of them cross to the opposite side in the pyramidal decussation. In the forebrain, midbrain and rhombencephalon, the corticospinal fibers are accompanied by corticopontine and corticobulbar pathways. In the echidna (Goldby, 1939) and in bats (Verhaart, 1962), the pyramidal decussation has been described as being less prominent and occurs in more rostral levels in the rhombencephalon. Similarly, in ungulates the medullary pyramids are small and the decussation is 'inconspicuous' (Phillips and Porter, 1977). It is interesting to note that in humans most spinal cords are asymmetric, the majority of cords being larger on the right side. More corticospinal fibers cross from right to left; apparently this is not due to handedness but is correlated with the observation that in 75% of decussations the crossing from left to right occurs more cranially than the opposite crossing (Nathan *et al.*, 1990).

Spinal course of the corticospinal tract

The crossed corticospinal tract of most mammals lies in the ventral part of the dorsal column of the spinal cord (Nudo and Masterton, 1990a). However, a significant number of crossed fibers assemble in the dorsolateral part of the lateral column. In primates, carnivores and lagomorphs, all or almost all of the crossed fibers are found in this lateral corticospinal tract and no significant dorsal corticospinal tract is found (Nudo and Masterton, 1990a). In anthropoid primates, the crossed lateral corticospinal tract is huge (Hefner and Masterton, 1975). Phillips and Porter (1977), suggesting that predominantly contralateral corticospinal tract projections are the rule in mammals whose behavior is dominated by vision. An uncrossed ventral corticospinal tract is also found in many ungulates, insectivores, edentates, the hyrax, and the elephant (Verhaart, 1967; Phillips and Porter, 1977). However, many of these uncrossed fibers may cross near their termination to reach the contralateral ventral horn (Brösamle and Schwab, 1997).

In most mammals, the corticospinal tracts descend to sacral levels (Masson *et al.*, 1991), but in some, including marsupials (Martin and Fisher, 1968; Watson, 1971; Frost *et al.*, 2000), the elephant (Verhaart, 1963), a tree shrew (Jane *et al.*, 1969) and an insectivore (Michaloudi *et al.*, 1988), the tract does not extend below mid-thoracic levels, with most fibers ending in the brachial enlargement. Unlike the strict topographical organization of the cuneate and gracile fasciculi, the corticospinal fibers in the rat do not maintain recognizable topographical relationship when they descend in the cervical spinal cord (Jeffrey and Fitzgerald, 1999).

The main area of termination of the corticospinal tract in most mammals is the medial parts of the base of the dorsal horn and the intermediate gray matter (Rexed's laminae 3, 4, 5, and 6). However, in some non-primate mammals direct projections to lamina 9 have been reported (Buxton and Goodman, 1967). These projections to lamina 9, presumed mostly to be direct corticomotoneuronal connections, are best developed in anthropoid primates, particularly in humans (Kuypers, 1981). In humans, up to 20% of the corticospinal fibers may terminate directly on motoneurons, compared to about 5% in macaques (see Figure 11.2).

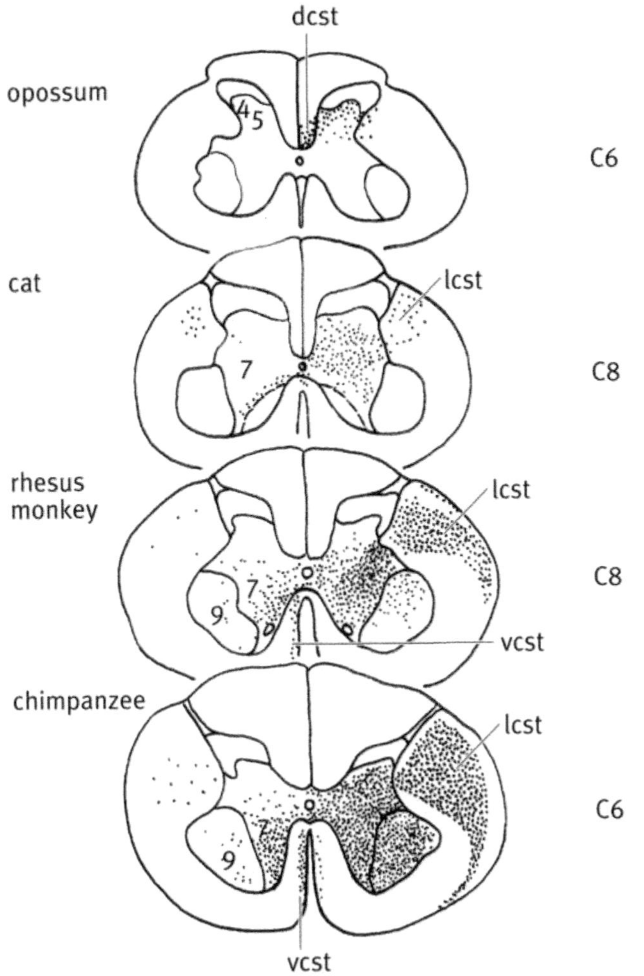

Figure 11.2 Corticospinal tract termination in the spinal cord of four different mammals

This diagram shows the position of the corticospinal tract fibres and their termination in lower cervical segments in four different mammals. In the opossum the corticospinal tract lies in the contralateral dorsal funiculus (dcst), and terminates in laminae 4, 5, and 6. In the cat the main corticospinal tract is in the dorsal part of the controlateral lateral funiculus (lcst), and the fibres terminate chiefly in laminae 5, 6, and 7. In the rhesus monkey the main corticospinal tract is in the dorsal part of the controlateral lateral funiculus (lcst), and a small ipsilateral ventral tract is seen (vcst). The fibres terminate chiefly in laminae 7, but there is a significant termination in lamina 9. In the chimpanzee the main corticospinal tract occupies a substantial area of the controlateral lateral funiculus (lcst) and a small ipsilateral ventral tract is seen (vcst). The fibres terminate in laminae 6 and 7, and there is a massive termination in lamina 9.

Hypothalamic and diencephalic projections to the spinal cord

Although the hypothalamus has been traditionally considered to be part of the diencephalon, developmental and gene expression data show it to be a separate and distinct part of the prosencephalon (see Puelles *et al.*, 2007 for a detailed discussion of this issue). The diencephalon should be considered to include the pretectum (prosomere 1), thalamus (prosomere 2), and prethalamus (prosomere 3) (Puelles *et al.*, 2007).

The ipsilateral paraventricular hypothalamic nuclear complex projects to the spinal cord (Basbaum and Fields, 1979; ten Donkelaar *et al.*, 1980), and there are also spinal projections from the lateral hypothalamus and parafornical area in mammals (Basbaum and Fields, 1979; Leong *et al.*, 1984; Holstege, 1987). Hallbeck and Blomqvist (1999) showed that in rats the spinal projection arises from a number of subnuclei of the paraventricular complex – mainly the lateral parvocellular nucleus, the ventral part of the medial parvocellular nucleus, and dorsal parvocellular nucleus, while only a few projecting neurons were found in the magnocellular nucleus. The magnocellular part of the paraventricular nucleus in the pigeon projects to the spinal cord, extending as far caudally as upper thoracic levels (Berk and Finkelstein, 1983). Hypothalamic projections to the spinal cord terminate mostly in lamina 1 but also in the preganglionic sympathetic and parasympathetic cell columns (Tracey, 2004). These pathways are involved in numerous functions, including blood pressure regulation and responses to stress.

The prethalamus (prosomere 3) projects to the spinal cord. There is a dopaminergic projection to the spinal cord from the A11 (Skagerberg and Lindvall, 1985) and a spinal projection from the zona incerta (Romanowski *et al.*, 1985), terminating in laminae 4, 5 and 10, particularly to cervical and lumbar regions (Tracey, 2004).

Two nuclei of the pretectal region (prosomere 1), project to the spinal cord. They are the interstitial nucleus of Cajal (InC) and nucleus of Darkschewitsch (Dk), which are often mistakenly considered to be located in the mesencephalon (see discussion in Puelles *et al.*, 2007). In the cat, InC was found to project bilaterally in the ventral funiculus as far as sacral spinal levels (Nyberg-Hansen, 1966), but other researchers found positive neurons only on the ipsilateral side (Castiglioni *et al.*, 1978; Satoda *et al.*, 2002). Spinal projections from InC and Dk were found in the rat (Leong *et al.*, 1984) and monkey (Carlton *et al.*, 1985). A spinal projection from InC and Dk was also found in the chick (Gross and Oppenheim, 1985) and barn owl (Masino and Knudsen, 1992). In elasmobranch fishes, only a few neurons in the area of InC were found positive, and no labeled cells were found in Dk after spinal injections of HRP (Cruce *et al.*, 1999).

The rubrospinal tract

The rubrospinal tract is prominent in many vertebrates that use limbs or pectoral fins for locomotion (ten Donkelaar, 1988b). As well as influencing general locomotion, the rubrospinal tract in at least some mammalian species has a cooperative role with corticospinal projections in controlling more skilled motor tasks (Whishaw *et al.*, 1998). The tract arises from the red nucleus, principally from the caudal magnocellular region (Massion, 1967; Murray and Gurule, 1979; Wild *et al.*, 1979; Strominger *et al.*, 1987). In mammals and birds, the tract crosses in the ventral tegmental decussation and descends in a position ventral to the spinal trigeminal tract and lateral to the superior olive and facial nucleus. It continues on to reach the dorsal part of the lateral column of the spinal cord. In animals with a prominent lateral corticospinal tract, the rubrospinal tract abuts the ventral border of that tract. The axons terminate chiefly in cervical and lumbosacral enlargements in lamina 5 and 6, but in many species there is a significant terminal field in lamina 7 (Martin and Dom, 1970; Warner and Watson, 1972; Brown, 1974; Wild *et al.*, 1979; Jankowska, 1988; Antal *et al.*, 1992; Küchler *et al.*, 2002). There is evidence that rubrospinal fibers make monosynaptic contact with last order interneurons in lamina 7 (Hongo *et al.*, 1969; Baldissera *et al.*, 1971). A direct projection to lamina 9 forepaw motoneurons has been reported in cats (McCurdy *et al.*, 1987) and rats (Küchler *et al.*, 2002).

In most species, some rubrospinal fibers do not cross and travel in the ipsilateral tract (Martin and Dom, 1970; Warner and Watson, 1972; Antal *et al.*, 1992; Küchler *et al.*, 2002).

A topographic organization of the cells of origin of the tract has been described in some mammals; the fibers terminating in the cervical enlargement arise from the dorsomedial sector of the caudal part of the red nucleus, and those terminating in the lumbosacral enlargement arise from the ventrolateral sector (Murray and Gerule, 1979; Larsen and Yumiya, 1980; Strominger *et al.*, 1987). Such a topographical organization was not found in the pigeon (Wild *et al.*, 1979).

The tectospinal tract

In mammals the tectospinal tract arises from cells in the superior colliculus and projects to cervical spinal cord (Nyberg-Hansen, 1964; Nudo and Masterton, 1989), and is primarily involved in the control of head and neck movements. The majority of fibers cross in the dorsal tegmental decussation and join the predorsal bundle as they travel caudally (Redgrave *et al.*, 1987; Harting, 1977; Dean *et al.*, 1989). The fibers arise from cells in the intermediate layers of the superior colliculus (Castiglioni *et al.*, 1978; Rhoades *et al.*, 1987; Nudo *et al.*, 1993), largely from the regions that receive input from the lower visual field (Nudo and Masterton, 1989). The tract is found in the ventral column of the spinal cord (Castiglioni *et al.*, 1978). Tectospinal fibers travel to upper cervical levels (Yasui *et al.*, 1998; Satoda *et al.*, 2002) and in rat terminate in layers 5, 7 and 8 (Yasui *et al.*, 1998), and in opossums terminate on the ventral medial nucleus and the accessory nucleus of lamina 9 (Martin, 1969). A small ipsilateral tectospinal projection has been shown to exist (Redgrave *et al.*, 1987; Olivier *et al.*, 1994). In rodents the uncrossed and crossed projections derive from different and segregated cell populations and mediate avoidance and defence, and orienting and approach, behaviors respectively (Dean *et al.*, 1989; Sefton *et al.*, 2004).

Nudo and Masterton (1989) found that the number of fibers in the tectospinal tract in mammals is surprisingly small. Carnivores had the largest number of fibers in the contralateral spinal cord (628 in the raccoon and 909 in the cat), but in 7 species of primates studied the number of fibers averaged only 220. The average for 23 non-carnivore mammals studied was 243. Harting (1977) and Rose *et al.*, (1991) also found only a small number of tectal fibers projecting to cervical spinal cord. Nudo and Masterton (1989) suggested that the influence of the tectum on neck movement may be chiefly mediated by tectal projections to hindbrain nuclei, which in turn project to the cervical spinal cord. ten Donkelaar (1976) did not find tectospinal projections in three reptile species.

Cerebellospinal projections

The deep cerebellar nuclei project to the spinal cord in amphibian and amniote vertebrates. When retrograde tracer was applied to the lateral funiculus in amphibians, many labeled cells were found in the cerebellar nucleus (which is embedded in the cerebellar peduncle) or scattered in the granule cell layer of the cerebellum (Larson-Prior and Cruce, 1992; ten Donkelaar *et al.*, 1998a). In pigeons, the cerebellospinal tract was shown to arise from the contralateral and ispilateral medial and lateral cerebellar nuclei, the fibers from the medial cerebellar nucleus terminated at layer VII of the cervical spinal cord down to cervical segment 8-9 (Arends and Zeigler, 1991). In the hatchling chick, spinally projecting neurons were found in the main and ventrolateral divisions of the medial (fastigial) nucleus and the border area between the medial and interpositus nuclei, and the terminals from the medial cerebellar nucleus did not extend caudal to mid-cervical levels (Gross and Oppenheim, 1985). In mammals, the cerebellospinal pathway originates mainly from the contralateral medial (fasitigial) nucleus and to a lesser extent from the lateral nucleus (Batton *et al.*, 1977). Some caudally projecting cells were located in the deep portion of the granule cell layer of the cerebellar cortex (Bangma *et al.*, 1984; Sánchez-Camacho *et al.*, 2001). In newborn rats, only a few neurons in the deep cerebellar nuclei were labeled and no definite pattern of distribution could be seen. The adult pattern of distribution was not seen until 10 days postnatally. The labeled neurons in the ventrolateral part of the medial portion of the medial cerebellar nucleus were continuous with those in the ventromedial portion of the posterior interpositus nucleus (Leong *et al.*, 1984).

The reticulospinal tracts

Reticulospinal projections are involved in preparatory and movement-related activities, postural control and modulation of some sensory and autonomic functions (Buford and Davidson, 2004; Tracey 2004; Deumans *et al.*, 2005). The medial and lateral reticulospinal tracts mainly arise in the medial reticular formation of the rhombencephalon (Peterson, 1979). Both arise from the area of reticular formation that is characterized by large cells (pontine caudal reticular nucleus and gigantocellular reticular nucleus), but large and small cells contribute axons to the tracts (Brodal, 1957; Torvik and Brodal, 1957; Nyberg-Hansen, 1965, 1966; Sirkin and Feng, 1987). The cells of origin of the medial reticulospinal tract are rostral to those giving rise to the lateral reticulospinal tract. The medial reticulospinal tract is found in the medial and ventral regions of the ventral column of the spinal cord, and the lateral reticulospinal tract is found in the ventral part of the lateral column close to the ventral horn. The medial tract is largely ipsilateral, but the lateral tract is made up of ipsilateral and contralateral components (Torvik and Brodal, 1957; Nyberg-Hansen, 1965, 1966; Peterson *et al.*, 1975). Although these reticulospinal projections are routinely described as forming tracts, they do not form a distinct bundle in the spinal cord; in humans reticulospinal fibers are often found to be scattered in the ventral and lateral columns and intermingled with fibers from other sources (Nathan *et al.*, 1996).

The medial (rostral or pontine) reticulospinal tract arises chiefly from neurons in the ipsilateral rostral gigantocellular reticular nucleus and the adjacent pontine caudal reticular nucleus (Petersen *et al.*, 1975). The fibers descend in or next to the medial longitudinal fasciculus in the caudal brain stem and reach the medial and ventral parts of the ventral column of the spinal cord (Nyberg-Hansen, 1965). The fibers terminate in laminae 6-9 at all spinal levels of the spinal cord with a predominance in more ventral laminae (Nyberg-Hansen, 1965).

The lateral (caudal or medullary) reticulospinal tract has ipsilateral and contralateral components. The tracts arise in the medial part of the gigantocellular reticular nucleus of the caudal rhombencephalon (Petersen *et al.*, 1975). They travel caudally in the ventral part of the lateral column of the spinal cord close to the ventral horn (Nyberg-Hansen, 1965). They send terminals to all levels of the spinal cord, ending principally in laminae 5 and 6, but also in laminae 7 to 9 (Nyberg-Hansen, 1965).

A distinct component of the ipsilateral reticulospinal tract projects specifically to the cervical spinal cord. These fibers arise from the dorsal region of the rostral gigantocellular reticular nucleus, near the abducens nucleus (Petersen, 1979).

Descending trigeminal and dorsal column nuclei projections

The spinal and principal trigeminal nuclei, and the gracile and cuneate dorsal column nuclei have all been reported to send axons to the spinal cord, terminating predominantly in the dorsal horn (Leong *et al.*, 1984; Tracey, 2004), many at cervical levels but also apparently descending to more caudal locations (Ruggiero *et al.*, 1981).

The vestibulospinal tracts

The major projections of the vestibular complex to the spinal cord are the lateral vestibulospinal tract, which arises in the lateral vestibular nucleus (Brodal *et al.*, 1962), and the medial vestibulospinal tract, which arises in the medial and spinal vestibular nuclei (Brodal *et al.*, 1962; Kneisley *et al.*, 1978; Peterson *et al.*, 1978). The vestibulospinal tracts are the main initiators of coordinated postural extensor activity in the limbs and trunk (Pompeiano, 1972). Given the similarity of organization of the main vestibular nuclei in mammalian, avian and reptilian brains, it is likely that the pattern of organization of the vestibulospinal tracts is consistent among amniote vertebrates.

The lateral vestibulospinal tract, primarily driven by otolith signals, arises from the ispilateral lateral vestibular nucleus. The fibers arise from large and small cells in the lateral vestibular nucleus (Peterson and Coulter, 1977) travels in the ventrolateral margin of the ventral column to cervical, thoracic, and lumbar levels of spinal cord. It is responsible for initiating extensor tone in the limbs (Pompeiano, 1972). A number of studies (Peterson and Coulter, 1977; Kneisley *et al.*, 1978; Shamboul, 1980) found that the larger lumbosacral component of the lateral vestibular tract arose in the dorsocaudal two thirds of the lateral vestibular nucleus, whereas the smaller thoracic and cervical components arose in the rostroventral parts of the nucleus.

Hayes and Rustioni (1981) confirmed that vestibulospinal fibers arising in the lateral vestibular nucleus were almost all ipsilateral, but those arising in the medial and spinal vestibular nuclei projected to both sides of the spinal cord. Boyle (2000) showed that most fibers of the lateral vestibulospinal tract give off very few collaterals before they reach lumbosacral levels, and they therefore constitute a kind of 'private line' from the lateral vestibular nucleus to the extensor muscles of the lower limb and tail.

The medial vestibulospinal tract, primarily driven by input from the semicircular canals, arises in the ipsilateral and contralateral medial and spinal vestibular nuclei, and travels in the ventral column in the area known as the sulcomarginal fasciculus (Nyberg-Hansen, 1966). The fibers terminate in laminae 7 and 8 in cervical and thoracic spinal cord segments (Wilson *et al.*, 1967). It coordinates head position with the position of the body in space and mediates the vestibulocollic reflex (Wilson *et al.*, 1995).

The Peterson group (Peterson and Coulter, 1977; Peterson *et al.*, 1978) have described a third vestibulospinal tract, which they call the caudal vestibulospinal tract. Bankoul and Neuhuber (1992) have also identified a tract originating in the caudal part of the medial vestibular nucleus. The fibers of the caudal vestibulospinal tract of Peterson *et al.*, (1978) arise in the caudal parts of the spinal, medial, and group F nuclei of the vestibular nuclear complex. The fibers travel in the ventral and ventrolateral columns of both sides of the spinal cord. They travel as far caudally as lumbar levels of the cord. The fibers of this tract have slow conduction velocities compared to fibers of the medial and lateral vestibulospinal tracts.

Raphespinal and coeruleospinal tracts

Descending raphespinal and coeruleospinal projections have neuromodulatory influences on motor function and are also involved in modulating some autonomic, reproductive and excretory functions, as well as modulating perception to pain (e.g. Ono and Fukuda, 1995; Mason 1999; Deumans *et al.*, 2005). Medullary raphe projections derive from raphe magnus, raphe obscurus and raphe pallidus nuclei (Skagerberg and Bjorklund, 1985). Neurons in the raphe magnus and part of the gigantocellular reticular formation project to the spinal cord in the dorsolateral funiculus, terminating mainly in the dorsal horn. This pathway contains a non-serotinergic component pathway and plays an important role in gating pain (Mason, 1999). Serotinergic neurons in the raphe obscurus and raphe pallidus nuclei project in the ventrolateral white matter and terminate in the intermediate gray and on motoneurons in ventral horn (Tracey, 2004), especially at thoracolumbar and upper sacral levels (Skagerberg and Bjorklund, 1985). The noradrenergic spinal projections from locus coeruleus travel mostly in the ipsilateral ventral funiculus and project into both dorsal and ventral horns. Descending fibers also arise from nucleus subcoeruleus (A7 group). There are intriguing strain-dependent differences in these various projections in the rat (Sluka and Westlund, 1992).

The solitariospinal tract

The solitariospinal tract is a small fiber bundle that arises from the ipsilateral nucleus of the solitary tract. The fibers terminate around the phrenic nucleus (C4-6) and in the ventral horn in thoracic levels (Loewy and Burton, 1978; Kneisley *et al.*, 1978; Mtui *et al.*, 1993). Although most solitariospinal fibers terminate in cervical and thoracic levels, some travel the whole length of the spinal cord. The projection to the phrenic nucleus and intercostal muscles is consistent with the role of the nucleus of the solitary tract in respiratory functions, as well as in the control of vomiting and possibly micturition behaviors.

Projection from the retroambiguus nucleus to the spinal cord

A pathway from the retroambiguus nucleus to the spinal cord mediates vocalization and copulatory behavior in the cat (Holstege *et al.*, 1997; Holstege and Georgiadis, 2004). A similar projection is present in the rat (Holstege *et al.*, 1997). The retroambiguus nucleus receives afferents from the midbrain periaqueductal gray.

References

Antal M, Sholomenko GN, Moschovakis AK, Storm-Mathisen J, Heizmann CW, Hunziker W (1992) The termination pattern and polysynaptic targets of rubrospinal fibers in the rat spinal cord. A light and electron microscope study. J Comp Neurol, 325, 22-37.

Aravamuthan BR, Muthusamy KA, Stein JF, Aziz TZ, Johansen-Berg H (2007) Topography of cortical and subcortical connections of the human pedunculopontine and subthalamic nuclei. Neuroimage, 37, 694-705.

Arends JJ, Zeigler HP (1991) Organization of the cerebellum in the pigeon *(Columba livia)*: 11. Projections of the Cerebellar Nuclei. J Comp Neurol 306, 245-272.

Baldissera F, Ten Bruggencate G, Lundberg A (1971) Rubrospinal monosynaptic connections with last-order interneurons of polysynaptic reflex paths. Brain Res. 27, 390-392.

Bangma GC, ten Donkelaar HJ, Dederen PJW, de Boer-van Huizen R (1984) Cerebellar efferents in the lizard *Varanus exanthematicus*. II. Projections of the cerebellar nuclei. J Comp Neurol 230, 218-230.

Bankoul S, Neuhuber WL (1992) A direct projection from the medial vestibular nucleus to the cervical spinal dorsal horn of the rat, as demonstrated by anterograde and retrograde tracing. Anat Embryol 185, 77-85.

Basbaum AI, Fields HL (1979) The origin of descending pathways in the dorsolateral funiculus of the spinal cord of the cat and rat: further studies on the anatomy of pain modulation. J Comp Neurol 187, 513-531.

Batton RR, Jayaraman A, Ruggiero D, Carpenter MB (1977) Fastigial efferent projections in the monkey: An autoradiographic study. J Comp Neurol 174, 281-305.

Berk ML, Finkelstein JA (1983) Long descending projections of the hypothalamus in the pigeon, *Columba livia.* J Comp Neurol 220, 127-136.

Bortoff GA, Strick PL (1993) Corticospinal terminations in two new-world primates: further evidence that corticomotoneuronal connections provide part of the neural substrate for manual dexterity. J Neurosci 13, 5105-5118.

Boyle R (2000) Morphology of lumbar-projecting lateral vestibulospinal neurons in the brainstem and cervical spinal cord in the squirrel monkey. Arch Ital Biol. 138, 107-122

Brodal A (1957) *The Reticular Formation of the Brain Stem. Anatomical aspects and functional correlations.* Charles C Thomas, Springfield.

Brodal A (1981). *Neurological Anatomy in Relation to Clinical Medicine.* 3rd edn. Oxford University Press, New York.

Brodal A, Pompeiano O, Walberg F (1962) *The Vestibular Nuclei and their Connections, Anatomy and Functional Correlations.* Oliver and Boyd, Edinburgh.

Brösamle C, Schwab ME (1997) Cells of origin, course, and termination patterns of the ventral, uncrossed component of the mature rat corticospinal tract. J Comp Neurol 386, 293-303

Brown LT (1971) Projections and termination of the corticospinal tract in rodents. Exp Brain Res 13, 432-450.

Brown LT (1974) Rubrospinal projections in the rat. J Comp Neurol 154, 169-187.

Buford JA, Davidson AG (2004) Movement-related and preparatory activity in the reticulospinal system of the monkey. Exp Brain Res 159, 284-300.

Buxton DF, Goodman DC (1967) Motor function and the corticospinal tracts in the dog and raccoon. J Comp Neurol 129, 341-60.

Carlton SM, Chung JM, Leonard R.B and Willis WD (1985) Funicular trajectories of brainstem neurons projecting to the lumbar spinal cord in the monkey *(Macaca fascicularis):* A retrograde labeling study. J Comp Neurol 241, 382-404.

Castiglioni AJ, Gallaway MC, Coulter JD (1978) Spinal projections from the midbrain in monkey. J Comp Neurol 178, 329-345.

Cruce WLR, Stuesse SL, Northcutt RG (1999) Brainstem neurons with descending projections to the spinal cord of two elasmobranch fishes: thornback guitarfish, *Platyrhinoidis triseriata,* and horn shark, *Heterodontus francisci.* J Comp Neurol 403, 534-560.

Dean P, Redgrave P, Westby GWM (1989) Event or emergency? Two response systems in the mammalian superior colliculus. Trends Neurosci 12, 137-147.

Deumans R, Koopmans GC, Joosten EAJ (2005) Regeneration of descending axon tracts after spinal cord injury. Prog Neurobiol 77, 57-89.

Frost SB, Milliken GW, Plautz EJ, Masterton RB, Nudo RJ (2000) Somatosensory and motor cortex representations in cerebral cortex of a primitive mammal *(Monodelphis domestica):* a window into the early evolution of sensorimotor cortex. J Comp Neurol 421, 29-51.

Goldby F (1939) An experimental investigation of the motor cortex and pyramidal tract of *Echidna aculeata.* J Anat 73, 509-524.

Gross GH, Oppenheim RW (1985) Novel sources of descending input to the spinal cord of the hatchling chick. J Comp Neurol 232, 162-179.

Hallbeck M, Blomqvist A (1999) Spinal cord-projecting vasopressinergic neurons in the rat paraventricular hypothalamus. J Comp Neurol 411, 201-211.

Harting JK (1977) Descending pathways from the superior colliculus: An autoradiographic analysis in the rhesus monkey (*Macaca mulatta*). J Comp Neurol 173, 583-612.

Hayes NL, Rustioni A (1981) Descending projections from brainstem and sensorimotor cortex to spinal enlargements in the cat. Exp Brain Res. 41, 89-107.

Heffner RS, Masterton RB (1975) Variation in form of the pyramidal tract and its relationship to digital dexterity. Brain Behav Evol 12, 161-200

Heffner RS, Masterton RB (1983) The role of the corticospinal tract in the evolution of human digital dexterity. Brain Behav Evol 23, 165-183

Holstege G (1987) Some anatomical observations on the projections from the hypothalamus to brainstem and spinal cord: an HRP and autoradiographic tracing study in the cat. J Comp Neurol 260, 98-126.

Holstege G, Kerstens L, Moes MC, VanderHorst VGJM (1997) Evidence for a periaqueductal gray – nucleus retroambiguus – spinal cord pathway in the rat. Neuroscience 80, 587-598.

Holstege G, Georgiadis JR (1998) The emotional brain: neural correlates of cat sexual behavior and human male ejaculation. Prog Brain Res 143, 39-42.

Hongo R, Jankowska E, Lundberg A (1969) The Rubrospinal tract. II. Facilitation of neuronal transmission in reflex paths to motoneurons. Exp Brain Res 7, 365-391.

Jane JA, Campbell CBG, Yashon D (1969) The origin of the corticospinal tract in the tree shrew *(Tupaia glis)* with observations on its brain stem and spinal terminations. Brain Behav Evol 2, 160-182.

Jankowska E (1988) Target cells of rubrospinal tract fibers within the lumbar spinal cord. Behav Brain Res 28, 91-96.

Jeffrey ND, Fitzgerald M (1999) Lack of topographical organisation of the corticospinal tract in the cervical spinal cord of the adult rat. Brain Res 833, 315-318.

Joosten EA, Gribnau, AA, Dederen PJ (1987) An anterograde tracer study of the developing corticospinal tract in the rat: three components. Brain Res 433, 121-130

Kneisley LW, Biber MP, LaVail JH (1978) A study of the origin of brain stem projections to monkey spinal cord using the retrograde transport method. Exp Neurol 60 116-139.

Kuchler M, Fouad K, Weinmann O, Schwab MF, Raineteau O (2002) Red nucleus projections to distinct motor neuron pools in the rat spinal cord. J Comp Neurol 448, 349-359.

Kuypers HGJM (1981) Anatomy of descending pathways. In: *Handbook of Sensory Physiology: the Nervous System.* Vol 2, part 1. Brookhart JM, Mountcastle VB (eds). Williams and Wilkins, USA.

Larsen KD, Yamiya H (1980) The red nucleus of the monkey. Exp Brain Res 40, 393-404.

Larson-Prior LJ, Cruce WL (1992) The red nucleus and mesencephalic tegmentum in a ranid amphibian: a cytoarchitectonic and HRP connectional study. Brain Behav Evol 40, 273-286.

Lende RA (1963) Cerebral cortex: a sensorimotor amalgam in the marsupilia. Science 141, 730-732.

Leong SK, Shieh JY, Wong WC (1984) Localizing spinal-cord-projecting neurons in adult albino rats. J Comp Neurol 228, 1-17.

Li, XG, Florence SL, Kaas JH (1990) Areal distributions of cortical neurons projecting to different levels of the caudal brain stem and spinal cord in rats. Somatosens Mot Res 7, 315-335.

Linowiecki AJ (1914) The comparative anatomy of the pyramidal tract. J Comp Neurol 24, 509-530.

Loewy A, Burton F (1978) Nuclei of the solitary tract: efferent projections to the lower brain stem and spinal cord of the cat. J Comp Neurol 181, 421-449.

Martin GF (1969) Efferent tectal pathways of the opossum *(Didelphis virginiana).* J Comp Neurol 135, 209-224.

Martin GF, Dom R (1970). The rubrospinal tract of the opossum *(Didelphis virginiana).* J Comp Neurol 138, 19-30.

Martin GF, Fisher AM (1968) A further evaluation of the origin, course, and termination of the opossum cortico-spinal tract. J Neurol Sci 7, 177-189.

Martin GF, Megirian D, Connor JB (1972). The corticospinal tract of the Tasmanian pororoo *(Potorous apicalis).* J Anat 111, 263-281.

Masino T, Knudsen EI (1992) Anatomical pathways from the optic tectum to the spinal cord subserving orienting movements in the barn owl. Exp Brain Res 92, 194-208.

Mason P (1999) Central mechanisms of pain modulation. Curr Opin Neurobiol 9, 436-441.

Massion J (1967) The mammalian red nucleus. Physiol Rev 47, 383-436.

Masson RL Jr, Sparkes ML, Ritz LA (1991) Descending projections to the rat sacrocaudal spinal cord. J Comp Neurol 307, 120-130.

McCurdy ML, Hansma DI, Houk JC, Gibson AR (1987) Selective projections from the cat red nucleus to digit motor neurons. J Comp Neurol 265, 367-379.

Michaloudi H, Donopoulos A, Karamanlidis AN, Papadopoulos GC, Antonopoulos J. (1988) Cortical and brain stem projections to the spinal cord of the hedgehog *(Erinaceus europaeus).* Anat Embryol 178, 259-270.

Miller MW (1987) The origin of corticospinal projection neurons in the rat. Exp Brain Res 67, 339-351.

Mtui EP, Anwar M, Gomez R, Reis DJ, Ruggiero DA (1993) Projections from the nucleus tractus solitarii to the spinal cord. J Comp Neurol 337, 231-252.

Murray HM, Gurule ME (1979) Origin of the rubrospinal tract of the rat. Neurosci Lett 14, 19-23.

Nathan PW, Smith MC, Deacon P (1990) The corticospinal tracts in man. Course and location of fibers at different segmental levels. Brain 113, 303-324.

Nathan PW, Smith MC, Deacon P (1996) Vestibulospinal, reticulospinal and descending propriospinal nerve fibers in man. Brain 119, 1809-1833.

Nudo RJ, Frost SB (2007) The evolution of motor cortex and motor systems. In *Evolution of Nervous Systems.* Volume 3 – Mammals. Kaas JH and Krubitzer LH (eds). Elsevier-Academic Press, San Diego.

Nudo RJ, Masterton RB (1988) Descending pathways to the spinal cord: a comparative study of 23 mammals. J Comp Neurol 277, 11, 53-79.

Nudo RJ, Masterton RB (1989) Descending pathways to the spinal cord, II: Quantitative study of the tectospinal tract in 23 mammals. J Comp Neurol 286, 96-119.

Nudo RJ, Masterton RB (1990a) Descending pathways to the spinal cord, III: Sites of origin of the corticospinal tract. J Comp Neurol 296, 559-583.

Nudo RJ, Masterton RB (1990b) Descending pathways to the spinal cord, IV: Some factors related to the amount of cortex devoted to the corticospinal tract. J Comp Neurol 296, 584-597.

Nudo RJ, Sutherland DP, Masterton RB (1993) Inter- and intra-laminar distribution of tectospinal neurons in 23 mammals. Brain Behav Evol 42, 1-23.

Nyberg-Hansen R (1964) The location and termination of tectospinal fibers in the cat. Exp Neurol 9, 212-227.

Nyberg-Hansen R (1965) Sites and mode of termination of reticulospinal fibers in the cat. An experimental study with silver impregnation methods. J Comp Neurol 124, 71-99.

Nyberg-Hansen R (1966) Functional organization of descending supraspinal fiber systems to the spinalcord. Anatomical observations and physiological correlations. Embryol Cell Biol 39, 3-48.

O'Leary DD, Stanfield BB (1986) A transient pyramidal tract projection from the visual cortex in the hamster and its removal by selective collateral elimination Brain Res 392, 87-89.

Olivier E, Katama T, Grantyn A (1994) Anatomical evidence for ipsilateral collicular projections to the spinal cord in the cat. Exp Brain Res 100, 160-164.

Ono H, Fukuda H (1995) Pharmacology of descending noradrenergic systems in relation to motor function. Pharmacol Ther 68, 105-112.

Oudega M, Varon S, Hagg T (1994) Distribution of corticospinal motor neurons in the postnatal rat. Quantitative evidence for massive collateral elimination and modest cell death. J Comp Neurol 347, 115-126.

Peterson BW (1979) Reticulospinal projections to spinal motor nuclei. Ann Rev Physiol 41, 127-140.

Peterson BW, Coulter JD (1977) A new long spinal projection from the vestibular nuclei in the cat. Brain Res 122, 357-356.

Peterson BW, Maunz RA, Fukushima K (1978) Properties of a new vestibulospinal projection, the caudal vestibulospinal tract. Exp Brain Res 32, 287-292.

Peterson BW, Maunz RA, Pitts NG, Mackel RG (1975) Patterns of projection and branching of reticulospinal neurons. Exp Brain Res 23, 331-353.

Petras JM (1969). Some efferent connections of the motor and somatosensory cortex of simian primates and felid, canid and procyonid carnivores. Ann NY Acad Sci 167, 469-505.

Phillips CG, Porter R (1977) *Corticospinal Neurons: their role in movement.* Monographs of the Physiological Society #34. Academic Press, London.

Pompeiano O (1972) Spinovestibular relations. Anatomical and physiological aspects. Prog Brain Res 37, 263-291

Puelles L, Martinez-de-la-Torre M, Paxinos G, Watson C, Martinez S (2007) *The Chick Brain in Stereotaxic Coordinates.* Elsevier-Academic Press, Oxford.

Redgrave P, Mitchell IJ, Dean P (1987) Descending projections from the superior colliculus in rat: a study using orthograde transport of wheatgerm-agglutinin conjugated horseradish peroxidase. Exp Brain Res 68, 147-167.

Rhoades RW, Mooney RD, Klein BG, Jacquin MF, Szczepanik AM, Chiaia NL (1987) The structural and functional characteristics of tectospinal neurons in the golden hamster. J Comp Neurol 55, 451-465.

Romanowski CA, Mitchell IJ, Crossman AR (1985) The organisation of the efferent projections of zona incerta. J Anat 143, 75-95.

Rose PK, MacDonald J, Abraham VC (1991) Projections of the tectospinal tract to the upper cervical spinal cord of the cat: A study with the anterograde tracer PHA-L. J Comp Neurol 314, 91-105.

Ruggiero DA, Ross CA, Reis DJ (1981) Projections from the spinal trigeminal nucleus to the entire length of the spinal cord in the rat. Brain Res 225, 225-233.

Sánchez-Camacho C, Marin O, ten Donkelaar HJ, Lez AG (2001) Descending supraspinal pathways in amphibians. I. A dextran amine tracing study of their cells of origin. J Comp Neurol 434, 186-208.

Satoda T, Matsumoto H, Zhou L, Rose P, Richmond, F (2002) Mesencephalic projections to the first cervical segment in the cat. Exp Brain Res 144, 397-413.

Schreyer DJ, Jones EG (1988) Topographic sequence of outgrowth of corticospinal axons in the rat: a study using retrograde axonal labelling with fast blue. Brain Res 38, 89-101.

Sefton A, Dreher B, Harvey AR (2004). Visual system. In: *The Rat Nervous System*, 3rd edn. Paxinos, G. (Ed). Academic Press, San Diego.

Sirken DW, Feng AS (1987) Autoradiographic study of descending pathways from the pontine reticular formation and mesencephalic trigeminal nucleus in the rat. J Comp Neurol 256, 483-493.

Shamboul KM (1980) Lumbosacral predominance of vestibulospinal fiber projection in the rat. J Comp Neurol 192, 519-530.

Skagerberg G, Björklund A (1985) Topographic principles in the spinal projections of serotonergic and non-serotonergic brainstem neurons in the rat. Neuroscience 15, 445-480.

Skagerberg G, Lindvall O (1985) Organization of diencephalic dopamine neurones projecting to the spinal cord in the rat. Brain Res 342, 340-351.

Sluka KA, Westlund KN (1992) Spinal projections of the locus coeruleus and the nucleus subcoeruleus in the Harlan and the Sasco Sprague-Dawley rat. Brain Res 579, 67-73.

Strominger NL (1969) A comparison of the pyramidal tracts in two species of edentates. Brain Res 15, 259-262.

Strominger RN, McGiffen JE, Strominger NL (1987) Morphometric and experimental studies of the red nucleus in the albino rat. Anat Rec 219, 420-428.

ten Donkalaar HJ (1976) Descending pathways from the brain stem to the spinal cord in some reptiles. II. Course and site of termination. J Comp Neurol 167, 443-463.

ten Donkelaar HJ (1998a) Anurans. In: *The Central Nervous System of Vertebrates.* Nieuwenhuys R, ten Donkelaar HJ, Nicholson C (eds). Springer-Verlag, Berlin.

ten Donkelaar HJ (1988b) Evolution of the red nucleus and rubrospinal tract. Behav Br Res 28, 9-20.

ten Donkelaar HJ, Kusuma A, DeBoer-Van Huizen R (1980) Cells of origin of pathways descending to the spinal cord in some quadrupedal reptiles. J Comp Neurol 192, 827-851.

Tracey D (2004). Ascending and descending tracts in the spinal cord. In: *The Rat Nervous System*, 3rd edn. Paxinos, G. (Ed). Academic Press, San Diego.

Torvik A, Brodal A (1957) The origin of reticulospinal fibers in the cat; an experimental study. Anat Rec 128, 113-137.

Verhaart WJC (1962) The pyramidal tract. Structure and function in man and animals. World Neurol 3, 43-53.

Verhaart WJC (1963) Pyramidal tract in the cord of the elephant. J Comp Neurol 121, 45-49.

Verhaart WJC (1967) The non-crossing of the pyramidal tract in Procavia capensis (Storr) and other instances of absence of the pyramidal crossing J Comp Neur 131, 387-392.

Warner G, Watson CRR (1972) The rubrospinal tract in a diprotodont marsupial *(Trichosurus vulpecula)*. Brain Res 28, 9-20.

Watson CRR (1971) The corticospinal tract of the quokka wallaby *(Setonix brachyurus)*. J Anat 109, 127-133.

Watson CRR, Freeman BW (1977) The corticospinal tract in the kangaroo. Brain Behav Evol 14, 341-351.

Webster DM, Rogers LJ, Pettigrew JD, Steeves JD (1990) Origins of descending spinal pathways in prehensile birds: do parrots have a homologue to the corticospinal tract of mammals? Brain Behav Evol, 36, 216-226.

Whishaw IQ, Gorny B, Sarna J (1998) Paw and limb use in skilled and spontaneous reaching after pyramidal tract, red nucleus and combined lesions in the rat: behavioral and anatomical dissociations. Behav Brain Res 93, 167-183.

Wild JM, Williams MN (2000) Rostral Wulst in passerine birds. I. Origin, course, and terminations of an avian pyramidal tract. J Comp Neurol 416, 429-450.

Wild JM, Cabot JB, Cohen DH, Karten HJ (1979) Origin, course and terminations of the rubrospinal tract in the pigeon *(Columbia livia)*. J Comp Neurol 187, 639-654.

Wilson VJ, Boyle R, Fukushima K, Rose PK, Shinoda Y, Shugiuchi Y, Uchino Y (1995) The vestibulocolic reflex. J Vestib Res 5, 147-170.

Wilson VJ, Wylie RM, Marco LA (1967) Projection of the spinal cord from the medial and descending vestibular nuclei of the cat. Nature 215, 419-430.

Yasui Y, Ono K, Tsumori T, Yokota S, Kishi T (1998) Tectal projections to the parvicellular reticular formation and the upper cervical spinal cord in the rat, with special reference to axon collateral innervation. Brain Res 804, 149-154.

12 The Propriospinal System

Amanda C Conta and Dennis J Stelzner

The propriospinal system: definition and overall function

In broad terms, the word 'propriospinal' refers to neurons that are intrinsic to the spinal cord and whose axons terminate within its boundaries (Chung and Coggeshall, 1983). This is in contrast to 'supraspinal' projections, whose cell bodies of origin are located outside, 'above', the spinal cord, and whose axons project to spinal regions (e.g. corticospinal, rubrospinal, vestibulospinal projections). Propriospinal fibers (PS) constitute a large proportion of the spinal cord white matter. Likewise, the number of propriospinal neurons in spinal cord gray matter is believed to be greater than the number of motoneurons (Chung *et al.*, 1984). Studies by Chung and colleagues suggest that 25% and 44% of axons in dorsal and dorsolateral funiculi, respectively, are of propriospinal origin (Chung *et al.*, 1987), as are 60% of the axons in sacral spinal cord segments (Chung *et al.*, 1984; Chung and Coggeshall, 1988). In more specific terms, the word 'propriospinal' is often used in reference to spinal interneurons that are rostral, and/or have direct input, to motoneuron (MN) pools and to other spinal interneurons. In this way, propriospinal projections participate in a variety of physiological and behavioral processes, such as modulation of afferent and descending input to the central pattern generators (CPG) for locomotion and respiration, as well as autonomic functions like visceroreception and pain perception. Within the cervical enlargement, for example, short-axon propriospinal neurons (also referred to as premotoneurons) mediate and modulate corticospinal and sensory input to upper/forelimb MNs, thereby playing key roles in reaching and grasping tasks (Alstermark and Sasaki, 1986a,b; Alstermark *et al.*, 1987a; Pierrot-Deseilligny, 2002). In a similar manner, propriospinal networks within the lumbosacral enlargement modulate input to lower/hindlimb MN pools (Kostyuk *et al.*, 1971; Kostyuk and Maisky, 1972; Kostyuk and Vasilenko, 1978; Gerasimenko *et al.*, 2002; Jordan and Schmidt, 2002). These limb segment propriospinal networks are regarded as cervical and lumbar spinal locomotor pattern generators, as they are intimately involved in the control and coordination of flexor and extensor MNs of the shoulder/forelimb and pelvic girdle/hindlimb regions, respectively (Kjaerulff and Kiehn, 1996; Ballion *et al.*, 2001; Juvin *et al.*, 2005). Cervical and lumbar spinal cord are further interconnected by ascending and descending long-axon propriospinal projections that coordinate the activity between cervical and lumbosacral enlargements during locomotion (Miller, 1970; Miller *et al.*, 1971, 1973; Miller SvdB, 1973; Matsushita *et al.*, 1979).

Respiratory propriospinal neuron groups relay descending input to MN pools that innervate primary inspiratory and expiratory muscles. High cervical inspiratory neurons modulate input from supraspinal regions onto phrenic and intercostal motor neurons (Hoskin *et al.*, 1988; Winslow and Rozovsky, 2003). In addition, they exert inhibitory effects on expiratory motoneurons from low thoracic and lumbar regions innervating abdominal muscle groups (Miller *et al.*, 1998). Thoracic respiratory interneurons are believed to be 'driven by supraspinal respiratory activity' and thereby modulate respiratory input to intercostal MN pools (Kirkwood *et al.*, 1988; Lu *et al.*, 2004). In addition, a subset of upper thoracic respiratory interneurons receive non-respiratory input, namely afferent viscerosomatic input, such as that resulting from esophageal distension, chemical stimulation of cardiac afferents and mechanical stimulation of somatic fields (Qin *et al.*, 2002a,b). This implies a role for these PS neurons in 'integrating nociceptive and nonnociceptive information from visceral organs and somatic fields' (Kirkwood *et al.*, 1988; Lundberg, 1979; Qin *et al.*, 2002a).

Propriospinal networks involved in autonomic function are located within superficial dorsal horn regions (laminae 2-4) as well as within lamina 10 (around the central canal), and constitute ipsi- and contralateral projecting circuits (Fitzgerald, 1982, 1983; Matsushita, 1998; Petko and Antal, 2000). Propriospinal projections that cross via the anterior and posterior commissures (a.k.a. commissural propriospinal connections; Petko *et al.*, 2004) provide the anatomical substrate that connects interneurons in laminae 3-4 of the dorsal horn on both sides of the spinal cord (Petko and Antal, 2000; Petko *et al.*, 2004). These reciprocal circuits, for instance, have been proposed to mediate the responses of spinal dorsal horn interneurons to mechanical or noxious stimulation in the contralateral hindlimb or tail (Fitzgerald, 1982, 1983; Petko *et al.*, 2004). A large proportion of propriospinal neurons located dorsal to the central canal (sometimes referred to as the dorsal commissural nucleus, or 'area 10 neurons'; Matsushita, 1998) are believed to modulate nociceptive and visceroreceptive information to and from the spinal cord. Different subsets of dorsal commissural propriospinal neurons are driven by nociceptive input, unlike others, which receive visceral (bowel and bladder) and somatic (cutaneous and subcutaneous) input (Honda and Lee, 1985; Matsushita, 1998). These propriospinal axons travel with supraspinal projections within the dorsal and dorsolateral funiculi, and project to interneurons in thoracic spinal cord, which, in turn, activate sympathetic preganglionic neurons of the intermediolateral cell column resulting in increased blood pressure (Petko and Antal, 2000; Cameron *et al.*, 2006).

As a result of their location, propriospinal networks in general exist in what could be considered a strategic position, by virtue of which they integrate afferent and descending input to the spinal cord and mediate essential motor, locomotor and respiration-related functions, as well as autonomic, posture and balance control mechanisms (Anderson, 1963; Giovanelli Barilari and Kuypers, 1969; Vasilenko, 1975; Vasilenko and Kostyukov, 1976; Kuypers and Maisky, 1977; Jordan and Schmidt, 2002).

Subgroups of propriospinal networks

Propriospinal projections can be classified into short and long-axon PS neuron subgroups. Short-axon networks usually have projections that extend up to six segments, and include among others: limb segment (i.e. cervical and lumbosacral enlargements) PS networks, short thoracic PS projections, and segmental inhibitory (thoracic) propriospinal circuits. Long-axon networks refer to those projecting beyond five or six spinal segments, for example, PS fibers interconnecting the cervical and lumbosacral enlargements (i.e. long descending and long ascending propriospinal tract projections) and upper cervical inspiratory interneurons.

Short-axon propriospinal projections

Limb segment propriospinal projections: cervical and lumbosacral

Short PS projections within the cervical and lumbosacral enlargements have their cell bodies of origin located throughout the gray matter, excluding lamina 9 motor neuron pools, and their axons extend intra- and intersegmentally for short distances (one or two segments) before synapsing onto motoneurons within lamina 9 or onto other neighboring interneurons.

Cervically located PS neurons have been characterized in great detail in the cat, by Alstermark and colleagues (Alstermark *et al.*, 1987a; Alstermark *et al.*, 2007; Alstermark and Kummel, 1986), and although the main circuitry is different from that seen in other mammals (including primates and humans), their role/function remains the same. Often referred to as the C3-C4 premotoneuronal system, short-axon cervical PS neurons are key modulators of reaching and grasping tasks. As such, C3-C4 PS neurons mediate descending input to the forelimb musculature. For example, in the cat there are no direct monosynaptic cortico-motoneuronal projections. All descending motor commands are relayed to relevant MN pools via spinally located C3-C4 PS neurons (Alstermark and Kummel, 1986; Alstermark *et al.*, 1987a; Alstermark *et al.*, 2007; Alstermark and Isa *et al.*, 2002). There is increasing evidence supporting the existence of a comparable C3-C4 propriospinal system in humans that is similarly responsible for mediating corticospinal input to the upper extremities (Dimitrijevic *et al.*, 1983, 2005; Mazevet and Pierrot-Deseilligny, 1994; Nicolas *et al.*, 2001; Gerasimenko *et al.*, 2002; Pierrot-Deseilligny, 2002; Pierrot-Deseilligny and Marchand-Pauvert, 2002; Stinear and Byblow, 2004). While humans, unlike cats, do have direct monosynaptic corticomotoneuronal connections, essential for the execution of finer, more complex motor tasks, a large part of corticospinal control of simpler motor tasks is believed to be the result of CST projections synapsing onto cervical premotorneurons, thus exerting their function indirectly onto MNs, via disynaptic or multisynaptic connections (Nicolas *et al.*, 2001). Descending input to C3-C4 PS neurons is further modulated by excitatory and inhibitory sensory peripheral/afferent information. In this manner, PS networks serve both as a gain control mechanism and as a lateral inhibition mechanism, enabling the finer control of the force of movement and preventing unnecessary activation of neighboring PS neurons not needed for a particular movement, respectively (Nicolas *et al.*, 2001; Pierrot-Deseilligny, 2002; Pierrot-Deseilligny and Marchand-Pauvert, 2002).

A similar propriospinal circuitry to that described in the cervical enlargement has also been described within the lumbosacral cord. These networks modulate descending and afferent input to lower/hindlimb motoneurons (Szentagothai, 1951; Sterling and Kuypers, 1968; Rustioni *et al.*, 1971). In their 1971 study, Rustioni and colleagues inferred the location of short-axon propriospinal neurons from Nauta and Fink-Heimer silver impregnation experiments. They suggested that lumbosacral PS neurons are differentially located within the intermediate gray matter, depending on whether they project to MNs of the pelvic girdle and thigh musculature (axial MNs) or to MNs of the distal hindlimb muscles. The former are believed to reside in medial laminae 7 and 8, and receive direct input from reticulospinal and vestibulospinal tracts traveling within the ventral white matter. The latter are more laterally placed in laminae 5-7, receiving input from rubrospinal and corticospinal tracts traveling within lateral, dorsolateral and dorsal white matter regions (Rustioni *et al.*, 1971). In this manner, subsets of short-axon lumbosacral PS projections play a role in modulating brain stem input to the lumbar CPG for locomotion, both in bipeds and quadrupeds (Grillner *et al.*, 1981; Cazalets *et al.*, 1995; Cowley and Schmidt, 1997, 2000; Zaporozhets, 2001; Jordan and Schmidt, 2002).

Much of the evidence for the existence of a CPG for locomotion in man comes from studies on individuals with incomplete spinal cord injuries. In these scenarios, injured subjects exhibit involuntary stepping movements, alternating right and left lower limbs, but only when in a supine position following hip extension and strong cutaneous input (Calancie *et al.*, 1994, Dobkin *et al.*, 1995, Duysens and Van de Crommert, 1998). Similar to what is reported in incomplete spinal cord injured individuals, involuntary stepping patterns can also be elicited via relatively strong electrical stimulation of lumbar spinal cord in individuals with complete cord transections at the cervical and thoracic spinal levels (Gerasimenko *et al.*, 2002; Jilge *et al.*, 2004; Minassian *et al.*, 2004). These observations imply that a spinally located interneuronal circuit may be responsible for modulating such peripheral stimuli and further integrating it with the local reflex circuitry to yield muscle contractions necessary for the elicitation of stepping patterns (Duysens and Van de Crommert, 1998).

Short thoracic propriospinal projections

Short thoracic PS neurons are distributed throughout gray matter regions, except lamina 9, and their axons either ascend or descend for a few (1-5) segments throughout the cord before terminating within gray matter regions and synapsing onto other local interneurons throughout laminae 3-8 and 10, as well as within motoneuron pools (lamina 9) (Giovanelli Barilari and Kuypers, 1969; Kostyuk *et al.*, 1971; Miller SvdB, 1973; Delwaide *et al.*, 1977; Kostyuk and Vasilenko, 1979; Molenaar and Kuypers, 1978; Szentagothai, 1951). Anterograde labeling of short thoracic PS fibers whose cell bodies of origin are located in mid-thoracic cord (T6-T8) indicates that these axons extend all the way to lumbosacral cord, terminating within motoneuron pools in the lumbar enlargement (Conta and Stelzner, 2004). In regards to their function, short thoracic PS neurons have been shown to be important for control of axial musculature and in postural mechanisms (Anderson, 1963; Vasilenko, 1975).

Thoracic respiratory interneurons

Thoracic respiratory propriospinal neurons are believed to be mainly involved in modulating/relaying descending respiratory drive from the medulla to intercostal and abdominal MN pools (Merrill and Lipski, 1987). Studies by Qin and colleagues, however, have provided evidence in support of the hypothesis that a subgroup of thoracic respiratory interneurons throughout the thoracic cord also receives and processes noxious somatic and visceral input (Qin *et al.*, 2002a,b). This is supported by studies showing that thoracic respiratory interneurons are stimulated/activated under experimental conditions following esophageal distension, chemical stimulation of cardiac afferents and mechanical stimulation of somatic fields.

Long-axon propriospinal projections

Long descending propriospinal tract projections

Long descending propriospinal tract neuron cell bodies are located in the deeper laminae of the dorsal horn (laminae 4 and 5) and intermediate gray matter (lamina 7 and 8), as well as dorsal to the central canal (lamina 10) of the cervical enlargement (C4-C6) (Giovanelli Barilari and Kuypers, 1969; Delwaide *et al.*, 1977; Matsushita *et al.*, 1979; Skinner *et al.*, 1979; Menetrey *et al.*, 1985, 1989; Conta and Stelzner, 2004). These neurons are arranged in a 'butterfly' pattern in these medial regions of the gray matter as well as around the central canal (lamina 10). Their axons extend caudally for long distances and the main projection terminates within comparable gray matter regions in lumbosacral spinal cord (i.e. deep laminae of dorsal horn and intermediate gray matter), thus interconnecting both enlargements and the short-axon PS circuits contained therein (Conta and Stelzner, 2004; Reed *et al.*, 2006). Long descending propriospinal tract (LDPT) projections are regarded as an 'intraspinal coordination system that plays key roles in mediating reflex control and in coordination during locomotion…' (Jankowska *et al.*, 1974; Delwaide *et al.*, 1977; Stelzner and Cullen, 1991; Cowley and Schmidt, 2000; Jordan and Schmidt, 2002; Conta and Stelzner, 2004; Reed *et al.*, 2006). In particular, LDPT circuits are considered key mediators of forelimb and hindlimb coordination during quadrupedal locomotion in rat and cat (Alstermark *et al.*, 1987b; Delwaide *et al.*, 1977; Stelzner and Cullen, 1991; Cowley and Schmidt, 2000), and have further been proposed to be a part of the CPG for locomotion (Jordan and Schmidt, 2002).

Long ascending propriospinal tract projections

Long ascending propriospinal tract (LAPT) projections interconnect the lumbosacral and cervical spinal cord enlargements, reciprocal to the LDPT projections. Ascending long-axon propriospinal circuits are believed to facilitate and/or inhibit afferent hindlimb input to forelimb reflexes. Such hindlimb input is carried in a caudo-rostral direction by LAPT fibers located within the 'central portions of the lateral funiculus' (Miller *et al.*, 1973). Recent studies by Juvin *et al.*, (2005) suggest that ascending excitatory input to cervical pattern generating centers plays a crucial role in both lumbocervical coupling and right-left forelimb alternation (Juvin *et al.*, 2005). In these studies, the authors chemically-induced fictive locomotor patterns in an isolated rat spinal cord preparation. These locomotor patterns were characterized by in-phase firing of ipsilateral C8 and L2 ventral roots and out-of-phase firing of C8 right and left ventral roots. Following a midsagittal cord section from C1 to T7, in an attempt to abolish the alternating left-right C8 ventral root activity, both in-phase and out-of-phase firing patterns persisted. This and other complementary experiments carried out in the same study (Juvin *et al.*, 2005) provide compelling evidence for the possibility that cervical and lumbar pattern generators are independent of each other, since they each possess bilateral alternating activity, a characteristic cycle frequency and burst duration (Juvin *et al.*, 2005). However, they are also coupled, and such coupling is necessary for the elicitation of locomotor patterns. Most importantly, 'cervical generators are unable to impose their rhythm patterns on their lumbar counterparts' (Juvin *et al.*, 2005). Therefore, although ascending and descending long-axon propriospinal projections act, for the most part, in a reciprocal manner to coordinate the limbs during quadrupedal locomotion (Miller SvdB, 1973), LAPT circuits may very likely exert a stronger influence on cervical propriospinal networks than the other way around, i.e. LDPT projections on lumbar propriospinal circuitry (Reed *et al.*, 2006; Viala and Vidal, 1978).

Indirect evidence for the existence of long-axon PS projections that interconnect the spinal enlargements in humans comes from studies in cervical spinal cord injured individuals, where electrical stimulation of the tibial nerve in the ankle causes an 'interlimb reflex response' in muscles in the forearm and hand (Calancie *et al.*, 1996). Also, upper and lower limb movements appear to be coordinated during functions like swimming or crawling, further suggesting that circuits controlling arm and leg functions are coupled (Wannier *et al.*, 2001; Dietz, 2002).

Upper cervical inspiratory projections

Upper cervical inspiratory neurons are regarded as propriospinal neurons that are located in high cervical (C1-C2) spinal regions and modulate brain stem input to respiratory MN pools (i.e. intercostal and abdominal MNs, and, to a lesser extent phrenic MNs) (Lipski *et al.*, 1993). Tracing studies (Miller *et al.*, 1998) and electrophysiological experiments (Hoskin *et al.*, 1988) suggest that collaterals from C1-C2 PS axons project to phrenic MN regions in C3-C5 spinal cord as they continue in their course to rostral as well as to caudal thoracic and upper lumbar cord regions, where inspiratory (intercostal) and expiratory (abdominal) MNs reside, respectively (Hoskin *et al.*, 1988; Winslow and Rozovsky, 2003). Subsets of C1-C2-originating propriospinal inspiratory projections have been shown to send axons as far as the lower lumbar and sacral spinal cord (L5-S1) (Miller *et al.*, 1998). Although upper cervical inspiratory projections most likely act in a monosynaptic manner on phrenic MNs, they are believed to act indirectly, in a disynaptic or multisynaptic manner on thoracic and lumbar MNs, by virtue of their interaction with more local thoracic respiratory propriospinal networks. In some instances, upper cervical inspiratory neurons 'excite inspiratory [i.e. intercostal] MNs via excitatory segmental/thoracic interneurons' (Lipski and Duffin, 1986); in other, complementary/reciprocal instances, the same C1-C2 propriospinal neurons constitute 'excitatory interneurons that inhibit expiratory [i.e. abdominal] MNs via segmental inhibitory/thoracic interneurons' (Hoskin *et al.*, 1988).

In addition to their effect on thoracic respiratory PS neurons, upper cervical inspiratory PS neurons exert descending inhibition on visceroreceptive thoracic interneurons. The increased activity elicited in thoracic or lumbosacral PS cells following cardiac afferent stimulation, esophageal distension and splanchnic stimulation or colorectal distension, respectively, was inhibited following glutamate application to upper cervical cord regions (Poree and Schramm, 1992; Qin *et al.*, 1999, 2004; Lu *et al.*, 2004;). Interestingly, this descending inhibition of thoracic interneurons following chemical activation of C1-C2 PS neurons did not affect the phasic respiratory drive of thoracic respiratory PS neurons to intercostal MNs (Lu *et al.*, 2004).

Propriospinal networks and neurotransmitters

Most of what is known today regarding the neurotransmitter characteristics of propriospinal neuron subgroups stems from stimulation studies of the in vitro spinal cord preparation (Cowley and Schmidt, 2000; Jordan and Schmidt, 2002; Juvin *et al.*, 2005). It is fairly well accepted that serotonin (5-HT) plays a key role in the participation of propriospinal circuits during motor and locomotor functions (Steeves *et al.*, 1980; Schmidt and Jordan, 2000; Jordan and Schmidt, 2002), as do N-methyl-D,L-aspartate (NMDA) and dopamine (DA) (Cowley and Schmidt, 2000; Jordan and Schmidt, 2002; Juvin *et al.*, 2005). Studies by Privat and colleagues suggest that exogenous delivery of 5-HT via transplanted embryonic raphe cells is capable of rescuing locomotor rhythmic activity following SCI (Feraboli-Lohnherr *et al.*, 1997; Ribotta *et al.*, 2000). Similarly, pharmachological interventions targeting serotonin receptors were able to elicit locomotor-like patterns in the spinal rat (Antri *et al.*, 2002, 2005). These findings support the possibility that serotonin can drive the isolated spinal cord of adult cat or rat and can reelicit locomotor behavior, following complete transection injury, further suggesting that PS neurons respond to 5-HT as a neurotransmitter. Recording experiments done on isolated neonatal rat spinal cord preparations showed that the spinal cord was usually inactive under control conditions, within a saline bath. Only when a mix of NMDA, 5-HT and DA was added to the bath was fictive locomotion elicited; consistent, coordinated and synchronous recordings were obtained from the ipsilateral ventral motor roots of cervical extensor and lumbar flexor MNs (Barriere *et al.*, 2004; Juvin *et al.*, 2005).

Interestingly, such alternating, yet coupled, locomotor-like rhythms elicited under bath application of the neurotransmitter cocktail became independent of each other following reversible axonal blockade at the level of the thoracic spinal cord (Juvin *et al.*, 2005). A similar finding was reported when the rhythmic coupling between cervical and lumbar enlargement regions was abolished following atropine application to the mid-thoracic spinal cord (Cowley and Schmidt, 2000), providing further evidence for the coordinating role of LDPT and LAPT projections during locomotion. In line with this latter observation, subsets of propriospinal neurons have been shown to be a part of what has been termed a 'cholinergic propriospinal innervation' of the spinal cord. This subset is consisted of short-axon propriospinal neurons from two specific groups: 1) those located primarily around the central canal (lamina 10 propriospinal cells), and 2) those found in lamina 7 ('partition' cells) (Sherriff and Henderson, 1994). In these experiments, retrogradely labeled PS cells were also immunoreactive for choline acetyltransferase (ChAT).

Upper cervical inspiratory propriospinal neurons are activated following glutamate application to the spinal cord surface (Qin *et al.*, 2002b, Lu *et al.*, 2004). This 'glutamate-activated cervical network' appears to play a key role in descending inhibition of activated viscerosomatic thoracic PS neurons (Poree and Schramm, 1992). This inhibitory effect is still evident following a complete C1 transection, confirming the propriospinal nature of this process (Poree and Schramm, 1992; Lu *et al.*, 2004).

Despite the paucity of detailed information regarding the inhibitory neurotransmitter characteristics of propriospinal circuits, the reciprocal firing patterns of MNs within the locomotor and respiratory CPGs is the result of alternating inhibition and excitation within the spinal cord. Preliminary studies in our laboratory indicate that many cervical and thoracic propriospinal neurons retrogradely labeled from the upper lumbar enlargement with fluorogold are immunoreactive for the GABA (A) α 1 receptor, and therefore are likely to use GABA as a neurotransmitter (Stelzner, unpublished observations).

Locomotor propriospinal system across species

It has been suggested that both bipedal and quadrupedal locomotor functions share similarities in the way upper and lower extremity muscles are coordinated (i.e. forelimb-hindlimb coordination in quadrupeds), such that the neuronal circuitry underlying interlimb coordination and reflex control during locomotion may be conserved from quadrupeds to primates to humans (Dietz V 2002). The main feature of such a neuronal circuitry for locomotion is a propriospinally-mediated integration of the cortical command and peripheral input during the execution of a locomotor function. That said, bipedal locomotion is considered the 'evolutionarily newer neuronal control mechanism' (Dietz, 2002) compared to a 'phylogenetically older quadrupedal circuitry' (Dietz, 2002, 2003). The difference between 'newer' and 'older' mechanisms is twofold and includes, for a bipedal circuitry: a) a specialization of the upper extremities in primates and humans, whereby the hand is now able to perform very discrete and skilled movements, and b) the eventual acquirement of upright stance and gait (Nakajima *et al.*, 2000; Dietz, 2002).

The underlying neuronal circuitry that gives rise to and enables skilled, fine motor control of hand movements in higher primates and man, in particular the monosynaptic cortico-motoneuronal projections, has been hypothesized to coexist with the evolutionarily older propriospinal circuitry that appears to mediate bipedal locomotion in humans (Dietz, 2002). Specifically, during locomotor functions that require coupling between cervical and lumbosacral arm and leg circuitry, indirect, propriospinally-mediated cortical input takes over. Alternatively, during more specific, skilled hand movement tasks, this indirect propriospinal pathway is inhibited, allowing for direct, monosynaptic cortico-motoneuronal control (Dietz, 2002, 2003). Evidence for this comes from stimulation studies in normal, healthy individuals showing that upper limb EMG responses could be elicited following a mechanical stimulus to one leg, but only when walking. Further, such interlimb reflex responses were absent when the test subject was standing and/or sitting while writing (i.e. carrying out a task that requires skilled hand movements) (Dietz *et al.*, 2001).

Propriospinal projections and experimental spinal cord injury

Studies seeking to promote regeneration following traumatic spinal cord injury (SCI) have been consistently unable to achieve long-distance, functional regrowth of axotomized supraspinal projections following SCI and a variety of therapeutic interventions. Only recently have propriospinal projections come into the picture as potential targets for regenerative and neuroprotective therapies post-SCI. Studies by Bareyre and colleagues (2004), for example, suggested that it may not be critical to achieve long distance regeneration of damaged supraspinal projections in order to promote functional recovery, since these damaged descending axons sprout collaterals onto subsets of local, intraspinal neurons that are not themselves severely affected by SCI. These local projections normally project to lumbar spinal cord intermediate gray matter as part of the LDPT subgroup of propriospinal fibers. Following SCI, they appear capable of sprouting into areas within lumbosacral cord that they would not normally target, i.e. motoneuron pools in lamina 9, thereby 'bridging' the lesion site and functioning as a 'detour circuit' (Bareyre *et al.*, 2004; Courtine *et al.*, 2008). As well as spared LDPT axons being important for recovery from SCI, damaged short PS projections may be a better target for therapies to improve axonal regeneration.

Experiments indicate that neurons whose axons are severed close to the cell body have a greater intrinsic capacity to mount a regenerative response (David and Aguayo, 1981; Benfey and Aguayo, 1982; Doster *et al.*, 1991) than long tract axons damaged in the spinal cord. Therefore, these propriospinal neurons are likely to be better candidates for such regeneration studies than, for example, the farther-located cell bodies of origin of the corticospinal tract or other brain stem projections. A handful of studies provide some evidence for this possibility by linking the reported functional recovery following spinal cord hemisections to the preferential regeneration of propriospinal projections (Xu *et al.*, 1995; Bregman *et al.*, 1997, 2002; Guest *et al.*, 1997; Iannotti *et al.*, 2003; Blesch *et al.*, 2004).

Interestingly, very few studies have been carried out in which the fate of propriospinal networks was assessed post-SCI. Recent studies from our laboratory have shown that propriospinal projections have a differential vulnerability to SCI and that, although a large proportion of long and short axon propriospinal neurons are lost within the first two weeks post-SCI, a small subset is still amenable to retrograde tracing, even at chronic survival times post-SCI. These results suggest that propriospinal networks may still be viable following different injury severities, and therefore may constitute good candidate substrates for regeneration and neuroprotection studies.

Differential vulnerability of propriospinal tract neurons to SCI in rat

Propriospinal tract neurons and their axons have a differential vulnerability to SCI, such that a subset of LDPT projections is spared following mild, moderate and severe SCI, whereas most short thoracic PS neurons appear to be axotomized by the same injuries (Conta and Stelzner, 2004; Conta and Stelzner, 2006, unpublished findings). In these experiments, even after the most severe of contusion injuries, a small number of LDPT neurons could be retrogradely labeled when fluorogold (FG) was injected caudal to the SCI. Conversely, very few short thoracic PS axons located near the same injury were spared, as indicated by the absence of retrogradely labeled cell bodies, from the same injections of FG. Interestingly, prelabeling the cell bodies of origin of these short thoracic PS axons prior to SCI showed that a significant proportion (~ 20%) of these neurons were still intact 2 weeks post-injury, compared to later time points (Conta and Stelzner, unpublished findings). These findings are most likely explained by the differential projection patterns of long- and short-axon PS fibers (Matsushita, 1970; Kostyuk and Maisky, 1972; Molenaar *et al.*, 1974; Molenaar and Kuypers, 1978; Conta and Stelzner 2004). At mid-low thoracic levels (T9-T10) LDPT axons are localized to the lateral-most portions of the lateral and ventrolateral funiculi, whereas short thoracic PS fibers travel more medially, within lateral white matter regions abutting the gray-white matter boundaries (Conta and Stelzner, 2004). In light of the dorsal approach that characterizes contusion lesions, as well as the centrifugal progression of these types of lesions (evolving from the center outwards and most severely affecting gray matter areas) the majority of the more medially located short PS projections are likely to be axotomized by this injury, compared to more laterally located LDPT fibers. An alternative explanation for our findings is that LDPT neurons are less vulnerable to SCI than their thoracic counterparts because they are located farther from the lesion site.

Conclusion

In summary, propriospinal neurons and their axons are contained within the boundaries of the spinal cord, making up an important proportion of the gray and white matter, respectively. They exert a variety of functions, including the modulation of descending and afferent input to locomotor, respiratory and autonomic centers throughout the spinal cord. Short-axon propriospinal projections within thoracic regions play important roles in axial musculature control and balance mechanisms. Short thoracic inspiratory propriospinal circuits innervate intercostal and abdominal muscles and mediate descending (brainstem) input to the corresponding motoneurons involved in respiration. Long ascending and descending propriospinal tracts connect cervical and lumbosacral enlargements and the short-axon circuits contained therein, and play key roles in interlimb coordination during quadrupedal locomotion. High cervical inspiratory interneurons originate at C1-C2 spinal levels and have axons that project caudally for long distances. This subpopulation modulates brainstem input to thoracic respiratory interneurons and exerts a descending inhibitory effect on visceroreceptive thoracic interneurons. In this way propriospinal networks appear to be essential for the transmission and processing of information pertaining to vital/basic/everyday functions like walking and breathing.

Their disruption following injury may very likely contribute to the observed overall dysfunction, and targeting propriospinal neurons and/or their axons may constitute an important aspect of therapeutic and pharmacological interventions seeking to promote significant functional recovery.

References

Alstermark B, Isa T. 2002. Premotoneuronal and direct corticomotoneuronal control in the cat and macaque monkey. Adv Exp Med Biol 508, 281-297.

Alstermark B, Isa T, Pettersson LG, Sasaki S (2007) The C3-C4 propriospinal system in the cat and monkey: a spinal pre-motoneuronal centre for voluntary motor control. Acta Physiol 189, 123-140.

Alstermark B, Kummel H (1986) Transneuronal labelling of neurones projecting to forelimb motoneurones in cats performing different movements. Brain Res 376, 387-391.

Alstermark B, Kummel H, Pinter MJ, Tantisira B (1987b) Branching and termination of C3-C4 propriospinal neurones in the cervical spinal cord of the cat. Neurosci Lett 74, 291-296.

Alstermark B, Lundberg A, Pinter M, Sasaki S (1987b) Long C3-C5 propriospinal neurones in the cat. Brain Res 404, 382-388.

Alstermark B, Sasaki, S (1986a) Integration in descending motor pathways controlling the forelimb in the cat. 14. Differential projection to fast and slow motoneurones from excitatory C3-C4 propriospinal neurones. Exp Brain Res 63, 530-542.

Alstermark B, Sasaki, S (1986b) Integration in descending motor pathways controlling the forelimb in the cat. 15. Comparison of the projection from excitatory C3-C4 propriospinal neurones to different species of forelimb motoneurones. Exp Brain Res 63, 543-556.

Anderson FD (1963) The structure of a chronically isolated segment of the cat spinal cord. J Comp Neurol 120, 297-315.

Antri M, Orsal D, Barthe JY (2002) Locomotor recovery in the chronic spinal rat: effects of long-term treatment with a 5-HT2 agonist. Eur J Neurosci 16, 467-476.

Antri M, Barthe JY, Mouffle C, Orsal D (2005) Long-lasting recovery of locomotor function in chronic spinal rat following chronic combined pharmacological stimulation of serotonergic receptors with 8-OHDPAT and quipazine. Neurosci Lett 384(1-2), 162-167.

Ballion B, Morin D, Viala D (2001) Forelimb locomotor generators and quadrupedal locomotion in the neonatal rat. Eur J Neurosci 14(10), 1727-1738.

Bareyre FM, Kerschensteiner M, Raineteau O, Mettenleiter TC, Weinmann O, Schwab ME (2004) The injured spinal cord spontaneously forms a new intraspinal circuit in adult rats. Nat Neurosci 7, 269-277.

Barriere G, Mellen N, Cazalets JR (2004) Neuromodulation of the locomotor network by dopamine in the isolated spinal cord of newborn rat. Eur J Neurosci 19, 1325-1335.

Benfey M, Aguayo AJ (1982) Extensive elongation of axons from rat brain into peripheral nerve grafts. Nature 296(5853), 150-152.

Blesch A, Yang H, Weidner N, Hoang A, Otero, D (2004) Axonal responses to cellularly delivered NT-4/5 after spinal cord injury. Mol Cell Neurosci 27, 190-201.

Bregman BS, McAtee M, Dai HN, Kuhn PL (1997) Neurotrophic factors increase axonal growth after spinal cord injury and transplantation in the adult rat. Exp Neurol 148, 475-494.

Bregman BS, Coumans JV, Dai HN, Kuhn PL, Lynskey J, McAtee M, Sandhu F (2002) Transplants and neurotrophic factors increase regeneration and recovery of function after spinal cord injury. Prog Brain Res 137, 257-273.

Calancie B, Needham-Shropshire B, Jacobs P, Willer K, Zych G, Green BA (1994) Involuntary stepping after chronic spinal cord injury. Evidence for a central rhythm generator for locomotion in man. Brain 117 (Pt 5), 1143-1159.

Calancie B, Lutton S, Broton JG (1996) Central nervous system plasticity after spinal cord injury in man: interlimb reflexes and the influence of cutaneous stimulation. Electroencephalogr Clin Neurophysiol 101, 304-315.

Cameron AA, Smith GM, Randall DC, Brown DR, Rabchevsky AG (2006) Genetic manipulation of intraspinal plasticity after spinal cord injury alters the severity of autonomic dysreflexia. J Neurosci 26(11), 2923-2932.

Cazalets JR, Borde M, Clarac F (1995) Localization and organization of the central pattern generator for hindlimb locomotion in newborn rat. J Neurosci 15(7 Pt 1), 4943-4951.

Chung K, Coggeshall, RE (1983) Propriospinal fibers in the rat. J Comp Neurol 217, 47-53.

Chung, K, Coggeshall RE (1988) Propriospinal fibers in the white matter of the cat sacral spinal cord. J Comp Neurol 269, 612-617.

Chung K, Kevetter GA, Willis WD, Coggeshall RE (1984) An estimate of the ratio of propriospinal to long tract neurons in the sacral spinal cord of the rat. Neurosci Lett 44, 173-177.

Chung K, Langford LA, Coggeshall RE (1987) Primary afferent and propriospinal fibers in the rat dorsal and dorsolateral funiculi. J Comp Neurol 263, 68-75.

Conta A, Stelzner DJ (2006) Unpublished findings. SUNY Upstate Medical University.

Conta AC, Stelzner DJ (2004) Differential vulnerability of propriospinal tract neurons to spinal cord contusion injury. J Comp Neurol 479, 347-359.

Courtine G, Song B, Roy RR, Zhong H, Herrmann JE, Ao Y, Qi J, Edgerton VR, Sofroniew MV (2008) Recovery of supraspinal control of stepping via indirect propriospinal relay connections after spinal cord injury. Nat Med 14, 69-74.

Cowley CK, Schmidt, BJ (2000) Characterization of propriospinal coupling between the cervical and lumbar regions during rhythmic motor activity in the in vitro neonatal spinal cord. Soc Neurosci Abstr 26, 60.2.

Cowley KC, Schmidt BJ (1997) Regional distribution of the locomotor pattern-generating network in the neonatal rat spinal cord. J Neurophysiol 77, 247-259.

David S, Aguayo AJ (1981) Axonal elongation into peripheral nervous system "bridges" after central nervous system injury in adult rats. Science 214, 931-933.

Delwaide PJ, Figiel C, Richelle C (1977) Effects of postural changes of the upper limb on reflex transmission in the lower limb. Cervicolumbar reflex interactions in man. J Neurol Neurosurg Psychiatry 40, 616-621.

Dietz V (2002) Do human bipeds use quadrupedal coordination? Trends Neurosci 25, 462-467.

Dietz V (2003) Spinal cord pattern generators for locomotion. Clin Neurophysiol 114, 1379-1389.

Dietz V, Fouad K, Bastiaanse CM (2001) Neuronal coordination of arm and leg movements during human locomotion. Eur J Neurosci 14(11), 1906-1914.

Dimitrijevic MR, Faganel J, Lehmkuhl D, Sherwood A (1983) Motor control in man after partial or complete spinal cord injury. Adv Neurol 39, 915-926.

Dimitrijevic MR, Persy I, Forstner C, Kern H, Dimitrijevic MM (2005) Motor control in the human spinal cord. Artif Organs 29, 216-219.

Dobkin BH, Harkema S, Requejo P, Edgerton, VR (1995) Modulation of locomotor-like EMG activity in subjects with complete and incomplete spinal cord injury. J Neurol Rehabil 9, 183-190.

Doster SK, Lozano AM, Aguayo AJ, Willard MB (1991) Expression of the growth-associated protein GAP-43 in adult rat retinal ganglion cells following axon injury. Neuron 6, 635-647.

Duysens J, Van de Crommert HW (1998) Neural control of locomotion; The central pattern generator from cats to humans. Gait Posture 7, 131-141.

Feraboli-Lohnherr D, Orsal D, Yakovleff A, Gimenez y Ribotta M, Privat A (1997) Recovery of locomotor activity in the adult chronic spinal rat after sublesional transplantation of embryonic nervous cells: specific role of serotonergic neurons. Exp Brain Res 113, 443-454.

Fitzgerald M (1982) The contralateral input to the dorsal horn of the spinal cord in the decerebrate spinal rat. Brain Res 236, 275-287.

Fitzgerald M (1983) Influences of contralateral nerve and skin stimulation on neurones in the substantia gelatinosa of the rat spinal cord. Neurosci Lett 36, 139-143.

Gerasimenko YP, Makarovskii AN, Nikitin OA (2002) Control of locomotor activity in humans and animals in the absence of supraspinal influences. Neurosci Behav Physiol 32, 417-423.

Giovanelli Barilari M, Kuypers HG (1969) Propriospinal fibers interconnecting the spinal enlargements in the cat. Brain Res 14, 321-330.

Grillner S, McClellan A, Sigvardt K, Wallen P, Wilen M (1981) Activation of NMDA-receptors elicits 'fictive locomotion' in lamprey spinal cord in vitro. Acta Physiol Scand 113, 549-551.

Guest JD, Hesse D, Schnell L, Schwab ME, Bunge MB, Bunge RP (1997) Influence of IN-1 antibody and acidic FGF-fibrin glue on the response of injured corticospinal tract axons to human Schwann cell grafts. J Neurosci Res 50, 888-905.

Honda CN, Lee CL (1985) Immunohistochemistry of synaptic input and functional characterizations of neurons near the spinal central canal. Brain Res 343, 120-128.

Hoskin RW, Fedorko LM, Duffin, J (1988) Projections from upper cervical inspiratory neurons to thoracic and lumbar expiratory motor nuclei in the cat. Exp Neurol 99, 544-555.

Iannotti C, Li H, Yan P, Lu X, Wirthlin L, Xu XM (2003) Glial cell line-derived neurotrophic factor-enriched bridging transplants promote propriospinal axonal regeneration and enhance myelination after spinal cord injury. Exp Neurol 183, 379-393.

Jankowska E, Lundberg A, Roberts WJ, Stuart D (1974) A long propriospinal system with direct effect on motoneurones and on interneurones in the cat lumbosacral cord. Exp Brain Res 21, 169-194.

Jilge B, Minassian K, Rattay F, Pinter MM, Gerstenbrand F, Binder H, Dimitrijevic MR (2004) Initiating extension of the lower limbs in subjects with complete spinal cord injury by epidural lumbar cord stimulation. Exp Brain Res 154, 308-326.

Jordan LM, Schmidt BJ (2002) Propriospinal neurons involved in the control of locomotion, potential targets for repair strategies? Prog Brain Res 137, 125-139.

Juvin L, Simmers J, Morin D (2005) Propriospinal circuitry underlying interlimb coordination in mammalian quadrupedal locomotion. J Neurosci 25(25), 6025-6035.

Kirkwood PA, Munson JB, Sears TA, Westgaard RH (1988) Respiratory interneurones in the thoracic spinal cord of the cat. J Physiol 395, 161-192.

Kjaerulff O, Kiehn O (1996) Distribution of networks generating and coordinating locomotor activity in the neonatal rat spinal cord in vitro: a lesion study. J Neurosci 16(18), 5777-5794.

Kostyuk PG, Maisky VA (1972) Propriospinal projections in the lumbar spinal cord of the cat. Brain Res 39, 530-535.

Kostyuk PG, Vasilenko DA (1978) Propriospinal neurones as a relay system for transmission of cortico-spinal influences. J Physiol (Paris) 74, 247-250.

Kostyuk PG, Vasilenko DA (1979) Spinal interneurons. Annu Rev Physiol 41, 115-126.

Kostyuk PG, Vasilenko DA, Lang, E (1971) Propriospinal pathways in the dorsolateral funiculus and their effects on lumbosacral motoneuronal pools. Brain Res 28, 233-249.

Kuypers HG, Maisky VA (1977) Funicular trajectories of descending brain stem pathways in cat. Brain Res 136, 159-165.

Lipski J, Duffin J (1986) An electrophysiological investigation of propriospinal inspiratory neurons in the upper cervical cord of the cat. Exp Brain Res 61, 625-637.

Lipski J, Duffin J, Kruszewska B, Zhang X (1993) Upper cervical inspiratory neurons in the rat: an electrophysiological and morphological study. Exp Brain Res 95, 477-487.

Lu F, Qin C, Foreman RD, Farber JP (2004) Chemical activation of C1-C2 spinal neurons modulates intercostal and phrenic nerve activity in rats. Am J Physiol Regul Integr Comp Physiol 286, 1069-1076.

Lundberg A (1979) Multisensory control of spinal reflex pathways. Prog Brain Res 50, 11-28.

Matsushita, M (1970) The axonal pathways of spinal neurons in the cat. J Comp Neurol 138, 391-417.

Matsushita M (1998) Ascending propriospinal afferents to area X (substantia grisea centralis) of the spinal cord in the rat. Exp Brain Res 119, 356-366.

Matsushita M, Ikeda M, Hosoya Y (1979) The location of spinal neurons with long descending axons (long descending propriospinal tract neurons) in the cat: a study with the horseradish peroxidase technique. J Comp Neurol 184, 63-80.

Mazevet D, Pierrot-Deseilligny E (1994) Pattern of descending excitation of presumed propriospinal neurones at the onset of voluntary movement in humans. Acta Physiol Scand 150, 27-38.

Menetrey D, de Pommery J, Roudier F (1985) Propriospinal fibers reaching the lumbar enlargement in the rat. Neurosci Lett 58, 257-261.

Merrill EG, Lipski J (1987) Inputs to intercostal motoneurons from ventrolateral medullary respiratory neurons in the cat. J Neurophysiol 57, 1837-1853.

Miller KE, Douglas VD, Richards AB, Chandler MJ, Foreman, RD (1998) Propriospinal neurons in the C1-C2 spinal segments project to the L5-S1 segments of the rat spinal cord. Brain Res Bull 47, 43-47.

Miller S (1970) Excitatory and inhibitory propriospinal pathways from lumbo-sacral to cervical segments in the cat. Acta Physiol Scand 80, 25A-26A.

Miller S, Reitsma DJ, van der Meche FG (1971) Excitatory ascending propriospinal actions between lumbosacral and cervical segments in the cat. J Physiol 218, 76P-77P.

Miller S, Reitsma DJ, van der Meche FG (1973) Functional organization of long ascending propriospinal pathways linking lumbo-sacral and cervical segments in the cat. Brain Res 62, 169-188.

Miller SvdB, J (1973) *The function of long propriospinal pathways in the co-ordination of quadrupedal stepping in the cat. Control of posture and locomotion.* Stein RD, Redford JD (eds). Plenum Press, New York.

Minassian K, Jilge B, Rattay F, Pinter MM, Binder H, Gerstenbrand F, Dimitrijevic MR (2004) Stepping-like movements in humans with complete spinal cord injury induced by epidural stimulation of the lumbar cord: electromyographic study of compound muscle action potentials. Spinal Cord 42, 401-416.

Molenaar, I, Kuypers, HG (1978) Cells of origin of propriospinal fibers and of fibers ascending to supraspinal levels. A HRP study in cat and rhesus monkey. Brain Res 152, 429-450.

Molenaar, I, Rustioni, A, Kuypers, HG (1974) The location of cells of origin of the fibers in the ventral and the lateral funiculus of the cat's lumbo-sacral cord. Brain Res 78, 239-254.

Nakajima K, Maier MA, Kirkwood PA, Lemon RN (2000) Striking differences in transmission of corticospinal excitation to upper limb motoneurons in two primate species. J Neurophysiol 84, 698-709.

Nicolas G, Marchand-Pauvert V, Burke D, Pierrot-Deseilligny E (2001) Corticospinal excitation of presumed cervical propriospinal neurones and its reversal to inhibition in humans. J Physiol 533(Pt 3), 903-919.

Petko M, Antal M (2000) Propriospinal afferent and efferent connections of the lateral and medial areas of the dorsal horn (laminae I-IV) in the rat lumbar spinal cord. J Comp Neurol 422, 312-325.

Petko M, Veress G, Vereb G, Storm-Mathisen J, Antal M (2004) Commissural propriospinal connections between the lateral aspects of laminae III-IV in the lumbar spinal cord of rats. J Comp Neurol 480, 364-377.

Pierrot-Deseilligny E (2002) Propriospinal transmission of part of the corticospinal excitation in humans. Muscle Nerve 26, 155-172.

Pierrot-Deseilligny E, Marchand-Pauvert V (2002) A cervical propriospinal system in man. Adv Exp Med Biol 508, 273-279.

Poree LR, Schramm LP (1992) Role of cervical neurons in propriospinal inhibition of thoracic dorsal horn neurons. Brain Res 599, 302-308.

Qin C, Chandler MJ, Miller KE, Foreman RD (1999) Chemical activation of cervical cell bodies: effects on responses to colorectal distension in lumbosacral spinal cord of rats. J Neurophysiol 82, 3423-3433.

Qin C, Chandler MJ, Foreman RD, Farber JP (2002a) Upper thoracic respiratory interneurons integrate noxious somatic and visceral information in rats. J Neurophysiol 88, 2215-2223.

Qin C, Farber JP, Chandler MJ, Foreman RD (2002b) Chemical activation of C(1)-C(2) spinal neurons modulates activity of thoracic respiratory interneurons in rats. Am J Physiol Regul Integr Comp Physiol 283, R843-852.

Qin C, Kranenburg A, Foreman RD (2004) Descending modulation of thoracic visceroreceptive transmission by C1-C2 spinal neurons. Auton Neurosci 114, 11-16.

Reed WR, Shum-Siu A, Onifer SM, Magnuson DS (2006) Inter-enlargement pathways in the ventrolateral funiculus of the adult rat spinal cord. Neurosci 142, 1195-1207.

Ribotta MG, Provencher J, Feraboli-Lohnherr D, Rossignol S, Privat A, Orsal D (2000) Activation of locomotion in adult chronic spinal rats is achieved by transplantation of embryonic raphe cells reinnervating a precise lumbar level. J Neurosci 20, 5144-5152.

Rustioni A, Kuypers HG, Holstege G (1971) Propiospinal projections from the ventral and lateral funiculi to the motoneurons in the lumbosacral cord of the cat. Brain Res 34, 255-275.

Schmidt BJ, Jordan LM (2000) The role of serotonin in reflex modulation and locomotor rhythm production in the mammalian spinal cord. Brain Res Bull 53, 689-710.

Sherriff FE, Henderson Z (1994) A cholinergic propriospinal innervation of the rat spinal cord. Brain Res 634, 150-154.

Skinner RD, Coulter JD, Adams RJ, Remmel RS (1979) Cells of origin of long descending propriospinal fibers connecting the spinal enlargements in cat and monkey determined by horseradish peroxidase and electrophysiological techniques. J Comp Neurol 188, 443-454.

Skinner RD, Nelson R, Griebel M, Garcia-Rill E (1989) Ascending projections of long descending propriospinal tract (LDPT) neurons. Brain Res Bull 22, 253-258.

Steeves JD, Schmidt BJ, Skovgaard BJ, Jordan LM (1980) Effect of noradrenaline and 5-hydroxytryptamine depletion on locomotion in the cat. Brain Res 185, 349-362.

Stelzner DJ, Cullen JM (1991) Do propriospinal projections contribute to hindlimb recovery when all long tracts are cut in neonatal or weanling rats? Exp Neurol 114, 193-205.

Sterling P, Kuypers HG (1968) Anatomical organization of the brachial spinal cord of the cat. 3. The propriospinal connections. Brain Res 7, 419-443.

Stinear JW, Byblow WD (2004) The contribution of cervical propriospinal premotoneurons in recovering hemiparetic stroke patients. J Clin Neurophysiol 21, 426-434.

Szentagothai J (1951) Short propriospinal neurons and intrinsic connections of the spinal gray. Acta Morph. Acad. Sci. Hung. 1, 81-94.

Vasilenko DA (1975) Propriospinal pathways in the ventral funicles of the cat spinal cord: their effects on lumbosacral motoneurones. Brain Res 93, 502-506.

Vasilenko DA, Kostyukov AI (1976) Brain stem and primary afferent projections to the ventromedial group of propriospinal neurones in the cat. Brain Res 117, 141-146.

Viala D, Vidal C (1978) Evidence for distinct spinal locomotion generators supplying respectively fore and hindlimbs in the rabbit. Brain Res 155, 182-186.

Wannier T, Bastiaanse C, Colombo G, Dietz V (2001) Arm to leg coordination in humans during walking, creeping and swimming activities. Exp Brain Res 141, 375-379.

Winslow C, Rozovsky J (2003) Effect of spinal cord injury on the respiratory system. Am J Phys Med Rehabil 82, 803-814.

Xu XM, Guenard, V, Kleitman N, Aebischer P, Bunge MB (1995) A combination of BDNF and NT-3 promotes supraspinal axonal regeneration into Schwann cell grafts in adult rat thoracic spinal cord. Exp Neurol 134, 261-272.

Zaporozhets EaS BJ (2001) Propriospinal connections mediate bulbospinal activation of locomotor-like activity in the in vitro neonatal rat spinal cord. Soc Neurosci Abstr 27, 297.11.

13 Spinal Cord Transmitter Substances

Claire Heise and Gulgun Kayalioglu

Introduction

The spinal cord, like the brain, is a neurochemically diverse structure. Neurons in the spinal cord use a variety of neurotransmitters to carry out their functions, and they in turn receive a variety of neurotransmitter inputs. For the most part these neurotransmitters show distinct patterns of distribution within the spinal cord that are similar across different species. Furthermore, specific neurotransmitter systems in the spinal cord are often associated with specific functions. This chapter focuses on the distribution of acetylcholine (ACh) and substance P (SP) within the cord, and examines the serotoninergic (5-HT), noradrenergic (NA), and dopaminergic (DA) projections from the brain to the spinal cord.

Cholinergic neurons

Over the last several decades, many studies have shed light on the morphology and function of the cholinergic neurons in the central nervous system of several mammalian species. Three enzymes are commonly used as markers of cholinergic cells and processes: choline acetyltransferase (ChAT), which synthesizes ACh; acetylcholinesterase (AChE), the hydrolyzing enzyme of acetylcholine; and vesicular acetylcholine transporter protein (VAChT), which is responsible for the transportation of acetylcholine from the cytoplasm into the synaptic vesicles (Kimura *et al.*, 1981; Houser *et al.*, 1983; Barber *et al.*, 1984; Borges and Iversen, 1986; Davidoff *et al.*, 1989; Arvidsson *et al.*, 1997; Kobayashi *et al.*, 2002; VanderHorst and Ulfhake, 2006). In the spinal cord, the distribution of neurons labeled with AChE histochemistry and immunohistochemistry for ChAT shows an identical distribution (Davidoff *et al.*, 1989). These neurons are found throughout the longitudinal extent of the spinal cord from cervical to coccygeal segments. In most species, they are localized to the gray matter of the spinal cord (Kimura *et al.*, 1981; Houser *et al.*, 1983; Barber *et al.*, 1984; Borges and Iversen, 1986; Davidoff *et al.*, 1989; Kobayashi *et al.*, 2002). Within the gray matter, cholinergic neurons and their processes are found distributed similarly in the dorsal horn, the intermediate gray and the ventral horn in all mammals studied thus far (rabbit – Kan *et al.*, 1978; cat – Aquilonius *et al.*, 1981; Kimura *et al.*, 1981; rat – Aquilonius *et al.*, 1981; Houser *et al.*, 1983; Barber *et al.*, 1984; Borges and Iversen, 1986; monkey – Kobayashi *et al.*, 2002; guinea pig – Davidoff *et al.*, 1989; human – Aquilonius *et al.*, 1981; Kobayashi *et al.*, 2002).

Dorsal horn

Cell bodies immunoreactive for ChAT are present in the dorsal horn, and those located in laminae 3-5 extend dendrites dorsally into a longitudinal plexus within lamina 3, which shows intense ChAT immunoreactivity (Figure 13.1; Kimura *et al.*, 1981; Houser *et al.*, 1983; Barber *et al.*, 1984; Borges and Iversen, 1986; Kobayashi, 2002). The majority of this ChAT immunoreactivity is associated with a dense array of fibers and terminals. However, a few bipolar ChAT immunoreactive neurons are also present in the dorsal horn, particularly in laminae 3, 4 and 5 (Kimura *et al.*, 1981; Houser *et al.*, 1983; Barber *et al.*, 1984; Borges and Iversen, 1986; Kobayashi *et al.*, 2002). Sparser ChAT immunoreactivity is seen in the other laminae of the dorsal horn; lamina 1 as well as the inner portion of lamina 2 contains ChAT immunoreactive terminals

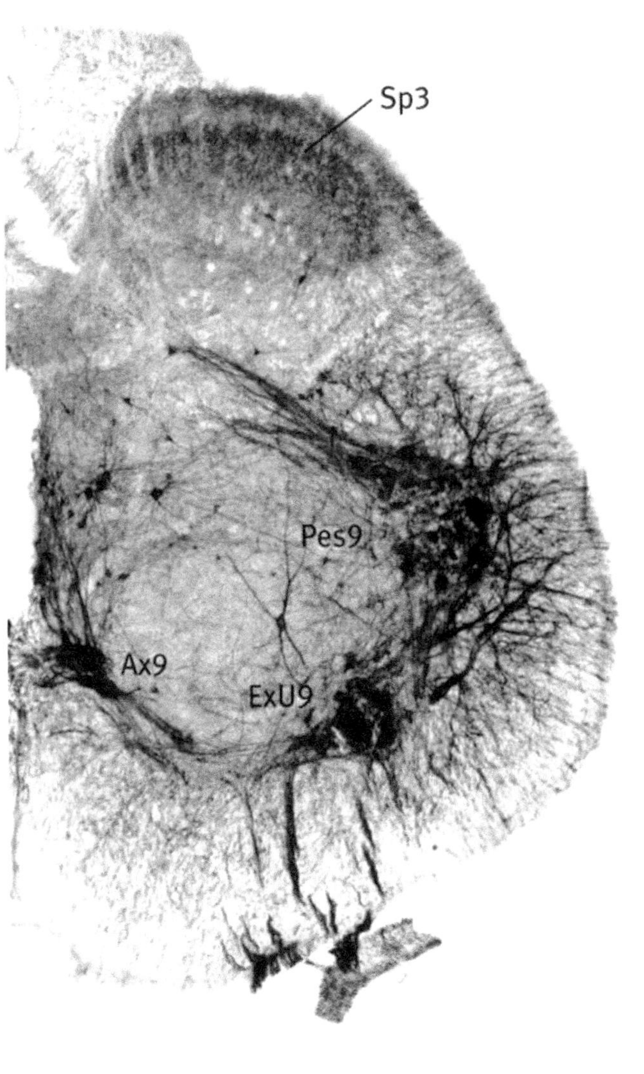

Figure 13.1 ChAT reactivity in lumbar spinal cord

This photomicrograph shows prominent ChAT immuno-staining in the dorsal horn, most prominently in lamina 3 (Sp3). It also shows dense reactivity in three lamina 9 motoneuron groups (Ax9, ExU9, and Pes9).
This figure is reproduced in the color insert.

(Houser *et al.*, 1983; Barber *et al.*, 1984; Borges and Iversen, 1986). The majority of ChAT immunoreactive fibers in the dorsal horn are thought to arise from cholinergic neurons located within the spinal gray matter (Houser *et al.*, 1983; Barber *et al.*, 1984; Borges and Iversen, 1986). Indeed, the ChAT immunoreactive cell bodies located in laminae 3-5 have been shown to contribute to the plexus of ChAT immunoreactive axon terminals seen in laminae 1, 3 and the inner portion of lamina 2 (Barber *et al.*, 1984; Borges and Iversen, 1986). In addition, some of these cells have processes extending into laminae 4, 7 and 9 (Borges and Iversen, 1986). Neurons immunoreactive for ChAT are presynaptic to primary sensory fibers and dorsal horn neurons (Ribeiro-da-Silva and Cuello, 1990). The majority of ChAT immunoreactive lamina 3 neurons also show GABA immunoreactivity (Todd, 1991).

Functionally, cholinergic neurons in the dorsal horn are thought to be a part of an endogenous spinal antinociceptive system (Yaksh *et al.*, 1985; Zhuo and Gebhart, 1991; Detweiler *et al.*, 1993; Yaksh and Malmberg, 1994; Naguib and Yaksh, 1997; Xu *et al.*, 1997; Bannon *et al.*, 1998; Pan *et al.*, 1999).

Cholinergic mediated antinociception appears to be generated in two main ways:

1. By direct inhibition of excitatory transmission. Presynaptic ACh receptors have been found on primary afferent terminals in the spinal dorsal horn and activation of these receptors (in particular, muscarinic receptors) is thought to inhibit glutamate release from these terminals (Gillberg and Wiksten, 1986; Riberio-da-Silva and Cuello, 1990; Pan *et al.*, 1999; Li *et al.*, 2002). In addition, activation of ACh receptors located postsynaptically on dorsal horn neurons inhibits excitatory sensory transmission in these neurons (Riberio-da-Silva and Cuello, 1990; Iwamoto and Marion, 1993; Travagli 1996; Li and Zhuo, 2001).
2. By facilitating inhibitory transmission. Activation of muscarinic and nicotinic receptors located on inhibitory interneurons in the dorsal horn increases GABA and glycine release, respectively (Todd and Spike, 1993; Riberio-da-Silva, 1995; Baba *et al.*, 1998; Hwang *et al.*, 2001; Kiyosawa *et al.*, 2001; Li *et al.*, 2002). These interneurons synapse on dorsal horn neurons, thereby, inhibiting sensory transmission. Further, an increase in GABA may have a direct influence on primary afferent transmission as activation of presynaptic GABA receptors decreases glutamate release from primary afferent terminals in the spinal dorsal horn (Iyadomi *et al.*, 2000).

Intermediate gray

The preganglionic cells of the spinal autonomic nervous system are found in the intermediate gray matter of the spinal cord. These cells are cholinergic, and for the most part are found in clearly defined columns. In thoracic and lumbar regions of the spinal cord, ChAT immunoreactive neurons form the intermediolateral column of the intermediate gray (Figure 13.1; Kimura *et al.*, 1981; Houser *et al.*, 1983; Barber *et al.*, 1984; Borges and Iversen, 1986; Kobayashi *et al.*, 2002). The intermediolateral nucleus (IML) contains preganglionic sympathetic neurons that extend from the first thoracic segment to upper lumbar segments. Both the principal part of the IML (IMLp) and the funicular IML (IMLf) contain ChAT immunoreactive neurons. The parasympathetic preganglionic neurons constitute the sacral parasympathetic nucleus, located in the upper sacral segments (Barber *et al.*, 1984). In addition to the IML, sympathetic preganglionic neurons immunoreactive for ChAT are found in the central autonomic cell area (CA), named also the paraependymal part of the intercalated nucleus (ICLPe) located dorsal to the central canal,

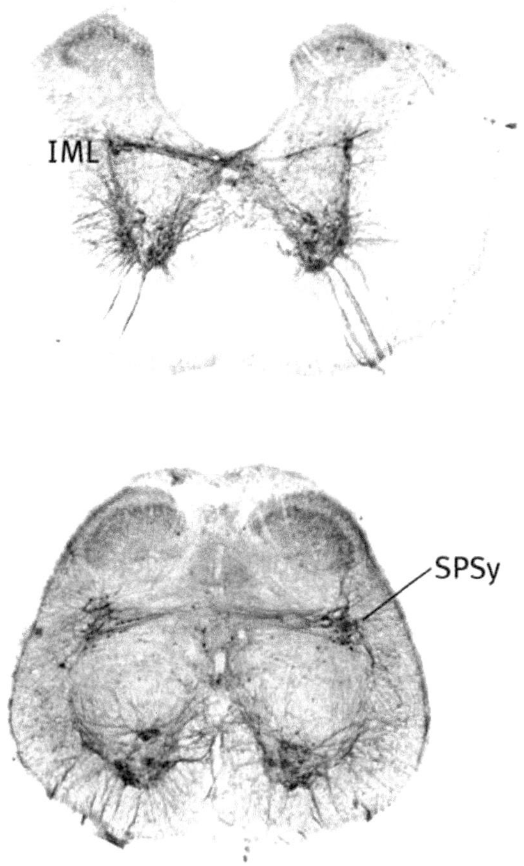

Figure 13.2 ChAT reactivity in sympathetic and parasympathetic preganlionic neurons

This photomicrograph shows prominent ChAT immuno-staining in the thoracic intermediolateral group (IML) and in the sacral parasympathetic group (SPSy). *This figure is reproduced in the color insert.*

and in the intercalated nucleus located between the IMLp and CA. The dendrites of IMLp and CA neurons form substantial longitudinal bundles within each group, as well as transverse bundles between the groups that resemble the rungs of a ladder (Figures 13.1 and 13.2; Barber *et al.*, 1984; Borges and Iversen, 1986).

The intermediate gray also contains cell bodies immunoreactive for ChAT in lamina 10 (Barber *et al.*, 1984). In general, these cells are small- to medium-sized, multipolar neurons classified as the central canal cluster cells by Barber *et al.*, (1984). These are present at all levels of the spinal cord and most numerous at the lumbar enlargement in the rat (Borges and Iversen, 1986). The cholinergic cluster cells have been reported to project to the dorsal horn, the CA, the contralateral lamina 10 and the ventral horn (Barber *et al.*, 1984; Borges and Iversen, 1986). The location of the cholinergic cells of lamina 10 appears to be ideal to integrate sensory, motor, and autonomic information in the spinal cord. However, the exact functional role of these neurons remains to be fully elucidated (Borges and Iversen, 1986).

Ventral horn

In the ventral horn of the spinal cord, ChAT immunoreactivity identifies the ventral somatic motoneurons of lamina 9 (Kimura *et al.*, 1981; Houser *et al.*, 1983; Satoh *et al.*, 1983; Barber *et al.*, 1984; Borges and Iversen, 1986; Kobayashi *et al.*, 2002). The somatic motoneurons have larger cells bodies than other ChAT neurons in the central nervous system. The cholinergic motoneurons in the ventral horn are aggregated into several distinct columns that extend rostrocaudally (see Chapter 7; Barber *et al.*, 1984). The majority of cholinergic neurons in the ventral horn are large rounded cells that correspond to α motoneurons (Kimura *et al.*, 1981; Houser *et al.*, 1983; Barber *et al.*, 1984; Borges and Iversen, 1986; Kobayashi *et al.*, 2002). In general, it is the smaller cells that stain most intensely for ChAT (Borges and Iversen, 1986; Ichikawa and Shimizu, 1998; Kobayashi *et al.*, 2002). These smaller cholinergic cells represent the γ motoneurons (Bryan *et al.*, 1972; Westbury 1982). As in the IML, dendritic bundling is a distinctive feature of motor columns in the spinal cord (Figure 13.3; Barber *et al.*, 1984). Longitudinal and transverse bundles link together ChAT positive dendrites of the same and separate columns (Barber *et al.*, 1984). The functional significance of this dendritic bundling is not fully understood, but probably aid in the synchronization and coordination of muscle activity (Scheibel and Scheibel, 1970; Schoenen, 1982). VAChT immunoreactive cell bodies were also found both in large and small neurons of the ventral horn, presumably representing α and γ motoneurons (Arvidsson *et al.*, 1997).

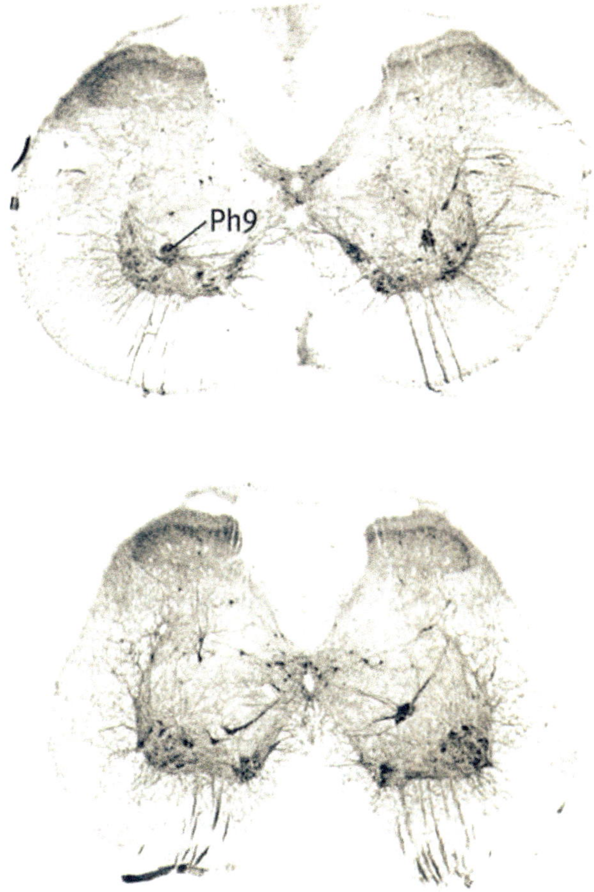

Figure 13.3 ChAT reactivity in cervical and lumbar motoneurons

This photomicrograph shows prominent ChAT immuno-staining in motoneurons at C4 (top) and at L3 (above). Dendritic bundling is a feature of the phrenic nucleus (Ph9). *This figure is reproduced in the color insert.*

Substance P

Substance P (SP), a member of the tachykinin neuropeptide family, is an important signaling molecule that has been associated with functionally distinct spinal cord systems. SP is involved in sensory, autonomic, and motor functions of the spinal cord. The distribution of SP in the spinal cord has been described for several species including the rat (Cuello and Kanazawa, 1978; Ljungdahl *et al.*, 1978; Barber *et al.*, 1979; Gibson *et al.*, 1981; Hunt *et al.*, 1981), cat (Hökfelt *et al.*, 1975), monkey (De Lanerolle and LaMotte, 1982b; Chung *et al.*, 1989), and human (De Lanerolle and LaMotte, 1982a,b). In general, the distribution of SP in the spinal cord of these species is similar. SP immunoreactivity is present in neurons of the dorsal horn (Gibson *et al.*, 1981; Hunt *et al.*, 1981; Danks *et al.*, 1986; Ribeiro-da-Silva *et al.*, 1991), small unmyelinated or thinly myelinated sensory fibers (Hökfelt *et al.*, 1975; DiFiglia *et al.*, 1982; De Lanerolle and LaMotte, 1983), and descending fibers from the brain stem to the spinal cord (Gilbert *et al.*, 1982;

Menetrey and Basbaum, 1987; Sasek and Helke, 1989). SP-immunoreactive cell bodies are sparse and found in the dorsal horn only, in laminae 1-3 and 5 (Gibson *et al.*, 1981; Danks *et al.*, 1986). In contrast, terminals immunoreactive for SP are found throughout the gray matter with the densest labeling in lamina 1, the outer portion of lamina 2 and lamina 5 (Gibson *et al.*, 1981; Chung *et al.*, 1989; Ribeiro-da-Silva *et al.*, 1989). Fibers and terminals immunoreactive for SP have also been reported in the dorsolateral part of the lateral fasciculus, the intermediolateral nucleus, lamina 10 and the ventral horn (Gibson *et al.*, 1981; Davis *et al.*, 1984; Chung *et al.*, 1989; LaMotte and Shapiro, 1991). SP is colocalized with enkephalin, glutamate, neurokinin A and NMDA receptors in the spinal cord (Dalsgaard *et al.*, 1985; De Biasi and Rustioni, 1988; Senba *et al.*, 1988; Ribeiro-da-Silva *et al.*, 1991; Benoliel *et al.*, 2000). Some SP-immunoreactive neurons also contain calcitonin gene-related peptide, galanin and somatostatin (Ju *et al.*, 1987).

Dorsal horn

In the dorsal horn, SP is released along with glutamate from small-diameter sensory fibers that terminate principally in laminae 1 and 2 of the dorsal horn, the location of the spinal nociceptive neurons (Hökfelt *et al.*, 1975, 1977; Ljungdahl *et al.*, 1978; Barber *et al.*, 1979; Gulbenkian *et al.*, 1986; McCarthy and Lawson, 1989). It has been demonstrated that both peripheral inflammation and noxious stimulation lead to release of SP in the dorsal horn (Oku *et al.*, 1987; Duggan *et al.*, 1988; Schaible *et al.*, 1990). Furthermore, SP applied iontophoretically to the spinal cord has been shown to excite nociceptive, but not non-nociceptive neurons (Radhakrishnan and Henry, 1991). In animals with behavioral hyperalgesia, SP release in the dorsal horn can be produced by normally ineffective innocuous stimuli (Schaible *et al.*, 1990). These results led researchers to suggest that SP was the primary pain transmitter in the spinal cord. However, Cao *et al.*, (1998), in a study of mice with a disruption to the gene encoding SP, found that SP significantly reduced nociceptive pain responses to moderate to intense stimuli, but did not alter responses to mildly painful stimuli, indicating a subtler role for SP in pain transmission.

Intermediate gray

SP-immunoreactive fibers make synaptic contact with sympathetic preganglionic neurons within the intermediolateral nucleus (De Lanerolle and LaMotte, 1982; Helke *et al.*, 1982; Poulat *et al.*, 1992). The SP innervation of the intermediolateral nucleus arises from the ventral medulla and appears to be excitatory (Johansson *et al.*, 1981; Helke at al, 1982; Backman and Henry, 1984; Menetrey and Basbaum, 1987; Reddy *et al.*, 1990; Sasek *et al.*, 1990; Nevin *et al.*, 1994). Iontophoresis of SP evokes a moderate increase in the firing rate of intermediolateral nucleus neurons (Backman *et al.*, 1990). Behavioral studies in rats show that an increase in heart rate and blood pressure produced by electrical stimulation of the ventral medulla can be blocked by intrathecal administration of SP antagonists (Loewy and Sawyer, 1982; McKitrick and Calaresu, 1991). SP immunoreactivity was also observed in neurons and fibers in the sacral parasympathetic nucleus and the dorsal gray commissure (Sasek *et al.*, 1984).

Ventral horn

Immunohistochemical studies show SP immunoreactivity associated with fibers and terminals in the ventral horn (Hokfelt *et al.*, 1975; Cuello *et al.*, 1976; Cuello and Kanazawa, 1978; Barber *et al.*, 1979; Kishida *et al.*, 1979 De Lanerolle and LaMotte, 1982b), and electron microscope studies have identified SP immunoreactivity in the ventral horn was contained predominately on dendrites in close apposition to motoneurons (De Lanerolle and LaMotte, 1982b; Vacca *et al.*, 1982). The majority of the SP innervation of the motoneurons appears to arise from the raphe nuclei of the ventral medulla (Chan-Palay *et al.*, 1978; Hökfelt *et al.*, 1978; Helke *et al.*, 1982). However, there is also evidence indicating that a small SP innervation of the ventral horn arises from primary afferents as well as intraspinal SP-immunoreactive neurons (Jessell *et al.*, 1979). The application of SP to the spinal cord in the rat has been reported to cause a depolarization of spinal motoneurons as well as a potentiation of monosynaptic reflexes (Konishi and Otsuka, 1974 a,b). However, in a study on cats, Krivoy *et al.*, (1980) showed iontrophoretic application of SP onto motoneurons at different strengths did not cause the motoneuron to discharge spontaneously, but produced gradual and prolonged changes in synaptic excitability of motoneurons, suggesting SP may play more of a modulatory role on the activity of spinal motoneurons.

Noradrenergic projections to the spinal cord

The locus coeruleus (LC) (A6 of Dahlström and Fuxe, 1964, 1965), the largest of the noradrenergic (NA) nuclei in mammalian brains, is a clearly defined cluster of neurons located adjacent to the fourth ventricle in the pontine brainstem. NA containing fibers to the spinal cord originate from the LC, and other NA cell groups, namely the locus subcoeruleus (SC) and the A5 and A7 cell groups of the lateral tegmentum. All NA in the spinal cord comes from these four areas (Blessing *et al.*, 1981; Martin *et al.*, 1982; Sawchenko and Swanson, 1982; Westlund *et al.*, 1982, 1983, 1984, 1991; Lyons and Grzanna, 1988; Fritschy and Grzanna, 1990; Kwiat and Basbaum, 1992; Hökfelt *et al.*, 2000). About 91% of LC and

86% of SC neurons projecting to the spinal cord contain NA (Sluka and Westlund, 1992). The NA projection targets all levels of the spinal gray with highest density in the intermediolateral nucleus, ventral horn, lamina 10, and in superficial laminae of the dorsal horn (Carlsson *et al.*, 1964; Dalhstrom and Fuxe, 1965; Crutcher and Bingham, 1978; Glazer and Ross, 1980; Lackner, 1980; Mizukawa, 1980; McLachlan and Oldfield, 1981; Satoh *et al.*, 1982; Westlund *et al.*, 1983, 1984; Kojima *et al.*, 1984; Schrøder and Skagerberg, 1985). The axons of LC and SC neurons travel through the ipsilateral ventral and ventrolateral funiculi, respectively (Clarke and Proudfit, 1991). However, Fritschy and Grzanna (1990) found the LC axons descend the length of the spinal cord within laminae 1-2. Axons of A5 and A7 neurons travel in the ventral and dorsolateral funiculi (Fritschy and Grzanna, 1990; Clark and Proudfit, 1991, 1993). These terminations are mainly ipsilateral (Clarke and Proudfit, 1991); however, there are reports of a bilateral projection (Karoum *et al.*, 1980; Commissiong, 1981; Fritschy and Grzanna 1990).

There is some disagreement over the exact organization of the origin of NA inputs to the spinal cord. Fritschy *et al.*, (1987) and Sluka and Westlund (1992) showed, in Harlan Sprague-Dawley rats, that LC neurons project primarily to the dorsal horn, whereas A5 and A7 neurons project to somatic and preganglionic motoneurons. However, Clark and Proudfit (1991) and Sluka and Westlund (1992), using similar methods in Sasco Sprague-Dawley rats, found the LC neurons project to the ventral horn and A7 and A5 neurons project to the dorsal horn and the intermediate zone, respectively. This difference was later confirmed to be due to differences in the two substrains of rats. The differential LC spinal projections were also found to be associated with different physiological functions in the two substrains, antinociception in Harlan rats but not in Sasco rats (Clark *et al.*, 1991; Clark and Proudfit, 1992; West *et al.*, 1993).

Dorsal horn

LC neurons provide a moderately dense innervation to the ventral part of the dorsal horn, but only a very sparse innervation to the superficial dorsal horn (Proudfit and Clark, 1991). Axons of A5 neurons terminate primarily in the ipsilateral deep dorsal horn, in laminae 4-6 (Clark and Proudfit, 1993). Axons of neurons of A7 cell group terminate most heavily in the laminae 1-4 of the dorsal horn (Clark and Proudfit, 1991). Stimulation of LC causes an increase in NA release, which inhibits nociceptive transmission in the dorsal horn through α-2 adrenergic receptors (Crawley *et al.*, 1979; Westlund *et al.*, 1983; Jones and Gebhart, 1987; Pan *et al.*, 2002; Sonohata *et al.*, 2003). Furthermore, iontophoretic application of NA induces a reduction of single unit responses to noxious stimuli in dorsal horn neurons of anaesthetized rats (Headley *et al.*, 1978; Howe and Zieglgansberger, 1987). Intracellular (North and Yoshimura, 1984) and whole-cell patch-clamp (Yajiri *et al.*, 1997; Baba *et al.*, 2000) recordings have revealed that NA application produces hyperpolarization in lamina 2 neurons in transverse spinal cord slices. Behaviorally, intrathecal administration of NA has antinociceptive effects when assessed by tail-flick test (Vaida *et al.*, 1986). Spinal application of NA (Yaksh *et al.*, 1985; Eisenach *et al.*, 1996; Honoré *et al.*, 1996; Supowit *et al.*, 1998) and electrical stimulation of A5, A6 and A7 cell groups all elicit robust antinociception (Stamford, 1995; Zhao *et al.*, 1999; Nuseir and Proudfit, 2000). In rats, the antinociceptive effect of NA is mediated by presynaptic inhibitions of primary afferents, thereby reducing noxious stimuli induced excitation, and through postsynaptic lamina 2 neurons (Sonohata *et al.*, 2003). In addition, NA can also activate GABAergic and/or glycinergic interneurons, which are present in alpha-1 adrenoceptors in lamina 2 (Todd and McKenzie, 1989).

Intermediate gray

Axons of NA A5 neurons terminate in lamina 7 and in the intermediolateral nucleus in thoracic spinal cord levels (Clark and Proudfit, 1993; Fritschy and Grzanna, 1990). Axons of A7 neurons terminate most heavily in the medial part of lamina 7 and in lamina 10 (Proudfit and Clark, 1991). Axons of SC neurons end in a diffuse area of termination in the intermediate gray of the spinal cord (Proudfit and Clark, 1991). Stimulation at sites near A5 neurons produces changes in cardiovascular reflexes (Loewy *et al.*, 1979 a, b; 1986; Drye *et al.*, 1990). Furthermore, NA appears to be involved in the interplay between cardiovascular spinal systems (through the intermediolateral nucleus) and somatosensory spinal systems (Thurston and Randich, 1990; Taylor *et al.*, 2001). For example, alterations in the activity of descending NAergic pathways have also been implicated in the mediation of antinociception elicited by acute hypertension, and in alteration of nociception displayed by spontaneously-hypertensive rats (Thurston and Randich, 1990; Taylor *et al.*, 2001).

Ventral horn

Axons of LC and SC neurons terminate in lamina 8 and in the motoneuron groups of lamina 9 of the ventral horn (Proudfit and Clark, 1991). There are also projections from A5 and A7 neurons to the ventral horn (Fritsch and Grzanna, 1990; Clark and Proudfit, 1993). The role of NA in the ventral horn is generally considered to be excitatory, though results from experimental studies vary (Engberg and Marshall, 1971;

Strahlendorf *et al.*, 1980; Fung and Barnes, 1981; Pun *et al.*, 1985; Chan *et al.*, 1986; Hino *et al.*, 1987; Fung *et al.*, 1991; Ono *et al.*, 1991). Electrical stimulation of the LC and SC results in excitation of motoneurons, and antagonists of NA block these facilitating effects (Strahlendorf *et al.*, 1980; Fung and Barnes, 1981; Chan *et al.*, 1986; Hino *et al.*, 1987; Fung *et al.*, 1991; Ono *et al.*, 1991).

Serotoninergic projections from the raphe nuclei

Serotonin (5-hydroxytryptamine) is an indolamine which is found in the midline raphe nuclei of the brain stem (Dahlström and Fuxe, 1964, 1965). Jones and Light (1992) showed that approximately 61% of neurons in the raphe nuclei project to the spinal cord and 47% of these raphespinal neurons are serotoninergic. Many of the raphe neurons also contain other peptides such as enkephalin (Tashiro *et al.*, 1988), galanin (Arvidsson *et al.*, 1991), glutamate (Conte *et al.*, 2005), substance P (Chan-Palay *et al.*, 1978; Bowker *et al.*, 1983; Wessendorf and Elde, 1987), and thyrotropin-releasing hormone (Arvidsson *et al.*, 1990). The serotoninergic projections from ventral medullary raphe nuclei arise from raphe magnus (RMg), raphe pallidus (RPa), and raphe obscurus (ROb) nuclei; they terminate in all laminae of the spinal cord (Bowker *et al.*, 1983; Bullitt and Light, 1989; Jones and Light, 1992; Kwiat and Basbaum, 1992). In the spinal cord, serotoninergic receptors are located principally on axon terminals (Kidd *et al.*, 1993; Kia *et al.*, 1995). However, rhizotomy leads to a marked decrease, but not a complete abolition, of binding sites and immunoreactivity for serotonin, indicating the presence of serotoninergic receptors in some dorsal horn neurons (Kidd *et al.*, 1993; Kia *et al.*, 1995); this has been confirmed by immunocytochemical and in situ hybridization studies (Kia *et al.*, 1995).

Serotoninergic terminals are found in lamina 1, the outer portion of lamina 2, lamina 10, the intermediolateral nucleus, and in the ventral horn (Mizukawa *et al.*, 1986; LaMotte, 1988; Anderson *et al.*, 1989). There are also serotoninergic projections to ependymal cells lining the spinal central canal (Azmitia, 1999). Retrograde and anterograde labeling studies showed that the NRM projects mainly to the dorsal horn via the dorsolateral funiculus, whereas the more caudally located RPa and ROb project mainly to the intermediate zone and the motoneuronal cell groups via the deeper fiber bundles located in the lateral and ventral funiculi (Dahlström and Fuxe, 1965; Basbaum and Fields, 1978; Skagerberg and Björklund, 1985). These serotoninergic terminals are also found to contain the neuropeptides mentioned above. It has been shown that 99% of raphespinal axons projecting to ventral horn contain substance P, 50% of raphespinal axons projecting to intermediolateral nucleus contain substance P, but only 3% of raphespinal axons projecting to the dorsal horn contain substance P (Wessendorf and Elde, 1987). In the dorsal horn, serotoninergic terminals that contain substance P, thyrotrophin-releasing hormone, and/or galanin are found to specifically project to lamina 5 (Wessendorf and Elde, 1987; Bowker and Abbott, 1990; Maxwell *et al.*, 1996), whereas the serotoninergic terminals in the superficial laminae do not contain these neuropeptides (Millhorn *et al.*, 1988; Stamp and Semba, 1995; Maxwell *et al.*, 1996).

Dorsal horn

Lamina 1 and the outer portion of lamina 2 of the dorsal horn contain a high density of serotoninergic terminals (Ruda *et al.*, 1982; LaMotte and De Lanerolle, 1983; Light *et al.*, 1983; Anderson *et al.*, 1989; Marlier *et al.*, 1991). Some serotoninergic terminals are also found in laminae 3 and 4 (Marlier *et al.*, 1991). In the dorsal horn, descending serotoninergic input is thought to provide key modulation of spinal pain systems. Stimulation of the medullary raphe nuclei inhibits spinal nociceptive reflexes (Mayer and Price, 1976; Besson *et al.*, 1981; Hammond, 1986) and nociceptive response in dorsal horn neurons (see Willis, 1982 and Gebhart, 1986 for reviews). Furthermore, serotoninergic receptor antagonists attenuate the stimulation-produced descending inhibition from the medullary nuclei (Satoh *et al.*, 1980, 1983; Hammond and Yaksh, 1984; Jensen and Yaksh, 1984, 1986; Aimone *et al.*, 1987; Barbaro *et al.*, 1985). This inhibitory role seems to be particularly significant in stressed animals, in which it may be tonically activated; serotonin may have only a minimal effect on nociception in resting animals (Milne and Gamble, 1990; Mitchell *et al.*, 1998). Recent studies have shown that serotonin may also have pronociceptive effects (Millan *et al.*, 1996; Green *et al.*, 2000). These opposing actions of serotonin are most likely due to the actions of different classes of serotonin receptors or different concentration of serotonin levels (Li and Zhuo, 1998) that either functionally enhance or decrease neuronal activity. Jankowska *et al.*, (1998) have shown that serotonin and raphespinal stimulation can exert a differential control over primary afferents of different modalities.

Intermediate gray

Serotoninergic fibers are found in the intermediolateral nucleus (Kojima *et al.*, 1983; Sur *et al.*, 1996). Here, an excitatory serotoninergic drive upon preganglionic neurons is thought to be important in maintaining vasomotor tone and arterial pressure (Loewy, 1990; Vera *et al.*, 1990; McCall and Clement, 1994).

Ventral horn

Large motoneurons of the ventral horn are surrounded by fine networks of serotoninergic fibers and terminals (Kojima *et al.*, 1982; Kojima and Sano, 1983). Electron microscopic studies have shown that the vast majority of serotoninergic boutons are apposed to dendritic branches rather than to somatic surface of motoneurons (Alvarez *et al.*, 1998). In the ventral horn, several studies, both in vitro and in vivo, showed that application of serotonin or serotonin precursors has an excitatory effect on spinal motoneurons (Neuman, 1985; Connell and Wallis, 1988; Elliott and Wallis, 1990; Takahashi and Berger, 1990; Huangfu *et al.*, 1994). Electrical stimulation of the raphe nuclei also produces excitation of spinal motoneurons (Fung and Barnes, 1989). It is generally thought that serotonin raphe innervation of the spinal motoneurons acts as a gain setting system by enhancing the overall responsiveness of motoneurons (Jacobs and Fornal, 1993). Consistent with this hypothesis, iontophoretic application of serotonin on motoneurons in vivo is found to enhance the excitability of motoneurons for glutamate or aspartate (White, 1985).

Dopaminergic projections to the spinal cord

Dopamine in the spinal cord has been implicated in sensory, autonomic and motor control. Anatomically, dopaminergic fibers, and presumably terminals, are found in the dorsal horn, the intermediolateral nucleus, the lateral spinal nucleus, lamina 10, and the ventral horn (Skagerberg *et al.*, 1982; Skagerberg and Lindvall, 1985; Scatton *et al.*, 1986; Holstege *et al.*, 1996; van Dijken *et al.*, 1996; Levant and McCarson, 2001). The strongest dopamine immunoreactivity is observed in the intermediolateral nucleus (Holstege *et al.*, 1996). Dopamine immunoreactivity is also found in the terminals within the motoneuron cell groups, with the densest innervation in the cremaster nucleus, Onuf's nucleus (in the cat and monkey), and its homolog in the rat (Holstege *et al.*, 1996). This pattern of dopaminergic innervation is similar across all mammals studied thus far (rat, cat, monkey – Holstege *et al.*, 1996; mouse – Qu *et al.*, 2006; cat – Fleetwood-Walker and Coote, 1981; human – Commissiong and Sedgwick, 1975).

The majority of, if not all, dopaminergic innervation of the spinal cord arises from the multipolar neurons in the A11 cell group of Dahlström and Fuxe (Dahlström and Fuxe, 1964; Commissiong *et al.*, 1978; Commissiong and Neff, 1979; Swanson and Kuypers, 1980; Skagerberg *et al.*, 1982; Skagerberg and Lindvall, 1985; Holstege and Kuypers, 1987; Hosoya *et al.*, 1991,1988; Qu *et al.*, 2006). The A11 cell group is conventionally described as being located in the posterior and dorsal hypothalamic areas and in the periventricular gray matter of the caudal thalamus (Skagerberg and Lindvall, 1985; Hosoya *et al.*, 1991). A projection from the paraventricular nucleus has also been described by Swanson *et al.*, in 1981. The dopaminergic component forms only about 4-6% of the total hypothalamic projection to the spinal cord (Skagerberg and Lindvall, 1985). The dopaminergic cells are distributed evenly throughout A11 without indication of topography, and only account for a small number of the total A11 neuronal population (Skagerberg and Lindvall, 1985). Thus, only few A11 neurons provide the bulk of the dopaminergic innervation of the entire spinal cord. This is achieved through a highly collateralized system with most neurons projecting to most of the length of the spinal cord (Skagerberg and Lindvall, 1985). The descending dopaminergic fibers travel to the spinal cord within the dorsolateral funiculus and partly along the central canal. The majority of dopaminergic neurons project ipsilaterally to the spinal cord, but there is evidence for a minor crossed projection in the rat (Skagerberg and Lindvall, 1985). In addition to the A11 cell group, dopaminergic projections to the spinal cord from the diencephalic A10 cell group have recently been found in mice (Qu *et al.*, 2006). Dopaminergic cells have also been identified in the dorsal root ganglia of the rat (Price and Mudge, 1983; Weil-Fugazza *et al.*, 1993; Bertrand and Weil-Fugazza, 1995). Finally, a population of tyrosine hydroxylase-immunoreactive cells have been described in the upper cervical cord and in the sacral cord (Mouchet *et al.*, 1986). These cells may be caudal equivalents of the medullary A1 and A2 cell groups. Intrinsic spinal dopaminergic cells may therefore be a source of dopamine in the spinal cord, but the significance of these projections is not known.

Dorsal horn

Dopamine receptors are localized predominately in the superficial laminae of the cervical and lumbar segments of the spinal cord, with the exception of lamina 2, which has a scarce dopamine content (Holstege *et al.*, 1996; Levant and McCarson, 2001). Dopamine receptors are also present in the deep dorsal horn (Holstege *et al.*, 1996; van Dijken *et al.*, 1996; Levant and McCarson, 2001). In the dorsal horn, dopamine plays a role in the transmission of sensory or nociceptive information. Iontophoretically administered dopamine, as well as electrical stimulation of A11, causes inhibition of responses to noxious stimuli in nociceptive neurons in the dorsal horn (Grossmann, *et al.*, 1973; Tulunay *et al.*, 1975; Fleetwood-Walker *et al.*, 1988). Intrathecally administered apomorphine and dopamine have been reported to increase the tail-flick latency in spinal rats, which is counteracted by previous application of dopamine receptor antagonists (Jensen and Smith, 1982, 1983; Jensen and Yaksh, 1984b; Barasi and Duggal, 1985). Furthermore, the administration of D2 receptor agonists generally elicits antinociception (Barasi and Duggal,

1985; Tamae *et al.*, 2005). The reported actions of dopamine in the dorsal horn may take place on primary afferents themselves, as dopamine can hyperpolarize dorsal root ganglion neurons (Gallagher *et al.*, 1980) via D2 receptors (Abramets and Samoilovich, 1991). Dopamine has also been shown to depress Ca^{2+} channel currents in isolated dorsal root ganglion neurons (Formenti *et al.*, 1998). Thus, dopamine in the spinal dorsal horn appears to exert an inhibitory effect on nociceptive transmission through D2 receptors.

Intermediate gray

Strong dopamine immunoreactivity has been shown in the intermediolateral nucleus and the sacral parasympathetic nucleus of the intermediate gray (Holstege *et al.*, 1996; van Dijken *et al.*, 1996). Sparse dopamine reactivity is present in laminae 6-7 (Holstege *et al.*, 1996). In the intermediolateral nucleus, iontophoretic application of dopamine onto preganglionic sympathetic neurons results in an increase in firing rate of these neurons (Lewis and Coote, 1990). Spinal application of dopaminergic receptor agonists produces hypotension and bradycardia (Pellissier and Demenge, 1991). Further, Simon and Schramm (1983) found that dopamine increased spontaneous renal nerve activity and that depletion of spinal dopamine reduced renal nerve excitation elicited by electrical stimulation in L4-L7 spinal cord segments. Taken together, these results suggest that dopamine input to the intermediolateral nucleus of the spinal cord may be part of the sympathoexcitatory system.

Ventral horn

In the ventral horn of the lumbosacral spinal cord, the sexually dimorphic cremaster nucleus and Onuf's nucleus (and its rodent homolog) show much stronger immunoreactivity than other motoneuron nuclei (Holstege *et al.*, 1996; van Dijken *et al.*, 1996). The effects of dopamine on somatic motoneurons in the ventral horn are less than clear, and both excitatory and inhibitory actions of dopamine have been reported. Barasi and Roberts (1977) found, in the rat, that dopamine directly increases the excitability of motorneurons. In contrast, rats with lesions to A11 cell group and D3 receptor knockout mice show excessive locomotor activity (Accili *et al.*, 1996; Ondo *et al.*, 2000; Clemens and Hochman, 2004). Furthermore, dopamine agonists are used in the treatment of restless legs syndrome, a disorder of dysfunction in spinal excitability (Bara-Jimenez *et al.*, 2000). These studies suggest that dopamine also has an inhibitory action on the activity of spinal cord motoneurons.

References

Abramets II, Samoilovich IM (1991) Analysis of two types of dopaminergic responses of neurons of the spinal ganglia of rats. Neurosci Behav Physiol 21, 435-440.

Accili D, Fishburn CS, Drago J, Steiner H, Lachowicz JE, Park BH, Gauda EB,Lee EJ, Cool MH, Sibley DR, Gerfen CR, Westphal H, Fuchs S (1996) A targeted mutation of the D3 dopamine receptor gene is associated with hyperactivity in mice. Proc Natl Acad Sci USA 93, 1945-1949.

Aimone LD, Jones SL, Gebhart GF (1987) Stimulation-produced descending inhibition from the periaqueductal gray and nucleus raphe magnus in the rat: mediation by spinal monoamines but not opioids. Pain 31, 123-136.

Alvarez FJ, Pearson JC, Harrington D, Dewey D, Torbeck L, Fyffe RE (1998). Distribution of 5-hydroxytryptamine-immunoreactive boutons on alpha-motoneurons in the lumbar spinal cord of adult cats. J Comp Neurol 393, 69-83.

Anderson CR, McLachlan EM, Srb-Christie O (1989) Distribution of sympathetic preganglionic neurons and monoaminergic nerve terminals in the spinal cord of the rat. J Comp Neurol 283, 269-284.

Aquilonius SM, Eckernäs SA, Gillberg PG (1981) Topographical localization of choline acetyltransferase within the human spinal cord and a comparison with some other species. Brain Res 211, 329-340

Arvidsson U, Cullheim S, Ulfhake B, Bennett GW, Fone KC, Cuello AC, Verhofstad AA, Visser TJ, Hökfelt T (1990) 5 Hydroxytryptamine, substance P, and thyrotropin-releasing hormone in the adult cat spinal cord segment L7: immunohistochemical and chemical studies. Synapse 6, 237-270.

Arvidsson U, Ulfhake B, Cullheim S, Bergstrand A, Theodorson E, Hokfelt T (1991) Distribution of 125I-galanin binding sites, immunoreactive galanin, and its coexistence with 5-hydroxtryptamine in the cat spinal cord: biochemical, histochemical, and experimental studies at the light and electron microscopic level. J Comp Neurol 308, 115-138.

Arvidsson U, Riedl M, Elde R, Meister B (1997) Vesicular acetylcholine transporter (VAChT) protein: a novel and unique marker for cholinergic neurons in the central and peripheral nervous systems. J Comp Neurol 378, 454-467.

Azmitia EC (1999) Serotonin neurons, neuroplasticity, and homeostasis of neural tissue. Neuropsychopharmacology 21, 33S-45S.

Baba H, Kohno T, Okamoto M, Goldstein PA, Shimoji K, Yoshimura M (1998) Muscarinic facilitation of GABA release in the substantia gelatinosa of the rat spinal dorsal horn. J Physiol 508, 83-93.

Baba H, Shimoji K, Yoshimura M (2000) Norepinephrine facilitates inhibitory transmission in substantia gelatinosa of adult rat spinal cord (part 1): effects on axon terminals of GABAergic and glycinergic neurons. Anesthesiology, 92, 473-484.

Backman SB, Henry JL (1984) Effects of substance P and thyrotropin-releasing hormone on sympathetic preganglionic neurones in the upper thoracic intermediolateral nucleus of the cat. Can J Physiol Pharmacol 62, 248-251.

Backman SB, Sequeira-Martinho H, Henry JL (1990) Adrenal versus nonadrenal sympathetic preganglionic neurones in the lower thoracic intermediolateral nucleus of the cat: effects of serotonin, substance P, and thyrotropin-releasing hormone. Can J Physiol Pharmacol 68, 1108-1118.

Bannon AW, Decker MW, Holladay MW, Curzon P, Donnelly-Roberts D, Puttfarcken PS, Bitner RS, Diaz A, Dickenson AH, Porsolt RD, Williams M, Arneric SP (1998) Broad-spectrum, non-opioid analgesic activity by selective modulation of neuronal nicotinic acetylcholine receptors. Science 279, 77-81.

Bara-Jimenez W, Aksu M, Graham B, Sato S, Hallett M (2000) Periodic limb movements in sleep: state-dependent excitability of the spinal flexor reflex. Neurology 54, 1609-1616.

Barasi S, Duggal KN (1985) The effect of local and systemic application of dopaminergic agents on tail flick latency in the rat. Eur J Pharmacol 117, 287-294.

Barasi S, Roberts MHT (1977) Responses of motoneurones to electrophorectically applied dopamine. Br J Pharmacol 60, 29-34.

Barbaro NM, Hammond DL, Fields HL (1985) Effects of intrathecally administered methysergide and yohimbine on microstimulation-produced antinociception in the rat. Brain Res 343, 223-229.

Barber RP, Vaughn JE, Slemmon JR, Salvaterra, PM, Roberts E, Leeman SE (1979) Origin, distribution and synaptic relationships of substance P axons in rat spinal cord. J Comp Neurol 184, 331-351.

Barber RP, Phelps PE, Houser CR, Crawford GD, Salvaterra PM, Vaughn JE (1984) The morphology and distribution of neurons containing choline acetyltransferase in the adult rat spinal cord: an immunocytochemical study. J Comp Neurol 229, 329-346.

Basbaum AI, Fields HL (1978) Endogenous pain control mechanisms: review and hypothesis. Ann Neurol 4, 451-462.

Benoliel R, Tanaka M, Caudle RM, Iadarola MJ (2000) Co-localization of N-methyl-D-aspartate receptors and substance P (neurokinin-1) receptors in rat spinal cord. Neurosci Lett 291, 61-64.

Bertrand A, Weil-Fugazza J (1995) Sympathectomy does not modify the levels of dopa or dopamine in the rat dorsal root ganglion. Brain Res 681, 201-204.

Besson JM, Oliveras JL, Chaouch A, Rivot JP (1981) Role of the raphe nuclei in stimulation producing analgesia. Adv Exp Med Biol 133, 153-176.

Blessing WW, Goodchild AK, Dampney RAL, Chalmers JP (1981) Cell groups in the lower brain stem of the rabbit projecting to the spinal cord, with special reference to catecholamine-containing neurons. Brain Res 221, 35-55.

Borges LF, Iversen SD (1986) Topography of choline acetyltransferase immunoreactive neurons and fibers in the rat spinal cord. Brain Res 362, 140-148.

Bowker RM, Abbott LC (1990) Quantitative re-evaluation of descending serotonergic and non-serotonergic projections from the medulla of the rodent: evidence for extensive co-existence of serotonin and peptides in the same spinally projecting neurons, but not from the nucleus raphe magnus. Brain Res 512, 15-25.

Bowker RM, Westlund KN, Sullivan MC, Wilber JF, Coulter JD (1983) Descending serotonergic, peptidergic and cholinergic pathways from the raphe nuclei: a multiple transmitter complex. Brain Res 288, 33-48.

Bryan RN, Trevino DL, Willis WD (1972) Evidence for a common location of alpha and gamma motoneurons. Brain Res 38 193-196.

Bullitt E, Light AR (1989) Intraspinal course of descending serotoninergic pathways innervating the rodent dorsal horn and lamina X. J Comp Neurol 286, 231-242.

Cao YQ, Manthy PW, Carlson EJ, Gillespie AM, Epstein CJ, Basbaum AI (1998) Primary afferent tachykinins are required to experience moderate to intense pain. Nature 392, 390-394.

Carlsson A, Falck B, Fuxe K, Hillarp NÅ (1964) Cellular localization of monoamines in the spinal cord. Acta Physiol Scand 60, 112-119.

Chan JYH, Fung SJ, Chan SH, Barnes CD (1986) Facilitation of lumbar monosynaptic reflexes by locus coeruleus in the rat. Brain Res 369, 103-109.

Chan-Palay V, Jonsson G, Palay SL (1978) Serotonin and substance P coexist in neurons of the rat's central nervous system. Proc Natl Acad Sci U S A 75, 1582-1586.

Chung K, Briner RP, Carlton SM, Westlund KN (1989) Immunohistochemical localization of seven different peptides in the human spinal cord. J Comp Neurol 280, 158-170.

Clark FM, Proudfit HK (1991) The projection of locus coeruleus neurons to the spinal cord in the rat determined by anterograde tracing combined with immunocytochemistry. Brain Res 538, 231-245.

Clark FM, Proudfit HK (1992) Anatomical evidence for genetic difference in the innervation of the rat spinal cord by noradrenergic locus coeruleus neurons. Brain Res 591, 44-53.

Clark FM, Proudfit HK (1993) The projections of noradrenergic neurons in the A5 catecholamine cell group to the spinal cord in the rat: anatomical evidence that A5 neurons modulate nociception. Brain Res 616, 200-210.

Clark FM, Yeomans DC, Proudfit HK (1991) The noradrenergic innervation of the spinal cord: differences between two substrains of Sprague-Dawley rats determined using retrograde tracers combined with immunocytochemistry. Neurosci Lett 125, 155-158.

Clemens S, Hochman S (2004) Conversion of the modulatory actions of dopamine on spinal reflexes from depression to facilitation in D3 receptor knock-out mice. J Neurosci 24, 11337-11345.

Commissiong JW (1981) Evidence that the noradrenergic coerulospinal projection decussates at the spinal level. Brain Res 212, 145-151.

Commissiong JW, Galli CL, Neff NH (1978) Differentiation of dopaminergic and noradrenergic neurons in rat spinal cord. J Neurochem 30, 1095-1099.

Commissiong JW, Neff NH (1979) Current status of dopamine in the mammalian spinal cord. Biochem Pharmacol 28, 1569-1573.

Commissiong JW, Sedgwick EM (1975) Letter: Dopamine and noradrenaline in human spinal cord. Lancet 1, 347.

Connell LA, Wallis DI (1988) Responses to 5-hydroxytryptamine evoked in the hemisected spinal cord of the neonate rat. Br J Pharmacol 94, 1101-1114.

Conte D, Legg ED, McCourt AC, Silajdzic E, Nagy GG, Maxwell DJ (2005) Transmitter content, origins and connections of axons in the spinal cord that possess the serotonin (5-hydroxytryptamine) 3 receptor. Neuroscience 13, 165-173.

Crawley JN, Laverty R, Roth RH (1979) Clonidine reversal of increased norepinephrine metabolite levels during morphine withdrawal. Eur J Pharmacol 57, 247-250.

Crutcher KA, Bingham WG Jr (1978) Descending monaminergic pathways in the primate spinal cord. Am J Anat 153,159-164.

Cuello AC, Kanazawa I (1978) The distribution of substance P immunoreactive fibers in the rat central nervous system. J Comp Neurol 178, 129-156.

Cuello AC, Polack JM, Pearse A G E (1976) Substance P: a naturally occurring transmitter in human spinal cord. Lancet 2, 1054-1056.

Dahlström A, Fuxe K (1964) Evidence for the existence of monoamine-containing neurons in the central nervous system I. Demonstration of monoamines in the cell bodies of brainstem neurons. Acta Physiol Scand 62 (Suppl 232), 1-55.

Dahlstrom A, Fuxe K (1965) Evidence for the existence of monoamine neurons in the central nervous system. II. Experimentally induced changes intraneuronal amine levels of bulbospinal neuron systems. Acta Physiol Scand 64 (Suppl 247), 1-36.

Dalsgaard CJ, Haegerstrand A, Theodorsson-Norheim E, Brodin E, Hokfelt T (1985) Neurokinin A-like immunoreactivity in rat primary sensory neurons; coexistence with substance P. Histochemistry 83, 37-39.

Danks JA, Rothman RB, Cascieri MA, Chicchi GG, Liang T, Herkenham M (1986) A comparative autoradiographic study of the distributions of substance P and eledoisin binding sites in rat brain. Brain Res 385, 273-281.

Davidoff MS, Galabov PG, Bergmann M (1989) The vegetative network in the thoracolumbar spinal cord of the guinea pig: a comparison of the distribution of AChE-enzyme activity and choline acetyltransferase-like immunoreactivity. J Hirnforsch 30, 707-717.

Davis BM, Krause JE, McKelvy JF, Cabot JB (1984) Effects of spinal lesions on substance P levels in the rat sympathetic preganglionic cell column: evidence for local spinal regulation. Neuroscience 13, 1311-1326.

De Biasi S, Rustioni A (1988) Glutamate and substance P coexist in primary afferent terminals in the superficial laminae of spinal cord. Proc Natl Acad Sci USA 85, 7820-7824.

De Lanerolle NC, LaMotte CC (1983) Ultrastructure of chemically defined neuron systems in the dorsal horn of the monkey. I. Substance P immunoreactivity. Brain Res 274, 31-49.

De Lanerolle NC, LaMotte CC (1982a) The human spinal cord: substance P and methionine-enkephalin immunoreactivity. J Neurosci 2, 1369-1386.

De Lanerolle NC, LaMotte CC (1982b) The morphological relationships between substance P immunoreactive processes and ventral horn neurons in the human and monkey spinal cord. J Comp Neurol 207, 305-313.

Detweiler DJ, Eisenach JC, Tong C, Jackson C (1993) A cholinergic interaction in alpha 2 adrenoreceptor-mediated antinociception in sheep, J Pharmacol Exp Ther 256, 536-542.

DiFiglia M, Aronin N, Leeman SE (1982) Light microscopic and ultrastructural localization of immunoreactive substance P in the dorsal horn of monkey spinal cord. Neuroscience 7, 1127-1139.

Drye RG, Baisden RH, Whittington DL, Woodruff ML (1990) The effects of stimulation of the A5 region on blood pressure and heart rate in rabbits. Brain Res Bull 24, 33-39.

Duggan AW, Hendry IA, Morton CR, Hutchison WD, Zhao ZQ (1988) Cutaneous stimuli releasing immunoreactive substance P in the dorsal horn of the cat. Brain Res 451, 261-273.

Eisenach JC, De Kock M, Klimscha W (1996) Alpha(2)-adrenergic agonists for regional anesthesia. A clinical review of clonidine (1984-1995) Anesthesiology 85, 655-674.

Elliott P, Wallis DI (1990) The actions of 5-hydroxytryptamine on lumbar motoneurones in neonatal rat spinal cord in vitro. J Physiol 426, 54P.

Engberg I, Marshall KC (1971) Mechanism of noradrenaline hyperpolarization in spinal cord motoneurones of the cat. Acta Physiol Scand 83, 142-144.

Fleetwood-Walker SM, Coote JH (1981) Contribution of noradrenaline-, dopamine- and adrenaline-containing axons to the innervation of different regions of the spinal cord of the cat. Brain Res 206, 95-106.

Fleetwood-Walker SM, Hope PJ, Mitchell R (1988) Antinociceptive actions of descending dopaminergic tracts on cat and rat dorsal horn somatosensory neurones. J Physiol 399, 335-348.

Formenti A, Martina M, Plebani A, Mancia M (1998) Multiple modulatory effects of dopamine on calcium channel kinetics in adult rat sensory neurons. J Physiol 509, 395-409.

Fritschy JM, Grzanna R (1990) Demonstration of two separate descending noradrenergic pathways to the rat spinal cord: evidence for an intragriseal trajectory of locus coeruleus axons in the superficial layers of the dorsal horn. J Comp Neurol 291, 553-582.

Fritschy JM, Lyons WE, Mullen CA, Kosofsky BE, Molliver ME, Grzanna R (1987) Distribution of locus coeruleus axons in the rat spinal cord: a combined anterograde transport and immunohistochemical study. Brain Res 437, 176-180.

Fung SJ, Barnes CD (1981) Evidence of facilitatory coerulospinal action in lumbar motoneurons of cats. Brain Res 216, 299-311.

Fung SJ, Barnes CD (1989) Raphé-produced excitation of spinal cord motoneurons in the cat. Neurosci Lett 103, 185-190.

Fung SJ, Manzoni D, Chan JY, Pompeiano O, Barnes CD (1991) Locus coeruleus control of spinal motor output. Prog Brain Res 88, 395-409.

Gallagher JP, Inokuchi H, Shinnick-Gallagher P (1980) Dopamine depolarisation of mammalian primary afferent neurones. Nature 283, 770-772.

Gebhart GF (1986) Modulatory effects of descending systems on spinal dorsal horn neurons. In: *Spinal Afferent Processing.* Yaksh TL (ed) New York: Plenum, 363-390.

Gibson SJ, Polak JM, Bloom SR, Wall PD (1981) The distribution of nine peptides in rat spinal cord with special emphasis on the substantia gelatinosa and on the area around the central canal (lamina X). J Comp Neurol 201, 65-79.

Gilbert RF, Emson PC, Hunt SP, Bennett GW, Marsden CA, Sandberg BE, Steinbusch HW, Verhofstad AA (1982). The effects of monoamine neurotoxins on peptides in the rat spinal cord. Neuroscience 7, 69-87.

Gillberg PG, Wiksten B (1986) Effects of spinal cord lesions and rhizotomies on cholinergic and opiate receptor binding sites in rat spinal cord. Acta Physiol Scand 126, 575-582.

Glazer EJ, Ross LL (1980) Localization of noradrenergic terminals in sympathetic preganglionic nuclei of the rat: demonstration by immunocytochemical localization of dopamine-beta-hydroxylase. Brain Res 185, 39-49.

Green GM, Scarth J, Dickenson A (2000) An excitatory role for 5-HT in spinal inflammatory nociceptive transmission; state-dependent actions via dorsal horn 5-HT(3) receptors in the anaesthetized rat. Pain 89, 81-88.

Grossmann W, Jurna I, Nell T, Theres C (1973) The dependence of the anti-nociceptive effect of morphine and other analgesic agents on spinal motor activity after central monoamine depletion. Eur J Pharmacol 24, 67-77.

Gulbenkian S, Merighi A, Wharton J, Varndell IM, Polak JM (1986) Ultrastructural evidence for the coexistence of calcitonin gene-related peptide and substance P in secretory vesicles of peripheral nerves in the guinea pig. J Neurocytol 15, 535-542.

Hammond DL (1986) Control systems for nociceptive afferent processing: the descending inhibitory pathways. In: *Spinal Afferent Processing*. Yaksh TL (ed). Plenum New York.

Hammond DL, Yaksh TL (1984) Antagonism of stimulation-produced antinociception by intrathecal administration of methysergide or phentolamine. Brain Res 298, 329-337.

Headley PM, Duggan AW, Griersmith BT (1978) Selective reduction by noradrenaline and 5-hydoxytryptamine of nociceptive responses of cat dorsal horn neurones. Brain Res 145, 185-189.

Helke CJ, Neil JJ, Massari VJ, Loewy AD (1982) Substance P neurons project from the ventral medulla to the intermediolateral cell column and ventral horn in the rat. Brain Res 243, 147-152.

Hino M, Ono H, Fukuda H (1987) Involvement of noradrenergic systems in the effects of conditioning stimulation of the lower brain stem on the lumbar spinal reflex in rats. Gen Pharmacol 18, 41-45.

Hokfelt T, Kellerth JO, Nilsson G, Pernow B (1975) Substance P: localization in the central nervous system and in some primary sensory neurons. Science 190, 889-890.

Hökfelt T, Ljungdahl A, Terenius L, Elde R, Nilsson G (1977) Immunohistochemical analysis of peptide pathways possibly related to pain and analgesia: enkephalin and substance P. Proc Natl Acad Sci USA 74, 3081-3085.

Hökfelt T, Ljungdahl A, Steinbusch H, Verhofstad A, Nilsson G, Brodin E, Pernow B, Goldstein M (1978) Immunohistochemical evidence of substance P-like immunoreactivity in some 5-hydroxytryptamine-containing neurons in the rat central nervous system. Neuroscience 3, 517-538.

Hökfelt T, Broberger C, Xu ZQ, Sergeyev V, Ubink R, Diez M (2000) Neuropeptides – an overview. Neuropharmacology 39, 1337-1356.

Holstege JC, Kuypers, HG (1987) Brainstem projections to spinal motoneurons: an update. Neuroscience 23, 809-821.

Holstege JC, Van Dijken H, Buijs RM, Goedknegt H, Gosens T, Bongers CM (1996) Distribution of dopamine immunoreactivity in the rat, cat and monkey spinal cord. J Comp Neurol 376, 631-652.

Honoré P, Chapman V, Buritova J, Besson JM (1996) To what extent do spinal interactions between an alpha-2 adrenoceptor agonist and a mu opioid agonist influence noxiously evoked c-Fos expression in the rat? A pharmacological study. J Pharmacol Exp Ther 278, 393-403.

Hosoya Y, Sugiura Y, Okado N, Loewy AD, Kohno K (1991) Descending input from the hypothalamic paraventricular nucleus to sympathetic preganglionic neurons in the rat. Exp Brain Res 85, 10-20.

Houser CR, Crawford GD, Barber RP, Salvaterra PM, Vaughn JE (1983) Organization and morphological characteristics of cholinergic neurons: an immunocytochemical study with a monoclonal antibody to choline acctyltransferase. Brain Res 266, 97-119.

Howe JR, Zieglgänsberger W (1987) Responses of rat dorsal horn neurons to natural stimulation and to iontophoretically applied norepinephrine. J Comp Neurol 255, 1-17.

Huangfu D, Hwang LJ, Riley TA, Guyenet PG (1994) Role of serotonin and catecholamines in sympathetic responses evoked by stimulation of rostral medulla. Am J Physiol 266, R338-352.

Hunt SP, Kelly JS, Emson P, Kimmel JR, Miller RJ, Wu JY (1981) An immunohistochemical study of neuronal populations containing neuropeptides or gamma-aminobutyrate within the superficial layers of the rat dorsal horn. Neuroscience 6, 1883-1898.

Hwang JH, Kim YK, Park JY, Lee EJ (2001) The mechanism of antiallodynic effect of intrathecal morphine in neuropathic pain induced by spinal nerve ligation. Korean J Anesthesiol 40, 244-251.

Ichikawa T, Shimizu T (1998) Organization of choline acetyltransferase-containing structures in the cranial nerve motor nuclei and spinal cord of the monkey. Brain Res 779, 96-103.

Iwamoto ET, Marion L (1993) Characterization of the antinociception produced by intrathecally administered muscarinic agonists in rats. J Pharmacol Exp Ther 266, 329-338.

Iyadomi M, Iyadomi I, Kumamoto E, Tomokuni K, Yoshimura M (2000) Presynaptic inhibition by baclofen of miniature EPSCs and IPSCs in substantia gelatinosa neurons of the adult rat spinal dorsal horn. Pain 85, 385-393.

Jacobs BL, Fornal CA (1993) 5-HT and motor control: a hypothesis. Trends Neurosci 16, 346-352.

Jankowska E, Simonsberg IH, Chojnicka B (1998) Modulation of information forwarded to feline cerebellum by monoamines. Ann N Y Acad Sci 860, 106-119.

Jensen TS, Smith DF (1982) Dopaminergic effects on tail-flick response in spinal rats. Eur J Pharmacol 79, 129-133.

Jensen TS, Smith DF (1983) Role of 5-HT and NA in spinal dopaminergic analgesia. Eur J Pharmacol 86, 65-70.

Jensen TS, Yaksh TL (1984a) Spinal monoamine and opiate systems partially mediate the antinociceptive effects produced by glutamate at brainstem sites. Brain Res 321, 287-297.

Jensen TS, Yaksh TL (1984b) Effects of an intrathecal dopamine agonist, apomorphine, on thermal and chemical evoked noxious responses in rats. Brain Res 296, 285-293.

Jensen TS, Yaksh TL (1986) Examination of spinal monoamine receptors through which brainstem opiate-sensitive systems act in the rat. Brain Res 363, 114-127.

Jessell TM, Tsunoo A, Kanazawa I, Otsuka M (1979) Substance P: depletion in the dorsal horn of rat spinal cord after section of the peripheral processes of primary sensory neurons. Brain Res 168, 247-259.

Johansson O, Hokfelt T, Pernow B, Jeffcoate SL, White N, Steinbusch HW, Verhofstad AA, Emson PC, Spindel E (1981) Immunohistochemical support for three putative transmitters in one neuron: coexistence of 5-hydroxytryptamine, substance P- and thyrotropin releasing hormone-like immunoreactivity in medullary neurons projecting to the spinal cord. Neuroscience 6, 1857-1881.

Jones SL, Gebhart GF (1987) Spinal pathways mediating tonic, coeruleospinal and raphe-spinal descending inhibition in the rat. J Neurophysiol 58, 138-159.

Jones SL, Light AR (1992) Serotoninergic medullary raphespinal projection to the lumbar spinal cord in the rat: a retrograde immunohistochemical study. J Comp Neurol 322, 599-610.

Ju G, Hökfelt T, Brodin E, Fahrenkrug J, Fischer JA, Frey P, Elde RP, Brown JC (1987) Primary sensory neurons of the rat showing calcitonin gene-related peptide immunoreactivity and their relation to substance P-, somatostatin-, galanin-, vasoactive intestinal polypeptide- and cholecystokinin-immunoreactive ganglion cells. Cell Tissue Res 247, 417-431.

Kan, KS, Chao LP, Eng LF (1978) Immunohistochemical localization of choline acetyltransferase in rabbit spinal cord and cerebellum. Brain Res 146, 221-229.

Karoum F, Commissiong JW, Neff NH, Wyatt RJ (1980) Biochemical evidence for uncrossed and crossed locus coeruleus projections to the spinal cord. Brain Res 196, 237-241.

Kia HK, Miquel MC, McKernan RM, Laporte AM, Lombard MC, Bourgoin S, Hamon M, Vergé D (1995) Localization of 5-HT3 receptors in the rat spinal cord: immunohistochemistry and in situ hybridization. Neuroreport 6, 257-261.

Kidd EJ, Laporte AM, Langlois X, Fattaccini CM, Doyen C, Lombard MC, Gozlan H, Hamon M (1993) 5-HT3 receptors in the rat central nervous system are mainly located on nerve fibres and terminals. Brain Res 612, 289-298.

Kimura H, McGeer PL, Peng JH, McGeer EG (1981) The central cholinergic system studied by choline acetyltransferase immunohistochemistry in the cat. J Comp Neurol 200, 151-201.

Kishida T, Kito S, Itoga E, Yanaihara N, Ogana N, Wakabayashi I (1979) Immunohistochemical studies on the distribution of biologically active oligopeptides in the rat central nervous system. Acta Histochem Cytochem.12, 301-324.

Kiyosawa A, Katsurabayashi S, Akaike N, Pang ZP, Akaike N (2001) Nicotine facilitates glycine release in the rat spinal dorsal horn. J Physiol 536, 101-110.

Kobayashi Y, Okuda T, Fujioka Y, Matsumura G, Nishimura Y, Haga T (2002) Distribution of the high-affinity choline transporter in the human and macaque monkey spinal cord. Neurosci Lett 317, 25-28.

Kojima M, Matsuura T, Kimura H, Nojyo Y, Sano Y (1984) Fluorescence histochemical study on the noradrenergic control to the anterior column of the spinal lumbosacral segments of the rat and dog, with special reference to motoneurons innervating the perineal striated muscles (Onuf's nucleus). Histochemistry 81, 237-241.

Kojima M, Sano Y (1983) The organization of serotonin fibers in the anterior column of the mammalian spinal cord. An immunohistochemical study. Anat Embryol (Berl) 167, 1-11.

Kojima M, Takeuchi Y, Goto M, Sano Y (1982) Immunohistochemical study on the distribution of serotonin fibers in the spinal cord of the dog. Cell Tissue Res 226, 477-491.

Kojima M, Takeuchi Y, Goto M, Sano Y (1983) Immunohistochemical study on the localization of serotonin fibers and terminals in the spinal cord of the monkey (Macaca fuscata). Cell Tissue Res 229, 23-36.

Konishi S, Otsuka M (1974a) Excitatory action of hypothalamic substance P on spinal motoneurones of newborn rats. Nature 252, 734-735.

Konishi S, Otsuka M (1974b) Proceedings: Action of substance P on the isolated spinal cord of the newborn rat. Jpn J Pharmacol 24, s:104.

Krivoy WA, Couch JR, Stewart JM, Zimmermann E (1980) Modulation of cat monosynaptic reflexes by substance P Brain Res 202, 365-372.

Kwiat GC, Basbaum AI (1992) The origin of brainstem noradrenergic and serotonergic projections to the spinal cord dorsal horn in the rat. Somatosens Mot Res 9, 157-173.

Lackner KJ (1980) Mapping of monoamine neurones and fibres in the cat lower brainstem and spinal cord. Anat Embryol (Berl) 161, 169-195.

LaMotte CC (1988) Lamina X of primate spinal cord: distribution of five neuropeptides and serotonin. Neuroscience 25, 639-658.

LaMotte CC, de Lanerolle NC (1983) Ultrastructure of chemically defined neuron systems in the dorsal horn of the monkey. III. Serotonin immunoreactivity. Brain Res 274, 65-77.

LaMotte CC, Shapiro CM (1991) Ultrastructural localization of substance P, met-enkephalin, and somatostatin immunoreactivity in lamina X of the primate spinal cord. J Comp Neurol 306, 290-306.

Levant B, McCarson KE (2001) D(3) dopamine receptors in rat spinal cord: implications for sensory and motor function. Neurosci Lett 303, 9-12.

Lewis DI, Coote JH (1990) Excitation and inhibition of rat sympathetic preganglionic neurones by catecholamines, Brain Res 530, 229-234.

Li DP, Chen SR, Pan YZ, Levey AI, Pan HL (2002) Role of presynaptic muscarinic and GABA(B) receptors in spinal glutamate release and cholinergic analgesia in rats. J Physiol 543, 807-818.

Li P, Zhuo M (1998) Silent glutamatergic synapses and nociception in mammalian spinal cord. Nature 393, 695-698.

Li P, Zhuo M (2001) Cholinergic, noradrenergic, and serotonergic inhibition of fast synaptic transmission in spinal lumbar dorsal horn of rat. Brain Res Bull 54, 639-647.

Light AR, Kavookjian AM, Petrusz P (1983) The ultrastructure and synaptic connections of serotonin-immunoreactive terminals in spinal laminae I and II. Somatosens Res 1, 33-50.

Ljungdahl A, Hokfelt T, Nilsson G (1978) Distribution of substance P-like immunoreactivity in the central nervous system of the rat- I. Cell bodies and nerve terminals. Neuroscience 3, 861-943.

Loewy AD (1990) Central autonomic pathways. In: *Central Regulation of Autonomic Functions.* Loewy AD and Spyer KM (eds). Oxford University Press, Oxford.

Loewy AD, Marson L, Parkinson D, Perry MA, Sawyer WB (1986) Descending noradrenergic pathways involved in the A5 depressor response. Brain Res 386, 313-324.

Loewy AD, Sawyer WB (1982) Substance P antagonist inhibits vasomotor responses elicited from ventral medulla in rat. Brain Res 245, 379-383.

Loewy AD, McKellar S, Saper CB (1979a) Direct projections from the A5 catecholamine cell group to the intermediolateral cell column. Brain Res 174, 309-314.

Loewy AD, Gregorie EM, McKellar S, Baker RP (1979b) Electrophysiological evidence that the A5 catecholamine cell group is a vasomotor center. Brain Res 178, 196-200.

Lyons WE, Grzanna R (1988) Noradrenergic neurons with divergent projections to the motor trigeminal nucleus and the spinal cord: a double retrograde neuronal labeling study. Neuroscience 26, 681-693.

Marlier L, Sandillon F, Poulat P, Rajaofetra N, Geffard M, Privat A (1991) Serotonergic innervation of the dorsal horn of rat spinal cord: light and electron microscopic immunocytochemical study. J Neurocytol 20, 310-322.

Martin GF, Cabana T, Humbertson AO Jr (1982) The brainstem origin of monoaminergic projections to the spinal cord of the North American opossum: a study using fluorescent tracers and fluorescence histochemistry. Brain Res Bull 9, 217-225.

Mayer DJ, Price DD (1976) Central nervous system mechanisms of analgesia. Pain 2, 379-404.

Maxwell L, Maxwell DJ, Neilson M, Kerr R. (1996) A confocal microscopic survey of serotoninergic axons in the lumbar spinal cord of the rat: co-localization with glutamate decarboxylase and neuropeptides. Neuroscience 75, 471-480.

McCall RB, Clement ME (1994) Role of serotonin1A and serotonin2 receptors in the central regulation of the cardiovascular system. Pharmacol Rev 46, 231-243.

McCarthy PW, Lawson SN (1989) Cell type and conduction velocity of rat primary sensory neurons with substance P-like immunoreactivity. Neuroscience 28, 745-753.

McKitrick DJ, Calaresu FR (1991). Cardiovascular responses to combined microinjection of substance P and acetylcholine in the intermediolateral nucleus of the rat. J Auton Nerv Syst 32, 69-75.

McLachlan EM, Oldfield BJ (1981) Some observations on the catecholaminergic innervation of the intermediate zone of the thoracolumbar spinal cord of the cat. J Comp Neurol 200, 529-544.

Menetrey D, Basbaum Al (1987) The distribution of substance P, enkephalin- and dynorphin-immunoreactive neurons in the medulla of the rat and their contribution to bulbospinal pathways. Neuroscience 23, 173-187.

Millan MJ, Seguin L, Honoré P, Girardon S, Bervoets K (1996) Pro- and antinociceptive actions of serotonin (5-HT)1A agonists and antagonists in rodents: relationship to algesiometric paradigm. Behav Brain Res 73, 69-77.

Millhorn DE, Hökfelt T, Seroogy K, Verhofstad AA (1988) Extent of colocalization of serotonin and GABA in neurons in the ventral medulla oblongata in rat. Brain Res 461, 169-174.

Milne RJ, Gamble GD (1990) Behavioural modification of bulbospinal serotonergic inhibition and morphine analgesia. Brain Res 521, 167-174.

Mitchell JM, Lowe D, Fields HL (1998) The contribution of the rostral ventromedial medulla to the antinociceptive effects of systemic morphine in restrained and unrestrained rats. Neuroscience 87, 123-133.

Mizukawa K (1980) The segmental detailed topographical distribution of monoaminergic terminals and their pathways in the spinal cord of the cat. Anat Anz 147, 125-144.

Mizukawa K, Otsuka N, Hattori T (1986) Serotonin-containing nerve fibers in the rat spinal cord: electron microscopic immunohistochemistry. Acta Med Okayama 40, 1-10.

Mouchet P, Manier M, Dietl M, Feuerstein C, Berod A, Arluison M, Denoroy L, Thibault J (1986) Immunohistochemical study of catecholaminergic cell bodies in the rat spinal cord. Brain Res Bull 16, 341-353.

Naguib M, Yaksh TL (1997) Characterization of muscarinic receptor subtypes that mediate antinociception in the rat spinal cord. Anesth Analg 85, 847-853.

Neuman RS (1985) Action of serotonin and norepinephrine on spinal motoneurones following blockade of synaptic transmission. Can J Physiol Pharmacol 63, 735-738.

Nevin K, Zhuo H, Helke CJ (1994) Neurokinin A coexists with substance P and serotonin in ventral medullary spinally projecting neurons of the rat. Peptides 15, 1003-1011.

North RA, Yoshimura M (1984). The actions of noradrenaline on neurones of the rat substantia gelatinosa in vitro. J Physiol 349, 43-55.

Nuseir K, Proudfit HK (2000) Bidirectional modulation of nociception by GABA neurons in the dorsolateral pontine tegmentum that tonically inhibit spinally projecting noradrenergic A7 neurons. Neuroscience 96, 773-783.

Oku R, Satoh M, Fujii N, Otaka A, Yajima H, Takagi H (1987) Calcitonin gene-related peptide promotes mechanical nociception by potentiating release of substance P from the spinal dorsal horn in rats. Brain Res 403, 350-354.

Ondo WG, He Y, Rajasekaran S, Le WD (2000) Clinical correlates of 6-hydroxydopamine injections into A11 dopaminergic neurons in rats: a possible model for restless legs syndrome? Mov Disord 15, 154-158.

Ono H, Hasebe Y, Satoh M, Nagao T, Ohta S, Hirobe M, Fukuda H (1991) Amphetamine antagonistic properties of 4-phenyl-1,2,3,4-tetrahydroisoquinoline: inhibition of spinal reflex-enhancing effects of methamphetamine, phenylethylamine and nomifensine. Brain Res 564, 319-322.

Pan HL, Chen SR, Eisenach JC (1999) Intrathecal clonidine alleviates allodynia in neuropathic rats: interaction with spinal muscarinic and nicotinic receptors. Anesthesiology 90, 509-514.

Pan YZ, Li DP, Pan HL (2002) Inhibition of glutamatergic synaptic input to spinal lamina II(o) neurons by presynaptic alpha(2)-adrenergic receptors. J Neurophysiol 87, 1938-1947.

Pellissier G, Demenge P (1991) Hypotensive and bradycardic effects elicited by spinal dopamine receptor stimulation: effects of D1 and D2 receptor agonists and antagonists. J Cardiovasc Pharmacol 18, 548-555.

Poulat P, Marlier L, Rajaofetra N, Privat A (1992) 5-Hydroxytryptamine, substance P and thyrotropin-releasing hormone synapses in the intermediolateral cell column of the rat thoracic spinal cord. Neurosci Lett 136, 19-22.

Pun RY, Marshall KC, Hendelman WJ, Guthrie PB, Nelson PG (1985) Noradrenergic responses of spinal neurons in locus coeruleus-spinal cord co-cultures. J Neurosci 5, 181-191.

Price J, Mudge AW (1983) A subpopulation of rat dorsal root ganglion neurons is catecholaminergic. Nature 301, 241-243.

Proudfit HK, Clark FM (1991) The projections of locus coeruleus neurons to the spinal cord. Prog Brain Res 88, 123-141.

Qu S, Ondo WG, Zhang X, Xie WJ, Pan TH, Le WD (2006) Projections of diencephalic dopamine neurons into the spinal cord in mice. Exp Brain Res 168, 152-156.

Radhakrishnan V, Henry JL (1991) Novel substance P antagonist, CP-99,345, blocks responses of cat spinal dorsal horn neurons to noxious cutaneous stimulation and to substance P Neurosci Lett 132, 39-43.

Reddy VK, Cassini P, Ho RH, Martin GF (1990) Origins and terminations of bulbospinal axons that contain serotonin and either enkephalin or substance-P in the North American opossum. J Comp Neurol 294, 96-108.

Ribeiro-da-Silva A (1995) Substantia gelatinosa of spinal cord. In: Paxinos G, editor. *The Rat Nervous System*, 3rd ed. Sydney: Academic Press.

Ribeiro-da-Silva A, Cuello AC (1990) Choline acetyltransferase-immunoreactive profiles are presynaptic to primary sensory fibers in the rat superficial dorsal horn. J Comp Neurol 295, 370-384.

Ribeiro-da-Silva A, Tagari P, Cuello AC (1989) Morphological characterization of substance P-like immunoreactive glomeruli in the superficial dorsal horn of the rat spinal cord and trigeminal subnucleus caudalis: a quantitative study. J Comp Neurol 281, 497-515.

Ribeiro-da-Silva A, Pioro EP, Cuello AC (1991) Substance P- and enkephalin-like immunoreactivities are colocalized in certain neurons of the substantia gelatinosa of the rat spinal cord: an ultrastructural double-labeling study. J Neurosci 11, 1068-1080.

Ruda MA, Coffield J, Steinbusch HW (1982) Immunocytochemical analysis of serotonergic axons in laminae I and II of the lumbar spinal cord of the cat. J Neurosci 2, 1660-1671.

Sasek CA, Helke CJ (1989) Enkephalin-immunoreactive neuronal projections from the medulla oblongata to the intermediolateral cell column: relationship to substance P-immunoreactive neurons. J Comp Neurol 287, 484-494.

Sasek CA, Seybold VS, Elde RP (1984) The immunohistochemical localization of nine peptides in the sacral parasympathetic nucleus and the dorsal gray commissure in rat spinal cord. Neuroscience 12, 855-873.

Sasek CA, Wessendorf MW, Helke CJ (1990) Evidence for co-existence of thyrotropin-releasing hormone, substance P and serotonin in ventral medullary neurons that project to the intermediolateral cell column in the rat. Neuroscience 35, 105-119.

Satoh M, Akaike A, Nakazawa T, Takagi H (1980) Evidence for involvement of separate mechanisms in the production of analgesia by electrical stimulation of the nucleus reticularis paragigantocellularis and nucleus raphe magnus in the rat. Brain Res 194, 525-529.

Satoh K, Kashiba A, Kimura H, Maeda T (1982) Noradrenergic axon terminals in the substantia gelatinosa of the rat spinal cord: an electron-microscopic study using glyoxylic acid-potassium parmanganate fixation. Cell Tissue Res 222, 359-378.

Satoh M, Oku R, Akaike A (1983) Analgesia produced by microinjection of L-glutamate into the rostral ventromedial bulbar nuclei of the rat and its inhibition by intrathecal alpha-adrenergic blocking agents. Brain Res 261, 361-364.

Sawchenko PE, Swanson LW (1982) The organization of noradrenergic pathways from the brainstem to the paraventricular and supraoptic nuclei in the rat. Brain Res 257, 275-325.

Scatton B, Dennis T, L'Heureux R, Monfort JC, Duyckaerts C, Javoy-Agid F (1986) Degeneration of noradrenergic and serotonergic but not dopaminergic neurones in the lumbar spinal cord of parkinsonian patients. Brain Res 380, 181-185.

Schaible HG, Jarrott B, Hope PJ, Duggan AW (1990) Release of immunoreactive substance P in the spinal cord during development of acute arthritis in the knee joint of the cat: a study with antibody microprobes. Brain Res 529, 214-223.

Scheibel ME, Scheibel AB (1970) Developmental relationship between spinal motoneuron dendrite bundles and patterned activity in the hind limb of cats. Exp Neurol 29, 328-335.

Schoenen J (1982) Dendritic organization of the human spinal cord: the motoneurons. J Comp Neurol 211, 226-247.

Schrøder HD, Skagerberg G (1985) Catecholamine innervation of the caudal spinal cord in the rat. J Comp Neurol 242, 358-368.

Senba E, Yanaihara C, Yanaihara N, Tohyama M (1988) Co-localization of substance P and Met-enkephalin-Arg6-Gly7-Leu8 in the intraspinal neurons of the rat, with special reference to the neurons in the substantia gelatinosa. Brain Res 453, 110-116.

Simon OR, Schramm LP (1983) Spinal superfusion of dopamine excites renal sympathetic nerve activity. Neuropharmacology 22, 287-293.

Skagerberg G, Björklund A (1985) Topographic principles in the spinal projections of serotonergic and non-serotonergic brainstem neurons in the rat. Neuroscience 15, 445-480.

Skagerberg G, Lindvall O (1985) Organization of diencephalic dopamine neurones projecting to the spinal cord in the rat. Brain Res 342, 340-351.

Skagerberg G, Björklund A, Lindvall O, Schmidt RH (1982) Origin and termination of the diencephalo-spinal dopamine system in the rat. Brain Res Bull 9, 237-244.

Skagerberg G, Meister B, Hökfelt T, Lindvall O, Goldstein M, Joh T, Cuello AC (1988) Studies on dopamine-, tyrosine hydroxylase and aromatic L-amino acid decarboxylase-containing cells in the rat diencephalon: comparison between formaldehyde-induced histofluorescence and immunofluorescence. Neuroscience 24, 605-620.

Sluka KA, Westlund KN (1992) Spinal projections of the locus coeruleus and the nucleus subcoeruleus in the Harlan and the Sasco Sprague-Dawley rat. Brain Res 579, 67-73.

Sonohata M, Furue H, Katafuchi T, Yasaka T, Doi A, Kumamoto E, Yoshimura M (2003) Actions of noradrenaline on substantia gelatinosa neurones in the rat spinal cord revealed by in vivo patch recording. J Physiol 555, 515-526.

Stamford JA (1995) Descending control of pain. Br J Anaesth, 75, 217-27.

Stamp JA, Semba K (1995) Extent of co-localization of serotonin and GABA in the neurons of the rat raphe nuclei. Brain Res 677, 39-49.

Strahlendorf JC, Strahlendorf HK, Kingsley RE, Gintautas J, Barnes CD (1980) Facilitation of the lumbar monosynaptic reflexes by locus coeruleus stimulation. Neuropharmacology 19, 225-230.

Supowit SC, Hallman DM, Zhao H, DiPette DJ (1998). Alpha 2-adrenergic receptor activation inhibits calcitonin gene-related peptide expression in cultured dorsal root ganglia neurons. Brain Res 782, 184-193.

Sur C, Betz H, Schloss P (1996) Localization of the serotonin transporter in rat spinal cord. Eur J Neurosci 8, 2753-2757.

Swanson LW, Kuypers HG (1980) The paraventricular nucleus of the hypothalamus: cytoarchitectonic subdivisions and organization of projections to the pituitary, dorsal vagal complex and spinal cord as demonstrated by retrograde fluorescence double labeling methods. J Comp Neurol 194, 555-570.

Swanson LW, Sawchenko PE, Bérod A, Hartman BK, Helle KB, Vanorden DE (1981) An immunohistochemical study of the organization of catecholaminergic cells and terminal fields in the paraventricular and supraoptic nuclei of the hypothalamus. J Comp Neurol 196, 271-285.

Takahashi T, Berger AJ (1990) Direct excitation of rat spinal motoneurones by serotonin. J Physiol 423, 63-76.

Tamae A, Nakatsuka T, Koga K, Kato G, Furue H, Katafuchi T, Yoshimura M (2005) Direct inhibition of substantia gelatinosa neurones in the rat spinal cord by activation of dopamine D2-like receptors. J Physiol 568, 243-253.

Tashiro T, Satoda T, Takahashi O, Matsushima R, Mizuno N. (1988) Distribution of axons exhibiting both enkephalin- and serotonin-like immunoreactivities in the lumbar cord segments: an immunohistochemical study in the cat. Brain Res 440, 357-362.

Taylor BK, Roderick RE, St Lezin E, Basbaum AI (2001) Hypoalgesia and hyperalgesia with inherited hypertension in the rat. Am J Physiol Regul Integr Comp Physiol 280, R345-R354.

Thurston CL, Randich A (1990) Acute increases in arterial blood pressure produced by occlusion of the abdominal aorta induces antinociception: peripheral and central substrates. Brain Res 519, 12-22.

Todd AJ (1991) Immunohistochemical evidence that acetylcholine and glycine exist in different populations of GABAergic neurons in lamina III of rat spinal dorsal horn. Neuroscience 44, 741-746.

Todd AJ, McKenzie J (1989) GABA-immunoreactive neurons in the dorsal horn of the rat spinal cord. Neuroscience 31, 799-806.

Todd AJ, Spike RC (1993) The localization of classical transmitters and neuropeptides within neurons in laminae I-III of the mammalian spinal dorsal horn. Prog Neurobiol 41, 609-645.

Travagli RA (1996) Muscarine receptor activation in the substantia gelatinosa of the spinal trigeminal nucleus of the guinea pig. J Neurophysiol 76, 3817-3822.

Tulunay FC, Sparber SB, Takemori AE (1975) The effect of dopaminergic stimulation and blockade on the nociceptive and antinociceptive responses of mice. Eur J Pharmacol 31, 65-70.

Vacca LL, Hobbs J, Abrahams S, Naftchi E (1982) Ultrastructural localization of substance P immunoreactivity in the ventral horn of the rat spinal cord. Histochemistry 76, 33-49.

Vaida GT, Moss P, Capan LM, Turndorf H (1986). Prolongation of lidocaine spinal anesthesia with phenylephrine. Anesth Analg 65, 781-785.

Van Dijken H, Dijk J, Voom P, Holstege JC (1996) Localization of dopamine D2 receptor in rat spinal cord identified with immunocytochemistry and in situ hybridization. Eur J Neurosci 8, 621-628.

VanderHorst VG, Ulfhake B (2006) The organization of the brainstem and spinal cord of the mouse: relationships between monoaminergic, cholinergic, and spinal projection systems. J Chem Neuroanat 31, 2-36.

Vera PL, Holets VR, Miller KE (1990) Ultrastructural evidence of synaptic contacts between substance P-, enkephalin-, and serotonin-immunoreactive terminals and retrogradely labelled sympathetic preganglionic neurons in the rat: a study using a double-peroxidase procedure. Synapse 6, 221-229.

Weil-Fugazza J, Onteniente B, Audet G, Philippe E (1993) Dopamine as trace amine in the dorsal root ganglia. Neurochem Res 18, 965-969.

Wessendorf MW, Elde R (1987) The coexistence of serotonin- and substance P-like immunoreactivity in the spinal cord of the rat as shown by immunofluorescent double labeling. J Neurosci 7, 2352-2363.

West WL, Yeomans DC, Proudfit HK (1993) The function of noradrenergic neurons in mediating antinociception induced by electrical stimulation of the locus coeruleus in two different sources of Sprague-Dawley rats. Brain Res 626, 127-135.

Westbury DR (1982) A comparison of the structures of alpha and gamma-spinal motoneurones of the cat. J Physiol 325, 79-91.

Westlund KN, Bowker RM, Ziegler MG, Coulter JD (1982) Descending noradrenergic projections and their spinal terminations. Prog Brain Res 57, 219-238.

Westlund IN, Bowker RM, Ziegler MG, Coulter JD (1983) Noradrenergic projections to the spinal cord of the rat. Brain Res, 263: 15-31.

Westlund KN, Bowker RM, Ziegler MG, Coulter JD (1984) Origins and terminations of descending noradrenergic projections to the spinal cord of monkey. Brain Res 292, 1-16.

Westlund KN, Zhang D, Carlton SM, Sorkin LS, Willis WD (1991) Noradrenergic innervation of somatosensory thalamus and spinal cord. Prog Brain Res 88, 77-88.

White SR (1985) A comparison of the effects of serotonin, substance P and thyrotropin-releasing hormone on excitability of rat spinal motoneurons in vivo. Brain Res 335, 63-70.

Willis WD (1982) Control of nociceptive transmission in the spinal cord. In: *Progress in Sensory Physiology,* vol 3 Ottoson D (ed). Springer, Berlin.

Xu Z, Tong C, Pan HL, Cerda SE, Eisenach JC (1997) Intravenous morphine increases release of nitric oxide from spinal cord by an alpha-adrenergic and cholinergic mechanism. J Neurophysiol 78, 2072-2078.

Yajiri Y, Yoshimura M, Okamoto M, Takahashi H, Higashi H (1997). A novel slow excitatory postsynaptic current in substantia gelatinosa neurons of the rat spinal cord in vitro. Neuroscience 76, 673-688.

Yaksh TL, Malmberg AB (1994) Central pharmacology of nociceptive transmission. In: *Textbook of Pain,* 3rd edn. Wall P, Melzack, R (eds.). Churchill Livingstone, Edinburgh.

Yaksh TL, Dirksen R, Harty GJ (1985) Antinociceptive effects of intrathecally injected cholinomimetic drugs in the rat and cat. Eur J Pharmacol 117, 81-88.

Zhao Y, Zhang C, Kang YM, Qiao JT, Dafny N (1999) Endogenous adenosine involved in the mediation of spinal antinociception produced by stimulating locus coeruleus. Life Sci 65, PL67-74.

Zhuo M, Gebhart GF (1991) Tonic cholinergic inhibition of spinal mechanical transmission. Pain 46, 211-222.

14 Spinal Cord Injury: experimental animal models and relation to human therapy

Stuart I Hodgetts, Giles W Plant
and Alan R Harvey

Introduction

Increasing use is being made of experimental animal models of spinal cord injury (SCI). The aims are to better understand the acute and chronic morphological, cellular and molecular consequences of SCI, and with this information develop and test new therapies to enhance anatomical repair and functional recovery. The ultimate aim is improved treatment and rehabilitation of patients suffering SCI. The majority of SCI events occur in the 18-35 year age group and the long-term personal, social and economic costs are enormous. From a therapeutic point of view, even partial restoration of damaged spinal tracts following SCI could result in improvements in respiratory function, upper/lower limb function (including tasks such as eating, drinking, reading and personal hygiene), and bowel/bladder/sexual function. As the goal for clinical therapies becomes increasingly reachable, associating these often broad behavioral outcomes with experimental strategies aimed at tissue repair/sparing, remyelination, neuronal protection, cellular replacement and axonal regeneration, as well as the activation of endogenous host stem/progenitor cell responses, has become more crucial but also more complex. There are many differences between the nervous systems of humans and animals commonly used in SCI studies. Issues such as size, gait, neuroanatomical, neurophysiological and behavioral differences as well as disparities in immunological and inflammatory responses following SCI, all point to potential limitations of animal models when assessing the efficacy and safety of possible SCI treatments in humans. Use of non-human primates as an intermediate step between rodents and humans may aid in translation of effective therapies to the clinic (Courtine *et al.*, 2007).

General pathophysiology of SCI

Human SCI often results from an initial ballistic insult to the spinal column and contents to produce a contusion injury. There is subsequent displacement of the spinal cord due to pressure from broken bone, disk fragments and hematoma and swelling inside the closed vertebral canal. Most human injuries do not compromise the dura mater or involve transection of the spinal cord, whereas in many rodent SCI models the pathological responses associated with opening the dura and transecting the spinal cord can be very different. In this context, the development of reproducible contusion models in mice and rats has been a major advance in recent years. Nonetheless there are some basic similarities in the pathophysiology of human SCI and experimental animal SCI models. The impact of SCI is characterized by a biphasic pattern. First there is the primary mechanical injury, typically focal and resulting from acute compression and contusion of the spinal cord via bone or disk displacement within the spinal column following fracture-dislocation or burst fracture of the spine (Sekhon and Fehlings, 2001). There is hemorrhage and the generation of a necrotic epicenter which is then followed by an expanding wave of secondary degeneration and cell death (extending millimeters in rodents vs centimeters in humans from the primary injury site) activated by a variety of detrimental pathophysiological events (Kwon *et al.*, 2004; Profyris *et al.*, 2004; Hagg and Oudega, 2006; Bramlett and Dietrich, 2007; Donnelly and Popovich, 2008). Typically, more peripheral portions of white matter are spared while there is extensive damage of gray matter and the gradual formation of cysts. Within damaged white matter, there is degeneration of ascending and descending axons, as well as progressive demyelination associated with loss of oligodendrocytes. Axonopathy leads to a loss of functional connections (denervation) and retraction of the proximal axon. There are also progressive changes in peripheral nerve excitability in SCI patients (Lin *et al.*, 2007).

In humans and animals there are variations in the extent of secondary injury pathology, and differences in the mechanisms of cell injury/death, immune responses and neural regeneration. During progressive secondary injury there is edema, disruption of blood flow and ischemia. Such processes lead to ionic imbalances (Nashmi and Fehlings, 2001) and further cell death via excitotoxicity, which occurs as a result of the accumulation of molecules such as glutamate in the extracellular fluid, leading to hyperactivation of neurotransmitter receptors (Park *et al.*, 2004; Baptiste and Fehlings, 2006). High levels of calcium enter the cell and activate enzymes (phospholipases, proteases, etc.) that go on to damage cell structures (Profyris *et al.*, 2004). This cell damage and other changes often lead to further induced cell death (apoptosis) and free radical-mediated lipid oxidation (Liu XZ *et al.*, 1997; Keane *et al.*, 2006). Reperfusion of tissue in the days following SCI causes further oxidative damage. After SCI it is thought that necrosis occurs with, and progresses after, the primary traumatic event, while apoptosis predominates over the more delayed stages of secondary injury (Casha *et al.*, 2001). Oligodendrocytes also appear to be more susceptible to apoptosis than neurons (Emery *et al.*, 1998; Casha *et al.*, 2001). Reactive astrocytes (activated immediately after SCI) produce extracellular matrix components such as chondroitin and keratan sulfate proteoglycans to eventually form a glial scar that encapsulates cystic regions and presents both a physical barrier and an inhibitory environment for axonal regeneration and remyelination (Fitch and Silver, 2008). As a result of these destructive processes, innate and adaptive (cellular) inflammatory events also play a role in both the early and chronic stages of spinal cord injury (Hausmann, 2003; Fleming *et al.*, 2006; Keane *et al.*, 2006; Donnelly and Popovich, 2008). The spatial extent of secondary injury events spreads radially and longitudinally along the spinal cord. Ultimately, the final outcome of SCI is related to the extent of the initial trauma and the secondary events that follow, resulting in neuronal and glial cell death and cavitation of central gray matter.

Types of spinal cord injury

Cervical vs thoracic vs lumbar vs sacral injury

Regardless of animal species, SCI results in severe functional loss, ranging from total paraplegia to specific loss of neural control to a particular organ, allodynia or even simple neuralgia. The extent of the neurological deficit in different animal species may vary depending on the destination of nerve branches entering or exiting the spinal cord below the level of the injury. Demographic studies reveal that in humans, cervical SCI is more frequent than thoracic, lumbar or sacral injuries. Experimental cervical SCI lesions facilitate the detailed study of manual dexterity and fine motor control recovery. Even limited sprouting or regeneration at the cervical level that affords some fine motor function improvement can be extremely beneficial to humans. Thoracic/lumbar/sacral SCI in nonhuman primates and other animal models facilitates the assessment of locomotory recovery and postural activities. Depending on the animal model used, and the type of injury (e.g. contusion, transection), the descending and ascending tracts in thoracic injuries may need to form new connections over much longer distances in order to become functional. The modest reports of limited regeneration in rodent models, for example, may require further refinement and improvement before they could contribute to regeneration over longer distances in humans (Fig. 14.1). It is also worth noting that in contrast to the human condition, experimental SCI in animals is performed under anesthesia and therapeutic regimes can be initiated immediately after, or even before, the injury. The suitability of such experimental paradigms should always be considered when interpreting the results in a clinical context.

Acute vs chronic injury

The extent and progression of SCI pathophysiology with time is defined by whether the injury is acute (immediate or short term) or chronic (persistent or long term). In animal models, an acute injury is usually represented by hours or days whilst the chronic injury is usually represented by weeks and months. In humans, the neuropathological profile resulting from an initial trauma follows a different temporal (and often spatial) distribution, depending on the type of injury (e.g. contusion vs section) (Bunge *et al.*, 1993; Kakulas, 1999). An early phase of spreading hemorrhagic necrosis and edema with secondary damage due to inflammatory cascades is followed by an intermediate phase of limited partial repair and tissue reorganization before ultimately reaching a chronic phase characterized by centrally located cysts or cavities within atrophic parenchyma and glial scar tissue. These progressive post-SCI events are generally similar in most animal models of contusion SCI, (Noble and Wrathall, 1985; Bresnahan *et al.*, 1987, 1991; Guízar-Sahagún *et al.*, 1994;), facilitating evaluation of neuroprotective strategies in the acute injury phase. Some species differences do, however, exist. For example, as in humans, rat contusive SCI leads to the formation of microscopic, small and large cavities and fluid-filled cysts, whereas in the mouse, a similar SCI results in the development of a dense connective tissue matrix that more closely resembles the long-term effects of some laceration and massive compression injuries in humans (Sroga *et al.*, 2003; Norenberg *et al.*, 2004). Such anatomical differences are likely to be related to observed differences in post-injury recovery in these species (Basso *et al.*, 2006). Depending on the species, a decision as to whether an injury is acute versus chronic depends on a variety of factors including the type of injury, rate of functional recovery (if any), anatomical features (e.g. cyst formation, scar tissue formation), profiles of neuronal marker expression, profiles of growth promoting and inhibitory molecule expression, and secondary tissue damage cascades (and inflammatory mediators).

Chronic injuries involve analysis of long term deficits that result from both the primary injury itself and secondary tissue damage. In humans an experimental treatment administered within a few days of SCI is currently viewed as an acute treatment. Exactly when the acute stage of human SCI becomes a chronic stage remains controversial, although reviewing the available data on the temporal progression of the pathophysiology of human SCI, it has been suggested that the chronic state is only attained 12 months after SCI and all time points in between are classified as subacute (Fawcett *et al.*, 2007). Current treatments (summarized later) may ultimately form part of combinatorial strategies, in particular attempting to revert chronic injuries back to a perhaps more treatable 'acute' state (e.g. via enzymatic digestion of the glial scar, or debrasion).

Complete vs incomplete injury

Complete SCI typically involves total transection through the spinal cord and disconnection of all ascending and descending tracts. In humans, complete transection injuries are not common, and it is widely believed that experimental contusion injury models more closely resemble the human condition. In animals such as mice, rats and cats, after complete spinal cord transection, maximum levels of postoperative care are usually required to ensure adequate feeding, prevent dehydration, control infection, pain management and bladder/bowel function.

Incomplete SCI in animal models involves partial transections, contusion injuries (mild, moderate or severe), compression injury (clips, forceps), focal myelotomy (incision), hypoxia/ischemia as well as induction by chemicals, temperature, laser and irradiation. Such incomplete (partial) injuries typically leave a percentage of motor and sensory pathways intact. Depending on the extent of each type of partial injury, the resulting neurologic deficit can range from mild (advantageous for postoperative animal care) to severe (Fehlings and Tator, 1995). Although more pertinent to most human SCI, analysis of incomplete injuries in animal models is complex due to differences in responses and recovery between, and even within, species (Basso *et al.*, 2006; Birdsall Abrams *et al.*, 2007), as well as difficulties in determining whether any functional improvements are a result of true regeneration of injured tracts or a consequence of functional compensation (e.g. behavioral), remyelination of damaged (but still intact) axons, or sprouting of other neural systems spared by the initial trauma.

Immune and inflammatory responses following SCI

SCI and secondary neuronal cell damage involves an inflammatory response (Hausmann, 2003; Sroga *et al.*, 2003). This rapid and robust response involves the participation of different cell types of the immune system as well as resident cells of the CNS and non-cellular components (such as adhesion molecules, cytokines and chemokines) (Dusart and Schwab, 1994; Popovich *et al.*, 1997; Zhang *et al.*, 1997; Schnell *et al.*, 1999;). Later phases include wound healing, extracellular matrix deposition, revascularization and scarring (Profyris *et al.*, 2004). During neuroinflammation, chemotaxis recruits immune cells to the CNS. Phagocytes (i.e. neutrophils and macrophages/monocytes) are the primary cellular components of the innate immune system, which is evolutionarily conserved and non-specific for antigens. By contrast, lymphocytes mediate adaptive immunity, activated in response to specific antigenic signals, and are considered secondary cellular components of the immune response. Macrophages in the injured spinal cord are derived from (i) resident microglia (Popovich *et al.*, 1999), which become activated within minutes to hours after SCI and transform into macrophages (Popovich *et al.*, 2002), and (ii) blood-borne monocyte/macrophages that infiltrate the lesion at 2 days, peak at 5-7 days, and can persist for weeks to months after SCI in rats (Carlson *et al.*, 1998; Popovich *et al.*, 1997). Depending upon the species and strain of animal (Sroga *et al.*, 2003), T-lymphocytes enter the injured spinal cord at different times (Jones *et al.*, 2005). Early inflammatory events, including expression of proinflammatory cytokines such as tumor necrosis factor alpha (TNF-α) and interleukins 1 and 6 at the site of injury, influence many cellular events after SCI and promote tissue damage (Carlson *et al.*, 1998; Streit *et al.*, 1998; Hayashi *et al.*, 2000; Popovich and Jones, 2003). Levels of TNF-α and TNF-α mRNA increase markedly within 1 hour after SCI in rats (Wang *et al.*, 1996; Streit *et al.*, 1998). Microglial cells (CNS resident macrophages) are considered the source of cytotoxic cytokines, such as TNF-α, that kill oligodendrocytes, and are also involved in the formation of reactive species during secondary SCI injury (Xu *et al.*, 2001).

Accumulation of debris in the CNS after injury is probably sufficient to initiate neutrophil and macrophage phagocytosis, as well as the release of neurotoxins and inflammatory mediators that can cause necrotic cavitation pathology and demyelination (Giulian *et al.*, 1993; Fitch *et al.*, 1999; Nguyen MD *et al.*, 2002; Popovich *et al.*, 2002; Stys, 2004). Secondary damage to the animal spinal cord mediated by neutrophils and macrophages is caused partly by oxidative enzymes (Taoka *et al.*, 1997; Bao *et al.*, 2004) and proteolytic enzymes such as matrix metalloproteinases (MMPs) (Noble *et al.*, 2002). Macrophages contribute to bystander damage by releasing pro-inflammatory cytokines, reactive oxygen species (ROS), nitric oxide (NO) (Conti *et al.*, 2007) and proteases (Popovich *et al.*, 1999, 2002), but also participate in the removal of injured tissue debris and the release of protective cytokines that promote neuronal regeneration, wound healing and tissue repair (Schwartz, 2003). The anatomical and functional outcomes after SCI in rats can be improved by pharmacological blockade/depletion of reactive cellular products (Taoka *et al.*, 1998; Gris *et al.*, 2004; Kwon *et al.*, 2004; Okada *et al.*, 2004;). Attenuation or depletion of monocyte and/or macrophage activity after acute SCI also decreases secondary demyelination and/or axon loss and improves neurological recovery (Giulian and Robertson, 1990; Popovich *et al.*, 1999; Mabon *et al.*, 2000).

Note however that under some circumstances there is evidence that macrophages can have a beneficial effect on neuronal survival and axonal regeneration (Dougherty *et al.*, 2000; Yin *et al.*, 2003; Crutcher *et al.*, 2006; Yin *et al.*, 2006). Blood monocytes promote axon regeneration if they are activated in vitro with pre-degenerated peripheral nerve before transplantation into injured spinal cord (Rapalino *et al.*, 1998). Grafts of activated macrophages have also been reported to promote regeneration (Prewitt *et al.*, 1997; Rapalino *et al.*, 1998; Bomstein *et al.*, 2003; Schwartz, 2003). This is a highly complex and controversial issue; the timing of macrophage activation seems to be critical (Dougherty *et al.*, 2000; Yin *et al.*, 2003, 2006) and in the spinal cord and other parts of the CNS the effects of lymphocyte and macrophage manipulation may be species and strain dependent (Popovich *et al.*, 1997; Kigerl *et al.*, 2006; Luo and Shi, 2007).

Compared to animal (especially rodent) models, the inflammatory response after human SCI is less easy to define, particularly immediately after injury (Schmitt *et al.*, 2000; Yang *et al.*, 2004; Chang, 2007). However a recent study has documented the temporal and spatial distribution of cellular infiltration as well as microglial activation after human SCI (Fleming *et al.*, 2006). Macrophages are phagocytic for weeks to months after an injury. Infiltrating neutrophils appeared as early as 4 hours after injury, peaked at 1-3 days and remained significantly increased for up to 10 days. Activated microglia appeared by day 1 after SCI, and areas with activated microglia and neutrophils were correlated with increased tissue damage. From a few months to a year after SCI, most remaining macrophages appeared to be no longer phagocytic. Note that interpretation of pathophysiology and inflammatory responses in human SCI is made difficult by individual variation in profile/severity of the SCIs, as well as not always accurate information about pre-existing diseases, drugs taken prior to the trauma, infections or other diseases that occurred after the trauma, and cardiovascular and surgical histories (Fleming *et al.*, 2006).

Autoimmunity in SCI

Low numbers of autoreactive lymphocytes that evade thymic elimination during immunological development can react with CNS proteins and be involved in neuropathology (Martin R *et al.*, 1992), as well as be activated in SCI (Popovich *et al.*, 1996; Jones *et al.*, 2002). Activation of myelin basic protein (MBP)-reactive T cells after nerve trauma (Olsson *et al.*, 1993) and in SCI patients (Kil *et al.*, 1999) provides evidence of an association between CNS trauma and the activation of CNS-autoreactive T cells. The extent to which such T cells participate in tissue injury after SCI remains controversial (Nguyen HH *et al.*, 2002). They have been shown to exacerbate axonal injury, demyelination and functional loss after SCI (Popovich *et al.*, 1996; Jones *et al.*, 2002); however reports that such destructive autoimmune reactions initiated by SCI were self-limiting and terminated by either regulatory cytokine networks or extensive induction of apoptosis of infiltrating T-cells (Popovich *et al.*, 1996; Bauer *et al.*, 1998) have been countered by the proposal that trauma-induced activation of myelin-reactive T cells is a physiological (c.f. pathological) consequence of SCI and should be boosted to achieve neuroprotection or 'protective autoimmunity' (Schwartz and Kipnis, 2001). This view is based on reported improvements in anatomical and functional outcomes after SCI in animals immunized with MBP, MBP-reactive T-cell lines and 'myelin-like' proteins (Moalem *et al.*, 1999; Hauben *et al.*, 2000, 2001, 2003). Genetic predisposition to autoimmune susceptibility within and between species may also play a role in determining the nature of immune responses to SCI and the effectiveness of various therapies in promoting recovery (Birdsall Abrams *et al.*, 2007; Limaye *et al.*, 2008).

Methods to induce spinal cord injury

Complete transection

Evaluation of the effectiveness of therapeutic interventions in promoting axonal regeneration and functional recovery are simpler in complete transection models. There are some reports of spontaneous recovery without therapeutic intervention (mainly in cats), although they include periods of gravity-assisted treadmill hindlimb training (Bélanger *et al.*, 1996; de Leon *et al.*, 1998; Cha *et al.*, 2007; Petruska *et al.*, 2007) or pharmacologic intervention (Chau *et al.*, 1998). There are reports describing varying amounts of axonal regeneration in complete transection studies in rats after transplantation of olfactory ensheathing glia (OEG) (Ramón-Cueto *et al.*, 1998; Lu *et al.*, 2001; Cao *et al.*, 2004; Fouad *et al.*, 2005), Schwann cells (Oudega *et al.*, 1997; Xu *et al.*, 1997; Plant *et al.*, 2001b) or brain-derived neurotrophic factor (BDNF)-transfected bone marrow stromal cells (BMSCs) (Koda *et al.*, 2007), but without substantial functional recovery after implantation.

Incomplete transection

Experimental models of incomplete spinal cord transection are commonly used and have proved useful in assessing the effects of treatments aimed at promoting axonal regeneration and/or sprouting from spared axons, and for correlating anatomical changes with specific behavioral improvements. Advantages of these models are that critical physiological functions (e.g. bladder and bowel functions) are preserved, and animals recover relatively extensive postural control quickly. Such spared postural and locomotor capability means less intensive postoperative animal care is needed. Dorsal or lateral hemisection partial injury models facilitate comparisons of the regenerative response in particular tracts with respective uninjured contralateral counterparts (Muir and Whishaw, 2000; Anderson *et al.*, 2007). Focal myelotomy (incisions) using a Vibraknife or needles also results in transection of individually identified axons or groups of axons within the spinal cord (Kerschensteiner *et al.*, 2005) which allows assessment of treatments to be more easily compared with specific functions and/or outcomes. As described earlier, disadvantages of these incomplete transection models include the possibility that axons of particular tracts may have escaped injury, and local and supraspinal plasticity of any remaining intact fibers and their reorganization could account for some, if not all, of any observed functional recovery. Ultimately, although partial transection models are subject to criticism for their lack of applicability to the majority of blunt spinal cord injuries in humans, they yield substantial value in axonal regeneration research and retain sufficient clinical relevance to be a model for testing the efficacy of many therapies.

Hypoxia/ischaemia

The behavioral and anatomical/physiological characteristics of isolated spinal cord explants and cell lines under hypoxic conditions have been studied in vitro in order to reveal the importance of physiological cascades leading to neuronal cell death. In in vivo animal models, ischemic/hypoxic injury is usually induced by compressive restriction (typically via forceps or aortic cross clamping) of blood vessels networking the spinal cord for prolonged time periods, followed by periods of reperfusion. Calcium influx into cells is responsible for initiating the 'final pathway' to cell death in neuronal tissue after traumatic or hypoxic injury, although the pathways involved in this cascade are complex. Hypoxia/ischemia induced injury is characterized by changes in glial-axonal signaling, and enhanced calcium channel activity with excessive accumulation of intracellular calcium in spinal cord dorsal column white matter tracts (Profyris *et al.*, 2004).

Contusion and compression models

Because most human SCI results from blunt trauma, models of contusive SCI are considered to resemble human SCI more closely than transection injuries. Even in humans with complete paraplegia after contusive SCI, the cord is rarely completely transected, with cysts interspersed between residual, normal-appearing cord parenchyma – usually located peripherally at the lesion site (Bunge *et al.*, 1993). Contusion SCI animal models produce similar lesions in which neuronal tissue remains intact peripherally around the lesion site, but in contrast to humans the amount of 'spared' tissue generally correlates with residual locomotor function (Bresnahan *et al.*, 1987; Basso *et al.*, 1995, 1996;).

Several models in various animal species have been developed to deliver a blunt contusive or compressive force to the spinal cord, and considerable research has been directed at optimizing the consistency, duration and precision of the applied biomechanical forces to obtain reproducible SCI. The ability to deliver such consistent injury parameters and discard animals that fall outside acceptable limits of each device facilitates accurate measurement of anatomical and functional outcomes. The open field locomotory BBB score (Basso *et al.*, 1995) is increasingly being used in conjunction with other methods to assess behavioral recovery (such as the Ratwalk system), strengthening the interpretation of treatment regimes designed to improve locomotory function after SCI. Contusion injuries allow detailed assessment of potential therapies and the efficacy of cell transplantation (replacement) regimes or materials to fill the lesion site (cavity or cyst), as well as investigate the role of environmental enrichment in functional recovery (Fischer and Peduzzi, 2007). As with incomplete transections, certain physiological functions are preserved, animals recover modest locomotory function quickly and less intensive postoperative animal care is required. A brief summary of some of these experimental contusion injury models is given below.

New York University (NYU) or MASCIS Impactor

This rodent spinal cord contusion model was developed in 1992 (Gruner, 1992) and has been widely used to deliver mild to severe contusion injuries in rats. The impactor is now in its third generation and there are improvements over earlier models. Under anaesthesia and after laminectomy, a 10 g rod is dropped from heights of 6.25 to 50 mm (generating mild to severe SCI respectively) onto the exposed dorsal surface of the spinal cord (typically at T9-T10). The vertebral processes rostral and caudal to the lesion site are clamped to reduce motion of the spinal column during the impact. This Multicenter Animal Spinal Cord Injury Study device produces consistent injuries in terms of the resulting neuropathology and functional deficits as evaluated on the BBB scale (Basso *et al.*, 1995, 1996).

Ohio State University Impactor (OSU)

Instead of using weight drop, the OSU device employs an electromechanical impact probe controlled by computer feedback (Bresnahan *et al.*, 1987; Noyes, 1987a, b). Refined to improve the consistency of SCI (Stokes, 1992), this model uses a probe that is slowly driven down to contact the dural surface and displaces the spinal cord by 30 μm with a force of approximately 3000 dynes as a consistent starting point from which to initiate SCI. The system is then triggered and the probe rapidly impacts the cord and displaces it by a predetermined amount. Transducers measure the impact force transmitted and the amount of cord displacement. The OSU impactor has been adapted for use in mice, which will allow for the evaluation of contusion SCI in a wide variety of transgenic models (Jakeman *et al.*, 2000).

Infinite Horizon Impactor

The Infinite Horizon (IH) spinal cord injury device (Scheff *et al.*, 2003) creates a reliable contusion SCI by rapidly applying a force-defined impact with a stainless steel-tipped impounder. An in-line force sensor connected to the impounder tip measures the force applied by a motor driven rack. When the user-defined force is reached the impounder tip is immediately withdrawn from the exposed spinal cord. Software displays the actual force applied to the spinal cord, the amount of spinal cord displacement, the velocity of the impounder at the peak force measurement, and plots a graph of changes in displacement and force as a function of time from sequence initiation.

Clip Compression

Rivlin and Tator (1978) introduced the 'clip compression' model of SCI in rats. This model compresses the spinal cord between the arms of a modified aneurysmal clip for variable time periods. Advantages of this model are that it reveals the relation between the severity of SCI and the duration of compression, produces consistent injuries in the rat, and some consider the compressive force applied to the cord as more representative of the human injury. Adjustment of the closing force of the clip (Fehlings and Tator, 1995) and/or the duration of compression produces different severities of SCI and facilitates study of pathophysiology after cord injury (Fehlings *et al.*, 1989), the timing of decompression (Guha *et al.*, 1987), and the efficacy of treatment regimes such as electrical stimulation (Fehlings and Tator, 1992), addition of neuroprotective agents (Schwartz and Fehlings, 2001), application of factors to reduce ischemic and secondary damage (Akiyama *et al.*, 2004; Sakanaka *et al.*, 2007), and transplantation of various cell types (Karimi-Abdolrezaee *et al.*, 2006; Moreno-Flores *et al.*, 2006). A forceps crush injury in mice has also been described (Plemel *et al.*, 2008).

Balloon Compression

Microballoon compression injuries typically involve pressure and volume controlled inflation of a microballoon in the epidural space. Originally described in the rat (Martin D *et al.*, 1992), treatments last typically from one minute resulting in a functional/behavioral 'reversible' injury (Nesathurai *et al.*, 2006) to 24 hrs (Lim *et al.*, 2007), which tend to be 'irreversible'. Overall, balloon compression injuries favour comparison to human SCI, although longer periods of compression can result in propagation of secondary injury processes resulting in limited functional recovery and more extensive anatomical damage (Carlson *et al.*, 2003). Animal models used in studies of balloon compression SCI include; rats (Martin D *et al.*, 1992; Bouhy *et al.*, 2006; Šedý *et al.*, 2007), rabbits, (Baydin *et al.*, 2007), dogs (Lim *et al.*, 2007) and primates (Nesathurai *et al.*, 2006).

Experimental SCI can also be induced by other methods. For example, a focused laser beam of appropriate wavelength interacting with an i.v. injected photosensitive dye results in reproducible vascular occlusion and ischemic SCI (Cameron *et al.*, 1990). The long-term consequences of photochemically induced SCI have been characterized anatomically (Bunge *et al.*, 1994) and functionally (Gárcía-Alias *et al.*, 2006). Irradiation-induced injuries typically employ x-ray or photons (often with ethidium bromide injections) delivered directly to the exposed spinal cord, and result in what is often considered a focal demyelinated lesion. In SCI studies, irradiation has been used to demonstrate Schwann cell-mediated enhanced axonal conduction in the rat (Blakemore and Crang, 1985; Honmou *et al.*, 1996; Kohama *et al.*, 2001), and remylination following i.v. delivery of bone marrow stromal cells (Akiyama *et al.*, 2002; Inoue *et al.*, 2003) or OEG (Sasaki *et al.*, 2006).

Assessing functional recovery in animal models of SCI

- The Basso, Beattie and Bresnahan (BBB) scale, is an open-field locomotor test initially developed for rats (Basso *et al.*, 1995). Based on the Tarlov scale (see below), it analyzes hindlimb movements and locomotor coordination in an open field using a 21-point non-linear scale incorporating parameters such as joint movement, weight support, limb coordination, foot placement, and gait stability, but not other movements requiring coordinated spinal cord activity or pain, bowel, bladder, or sexual function. Note that statistical analyses can be difficult to interpret because the scale is non-linear and higher BBB values represent a score consisting of an increasing numbers of variables that are often unrelated to the previous (lower) scores.
- The Basso Mouse Scale (BMS) is an open-field locomotor test developed for mice. Based on the rat BBB scale, this test also assesses hindlimb function and locomotion (Basso *et al.*, 2006).

- The Tarlov scale is a 5-point scale to assess upper and lower limb locomotion and uses scores ranging from 0 (paraplegia) to 4 (normal locomotory function) and incorporates motor evoked potential (MEP) and somatosensory evoked potential (SSEP) in muscle and nerves (Muir, 2004).
- The Durham scale incorporates a Tarlov scale, as well as functional task, bowel hygiene, and neck position, but does not assess pain, bowel, bladder, or sexual function. A useful predictor of myelopathy and survival in an animal model of surgical treatments for cervical spondylotic myelopathy, like the Tarlov scale it appears to be best suited for assessment of incomplete injuries (Muir, 2004).
- The Catwalk (or Ratwalk) gait analysis uses a glass walkway and (digital) video recording of injured animals as they cross. The animals need to be pretrained. Several versions of this analysis software now exist and provide objective analysis of multiple locomotion parameters (usually with millisecond time resolution) during walkway crossing, including interlimb coordination, swing duration, paw print area, stride length and base of support (Hamers *et al.*, 2001).
- Electrophysiological assessment of neuronal activity non-invasively assesses MEPs or SSEP and identifies functional connections and correlates data to the impairment of locomotor activity.
- Directed forepaw reaching requires rats to reach under a barrier and pick up food with forepaws and consists of a limited scale for assessing coordinated limb and muscle movement.
- Rotor rod examines sensory feedback, coordination, and muscle strength required for locomotion by placing the animal on a rotating bar and timing how long it takes for the animal to lose balance. It does not assess restoration of fine motor control or other complications associated with SCI (Muir, 2004).
- Grid walking assesses the ability of mice and rats to walk over a wire mesh grid as a measure of coordinated walking (Muir, 2004).
- The inclined plane test is performed by placing the injured animal on a ramp of a preset incline and examines sensory feedback, coordination, and muscle strength required for locomotion, but does not assess restoration of fine motor control or other complications associated with SCI (Muir, 2004).
- Grooming response uses water to elicit grooming with the rat's forelimbs and measures forelimb response as an indicator of function. It is a brain stem-mediated spontaneous reflex sensitive to the level and severity of the injury, but lacks the ability to discriminate between loss of communication with brain stem or damage to other parts of the nervous system
- The footprint test examines an injured animal's gait by analyzing prints of paw position and toe drags as a measure of recovery of locomotion but does not assess restoration of fine motor control (Muir, 2004).
- The forepaw withdrawal test measures the time taken to withdraw the animal's forelimb from a heat block, which indicates motor function as well as sensory heat perception. The Von Frey Hair test uses graded hairs to assess nociceptive behavior as an allodynic response to mechanical stimulation; this test can be performed in various models of SCI (Mills *et al.*, 2001).
- The Morris water-maze test is an open-field water-maze procedure in which rats learn to escape from opaque water onto a hidden platform (Morris, 1984). This test assesses different aspects of spatial memory (e.g. working memory) and nonspatial discrimination learning as a measure of cerebral and locomotory function.

Assessing human functional recovery

- The American Spinal Injury Association (ASIA) International Standards for Neurological Classification assesses an injury's effects on motor and sensory systems and scores alphabetically from A (sensorimotor complete) to B (motor complete, sensory incomplete), C and D (motor and sensory incomplete), E (normal). The analysis is relatively insensitive to small but significant changes in motor and sensory functions and only covers five spinal levels in the cervical and lumbar regions. The ASIA impairment scale does not assess pain, bowel, bladder or sexual function (Ditunno *et al.*, 1994; Fawcett *et al.*, 2007).
- The Modified Benzel Classification is based on the ASIA but has grades from 1 to 7. The main difference is the expansion of ASIA D into three separate Benzel grades (Geisler *et al.*, 2001a, b; Coleman and Geisler, 2004).
- The Walking Index for Spinal Cord Injury (WISCI) is a test limited to assessing walking after SCI and grades physical assistance and any apparatus required for walking using a scale that measures functional limitations and changes in functional capacity with rehabilitation (Ditunno *et al.*, 2000).
- The Spinal Cord Injury-Functional Ambulation Inventory is a limited assessment of walking after SCI incorporating gait parameters, use of assistive devices, and temporal-distance measurements

- The Functional Independence Measure (FIM), is not designed specifically for SCI but assesses general neurological, musculoskeletal dysfunction, using an 18-item, 7-level ordinal scale (Hall *et al.*, 1999). The scale has few cognitive, behavioral, and communication-related functional items and also forms the basis of the Functional Assessment Measure (FAM).
- A Spinal Cord Independence Measure (SCIM) analyzes respiration, mobility and sphincter management in SCI and is more sensitive than FIM or FAM. It provides a total SCIM score of 100 points, while all the sub-items (e.g. feeding, bathing) are weighted (scores from 0-15) dependent on their relevance (Itzkovich *et al.*, 2002).
- An electrophysiological assessment of neuronal activity, noninvasively assesses motor evoked potential (MEP) and somatosensory evoked potential (SSEP) in muscle and nerves as a correlation to impairment of locomotor activity.
- Lower-extremity motor scores uses a basic functional assessment of five key l muscle groups and scores from 0-5 (no movement to normal resistance).
- The Barthel Index assesses an individual's independence in mobility and functional ability, incorporating mobility, self-care, and bowel/bladder continence. The scale cannot measure sensitive but significant changes in function.
- The Visual Analog Scale (VAS) is a limited self-assessment of pain using a graphic rating scale.
- The Ashworth Scale is a 6-point scale to assess spasticity and muscle tone and has limited application to assessing small changes in functional recovery (Ashworth, 1964; Pandyan *et al.*, 1999).
- The Stroke Rehabilitation Assessment of Movement (STREAM) is designed to assess voluntary movement and basic upper and lower extremity mobility using 30 mobility items, and a 3- or 4-point scale.
- A timed 'up and go' (TUG) is a widely used test of functional mobility that measures the ability of patients to stand out of a chair stand and walk forward three meters.
- The Box and Block test requires patients to move individual blocks from one compartment to another in 60 seconds, as a measure of unilateral gross manual dexterity, muscle strength and control.
- Quantitative Sensory Testing (QST) is a sensory assessment of three types of sensory nerve fibers: fast $A\beta$ (touch, joint position, mild pressure, vibration), small $A\delta$ (cold sensation, pain), and C fibers (warmth sensation, pain).

Differences between animal models and humans and functional recovery after SC

Size differences

The importance of size differences between animal SCI models and humans is reflected in the distances over which axons are required to regenerate after injury (Courtine *et al.*, 2007) (Fig 14.1). This may affect comparisons between regeneration studies in rodents versus primates, especially with injuries to cervical and thoricac cord. In addition, species differences in weight bearing due to differences in body size and anatomy, as well as gait, will affect nervous system networks and associated musculature. In addition to quadrupedal locomotion, many primate species practice bipedal walking with characteristics similar to humans; bipedal stepping can be quantified (Nakajima *et al.*, 2004) and artificially assisted after SCI.

Some comments on comparative neuroanatomy

The primate motor cortex and its descending output, the corticospinal tract (CST) projects extensively to the brainstem and spinal cord. The CST of human and nonhuman primates has undergone major evolutionary changes related to the development of skilled motor capacities, especially in distal musculature (Bortoff and Strick, 1993; Rouiller *et al.*, 1996; Lemon *et al.*, 2004). The cortical origin of CST axons and their location in the spinal cord varies between species, with increasing emphasis on projections from frontal lobe motor cortical areas in primates (Nudo and Frost, 2007). In rodents, unlike primates, most CST axons are located in the ventral part of the dorsal funiculus (Fig 14.1) and project mainly to dorsal horn neurons (in laminae 3-6) and premotor spinal circuits (Tracey, 2004). More complex CST projections in many nonhuman primates involve direct and indirect projections that network in the ventral horn; some axons synapse directly

Figure 14.1 Comparative anatomy of the rat, cat and human spinal cord

A The relative size and shape of cross-sections of rat, cat and human spinal cord at the level of the cervical and lumbar enlargements. B. Approximate location of some ascending and descending tracts in each species Top row – rat and human, bottom row – cat Unlike the cross-sections shown in A, these diagrams are not drawn to scale Dotted lines indicate areas where tracts appear to overlap. Note that, while tracts are drawn as discrete entitities, in some cases (such as the reticulospinal and raphespinal projections), fibres can be quite widely scattered in white matter and the diagrams therefore indicate the position of most axons of a particular type. S = sensory, M = motor C. Low-power views of the dorsal surface of a complete rat spinal cord and about six segments of human spinal cord. The entire rat spinal cord is about 9 cm in length, in comparison, individual segments of human cervical spinal cord can be 1.5-1.6 cm in length The longest human spinal cord segment is T6, about 2 2 cm long (Ko *et al*, 2004) or almost 25% the length of the rat spinal cord *This figure is reproduced in the color insert.*

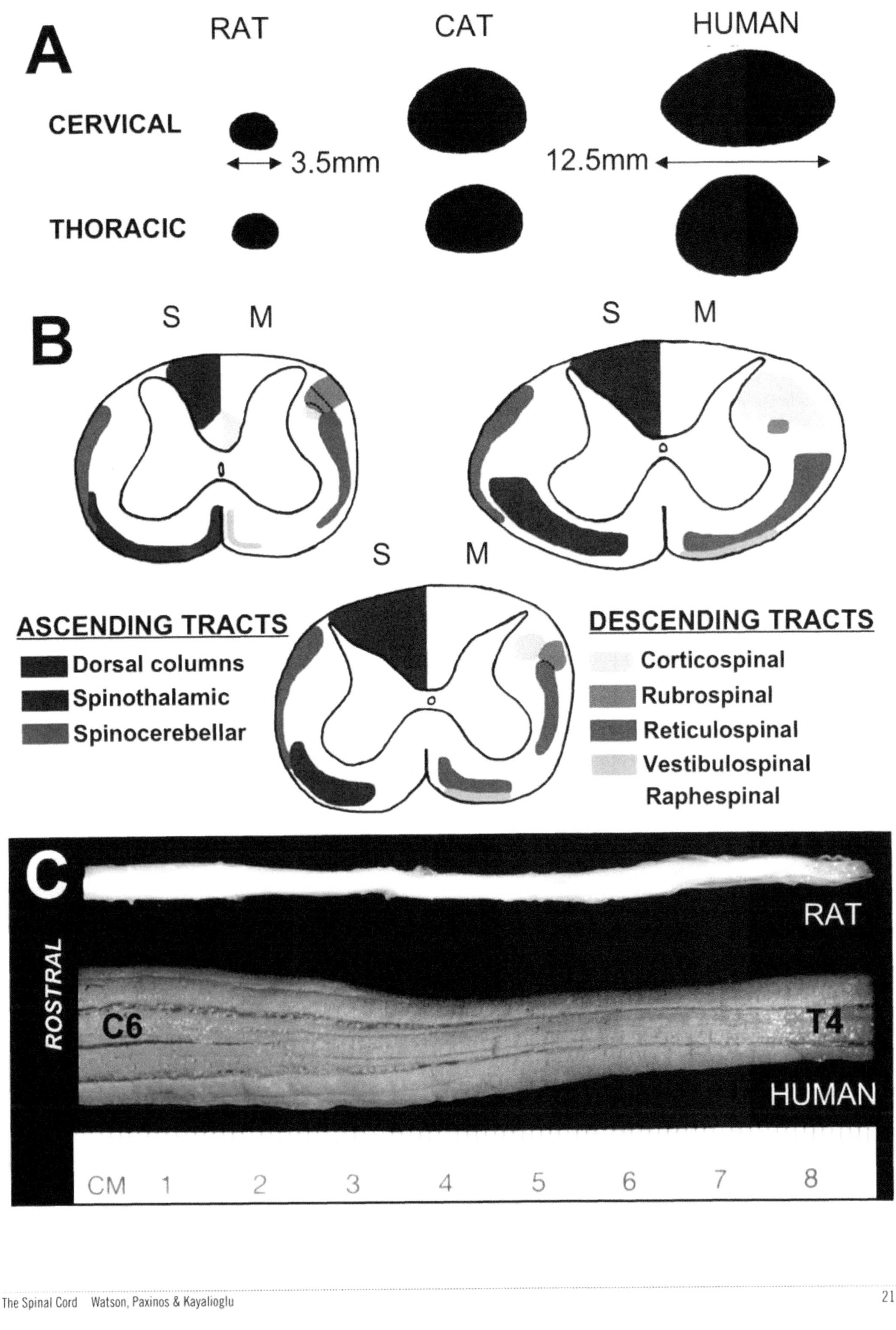
A
RAT
CAT
HUMAN
CERVICAL
THORACIC
3.5mm
12.5mm
B
S
M
ASCENDING TRACTS
Dorsal columns
Spinothalamic
Spinocerebellar
DESCENDING TRACTS
Corticospinal
Rubrospinal
Reticulospinal
Vestibulospinal
Raphespinal
C
ROSTRAL
RAT
C6
T4
HUMAN
CM 1 2 3 4 5 6 7 8

on motoneurons that innervate hand muscles (Rouiller *et al.*, 1996; Lemon *et al.*, 2004), a projection even more pronounced in humans (Kuypers, 1981; Heffner and Masterton, 1983). Motor responses following CST stimulation consequently differ between primates and rodents (Lemon and Griffiths, 2005), as well as between different primate species (Lemon *et al.*, 2004). A summary of other descending supraspinal projections is given in Chapter 11.

Functional behavior

The development of the enhanced ability to control hand musculature and unique manual dexterity functions such as the opposable thumb-finger grip in primates is facilitated by direct cortical projections to spinal motorneurons in cervical cord (Lemon *et al.*, 2004). CST lesions in primates cause major impairments in fine motor function of the hands and feet, and the extent of these deficits is most severe in humans (Nathan, 1994), whereas even though fine motor control of the forelimb in rodents is affected by CST lesions (Whishaw *et al.*, 1998; Anderson *et al.*, 2007), the sensitivity in digital control is much less developed in rodents than in primates and should be considered in any interpretation of results following SCI therapy. Such differences suggest that SCI research in nonhuman primates may provide a better prediction of the potential of therapies for SCI to mediate functional recovery (Courtine *et al.*, 2007). In rodents, CST projections to the lumbosacral enlargement are less than to cervical cord, and perhaps as a consequence CST lesions have minimal effects on stepping (Muir and Whishaw, 1999), whereas damage to the CST in nonhuman primates results in permanent deficits during stepping (Courtine *et al.*, 2005) and severely impairs independent walking in humans (Nathan, 1994; Courtine *et al.*, 2007). In primates, the intrinsic capacity to recover stepping and grasping after a SCI are still not fully understood (Courtine *et al.*, 2005). Rats and mice differ in their ability to recover from SCI, as shown by treadmill-assisted walking studies (Orsal *et al.*, 2002; Leblond *et al.*, 2003). Mice perform better on treadmills after injury alone compared to rats, which are unable to show patterned locomotion in the absence of other interventions.

Plasticity and evidence of spontaneous recovery

Any functional recovery that occurs after a CNS injury partly depends on reorganization of undamaged neural pathways (Tuszynski *et al.*, 1999; Maier and Schwab, 2006). Following SCI, the spinal circuitry and its pattern generator systems exhibit a degree of plasticity that is induced by both activity-dependent and injury-induced factors (Barbeau *et al.*, 2002; Cai *et al.*, 2006; Hagg and Oudega, 2006; Maier and Schwab, 2006; Scivoletto *et al.*, 2007; Bareyre, 2008). Limited functional improvements following SCI can also occur through the formation of new circuits via synaptic reorganization, and collateral sprouting (Bradbury and McMahon, 2006). Differences in the utilization of spinal and supraspinal motor systems between primates and other animal models of SCI may account for some of the different anatomical and functional outcomes observed following various therapeutic interventions. Enhancing reorganization at multiple sites (such as the cerebral cortex) after SCI as well as altering responses of non-CST supraspinal and intraspinal pathways might effectively improve motor recovery in response to incomplete injuries (Tuszynski *et al.*, 1999; Belci *et al.*, 2004; Schmidlin *et al.*, 2004; Kim *et al.*, 2006) and could be of benefit as part of a combinatorial SCI therapy (Lu and Tuszynski, 2008). In this context, there are several approaches to increasing plasticity of the spinal cord after injury. These include activity-dependent motor training incorporating proprioceptive and cutaneous input (Edgerton *et al.*, 2004); pharmacological intervention that modifies different neurotransmitter systems (de Leon *et al.*, 1999; Kim *et al.*, 2001; Rossignol *et al.*, 2001; Fong *et al.*, 2005); electrical stimulation (Dimitrijevic *et al.*, 1998; Mushahwar *et al.*, 2000); activity-dependent induction of neurotrophins (Gomez-Pinilla *et al.*, 2001) and environmental enrichment (Lankhorst *et al.*, 2001; Van Meeteren *et al.*, 2003; Fischer and Peduzzi, 2007). There is enhanced plasticity and regenerative capacity in the spinal cord after injury in immature animals (Fry *et al.*, 2003; Brown *et al.*, 2005).

The majority of SCI patients show some recovery of motor function below the initial injury level. Some sensory recovery occurs after SCI, with a similar time course as motor recovery. The rate of recovery in humans is rapid during the first three months and motor improvement is almost complete by 9 months, but ultimately plateaus at 12 to 18 months after SCI (Fawcett *et al.*, 2007). In patients with motor complete lesions (ASIA A and ASIA B) the majority of this functional return is likely to be limited to the zone of partial preservation just below the injury site, where some motor capacity remains. Spontaneous recovery of motor function in incomplete SCI patients (ASIA C and ASIA D) is more substantial and highly variable. Human postmortem studies report tangled masses of central axons found within human cerebral and spinal cord lesions that are claimed to be regeneration of central origin (Sung, 1981, 1987).

Strategies to treat SCI

Although limited return of function may occur in the weeks and months following SCI, there is a lack of extensive and functionally appropriate spontaneous regeneration of injured ascending and descending axons. This lack of long-distance regenerative growth is typical of the mature mammalian CNS and is a consequence of a number of factors including injury-induced loss of projection neurons; intrinsic changes in neuronal responsiveness that limit the potential for regenerative growth; formation of scar tissue by reactive astrocytes and other cells; and the presence of many types of growth inhibitory molecules. Several of these molecules are expressed by oligodendroglia and associated myelin, others such as chondroitin sulphate proteoglycans (CSPGs) and tenascin are found in the glial and meningeal scars that form in and around injuries. After SCI, therapeutic intervention is therefore needed to increase the viability of injured neurons and activate endogenous repair mechanisms, promote the regrowth and myelination of axons, maintain fiber growth within spinal cord tissue distal to the injury, and ensure that a significant proportion of regenerating axons re-innervate appropriate target regions and re-form functional synaptic connections. Reconstruction of pathways in the spinal cord after large lesions or in the chronic situation, where an injury has occurred some time previously, requires some form of tissue engineering to bypass or restore continuity across the traumatized region, giving regrowing axons access to 'normal' neuropil distal to the site of injury. It is beyond the scope of this chapter to document in detail all of the various strategies that are being tested in experimental SCI models, however a brief overview of some of these therapeutic approaches is now given.

Cell transplantation therapies

To date, a number of potentially clinically relevant cellular transplantation approaches have been used in attempts to treat SCI in animal models. These include fibroblasts (e.g. Nakahara *et al.*, 1996; Tuszynski *et al.*, 1996; Grill *et al.*, 1997a); peripheral nerve grafts and Schwann cell bridges (Richardson *et al.*, 1980; Xu *et al.*, 1997; Xu *et al.*, 1999; Plant *et al.*, 2001a, b; Bunge, 2002; Oudega and Xu, 2006; Pearse *et al.*, 2007); olfactory ensheathing glia (OEG) (Plant *et al.*, 2001a; Harvey and Plant, 2006; Ruitenberg *et al.*, 2006; Pearse *et al.*, 2007; Raisman and Li, 2007; Richter and Roskams, 2008); embryonic stem cells (McDonald *et al.*, 1999; Coutts and Keirstead, 2008); oligodendroglial progenitor cells (Blakemore *et al.*, 2000); adult neural progenitors (Kocsis *et al.*, 2004); neural stem cells (Zietlow *et al.*, 2008), autologous macrophages (Rapalino *et al.*, 1998); and bone marrow stromal cells (BMSCs) (Nandoe *et al.*, 2006; Sykova *et al.*, 2006a; Phinney and Prockop, 2007). Viability and integration of grafted cells within and adjacent to lesion sites is variable and depends on the type of cell that is used. Nonetheless, in rodent models, a majority of cell types that have been tested appear to enhance tissue sparing and promote endogenous repair processes (Mikami *et al.*, 2004; Okano *et al.*, 2007), and there are many reports of at least some functional improvements although unfortunately these are rarely correlated with anatomical evidence of large scale and targeted axonal regeneration. The effectiveness of grafts can also be enhanced by combining them with other therapeutic approaches such as combination with trophic factors (Xu *et al.*, 1995; Bregman *et al.*, 2002; Cheng *et al.*, 2004), and systemic administration of anti-inflammatory agents (Chen *et al.*, 1996; Nash *et al.*, 2002; Fu *et al.*, 2007) or immunosuppressive therapies (López-Vales *et al.*, 2006). Note however that the effectiveness of the latter approach may vary depending on the genetic and immunological background of the animals being treated (Cui *et al.*, 2007).

Gene therapy

Viral vectors can be used to introduce therapeutic genes into endogenous cells in the brain and/or spinal cord of animals with SCI, or they can be used to genetically modify donor cells ex vivo prior to transplantation. Recombinant adeno- or adeno-associated (AAV) viral vectors have been used to introduce genes into cells in the spinal cord in vivo (Liu Y *et al.*, 1997; Peel *et al.*, 1997; Romero and Smith, 1998; Boulis *et al.*, 1999; Blits and Bunge, 2006;) and have been shown to prevent apoptosis (Lou *et al.*, 1998), reduce neuropathic pain (Finegold *et al.*, 1999), and deliver growth factors such as neurotrophin-3 (NT-3) and brain-derived neurotrophic factor (BDNF) to brainstem, spinal and peripheral neurons to enhance protection and axonal outgrowth (Dijkhuizen *et al.*, 1997; Baumgartner and Shine, 1998; Zhang *et al.*, 1998; Blits *et al.*, 2003; Ruitenberg *et al.*, 2004; Kwon *et al.*, 2007; Nakajima *et al.*, 2007). Lentiviral (LV) vectors, which integrate into the host's genome resulting in stable and long-term gene expression in both dividing and quiescent cells, have also been trialed and LV vectors expressing BDNF prevent atrophy of rubrospinal neurons (Kwon *et al.*, 2007). Gene transfer to spinal cord scar tissue using LV vectors results primarily in transgene expression by astrocytes (Hendriks *et al.*, 2007).

Donor cells have been genetically modified ex vivo prior to transplantation using various transfection protocols, including retroviral and LV vectors (Tuszynski *et al.*, 1994, 1996; Nakahara *et al.*, 1996; Grill *et al.*, 1997a, b; Blesch *et al.*, 1998; Menei *et al.*, 1998; Liu *et al.*, 1999; Ruitenberg *et al.*, 2003, 2005, 2006; Cao *et al.*, 2004; Hu *et al.*, 2007a). Cells such as Schwann cells and fibroblasts have been modified to deliver a variety of neurotrophic factors to the injured spinal cord, including nerve growth factor (NGF) (Nakahara *et al.*, 1996; Tuszynski *et al.*, 1996; Grill *et al.*, 1997b); BDNF (Menei *et al.*, 1998; Liu *et al.*, 1999;); glial cell line-derived neurotrophic factor (GDNF) (Cao *et al.*, 2004) and NT-3 (Grill *et al.*, 1997a). LV efficiently transduces OEGs (Ruitenberg *et al.*, 2002), and after transplantation into contused rat spinal cord OEG that have been modified to express NT-3 promote tissue sparing and regeneration (Ruitenberg *et al.*, 2005). In human SCI it is envisaged that adult cells could be used to generate a variety of precursor cells as vehicles for gene therapy, including glial and neural restricted precursor cells. These cells would be modified ex vivo to secrete appropriate factors prior to transplantation into the injured CNS. The ability to safely regulate (switch on or off) transgene expression in grafted cells is likely to be an essential next step in this process.

Tissue engineering

Progress made in tissue engineering as the result of recent advances in biomaterial design has accelerated the search for new repair strategies for SCI. These generally involve the implantation of acellular, biocompatible scaffolds that provide a terrain for regenerating axons to grow across the site of injury and/or create an environment for endogenous repair. Bioactive molecules can be incorporated into polymer bridges, and scaffolds can also be used in combination with different growth factors or as hybrid structures incorporating cells for tissue replacement.

A variety of biomatrices have been tested in experimental SCI models. For example, collagen by itself or containing astrocytes promotes axonal regeneration into grafts from both proximal and distal ends and ameliorates functional deficits (Joosten *et al.*, 2004; Yoshii *et al.*, 2004). Impregnation of collagen with growth factors increases the regrowth of axons into implants (Houweling *et al.*, 1998) and impregnation of synthetic hydrogel guidance channels with NT-3 improves localized regeneration (Tsai *et al.*, 2006). In humans, functional recovery has been reported in a patient with SCI using collagen implantation and omental transposition (Goldsmith *et al.*, 2005). Fibrin has also be used to bridge SCIs with reported enhancement of neuronal survival, axonal regrowth and function, usually in combination with growth factors or growth-supporting cells such as Schwann cells and OEG (Guest *et al.*, 1997; Plant *et al.*, 2003; Meijs *et al.*, 2004; Taylor *et al.*, 2004). Alginate hydrogel (derived from brown seaweed) has also been tested (Suzuki *et al.*, 2002; Prang *et al.*, 2006) and used as a potential 'carrier' of OEG, Schwann cells and BMSCs (Wu *et al.*, 2001; Novikov *et al.*, 2002; Novikova *et al.*, 2006). Microencapsulation of BDNF-expressing fibroblasts in alginate-poly-L-ornithine promotes growth of regenerating axons and improved locomotory function in SCI rats (Tobias *et al.*, 2005). Implantation of Matrigel™ (which contains mainly laminin, fibronectin, and proteoglycans) does not appear to promote regeneration but in combination with bioactive molecules (Facchiano *et al.*, 2002; Iannotti *et al.*, 2003), or when used as a scaffold for OEG, Schwann cells or BMSCs (Bamber *et al.*, 2001; Pinzon *et al.*, 2001; Plant *et al.*, 2001b; Fouad *et al.*, 2005; Kamada *et al.*, 2005; Xiao *et al.*, 2005), some regrowth is seen. Other biologically derived bridges tested in rodent SCI models include matrices made from fibronectin (King *et al.*, 2003) and agarose (Jain *et al.*, 2006; Stokols and Tuszynski, 2006).

Many types of synthetic polymers, some biodegradable, are also being trialed (Nomura *et al.*, 2006). Synthetic hydrogels include nonbiodegradable poly [N-2-(hydroxypropyl) methacrylamide] (HPMA) (NeuroGel™) (Woerly *et al.*, 1999; Woerly, 2000) and poly(2-hydroxyethyl methacrylate-co-methyl methacrylate) (HEMA) hydrogels (Plant *et al.*, 1995). These consist of crosslinked networks of hydrophilic co-polymers that swell in water and provide a substrate for cell attachment and growth, as well as transport of small molecules (Woerly *et al.*, 1999, 2001a; Loh *et al.*, 2001; Bakshi *et al.*, 2004). It is possible to incorporate peptide or aminosugar sequences into HPMA hydrogels (Plant *et al.*, 1997) and in SCI this increases angiogenesis and axonal growth (Woerly *et al.*, 2001b). Poly (α-hydroxy acids) are synthetic polymers that are biodegradable (Wu and Ding, 2004). In SCI studies, variants of this polymer structure support axonal regeneration with little perturbation of the glial and inflammatory response (Gautier *et al.*, 1998; Patist *et al.*, 2004) although to date with little impact on function. Very recently, oriented self-assembling nanofiber structures have been reported to promote growth of descending and ascending fibers after SCI in mice, with some behavioural improvements (Tysseling-Mattiace *et al.*, 2008).

Pharmacological therapies

Pharmacological compounds that prevent necrosis and apoptosis, reduced glial scarring, improve electrical conduction of surviving axons, and promote the regrowth and remyelination of damaged axons represent emerging therapeutic targets for the management and treatment of SCI (Baptiste and Fehlings, 2006). Several preclinical studies since the early 1990s have reported potential therapeutic benefits of anti-apoptotic and/or neuroprotective approaches in animal models of SCI by targeting secondary injury mechanisms (Fawcett *et al.*, 2007) and a brief summary of some of the more (at present) promising approaches is given below.

Loss of ionic homeostasis due to membrane disruption causes neuronal degeneration and axonal dysfunction after traumatic injury (Stys, 2004). The Na^+ channel blocker Riluzole has been tested in attempts to reduce secondary pathophysiology after SCI (Fehlings and Agrawal, 1995; Schwartz and Fehlings, 2001, 2002; Wang *et al.*, 2004). The combined use of riluzole with methylprednisolone (MPS) may represent a beneficial therapy

(Mu *et al.*, 2000). N-methyl-D-aspartate (NMDA) and non-NMDA receptor antagonists are also potential therapeutic targets for SCI (Wrathall *et al.*, 1997; Hirbec *et al.*, 2001; Baptiste and Fehlings, 2006;). The therapeutic actions of the glucocorticoid MPS are controversial (Hurlbert, 2000; Fehlings, 2001) and have been evaluated in three major clinical trials of acute SCI (Bracken *et al.*, 1992, 1998). Use of the non-selective opioid receptor antagonist, naloxone for treatment of SCI is also controversial (Bracken *et al.*, 1992, 1998).

Rho and Rac are small intracellular GTPase-associated signaling proteins that, in their active forms (GTP-bound), mediate many of the growth-inhibitory effects that occur within the mature CNS (Yiu and He, 2006). Rho-GTP levels are elevated following experimental SCIs in rodents (Dubreuil *et al.*, 2003) and inactivation of Rho or downstream effectors like Rho kinase (ROCK) results in increased regeneration of axons and also reduces cell death at SCI sites (McKerracher and Higuchi, 2006). Animals treated with cell permeant forms of C3 show evidence of increased hindlimb recovery after SCI, and local application of the ROCK inhibitor Y-27632 at the lesion site promotes axonal regeneration and motor recovery in mice (Dergham *et al.*, 2002). In rats, intrathecal application of Y-27632 promotes regeneration of CST fibers and enhances the speed of recovery but not overall BBB outcome scores (Fournier *et al.*, 2003), and oral application over 10 days also did not improve BBB scores (Sung *et al.*, 2003). These studies highlight how differences in experimental SCI methodology, dosing regimes and delivery can lead to different functional outcomes. Note that Cethrin (BA-210), a C3 fusion protein, has been tested in phase I/IIa multinational clinical trials in patients with SCI (Baptiste and Fehlings, 2007). Remarkably, some commonly used nonsteroidal anti-inflammatory drugs (ibuprofen and indomethacin) also inhibit Rho and have recently been shown to stimulate axonal growth in vitro and promote axonal sprouting in vivo after SCI (Fu *et al.*, 2007). Improved locomotor activity was also reported after subcutaneous ibuprofen administration. Anti-Nogo (or IN-1 antigen) monoclonal antibody treatment neutralizes myelin-associated neurite outgrowth inhibition to promote CNS axon regeneration in vivo (Schnell and Schwab, 1990 Bregman *et al.*, 1995) and this approach is being tested in a multicenter clinical SCI trial.

Via a number of intracellular mechanisms, including altering neuronal responsiveness to growth-inhibitory molecules, elevation of cyclic AMP (cAMP) enhances the beneficial effects of neurotrophins on adult CNS neurons (Cui *et al.*, 2003; Lu *et al.*, 2004; Pearse *et al.*, 2004; Hu *et al.*, 2007b; Hannila and Filbin, 2008). Rolipram is a blood-brain-barrier permeable, phosphodiesterase type IV inhibitor that reduces breakdown of cAMP and has various neuroprotective properties including anti-inflammatory actions (Yoshikawa *et al.*, 1999; Beshay *et al.*, 2001; Zhang *et al.*, 2002) and inhibitory effects on excitatory amino acid toxicity (Block *et al.*, 2001). Rolipram used in association with cellular grafts promotes regeneration and enhances functional recovery after SCI (Gao *et al.*, 2004; Nikulina *et al.*, 2004; Pearse *et al.*, 2004).

Minocycline is a second-generation tetracycline derivative antibiotic that counteracts excitotoxicity and also moderates apoptosis and inflammation (Tikka and Koistinaho, 2001; Maier *et al.*, 2007). The drug has diverse neuroprotective effects after experimental SCI, with particular actions on oligodendrocytes (Wells *et al.*, 2003a; Stirling *et al.*, 2004; Yune *et al.*, 2007). Other methods to reduce oligodendrocyte loss (and demyelination) involve targeted blockade of transmembrane 'death' receptors (Fas) which results in improved sparing of axons and better functional outcomes after SCI (Demjen *et al.*, 2004; Ackery *et al.*, 2006; Baptiste and Fehlings, 2007). Thyrotropin-releasing hormone (TRH) and analogs have been found to reduce secondary damage cascades following SCI (Dumont *et al.*, 2001) with improved function after experimental SCI in rats (Hashimoto and Fukuda, 1991). A clinical trial assessed the safety and potential efficacy of TRH in patients and revealed improvements in neurological and sensory function for incomplete, but not complete SCI patients (Pitts *et al.*, 1995), however further follow-up studies appear not to have eventuated.

Cyclosporin-A reduces secondary damage due to lipid peroxidation (Diaz-Ruiz *et al.*, 2000) and may improve outcomes following SCI, however there are strain-specific differences in autoimmune susceptibility responses to this drug that may argue caution in its general use in the clinic (Cui *et al.*, 2007). Finally, reduction of glial scarring can also enhance regenerative growth potential; successful experimental examples include blockade of the receptor tyrosine kinase EphA4 (Goldshmit *et al.*, 2004; Fabes *et al.*, 2007) and digestion of the CSPG matrix within reactive spinal tissue (Fouad *et al.*, 2005; Bradbury and McMahon, 2006; Houlé *et al.*, 2006). In addition, matrix metalloproteinases (MMPs), which degrade extracellular matrix and proteins (Sternlicht and Werb, 2001) are up-regulated after compression SCI in mice (Wells *et al.*, 2003b) and improved functional recovery after SCI has been observed in MMP-12 and MMP-9 null mice, most likely due to a reduction of the lesion-induced permeability of the blood brain barrier and attenuation of inflammatory cell infiltration (Noble *et al.*, 2002). Studies have demonstrated the involvement of MMPs in experimental models of SCI although only a few correlative studies regarding the spatio-temporal distribution of MMPs in human SCI so far exist (Fleming *et al.*, 2006; Buss *et al.*, 2007).

Combinatorial strategies

Despite a wealth of information from animal studies that report enhanced axonal growth after SCI treatments, evidence indicates that experimental approaches that combine therapies to target distinct but complementary neural mechanisms achieve greater axonal growth, remyelination and repair than individual therapies alone (Lu *et al.*, 2004; Pearse *et al.*, 2004; Hu *et al.*, 2007b). It is also widely agreed that a combination of therapeutic strategies will be required to facilitate significant clinical improvements in SCI patients (Baptiste and Fehlings, 2006; Thuret *et al.*, 2006; Rossignol *et al.*, 2007; Lu and Tuszynski, 2008). The relative success of such combinations may influence the suitability of their application for acute vs chronic SCI. These combinations could include cell transplantation with delivery (virally or otherwise?) of neurotrophins and/or pharmaceuticals, tissue engineered scaffolds/polymers, anti-scarring agents and immunosuppressive regimes. Rehabilitation based on extant functional networks below an SCI may also be used in conjunction with these types of therapies, and further strategies may be required to enhance axonal excitability in peripheral nerves that connect below the site of injury (Lim *et al.*, 2007). Clearly, extensive structural and/or functional recovery after SCI, improvement of plasticity and axonal regeneration would need to be demonstrated in the most appropriate animal models before application to the clinical situation (Courtine *et al.*, 2007). It is also possible that proven combinations successful in acute SCI outcomes can only be applied to chronic SCI once the injury has been 'made acute' once more (e.g. via surgical debrading and/or enzymatic descarring).

Clinical trials

In recent years there have been an increasing number of small-scale trials testing experimental therapies in human SCI. Some of these trials (e.g. MPS and naloxone treatments) have been alluded to earlier in this chapter. OEGs are being tested in several SCI repair paradigms. These cells can be expanded from nasal biopsies to provide autologous cells for transplantation. Clinical transplantation of OEGs directly into a lesion has so far been reported to be safe (Féron *et al.*, 2005) and there are reports of functional improvement (Huang *et al.*, 2003; Lima *et al.*, 2006). A limited clinical study involving the injection of autologous hBMSCs into a small number of SCI patients with complete transections reported functional improvement within 3 weeks of administration (Park *et al.*, 2005). Another study used autologous BMC delivered either intra-arterially or intravenously in a Phase I/II clinical trial in patients (n=20) with transverse spinal cord lesions (Syková *et al.*, 2006b). It was reported that grafted cells survived without immunosuppression and functional improvements were seen at 3 and 4 weeks. These studies have been expanded to include a further 20 SCI patients with varying degrees of injury (Syková *et al.*, 2006a). Recently, it has been reported that autologous macrophage transplantation therapy is tolerated in patients with acute SCI (Knoller *et al.*, 2005). A range of pharmacological agents are also being, or are soon to be, tested including drugs designed to enhance the conduction of surviving intact demyelinated axons and antibodies to block Nogo signaling (Baptiste and Fehlings, 2007).

There is widespread agreement, as larger clinical trials are envisaged, that any potential therapies must be shown to be safe and reproducible using objective scientific and medical standards (Blight and Tuszynski, 2006; Thuret *et al.*, 2006). Careful recommendations and guidelines have been drawn up for the future conduct of clinical SCI trials (e.g. Fawcett *et al.*, 2007; Steeves *et al.*, 2007) which encompass clinical trial design, the types of patients to include in a given trial, assessment of spontaneous recovery, and a need for large enough cohorts and sufficient statistical power to allow proper assessment of outcomes after treatment. This is a difficult area, both practically and ethically, but the weight of recent scientific progress in experimental SCI, albeit with many caveats, suggests that these clinical guidelines will be in use, sooner rather than later.

References

Ackery A, Robins S, Fehlings MG (2006) Inhibition of Fas-mediated apoptosis through administration of soluble Fas receptor improves functional outcome and reduces posttraumatic axonal degeneration after acute spinal cord injury. J Neurotrauma 23, 604-616.

Akiyama Y, Radtke C, Kocsis JD (2002) Remyelination of the rat spinal cord by transplantation of identified bone marrow stromal cells. J Neurosci 22, 6623-6630.

Akiyama C, Yuguchi T, Nishio M, Tomishima T, Fujinaka T, Taniguchi M, Nakajima Y, Kohmura E, Yoshimine, T (2004) Src family kinase inhibitor PP1 reduces secondary damage after spinal cord compression in rats. J Neurotrauma 21, 923-931.

Anderson KD, Gunawan A, Steward O (2007) Spinal pathways involved in the control of forelimb motor function in rats. Exp Neurol 206, 318-331.

Ashworth B (1964) Preliminary Trial of Carisoprodol in Multiple Sclerosis. Practitioner 192, 540-542.

Bakshi A, Fisher O, Dagci T, Himes BT, Fischer I, Lowman A (2004) Mechanically engineered hydrogel scaffolds for axonal growth and angiogenesis after transplantation in spinal cord injury. J Neurosurg Spine 1, 322-329.

Bamber NI, Li H, Lu X, Oudega M, Aebischer P, Xu XM (2001) Neurotrophins BDNF and NT-3 promote axonal re-entry into the distal host spinal cord through Schwann cell-seeded mini-channels. Eur J Neurosci 13, 257-268.

Bao F, Chen Y, Dekaban GA, Weaver LC (2004) Early anti-inflammatory treatment reduces lipid peroxidation and protein nitration after spinal cord injury in rats. J Neurochem 88, 1335-1344.

Baptiste DC, Fehlings MG (2006) Pharmacological approaches to repair the injured spinal cord. J Neurotrauma 23, 318-334.

Baptiste DC, Fehlings MG (2007) Update on the treatment of spinal cord injury. Prog Brain Res 161, 217-233.

Barbeau H, Fung J, Leroux A, Ladouceur M (2002) A review of the adaptability and recovery of locomotion after spinal cord injury. Prog Brain Res 137, 9-25.

Bareyre FM (2008) Neuronal repair and replacement in spinal cord injury. J Neurol Sci 265, 63-72.

Basso DM, Beattie MS, Bresnahan JC (1995) A sensitive and reliable locomotor rating scale for open field testing in rats. J Neurotrauma 12, 1-21.

Basso DM, Beattie MS, Bresnahan JC (1996) Graded histological and locomotor outcomes after spinal cord contusion using the NYU weight-drop device versus transection. Exp Neurol 139, 244-256.

Basso DM, Fisher LC, Anderson A.J, Jakeman LB, McTigue DM, Popovich PG (2006) Basso Mouse Scale for locomotion detects differences in recovery after spinal cord injury in five common mouse strains. J Neurotrauma 23, 635-659.

Bauer J, Bradl M, Hickley WF, Forss-Petter S, Breitschopf H, Linington C, Wekerle H, Lassmann H (1998) T-cell apoptosis in inflammatory brain lesions: destruction of T-cells does not depend on antigen recognition. Am J Pathol 153, 715-724.

Baumgartner BJ, Shine HD (1998) Neuroprotection of spinal motoneurons following targeted transduction with an adenoviral vector carrying the gene for glial cell line-derived neurotrophic factor. Exp Neurol 153, 102-112.

Baydin A, Cokluk C, Aydin K (2007) A new minimally invasive experimental spinal cord injury model in rabbits. Min Invas Neurosurg 50, 170-172.

Bélanger M, Drew T, Provencher J, Rossignol S (1996) A comparison of treadmill locomotion in adult cats before and after spinal transection. J Neurophysiol 76, 471-491.

Belci M, Catley M, Husain M, Frankel HL, Davey NJ (2004) Magnetic brain stimulation can improve clinical outcome in incomplete spinal cord injured patients. Spinal Cord 42, 417-419.

Beshay E, Croze F, Prud'homme GJ (2001) The phosphodiesterase inhibitors pentoxifylline and rolipram suppress macrophage activation and nitric oxide production in vitro and in vivo. Clinical Immun 98, 272-279.

Birdsall Abrams M, Josephson A, Dominguez C, Oberg J, Diez M, Spenger C, Olson L, Piehl F, Lidman O. (2007) Recovery from spinal cord injury differs between rat strains in a major histocompatibility complex-independent manner. Eur J Neurosci 26, 1118-1127.

Blakemore WF, Crang AJ (1985) The use of cultured autologous Schwann cells to remyelinate areas of persistent demyelination in the central nervous system. J Neurol Sciences 70, 207-223.

Blakemore WF, Gilson JM, Crang AJ (2000) Transplanted glial cells migrate over a greater distance and remyelinate demyelinated lesions more rapidly than endogenous remyelinating cells. J Neurosci Res 61, 288-294.

Blesch A, Grill RJ, Tuszynski MH (1998) Neurotrophin gene therapy in CNS models of trauma and degeneration. Prog Brain Res 117, 473-484.

Blight AR, Tuszynski MH (2006) Clinical trials in spinal cord injury. J Neurotrauma 23, 586-593.

Blits B, Bunge MB (2006) Direct gene therapy for repair of the spinal cord. J Neurotrauma 23, 508-520.

Blits B, Oudega M, Boer GJ, Bartlett Bunge M, Verhaagen J (2003) Adeno-associated viral vector-mediated neurotrophin gene transfer in the injured adult rat spinal cord improves hind-limb function. Neuroscience 118, 271-281.

Block F, Schmidt W, Nolden-Koch M, Schwarz M (2001) Rolipram reduces excitotoxic neuronal damage. Neuroreport 12, 1507-1511.

Bomstein Y, Marder JB, Vitner K, Smirnov I, Lisaey G, Butovsky O, Fulga V, Yoles E (2003) Features of skin-coincubated macrophages that promote recovery from spinal cord injury. J Neuroimmunol 142, 10-16.

Bortoff GA, Strick PL (1993) Corticospinal terminations in two new-world primates: further evidence that corticomotoneuronal connections provide part of the neural substrate for manual dexterity. J Neurosci 13, 5105-5118.

Bouhy D, Malgrange B, Multon S, Poirrier AL, Scholtes F, Schoenen J, Franzen R (2006) Delayed GM-CSF treatment stimulates axonal regeneration and functional recovery in paraplegic rats via an increased BDNF expression by endogenous macrophages. Faseb J 20, 1239-1241.

Boulis NM, Bhatia V, Brindle TI, Holman HT, Krauss DJ, Blaivas M, Hoff J.T (1999) Adenoviral nerve growth factor and beta-galactosidase transfer to spinal cord: a behavioral and histological analysis. J Neurosurg 90, 99-108.

Bracken MB, Shepard MJ, Collins WF, Jr, Holford TR, Baskin DS, Eisenberg HM, Flamm E, Leo-Summers L, Maroon JC, Marshall LF, *et al.*, (1992) Methylprednisolone or naloxone treatment after acute spinal cord injury: 1-year follow-up data. Results of the second National Acute Spinal Cord Injury Study. J Neurosurgery 76, 23-31.

Bracken MB, Shepard MJ, Holford TR, Leo-Summers L, Aldrich EF, Fazl M, Fehlings, MG, Herr DL, Hitchon PW, Marshall LF, Nockels RP, Pascale V, Perot PL Jr, Piepmeier J, Sonntag VK, Wagner F, Wilberger JE, Winn HR, Young W (1998) Methylprednisolone or tirilazad mesylate administration after acute spinal cord injury: 1-year follow up. Results of the third National Acute Spinal Cord Injury randomized controlled trial. J Neurosurgery 89, 699-706.

Bradbury EJ, McMahon SB (2006) Spinal cord repair strategies: why do they work? Nat Rev Neurosci 7, 644-653.

Bramlett HM, Dietrich WD (2007) Progressive damage after brain and spinal cord injury: pathomechanisms and treatment strategies. Prog Brain Res 161, 125-141.

Bregman BS, Kunkel-Bagden E, Schnell L, Dai HN, Gao D, Schwab ME (1995) Recovery from spinal cord injury mediated by antibodies to neurite growth inhibitors. Nature 378, 498-501.

Bregman BS, Coumans JV, Dai HN, Kuhn PL, Lynskey J, McAtee M, Sandhu F (2002) Transplants and neurotrophic factors increase regeneration and recovery of function after spinal cord injury. Prog Brain Res 137, 257-273.

Bresnahan JC, Beattie MS, Todd FD 3rd, Noyes DH (1987) A behavioral and anatomical analysis of spinal cord injury produced by a feedback-controlled impaction device. Exp Neurol 95, 548-570.

Bresnahan JC, Beattie MS, Stokes BT, Conway KM (1991) Three-dimensional computer-assisted analysis of graded contusion lesions in the spinal cord of the rat. J Neurotrauma 8, 91-101.

Brown KM, Wolfe BB, Wrathall JR (2005) Rapid functional recovery after spinal cord injury in young rats. J Neurotrauma 22, 559-574.

Bunge MB (2002) Bridging the transected or contused adult rat spinal cord with Schwann cell and olfactory ensheathing glia transplants. Prog Brain Res 137, 275-282.

Bunge MB, Holets VR, Bates ML, Clarke TS, Watson BD (1994) Characterization of photochemically induced spinal cord injury in the rat by light and electron microscopy. Exp Neurol 127, 76-93.

Bunge RP, Puckett WR, Becerra JL, Marcillo A, Quencer RM (1993) Observations on the pathology of human spinal cord injury. A review and classification of 22 new cases with details from a case of chronic cord compression with extensive focal demyelination. Adv Neurol 59, 75-89.

Buss A, Pech K, Kakulas BA, Martin D, Schoenen J, Noth J, Brook, GA (2007) Matrix metalloproteinases their inhibitors in human traumatic spinal cord injury. BMC Neurology 7, 17.

Cai LL, Courtine G, Fong AJ, Burdick J.W, Roy RR, Edgerton VR (2006) Plasticity of functional connectivity in the adult spinal cord. Philos Trans R Soc Lond B Biol Sci 361, 1635-1646.

Cameron T, Prado R, Watson BD, Gonzalez-Carvajal M, Holets VR (1990) Photochemically induced cystic lesion in the rat spinal cord. I. Behavioral morphological analysis. Exp Neurol 109, 214-223.

Cao L, Liu L, Chen ZY, Wang LM, Ye JL, Qiu HY, Lu CL, He C (2004) Olfactory ensheathing cells genetically modified to secrete GDNF to promote spinal cord repair. Brain 127, 535-549.

Carlson GD, Gorden CD, Oliff HS, Pillai JJ, LaManna JC (2003) Sustained spinal cord compression: part I: time-dependent effect on long-term pathophysiology. J Bone Joint Surgery America 85-A, 86-94.

Carlson SL, Parrish ME, Springer JE, Doty K, Dossett L (1998) Acute inflammatory response in spinal cord following impact injury. Exp Neurol 151, 77-88.

Casha S, Yu WR, Fehlings MG (2001) Oligodendroglial apoptosis occurs along degenerating axons and is associated with FAS and p75 expression following spinal cord injury in the rat. Neuroscience 103, 203-218.

Cha J, Heng C, Reinkensmeyer DJ, Roy RR, Edgerton VR, De Leon RD (2007) Locomotor ability in spinal rats is dependent on the amount of activity imposed on the hindlimbs during treadmill training. J Neurotrauma 24, 1000-1012.

Chang, HT (2007) Subacute human spinal cord contusion: few lymphocytes and many macrophages. Spinal Cord 45, 174-182.

Chau C, Barbeau H, Rossignol S (1998) Effects of intrathecal alpha1- and alpha2-noradrenergic agonists and norepinephrine on locomotion in chronic spinal cats. J Neurophysiol 79, 2941-2963.

Chen A, Xu XM, Kleitman N, Bunge MB (1996) Methylprednisolone administration improves axonal regeneration into Schwann cell grafts in transected adult rat thoracic spinal cord. Exp Neurol 138, 261-276.

Cheng H, Liao KK, Liao SF, Chuang TY, Shih YH (2004) Spinal cord repair with acidic fibroblast growth factor as a treatment for a patient with chronic paraplegia. Spine 29, E284-288.

Coleman WP, Geisler FH (2004) Injury severity as primary predictor of outcome in acute spinal cord injury: retrospective results from a large multicenter clinical trial. Spine Journal 4, 373-378.

Conti A, Miscusi M, Cardali S, Germanò A, Suzuki H, Cuzzocrea S, Tomasello F (2007) Nitric oxide in the injured spinal cord: synthases cross-talk, oxidative stress and inflammation. Brain Res Rev 54, 205-218.

Courtine G, Roy RR, Raven J, Hodgson J, McKay H, Yang H, Zhong H, Tuszynski MH, Edgerton V.R. (2005) Performance of locomotion and foot grasping following a unilateral thoracic corticospinal tract lesion in monkeys (Macaca mulatta). Brain 128, 2338-2358.

Courtine G, Bunge MB, Fawcett JW, Grossman RG, Kaas JH, Lemon R, Maier I, Martin J, Nudo RJ, Ramón-Cueto A, Rouiller EM, Schnell L, Wannier T, Schwab ME, Edgerton VR (2007) Can experiments in nonhuman primates expedite the translation of treatments for spinal cord injury in humans? Nature Medicine 13, 561-566.

Coutts M, Keirstead HS (2008) Stem cells for the treatment of spinal cord injury. Exp Neurol, 209, 368-77.

Crutcher KA, Gendelman HE, Kipnis J, Perez-Polo JR, Perry VH, Popovich PG, Weaver LC (2006) Debate: is increasing neuroinflammation beneficial for neural repair? J Neuroimm Pharmacol 1, 195-211.

Cui Q, Pollett MA, Symons NA, Plant GW, Harvey AR (2003) A new approach to CNS repair using chimeric peripheral nerve grafts. J Neurotrauma 20, 17-31.

Cui Q, Hodgetts SI, Hu Y, Luo JM, Harvey AR (2007) Strain-specific differences in the effects of cyclosporin A and FK506 on the survival and regeneration of axotomized retinal ganglion cells in adult rats. Neuroscience 146, 986-999.

de Leon RD, Hodgson JA, Roy RR, Edgerton VR (1998) Locomotor capacity attributable to step training versus spontaneous recovery after spinalization in adult cats. J Neurophysiol 79, 1329-1340.

de Leon RD, Tamaki H, Hodgson JA, Roy RR, Edgerton VR (1999) Hindlimb locomotor and postural training modulates glycinergic inhibition in the spinal cord of the adult spinal cat. J Neurophysiol 82, 359-369.

Demjen D, Klussmann S, Kleber S, Zuliani C, Stieltjes B, Metzger C, Hirt UA, Walczak H, Falk W, Essig M, Edler L, Krammer PH, Martin-Villalba A (2004) Neutralization of CD95 ligand promotes regeneration and functional recovery after spinal cord injury. Nature Medicine 10, 389-395.

Dergham P, Ellezam B, Essagian C, Avedissian H, Lubell WD, McKerracher L. (2002) Rho signaling pathway targeted to promote spinal cord repair. J Neuroscience 22, 6570-6577.

Diaz-Ruiz A, Rios C, Duarte I, Correa D, Guizar-Sahagun G, Grijalva I, Madrazo I, Ibarra A (2000) Lipid peroxidation inhibition in spinal cord injury: cyclosporin-A vs methylprednisolone. Neuroreport 11, 1765-1767.

Dijkhuizen PA, Hermens WT, Teunis MA, Verhaagen J (1997) Adenoviral vector-directed expression of neurotrophin-3 in rat dorsal root ganglion explants results in a robust neurite outgrowth response. J Neurobiol 33, 172-184.

Dimitrijevic MR, Gerasimenko Y, Pinter MM (1998) Evidence for a spinal central pattern generator in humans. Annals of the NY Acad Sciences 860, 360-376.

Ditunno JF Jr, Young W, Donovan WH, Creasey G (1994) The international standards booklet for neurological and functional classification of spinal cord injury. American Spinal Injury Association. Paraplegia 32, 70-80.

Ditunno JF Jr, Ditunno PL, Graziani V, Scivoletto G, Bernardi M, Castellano V, Marchetti M, Barbeau H, Frankel HL, D'Andrea Greve JM, Ko HY, Marshall R, Nance, P (2000) Walking index for spinal cord injury (WISCI): an international multicenter validity and reliability study. Spinal Cord 38, 234-243.

Donnelly DJ, Popovich PG (2008) Inflammation and its role in neuroprotection, axonal regeneration and functional recovery after spinal cord injury. Exp Neurol 209, 378-388.

Dougherty KD, Dreyfus CF, Black IB (2000) Brain-derived neurotrophic factor in astrocytes, oligodendrocytes, and microglia/macrophages after spinal cord injury. Neurobiol Dis 7, 574-585.

Dubreuil CI, Winton MJ, McKerracher L. (2003) Rho activation patterns after spinal cord injury and the role of activated Rho in apoptosis in the central nervous system. J Cell Biol 162, 233-243.

Dumont RJ, Verma S, Okonkwo DO, Hurlbert RJ, Boulos PT, Ellegala DB, Dumont AS (2001) Acute spinal cord injury, part II: contemporary pharmacotherapy. Clin Neuropharmacol 24, 265-279.

Dusart I, Schwab ME (1994) Secondary cell death and the inflammatory reaction after dorsal hemisection of the rat spinal cord. Eur J Neurosci 6, 712-724.

Edgerton VR, Tillakaratne NJ, Bigbee, AJ, de Leon RD, Roy RR (2004) Plasticity of the spinal neural circuitry after injury. Ann Rev Neurosci 27, 145-167.

Emery E, Aldana P, Bunge MB, Puckett W, Srinivasan A, Keane RW, Bethea J, Levi AD (1998) Apoptosis after traumatic human spinal cord injury. J Neurosurg 89, 911-920.

Fabes J, Anderson P, Brennan C, Bolsover S. (2007) Regeneration-enhancing effects of EphA4 blocking peptide following corticospinal tract injury in adult rat spinal cord. Eur J Neurosci 26, 2496-2505.

Facchiano F, Fernandez E, Mancarella S, Maira G, Miscusi M, D'Arcangelo D, Cimino-Reale G, Falchetti M.L, Capogrossi MC, Pallini R. (2002) Promotion of regeneration of corticospinal tract axons in rats with recombinant vascular endothelial growth factor alone and combined with adenovirus coding for this factor. J Neurosurg 97, 161-168.

Fawcett JW, Curt A, Steeves JD, Coleman WP, Tuszynski MH, Lammertse D, Bartlett PF, Blight AR, Dietz V, Ditunno J, Dobkin BH, Havton LA, Ellaway PH, Fehlings MG, Privat A, Grossman R, Guest JD, Kleitman N, Nakamura M, Gaviria M, Short D (2007) Guidelines for the conduct of clinical trials for spinal cord injury as developed by the ICCP panel: spontaneous recovery after spinal cord injury and statistical power needed for therapeutic clinical trials. Spinal Cord 45, 190-205.

Fehlings MG (2001) Editorial: recommendations regarding the use of methylprednisolone in acute spinal cord injury: making sense out of the controversy. Spine 26, S56-57.

Fehlings MG, Agrawal S (1995) Role of sodium in the pathophysiology of secondary spinal cord injury. Spine 20, 2187-2191.

Fehlings MG, Tator CH (1992) The effect of direct current field polarity on recovery after acute experimental spinal cord injury. Brain Res 579, 32-42.

Fehlings MG, Tator CH (1995) The relationships among the severity of spinal cord injury, residual neurological function, axon counts, and counts of retrogradely labeled neurons after experimental spinal cord injury. Exp Neurol 132, 220-228.

Fehlings MG, Tator CH, Linden RD (1989) The relationships among the severity of spinal cord injury, motor and somatosensory evoked potentials and spinal cord blood flow. Electroencephalography in Clin Neurophysiol 74, 241-259.

Féron F, Perry C, Cochrane J, Licina P, Nowitzke A, Urquhart S, Geraghty T, Mackay-Sim A (2005) Autologous olfactory ensheathing cell transplantation in human spinal cord injury. Brain 128, 2951-2960.

Finegold AA, Mannes AJ, Iadarola MJ (1999) A paracrine paradigm for in vivo gene therapy in the central nervous system: treatment of chronic pain. Hum Gene Ther 10, 1251-1257.

Fischer FR, Peduzzi JD (2007) Functional recovery in rats with chronic spinal cord injuries after exposure to an enriched environment. J Spinal Cord Med 30, 147-155.

Fitch MT, Silver J (2008) CNS injury, glial scars, and inflammation: Inhibitory extracellular matrices and regeneration failure. Exp Neurol 209, 294-301.

Fitch MT, Doller C, Combs CK, Landreth GE, Silver J (1999) Cellular and molecular mechanisms of glial scarring and progressive cavitation: in vivo and in vitro analysis of inflammation-induced secondary injury after CNS trauma. J Neurosci 19, 8182-8198.

Fleming JC, Norenberg MD, Ramsay DA, Dekaban GA, Marcillo AE, Saenz AD, Pasquale-Styles M, Dietrich WD, Weaver LC (2006) The cellular inflammatory response in human spinal cords after injury. Brain 129, 3249-3269.

Fong AJ, Cai LL, Otoshi CK, Reinkensmeyer DJ, Burdick JW, Roy RR, Edgerton VR (2005) Spinal cord-transected mice learn to step in response to quipazine treatment and robotic training. J Neurosci 25, 11738-11747.

Fouad K, Schnell L, Bunge MB, Schwab ME, Liebscher T, Pearse DD (2005) Combining Schwann cell bridges and olfactory-ensheathing glia grafts with chondroitinase promotes locomotor recovery after complete transection of the spinal cord. J Neurosci 25, 1169-1178.

Fournier AE, Takizawa BT, Strittmatter SM (2003) Rho kinase inhibition enhances axonal regeneration in the injured CNS. J Neurosci 23, 1416-1423.

Fry EJ, Stolp HB, Lane MA, Dziegielewska KM, Saunders NR (2003) Regeneration of supraspinal axons after complete transection of the thoracic spinal cord in neonatal opossums (Monodelphis domestica). J Comp Neurol 466, 422-444.

Fu Q, Hue J, Li S. (2007) Nonsteroidal anti-inflammatory drugs promote axon regeneration via RhoA inhibition. J Neurosci 27, 4154-4164.

Gao Y, Deng K, Hou J, Bryson JB, Barco A, Nikulina E, Spencer T, Mellado W, Kandel ER, Filbin MT (2004) Activated CREB is sufficient to overcome inhibitors in myelin and promote spinal axon regeneration in vivo. Neuron 44, 609-621.

Garciá-Aliás G, Valero-Cabré A, López-Vales R, Forés J, Verdú E, Navarro X (2006) Differential motor and electrophysiological outcome in rats with mid-thoracic or high lumbar incomplete spinal cord injuries. Brain Research 1108, 195-204.

Gautier SE, Oudega M, Fragoso M, Chapon P, Plant GW, Bunge MB, Parel JM (1998) Poly(alpha-hydroxyacids) for application in the spinal cord: resorbability and biocompatibility with adult rat Schwann cells and spinal cord. J Biomed Mat Res 42, 642-654.

Geisler FH, Coleman WP, Grieco G, Poonian D (2001a) Measurements and recovery patterns in a multicenter study of acute spinal cord injury. Spine 26, S68-86.

Geisler FH, Coleman WP, Grieco G, Poonian D (2001b) The Sygen multicenter acute spinal cord injury study. Spine 26, S87-98.

Giulian D, Robertson C (1990) Inhibition of mononuclear phagocytes reduces ischemic injury in the spinal cord. Ann Neurol 27, 33-42.

Giulian D, Corpuz M, Chapman S, Mansouri M, Robertson C (1993) Reactive mononuclear phagocytes release neurotoxins after ischemic and traumatic injury to the central nervous system. J Neurosci Res 36, 681-693.

Goldshmit Y, Galea MP, Wise G, Bartlett PF, Turnley AM (2004) Axonal regeneration and lack of astrocytic gliosis in EphA4-deficient mice. J Neurosci 24, 10064-10073.

Goldsmith HS, Fonseca A Jr, Porter J (2005) Spinal cord separation: MRI evidence of healing after omentum-collagen reconstruction. Neurol Res 27, 115-123.

Gómez-Pinilla, F, Ying, Z, Opazo, P, Roy, R.R, Edgerton, V.R. (2001) Differential regulation by exercise of BDNF and NT-3 in rat spinal cord and skeletal muscle. Eur J Neurosci 13, 1078-1084.

Grill RJ, Murai K, Blesch A, Gage FH, Tuszynski MH (1997a) Cellular delivery of neurotrophin-3 promotes corticospinal axonal growth and partial functional recovery after spinal cord injury. J Neurosci 17, 5560-5572.

Grill RJ, Blesch A, Tuszynski MH (1997b) Robust growth of chronically injured spinal cord axons induced by grafts of genetically modified NGF-secreting cells. Exp Neurol 148, 444-452.

Gris D, Marsh DR, Oatway MA, Chen Y, Hamilton EF, Dekaban GA, Weaver LC (2004) Transient blockade of the CD11d/CD18 integrin reduces secondary damage after spinal cord injury, improving sensory, autonomic, and motor function. J Neurosci 24, 4043-4051.

Gruner JA (1992) A monitored contusion model of spinal cord injury in the rat. J Neurotrauma 9, 123-126; discussion 126-128.

Guest JD, Hesse D, Schnell L, Schwab ME, Bunge MB, Bunge RP (1997) Influence of IN-1 antibody and acidic FGF-fibrin glue on the response of injured corticospinal tract axons to human Schwann cell grafts. J Neurosci Res 50, 888-905.

Guha A, Tator CH, Endrenyi L, Piper I (1987) Decompression of the spinal cord improves recovery after acute experimental spinal cord compression injury. Paraplegia 25, 324-339.

Guízar-Sahagún G, Grijalva I, Madrazo I, Franco-Bourland R, Salgado H, Ibarra A, Oliva E, Zepeda A (1994) Development of post-traumatic cysts in the spinal cord of rats-subjected to severe spinal cord contusion. Surg Neurol 41, 241-249.

Hagg T, Oudega M (2006) Degenerative and spontaneous regenerative processes after spinal cord injury. J Neurotrauma 23, 263-280.

Hall KM, Cohen ME, Wright J, Call M, Werner P (1999) Characteristics of the Functional Independence Measure in traumatic spinal cord injury. Arch Phys Med Rehab 80, 1471-1476.

Hamers FP, Lankhorst AJ, van Laar TJ, Veldhuis WB, Gispen WH (2001) Automated quantitative gait analysis during overground locomotion in the rat: its application to spinal cord contusion and transection injuries. J Neurotrauma 18, 187-201.

Hannila SS, Filbin MT (2008) The role of cyclic AMP signaling in promoting axonal regeneration after spinal cord injury. Exp Neurol 209, 321-332.

Harvey AR, Plant GW (2006) Olfactory ensheathing glia and spinal cord injury: basic mechanisms to transplantation. Future Neurol 1, 453-463.

Hashimoto T, Fukuda N (1991) Effect of thyrotropin-releasing hormone on the neurologic impairment in rats with spinal cord injury: treatment starting 24 h and 7 days after injury. Eur J Pharmacol 203, 25-32.

Hauben E, Butovsky O, Nevo U, Yoles E, Moalem G, Agranov E, Mor F, Leibowitz-Amit R, Pevsner E, Akselrod S, Neeman M, Cohen IR, Schwartz M. (2000) Passive or active immunization with myelin basic protein promotes recovery from spinal cord contusion. J Neurosci 20, 6421-6430.

Hauben E, Agranov E, Gothilf A, Nevo U, Cohen A, Smirnov I, Steinman L, Schwartz M (2001) Posttraumatic therapeutic vaccination with modified myelin self-antigen prevents complete paralysis while avoiding autoimmune disease. J Clin Invest 108, 591-599.

Hauben E, Gothilf A, Cohen A, Butovsky O, Nevo U, Smirnov I, Yoles E, Akselrod S, Schwartz M (2003) Vaccination with dendritic cells pulsed with peptides of myelin basic protein promotes functional recovery from spinal cord injury. J Neurosci 23, 8808-8819.

Hausmann ON (2003) Post-traumatic inflammation following spinal cord injury. Spinal Cord 41, 369-378.

Hayashi M, Ueyama T, Nemoto K, Tamaki T, Senba E. (2000) Sequential mRNA expression for immediate early genes, cytokines, and neurotrophins in spinal cord injury. J Neurotrauma 17, 203-218.

Heffner RS, Masterton RB (1983) The role of the corticospinal tract in the evolution of human digital dexterity. Brain Behav Evol 23, 165-183.

Hendriks WT, Eggers R, Verhaagen J, Boer GJ (2007) Gene transfer to the spinal cord neural scar with lentiviral vectors: predominant transgene expression in astrocytes but not in meningeal cells. J Neurosci Res 85, 3041-3052.

Hirbec H, Gaviria M, Vignon J (2001) Gacyclidine: a new neuroprotective agent acting at the N-methyl-D-aspartate receptor. CNS Drug Review 7, 172-198.

Honmou O, Felts PA, Waxman SG, Kocsis JD (1996) Restoration of normal conduction properties in demyelinated spinal cord axons in the adult rat by transplantation of exogenous Schwann cells. J Neurosci 16, 3199-3208.

Houlé JD, Tom VJ, Mayes D, Wagoner G, Phillips N, Silver J (2006) Combining an autologous peripheral nervous system 'bridge' and matrix modification by chondroitinase allows robust, functional regeneration beyond a hemisection lesion of the adult rat spinal cord. J Neurosci 26, 7405-7415.

Houweling DA, Lankhorst AJ, Gispen WH, Bar PR, Joosten EA (1998) Collagen containing neurotrophin-3 (NT-3) attracts regrowing injured corticospinal axons in the adult rat spinal cord and promotes partial functional recovery. Exp Neurol 153, 49-59.

Hu Y, Arulpragasam A, Plant GW, Hendriks WT, Cui Q, Harvey AR (2007a) The importance of transgene and cell type on the regeneration of adult retinal ganglion cell axons within reconstituted bridging grafts. Exp Neurol 207, 314-328.

Hu, Y, Cui, Q, Harvey, A.R. (2007b) Interactive effects of C3, cyclic AMP and ciliary neurotrophic factor on adult retinal ganglion cell survival and axonal regeneration. Mol Cell Neurosci 34, 88-98.

Huang H, Chen L, Wang H, Xiu B, Li B, Wang R, Zhang J, Zhang F, Gu Z, Li Y, Song Y, Hao W, Pang S, Sun J (2003) Influence of patients' age on functional recovery after transplantation of olfactory ensheathing cells into injured spinal cord injury. Chin Med J (Engl) 116, 1488-1491.

Hurlbert RJ (2000) Methylprednisolone for acute spinal cord injury: an inappropriate standard of care. J Neurosurg 93, 1-7.

Iannotti C, Li H, Yan P, Lu X, Wirthlin L, Xu XM (2003) Glial cell line-derived neurotrophic factor-enriched bridging transplants promote propriospinal axonal regeneration and enhance myelination after spinal cord injury. Exp Neurol 183, 379-393.

Inoue M, Honmou O, Oka S, Houkin K, Hashi K, Kocsis JD (2003) Comparative analysis of remyelinating potential of focal and intravenous administration of autologous bone marrow cells into the rat demyelinated spinal cord. Glia 44, 111-118.

Itzkovich M, Tripolski M, Zeilig G, Ring H, Rosentul N, Ronen J, Spasser R, Gepstein R, Catz A (2002) Rasch analysis of the Catz-Itzkovich spinal cord independence measure. Spinal Cord 40, 396-407.

Jain A, Kim YT, McKeon RJ, Bellamkonda RV (2006) In situ gelling hydrogels for conformal repair of spinal cord defects, and local delivery of BDNF after spinal cord injury. Biomat 27, 497-504.

Jakeman LB, Guan Z, Wei P, Ponnappan R, Dzwonczyk R, Popovich PG, Stokes BT (2000) Traumatic spinal cord injury produced by controlled contusion in mouse. J Neurotrauma 17, 299-319.

Jones TB, Basso DM, Sodhi A, Pan JZ, Hart RP, MacCallum RC, Lee S, Whitacre CC, Popovich PG (2002) Pathological CNS autoimmune disease triggered by traumatic spinal cord injury: implications for autoimmune vaccine therapy. J Neurosci 22, 2690-2700.

Jones TB, Hart RP, Popovich PG (2005) Molecular control of physiological and pathological T-cell recruitment after mouse spinal cord injury. J Neurosci 25, 6576-6583.

Joosten EA, Veldhuis WB, Hamers FP (2004) Collagen containing neonatal astrocytes stimulates regrowth of injured fibers and promotes modest locomotor recovery after spinal cord injury. J Neurosci Res 77, 127-142.

Kakulas, B.A. (1999) The applied neuropathology of human spinal cord injury. Spinal Cord 37, 79-88.

Kamada T, Koda M, Dezawa M, Yoshinaga K, Hashimoto M, Koshizuka S, Nishio Y, Moriya H, Yamazaki M (2005) Transplantation of bone marrow stromal cell-derived Schwann cells promotes axonal regeneration and functional recovery after complete transection of adult rat spinal cord. J Neuropath Exp Neurol 64, 37-45.

Karimi-Abdolrezaee S, Eftekharpour E, Wang J, Morshead CM, Fehlings MG (2006) Delayed transplantation of adult neural precursor cells promotes remyelination and functional neurological recovery after spinal cord injury. J Neurosci 26, 3377-3389.

Keane RW, Davis AR, Dietrich WD (2006) Inflammatory and apoptotic signaling after spinal cord injury. J Neurotrauma 23, 335-344.

Kerschensteiner M, Schwab ME, Lichtman JW, Misgeld T (2005) In vivo imaging of axonal degeneration and regeneration in the injured spinal cord. Nature Medicine 11, 572-577.

Kigerl, K.A, McGaughy, V.M, Popovich PG (2006) Comparative analysis of lesion development and intraspinal inflammation in four strains of mice following spinal contusion injury. J Comp Neurol 494, 578-594.

Kil K, Zang YC, Yang D, Markowski J, Fuoco GS, Vendetti GC, Rivera VM, Zhang JZ (1999) T cell responses to myelin basic protein in patients with spinal cord injury and multiple sclerosis. J Neuroimmunol 98, 201-207.

Kim BG, Dai HN, McAtee M, Vicini S, Bregman BS (2006) Remodeling of synaptic structures in the motor cortex following spinal cord injury. Exp Neurol 198, 401-415.

Kim D, Murray M, Simansky KJ (2001) The serotonergic 5-HT(2C) agonist m-chlorophenylpiperazine increases weight-supported locomotion without development of tolerance in rats with spinal transections. Exp Neurol 169, 496-500.

King VR, Henseler M, Brown RA, Priestley JV (2003) Mats made from fibronectin support oriented growth of axons in the damaged spinal cord of the adult rat. Exp Neurol 182, 383-398.

Knoller N, Auerbach G, Fulga V, Zelig G, Attias J, Bakimer R, Marder JB, Yoles E, Belkin M, Schwartz M, Hadani M (2005) Clinical experience using incubated autologous macrophages as a treatment for complete spinal cord injury: phase I study results. J Neurosurg Spine 3, 173-181.

Ko HY, Park JH, Shin YB, Baek SY (2004) Gross quantitative measurements of spinal cord segements in humans. Spinal Cord 42, 35-40.

Kocsis JD, Akiyama Y, Radtke C (2004) Neural precursors as a cell source to repair the demyelinated spinal cord. J Neurotrauma 21, 441-449.

Koda M, Kamada T, Hashimoto M, Murakami M, Shirasawa H, Sakao S, Ino H, Yoshinaga K, Koshizuka S, Moriya H, Yamazaki M (2007) Adenovirus vector-mediated ex vivo gene transfer of brain-derived neurotrophic factor to bone marrow stromal cells promotes axonal regeneration after transplantation in completely transected adult rat spinal cord. Eur Spine J 16, 2206-2214.

Kohama I, Lankford KL, Preiningerova J, White FA, Vollmer TL, Kocsis JD (2001) Transplantation of cryopreserved adult human Schwann cells enhances axonal conduction in demyelinated spinal cord. J Neurosci 21, 944-950.

Kuypers HG (1981) Anatomy of descending pathways. In: *Handbook of Sensory Physiology: the Nervous System.* Volume 2, Part 1. Brookhart JM, Mountcastle VB (eds). Williams Wilkins, USA.

Kwon BK, Tetzlaff W, Grauer JN, Beiner J, Vaccaro AR (2004) Pathophysiology and pharmacologic treatment of acute spinal cord injury. Spine Journal 4, 451-464.

Kwon BK, Liu J, Lam C, Plunet W, Oschipok LW, Hauswirth W, Di Polo A, Blesch A, Tetzlaff W (2007) Brain-derived neurotrophic factor gene transfer with adeno-associated viral and lentiviral vectors prevents rubrospinal neuronal atrophy and stimulates regeneration-associated gene expression after acute cervical spinal cord injury. Spine 32, 1164-1173.

Lankhorst AJ, ter Laak MP, van Laar TJ, van Meeteren NL, de Groot JC, Schrama LH, Hamers FP, Gispen WH (2001) Effects of enriched housing on functional recovery after spinal cord contusive injury in the adult rat. J Neurotrauma 18, 203-215.

Leblond H, L'Esperance M, Orsal D, Rossignol S (2003) Treadmill locomotion in the intact and spinal mouse. J Neurosci 23, 11411-11419.

Lemon RN, Griffiths J (2005) Comparing the function of the corticospinal system in different species: organizational differences for motor specialization? Muscle Nerve 32, 261-279.

Lemon RN, Kirkwood PA, Maier MA, Nakajima K, Nathan P (2004) Direct and indirect pathways for corticospinal control of upper limb motoneurons in the primate. Prog Brain Res 143, 263-279.

Lim JH, Byeon YE, Ryu HH, Jeong Y.H Lee, YW, Kim WH, Kang KS, Kweon OK (2007) Transplantation of canine umbilical cord blood-derived mesenchymal stem cells in experimentally induced spinal cord injured dogs. J Vet Sci 8, 275-282.

Lima C, Pratas-Vital J, Escada P, Hasse-Ferreira A, Capucho C, Peduzzi JD (2006) Olfactory mucosa autografts in human spinal cord injury: a pilot clinical study. J Spinal Cord Med 29, 191-203; discussion 204-196.

Limaye N, Belobrajdic KA, Wandstrat AE, Bonhomme F, Edwards SV, Wakeland EK (2008) Prevalence and evolutionary origins of autoimmune susceptibility alleles in natural mouse populations. Genes Immunol 9, 61-68.

Lin CS, Macefield VG, Elam M, Wallin BG, Engel S, Kiernan MC (2007) Axonal changes in spinal cord injured patients distal to the site of injury. Brain 130, 985-994.

Liu XZ, Xu XM, Hu R, Du C, Zhang SX, McDonald J.W, Dong HX, Wu YJ, Fan GS, Jacquin, MF, Hsu CY, Choi DW (1997) Neuronal and glial apoptosis after traumatic spinal cord injury. J Neurosci 17, 5395-5406.

Liu Y, Himes BT, Moul J, Huang W, Chow SY, Tessler A, Fischer I (1997) Application of recombinant adenovirus for in vivo gene delivery to spinal cord. Brain Res 768, 19-29.

Liu Y, Kim D, Himes BT, Chow SY, Schallert T, Murray M, Tessler A, Fischer I (1999) Transplants of fibroblasts genetically modified to express BDNF promote regeneration of adult rat rubrospinal axons and recovery of forelimb function. J Neurosci 19, 4370-4387.

Loh NK, Woerly S, Bunt SM, Wilton SD, Harvey AR (2001) The regrowth of axons within tissue defects in the CNS is promoted by implanted hydrogel matrices that contain BDNF and CNTF producing fibroblasts. Exp Neurol 170, 72-84.

López-Vales R, Forés J, Verdú E, Navarro X. (2006) Acute and delayed transplantation of olfactory ensheathing cells promote partial recovery after complete transection of the spinal cord. Neurobiol Dis 21, 57-68.

Lou J, Lenke LG, Xu F, O'Brien M (1998) In vivo Bcl-2 oncogene neuronal expression in the rat spinal cord. Spine 23, 517-523.

Lu J, Féron F, Ho SM, Mackay-Sim A, Waite PM (2001) Transplantation of nasal olfactory tissue promotes partial recovery in paraplegic adult rats. Brain Res 889, 344-357.

Lu P, Tuszynski MH (2008) Growth factors and combinatorial therapies for CNS regeneration. Exp Neurol 209, 313-320.

Lu P, Yang H, Jones L.L, Filbin MT, Tuszynski MH (2004) Combinatorial therapy with neurotrophins and cAMP promotes axonal regeneration beyond sites of spinal cord injury. J Neurosci 24, 6402-6409.

Luo J, Shi R. (2007) Polyethylene glycol inhibits apoptotic cell death following traumatic spinal cord injury. Brain Res 1155, 10-16.

Mabon PJ, Weaver LC, Dekaban GA (2000) Inhibition of monocyte/macrophage migration to a spinal cord injury site by an antibody to the integrin alphaD: a potential new anti-inflammatory treatment. Exp Neurol 166, 52-64.

Maier IC, Schwab ME (2006) Sprouting, regeneration and circuit formation in the injured spinal cord: factors and activity. Philos Trans R Soc Lond B Biol Sci 361, 1611-1634.

Maier K, Merkler D, Gerber J, Taheri N, Kuhnert AV, Williams SK, Neusch C, Bähr M, Diem,R (2007) Multiple neuroprotective mechanisms of minocycline in autoimmune CNS inflammation. Neurobiol Dis 25, 514-525.

Martin D, Schoenen J, Delrée P, Gilson V, Rogister B, Leprince P, Stevenaert A, Moonen G (1992) Experimental acute traumatic injury of the adult rat spinal cord by a subdural inflatable balloon: methodology, behavioral analysis, and histopathology. J Neurosci Res 32, 539-550.

Martin R, McFarland HF, McFarlin DE (1992) Immunological aspects of demyelinating diseases. Ann Rev Immunol 10, 15-187.

McDonald JW, Liu XZ, Qu Y, Liu S, Mickey SK, Turetsky D, Gottlieb DI, Choi DW (1999) Transplanted embryonic stem cells survive, differentiate and promote recovery in injured rat spinal cord. Nature Medicine 5, 1410-1412.

McKerracher L, Higuchi H (2006) Targeting Rho to stimulate repair after spinal cord injury. J Neurotrauma 23, 309-317.

Meijs MF, Timmers L, Pearse DD, Tresco PA, Bates ML, Joosten EA, Bunge MB, Oudega M (2004) Basic fibroblast growth factor promotes neuronal survival but not behavioral recovery in the transected and Schwann cell implanted rat thoracic spinal cord. J Neurotrauma 21, 1415-1430.

Menei P, Montero-Menei C, Whittemore SR, Bunge R.P, Bunge MB (1998) Schwann cells genetically modified to secrete human BDNF promote enhanced axonal regrowth across transected adult rat spinal cord. Eur J Neurosci 10, 607-621.

Mikami Y, Okano H, Sakaguchi M, Nakamura M, Shimazaki T, Okano HJ, Kawakami Y, Toyama Y, Toda M (2004) Implantation of dendritic cells in injured adult spinal cord results in activation of endogenous neural stem/progenitor cells leading to de novo neurogenesis and functional recovery. J Neurosci Res 76, 453-465.

Mills CD, Hains BC, Johnson KM, Hulsebosch CE (2001) Strain and model differences in behavioral outcomes after spinal cord injury in rat. J Neurotrauma 18, 743-756.

Moalem G, Leibowitz-Amit R, Yoles E, Mor F, Cohen IR, Schwartz M (1999) Autoimmune T cells protect neurons from secondary degeneration after central nervous system axotomy. Nature Medicine 5, 49-55.

Moreno-Flores MT, Bradbury EJ, Martín-Bermejo MJ, Agudo M, Lim F, Pastrana E, Ávila J, Díaz-Nido J, McMahon SB, Wandosell F (2006) A clonal cell line from immortalized olfactory ensheathing glia promotes functional recovery in the injured spinal cord. Mol Ther 13, 598-608.

Morris R (1984) Developments of a water-maze procedure for studying spatial learning in the rat. J Neurosci Meth 11, 47-60.

Mu X, Azbill RD, Springer JE (2000) Riluzole and methylprednisolone combined treatment improves functional recovery in traumatic spinal cord injury. J Neurotrauma 17, 773-780.

Muir G (2004) *The Behavior of the Laboratory Rat: A Handbook With Tests.* Whishaw IQ, Kolb B (eds). Oxford University Press, Oxford.

Muir GD, Whishaw IQ (1999) Complete locomotor recovery following corticospinal tract lesions: measurement of ground reaction forces during overground locomotion in rats. Behav Brain Res 103, 45-53.

Muir GD, Whishaw IQ (2000) Red nucleus lesions impair overground locomotion in rats: a kinetic analysis. European J Neurosci 12, 1113-1122.

Mushahwar VK, Collins DF, Prochazka A (2000) Spinal cord microstimulation generates functional limb movements in chronically implanted cats. Exp Neurol 163, 422-429.

Nakahara Y, Gage FH, Tuszynski MH (1996) Grafts of fibroblasts genetically modified to secrete NGF, BDNF, NT-3, or basic FGF elicit differential responses in the adult spinal cord. Cell Transplant 5, 191-204.

Nakajima H, Uchida K, Kobayashi S, Inukai T, Horiuchi Y, Yayama T, Sato R, Baba H (2007) Rescue of rat anterior horn neurons after spinal cord injury by retrograde transfection of adenovirus vector carrying brain-derived neurotrophic factor gene. J Neurotrauma 24, 703-712.

Nakajima K, Mori F, Takasu C, Mori M, Matsuyama K, Mori S (2004) Biomechanical constraints in hindlimb joints during the quadrupedal versus bipedal locomotion of M. fuscata. Prog Brain Res 143, 183-190.

Nandoe RD, Hurtado A, Levi AD, Grotenhuis A, Oudega M. (2006) Bone marrow stromal cells for repair of the spinal cord: towards clinical application. Cell Transplant 15, 563-577.

Nash HH, Borke RC, Anders JJ (2002) Ensheathing cells and methylprednisolone promote axonal regeneration and functional recovery in the lesioned adult rat spinal cord. J Neurosci 22, 7111-7120.

Nashmi R, Fehlings MG (2001) Mechanisms of axonal dysfunction after spinal cord injury: with an emphasis on the role of voltage-gated potassium channels. Brain Research and Brain Res Rev 38, 165-191.

Nathan, PW (1994) Effects on movement of surgical incisions into the human spinal cord. Brain 117, 337-346.

Nesathurai S, Graham WA, Mansfield K, Magill D, Sehgal P, Westmoreland SV, Prusty S, Rosene DL, Sledge JB (2006) Model of traumatic spinal cord injury in Macaca fascicularis: similarity of experimental lesions created by epidural catheter to human spinal cord injury. J Med Primat 35, 401-404.

Nguyen HH, Jayasinha V, Xia B, Hoyte K, Martin PT (2002) Overexpression of the cytotoxic T cell GalNAc transferase in skeletal muscle inhibits muscular dystrophy in mdx mice. Proc Natl Acad Sci USA 99, 5616-5621.

Nguyen MD, Julien, JP, Rivest S (2002) Innate immunity: the missing link in neuroprotection and neurodegeneration? Nat Rev Neurosci 3, 216-227.

Nikulina E, Tidwell JL, Dai HN, Bregman BS, Filbin MT (2004) The phosphodiesterase inhibitor rolipram delivered after a spinal cord lesion promotes axonal regeneration and functional recovery. Proc Nat Acad Sci 101, 8786-8790.

Noble LJ, Wrathall JR (1985) Spinal cord contusion in the rat: morphometric analyses of alterations in the spinal cord. Exp Neurol 88, 135-149.

Noble LJ, Donovan F, Igarashi T, Goussev S, Werb Z (2002) Matrix metalloproteinases limit functional recovery after spinal cord injury by modulation of early vascular events. J Neurosci 22, 7526-7535.

Nomura H, Tator CH, Shoichet MS (2006) Bioengineered strategies for spinal cord repair. J Neurotrauma 23, 496-507.

Norenberg MD, Smith J, Marcillo A (2004) The pathology of human spinal cord injury: defining the problems. J Neurotrauma 21, 429-440.

Novikov LN, Novikova LN, Mosahebi A, Wiberg M, Terenghi G, Kellerth JO (2002) A novel biodegradable implant for neuronal rescue and regeneration after spinal cord injury. Biomat 23, 3369-3376.

Novikova LN, Mosahebi A, Wiberg M, Terenghi G, Kellerth JO, Novikov LN (2006) Alginate hydrogel and matrigel as potential cell carriers for neurotransplantation. J Biomed Mat Res A 77, 242-252.

Noyes DH (1987a) Correlation between parameters of spinal cord impact and resultant injury. Exp Neurol 95, 535-547.

Noyes DH (1987b) Electromechanical impactor for producing experimental spinal cord injury in animals. Med Biol Eng Comp 25, 335-340.

Nudo RJ, Frost SB (2007) The evolution of motor cortex and motor systems. In: *Evolution of Nervous Systems.* Volume 3 – Mammals. Kaas JH, Krubitzer LH (eds). Elsevier Academic Press, San Diego.

Okada S, Nakamura M, Mikami Y, Shimazaki T, Mihara M, Ohsugi Y, Iwamoto Y, Yoshizaki K, Kishimoto T, Toyama Y, Okano H (2004) Blockade of interleukin-6 receptor suppresses reactive astrogliosis and ameliorates functional recovery in experimental spinal cord injury. J Neurosci Res 76, 265276.

Okano H, Sakaguchi M, Ohki K, Suzuki N, Sawamoto K (2007) Regeneration of the central nervous system using endogenous repair mechanisms. J Neurochem 102, 1459-1465.

Olsson T, Sun JB, Solders G, Xiao BG, Höjeberg B, Ekre HP, Link H (1993) Autoreactive T and B cell responses to myelin antigens after diagnostic sural nerve biopsy. J Neurol Sci 117, 130-139.

Orsal D, Barthe JY, Antri M, Feraboli-Lohnherr D, Yakovleff A, Giménez y Ribotta M, Privat A, Provencher J, Rossignol S (2002) Locomotor recovery in chronic spinal rat: long-term pharmacological treatment or transplantation of embryonic neurons? Prog Brain Res 137, 213-230.

Oudega M, Xu XM (2006) Schwann Cell Transplantation for Repair of the Adult Spinal Cord. J Neurotrauma 23, 453-467.

Oudega M, Xu XM, Guénard V, Kleitman N, Bunge MB (1997) A combination of insulin-like growth factor-I and platelet-derived growth factor enhances myelination but diminishes axonal regeneration into Schwann cell grafts in the adult rat spinal cord. Glia 19, 247-258.

Pandyan AD, Johnson GR, Price CI, Curless RH, Barnes MP, Rodgers H (1999) A review of the properties and limitations of the Ashworth and modified Ashworth Scales as measures of spasticity. Clin Rehab 13, 373-383.

Park E, Velumian AA, Fehlings MG (2004) The role of excitotoxicity in secondary mechanisms of spinal cord injury: a review with an emphasis on the implications for white matter degeneration. J Neurotrauma 21, 754-774.

Park HC, Shim YS, Ha Y, Yoon SH, Park SR, Choi BH, Park HS (2005) Treatment of complete spinal cord injury patients by autologous bone marrow cell transplantation and administration of granulocyte-macrophage colony stimulating factor. Tiss Eng 11, 913-922.

Patist CM, Mulder MB, Gautier SE, Maquet V, Jérôme R, Oudega M (2004) Freeze-dried poly(D,L-lactic acid) macroporous guidance scaffolds impregnated with brain-derived neurotrophic factor in the transected adult rat thoracic spinal cord. Biomat 25, 1569-1582.

Pearse D.D, Pereira FC, Marcillo AE, Bates ML, Berrocal YA, Filbin MT, Bunge MB (2004) cAMP and Schwann cells promote axonal growth and functional recovery after spinal cord injury. Nature Medicine 10, 610-616.

Pearse DD, Sanchez AR, Pereira FC, Andrade CM, Puzis R, Pressman Y, Golden K, Kitay BM, Blits B, Wood PM, Bunge MB (2007) Transplantation of Schwann cells and/or olfactory ensheathing glia into the contused spinal cord: Survival, migration, axon association, and functional recovery. Glia 55, 976-1000.

Peel AL, Zolotukhin S, Schrimsher GW, Muzyczka N, Reier PJ (1997) Efficient transduction of green fluorescent protein in spinal cord neurons using adeno-associated virus vectors containing cell type-specific promoters. Gene Ther 4, 16-24.

Petruska JC, Ichiyama RM, Jindrich DL, Crown ED, Tansey KE, Roy RR, Edgerton VR, Mendell LM (2007) Changes in motoneuron properties and synaptic inputs related to step training after spinal cord transection in rats. J Neurosci 27, 4460-4471.

Phinney DG, Prockop DJ (2007) Concise review: mesenchymal stem/multipotent stromal cells: the state of transdifferentiation and modes of tissue repair – current views. Stem Cells 25, 2896-2902.

Pinzon A, Calancie B, Oudega M, Noga BR (2001) Conduction of impulses by axons regenerated in a Schwann cell graft in the transected adult rat thoracic spinal cord. J Neurosci Res 64, 533-541.

Pitts LH, Ross A, Chase GA, Faden AI (1995) Treatment with thyrotropin-releasing hormone (TRH) in patients with traumatic spinal cord injuries. J Neurotrauma 12, 235-243.

Plant GW, Ramón-Cueto A, Bunge M (2001a) Transplantation of Schwann cells and ensheathing glia to improve regeneration in adult spinal cord. In: *Axonal regeneration in the Central nervous system.* Ingoglia N, Murray M (eds). Informa Health Care, USA.

Plant GW, Bates ML, Bunge MB (2001b) Inhibitory proteoglycan immunoreactivity is higher at the caudal than the rostral Schwann cell graft-transected spinal cord interface. Mol Cell Neurosci 17, 471-487.

Plant GW, Harvey AR, Chirila TV (1995) Axonal growth within poly (2-hydroxyethyl methacrylate) sponges infiltrated with Schwann cells and implanted into the lesioned rat optic tract. Brain Res 671, 119-130.

Plant GW, Woerly S, Harvey AR (1997) Hydrogels containing peptide or aminosugar sequences implanted into the rat brain: influence on cellular migration and axonal growth. Exp Neurol 143, 287-299.

Plant GW, Christensen CL, Oudega M, Bunge MB (2003) Delayed transplantation of olfactory ensheathing glia promotes sparing/regeneration of supraspinal axons in the contused adult rat spinal cord. J Neurotrauma 20, 1-16.

Plemel JR, Duncan G, Chen KW, Shannon C, Park S, Sparling JS, Tetzlaff W (2008) A graded forceps crush spinal cord injury model in mice. J Neurotrauma 25, 350-370.

Popovich PG, Jones TB (2003) Manipulating neuroinflammatory reactions in the injured spinal cord: back to basics. Trends Pharmacol Sci 24, 13-17.

Popovich PG, Stokes BT, Whitacre CC (1996) Concept of autoimmunity following spinal cord injury: possible roles for T lymphocytes in the traumatized central nervous system. J Neurosci Res 45, 349-363.

Popovich PG, Wei P, Stokes BT (1997) Cellular inflammatory response after spinal cord injury in Sprague-Dawley and Lewis rats. J Comp Neurol 377, 443-464.

Popovich PG, Guan Z, Wei P, Huitinga I, van Rooijen N, Stokes BT (1999) Depletion of hematogenous macrophages promotes partial hindlimb recovery and neuroanatomical repair after experimental spinal cord injury. Exp Neurol 158, 351-365.

Popovich PG, Guan Z, McGaughy V, Fisher L, Hickey WF, Basso DM (2002) The neuropathological and behavioral consequences of intraspinal microglial/macrophage activation. J Neuropathol Exp Neurol 61, 623-633.

Prang P, Müller R, Eljaouhari A, Heckmann K, Kunz W, Weber T, Faber C, Vroemen M, Bogdahn U, Weidner N (2006) The promotion of oriented axonal regrowth in the injured spinal cord by alginate-based anisotropic capillary hydrogels. Biomat 27, 3560-3569.

Prewitt CM, Niesman IR, Kane CJ, Houlé JD (1997) Activated macrophage/microglial cells can promote the regeneration of sensory axons into the injured spinal cord. Exp Neurol 148, 433-443.

Profyris C, Cheema SS, Zang D, Azari MF, Boyle K, Petratos S (2004) Degenerative and regenerative mechanisms governing spinal cord injury. Neurobiol Dis 15, 415-436.

Raisman G, Li Y (2007) Repair of neural pathways by olfactory ensheathing cells. Nat Rev Neurosci 8, 312-319.

Ramón-Cueto A, Plant G.W, Avila J, Bunge MB (1998) Long-distance axonal regeneration in the transected adult rat spinal cord is promoted by olfactory ensheathing glia transplants. J Neurosci 18, 3803-3815.

Rapalino O, Lazarov-Spiegler O, Agranov E, Velan GJ, Yoles E, Fraidakis M, Solomon A, Gepstein R, Katz A, Belkin M, Hadani M, Schwartz M (1998) Implantation of stimulated homologous macrophages results in partial recovery of paraplegic rats. Nature Medicine 4, 814-821.

Richardson PM, McGuinness UM, Aguayo AJ (1980) Axons from CNS neurons regenerate into PNS grafts. Nature 284, 264-265.

Richter MW, Roskams AJ (2008) Olfactory ensheathing cell transplantation following spinal cord injury: Hype or hope? Exp Neurol 209, 353-367.

Romero MI, Smith GM (1998) Adenoviral gene transfer into the normal and injured spinal cord: enhanced transgene stability by combined administration of temperature-sensitive virus and transient immune blockade. Gene Ther 5, 1612-1621.

Rossignol S, Giroux N, Chau C, Marcoux J, Brustein E, Reader TA (2001) Pharmacological aids to locomotor training after spinal injury in the cat. J Physiol 533, 65-74.

Rossignol S, Schwab M, Schwartz M, Fehlings MG (2007) Spinal cord injury: time to move? J Neurosci 27, 11782-11792.

Rouiller EM, Moret V, Tanne J, Boussaoud D (1996) Evidence for direct connections between the hand region of the supplementary motor area and cervical motoneurons in the macaque monkey. Eur J Neurosci 8, 1055-1059.

Ruitenberg MJ, Plant GW, Christensen CL, Blits B, Niclou SP, Harvey AR, Boer GJ, Verhaagen J (2002) Viral vector-mediated gene expression in olfactory ensheathing glia implants in the lesioned rat spinal cord. Gene Ther 9, 135-146.

Ruitenberg MJ, Plant GW, Hamers FP, Wortel J, Blits B, Dijkhuizen PA, Gispen WH, Boer GJ, Verhaagen J (2003) Ex vivo adenoviral vector-mediated neurotrophin gene transfer to olfactory ensheathing glia: effects on rubrospinal tract regeneration, lesion size, and functional recovery after implantation in the injured rat spinal cord. J Neurosci 23, 7045-7058.

Ruitenberg MJ, Blits B, Dijkhuizen PA, te Beek ET, Bakker A, van Heerikhuize JJ, Pool CW, Hermens WT, Boer GJ, Verhaagen J (2004) Adeno-associated viral vector-mediated gene transfer of brain-derived neurotrophic factor reverses atrophy of rubrospinal neurons following both acute and chronic spinal cord injury. Neurobiol Dis 15, 394-406.

Ruitenberg MJ, Levison DB, Lee SV, Verhaagen J, Harvey AR, Plant GW (2005) NT-3 expression from engineered olfactory ensheathing glia promotes spinal sparing and regeneration. Brain 128, 839-853.

Ruitenberg MJ, Vukovic J, Sarich J, Busfield SJ, Plant GW (2006) Olfactory ensheathing cells: characteristics, genetic engineering, and therapeutic potential. J Neurotrauma 23, 468-478.

Sakanaka M, Zhu P, Zhang B, Wen TC, Cao F, Ma YJ, Samukawa K, Mitsuda N, Tanaka J, Kuramoto M, Uno H, Hata R (2007) Intravenous infusion of dihydroginsenoside Rb1 prevents compressive spinal cord injury and ischemic brain damage through upregulation of VEGF and Bcl-XL. J Neurotrauma 24, 1037-1054.

Sasaki M, Black JA, Lankford KL, Tokuno HA, Waxman SG, Kocsis JD (2006) Molecular reconstruction of nodes of Ranvier after remyelination by transplanted olfactory ensheathing cells in the demyelinated spinal cord. J Neurosci 26, 1803-1812.

Scheff SW, Rabchevsky AG, Fugaccia I, Main JA, Lumpp JE Jr (2003) Experimental modeling of spinal cord injury: characterization of a force-defined injury device. J Neurotrauma 20, 179-193.

Schmidlin E, Wannier T, Bloch J, Rouiller EM (2004) Progressive plastic changes in the hand representation of the primary motor cortex parallel incomplete recovery from a unilateral section of the corticospinal tract at cervical level in monkeys. Brain Res 1017, 172-183.

Schmitt AB, Buss A, Breuer S, Brook GA, Pech K, Martin D, Schoenen J, Noth J, Love S, Schroder JM, Kreutzberg GW, Nacimiento W (2000) Major histocompatibility complex class II expression by activated microglia caudal to lesions of descending tracts in the human spinal cord is not associated with a T cell response. Acta Neuropathol 100, 528-536.

Schnell L, Schwab ME (1990) Axonal regeneration in the rat spinal cord produced by an antibody against myelin-associated neurite growth inhibitors. Nature 343, 269-272.

Schnell, L, Fearn, S, Klassen, H, Schwab, M.E, Perry, V.H. (1999) Acute inflammatory responses to mechanical lesions in the CNS: differences between brain and spinal cord. Eur J Neurosci 11, 3648-3658.

Schwartz G, Fehlings MG (2001) Evaluation of the neuroprotective effects of sodium channel blockers after spinal cord injury: improved behavioral and neuroanatomical recovery with riluzole. J Neurosurg 94, 245-256.

Schwartz G, Fehlings MG (2002) Secondary injury mechanisms of spinal cord trauma: a novel therapeutic approach for the management of secondary pathophysiology with the sodium channel blocker riluzole. Prog Brain Res 137, 177-190.

Schwartz M (2003) Macrophages and microglia in central nervous system injury: are they helpful or harmful? J Cereb Bl Fl Metab 23, 385-394.

Schwartz M, Kipnis J (2001) Protective autoimmunity: regulation and prospects for vaccination after brain and spinal cord injuries. Trends Molec Med 7, 252-258.

Scivoletto G, Ivanenko Y, Morganti B, Grasso R, Zago M, Lacquaniti F, Ditunno J, Molinari M (2007) Plasticity of spinal centers in spinal cord injury patients: new concepts for gait evaluation and training. Neurorehab Neur Repair 21, 358-365.

Šedý J, Urdzíková L, Likavcanová K, Hejcl A, Jendelová P, Syková E (2007) A new model of severe neurogenic pulmonary edema in spinal cord injured rat. Neurosci Lett 423, 167-171.

Sekhon LH, Fehlings MG (2001) Epidemiology, demographics, and pathophysiology of acute spinal cord injury. Spine 26, S2-12.

Sroga JM, Jones TB, Kigerl KA, McGaughy VM, Popovich PG (2003) Rats and mice exhibit distinct inflammatory reactions after spinal cord injury. J Comp Neur 462, 223-240.

Steeves JD, Lammertse D, Curt A, Fawcett JW, Tuszynski MH, Ditunno JF, Ellaway PH, Fehlings MG, Guest JD, Kleitman N, Bartlett PF, Blight AR, Dietz V, Dobkin BH, Grossman R, Short D, Nakamura M, Coleman WP, Gaviria M, Privat A (2007) Guidelines for the conduct of clinical trials for spinal cord injury (SCI) as developed by the ICCP panel: clinical trial outcome measures. Spinal Cord 45, 206-221.

Sternlicht MD, Werb Z (2001) How matrix metalloproteinases regulate cell behavior. Ann Rev Cell Develop Biol 17, 463-516.

Stirling DP, Khodarahmi K, Liu J, McPhail LT, McBride CB, Steeves JD, Ramer MS, Tetzlaff W (2004) Minocycline treatment reduces delayed oligodendrocyte death, attenuates axonal dieback, and improves functional outcome after spinal cord injury. J Neurosci 24, 2182-2190.

Stokes BT (1992) Experimental spinal cord injury: a dynamic and verifiable injury device. J Neurotrauma 9, 129-131; discussion 131-124.

Stokols S, Tuszynski MH (2006) Freeze-dried agarose scaffolds with uniaxial channels stimulate and guide linear axonal growth following spinal cord injury. Biomaterials 27, 443-451.

Streit WJ, Semple-Rowland SL, Hurley SD, Miller RC, Popovich PG, Stokes BT (1998) Cytokine mRNA profiles in contused spinal cord and axotomized facial nucleus suggest a beneficial role for inflammation and gliosis. Exp Neurol 152, 74-87.

Stys PK (2004) White matter injury mechanisms. Cur Mol Med 4, 113-130.

Sung JH (1981) Tangled masses of regenerated central nerve fibers (non-myelinated central neuromas) in the central nervous system. J Neuropath Exp Neurol 40, 645-657.

Sung JH (1987) Tangled masses of central axons (central axonomas) in the brain stem: anatomical evidence for the regenerative growth of human central axons. J Neuropath Exp Neurol 46, 200-213.

Sung JK, Miao L, Calvert JW, Huang L, Louis Harkey H, Zhang JH (2003) A possible role of RhoA/Rho-kinase in experimental spinal cord injury in rat. Brain Res 959, 29-38.

Suzuki Y, Kitaura M, Wu S, Kataoka K, Suzuki K, Endo K, Nishimura Y, Ide C (2002) Electrophysiological and horseradish peroxidase-tracing studies of nerve regeneration through alginate-filled gap in adult rat spinal cord. Neurosci Lett 318, 121-124.

Syková E, Homola A, Mazanec R, Lachmann H, Konrádová SL, Kobylka P, Pádr R, Neuwirth J, Komrska V, Vávra V, Štulík J, Bojar M (2006a) Autologous bone marrow transplantation in patients with subacute and chronic spinal cord injury. Cell Transplant 15, 675-687.

Syková E, Jendelová P, Urdziková L, Lesný P, Hejcl A (2006b) Bone marrow stem cells and polymer hydrogels – two strategies for spinal cord injury repair. Cell Mol Neurobiol 26, 1113-1129.

Taoka Y, Okajima K, Uchiba M, Murakami K, Kushimoto S, Johno M, Naruo M, Okabe H, Takatsuki K (1997) Role of neutrophils in spinal cord injury in the rat. Neurosci 79, 1177-1182.

Taoka Y, Okajima K, Uchiba M, Murakami K, Harada N, Johno M, Naruo M (1998) Activated protein C reduces the severity of compression-induced spinal cord injury in rats by inhibiting activation of leukocytes. J Neurosci 18, 1393-1398.

Taylor SJ, McDonald JW 3rd, Sakiyama-Elbert SE (2004) Controlled release of neurotrophin-3 from fibrin gels for spinal cord injury. J Controlled Release 98, 281-294.

Thuret S, Moon LD, Gage FH (2006) Therapeutic interventions after spinal cord injury. Nat Rev Neurosci 7, 628-643.

Tikka TM, Koistinaho JE (2001) Minocycline provides neuroprotection against N-methyl-D-aspartate neurotoxicity by inhibiting microglia. J Immunol 166, 7527-7533.

Tobias CA, Han SS, Shumsky JS, Kim D, Tumolo M, Dhoot NO, Wheatley MA, Fischer I, Tessler A, Murray M (2005) Alginate encapsulated BDNF-producing fibroblast grafts permit recovery of function after spinal cord injury in the absence of immune suppression. J Neurotrauma 22, 138-156.

Tracey D (2004) Ascending and descending tracts in the spinal cord. In: *The Rat Nervous System*, 3rd edn. Paxinos G (ed). Academic Press, San Diego.

Tsai EC, Dalton PD, Shoichet MS, Tator CH (2006) Matrix inclusion within synthetic hydrogel guidance channels improves specific supraspinal and local axonal regeneration after complete spinal cord transection. Biomat 27, 519-533.

Tuszynski M, Edgerton R, Dobkin B (1999) Recovery of locomotion after experimental spinal cord injury: axonal regeneration or modulation of intrinsic spinal cord walking circuitry? J Spinal Cord Med 22, 143.

Tuszynski MH, Peterson DA, Ray J, Baird A, Nakahara Y, Gage FH (1994) Fibroblasts genetically modified to produce nerve growth factor induce robust neuritic ingrowth after grafting to the spinal cord. Exp Neurol 126, 1-14.

Tuszynski MH, Gabriel K, Gage FH, Suhr S, Meyer S, Rosetti A (1996) Nerve growth factor delivery by gene transfer induces differential outgrowth of sensory, motor, and noradrenergic neurites after adult spinal cord injury. Exp Neurol 137, 157-173.

Tysseling-Mattiace VM, Sahni V, Niece KL, Birch D, Czeisler C, Fehlings MG, Stupp SI, Kessler JA (2008) Self-assembling nanofibers inhibit glial scar formation and promote axon elongation after spinal cord injury. J Neurosci 28, 3814-3823.

Van Meeteren NL, Eggers R, Lankhorst AJ, Gispen WH, Hamers FP (2003) Locomotor recovery after spinal cord contusion injury in rats is improved by spontaneous exercise. J Neurotrauma 20, 1029-1037.

Wang CX, Nuttin B, Heremans H, Dom R, Gybels J (1996) Production of tumor necrosis factor in spinal cord following traumatic injury in rats. J Neuroimmunol 69, 151-156.

Wang SJ, Wang KY, Wang WC (2004) Mechanisms underlying the riluzole inhibition of glutamate release from rat cerebral cortex nerve terminals (synaptosomes). Neurosci 125, 191-201.

Wells JE, Hurlbert RJ, Fehlings MG, Yong VW (2003a) Neuroprotection by minocycline facilitates significant recovery from spinal cord injury in mice. Brain 126, 1628-1637.

Wells JE, Rice TK, Nuttall RK, Edwards DR, Zekki H, Rivest S, Yong VW (2003b) An adverse role for matrix metalloproteinase 12 after spinal cord injury in mice. J Neurosci 23, 10107-10115.

Whishaw IQ, Gorny B, Sarna J (1998) Paw and limb use in skilled and spontaneous reaching after pyramidal tract, red nucleus and combined lesions in the rat: behavioral and anatomical dissociations. Behav Brain Res 93, 167-183.

Woerly S (2000) Restorative surgery of the central nervous system by means of tissue engineering using NeuroGel™ implants. Neurosurg Rev 23, 59-77; discussion 78-59.

Woerly S, Petrov P, Syková E, Roitbak T, Simonova Z, Harvey AR (1999) Neural tissue formation within porous hydrogels implanted in brain and spinal cord lesions: ultrastructural, immunohistochemical, and diffusion studies. Tiss Eng 5, 467-488.

Woerly S, Doan VD, Sosa N, de Vellis J, Espinosa A (2001a) Reconstruction of the transected cat spinal cord following NeuroGel™ implantation: axonal tracing, immunohistochemical and ultrastructural studies. Int J Dev Neurosci 19, 63-83.

Woerly S, Pinet E, de Robertis L, Van Diep D, Bousmina M (2001b) Spinal cord repair with PHPMA hydrogel containing RGD peptides (NeuroGel™). Biomat 22, 1095-1111.

Wrathall JR, Teng YD, Marriott R (1997) Delayed antagonism of AMPA/kainate receptors reduces long-term functional deficits resulting from spinal cord trauma. Exp Neurol 145, 565-573.

Wu L, Ding J (2004) In vitro degradation of three-dimensional porous poly(D,L-lactide-co-glycolide) scaffolds for tissue engineering. Biomat 25, 5821-5830.

Wu S, Suzuki Y, Kitada M, Kitaura M, Kataoka K, Takahashi J, Ide C, Nishimura Y (2001) Migration, integration, and differentiation of hippocampus-derived neurosphere cells after transplantation into injured rat spinal cord. Neurosci Lett 312, 173-176.

Xiao M, Klueber KM, Lu C, Guo Z, Marshall CT, Wang H, Roisen FJ (2005) Human adult olfactory neural progenitors rescue axotomized rodent rubrospinal neurons and promote functional recovery. Exp Neurol 194, 12-30.

Xu J, Kim GM, Chen S, Yan P, Ahmed SH, Ku G, Beckman JS, Xu XM, Hsu CY (2001) iNOS and nitrotyrosine expression after spinal cord injury. J Neurotrauma 18, 523-532.

Xu XM, Guénard V, Kleitman N, Aebischer P, Bunge MB (1995) A combination of BDNF and NT-3 promotes supraspinal axonal regeneration into Schwann cell grafts in adult rat thoracic spinal cord. Exp Neurol 134, 261-272.

Xu XM, Chen A, Guénard V, Kleitman N, Bunge MB (1997) Bridging Schwann cell transplants promote axonal regeneration from both the rostral and caudal stumps of transected adult rat spinal cord. J Neurocytol 26, 1-16.

Xu XM, Zhang SX, Li H, Aebischer P, Bunge MB (1999) Regrowth of axons into the distal spinal cord through a Schwann-cell-seeded mini-channel implanted into hemisected adult rat spinal cord. Eur J Neurosci 11, 1723-1740.

Yang L, Blumbergs PC, Jones NR, Manavis J, Sarvestani GT, Ghabriel MN (2004) Early expression and cellular localization of proinflammatory cytokines interleukin-1beta, interleukin-6, and tumor necrosis factor-alpha in human traumatic spinal cord injury. Spine 29, 966-971.

Yin Y, Cui Q, Li Y, Irwin N, Fischer D, Harvey AR, Benowitz LI (2003) Macrophage-derived factors stimulate optic nerve regeneration. J Neurosci 23, 2284-2293.

Yin Y, Henzl MT, Lorber B, Nakazawa T, Thomas TT, Jiang F, Langer R, Benowitz LI (2006) Oncomodulin is a macrophage-derived signal for axon regeneration in retinal ganglion cells. Nature Neurosci 9, 843-852.

Yiu G, He Z (2006) Glial inhibition of CNS axon regeneration. Nat Rev Neurosci 7, 617-627.

Yoshii S, Oka M, Shima M, Taniguchi A, Taki Y, Akagi M (2004) Restoration of function after spinal cord transection using a collagen bridge. J Biomed Mat Res A 70, 569-575.

Yoshikawa M, Suzumura A, Tamaru T, Takayanagi T, Sawada M (1999) Effects of phosphodiesterase inhibitors on cytokine production by microglia. Multiple Sclerosis 5, 126-133.

Yune TY, Lee JY, Jung GY, Kim SJ, Jiang MH, Kim YC, Oh YJ, Markelonis GJ, Oh TH (2007) Minocycline alleviates death of oligodendrocytes by inhibiting pro-nerve growth factor production in microglia after spinal cord injury. J Neurosci 27, 7751-7761.

Zhang B, Yang L, Konishi Y, Maeda N, Sakanaka M, Tanaka J (2002) Suppressive effects of phosphodiesterase type IV inhibitors on rat cultured microglial cells: comparison with other types of cAMP-elevating agents. Neuropharmacol 42, 262-269.

Zhang Y, Dijkhuizen PA, Anderson PN, Lieberman AR, Verhaagen J (1998) NT-3 delivered by an adenoviral vector induces injured dorsal root axons to regenerate into the spinal cord of adult rats. J Neurosci Res 54, 554-562.

Zhang Z, Krebs CJ, Guth L (1997) Experimental analysis of progressive necrosis after spinal cord trauma in the rat: etiological role of the inflammatory response. Exp Neurol 143, 141-152.

Zietlow R, Lane EL, Dunnett SB, Rosser AE (2008) Human stem cells for CNS repair. Cell Tiss Res 331, 301-322.

15 Atlas of the Rat Spinal Cord

Charles Watson, George Paxinos, Gulgun Kayalioglu and Claire Heise

Introduction

There have been many excellent neuroanatomical studies of the spinal cord in laboratory animals over the past 60 years, but almost all have focused on a particular region, tract, or cell group. Exceptions are the pioneering work of Rexed on the cytoarchitecture of the spinal cord gray matter in the cat (1952) and the work of Sterling and Kuypers (1968), Sidman *et al.*, (1971), and Scheibel and Scheibel (1970). Chapters 6, 7, 8, 9, 10, 11, 12, and 13 of this book provide an account of the major studies that have been carried out on the mammalian spinal cord. However, despite this wealth of information on individual features of the spinal cord, there is no modern comprehensive atlas of the spinal cord. We have developed atlases of the rat and mouse spinal cord to help fill this gap.

We provide in this chapter the fundamental structural map of the spinal cord of the principal laboratory research animal, the rat.

The atlas of the rat spinal cord has the following features:

- Photographs of Nissl stained transverse sections from each of the spinal cord segments (obtained from a single unfixed spinal cord).
- Detailed diagrams of each of the spinal cord segments pictured, delineating the laminae of Rexed and all other significant neuronal groupings at each level.
- Photographs of additional sections displaying markers such as acetylcholinesterase (AChE), calbindin, calretinin, choline acetlytransferase, neurofilament protein (SMI 32), enkephalin, calcitonin gene-related peptide (CGRP), and neuronal nuclear protein (NeuN). These sections were obtained from a single perfused rat spinal cord.
- All the major motoneuron clusters are identified in relation to the individual muscles or muscle groups they supply.

The rat spinal cord atlas is based on a set of Nissl stained transverse sections from a single spinal cord (the master set), supplemented by additional sections at each level stained with a range of chemical marker stains. The Nissl photograph is located on the top left of the open book. We decided to use the unfixed tissue for the master Nissl set because Nissl staining is crisper in unfixed tissue and we found that the dorsal horns were less distorted in unfixed tissue. Directly below this Nissl photograph is the labeled drawing that corresponds to it.

The page opposite the Nissl master section and the diagram presents photographs of up to four sections stained with different markers taken from the same level but from a different rat (fixed tissue) from that on the left-hand side which is diagrammed. An exception is the AChE sections, which were always taken from fresh tissue from the same rat from which the Nissl sections were taken.

Methods

For the spinal cord that produced the Nissl master set of sections and the AChE sections (fresh, unfixed tissue), a male Wistar rat weighing approximately 150 g was decapitated. The vertebral column with the spinal cord in situ was dissected free. The vertebral column was cut in 4 pieces and quickly frozen with dry ice. Each piece was sectioned on a cryotome at a thickness of 40 μm. Collection of sections was greatly assisted by the use of the CryoJane Tape Transfer system. This system enables sections to be collected in a way that minimizes differential shrinkage within the section.

In the fresh spinal cord sections it was not always possible to firmly identify every spinal cord segment. Where any doubt existed, the identity of the segment was established by reference to a number of section sets taken from fixed specimens in which the segments were identified by dissection of spinal roots.

For the immunoreactions, we used perfused fixed tissue. A Wistar rat weighing approximately 150 g was deeply anesthetized with pentobarbital and a blunt needle was inserted into the left ventricle and advanced a few millimeters into the ascending aorta. After the left atrium was cut open, perfusion commenced at a constant flow using a peristaltic pump. The perfusate consisted of 250 ml of saline delivered over a period of 3 min, followed by 600 ml of a 4% paraformaldehyde solution in 0.1 M phosphate buffer (PB, pH 7.4) over a period of 25 min. The first 200 ml of the fixative was delivered at the same speed as the saline flush. The speed was then gradually reduced to pass the remaining 400 ml. The spinal cord was exposed by laminectomy and further dissected until all of the spinal nerves could be identified. The segments were separated by a series of transverse cuts and collected for later sectioning. The spinal cord was then removed and immersed in a sucrose solution (30% in PB) for 2 nights at 4°C.

The fixed spinal cord was maintained at -12°C and cut with a newly sharpened C-profile stainless steel blade. The section thickness was set at 40 μm and each section was collected for subsequent staining for calbindin, calretinin, choline acetlytransferase (ChAT), neurofilament protein (SMI 32), enkephalin, calcitonin-gene related peptide (CGRP), and neuronal nuclear protein (NeuN). Sections were collected individually with a paint brush and placed in sequence in a series of eight jars filled with cold PB and sitting on a tray of crushed ice to keep the solutions cold. Sections placed in the fourth jar were transferred to PB and immediately mounted on gelatine-coated slides; they were allowed to dry overnight and then stained for Nissl substance with cresyl violet. The most helpful stain for demonstrating major anatomical features of the spinal cord was ChAT, because it clearly shows the identity of the somatic motor neurons, the autonomic motoneurons and some of the laminae of the dorsal horn.

Cresyl violet staining and AChE histochemistry

See Paxinos and Watson (2007) or visit the website: www.powmri.edu.au/staff/paxinos.htm for a detailed protocol.

Immunohistochemical processing

All washes were done on an orbital shaker at room temperature. To minimize damage to the sections, tea strainers were used to lift and transfer the sections between washes. Jars, tea strainers, and paint brushes were cleaned in 50% alcohol and distilled water. The processing consisted of the following steps:

1. Wash in PB for 10 minutes.
2. Wash in 50% alcohol for 30 minutes.
3. Wash in 50% alcohol and 1% hydrogen peroxide for 30 minutes.
4. Wash in 5% normal horse serum in PB for 30 minutes.
5. Incubate with primary antibody in PBH [PB solution with normal horse serum (2%) and Triton X-100 (0.2%)]. The incubations were done in scintillation glass vials on the orbital shaker at 4ºC for 32 hours in a 3 ml volume. The antibodies and dilutions were:
   - anti-parvalbumin (mouse monoclonal IgG1; SWant Cat. No.235), 1:128,000;
   - anti-calbindin D28k (mouse monoclonal IgG1; SWant Cat. No.300), 1:20,000;
   - anti-calretinin (rabbit polyclonal; SWant Cat. No. 7696), 1:64,000;
   - SMI32 (mouse monoclonal IgG1 directed against nonphosphorylated epitopes in neurofilament B; Sternberger Monoclonal, Inc.), 1:32,000;
   - anti-tyrosine hydroxylase (mouse monoclonal IgG1; Incstar Cat. No. 22941), 1:16,000.

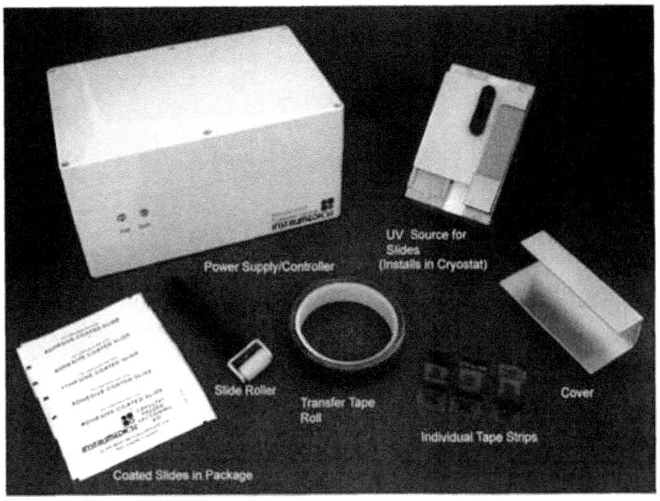

Figure 15.1 Components of the CryoJane Tape Transfer System

A Transfer Tape Strip is pressed onto the tissue face with the roller, and a section is then cut deep to the tape. The tape with section stuck to it is laid a coated slide. The slide carrying the section and the tape is exposed to UV light. The UV destroys the tape adhesive and so releases the section from the transfer tape. This process also activates adhesion of the section to the coated slide. The effect of the process is to minimize distortion due to shrinkage after cutting. The UV source is installed in the cryostat.

6. Wash four times in PB, 10 minutes each wash.
7. Incubate with biotinylated secondary antibody in PBH. The incubations were done in scintillation glass vials on the orbital shaker at room temperature for 1 hour in a 4 ml volume. The two antibodies used were:
   - Anti-mouse IgG (goat biotin-SP-conjugated AffiniPure; Jackson ImmunoResearch Laboratories), 1:200;
   - Anti-rabbit IgG (goat biotin-SP-conjugated AffiniPure, Jackson ImmunoResearch Laboratories, 1:200.
8. Wash four times in PB, 10 minutes each wash.
9. Incubate in ExtrAvidin-Peroxidase (Sigma E-2886) 1:1000 in scintillation glass vials on the orbital shaker at room temperature for 1 hour in a 4 ml volume.
10. Wash four times in 0.1 M Tris buffer (TB, pH 7.4) containing 0.004% nickel ammonium sulphate, 10 minutes each wash.
11. Incubate with 3,3'-diaminobenzidine (DAB) using Nickel enhancement and a glucose oxidase reaction to generate peroxide, as described by Llewellyn-Smith *et al.*, (1993). Briefly, the sections were incubated in 10 ml of a TB solution containing 0.04% nickel ammonium sulfate, 0.004% ammonium chloride, 0.2% d-glucose, and 0.05% DAB in a 10 ml volume. Three minutes later, 10 μl of glucose oxidase (1200 units/ml, Sigma G-6891) was added to the solution. The incubations were done in plastic jars, on the orbital shaker and at room temperature.

12. Stop the reaction by a wash in TB when specific immunostaining is judged to be optimal. Incubation times varied between 10 and 25 minutes.

13. Wash three times in TB, 10 minutes each wash.

Mounting

For the fresh spinal cord the tissue was mounted directly on special slides provided for the CryoJane Tape Transfer System. For the fixed spinal cord, the sections were mounted on gelatine-coated slides (4% gelatine) in 0.01 M Tris-buffered saline (pH 7.4). The slides were left to dry overnight. They were then cleared in 100% alcohol for 1 hr, followed by xylene for 1 hour, and finally were coverslipped with DPX.

Photography and diagrams

Photographs were taken through an Olympus Provis microscope with a 16 megapixel digital camera.

Sections sustained different degrees of shrinkage through different stains. We were guided by the Nissl standard because the Nissl sections were applied directly to the microscope slides through the CyroJane apparatus. We resized the images of the other stains to approximately match the size of the Nissls.

The scale bar refers to the diagrams. The photographs are slightly larger than the diagrams to permit easier detection of structures.

Nomenclature and abbreviations

The nomenclature used in this atlas is in most cases consistent with mainstream usage. Some new terms were coined and these are explained in the next paragraph. The abbreviations were constructed so as to be consistent with the master abbreviation set employed in *The Rat Brain in Stereotaxic Coordinates* (Paxinos and Watson, 2007) and *The Mouse Brain in Stereotaxic Coordinates* (Franklin and Paxinos, 2007). The use of consistent abbreviations for homologous structures in different species means that readers are not burdened with the task of learning different abbreviations for the same nucleus or region.

Basis of delineation of structures

We identified laminae in the spinal cord gray matter according to the schema developed by Rexed (1952). A description of the cytoarchitecture of these laminae is presented in Chapter 6. We used the prefix Sp to indicate each lamina so as to allow computer searching systems to distinguish between this set of digits and those used in other areas (e.g. layers of cerebral cortex).

Certain nuclei have been identified within the laminae of Rexed. Some of these are well known (such as the dorsal nucleus), but some others are less well recognized. In the latter category are the nuclei of origin of the ventral spinocerebellar tract. We have identified these as the lumbar and sacral precerebellar nuclei (LPrCb and SPrCb). See further discussion of these cell groups in Chapter 6.

The preganglionic nuclei of the sympathetic and parasympathetic nervous systems are clearly identified, and a description is presented in Chapter 8. We have labeled the sympathetic preganglionic cells as the intermediolateral cell group (IML), but we chose the name 'sacral parasympathetic nucleus' for the parasympathetic preganglionic neuron groups. We have not separately identified the lateral outlying parts of either of these nuclei, but it should be noted that some authors name these lateral outlying cells in the thoracic region as the funicular part of the IML.

We have attempted to resolve an area of nomenclatural confusion involving the lumbar and sacral intermediate gray area. The name 'dorsal commissural nucleus' has been applied by different groups of authors to two completely different nuclei, one lumbar and one sacral. We have named the former the lumbar dorsal commissural nucleus (LDCom) and the latter the sacral dorsal commissural nucleus (SDCom). See Chapters 6 and 8 for further discussion of these nuclei.

The motoneuron clusters in lamina 9 have been named for the muscles or muscle groups they supply. Chapter 7 gives a full account of the evidence for these designations. We have not attempted to subdivide these motonuclei into groups that supply individual muscles because there is considerable overlap and we felt that the drawings should be kept as clear as possible. Those who wish to identify the motoneurons supplying individual muscles should consult the references cited in Chapter 7.

We were not able to identify the majority of long tracts in these spinal cord sections. However, their presumed position can be located with reference to Chapters 10, 11, and 12. The tracts we were able to identify were the dorsolateral fasciculus, the gracile fasciculus, the cuneate fasciculus, the postsynaptic dorsal column pathway, the rubrospinal tract, and the dorsal corticospinal tract. See Chapters 10 and 11 for more detail on these tracts.

Naming of spinal cord segments

The spinal cord is customarily divided into a series of segments based on the emergence of spinal nerves. One segment ends where the last ventral and dorsal rootlets go to one spinal canal, and the first rootlets of the next segment diverge toward the next spinal canal. The spinal cord itself shows no evidence of segmentation, and may be said to have apparent segmentation imposed on it by the anatomy of the vertebrae.

The naming of the spinal nerves is not straightforward. The first nerve, emerging between the occipital bone and the atlas, is called the first cervical nerve (C1), and subsequent spinal nerves are named for the vertebra caudal to their exit. This works well until the cervico-thoracic junction. In the thoracic region, it is obvious that the spinal nerves that follow the ribs should be named for the thoracic vertebra above their exit (T1, T2 etc), but that leaves the last cervical nerve in no man's land. It cannot be named after the vertebra caudal to its exit (T1) because that name is already logically allocated to the nerve that follows the first rib. The solution has been to call this nerve the eighth cervical nerve, even though there are only seven cervical vertebrae.

Caudal to the T1 segment, all spinal nerves (and their parent segments) are named for the vertebra rostral to their exit (the L1 spinal nerve exits caudal to the first lumbar vertebra etc).

Identification of regions and segments of the spinal cord in the rat and mouse

Unfortunately, the traditional regional subdivision of the spinal cord into cervical, thoracic, lumbar, sacral, and coccygeal regions does not correlate well with the functional regions of the spinal cord. For example, the expanded ventral horn which houses the forelimb motoneurons begins at the C5 segment and ends in the T1 segment in rodents, and the region containing sympathetic preganglionic cells begins at the T2 segment and ends in the L1 segment. In both cases they disregard the vertebral transition point. On the other hand, the functional regions correlate well with regions of *Hox* gene expression in the limb enlargements and in the sympathetic preganglionic region (for a more detailed discussion of this issue, see Chapter 17).

The best way to view the functional subdivisions of the spinal cord is in ChAT or AchE sections, because they give a clear picture of motoneuron groups in the ventral horn as well as the sympathetic and parasympathetic preganglionic cell groups. This enables the identification of a number of segments or segment groups to aid localization of segments. These are:

- C3 (prominent trapezius/sternocleidomastoid nucleus in lateral position)
- C4 (phrenic nucleus)
- C5-C6 region (biceps and deltoid motoneuron groups)
- C7-C8 (triceps and forelimb motoneuron groups)
- T1 (manus motoneurons)
- T2-L1 (sympathetic intermediolateral horn)
- L2-:L3 (quadriceps and adductor motoneuron groups)
- L4-L5 (gluteal and crural motoneuron groups)
- L6 (pes motoneurons and rostral Onuf's nucleus)
- S1-S2 (sacral preganglionic cell group)
- S3-Co3 (tail motoneurons)

This schema can be applied with minor modification to all mammalian spinal cords. Further subdivision can be achieved by comparison of sections with those of the atlas master set.

References

Franklin KBJ, Paxinos G (2007) *The Mouse Brain in Stereotaxic Coordinates.* 3rd edn., Academic Press, San Diego.

Llewellyn-Smith IJ, Pilowsky P, Minson JB (1993) The tungstate-stabilized tetramethylbenzidine reaction for light and electron microscopic immunocytochemistry and for revealing biocytin-filled neurons. J Neurosci Methods 46, 27-40.

Paxinos G, Watson C (2007) *The Rat Brain in Stereotaxic Coordinates.* 6th edn., Academic Press, San Diego.

Rexed B (1952) The cytoarchitectonic organization of the spinal cord in the cat. J Comp Neurol 96, 414-495.

Scheibel ME, Scheibel AB (1970) Organization of spinal motoneuron dendrites in bundles. Exp Neurol 28, 106-112.

Sidman RL, Angevine J B Jr, Taber Pierce E (1971) *Atlas of the Mouse Brain and Spinal Cord.* Harvard University Press, Cambridge

Sterling P, Kuypers HG (1968) Anatomical organization of the brachial spinal cord in the cat. III. The propriospinal connections. Brain Res 7, 419-443.

List of structures

Long name	Abbreviation
accessory nerve nucleus	11N
spinal accessory nerve	11n
lamina 1 of the spinal gray	1Sp
lamina 2 of the spinal gray	2Sp
lamina 2 of the spinal gray, inner part	2SpI
lamina 2 of the spinal gray, outer part	2SpO
lamina 3 of the spinal gray	3Sp
lamina 4 of the spinal gray	4Sp
lamina 5 of the spinal gray	5Sp
lamina 5 of the spinal gray, lateral part	5SpL
lamina 5 of the spinal gray, medial part	5SpM
lamina 6 of the spinal gray	6Sp
lamina 6 of the spinal gray, lateral part	6SpL
lamina 6 of the spinal gray, medial part	6SpM
lamina 7 of the spinal gray	7Sp
lamina 8 of the spinal gray	8Sp
lamina 9 of the spinal gray	9Sp
lamina 10 of the spinal gray	10Sp
adductor motoneurons of lamina 9	Ad9
anterior spinal artery	asp
axial muscle motoneurons of lamina 9	Ax9
biceps motoneurons of lamina 9	Bi9
central canal	CC
cauda equina	CE
central cervical nucleus	CeCv
crural extensor motoneurons of lamina 9	CEx9
crural flexor motoneurons of lamina 9	CFl9
coccygeal vertebra	CoVert
cremaster motoneurons of lamina 9	Cr9
cuneate fasciculus	cu
dorsal nucleus (Clarke)	D
dorsal corticospinal tract	dcs
deltoid motoneurons of lamina 9	De9
dorsal funiculus	df
dorsolateral fasciculus (Lissauer)	dl
dorsal root	dr
dorsal root ganglion	DRG
external anal sphincter motoneurons of lamina 9	ExA9
external urethral sphincter motoneurons of lamina 9	ExU9
forearm extensor motoneurons of lamina 9	FEx9
forearm flexor motoneurons of lamina 9	FFl9
gluteal motoneurons of lamina 9	Gl9
gracile fasciculus	gr
hamstring motoneurons of lamina 9	Hm9
internal basilar nucleus	IB
intercalated nucleus	ICl
intercalated nucleus, paraependymal part	IClPe
intercostal muscle motoneurons of lamina 9	ICo9
infrahyoid muscle motoneurons of lamina 9	IH9
intermediolateral column	IML
intermediomedial column	IMM
iliopsoas motoneurons of lamina 9	IPs9
lateral cervical nucleus	LatC
latissimus dorsi motoneurons of lamina 9	LD9
lumbar dorsal commissural nucleus	LDCom
lateral funiculus	lf
lumbar precerebellar nucleus	LPrCb
levator scapulae motoneurons of lamina 9	LS9
lateral spinal nucleus	LSp
manus motoneurons of lamina 9	Man9
pectoral muscle motoneurons of lamina 9	Pec9
pes motoneurons of lamina 9	Pes9
phrenic motoneurons of lamina 9	Ph9
psoas motoneurons of lamina 9	Ps9
postsynaptic dorsal column pathway	psdc
quadriceps motoneurons of lamina 9	Q9
quadratus lumborum motoneurons of lamina 9	QL9
rhomboid muscle motoneurons of lamina 9	Rh9
rubrospinal tract	rs
scalene muscle motoneurons of lamina 9	Sc9
sacral dorsal commissural nucleus	SDCom
supraspinatus and infraspinatus motoneurons of lamina 9	SI9
sternomastoid motoneurons of lamina 9	SM9
spinous process of vertebra	SpPr
sacral precerebellar nucleus	SPrCb
sacral parasympathetic nucleus	SPSy
serratus anterior motoneurons in lamina 9	Sr9
tail muscle motoneurons of lamina 9	Tail9
thoracoabdominal wall muscle motoneurons of lamina 9	ThAb9
triceps motoneurons of lamina 9	Tr9
trapezius motoneurons of lamina 9	Tz9
trapezius and sternomastoid motoneurons of lamina 9	TzSM9
vertebral arch	VAr
vertebral body	VB
ventral funiculus	vf
ventral median fissure	vmf
ventral root	vr
ventral white commissure	vwc

Permission to reproduce figures

Acknowledgements

We are greatly indebted to Charles Scouten (Myneurolab.com, Leica Biosystems St. Louis LLC, 5918 Evergreen Blvd, St. Louis MO 63134) their donation of a CryoJane Tape Transfer System. This system enabled us to section the spinal cord with significantly less distortion than otherwise, and we recommend it to those who section brain or spinal cord. Financial support for the construction of *Atlas of the Rat Spinal Cord* was provided by the Office of Science and Medical Research of the government of New South Wales.

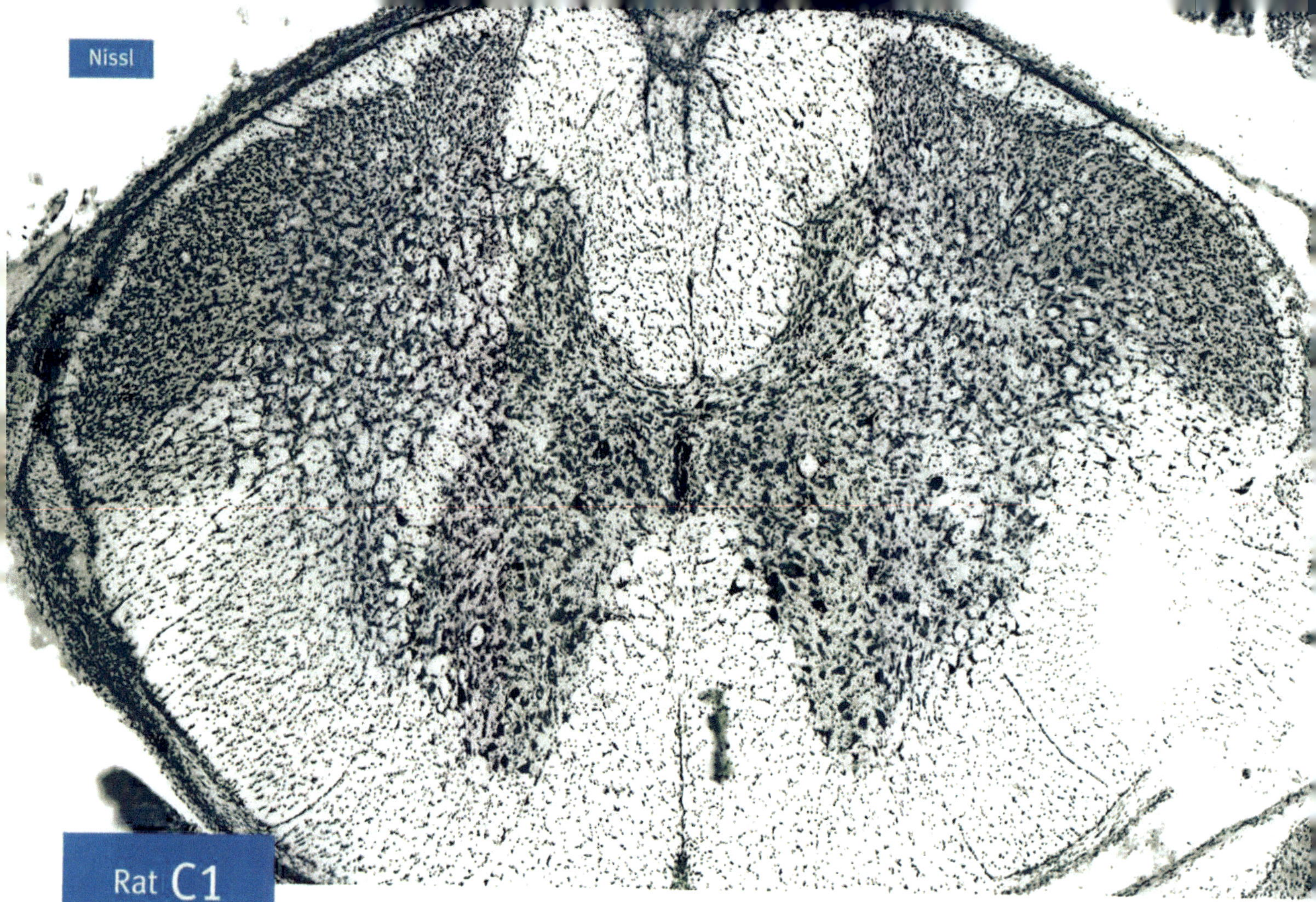
Nissl

Rat C1

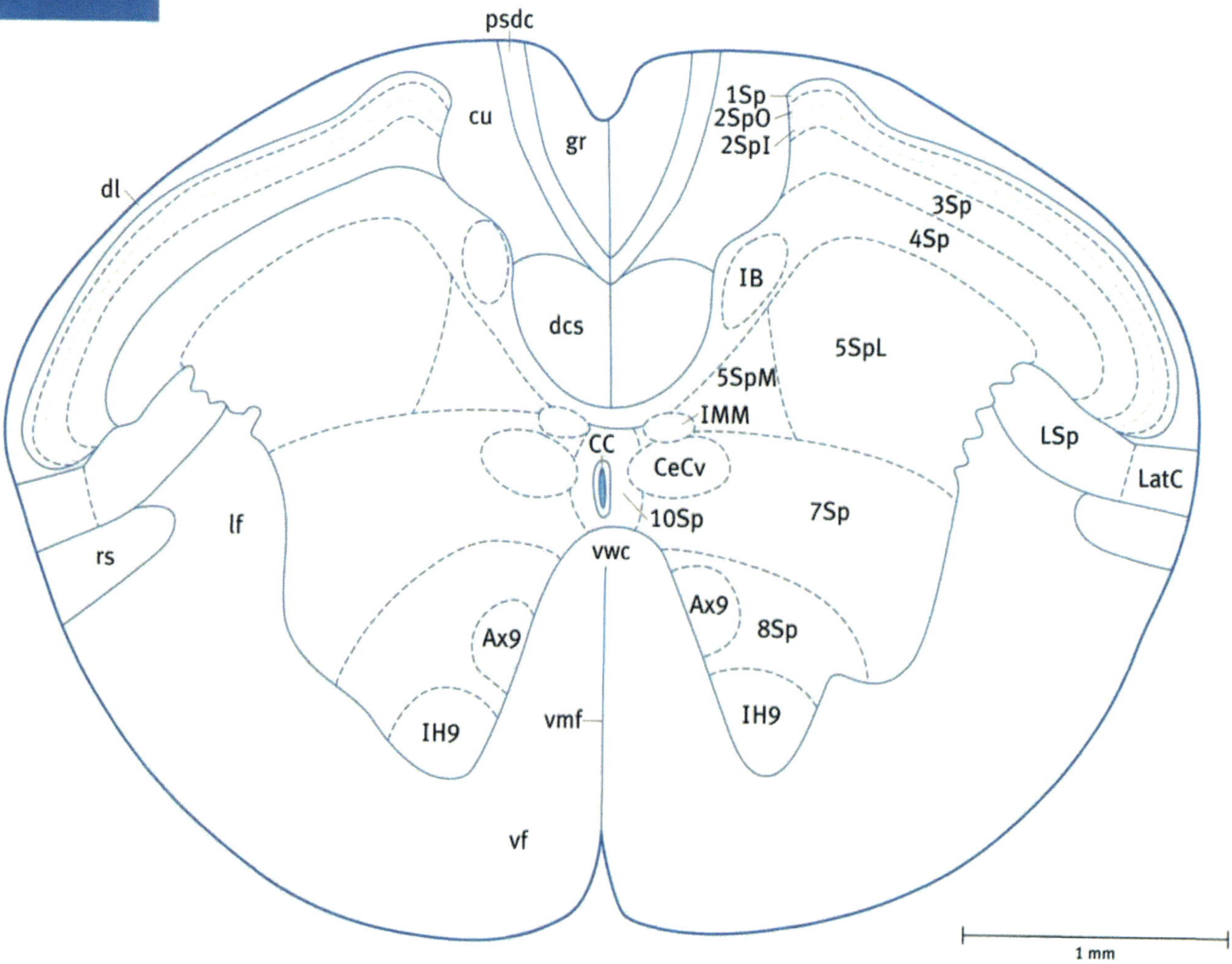
psdc
1Sp
2SpO
2SpI
cu
gr
dl
3Sp
4Sp
IB
dcs
5SpL
5SpM
IMM
CC
CeCv
LSp
LatC
10Sp
7Sp
lf
rs
vwc
Ax9
Ax9
8Sp
IH9
IH9
vmf
vf
1 mm

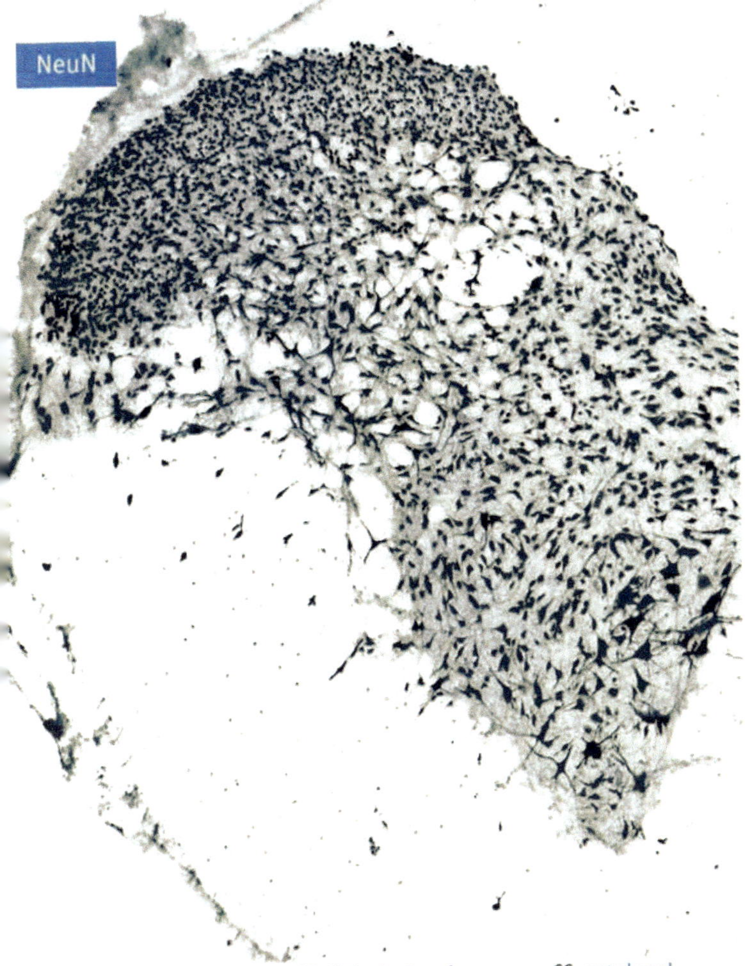

1Sp lamina 1
2SpI lamina 2, inner
2SpO lamina 2, outer
3Sp lamina 3
5SpL lamina 5, lat
5SpM lamina 5, med
7Sp lamina 7
8Sp lamina 8
10Sp lamina 10
Ax9 axial muscles
CC central canal
CeCv cent cervical nu
cu cuneate fasc
dcs dors corticosp
dl dorsolat fasc
gr gracile fascic
IB int basal nu
IH9 infrahyoid musc
IMM intermediomed
LatC lat cervical nu
LSp lat spinal nu
psdc postsyn dors col
rs rubrospinal tr
vmf vent med fiss
vwc vent white com

Rat C1

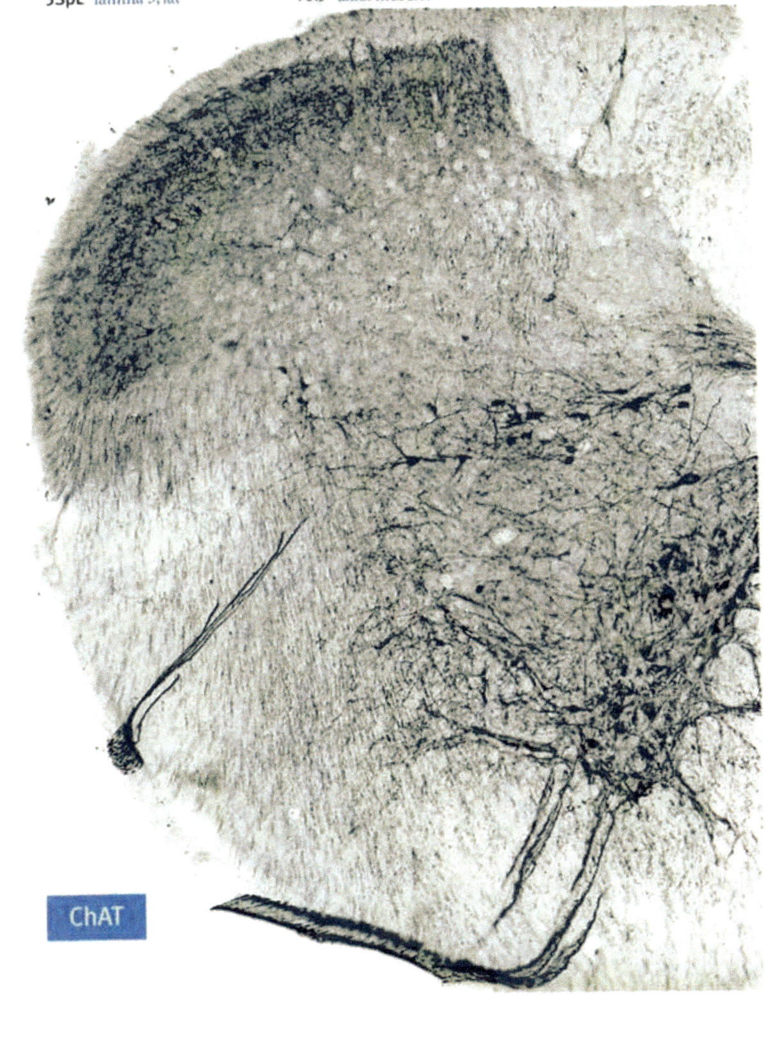

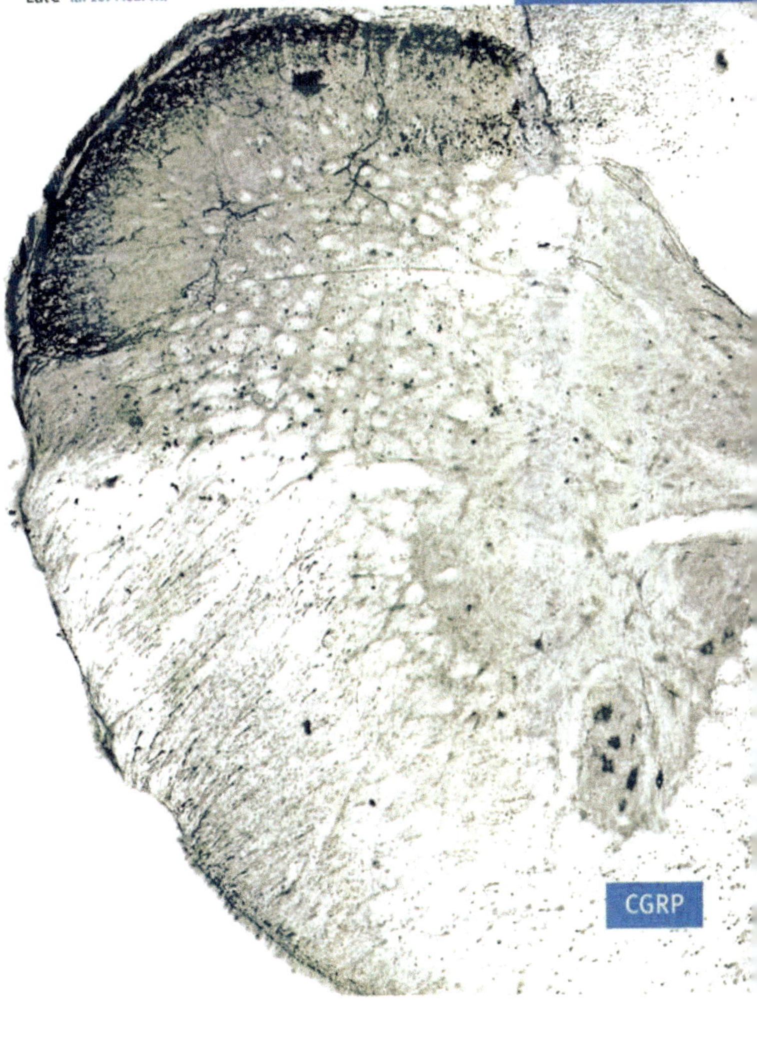

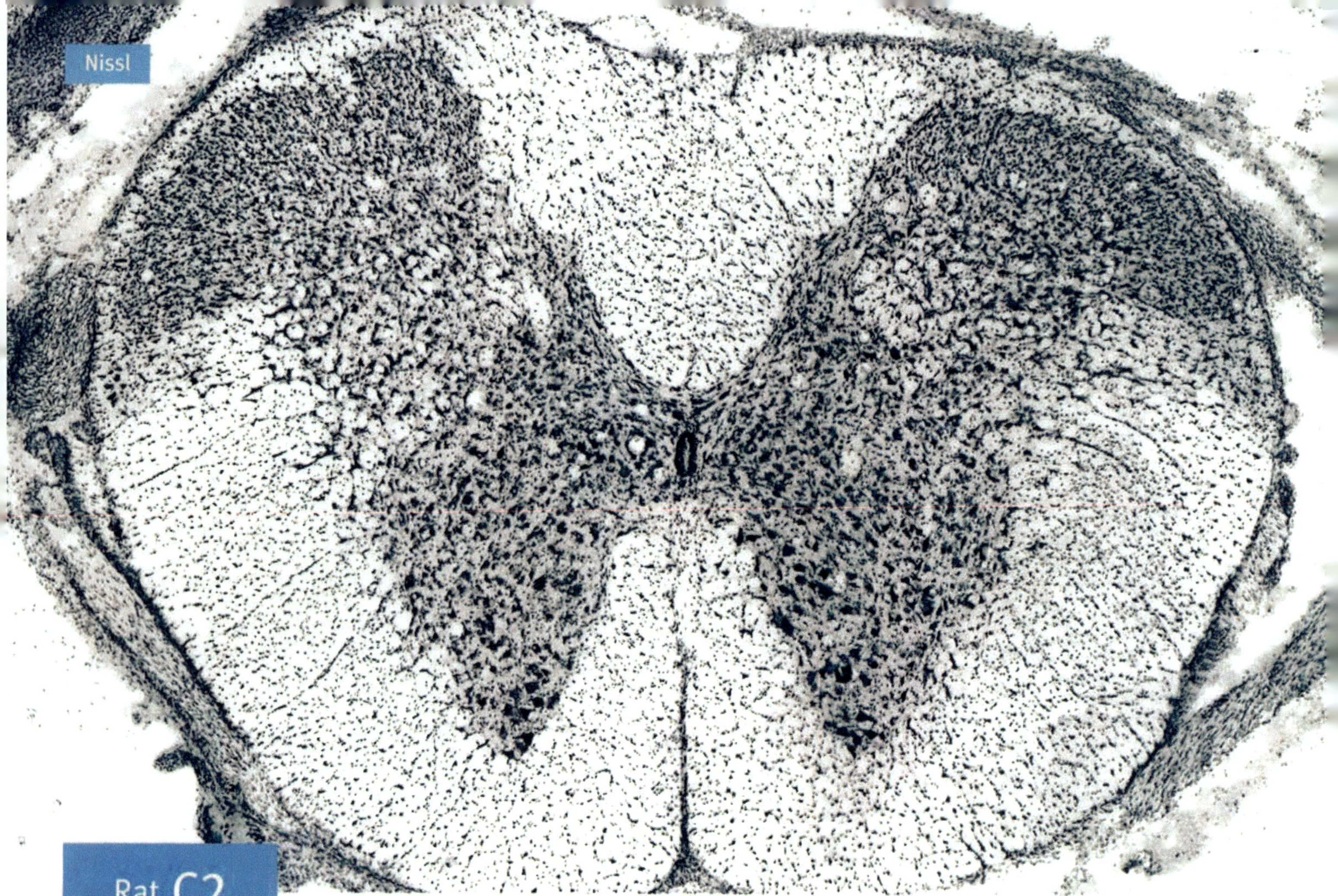
Nissl

Rat C2

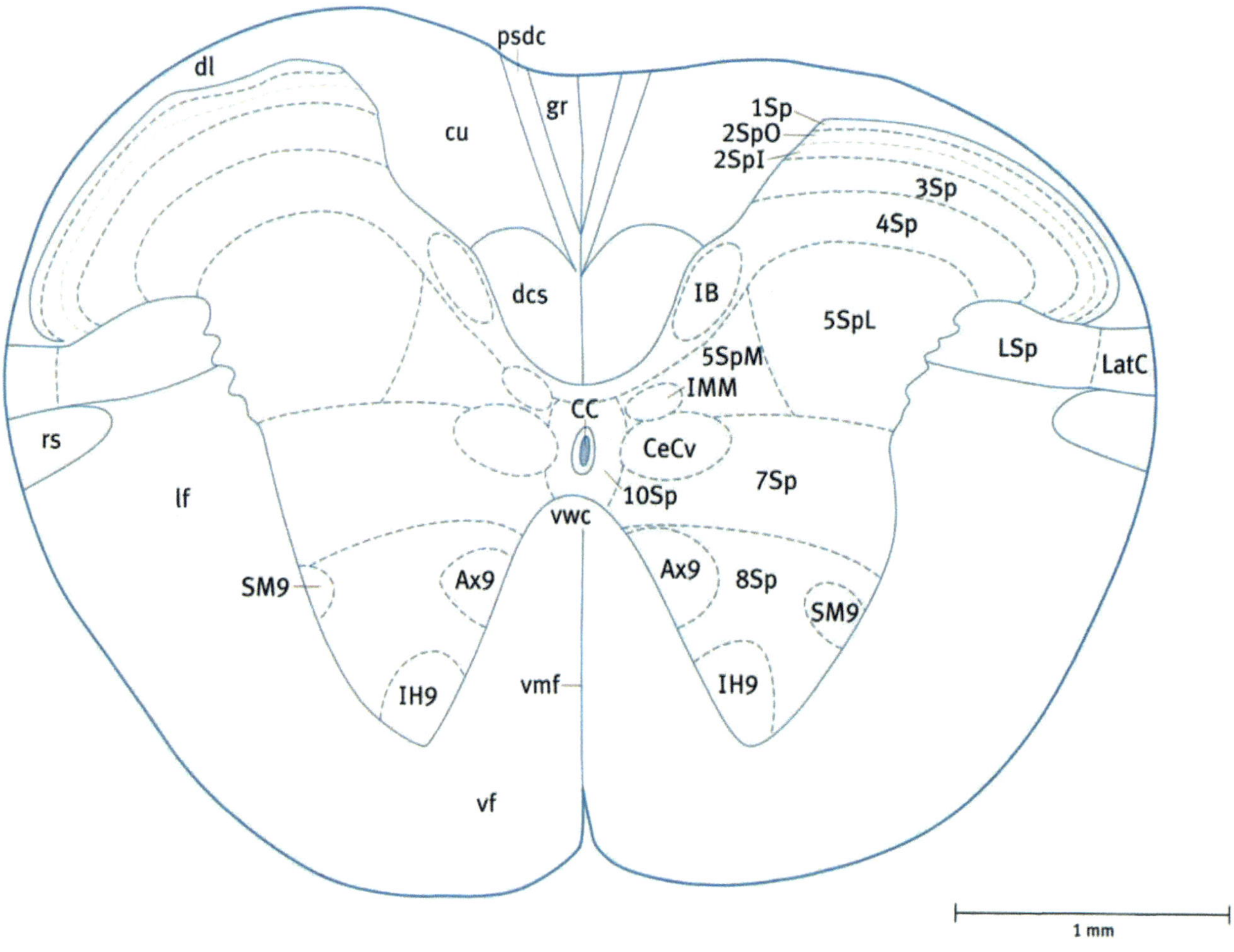
psdc
dl
gr
cu
1Sp
2SpO
2SpI
3Sp
4Sp
dcs
IB
5SpL
5SpM
LSp
LatC
IMM
CC
rs
CeCv
7Sp
10Sp
lf
vwc
SM9
Ax9
Ax9
8Sp
SM9
IH9
vmf
IH9
vf
1 mm

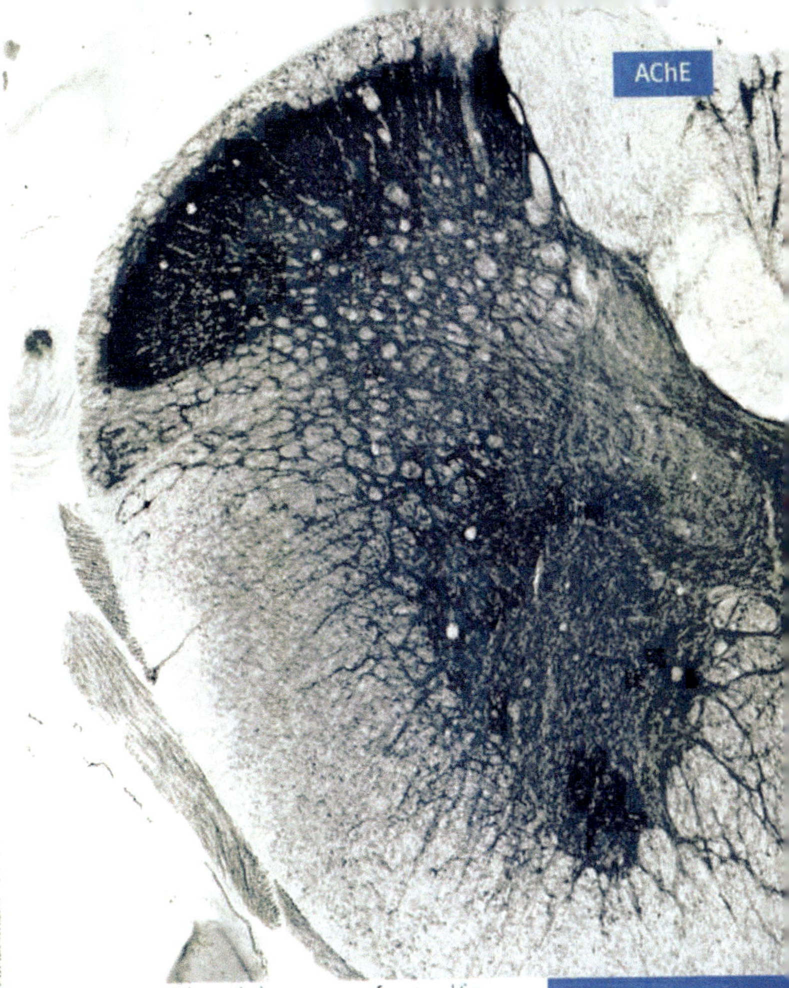

1Sp lamina 1
2SpI lamina 2, inner
2SpO lamina 2, outer
3Sp lamina 3
4Sp lamina 4
5SpL lamina 5, lat
5SpM lamina 5, med
7Sp lamina 7
8Sp lamina 8
10Sp lamina 10
Ax9 axial muscles
CC central canal
CeCv cent cervical nu
cu cuneate fasc
dcs dors corticosp
dl dorsolat fasc
gr gracile fascic
IB int basal nu
IH9 infrahyoid musc
IMM intermediomed
LatC lat cervical nu
LSp lat spinal nu
psdc postsyn dors col
rs rubrospinal tr
SM9 sternomastoid
vmf vent med fiss
vwc vent white com

Rat C2

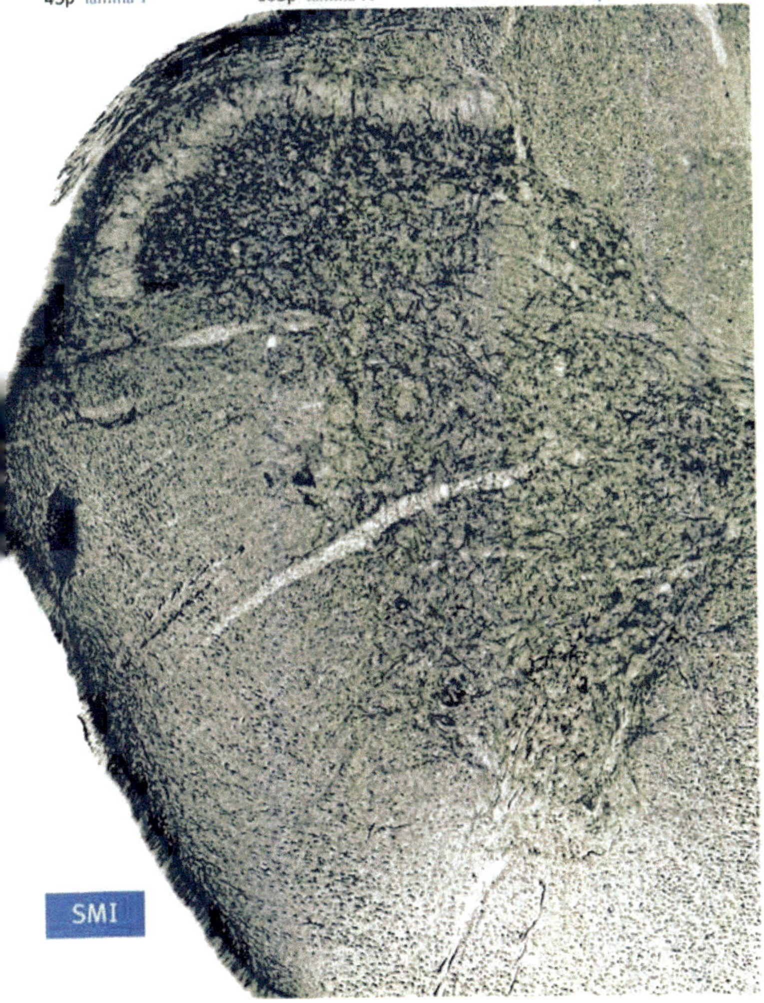

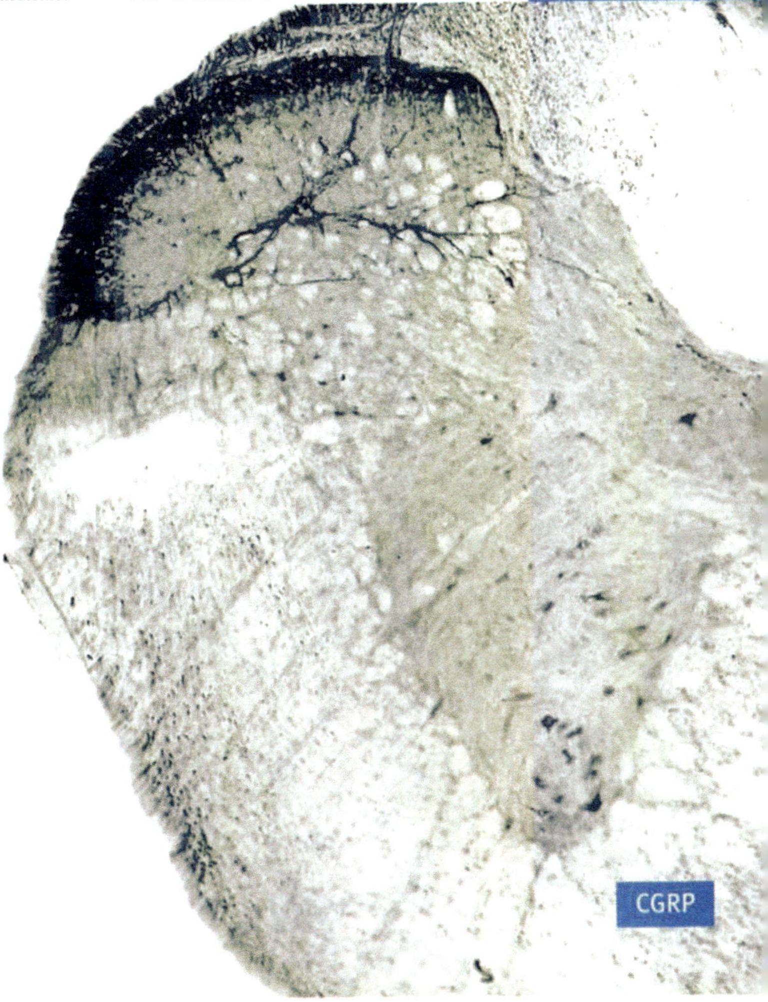

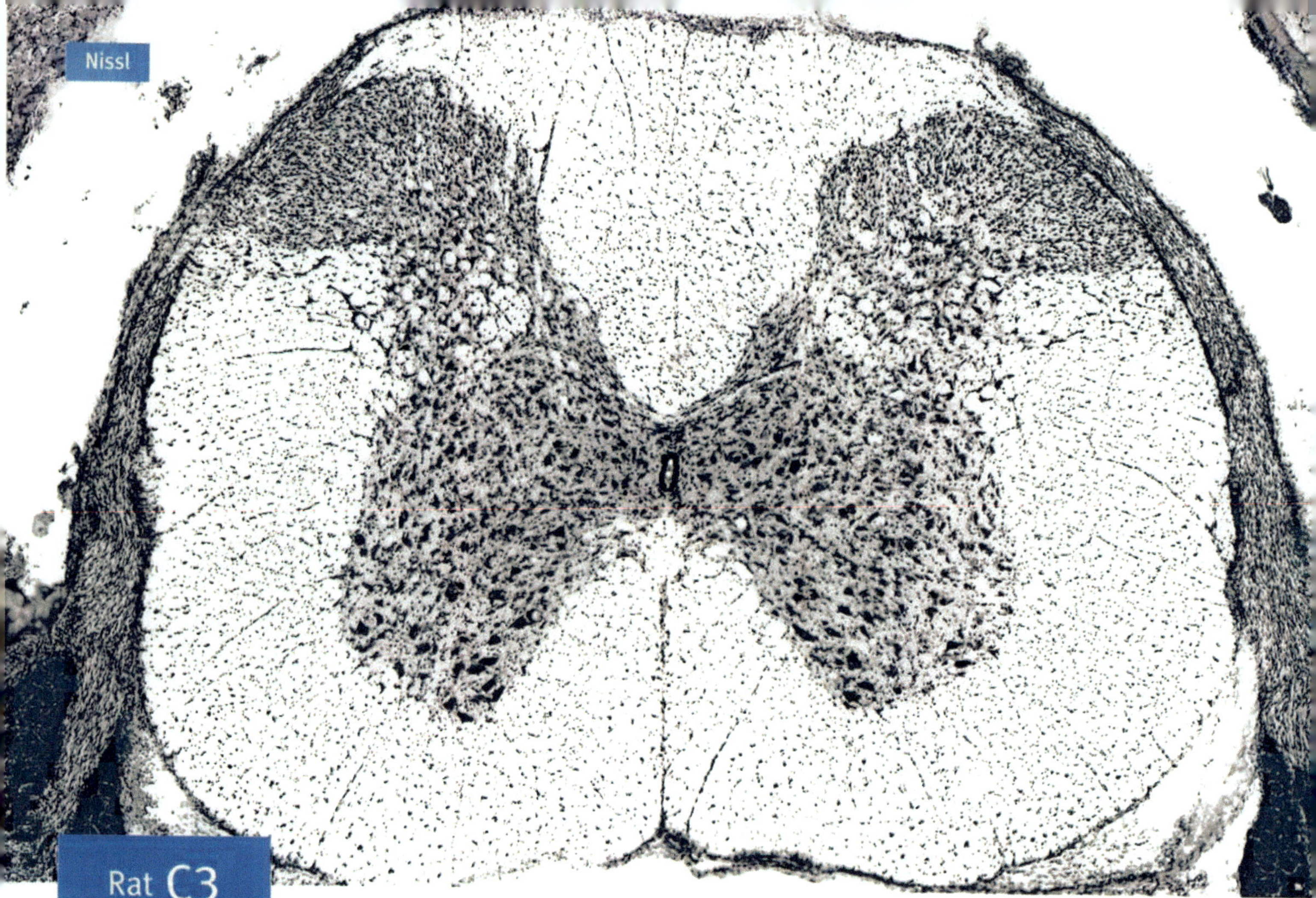
Nissl

Rat C3

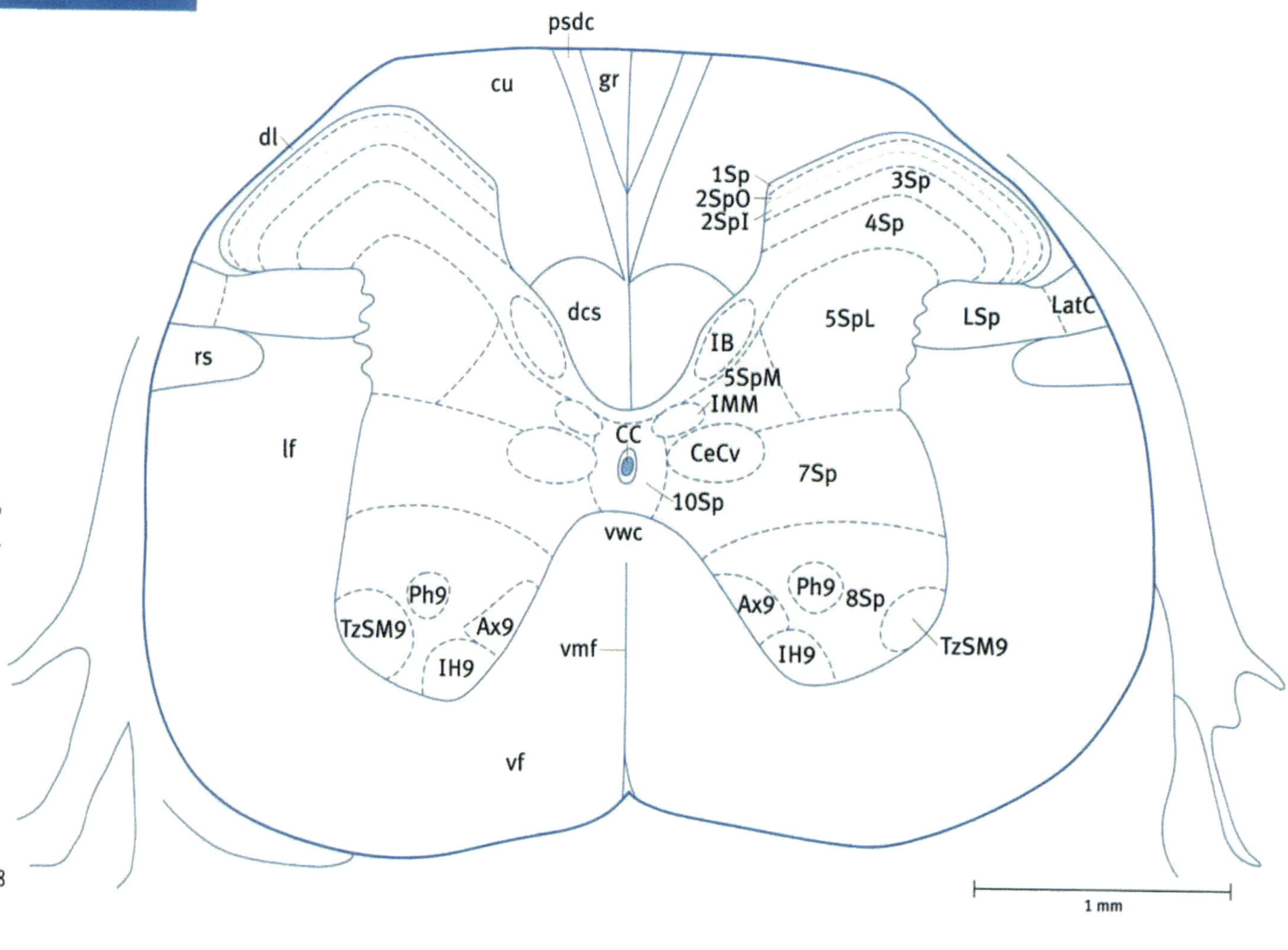
psdc
cu
gr
dl
1Sp
2SpO
2SpI
3Sp
4Sp
dcs
5SpL
LSp
LatC
rs
IB
5SpM
IMM
CC
CeCv
lf
7Sp
10Sp
vwc
Ph9
8Sp
Ph9
Ax9
TzSM9
Ax9
IH9
vmf
IH9
TzSM9
vf
1 mm

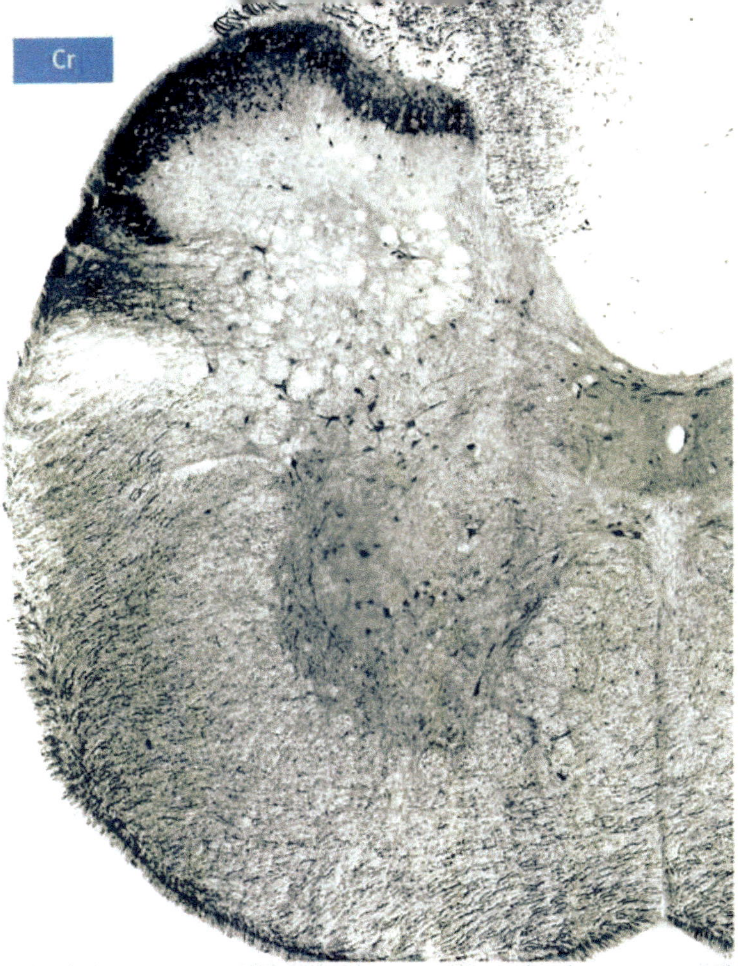

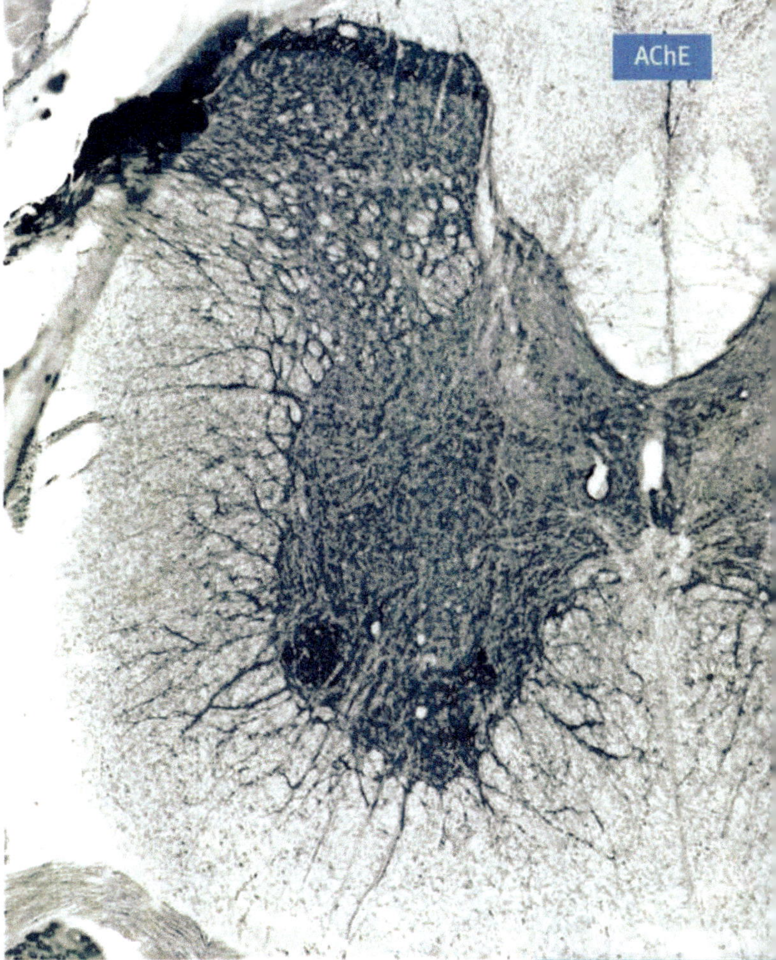

1Sp lamina 1
2SpI lamina 2, inner
2SpO lamina 2, outer
3Sp lamina 3
4Sp lamina 4
5SpL lamina 5, lat
5SpM lamina 5, med
7Sp lamina 7
10Sp lamina 10
Ax9 axial muscles
CC central canal
CeCv cent cervical nu
cu cuneate fasc
dcs dors corticosp
dl dorsolat fasc
gr gracile fascic
IB int basal nu
IH9 infrahyoid musc
IMM intermediomed
LatC lat cervical nu
LSp lat spinal nu
Ph9 phrenic
psdc postsyn dors col
rs rubrospinal tr
TzSM9 trap/sternom
vmf vent med fiss
vwc vent white com

Rat C3

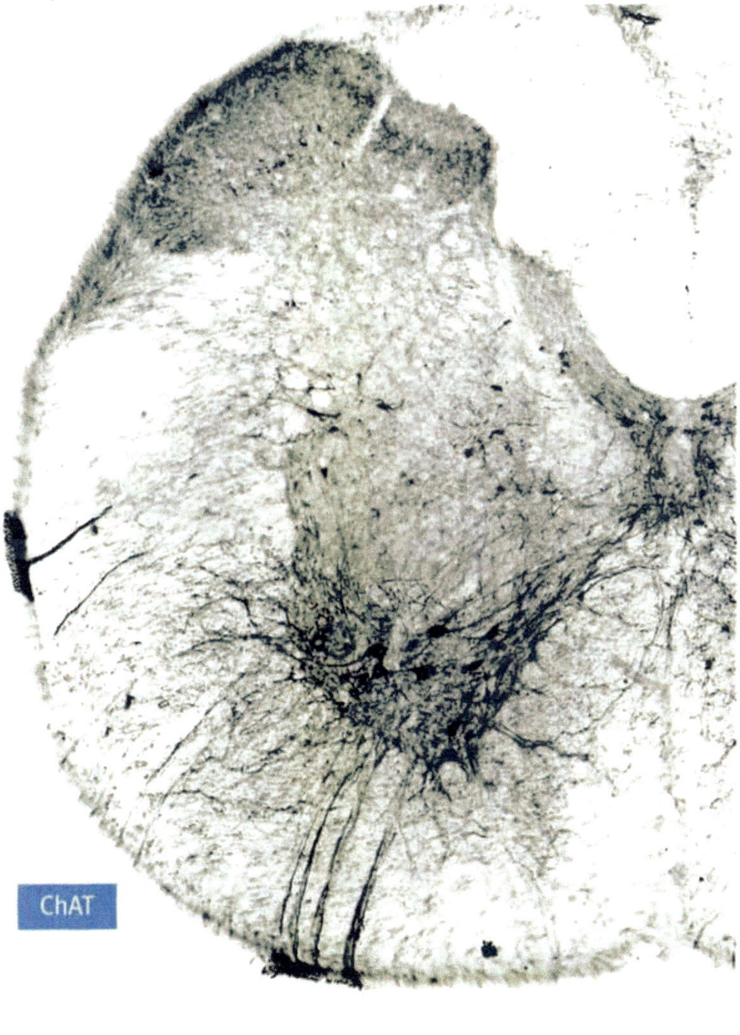

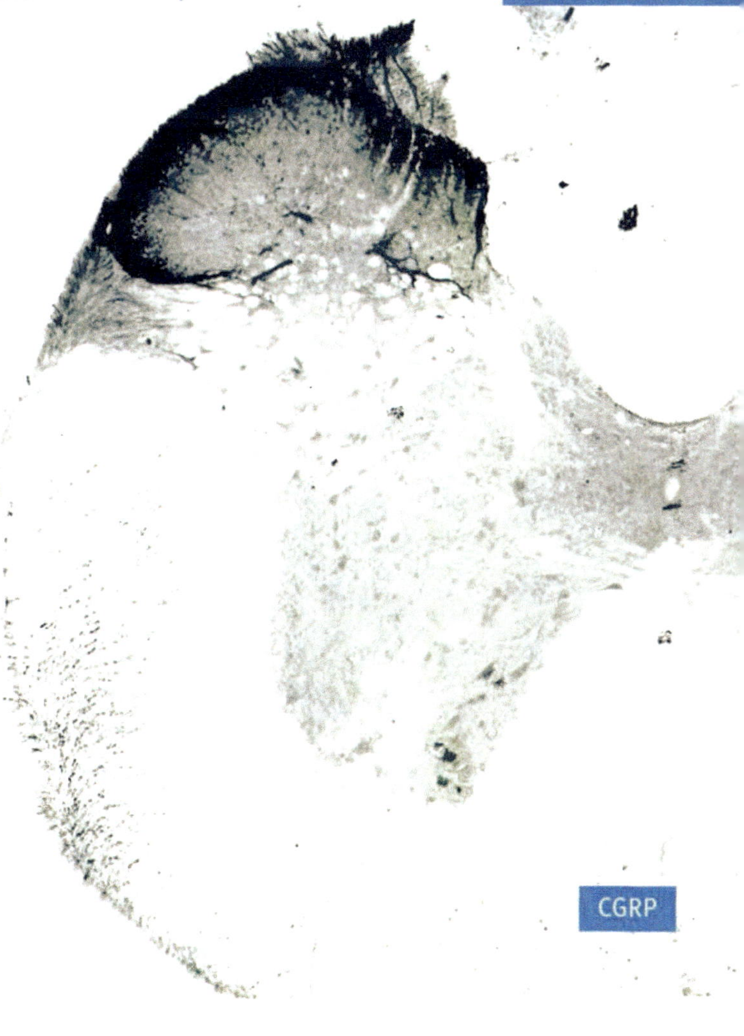

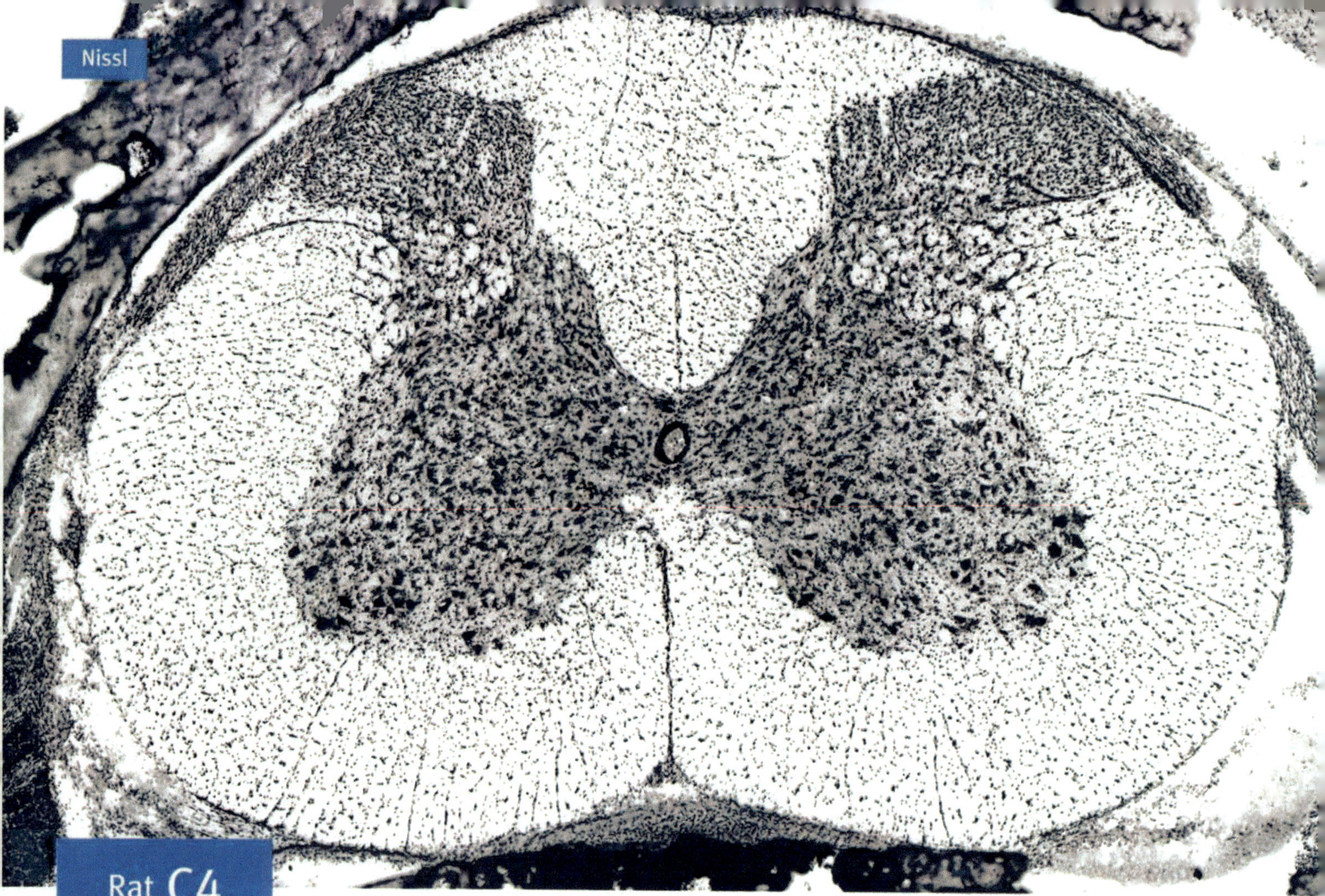
Nissl

Rat C4

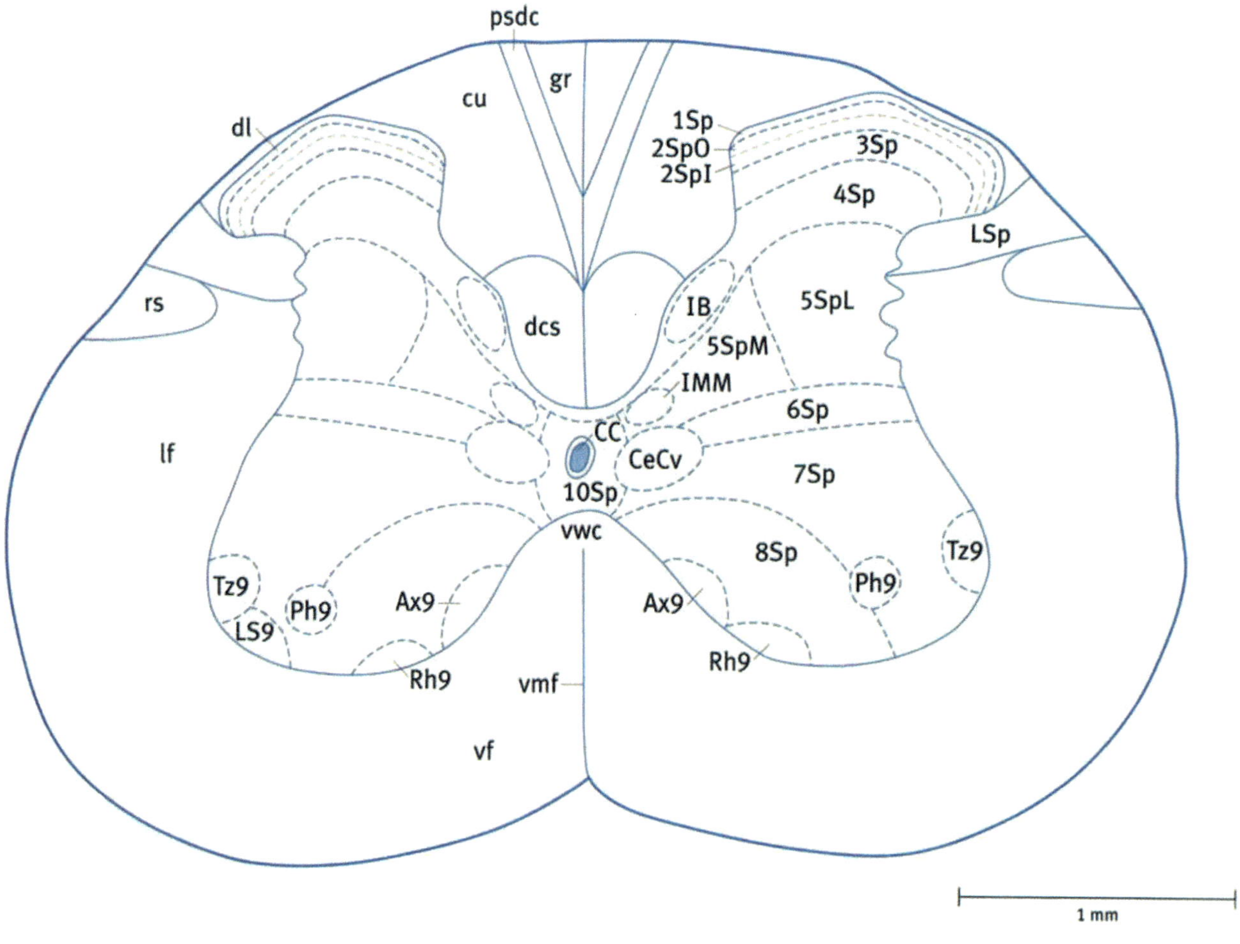
psdc
gr
cu
dl
1Sp
2SpO
2SpI
3Sp
4Sp
LSp
rs
dcs
IB
5SpL
5SpM
IMM
6Sp
CC
CeCv
10Sp
7Sp
lf
vwc
8Sp
Tz9
Ph9
Ax9
LS9
Rh9
vmf
vf
1 mm

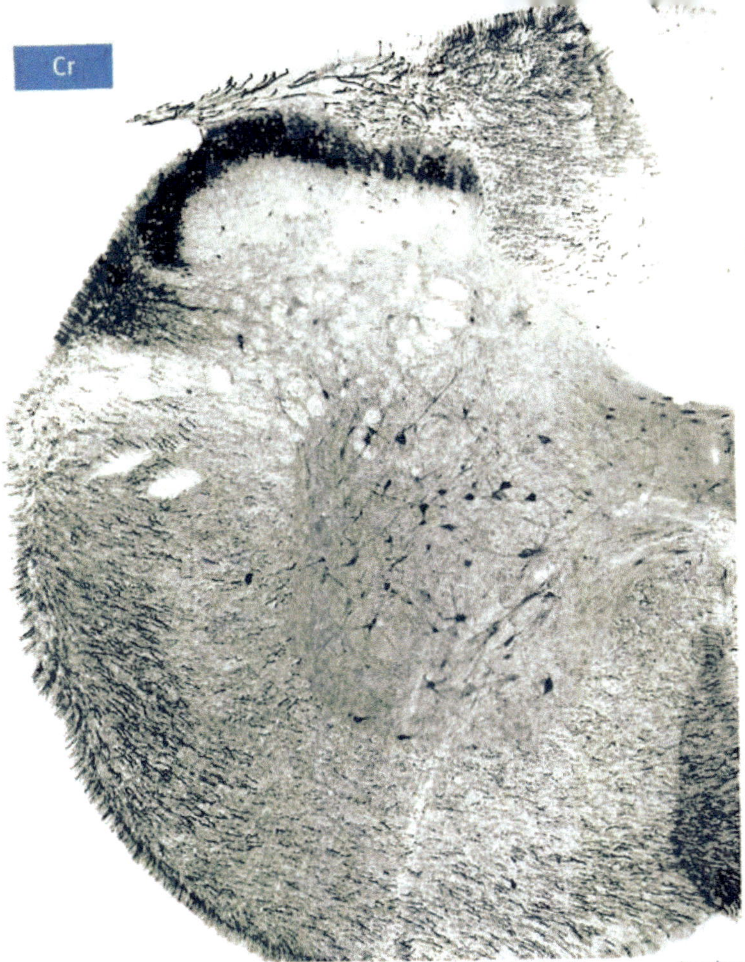

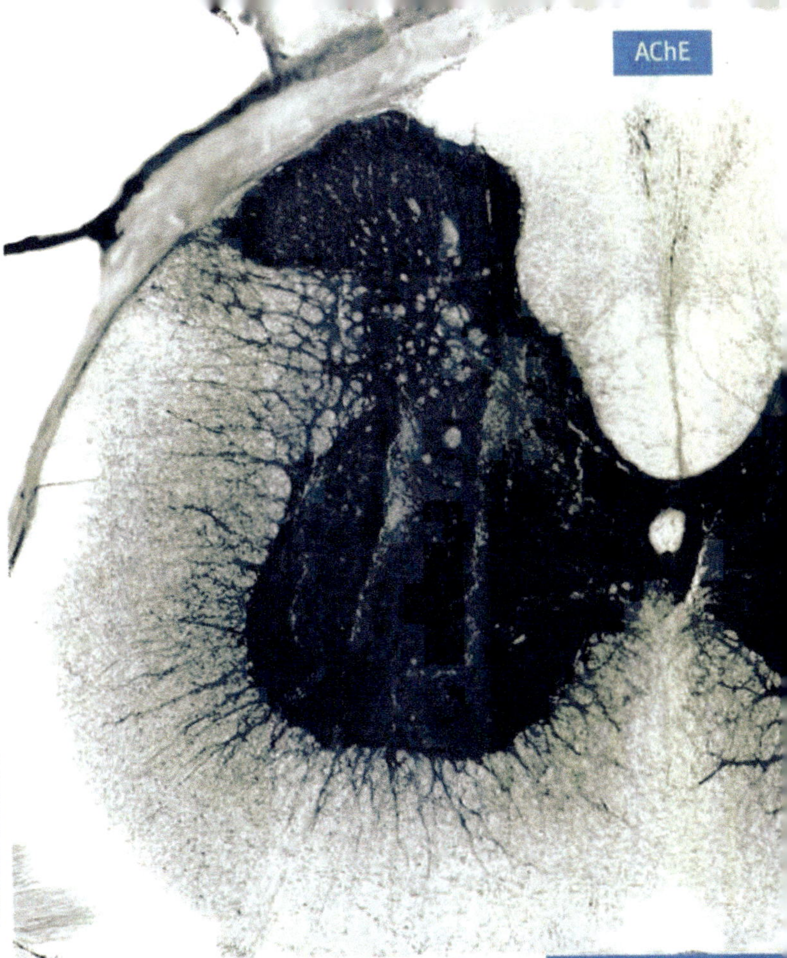

1Sp lamina 1
2SpI lamina 2, inner
2SpO lamina 2, outer
3Sp lamina 3
4Sp lamina 4
5SpL lamina 5, lat
5SpM lamina 5, med
6Sp lamina 6
7Sp lamina 7
8Sp lamina 8
10Sp lamina 10
Ax9 axial muscles
CC central canal
CeCv cent cervical nu
cu cuneate fasc
dcs dors corticosp
dl dorsolat fasc
gr gracile fascic
IB int basal nu
IMM intermediomed
LS9 levator scap
LSp lat spinal nu
Ph9 phrenic
psdc postsyn dors col
Rh9 rhomboid musc
rs rubrospinal tr
Tz9 trapezius
vmf vent med fiss
vwc vent white com

Rat C4

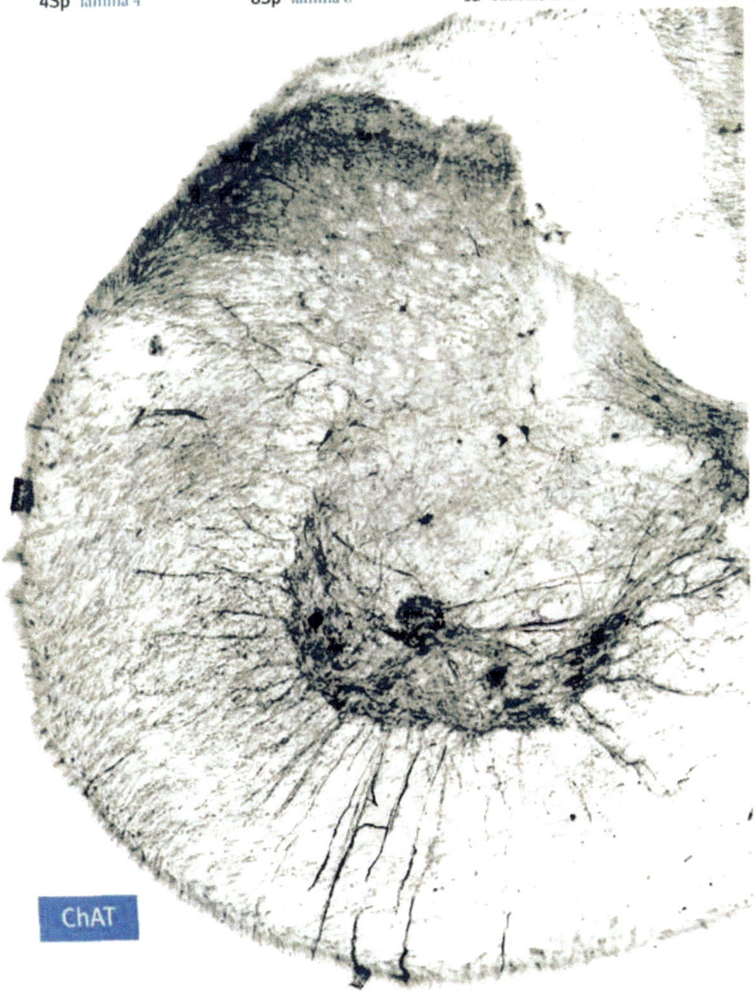

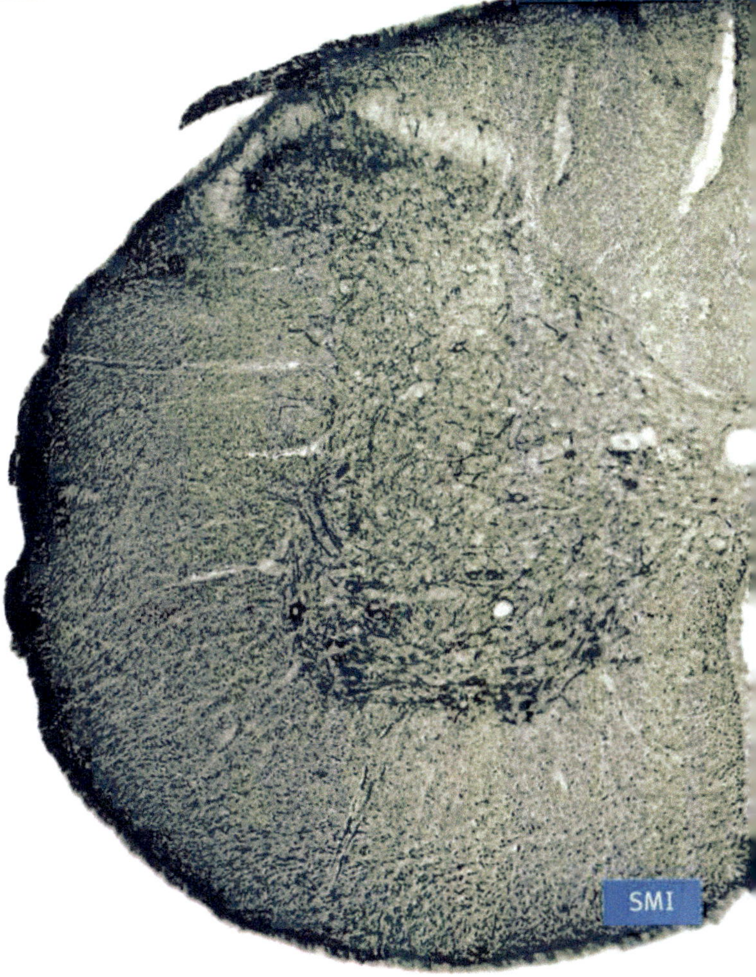

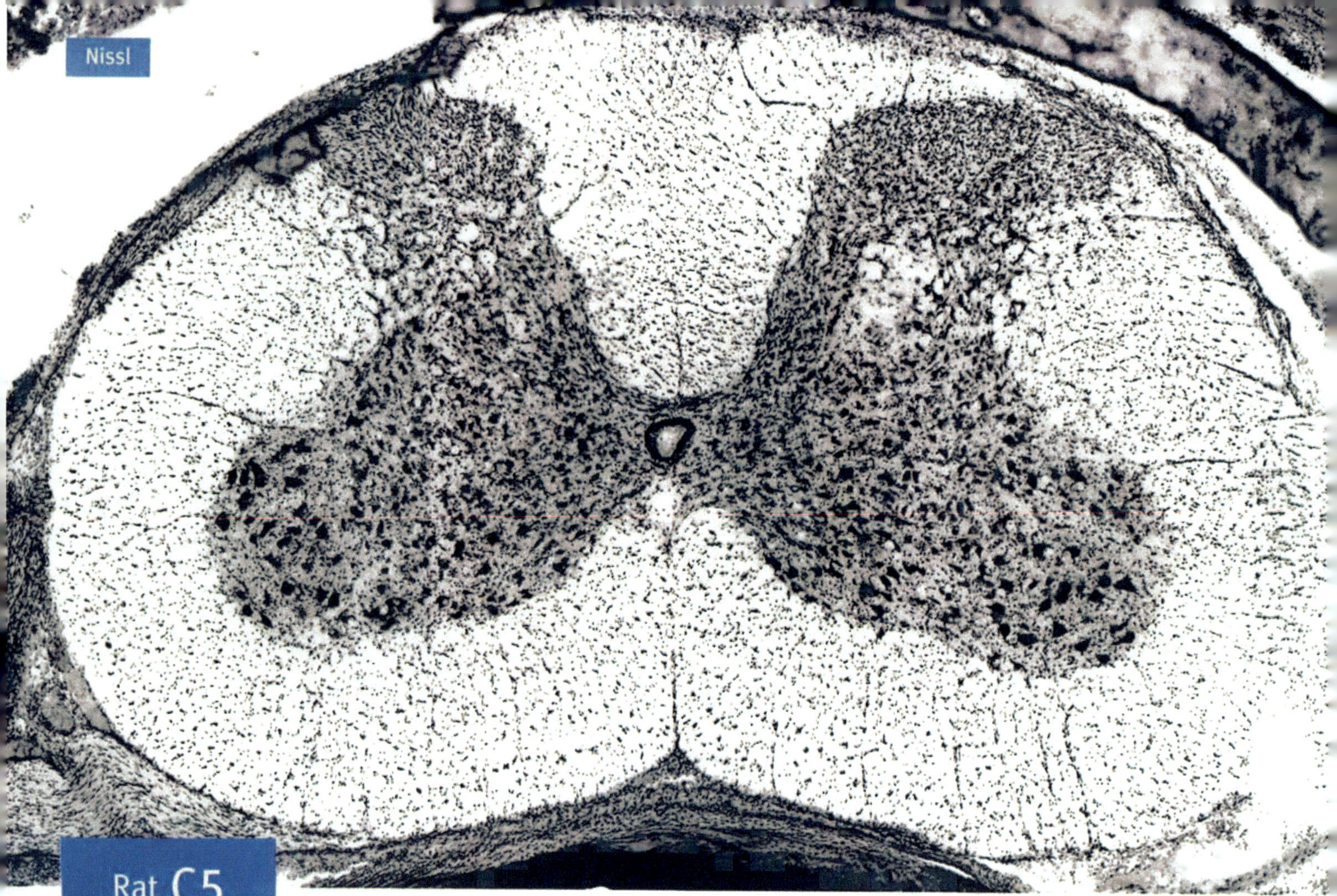
Nissl

Rat C5

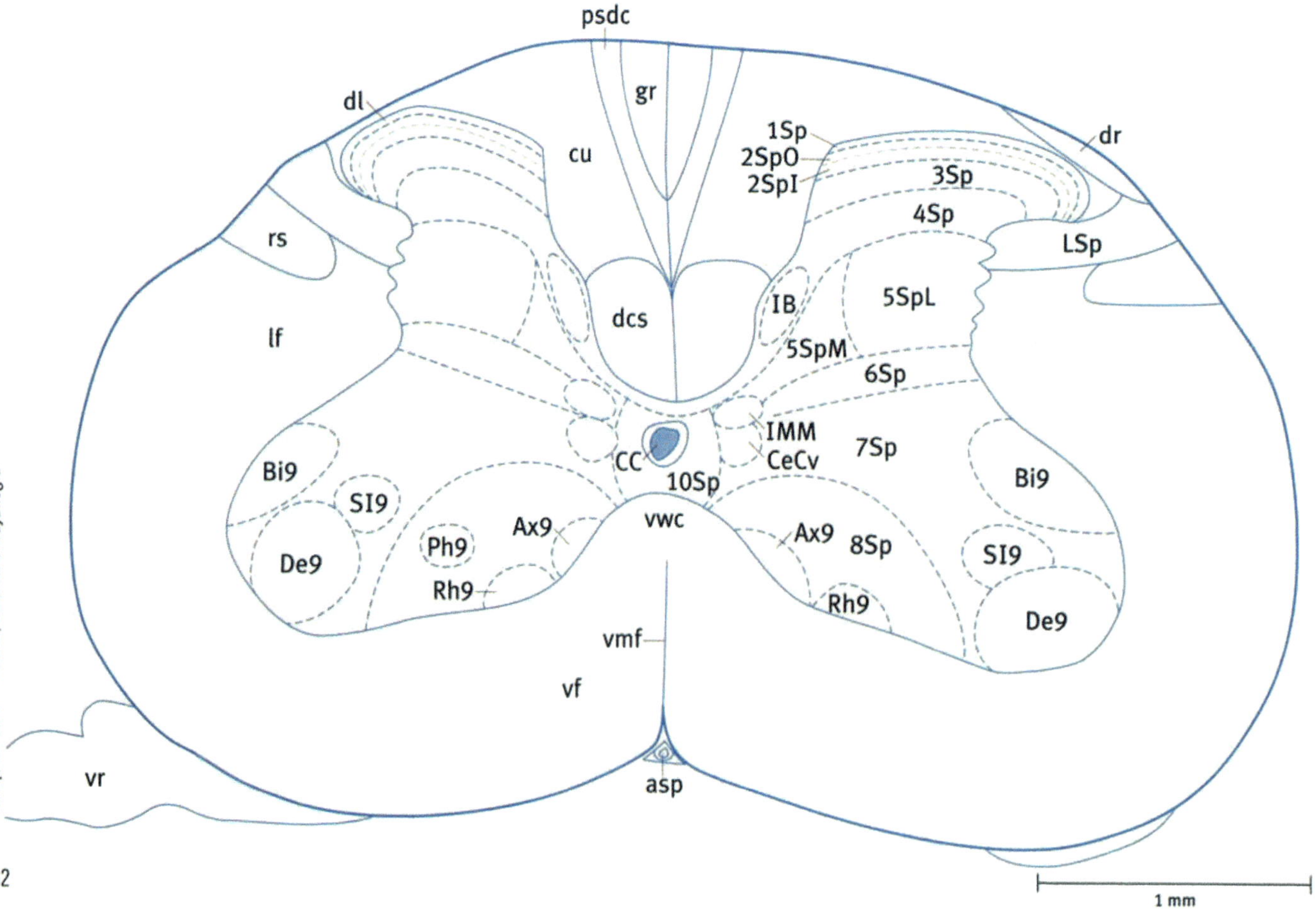
psdc
gr
dl
cu
1Sp
2SpO
2SpI
3Sp
dr
4Sp
rs
LSp
IB
5SpL
lf
dcs
5SpM
6Sp
IMM
CC
CeCv
7Sp
Bi9
10Sp
Bi9
SI9
vwc
Ax9
Ax9
8Sp
Ph9
SI9
De9
Rh9
Rh9
De9
vmf
vf
vr
asp
1 mm

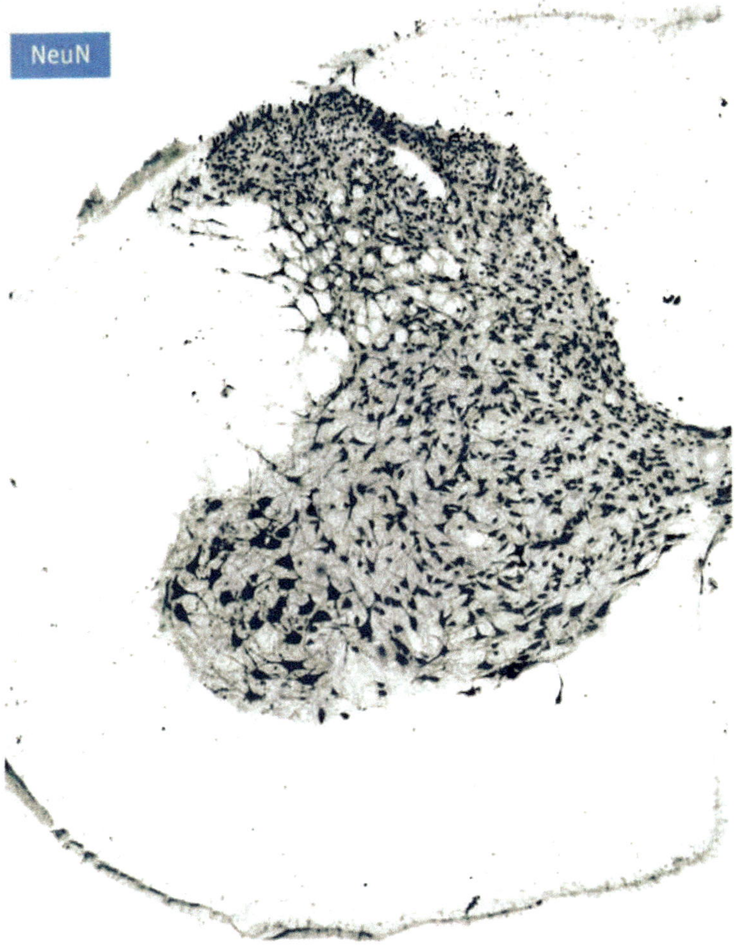

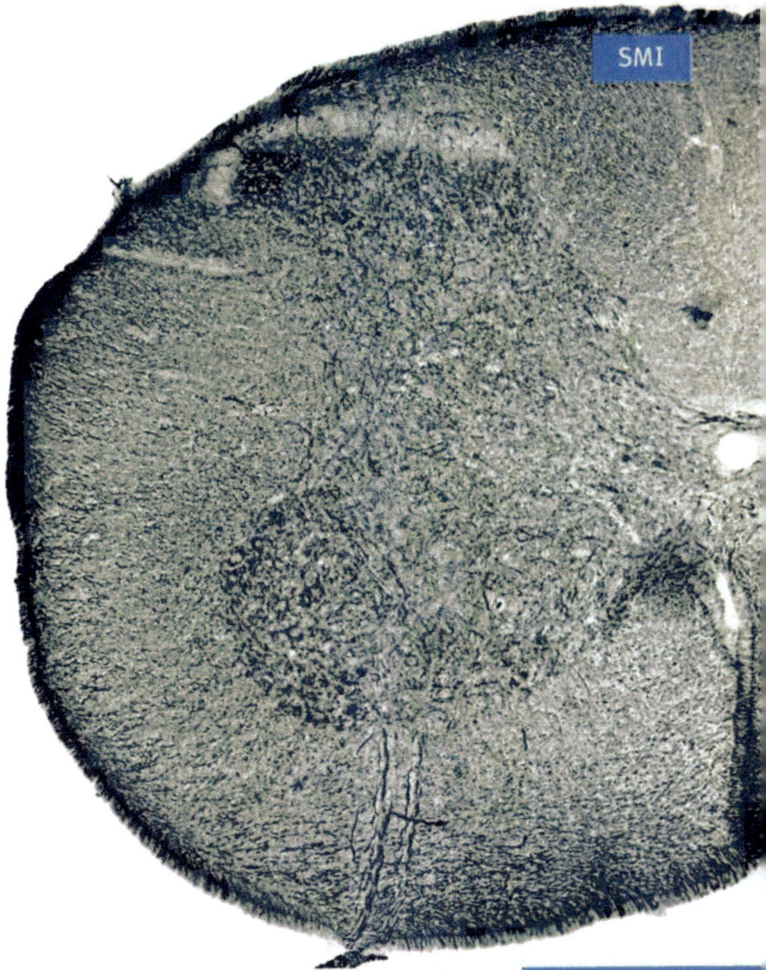

1Sp lamina 1
2SpI lamina 2, inner
2SpO lamina 2, outer
3Sp lamina 3
4Sp lamina 4
5SpL lamina 5, lat
6Sp lamina 6
7Sp lamina 7
8Sp lamina 8
asp ant spinal art
Ax9 axial muscles
Bi9 biceps
CC central canal
CeCv cent cervical nu
cu cuneate fasc
dcs dors corticosp
De9 deltoid
dl dorsolat fasc
dr dorsal root
gr gracile fascic
IB int basal nu
IMM intermediomed
LSp lat spinal nu
Ph9 phrenic
psdc postsyn dors col
Rh9 rhomboid musc
rs rubrospinal tr
SI9 supraspinatus
vmf vent med fiss
vr ventral root
vwc vent white com

Rat C5

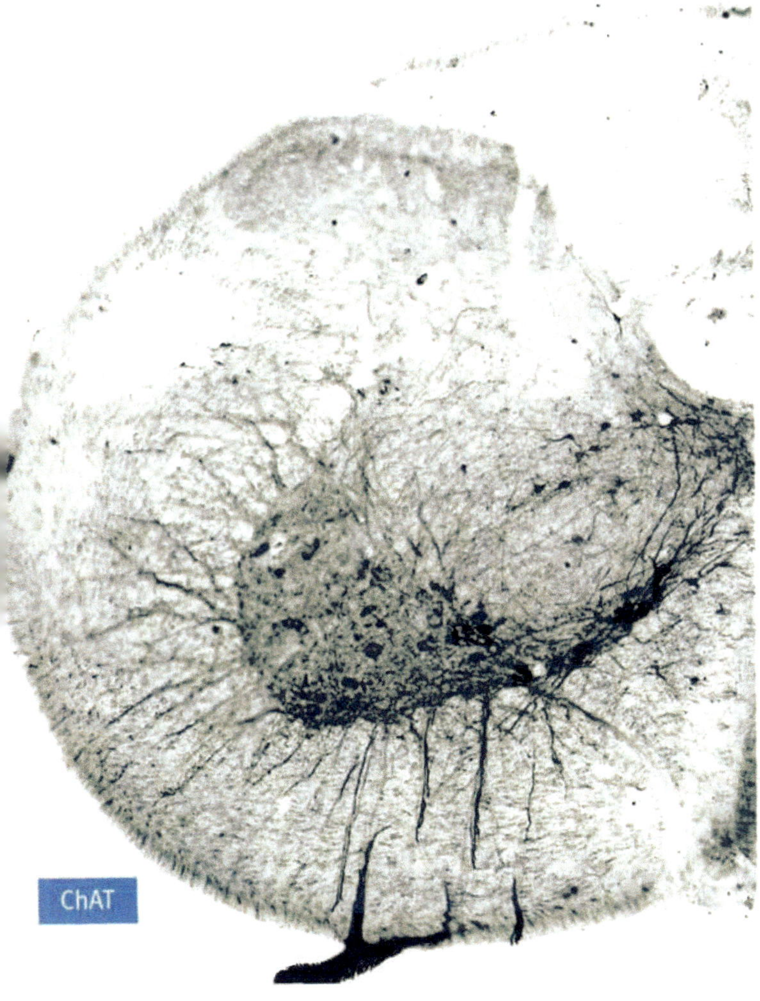

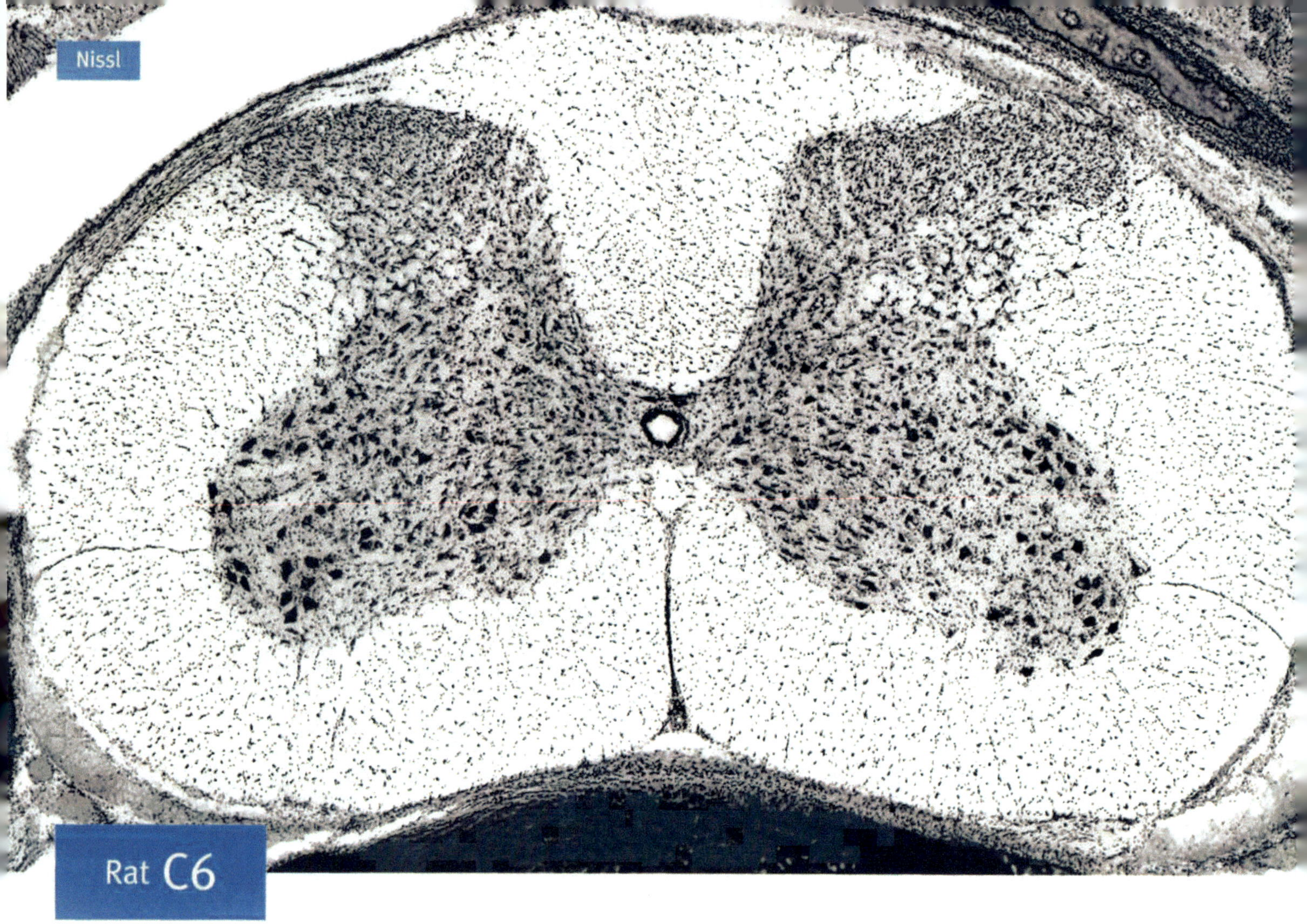
Nissl
Rat C6

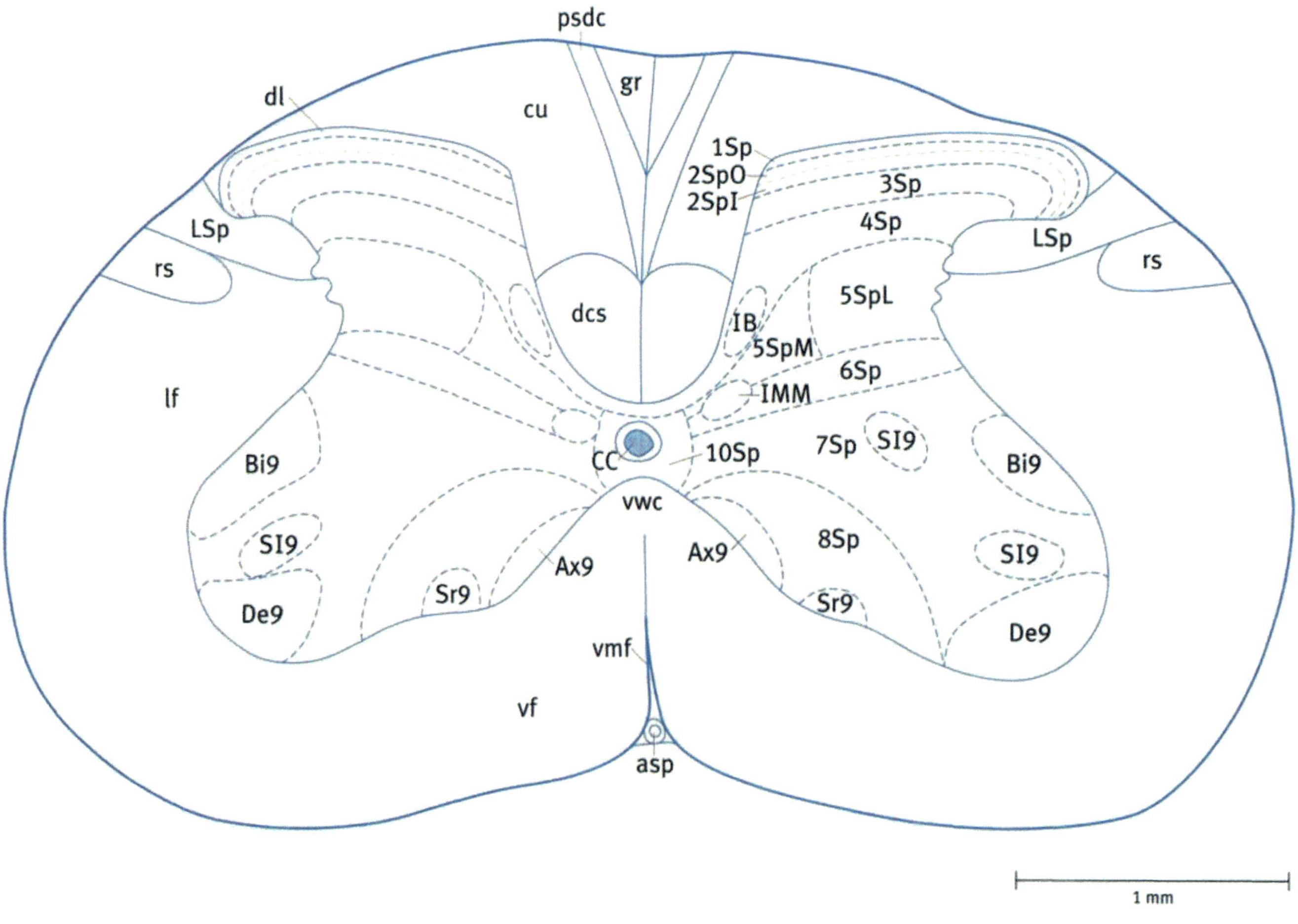
psdc
gr
dl
cu
1Sp
2SpO
2SpI
3Sp
4Sp
LSp
LSp
rs
rs
5SpL
dcs
IB
5SpM
6Sp
IMM
lf
SI9
7Sp
CC
10Sp
Bi9
Bi9
vwc
SI9
Ax9
Ax9
8Sp
SI9
Sr9
Sr9
De9
De9
vmf
vf
asp
1 mm

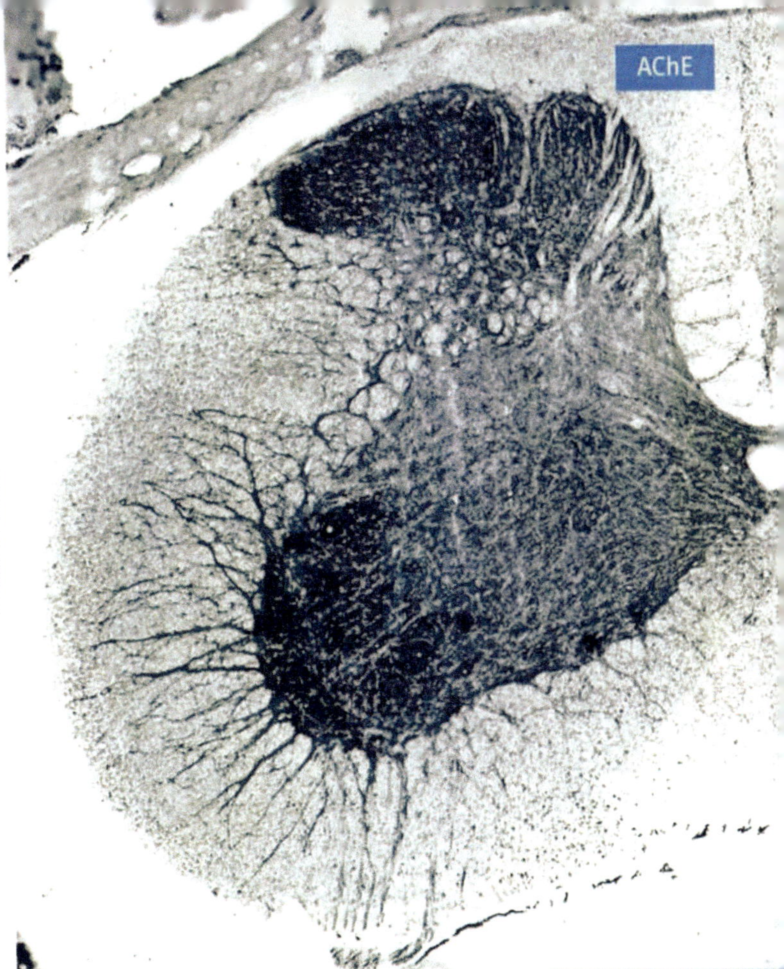

1Sp lamina 1
2SpI lamina 2, inner
2SpO lamina 2, outer
3Sp lamina 3
4Sp lamina 4
5SpL lamina 5, lat
5SpM lamina 5, med
6Sp lamina 6
7Sp lamina 7
8Sp lamina 8
10Sp lamina 10
asp ant spinal art
Ax9 axial muscles
Bi9 biceps
CC central canal
cu cuneate fasc
dcs dors corticosp
De9 deltoid
dl dorsolat fasc
gr gracile fascic
IB int basal nu
IMM intermediomed
LSp lat spinal nu
psdc postsyn dors col
rs rubrospinal tr
SI9 supraspinatus
Sr9 serratus ant
vmf vent med fiss
vwc vent white com

Rat C6

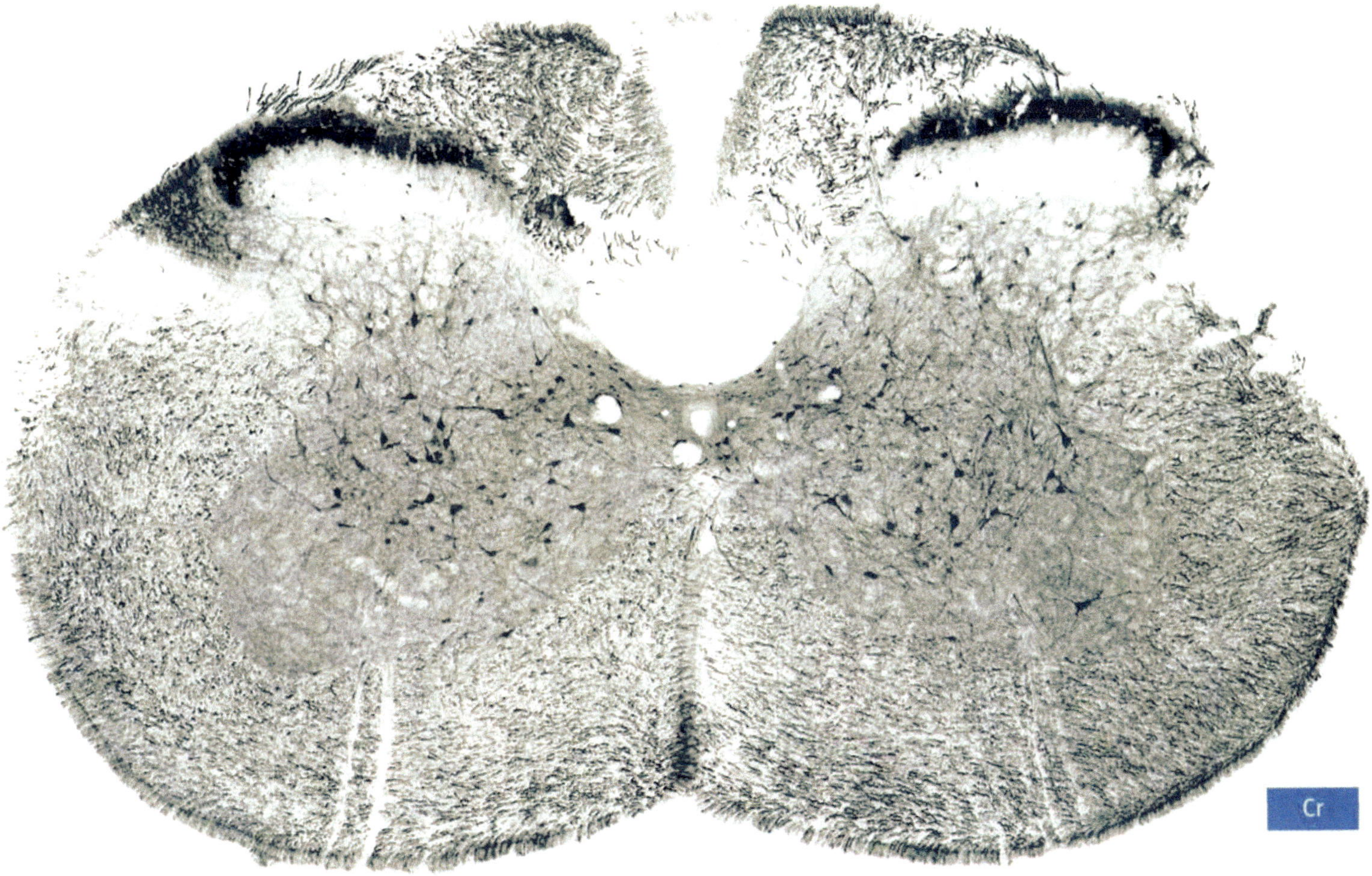

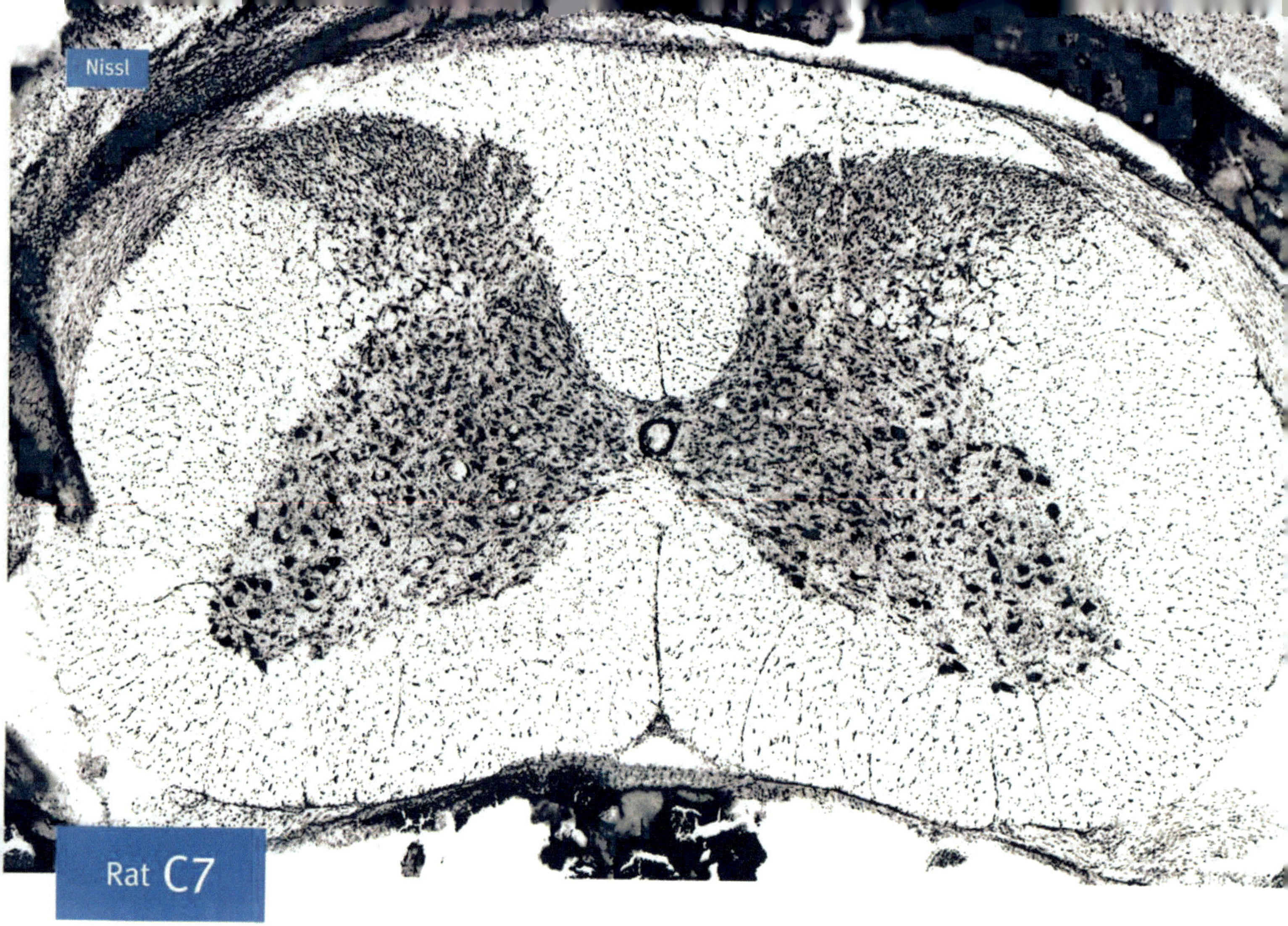
Nissl
Rat C7

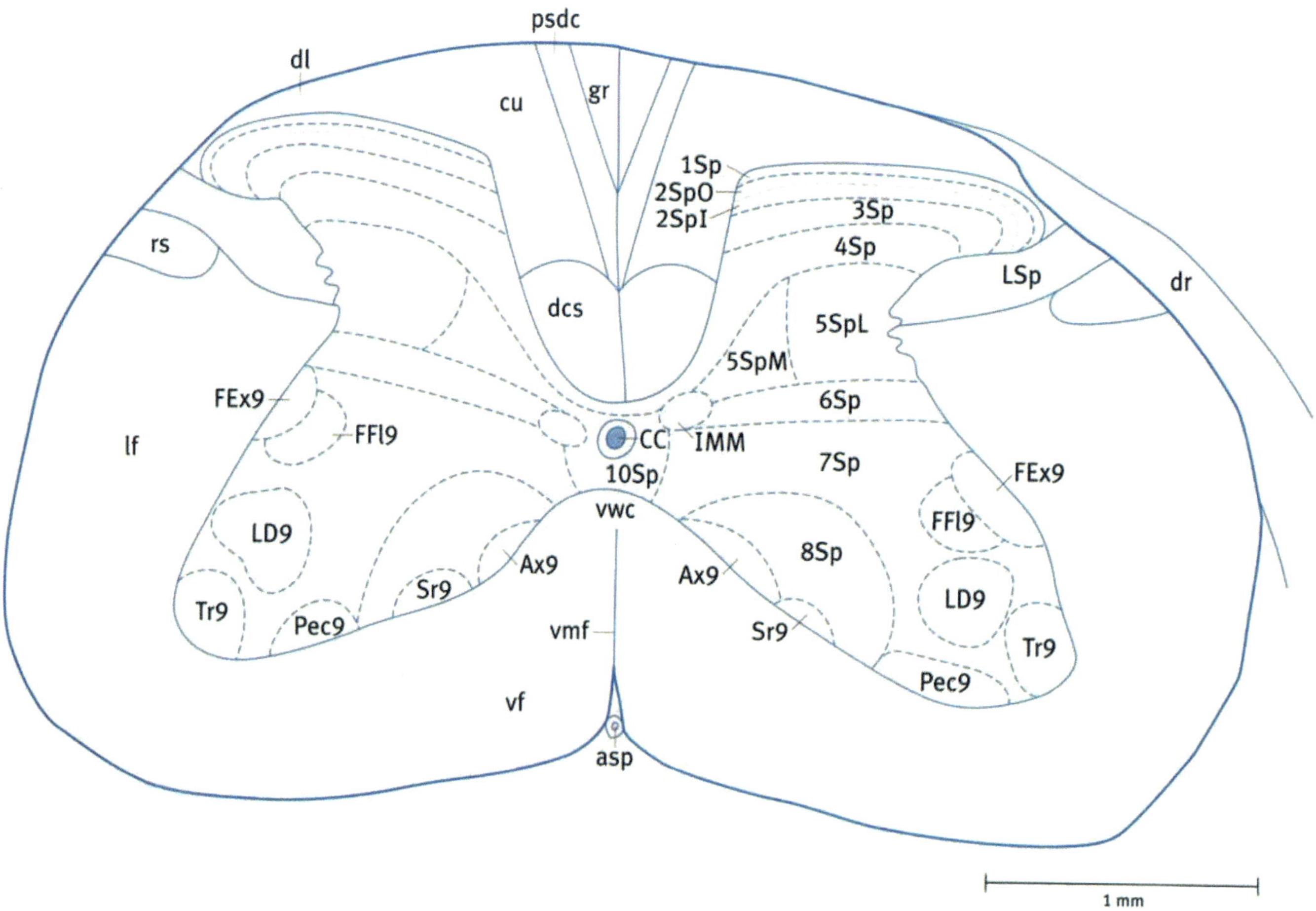
psdc
dl
cu
gr
1Sp
2SpO
2SpI
3Sp
4Sp
rs
LSp
dr
dcs
5SpL
5SpM
6Sp
FEx9
FFl9
lf
CC
IMM
10Sp
7Sp
FEx9
vwc
FFl9
LD9
Ax9
Ax9
8Sp
Sr9
LD9
Tr9
Pec9
vmf
Sr9
Tr9
Pec9
vf
asp
1 mm

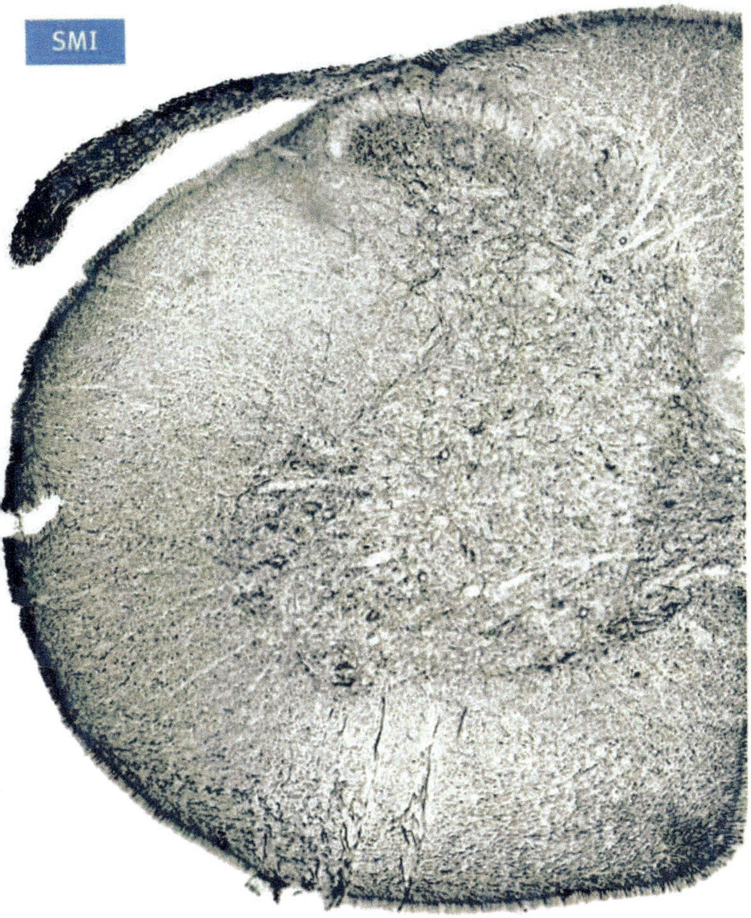

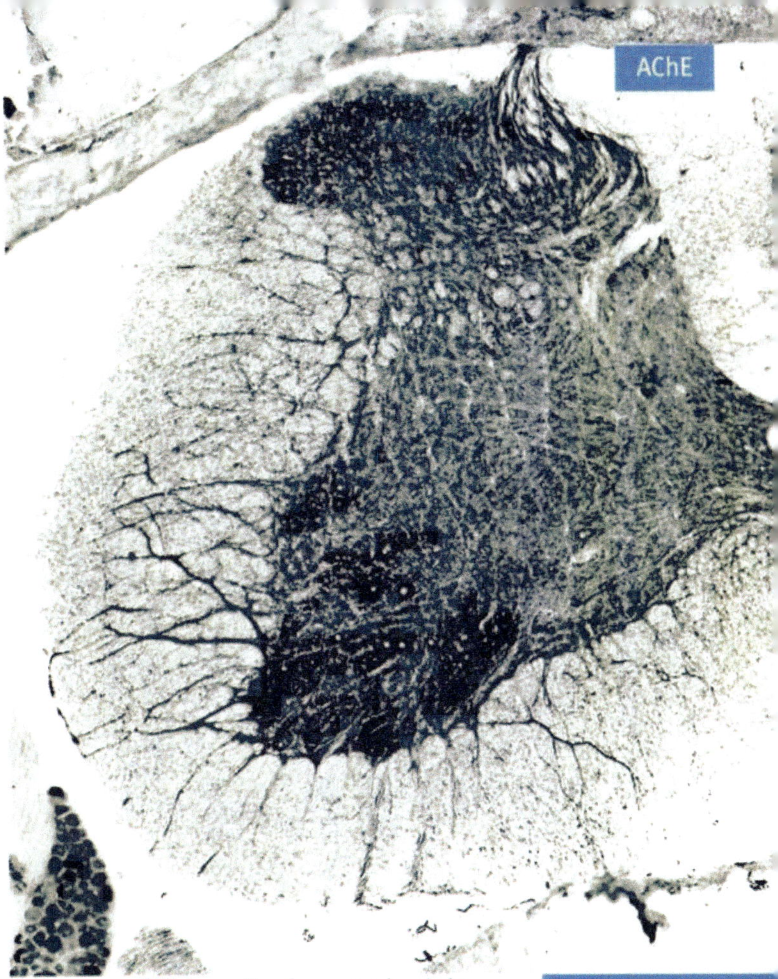

1Sp lamina 1
2SpI lamina 2, inner
2SpO lamina 2, outer
3Sp lamina 3
4Sp lamina 4
5SpL lamina 5, lat
5SpM lamina 5, med
6Sp lamina 6
7Sp lamina 7
8Sp lamina 8
10Sp lamina 10
asp ant spinal art
Ax9 axial muscles
CC central canal
cu cuneate fasc
dl dorsolat fasc
dr dorsal root
FEx9 forearm ext
FFl9 forearm flex
gr gracile fascic
IMM intermediomed
LD9 latiss dorsi
LSp lat spinal nu
Pec9 pectoral musc
psdc postsyn dors col
rs rubrospinal tr
Sr9 serratus ant
Tr9 triceps
vmf vent med fiss
vwc vent white com

Rat C7

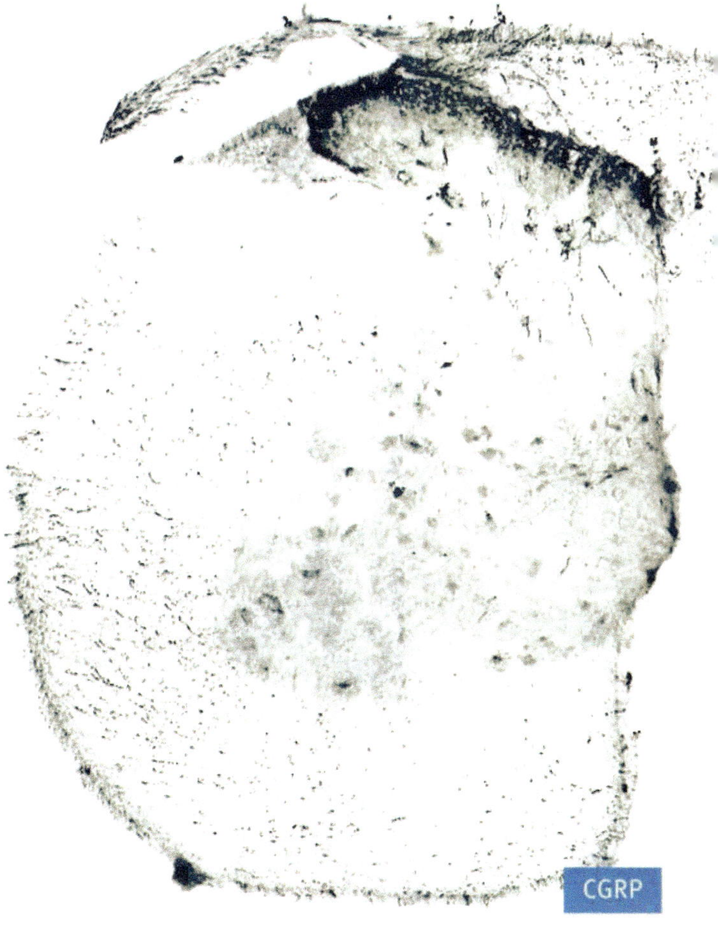

Rat C8

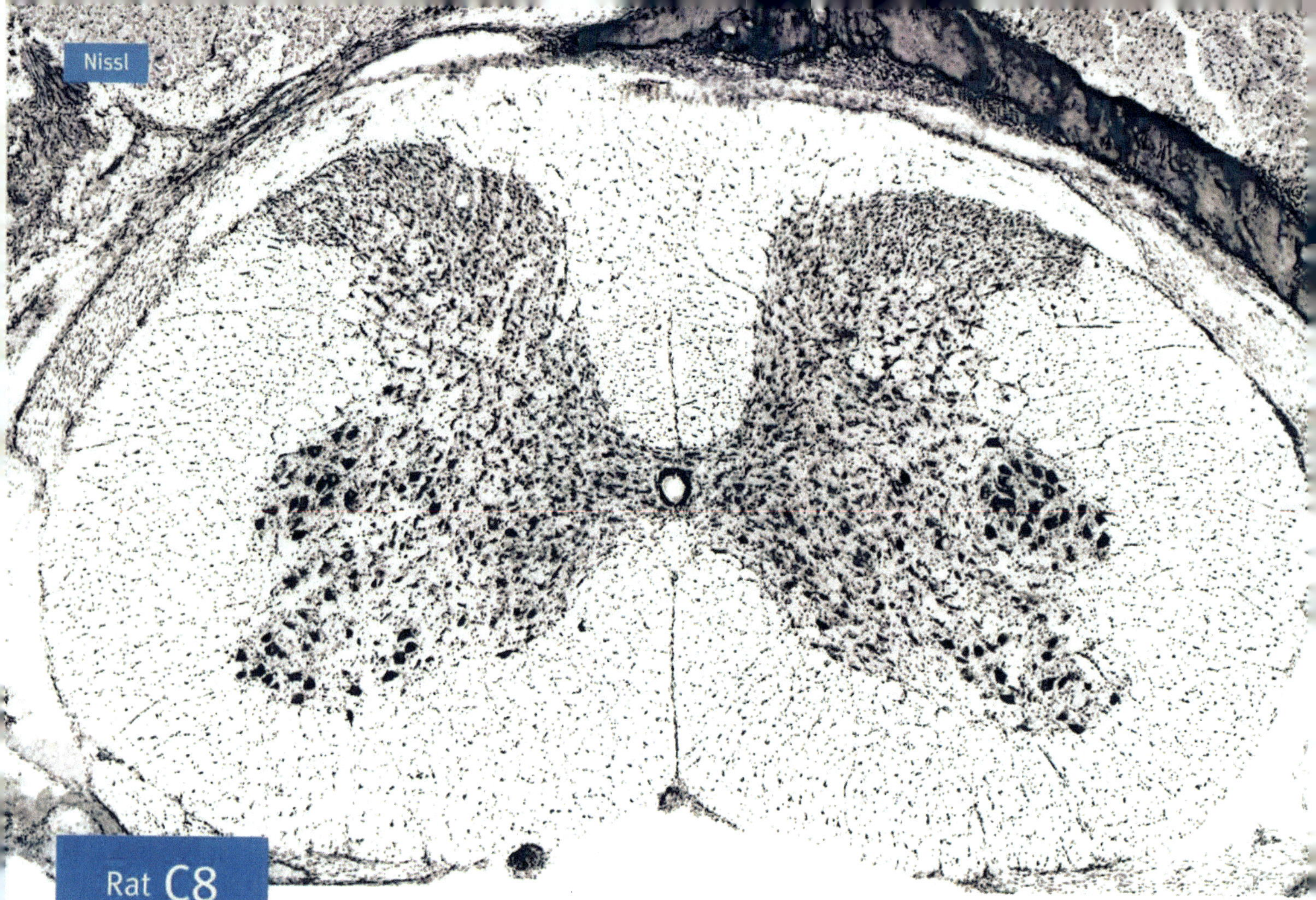
Nissl

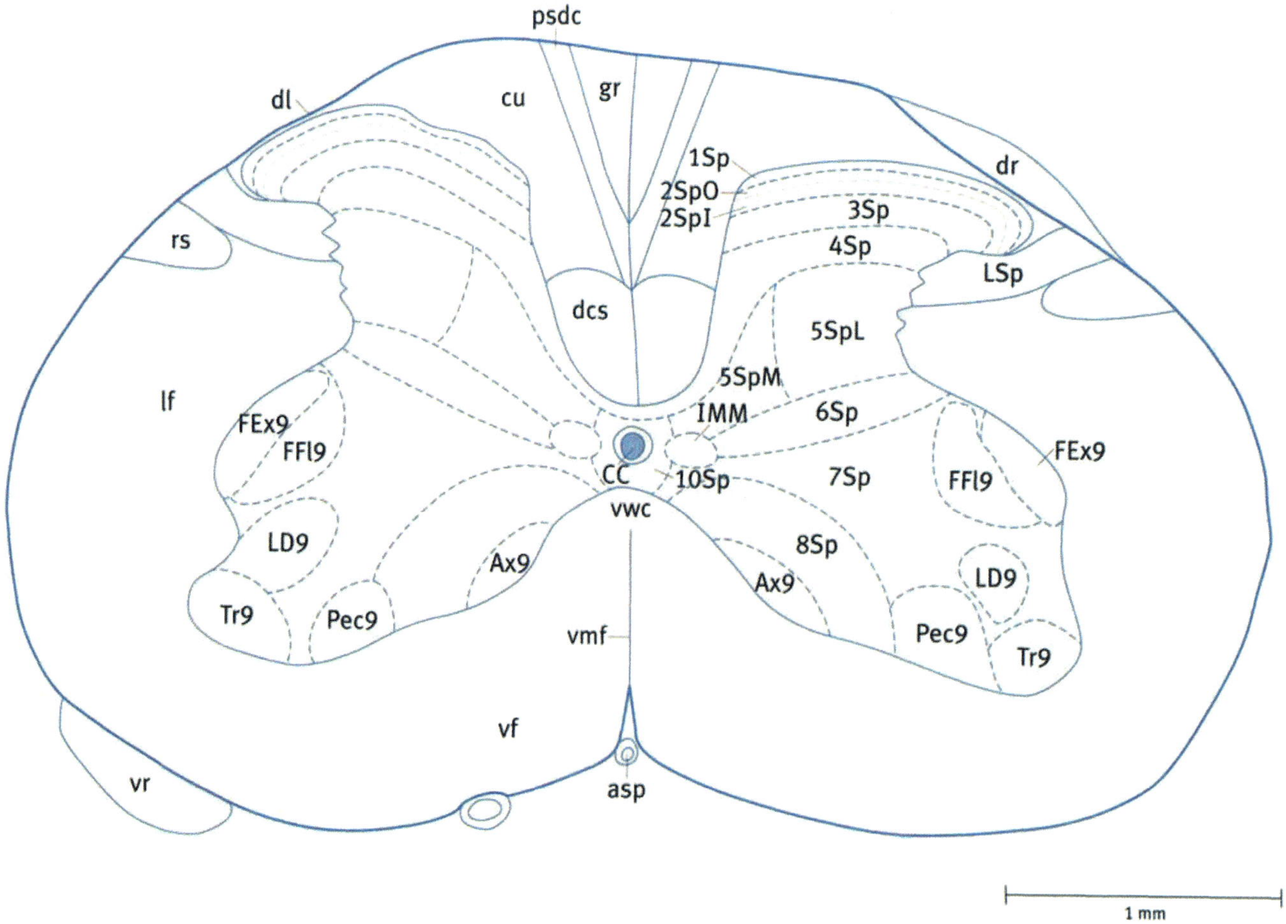
psdc
dl
cu
gr
1Sp
2SpO
2SpI
dr
3Sp
rs
4Sp
LSp
dcs
5SpL
5SpM
lf
IMM
6Sp
FEx9
FFl9
FEx9
CC
10Sp
7Sp
FFl9
vwc
LD9
8Sp
Ax9
Ax9
LD9
Tr9
Pec9
vmf
Pec9
Tr9
vf
vr
asp
1 mm

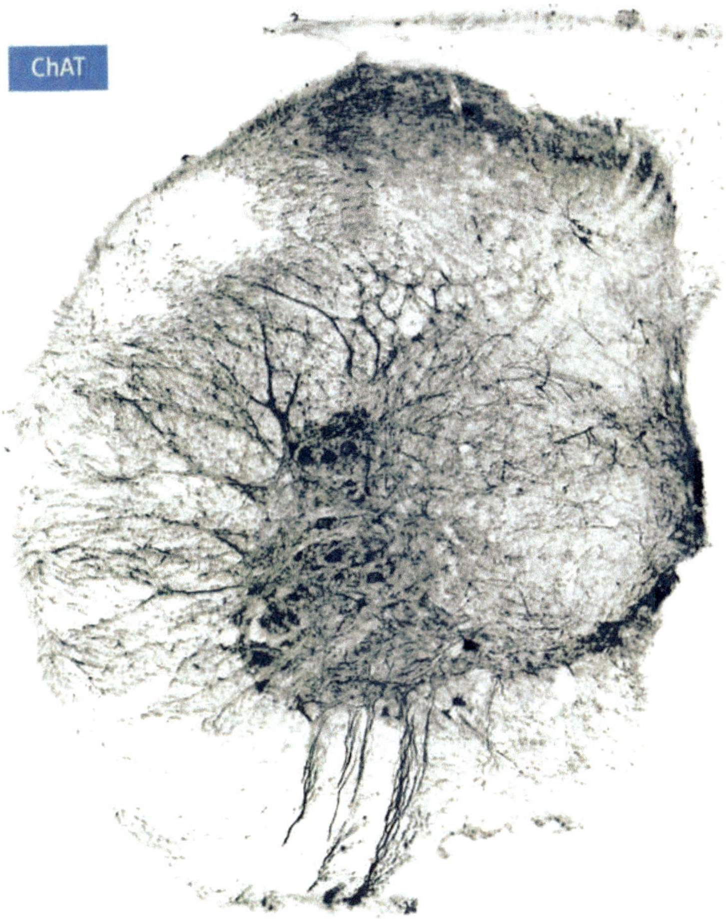

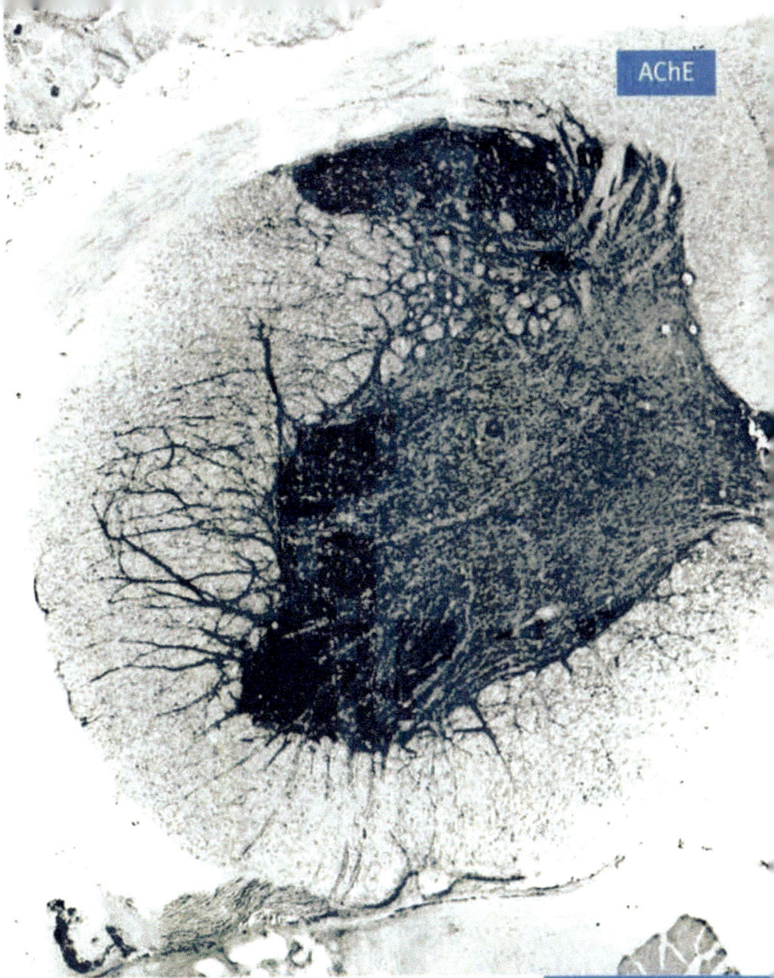

1Sp lamina 1
2SpI lamina 2, inner
2SpO lamina 2, outer
3Sp lamina 3
4Sp lamina 4
5SpL lamina 5, lat
5SpM lamina 5, med
6Sp lamina 6
7Sp lamina 7
8Sp lamina 8
10Sp lamina 10
asp ant spinal art
Ax9 axial muscles
CC central canal
cu cuneate fasc
dcs dors corticosp
dl dorsolat fasc
dr dorsal root
FEx9 forearm ext
FFl9 forearm flex
gr gracile fascic
IMM intermediomed
LD9 latiss dorsi
LSp lat spinal nu
Pec9 pectoral musc
psdc postsyn dors col
rs rubrospinal tr
Tr9 triceps
vmf vent med fiss
vr ventral root
vwc vent white com

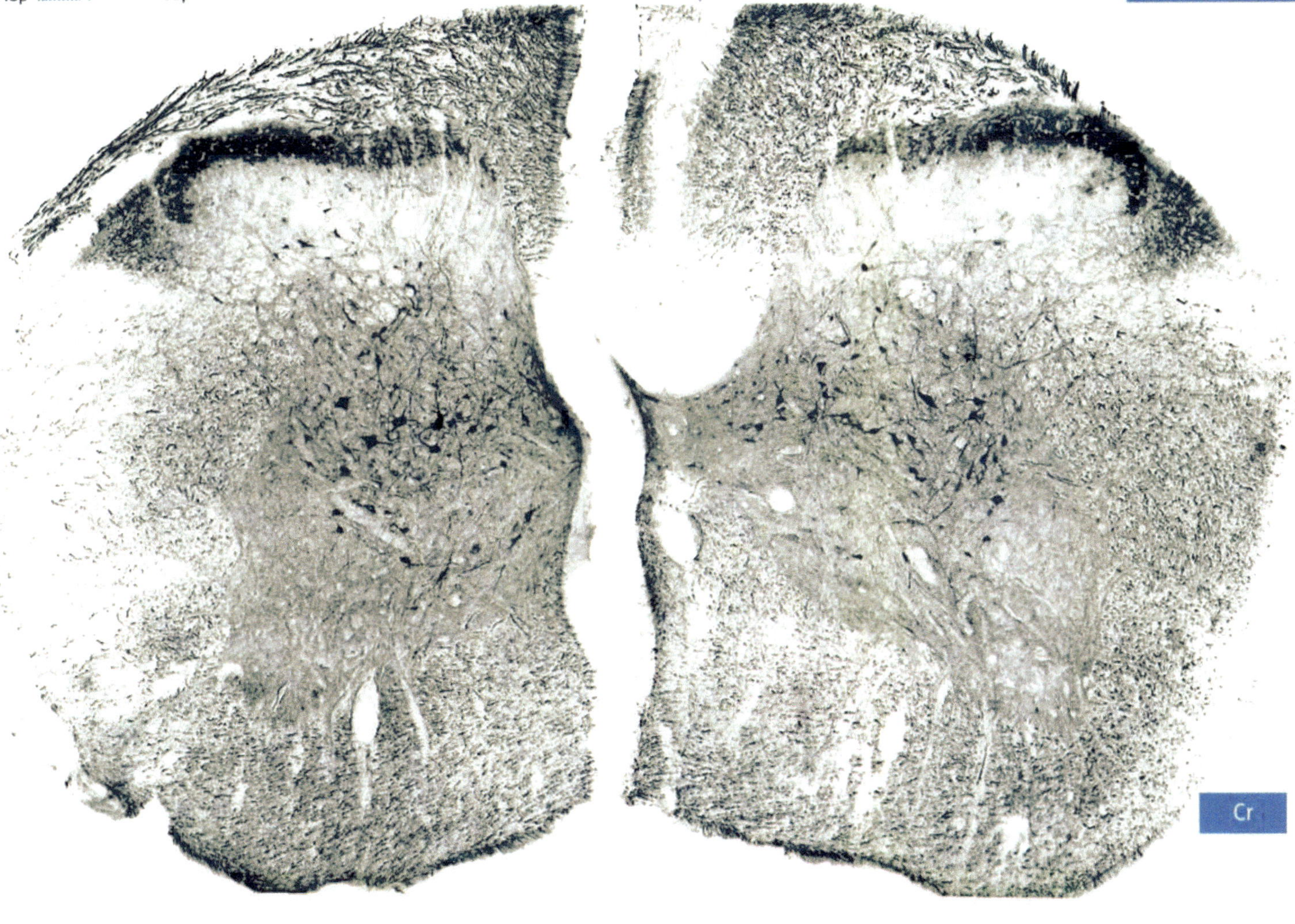

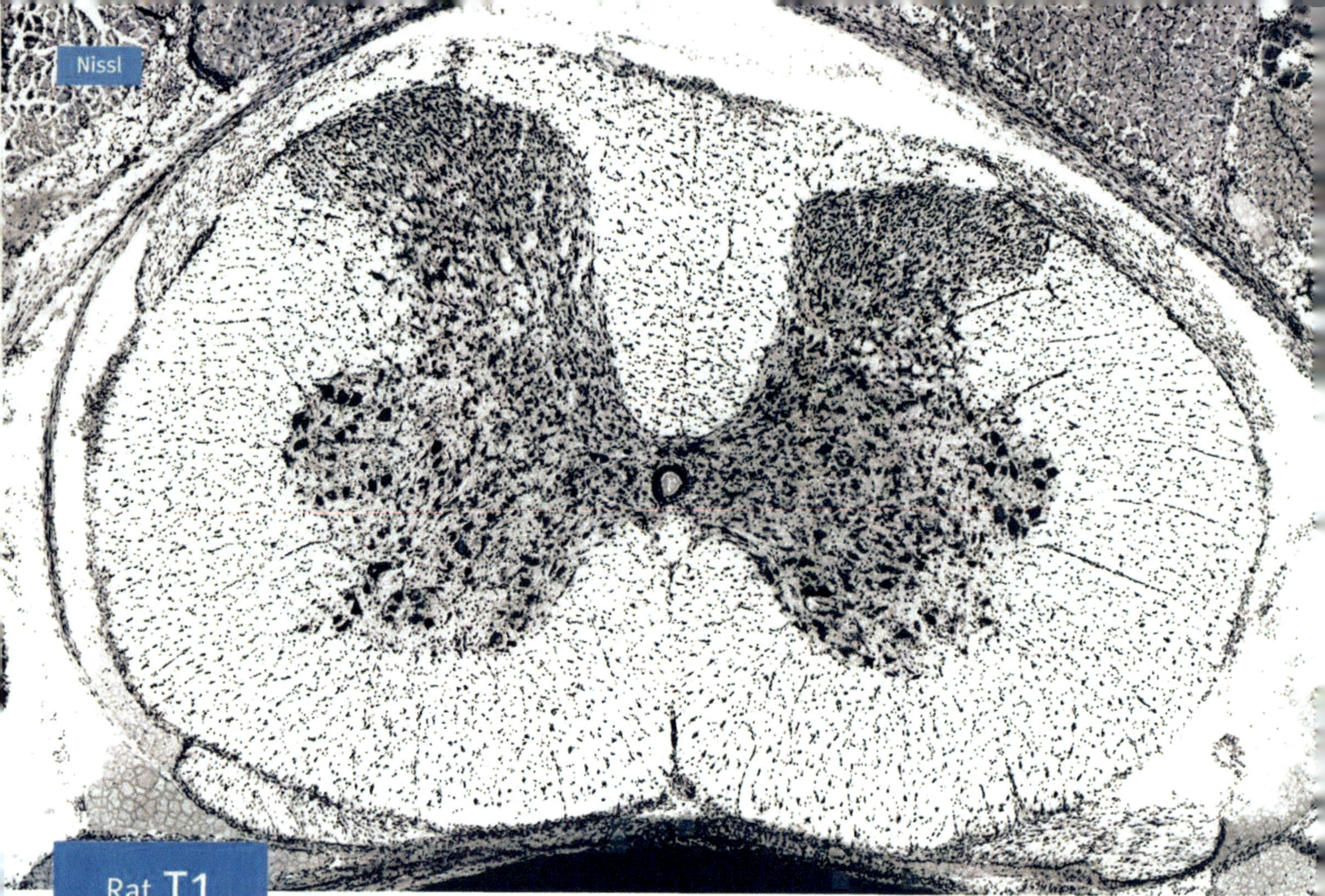
Nissl

Rat T1

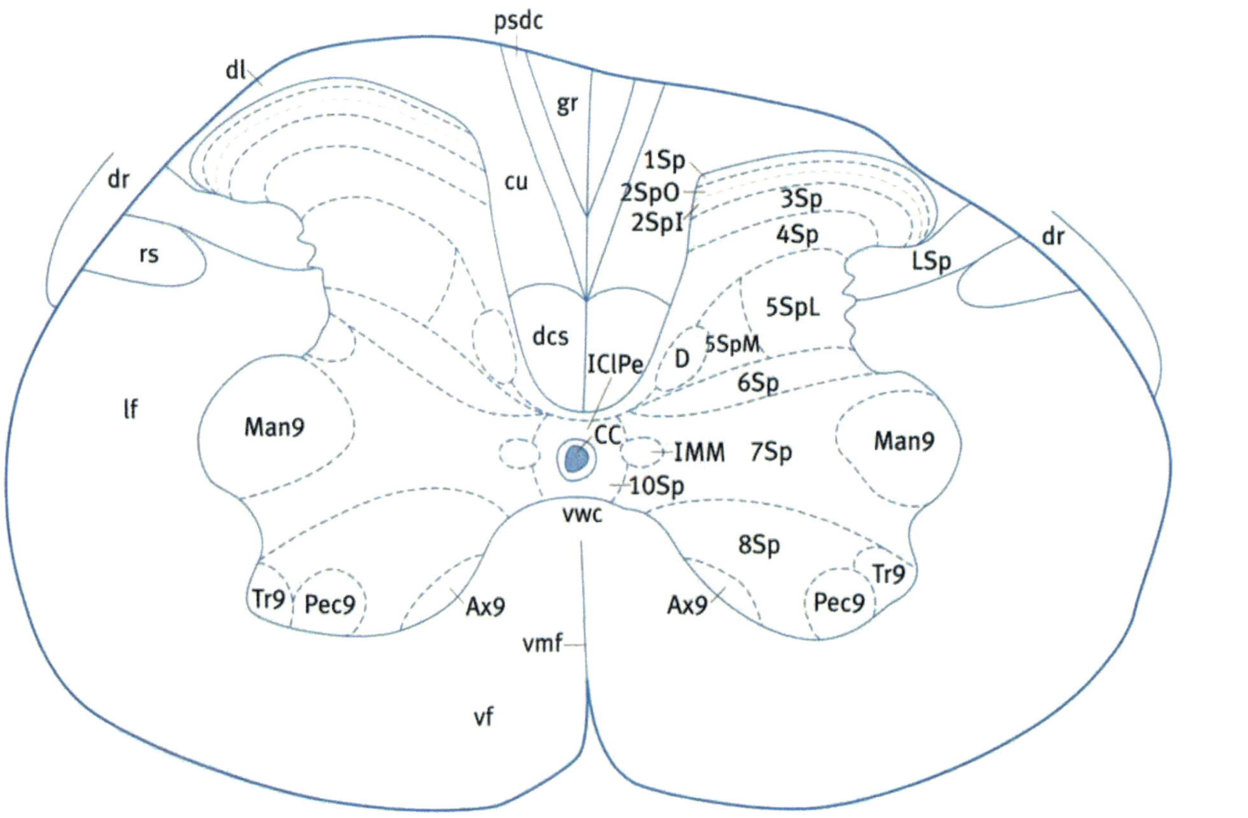
psdc
dl
gr
cu
1Sp
2SpO
2SpI
3Sp
4Sp
dr
dr
rs
LSp
5SpL
dcs
IClPe
D
5SpM
6Sp
lf
Man9
CC
IMM
7Sp
Man9
10Sp
vwc
8Sp
Tr9
Tr9
Pec9
Ax9
Ax9
Pec9
vmf
vf
1 mm

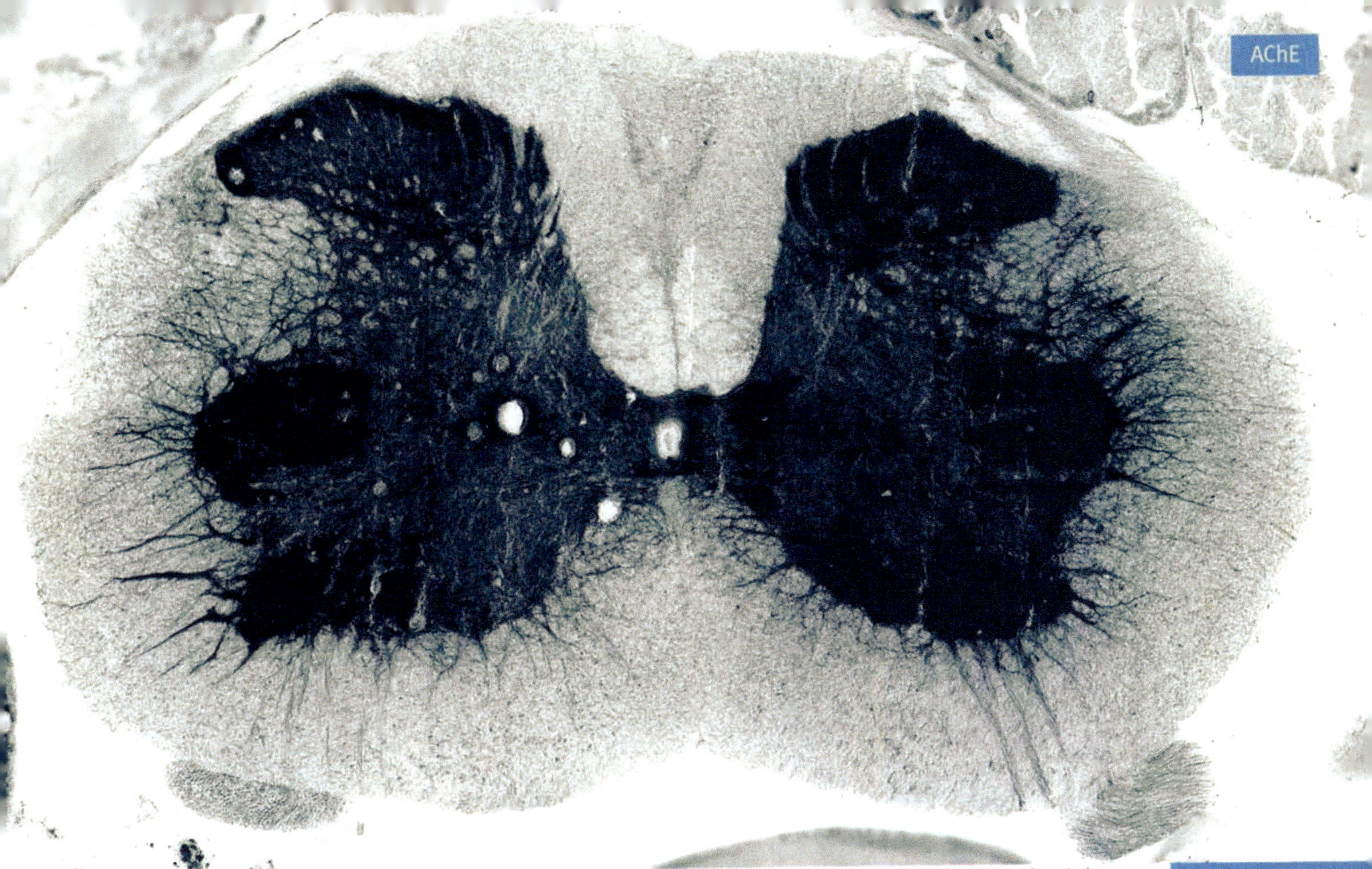

AChE

1Sp lamina 1
2SpI lamina 2, inner
2SpO lamina 2, outer
3Sp lamina 3
4Sp lamina 4
5SpL lamina 5, lat
5SpM lamina 5, med
6Sp lamina 6
7Sp lamina 7
8Sp lamina 8
10Sp lamina 10
Ax9 axial muscles
CC central canal
cu cuneate fasc
D dorsal nucleus
dcs dors corticosp
dl dorsolat fasc
dr dorsal root
gr gracile fascic
ICIPe intercal, periep
IMM intermediomed
LSp lat spinal nu
Man9 manus
Pec9 pectoral musc
psdc postsyn dors col
rs rubrospinal tr
Tr9 triceps
vmf vent med fiss
vwc vent white com

Rat T1

Nissl

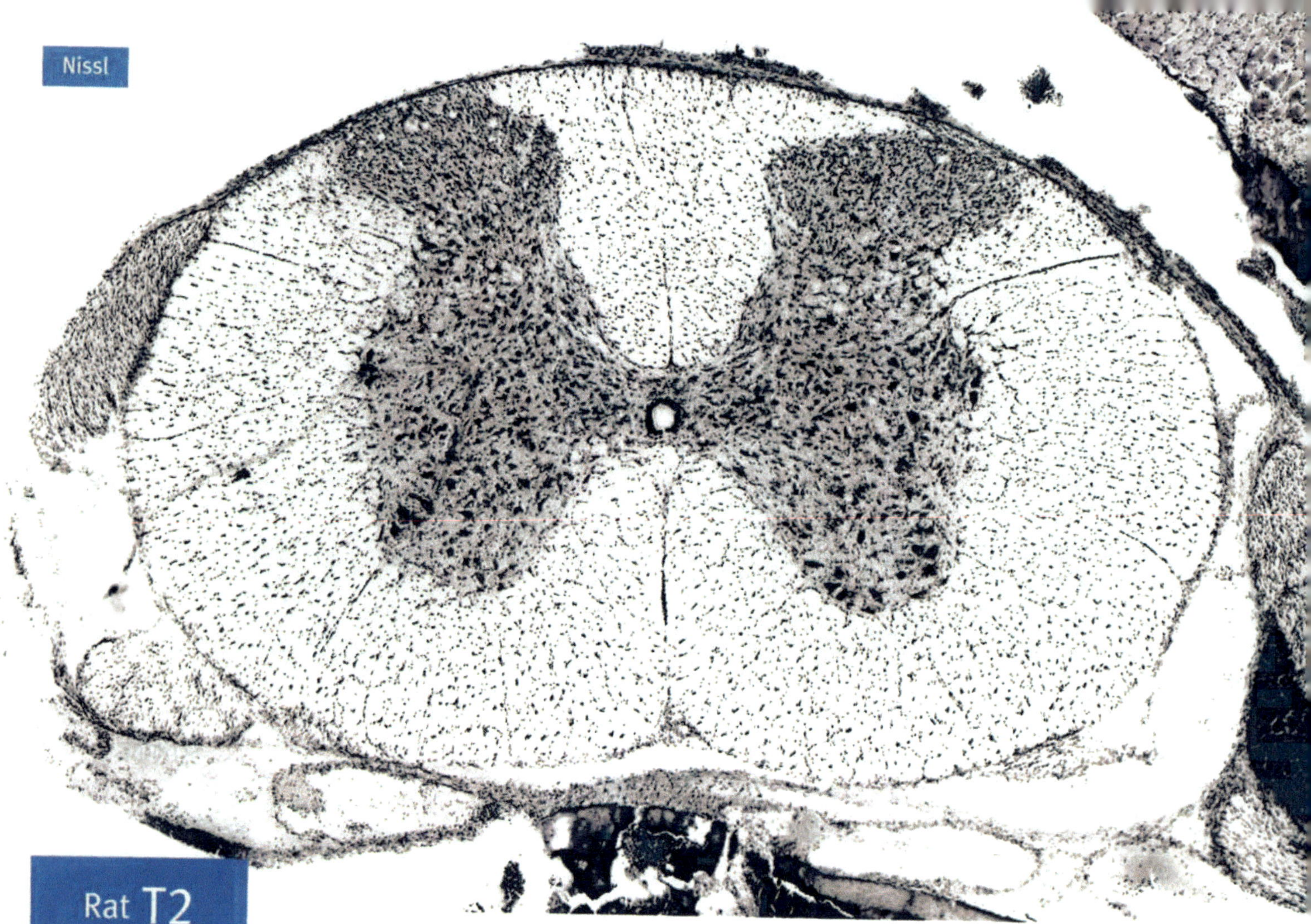

Rat T2

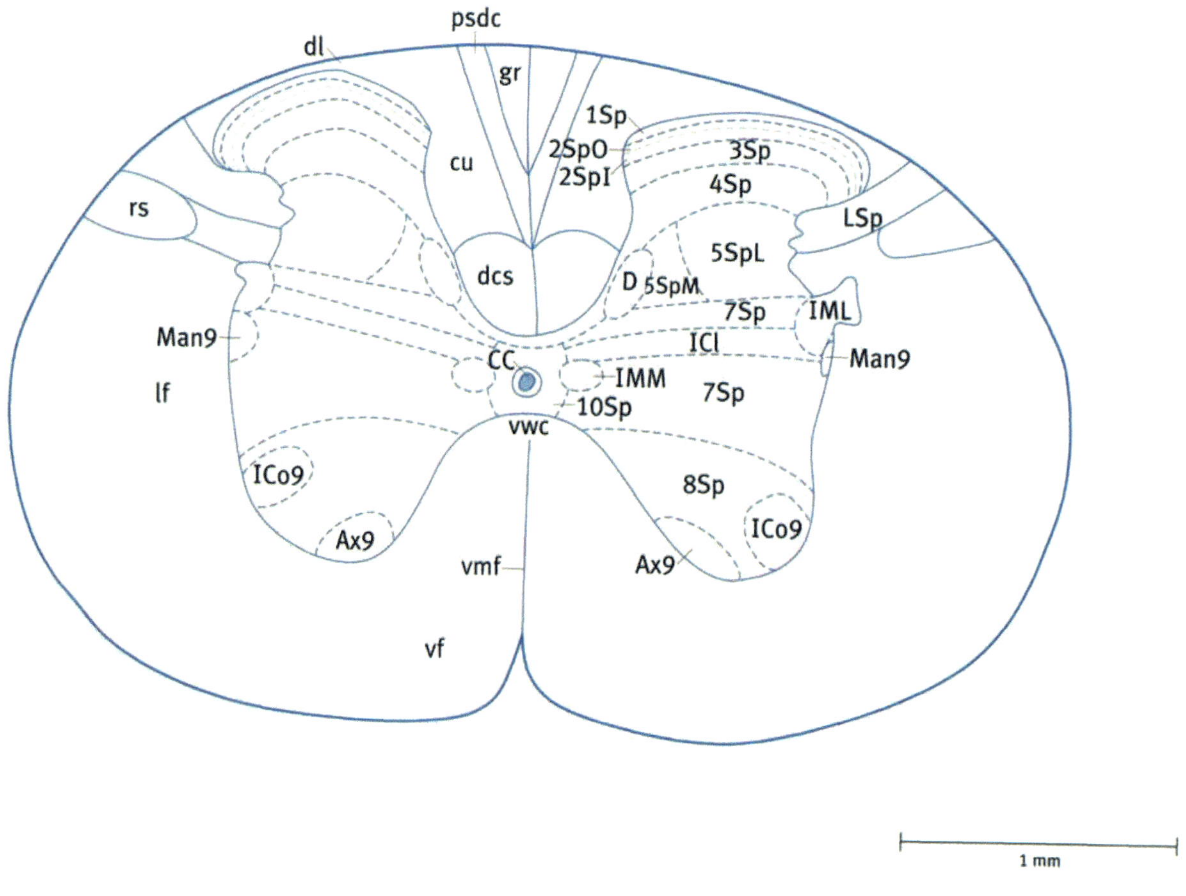
psdc
dl
gr
1Sp
2SpO
2SpI
3Sp
4Sp
cu
rs
LSp
5SpL
dcs
D
5SpM
7Sp
IML
Man9
ICl
CC
Man9
IMM
lf
7Sp
10Sp
vwc
8Sp
ICo9
ICo9
Ax9
Ax9
vmf
vf
1 mm

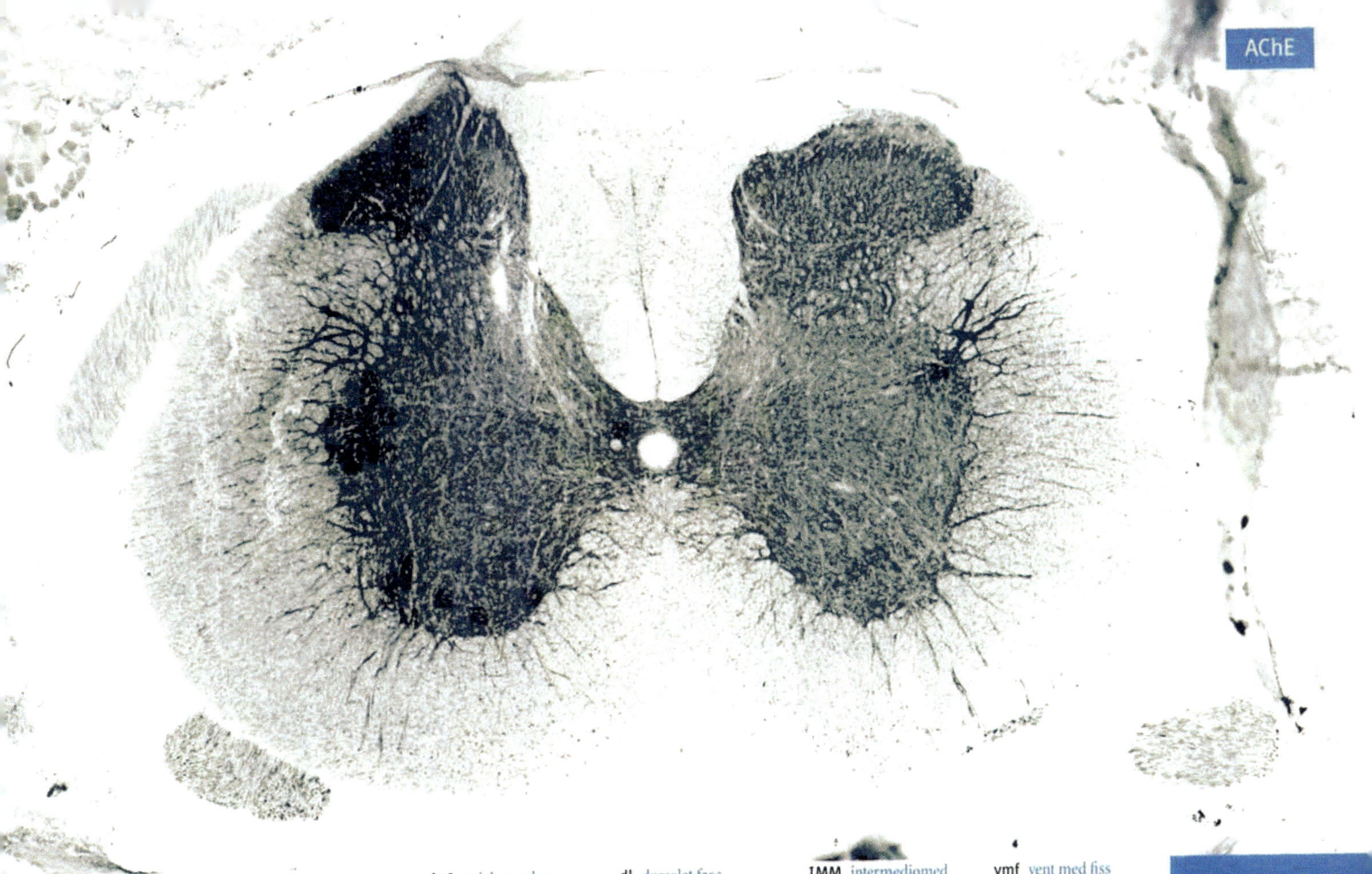

1Sp lamina 1
2SpI lamina 2, inner
2SpO lamina 2, outer
3Sp lamina 3
4Sp lamina 4
5SpL lamina 5, lat
5SpM lamina 5, med
7Sp lamina 7
8Sp lamina 8
10Sp lamina 10
Ax9 axial muscles
CC central canal
cu cuneate fasc
D dorsal nucleus
dcs dors corticosp
dl dorsolat fasc
gr gracile fascic
ICl intercalated nu
ICo9 intercostals
IML intermediolat
IMM intermediomed
LSp lat spinal nu
Man9 manus
psdc postsyn dors col
rs rubrospinal tr
vmf vent med fiss
vwc vent white com

Rat T2

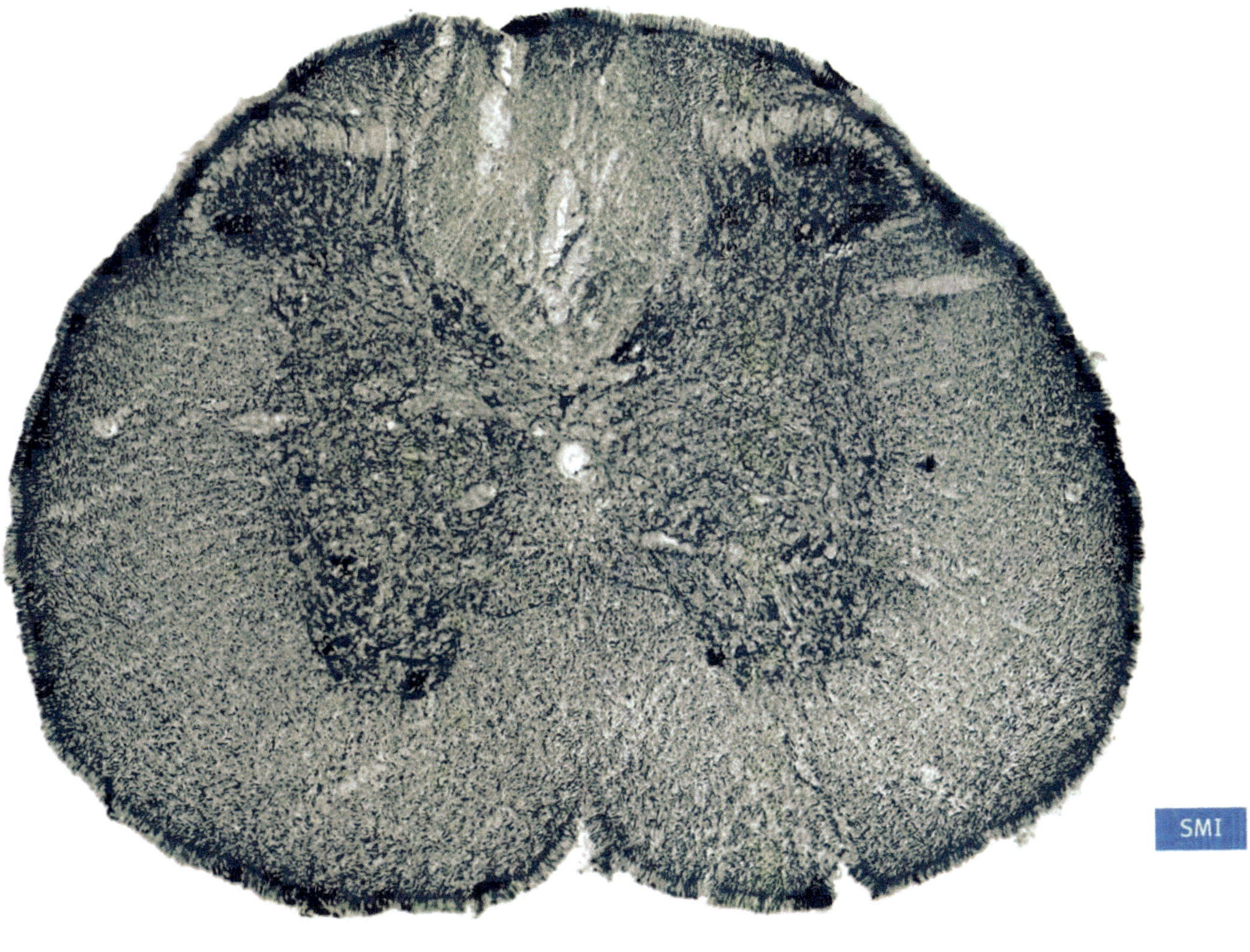

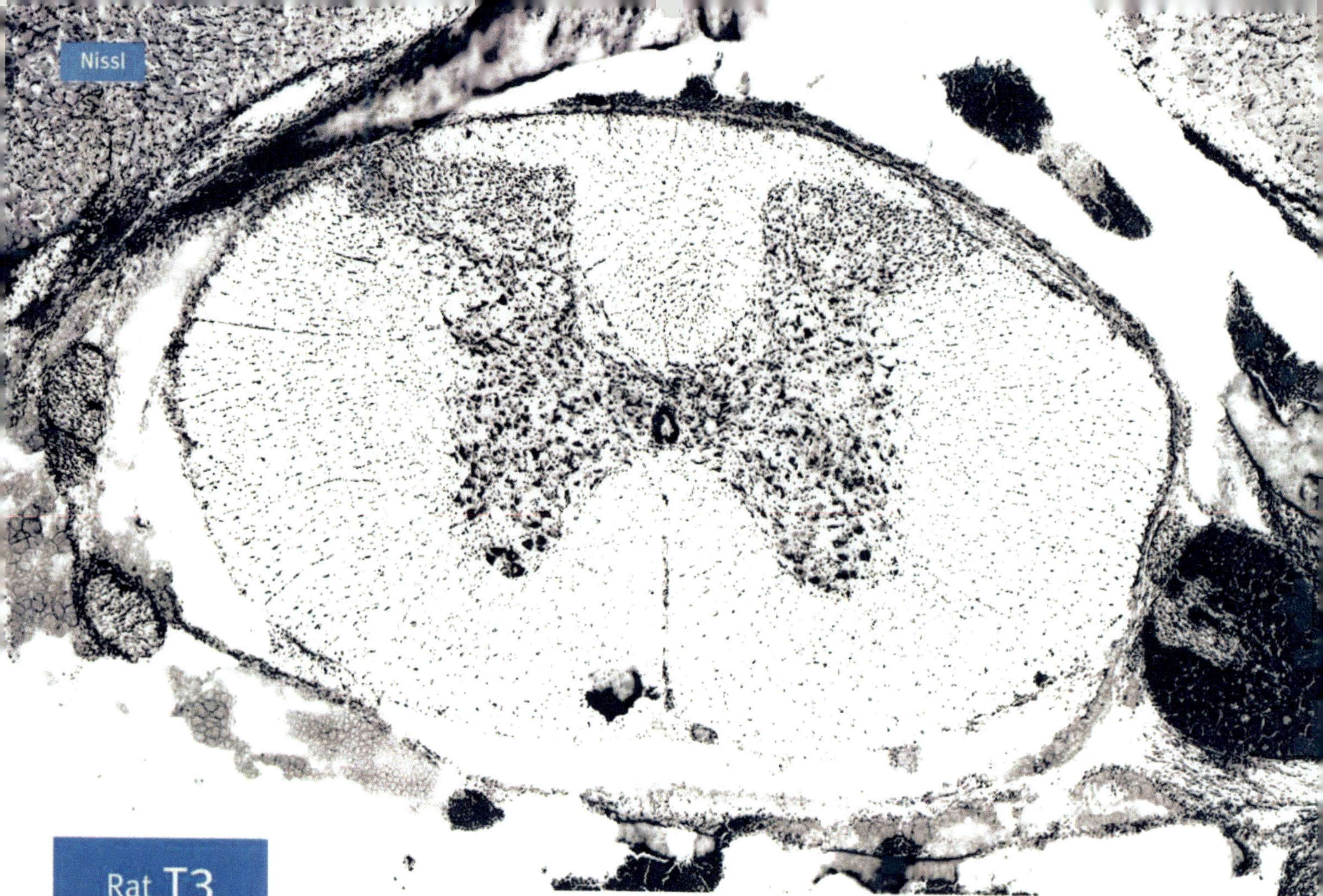
Nissl

Rat T3

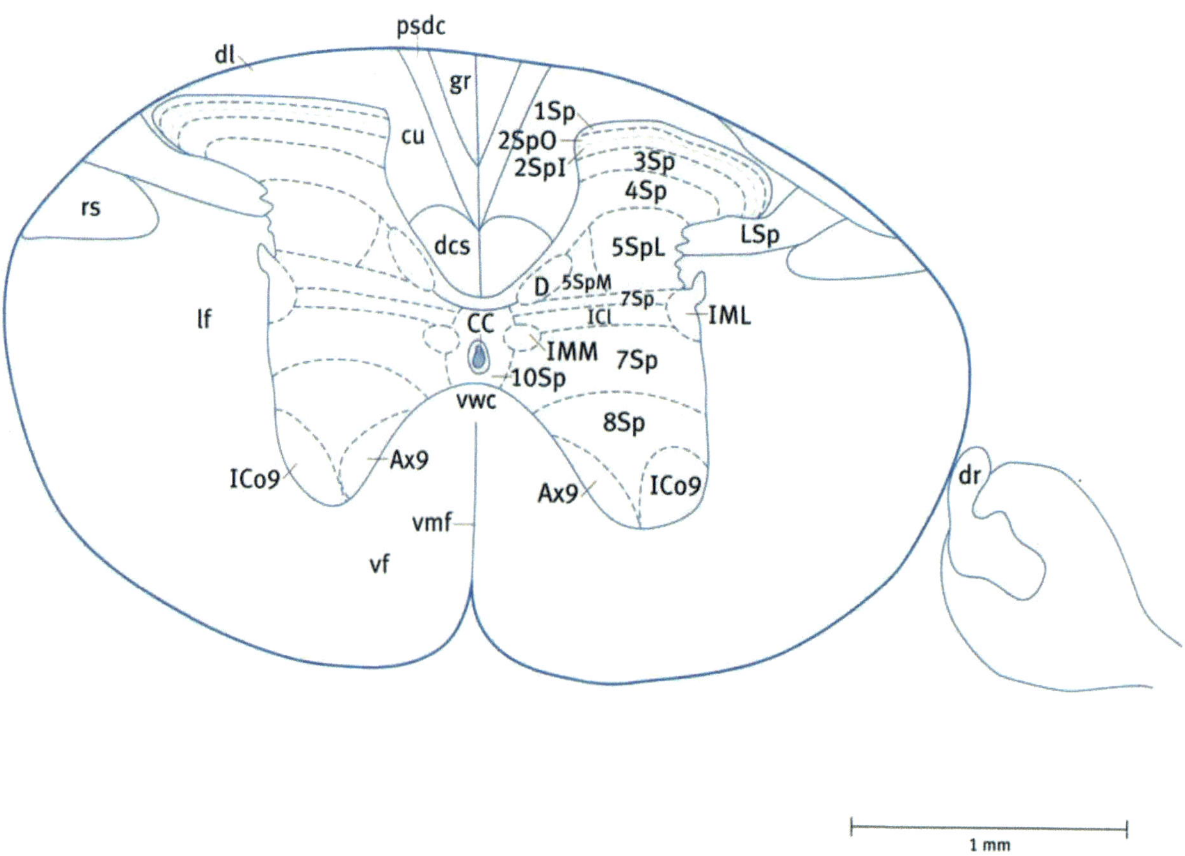
psdc
dl
gr
cu
1Sp
2SpO
2SpI
3Sp
4Sp
rs
LSp
dcs
5SpL
D
5SpM
7Sp
lf
CC
ICl
IML
IMM
7Sp
10Sp
vwc
8Sp
Ax9
ICo9
Ax9
ICo9
dr
vmf
vf
1 mm

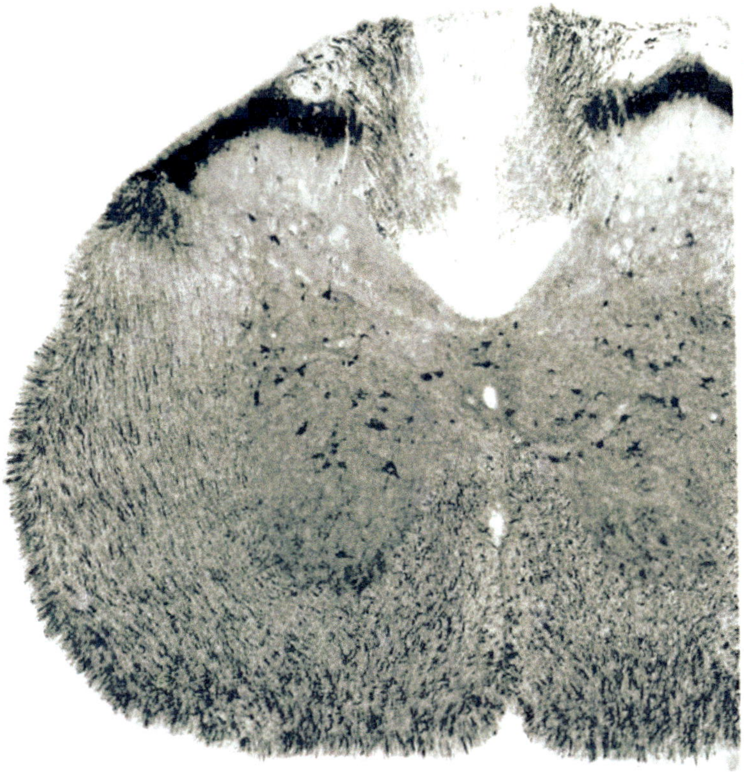

Cr

AChE

1Sp lamina 1
2SpI lamina 2, inner
2SpO lamina 2, outer
3Sp lamina 3
4Sp lamina 4
5SpL lamina 5, lat
5SpM lamina 5, med
7Sp lamina 7
8Sp lamina 8
10Sp lamina 10
Ax9 axial muscles
CC central canal
cu cuneate fasc
D dorsal nucleus
dcs dors corticosp
dl dorsolat fasc
dr dorsal root
gr gracile fascic
ICl intercalated nu
ICo9 intercostals
IML intermediolat
IMM intermediomed
LSp lat spinal nu
psdc postsyn dors col
rs rubrospinal tr
vmf vent med fiss
vwc vent white com

Rat T3

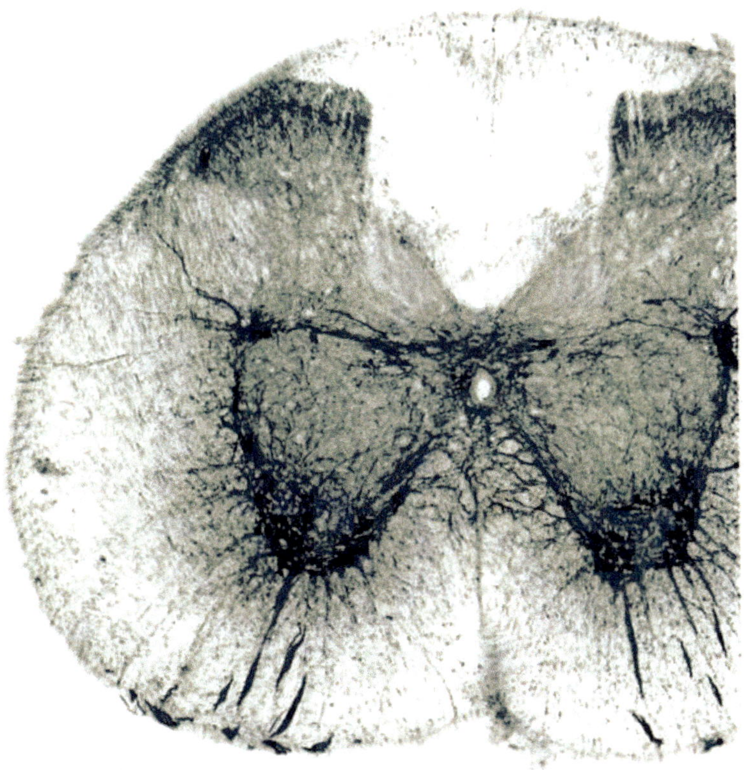

ChAT

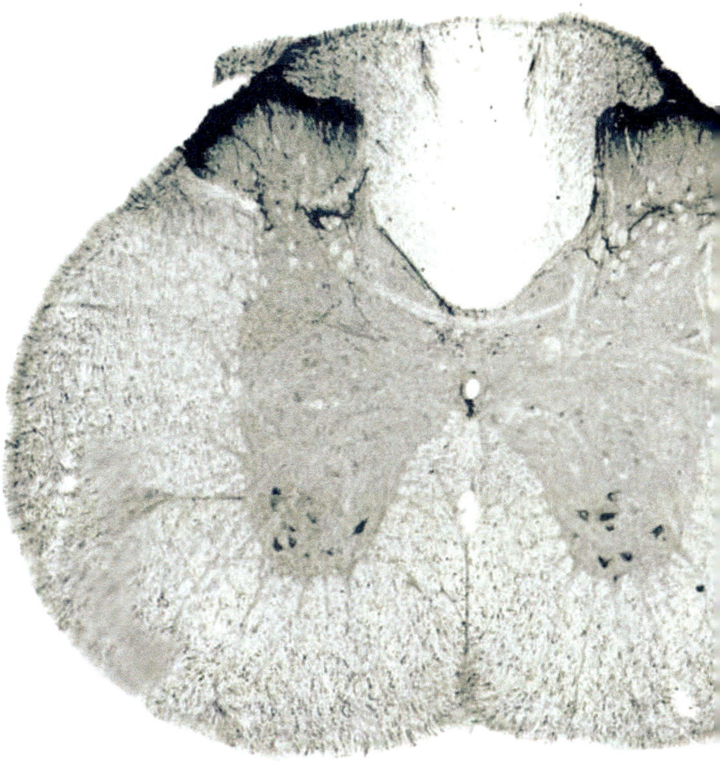

CGRP

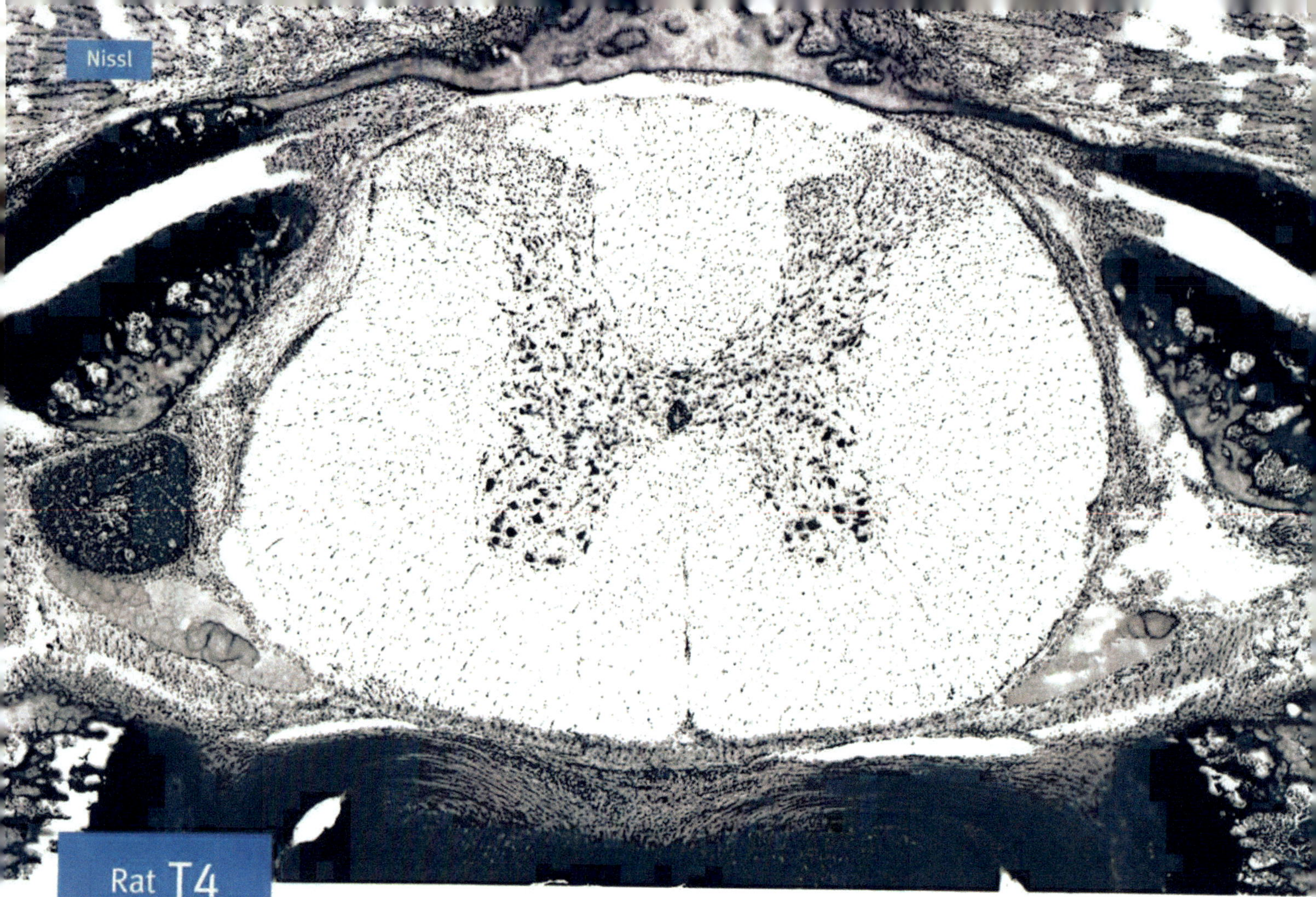
Nissl

Rat T4

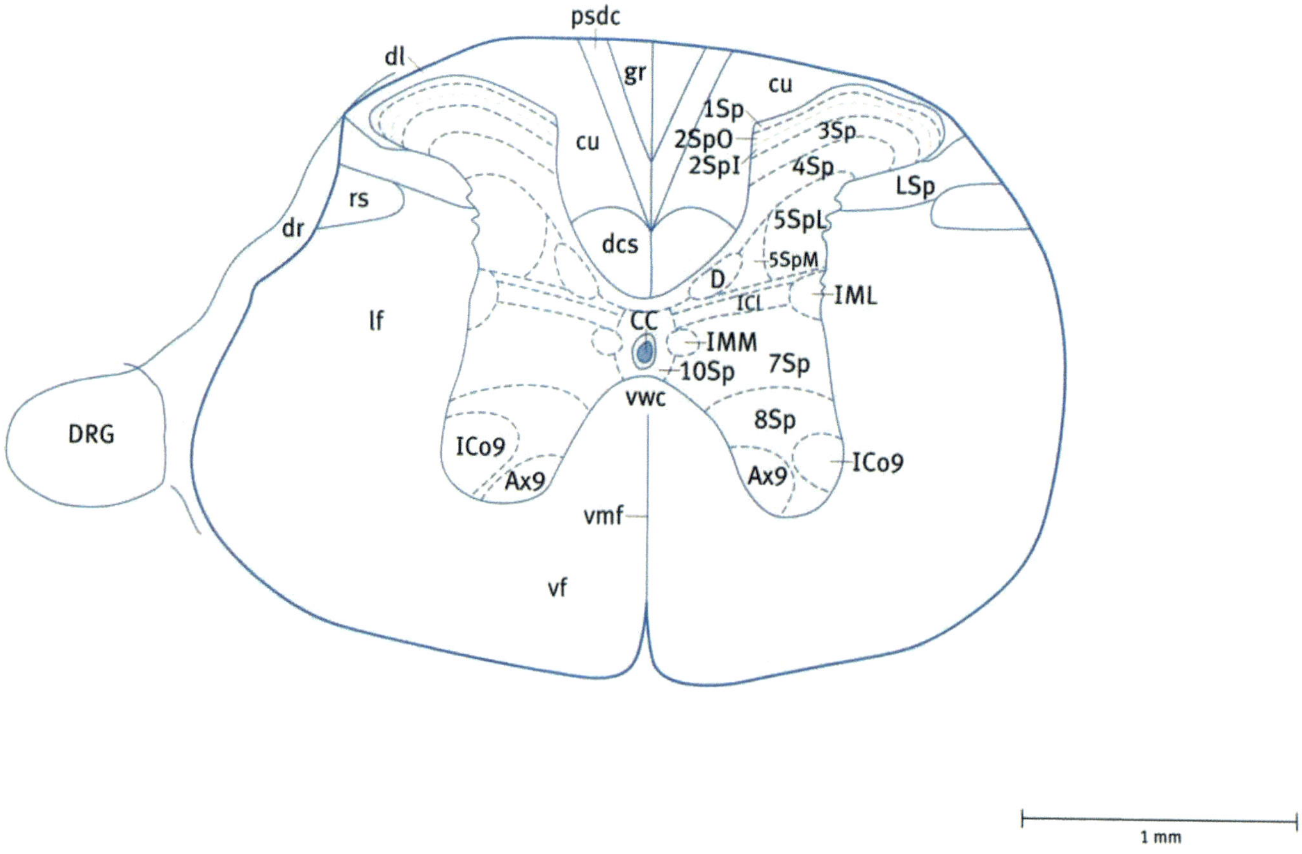
psdc
dl
gr
cu
1Sp
cu
2SpO
3Sp
2SpI
4Sp
rs
LSp
dr
5SpL
dcs
5SpM
D
IML
lf
ICl
CC
IMM
7Sp
10Sp
vwc
8Sp
DRG
ICo9
ICo9
Ax9
Ax9
vmf
vf
1 mm

NeuN

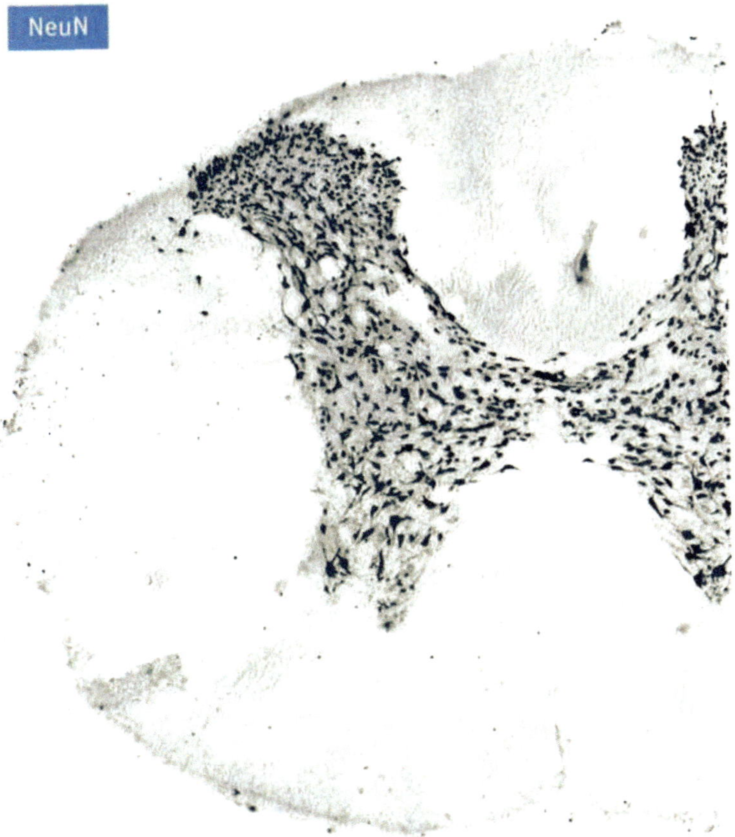

SMI

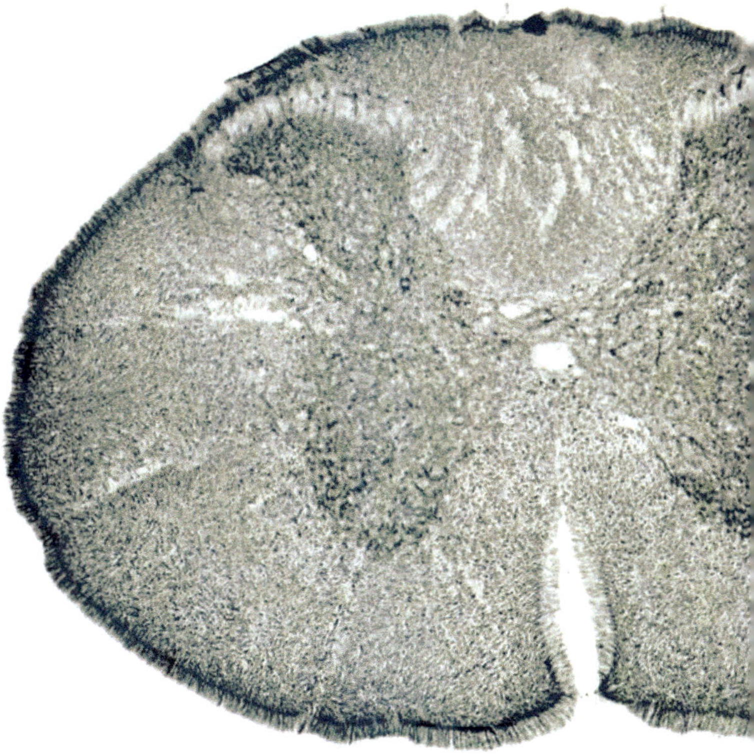

1Sp lamina 1
2SpI lamina 2, inner
2SpO lamina 2, outer
3Sp lamina 3
4Sp lamina 4
5SpL lamina 5, lat
5SpM lamina 5, med
7Sp lamina 7
8Sp lamina 8
10Sp lamina 10
Ax9 axial muscles
CC central canal
cu cuneate fasc
D dorsal nucleus
dcs dors corticosp
dl dorsolat fasc
dr dorsal root
DRG dors root gang
gr gracile fascic
ICl intercalated nu
ICo9 intercostals
IML intermediolat
IMM intermediomed
LSp lat spinal nu
psdc postsyn dors col
rs rubrospinal tr
vmf vent med fiss
vwc vent white com

Rat T4

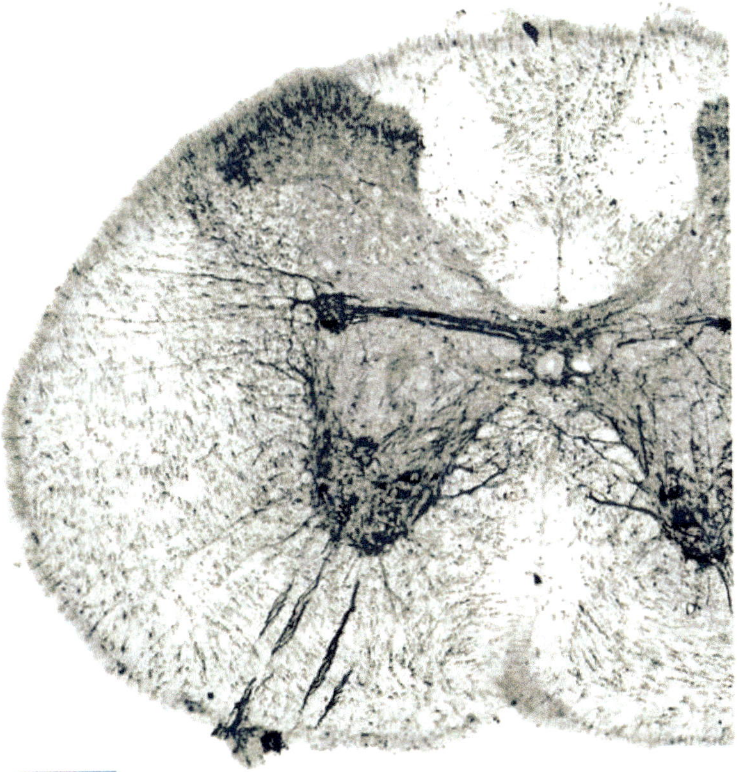

ChAT

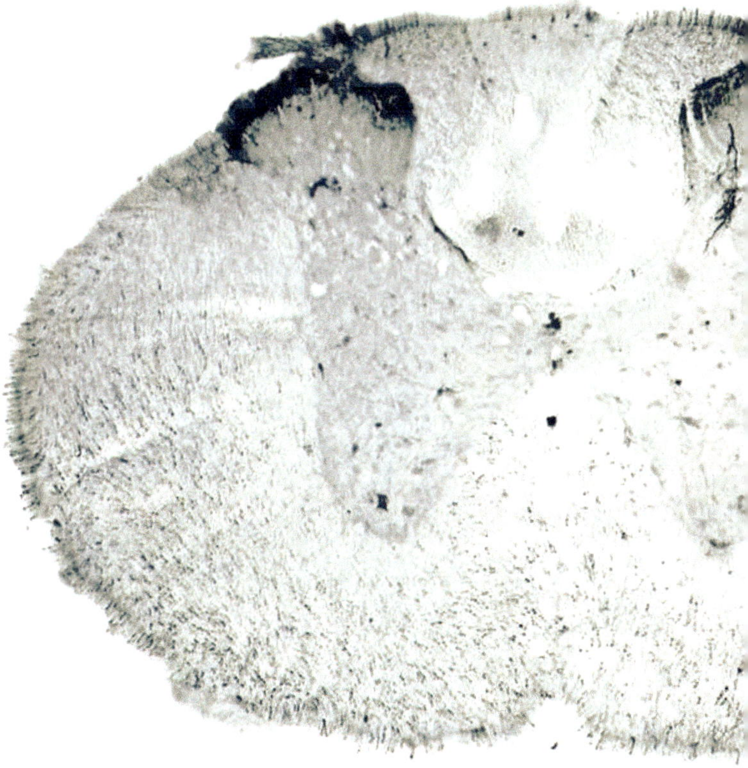

CGRP

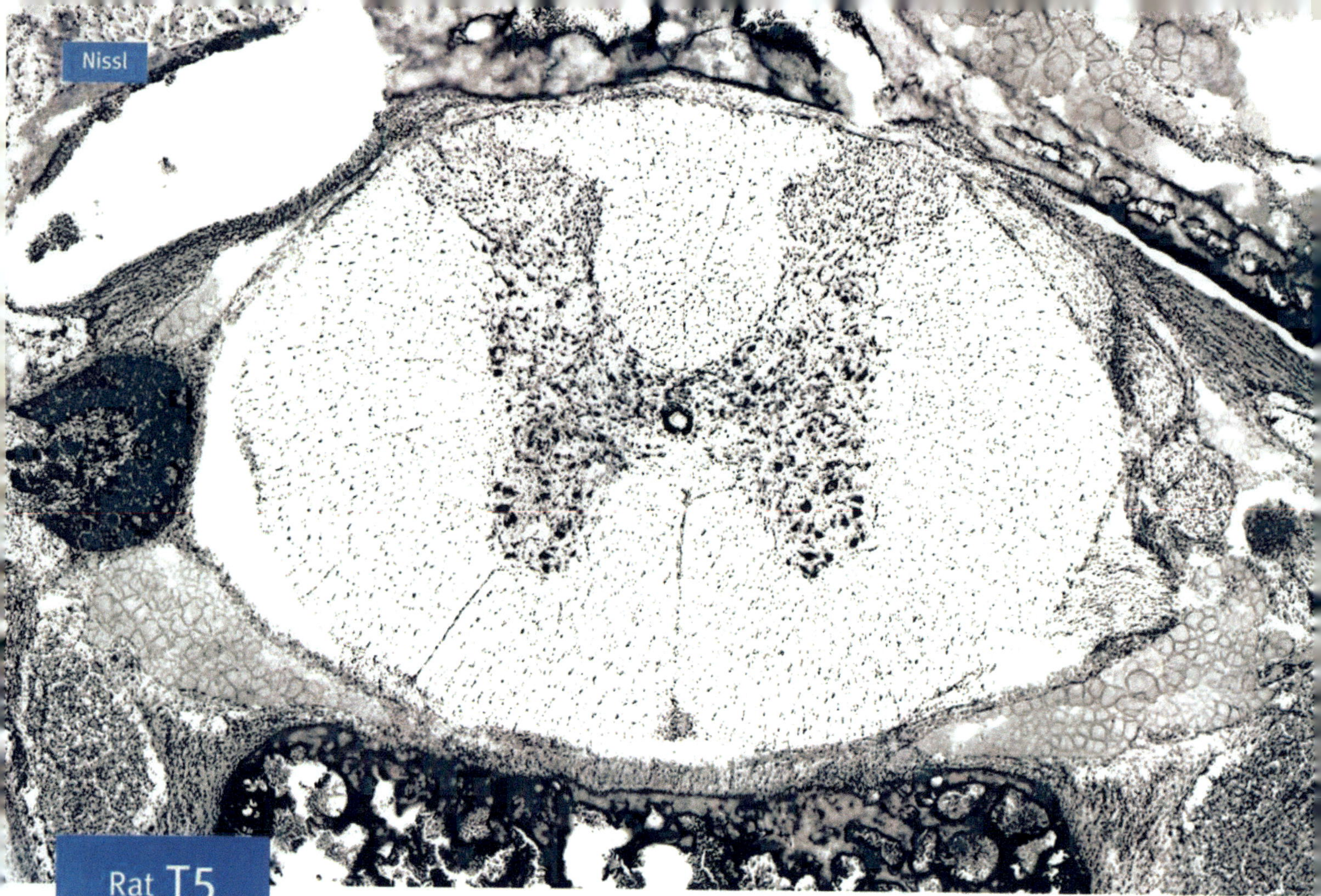
Nissl

Rat T5

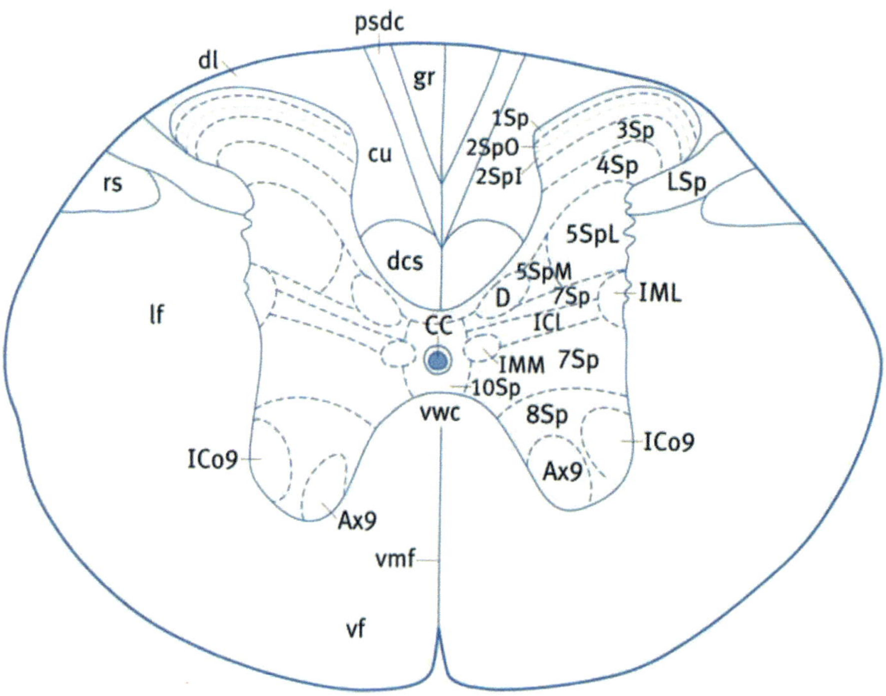
psdc
dl
gr
1Sp
3Sp
cu
2SpO
2SpI
4Sp
rs
LSp
5SpL
dcs
5SpM
D
7Sp
IML
lf
CC
ICl
IMM
7Sp
10Sp
vwc
8Sp
ICo9
ICo9
Ax9
Ax9
vmf
vf

1 mm

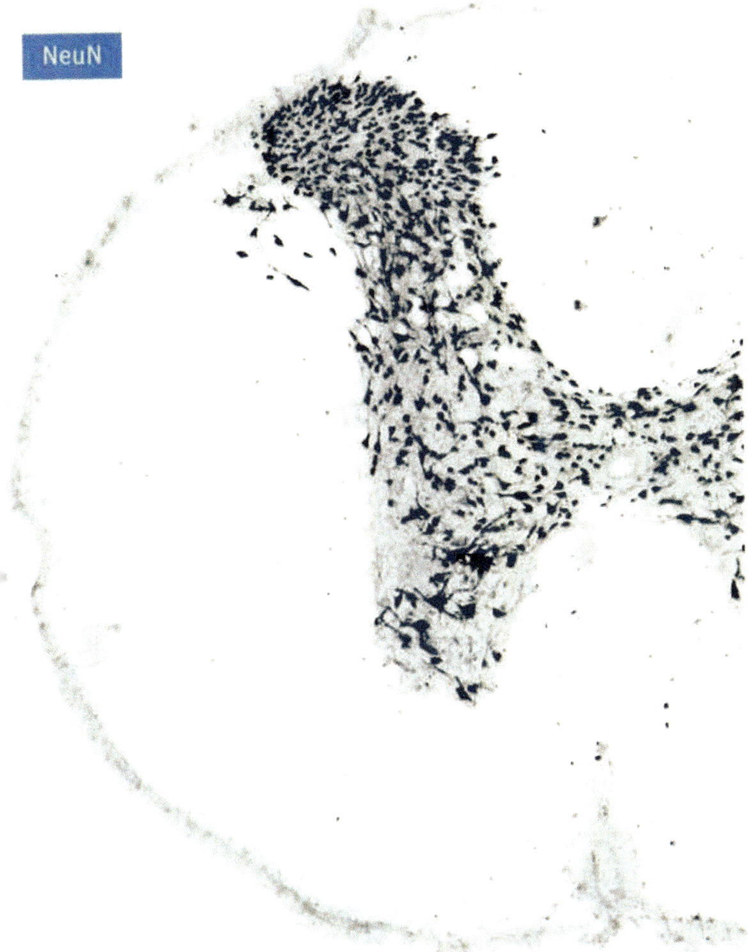

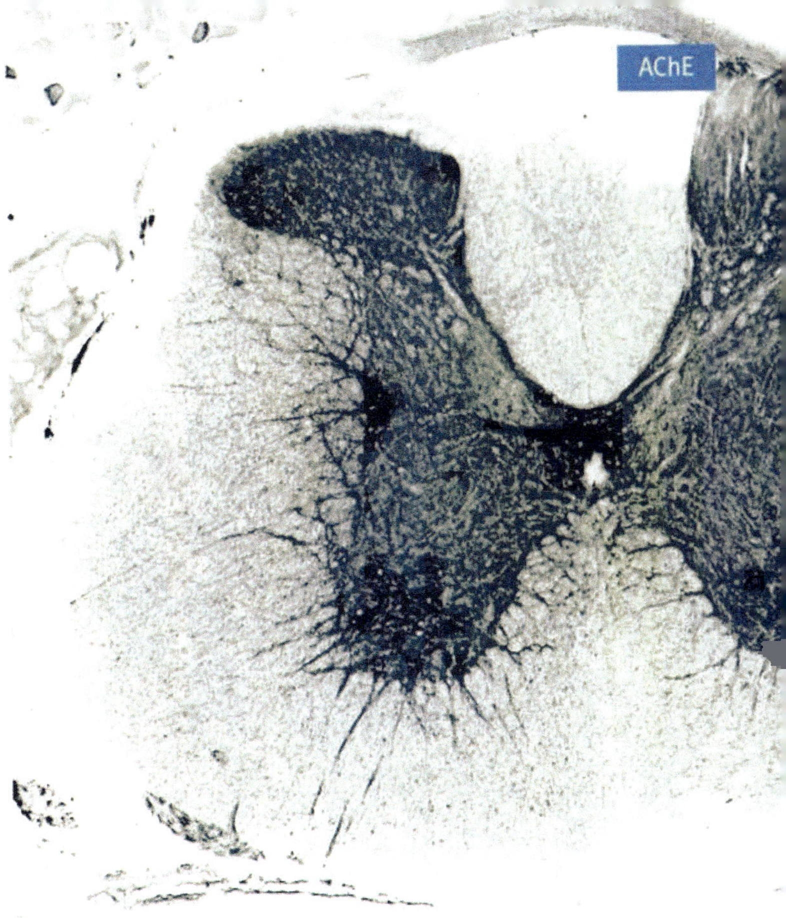

1Sp lamina 1
2SpI lamina 2, inner
2SpO lamina 2, outer
3Sp lamina 3
4Sp lamina 4
5SpL lamina 5, lat
5SpM lamina 5, med
7Sp lamina 7
8Sp lamina 8
10Sp lamina 10
Ax9 axial muscles
CC central canal
cu cuneate fasc
D dorsal nucleus
dcs dors corticosp
dl dorsolat fasc
gr gracile fascic
ICl intercalated nu
ICo9 intercostals
IML intermediolat
IMM intermediomed
LSp lat spinal nu
psdc postsyn dors col
rs rubrospinal tr
vmf vent med fiss
vwc vent white com

Rat T5

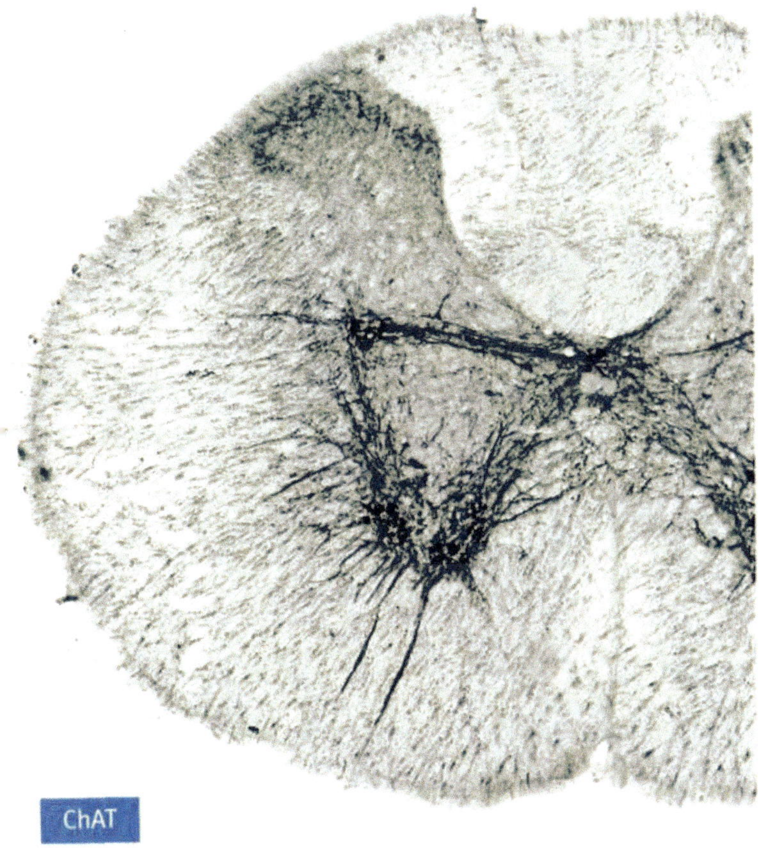

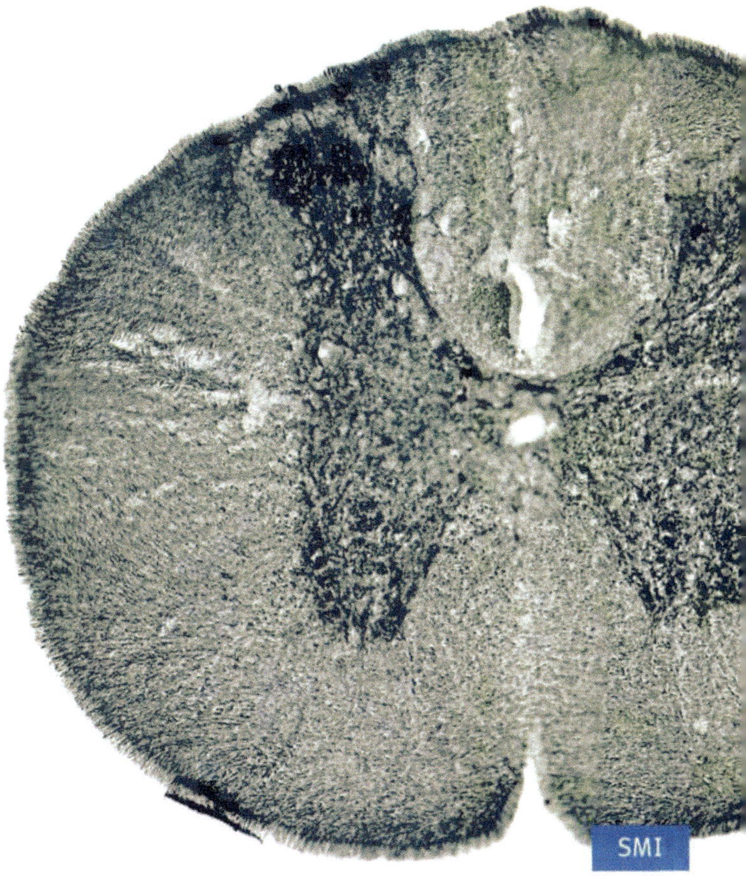

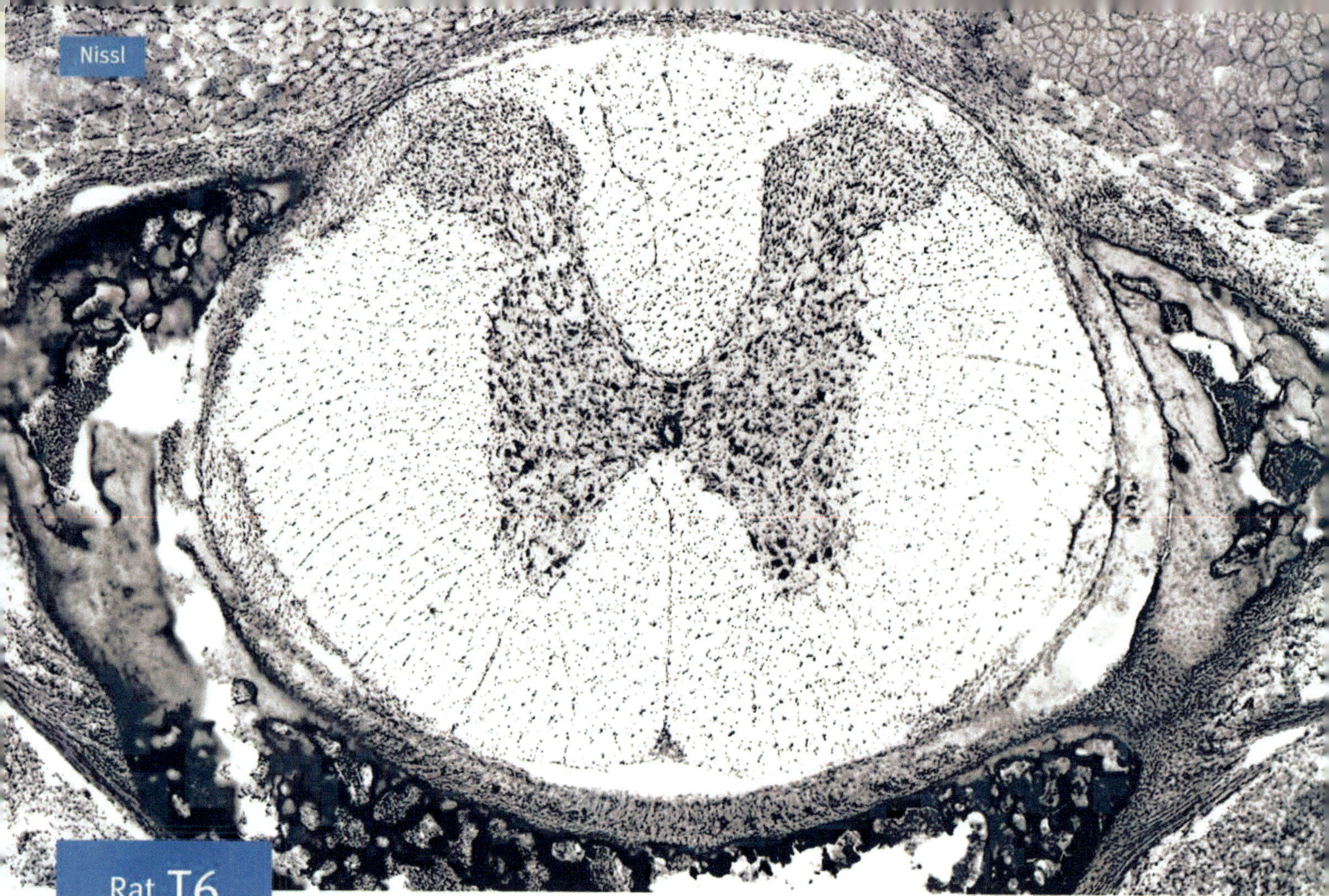
Nissl

Rat T6

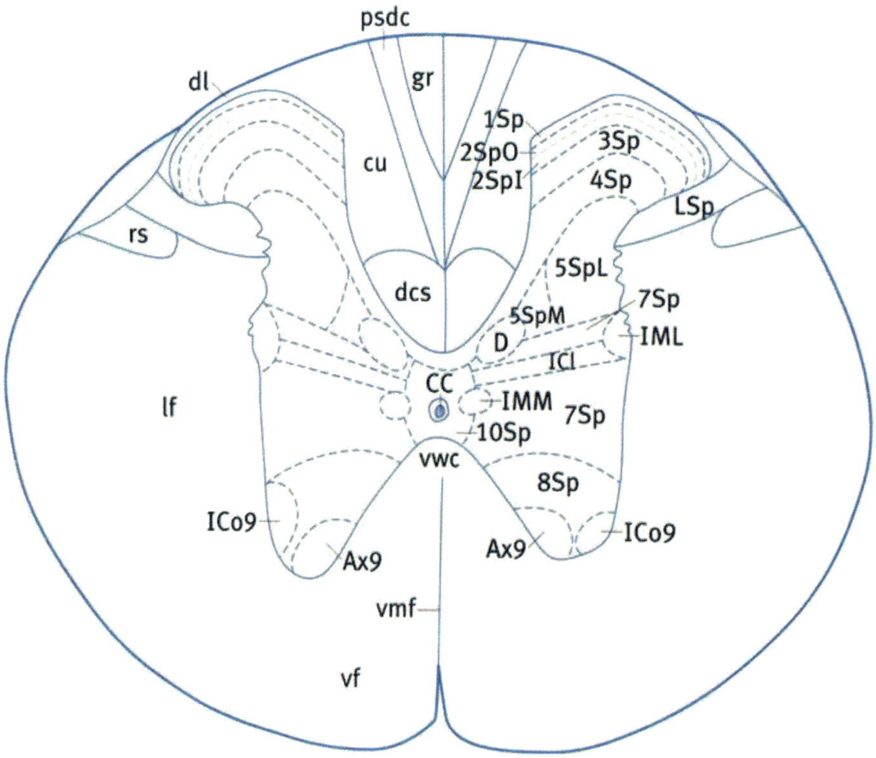
psdc
dl
gr
1Sp
cu
2SpO
3Sp
2SpI
4Sp
LSp
rs
5SpL
dcs
7Sp
5SpM
IML
D
ICl
CC
lf
IMM
7Sp
10Sp
vwc
8Sp
ICo9
ICo9
Ax9
Ax9
vmf
vf

1 mm

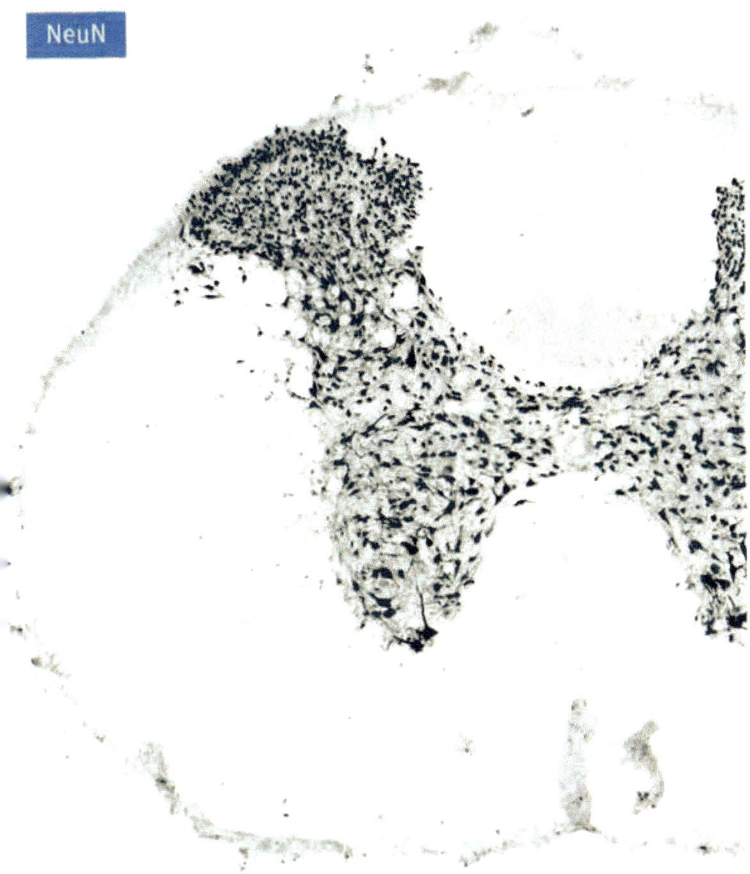

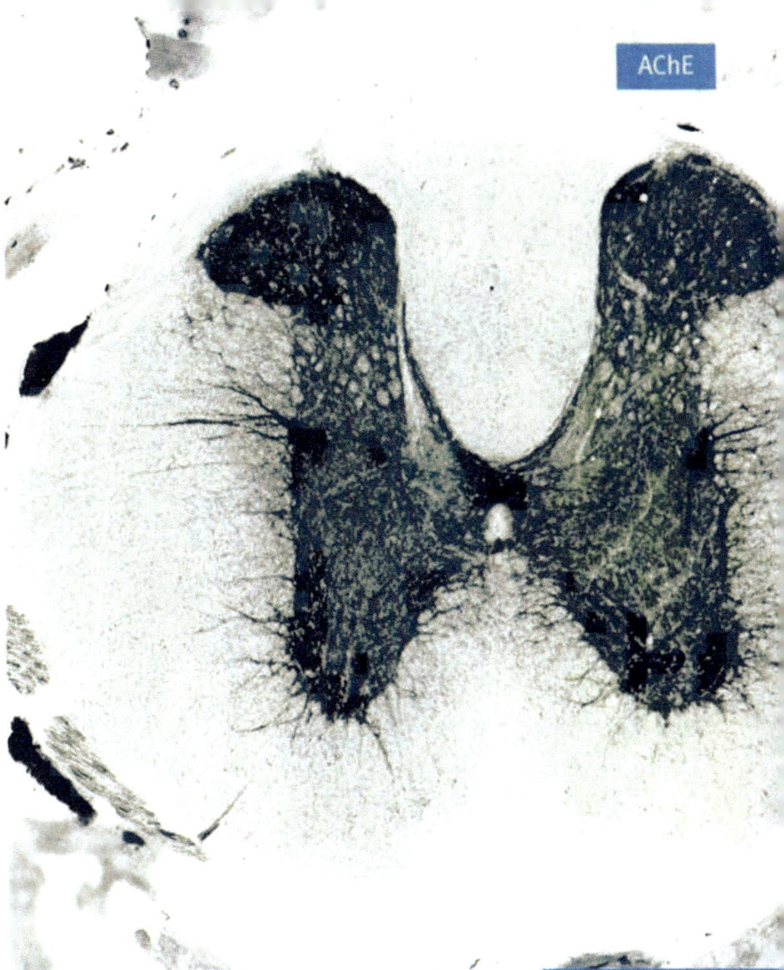

1Sp lamina 1
2SpI lamina 2, inner
2SpO lamina 2, outer
3Sp lamina 3
4Sp lamina 4
5SpL lamina 5, lat
5SpM lamina 5, med
7Sp lamina 7
8Sp lamina 8
10Sp lamina 10
Ax9 axial muscles
CC central canal
cu cuneate fasc
D dorsal nucleus
dcs dors corticosp
dl dorsolat fasc
gr gracile fascic
ICl intercalated nu
ICo9 intercostals
IML intermediolat
IMM intermediomed
LSp lat spinal nu
psdc postsyn dors col
rs rubrospinal tr
vmf vent med fiss
vwc vent white com

Rat T6

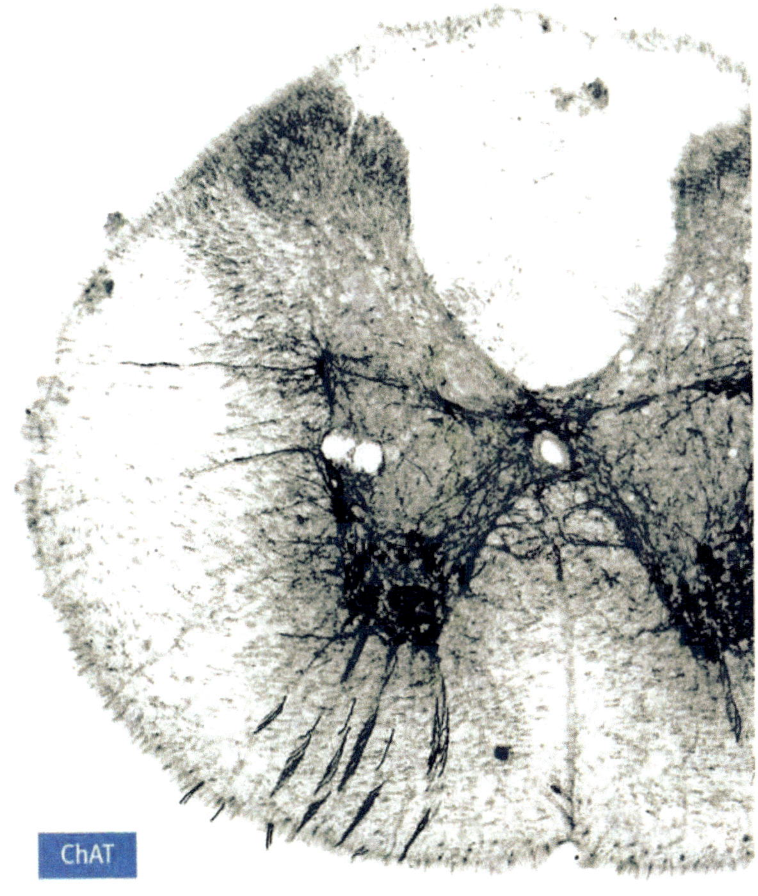

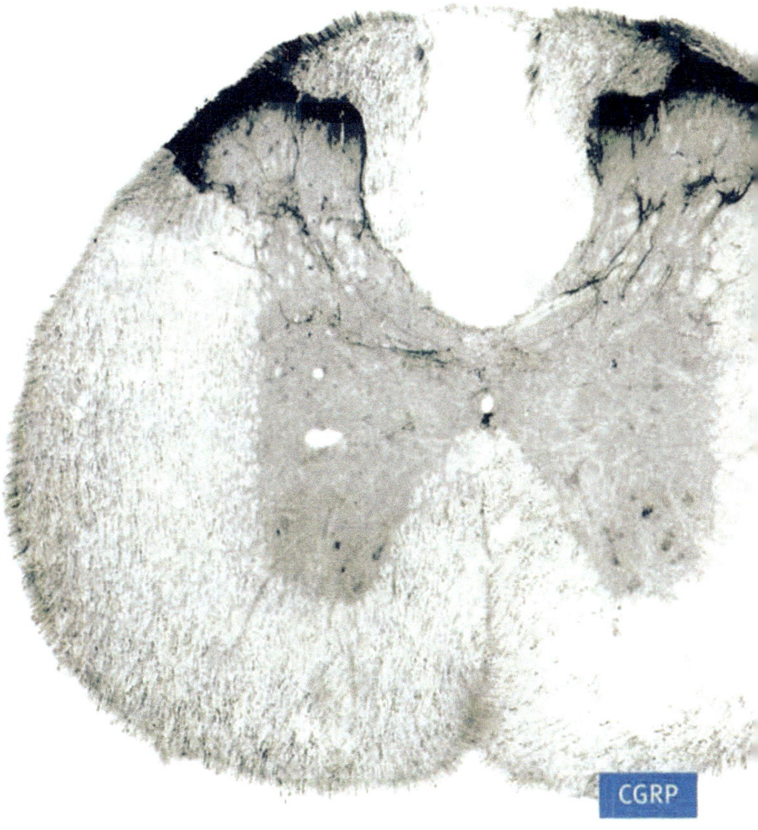

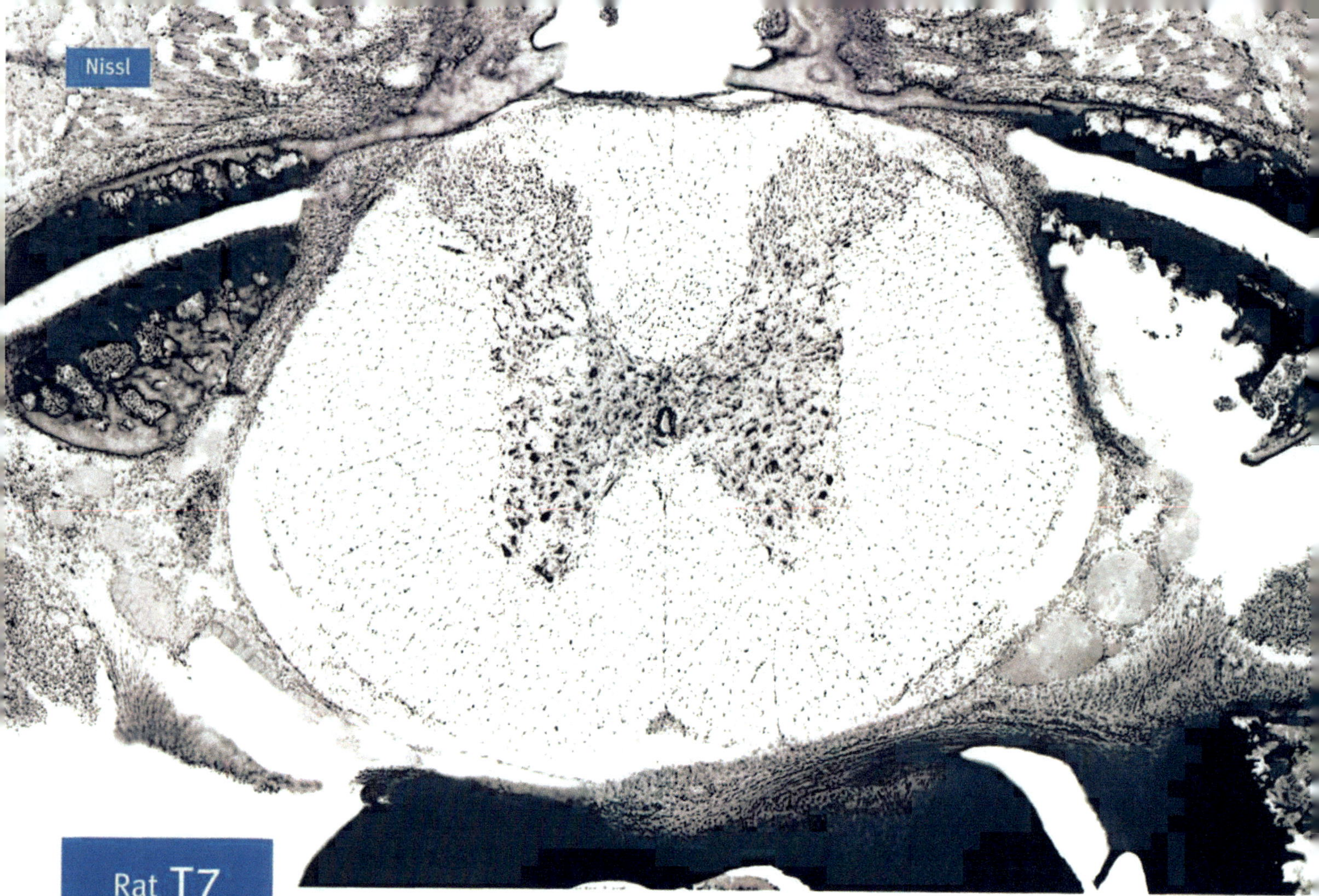
Nissl

Rat T7

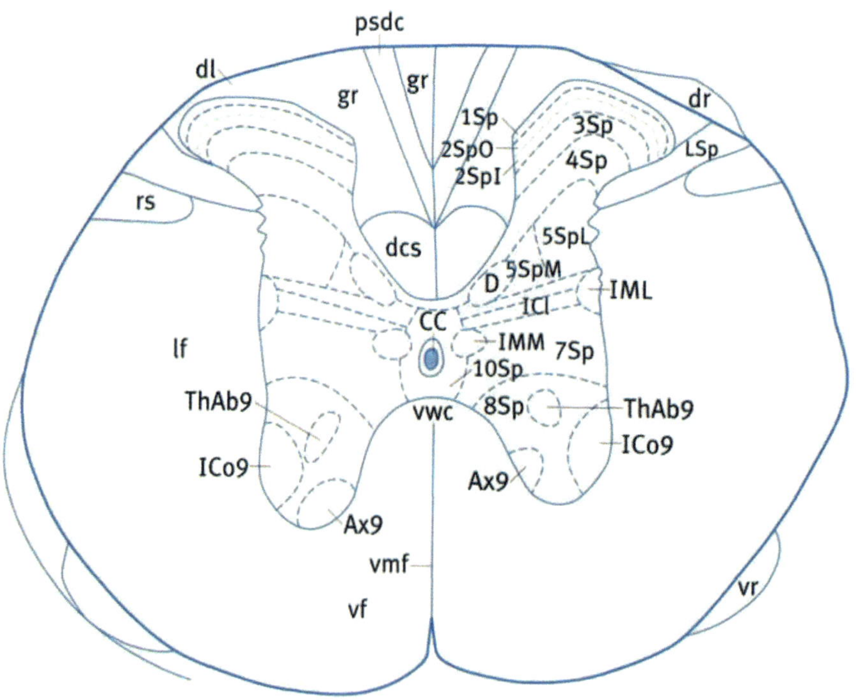
psdc
dl
gr
gr
dr
1Sp
2SpO
2SpI
3Sp
4Sp
LSp
rs
dcs
5SpL
5SpM
D
IML
ICl
CC
lf
IMM
7Sp
10Sp
ThAb9
8Sp
ThAb9
vwc
ICo9
ICo9
Ax9
Ax9
vmf
vf
vr

1 mm

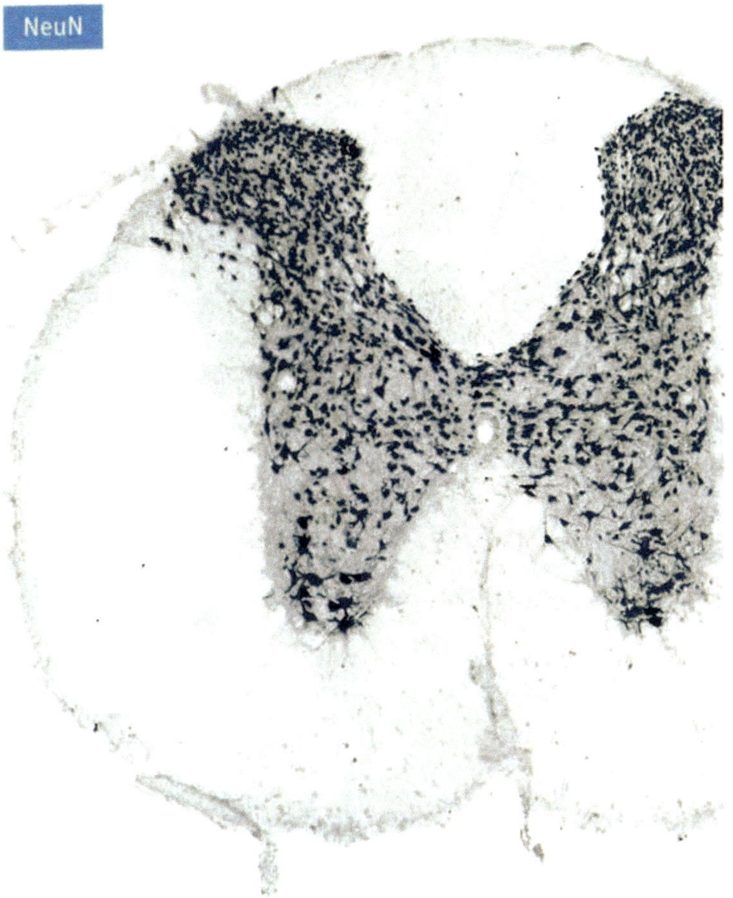

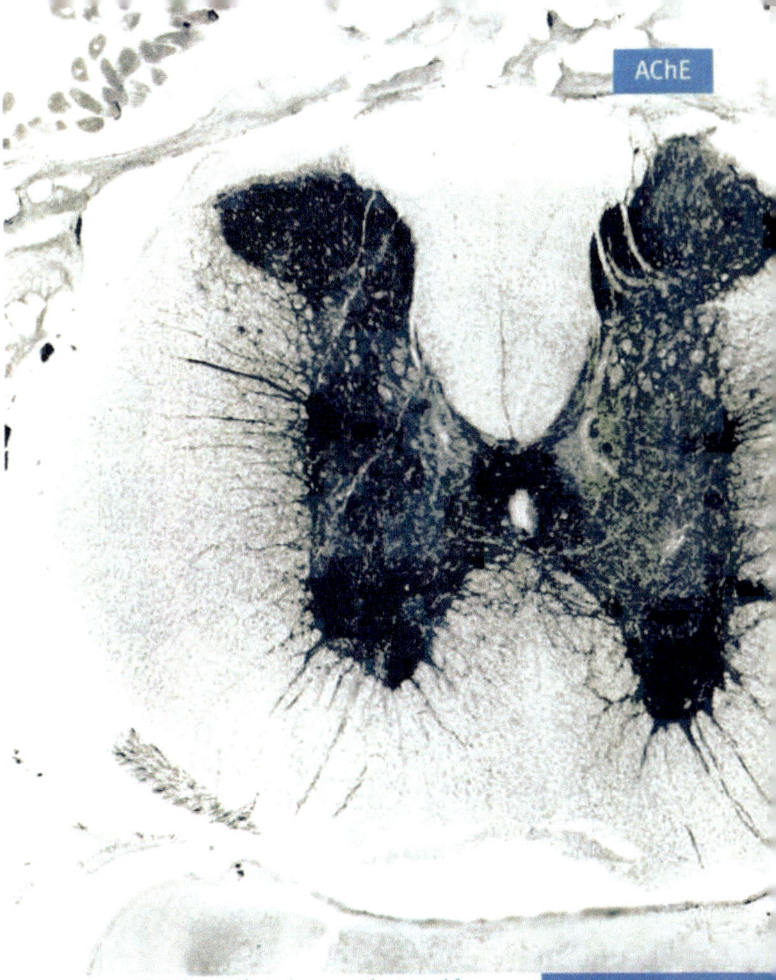

Rat T7

- **1Sp** lamina 1
- **2SpI** lamina 2, inner
- **2SpO** lamina 2, outer
- **3Sp** lamina 3
- **4Sp** lamina 4
- **5SpL** lamina 5, lat
- **5SpM** lamina 5, med
- **7Sp** lamina 7
- **8Sp** lamina 8
- **10Sp** lamina 10
- **Ax9** axial muscles
- **CC** central canal
- **D** dorsal nucleus
- **dcs** dors corticosp
- **dl** dorsolat fasc
- **dr** dorsal root
- **gr** gracile fascic
- **ICl** intercalated nu
- **ICo9** intercostals
- **IML** intermediolat
- **IMM** intermediomed
- **LSp** lat spinal nu
- **psdc** postsyn dors col
- **rs** rubrospinal tr
- **ThAb9** thor-abd wall
- **vmf** vent med fiss
- **vr** ventral root
- **vwc** vent white com

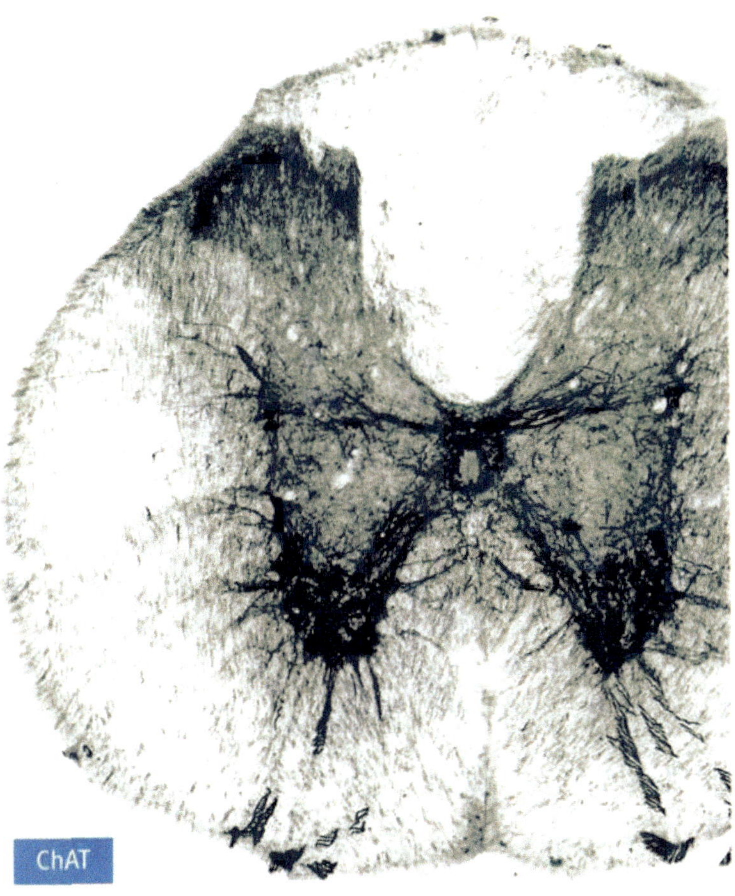

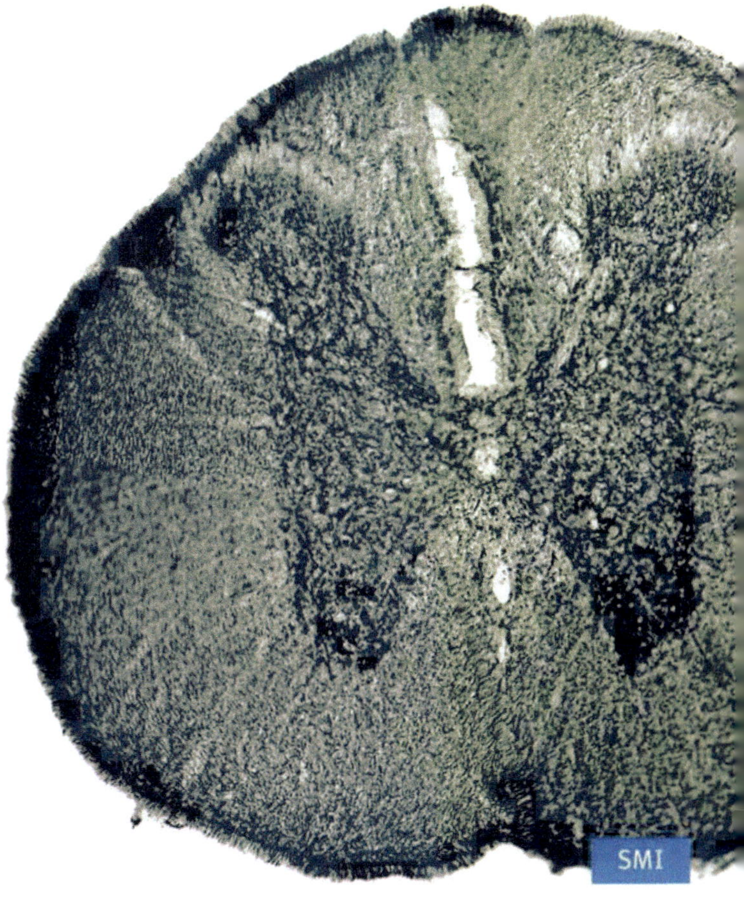

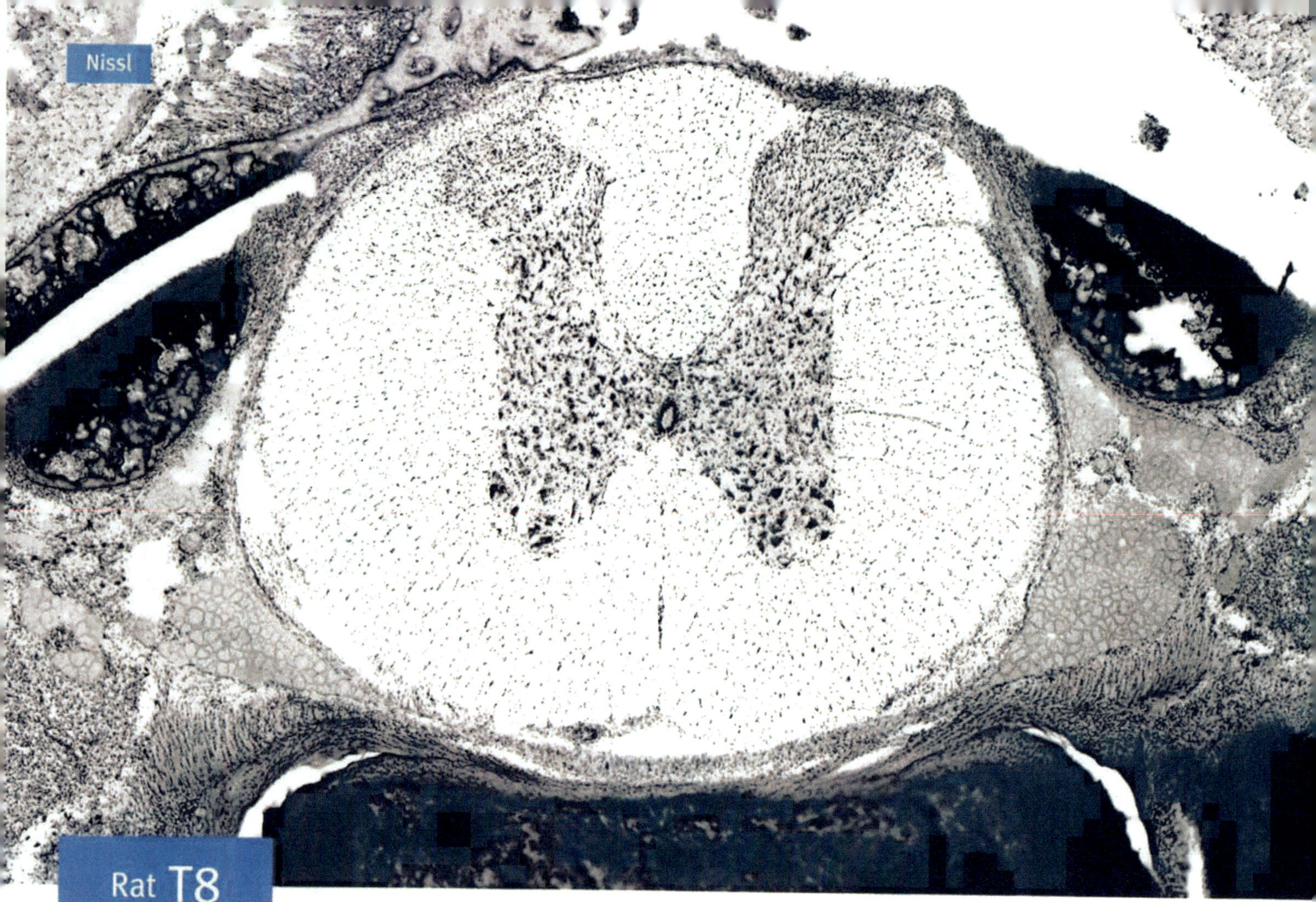
Nissl

Rat T8

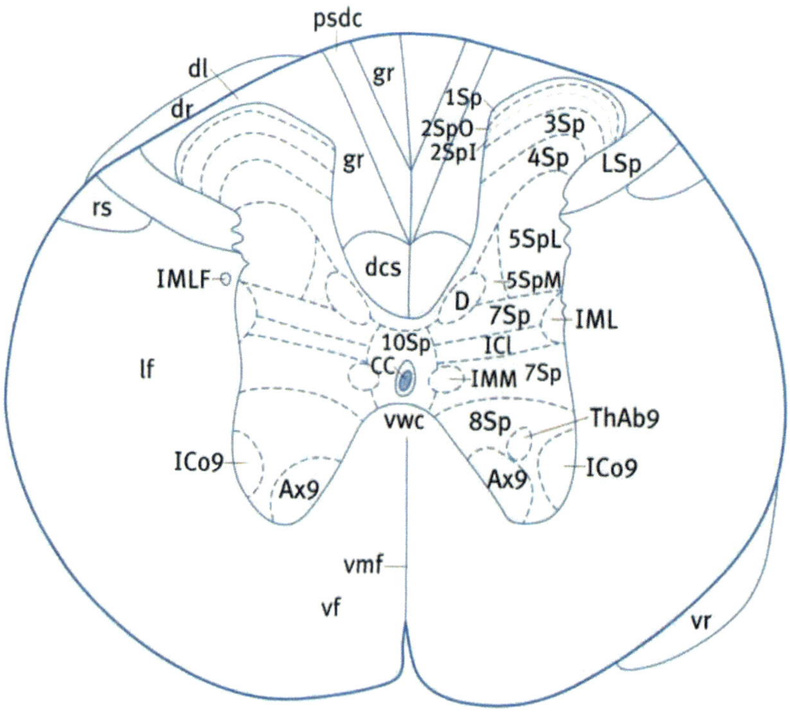
psdc
dl
dr
gr
gr
1Sp
2SpO
2SpI
3Sp
4Sp
LSp
rs
5SpL
dcs
5SpM
IMLF
D
7Sp
IML
10Sp
ICl
CC
IMM
7Sp
lf
vwc
8Sp
ThAb9
ICo9
Ax9
Ax9
ICo9
vmf
vf
vr

1 mm

Cr

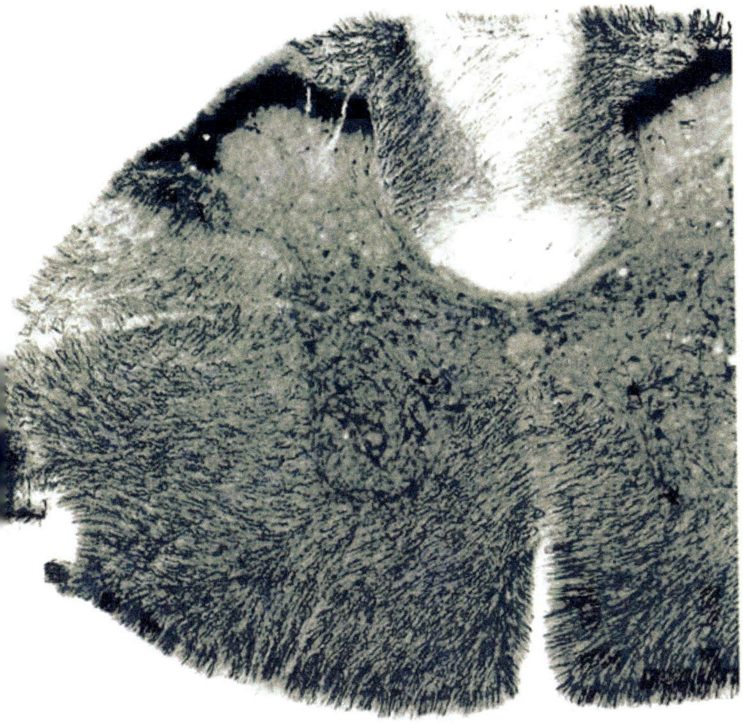

AChE

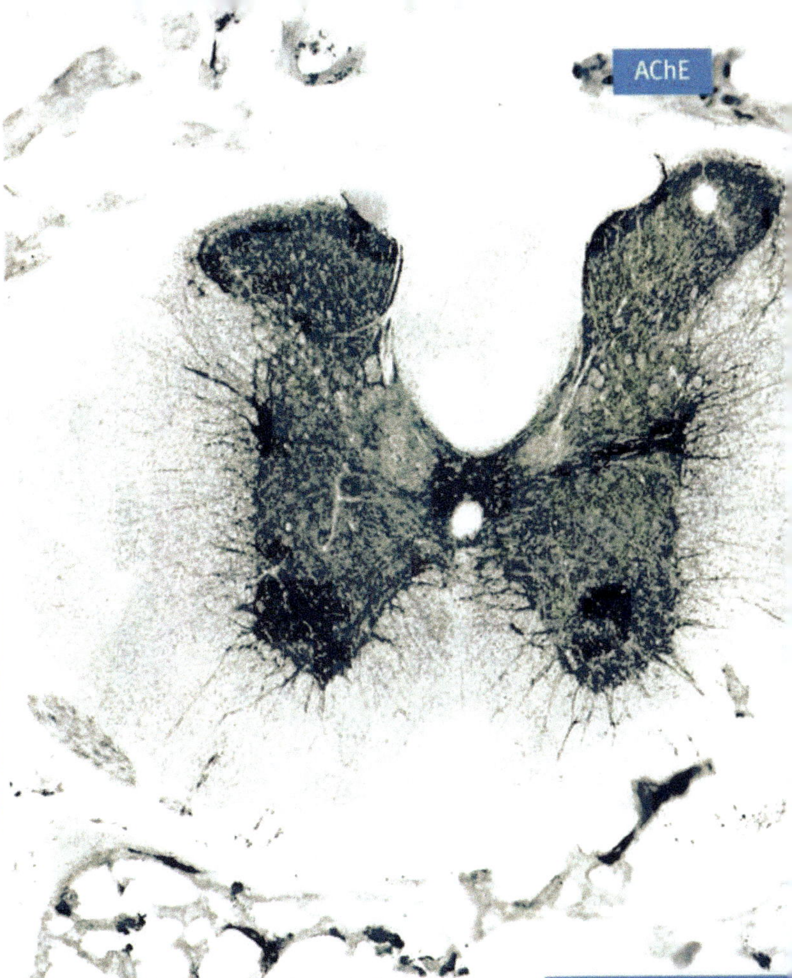

1Sp lamina 1
2SpI lamina 2, inner
2SpO lamina 2, outer
3Sp lamina 3
4Sp lamina 4
5SpL lamina 5, lat
5SpM lamina 5, med
7Sp lamina 7
8Sp lamina 8
10Sp lamina 10
Ax9 axial muscles
CC central canal
D dorsal nucleus
dcs dors corticosp
dl dorsolat fasc
dr dorsal root
gr gracile fascic
ICl intercalated nu
ICo9 intercostals
IML intermediolat
IMM intermediomed
LSp lat spinal nu
psdc postsyn dors col
rs rubrospinal tr
ThAb9 thor-abd wall
vmf vent med fiss
vr ventral root
vwc vent white com

Rat T8

NeuN

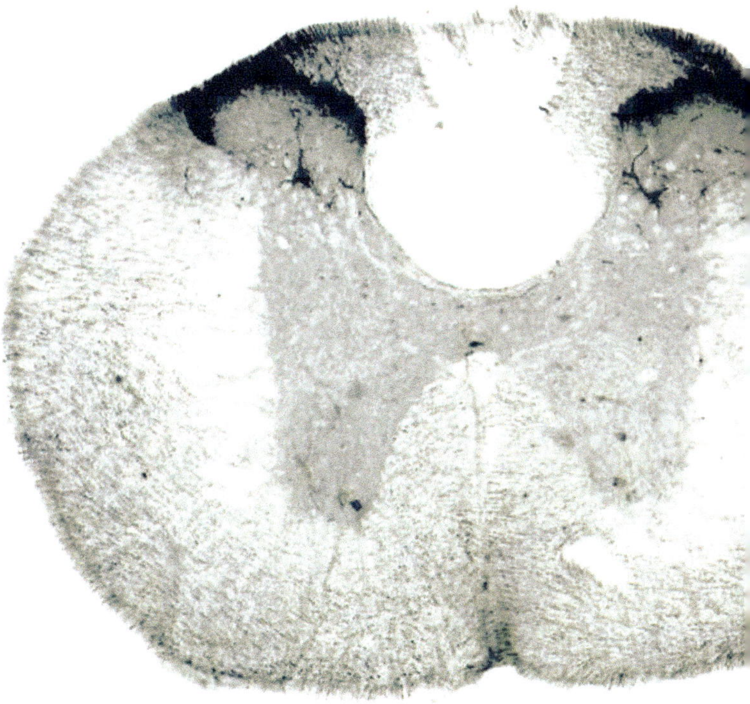

CGRP

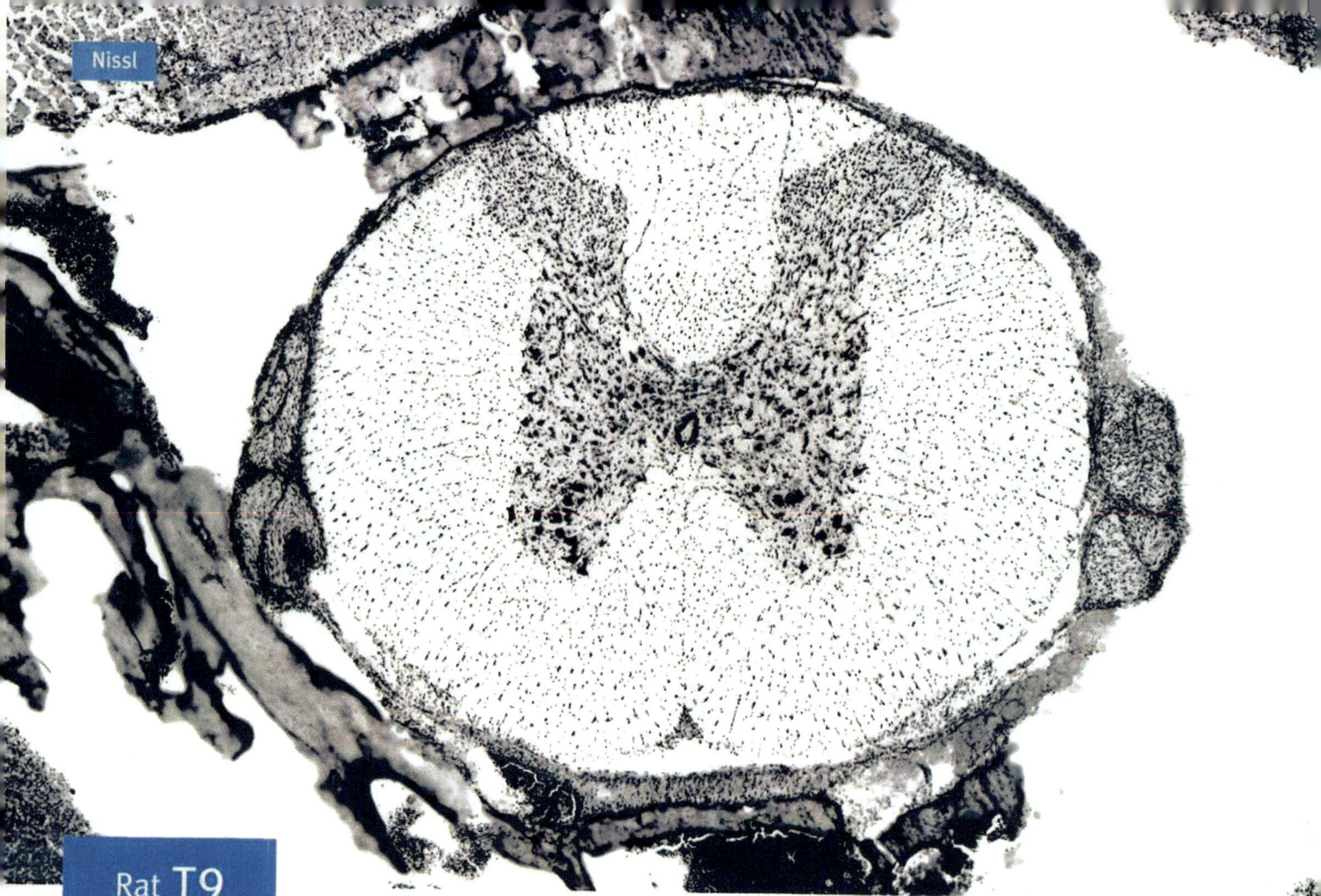
Nissl

Rat T9

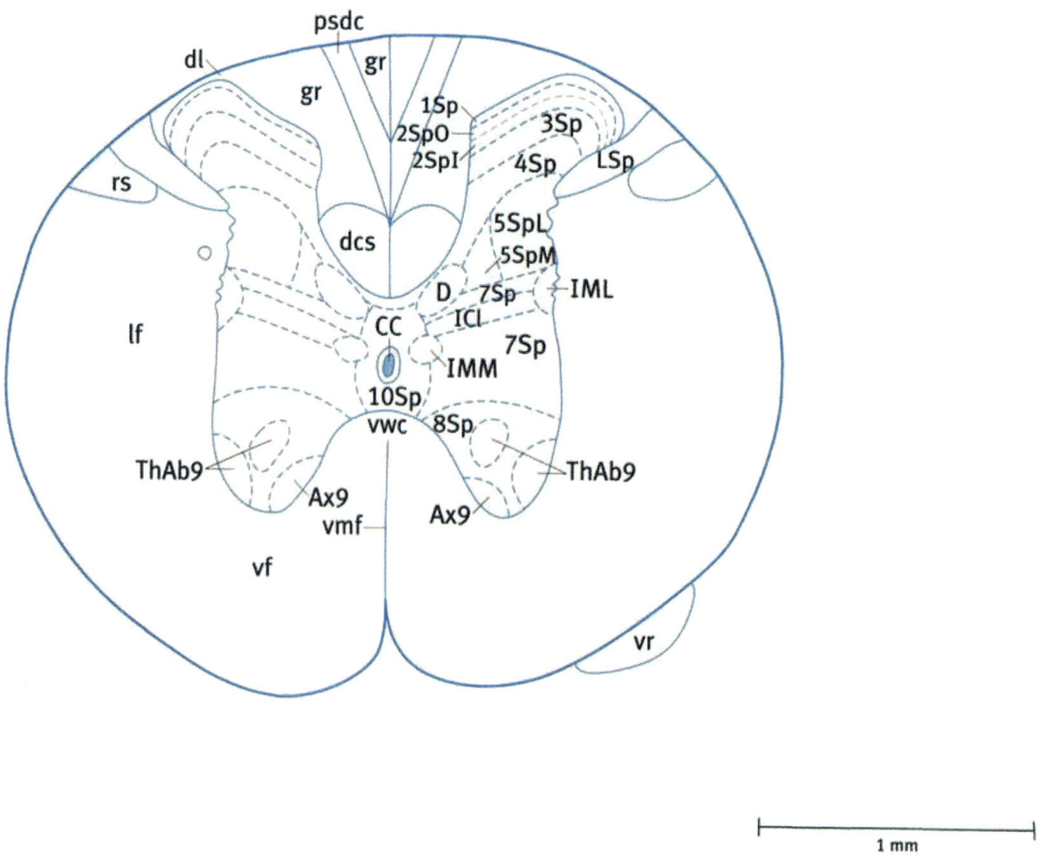
psdc
dl
gr
gr
1Sp
2SpO
2SpI
3Sp
4Sp
LSp
rs
dcs
5SpL
5SpM
D
7Sp
IML
ICl
CC
7Sp
lf
IMM
10Sp
vwc
8Sp
ThAb9
ThAb9
Ax9
Ax9
vmf
vf
vr
1 mm

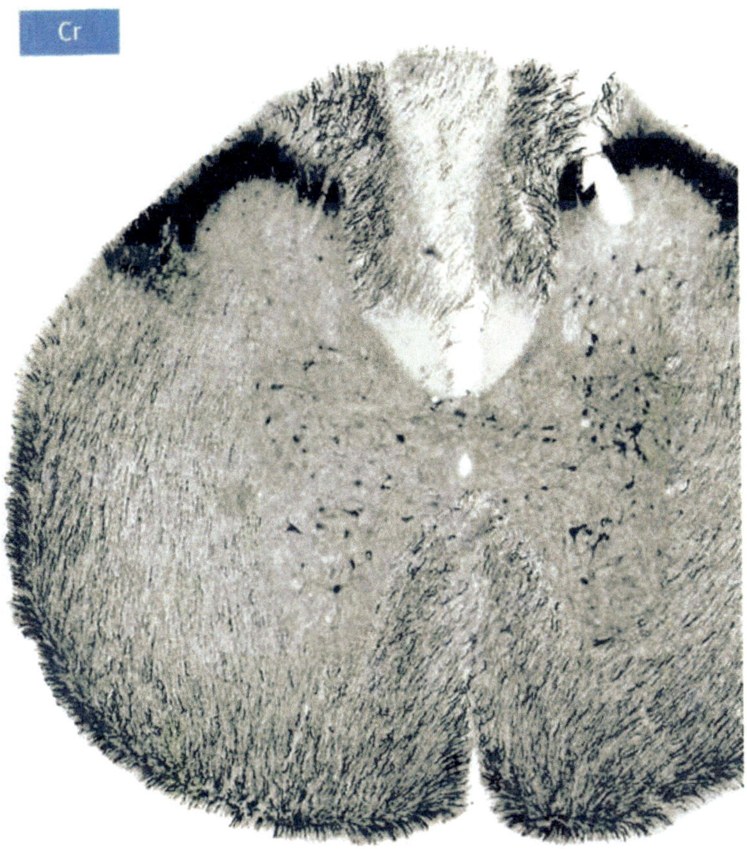

1Sp lamina 1
2SpI lamina 2, inner
2SpO lamina 2, outer
3Sp lamina 3
4Sp lamina 4
5SpL lamina 5, lat
5SpM lamina 5, med
7Sp lamina 7
8Sp lamina 8
10Sp lamina 10
Ax9 axial muscles
CC central canal
D dorsal nucleus
dcs dors corticosp
dl dorsolat fasc
gr gracile fascic
ICl intercalated nu
IML intermediolat
IMM intermediomed
LSp lat spinal nu
psdc postsyn dors col
rs rubrospinal tr
ThAb9 thor-abd wall
vmf vent med fiss
vr ventral root
vwc vent white com

Rat T9

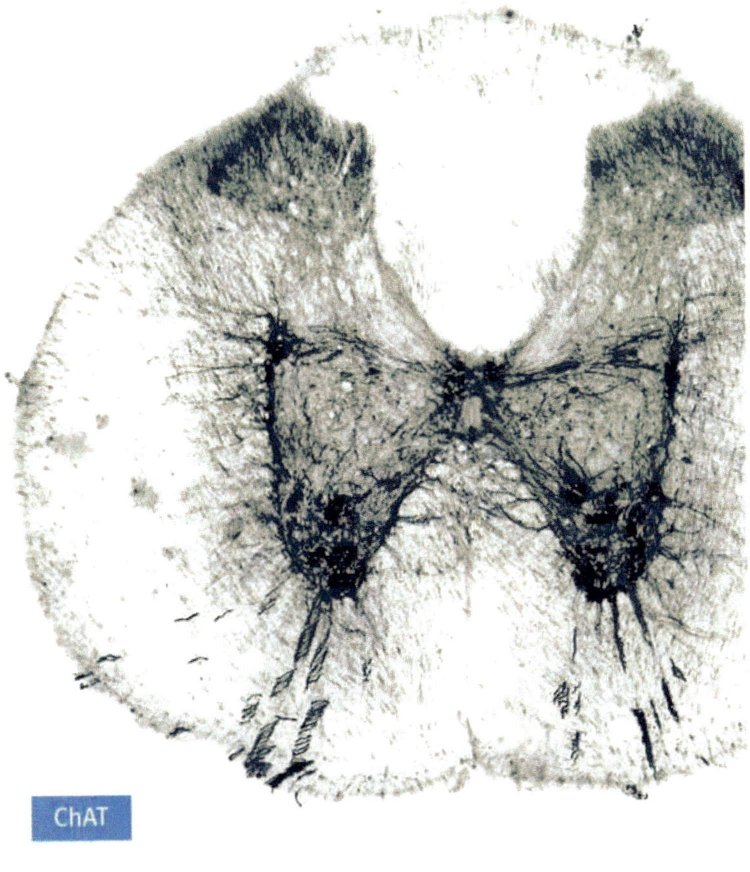

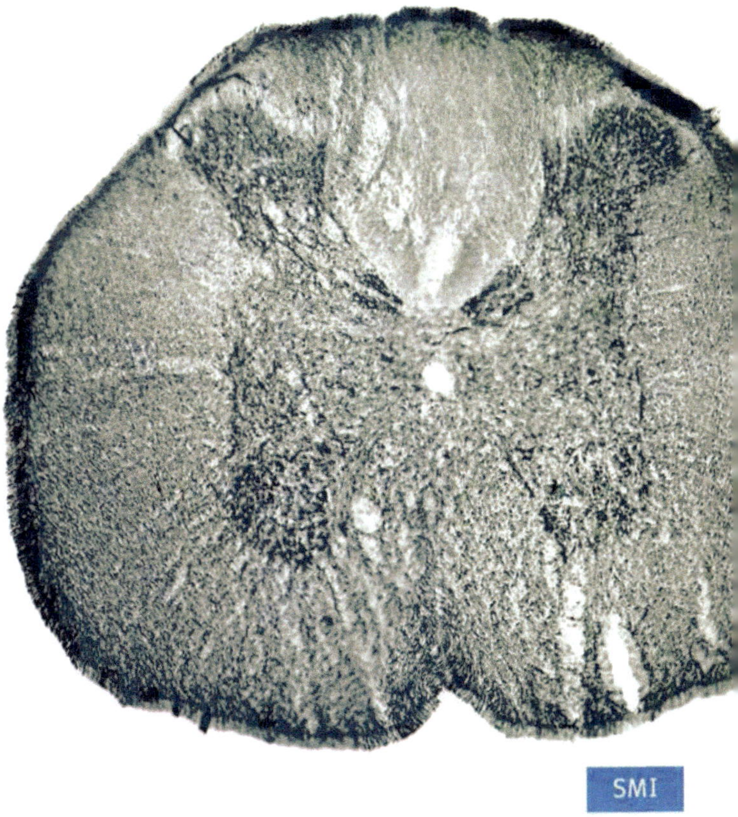

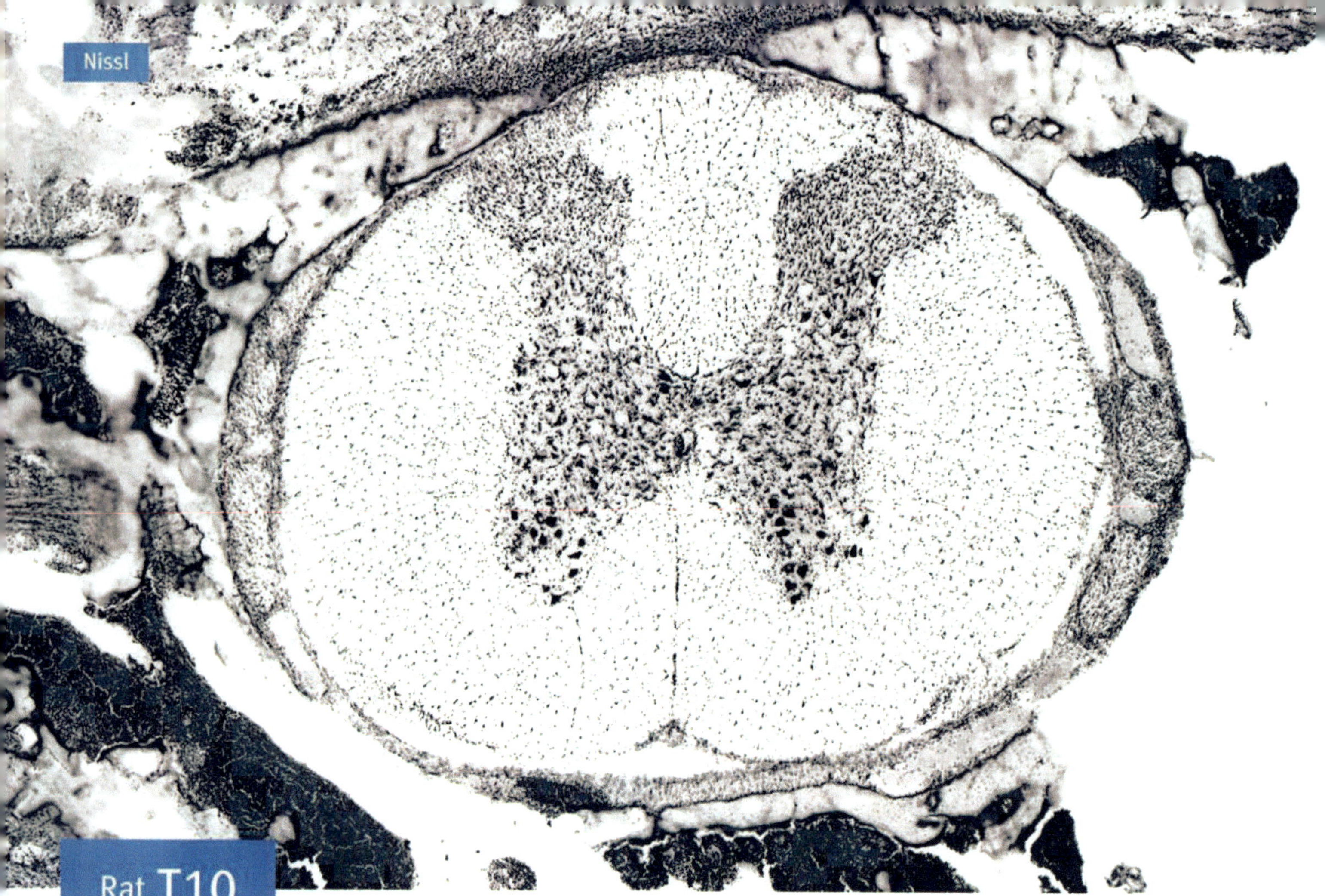
Nissl

Rat T10

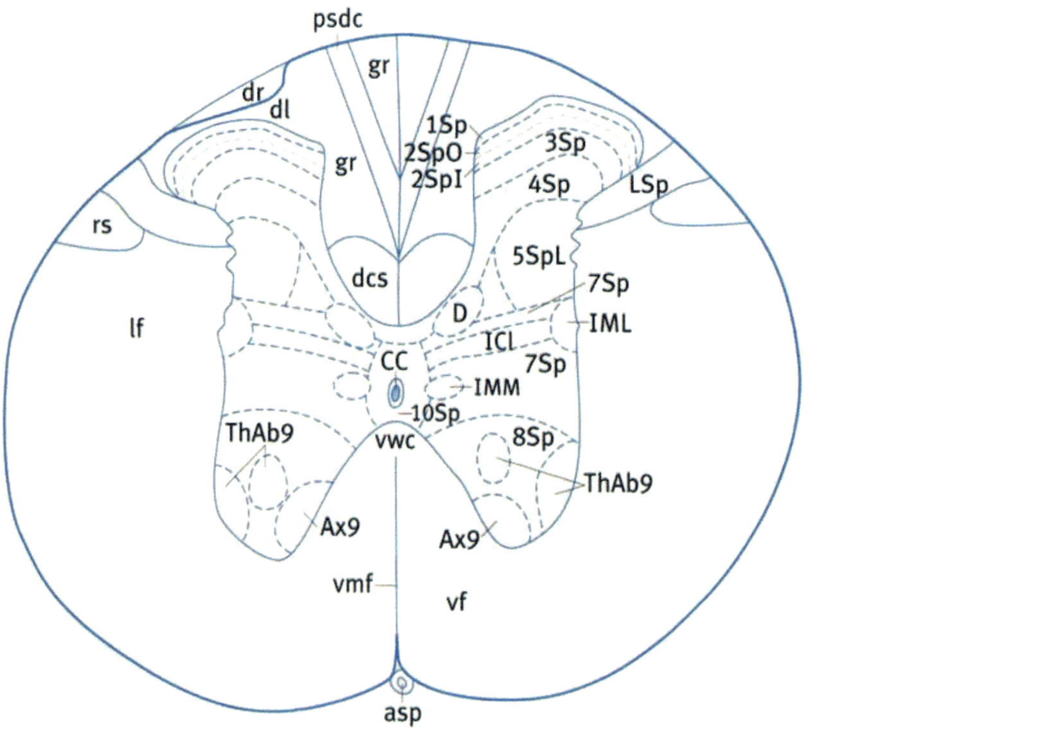
psdc
gr
dr
dl
gr
1Sp
2SpO
2SpI
3Sp
4Sp
LSp
rs
5SpL
7Sp
dcs
D
IML
lf
ICl
CC
7Sp
IMM
10Sp
ThAb9
vwc
8Sp
ThAb9
Ax9
Ax9
vmf
vf
asp
1 mm

NeuN

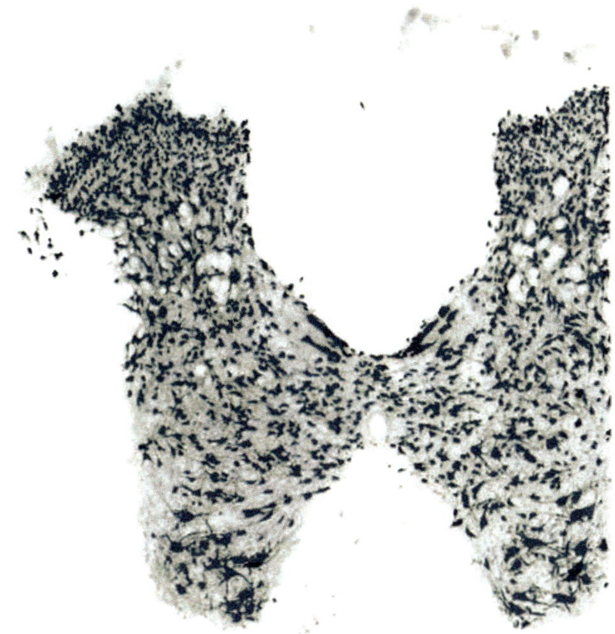

AChE

1Sp lamina 1
2SpI lamina 2, inner
2SpO lamina 2, outer
3Sp lamina 3
4Sp lamina 4
5SpL lamina 5, lat
7Sp lamina 7
8Sp lamina 8
10Sp lamina 10
asp ant spinal art
Ax9 axial muscles
CC central canal
D dorsal nucleus
dcs dors corticosp
dr dorsal root
gr gracile fascic
ICl intercalated nu
IML intermediolat
IMM intermediomed
LSp lat spinal nu
psdc postsyn dors col
rs rubrospinal tr
ThAb9 thor-abd wall
vmf vent med fiss
vwc vent white com

Rat T10

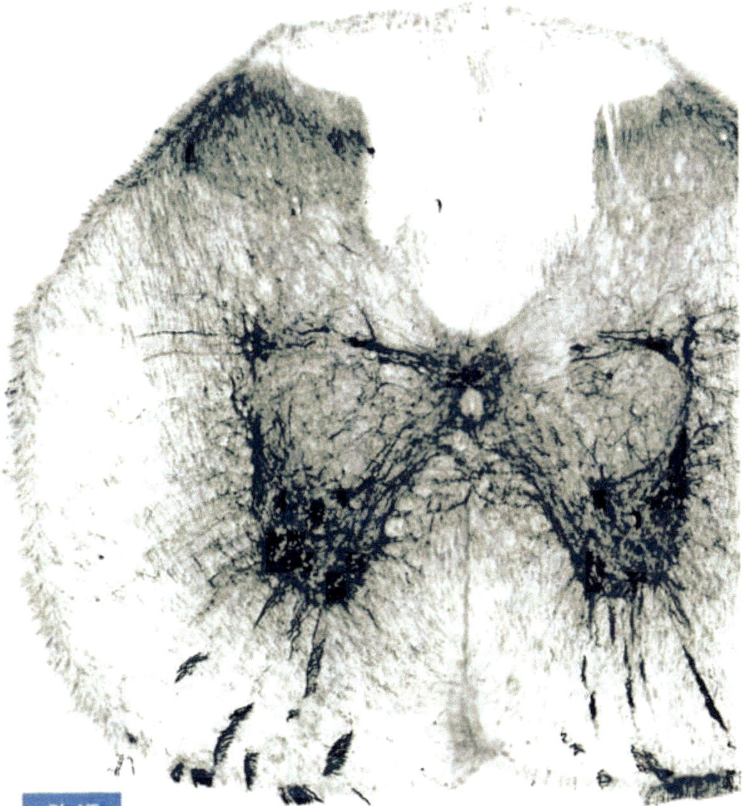

ChAT

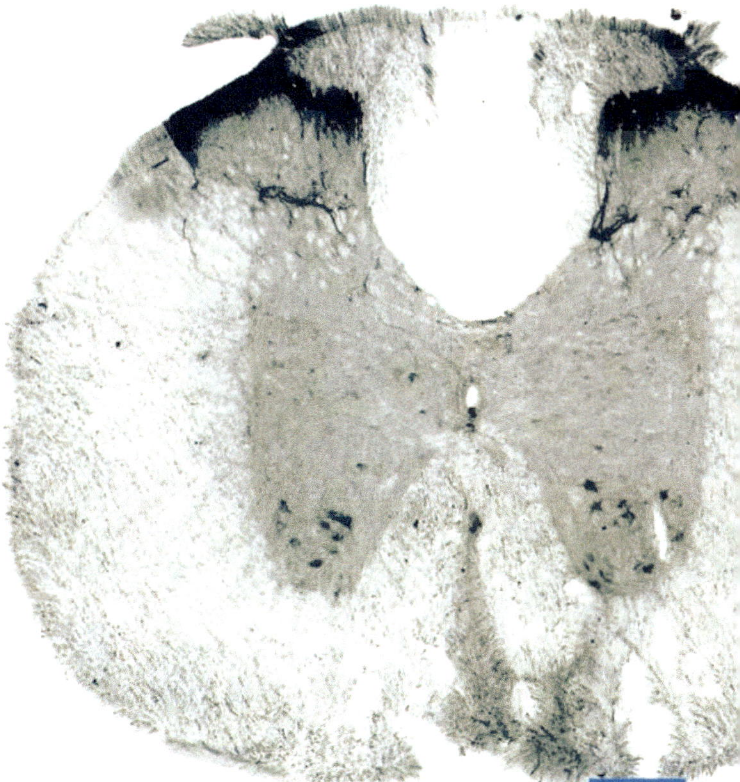

CGRP

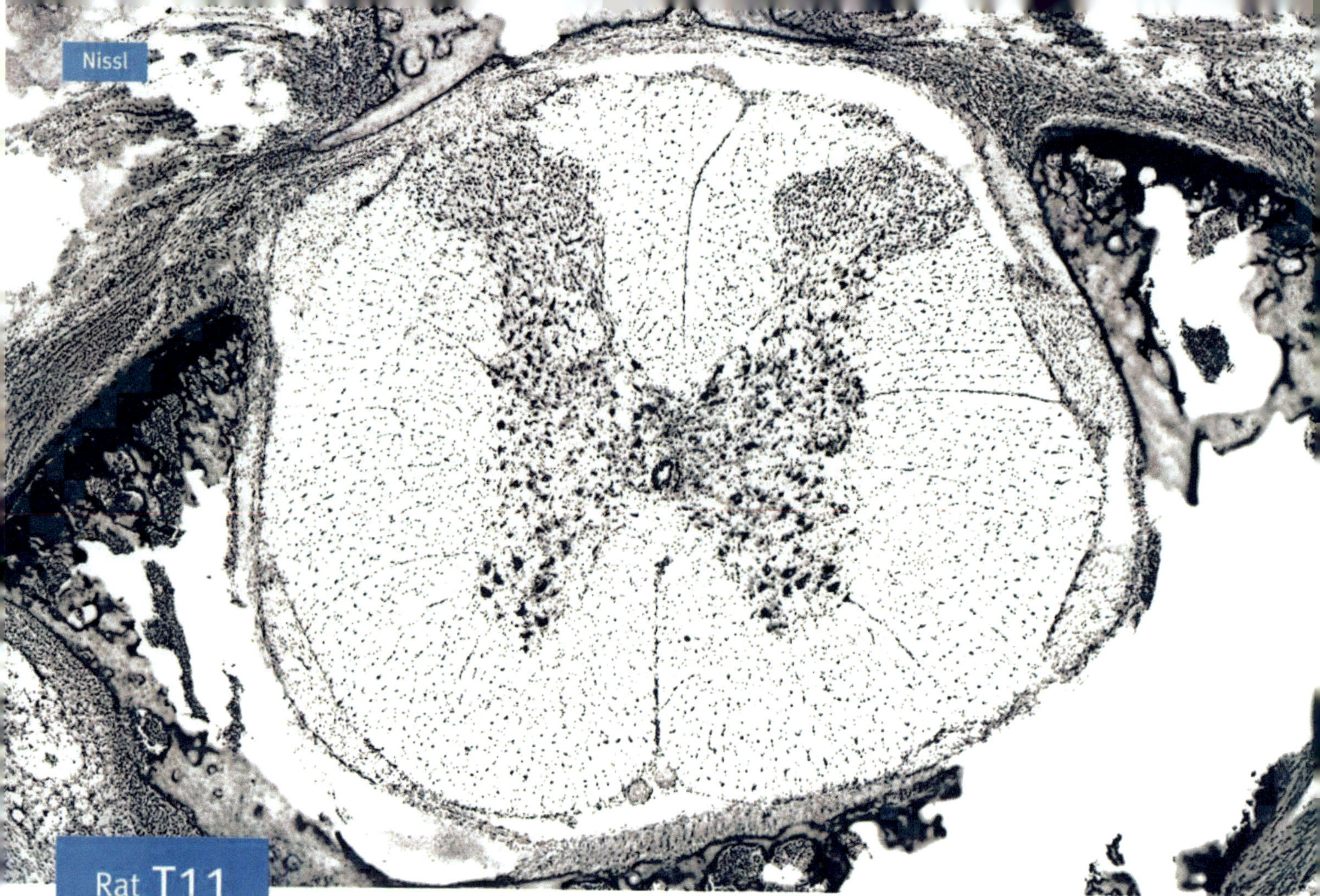
Nissl

Rat T11

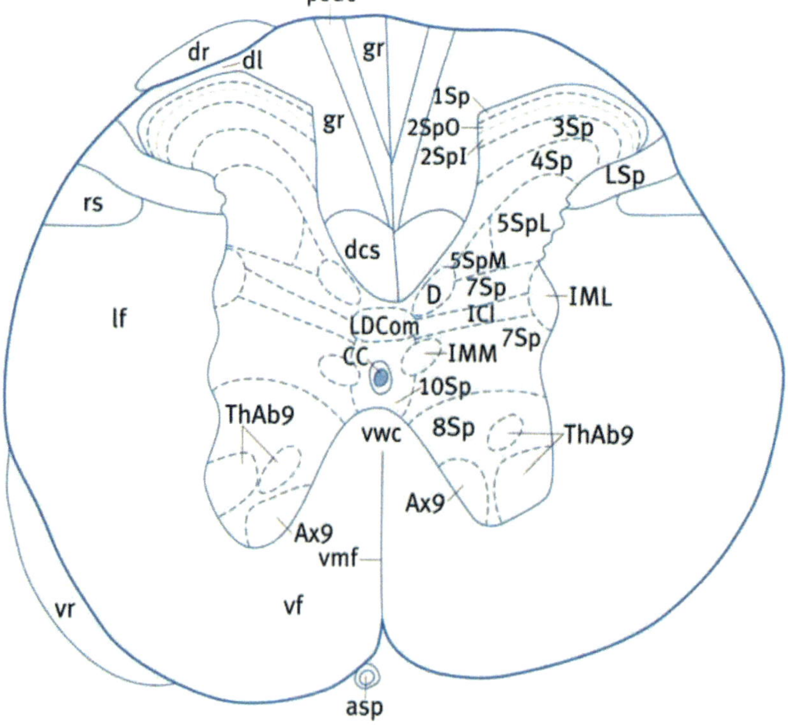
psdc
dr
dl
gr
gr
1Sp
2SpO
2SpI
3Sp
4Sp
LSp
rs
5SpL
dcs
5SpM
7Sp
D
IML
ICl
lf
LDCom
7Sp
CC
IMM
10Sp
ThAb9
vwc
8Sp
ThAb9
Ax9
Ax9
vmf
vr
vf
asp
1 mm

AChE

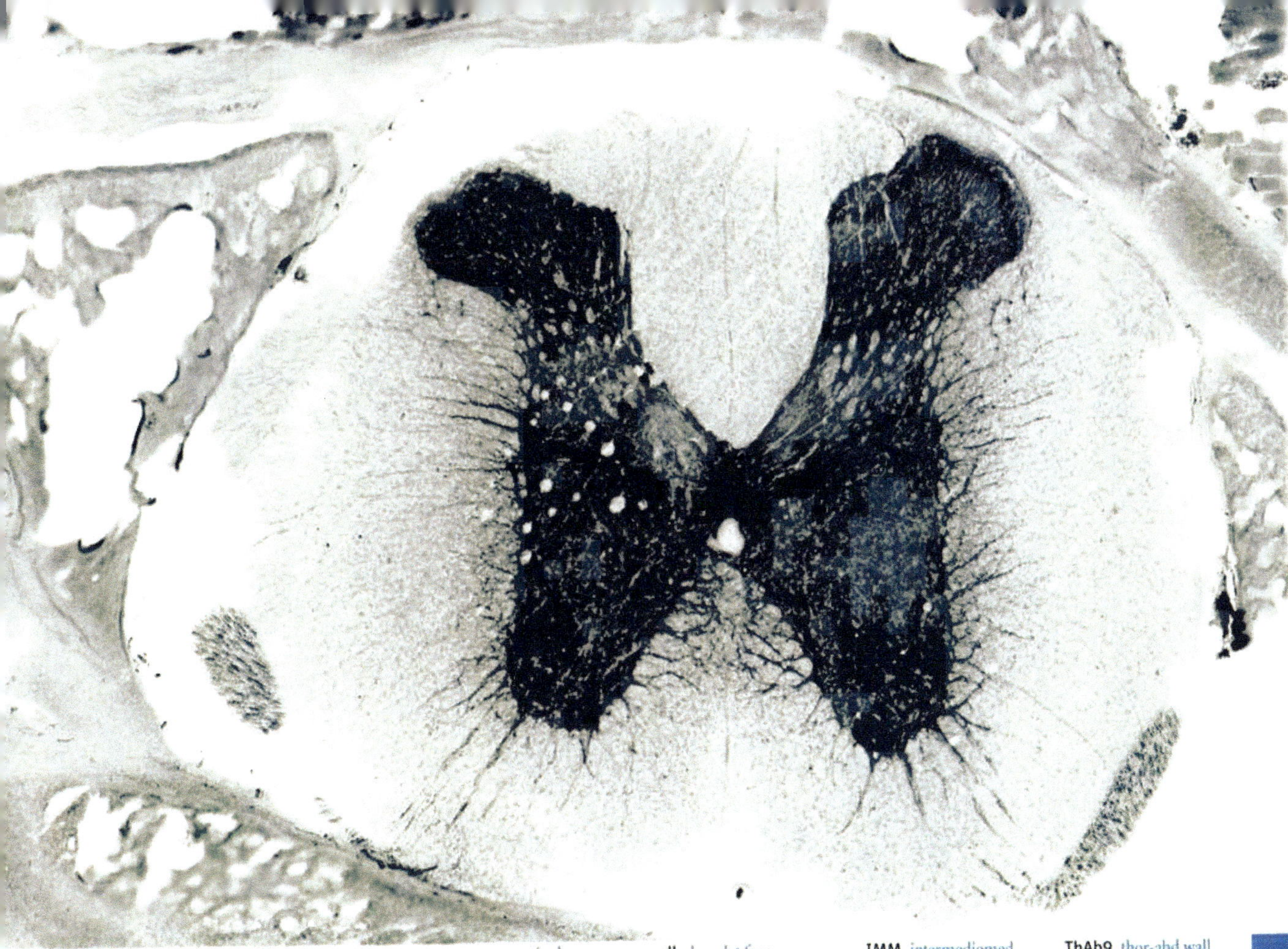

1Sp lamina 1
2SpI lamina 2, inner
2SpO lamina 2, outer
3Sp lamina 3
4Sp lamina 4
5SpL lamina 5, lat
5SpM lamina 5, med
7Sp lamina 7
8Sp lamina 8
10Sp lamina 10
asp ant spinal art
Ax9 axial muscles
CC central canal
D dorsal nucleus
dcs dors corticosp
dl dorsolat fasc
dr dorsal root
gr gracile fascic
ICl intercalated nu
IML intermediolat
IMM intermediomed
LDCom lum dors com
LSp lat spinal nu
psdc postsyn dors col
rs rubrospinal tr
ThAb9 thor-abd wall
vmf vent med fiss
vr ventral root
vwc vent white com

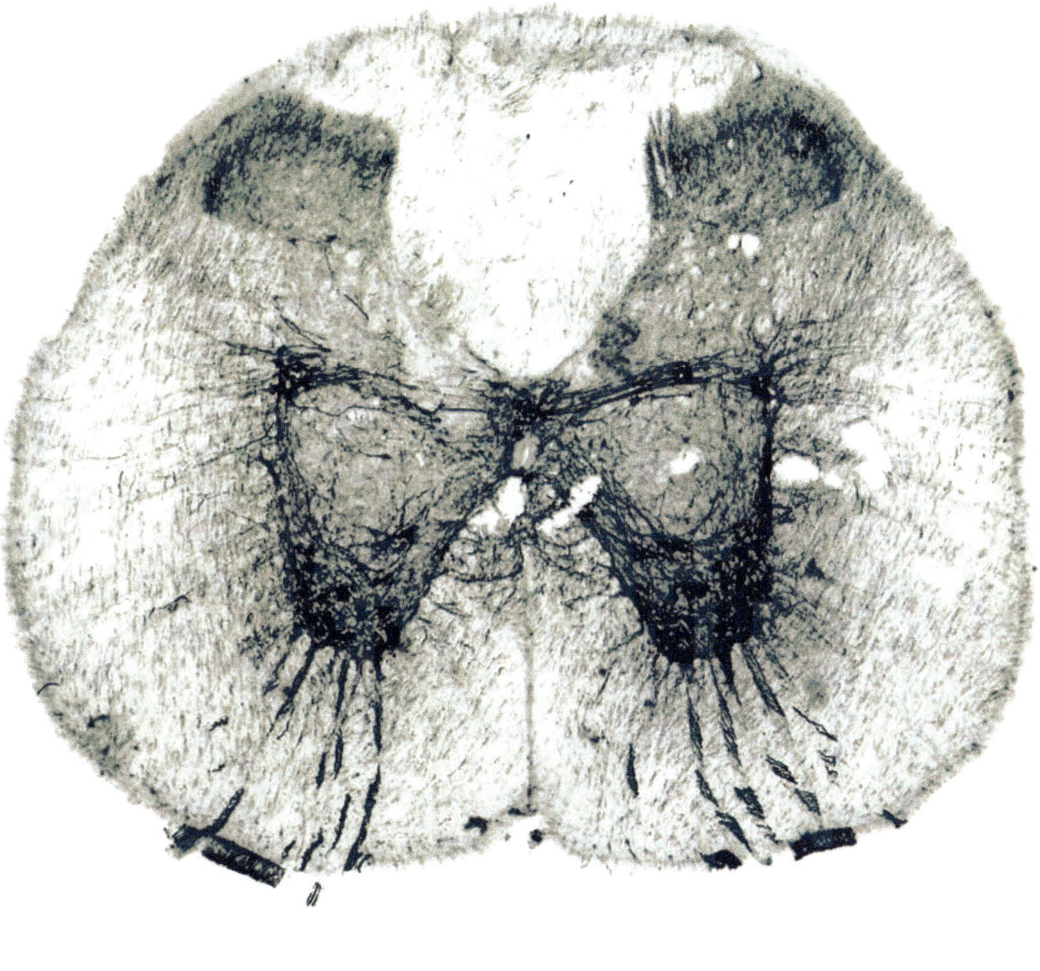

ChAT

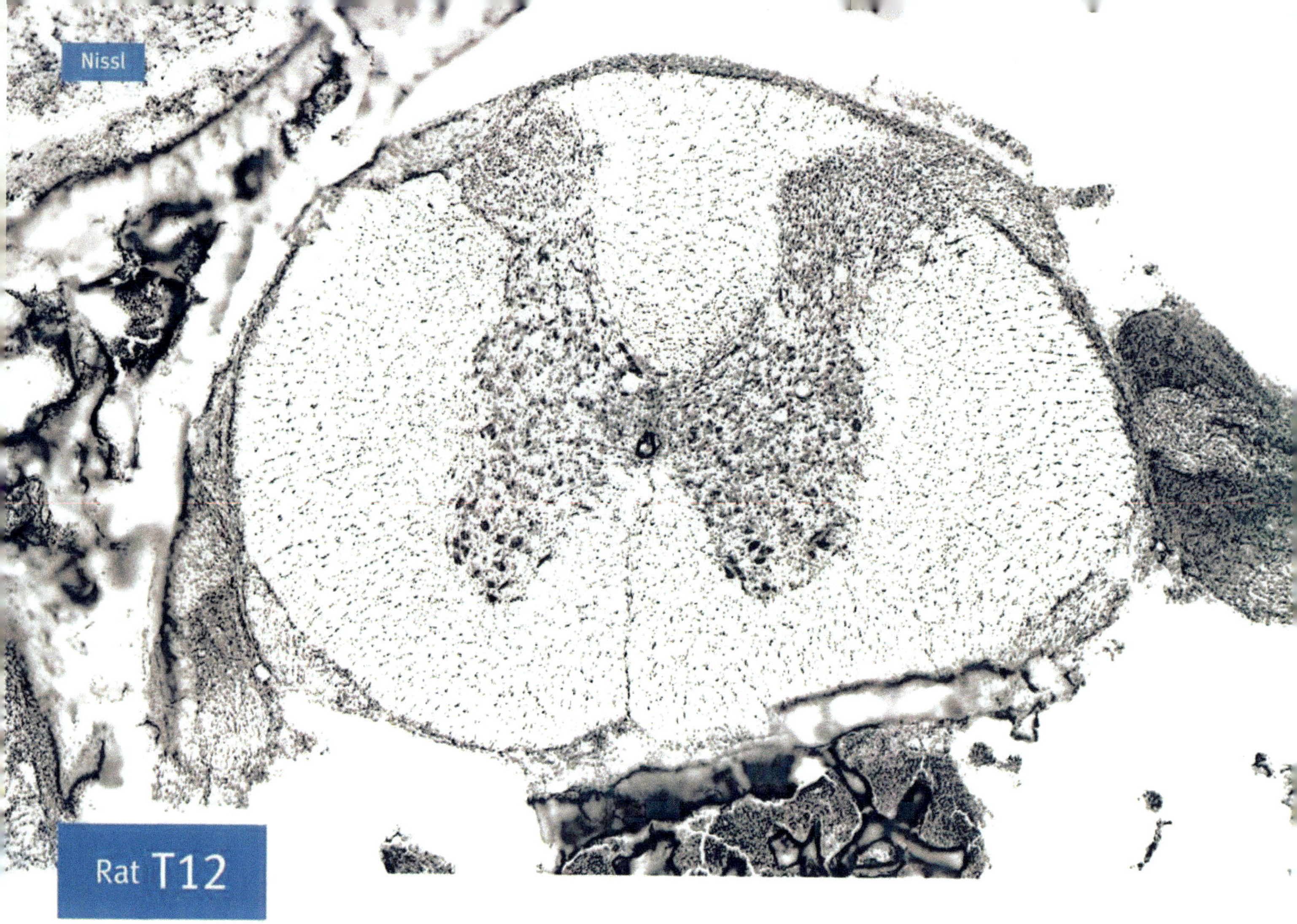
Nissl
Rat T12

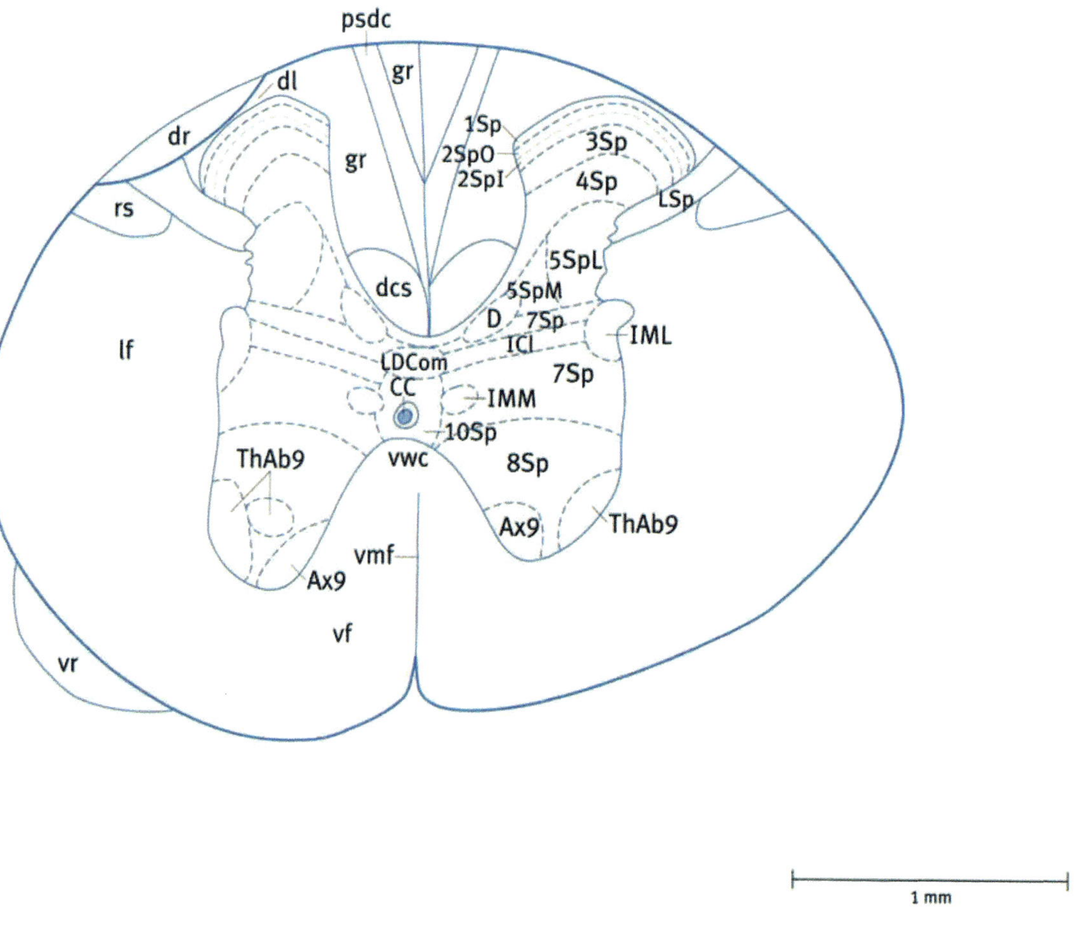
psdc
gr
dl
dr
gr
1Sp
2SpO
2SpI
3Sp
4Sp
LSp
rs
5SpL
dcs
5SpM
D
7Sp
ICl
IML
lf
LDCom
CC
7Sp
IMM
10Sp
ThAb9
vwc
8Sp
Ax9
ThAb9
vmf
Ax9
vf
vr
1 mm

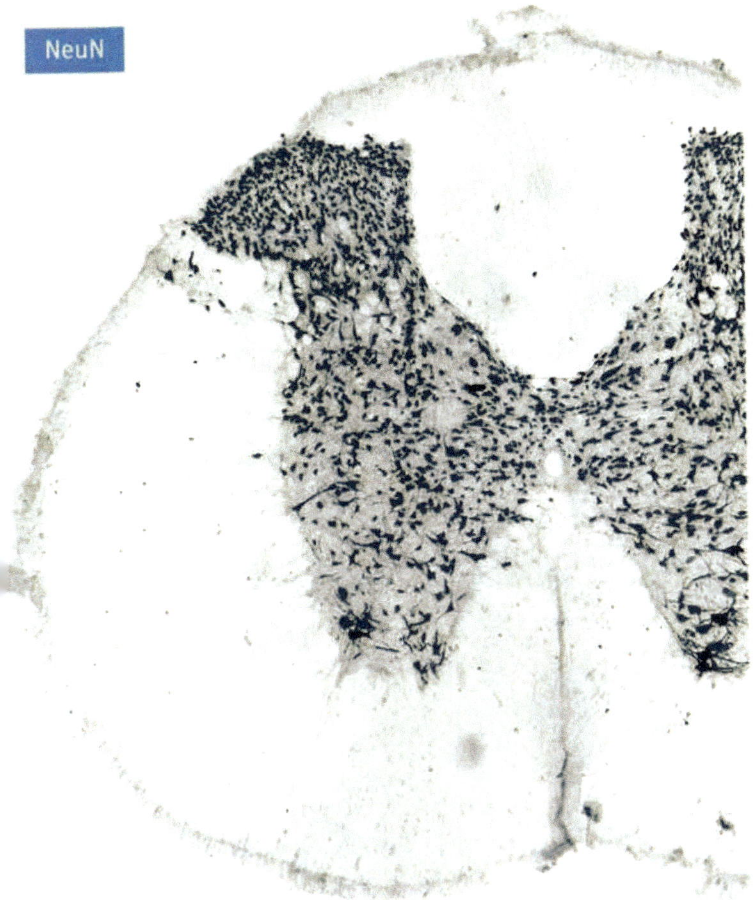

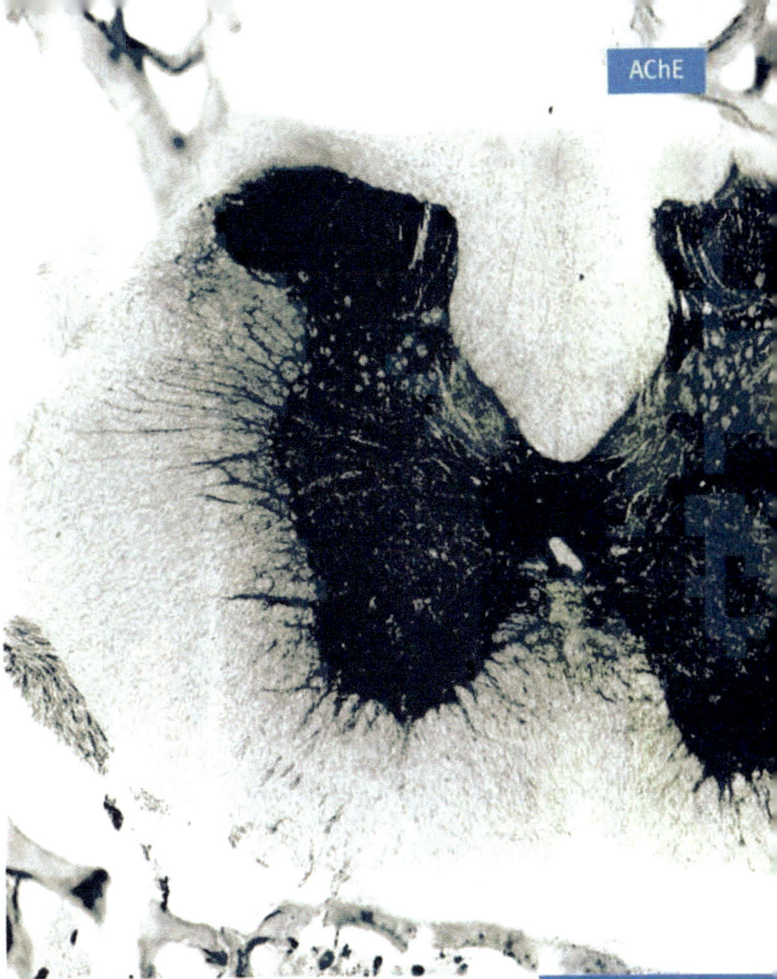

1Sp lamina 1
2SpI lamina 2, inner
2SpO lamina 2, outer
3Sp lamina 3
4Sp lamina 4
5SpL lamina 5, lat
5SpM lamina 5, med
7Sp lamina 7
8Sp lamina 8
10Sp lamina 10
Ax9 axial muscles
CC central canal
D dorsal nucleus
dcs dors corticosp
dl dorsolat fasc
dr dorsal root
gr gracile fascic
ICl intercalated nu
IML intermediolat
IMM intermediomed
LDCom lum dors com
LSp lat spinal nu
psdc postsyn dors col
rs rubrospinal tr
ThAb9 thor-abd wall
vmf vent med fiss
vr ventral root
vwc vent white com

Rat T12

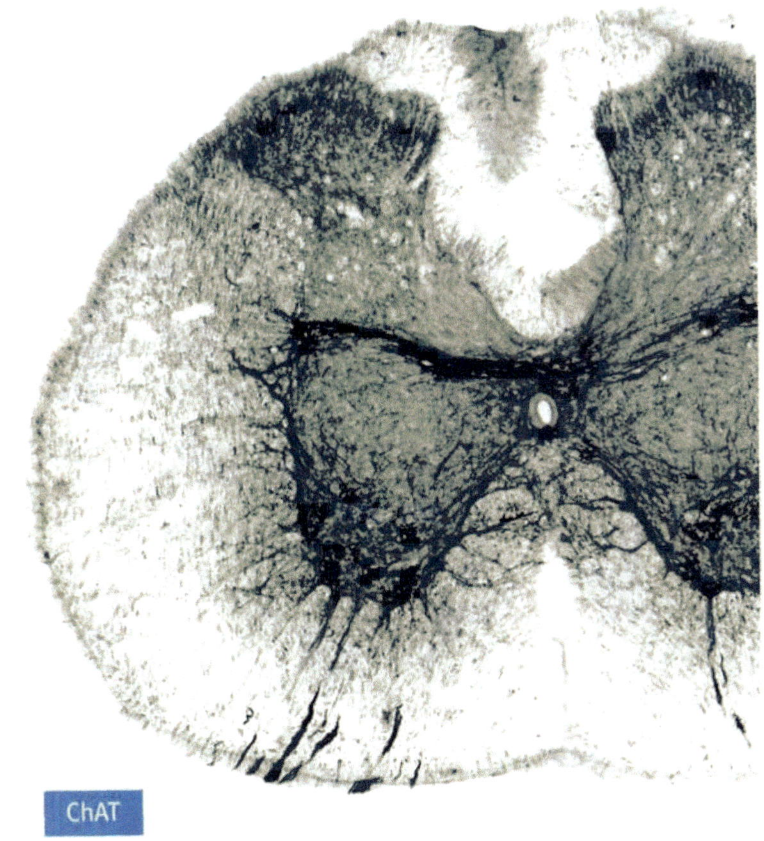

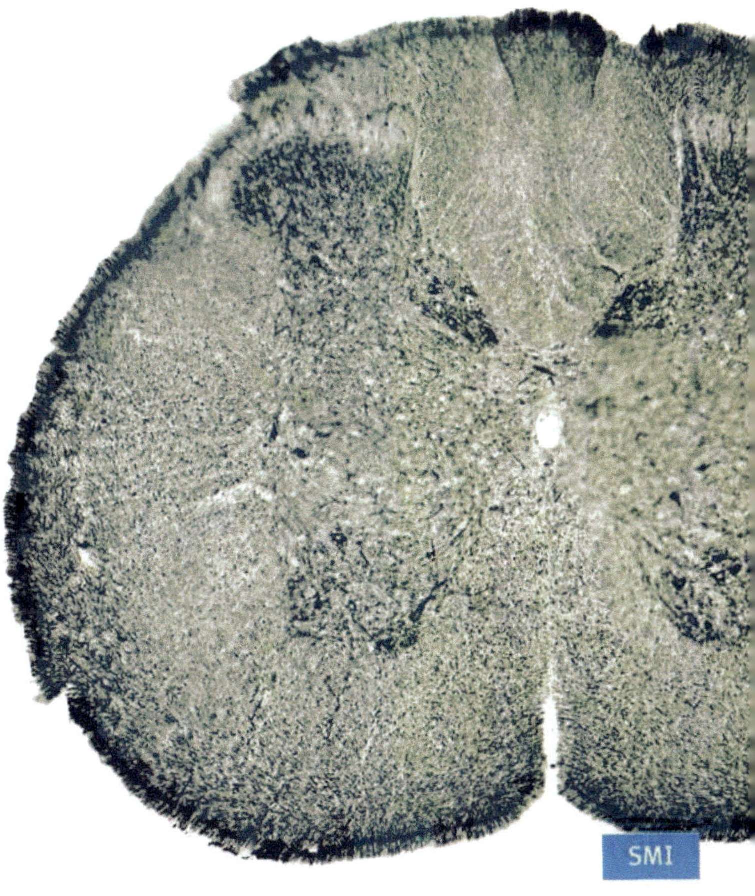

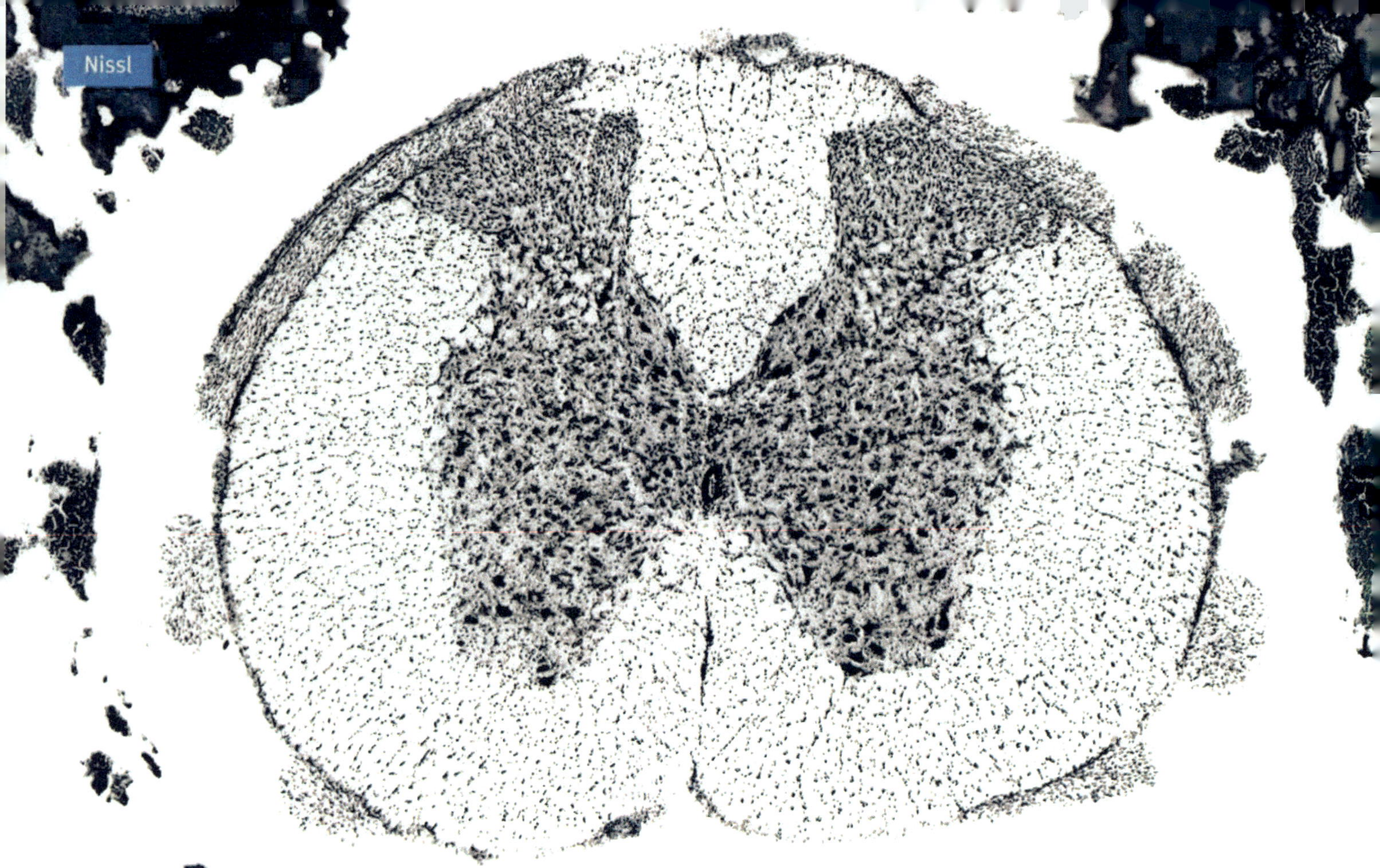
Nissl

Rat T13

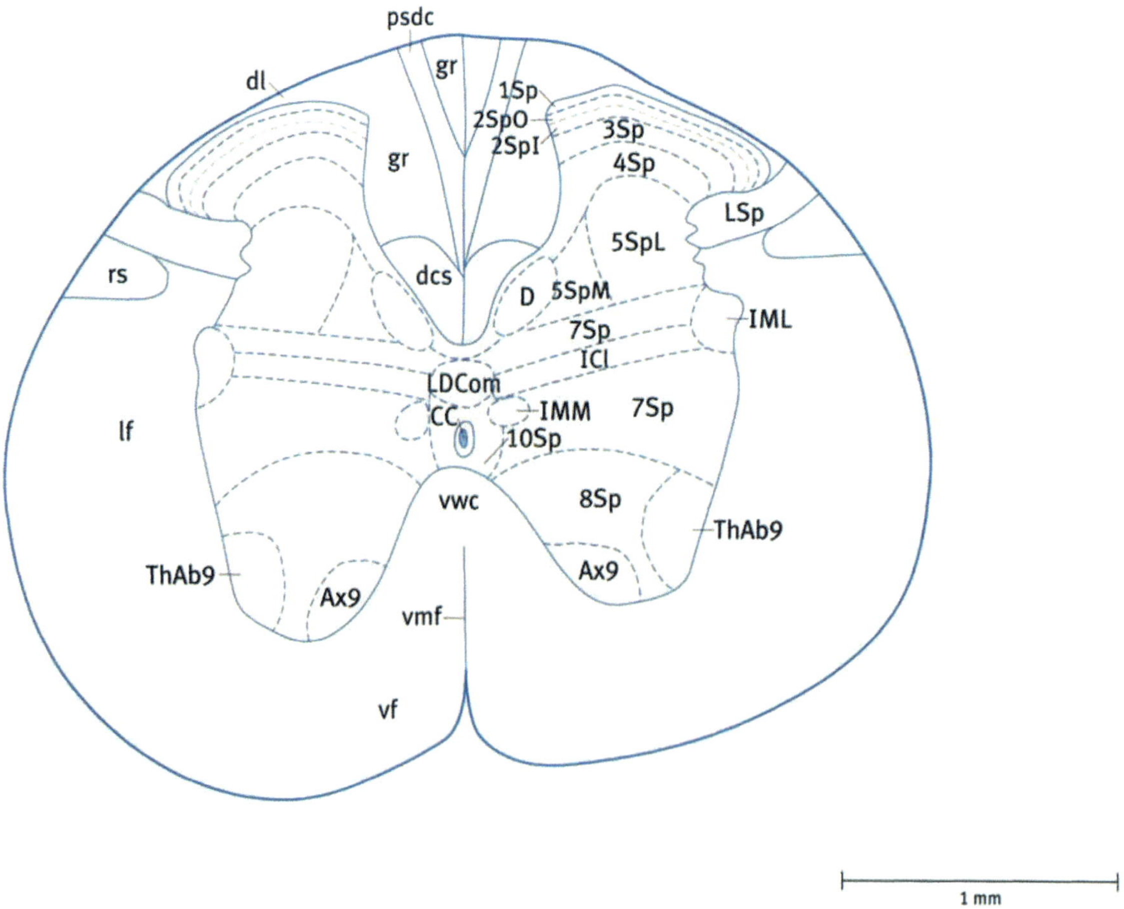
psdc
gr
dl
1Sp
2SpO
3Sp
2SpI
gr
4Sp
LSp
5SpL
rs
dcs
D
5SpM
IML
7Sp
ICl
LDCom
lf
CC
IMM
7Sp
10Sp
vwc
8Sp
ThAb9
ThAb9
Ax9
Ax9
vmf
vf
1 mm

1Sp lamina 1
2SpI lamina 2, inner
2SpO lamina 2, outer
3Sp lamina 3
4Sp lamina 4

5SpL lamina 5, lat
5SpM lamina 5, med
7Sp lamina 7
8Sp lamina 8
10Sp lamina 10

Ax9 axial muscles
CC central canal
D dorsal nucleus
dcs dors corticosp
gr gracile fascic

ICl intercalated nu
IML intermediolat
IMM intermediomed
LDCom lum dors com
LSp lat spinal nu

psdc postsyn dors col
rs rubrospinal tr
ThAb9 thor-abd wall
vmf vent med fiss
vwc vent white com

Rat T13

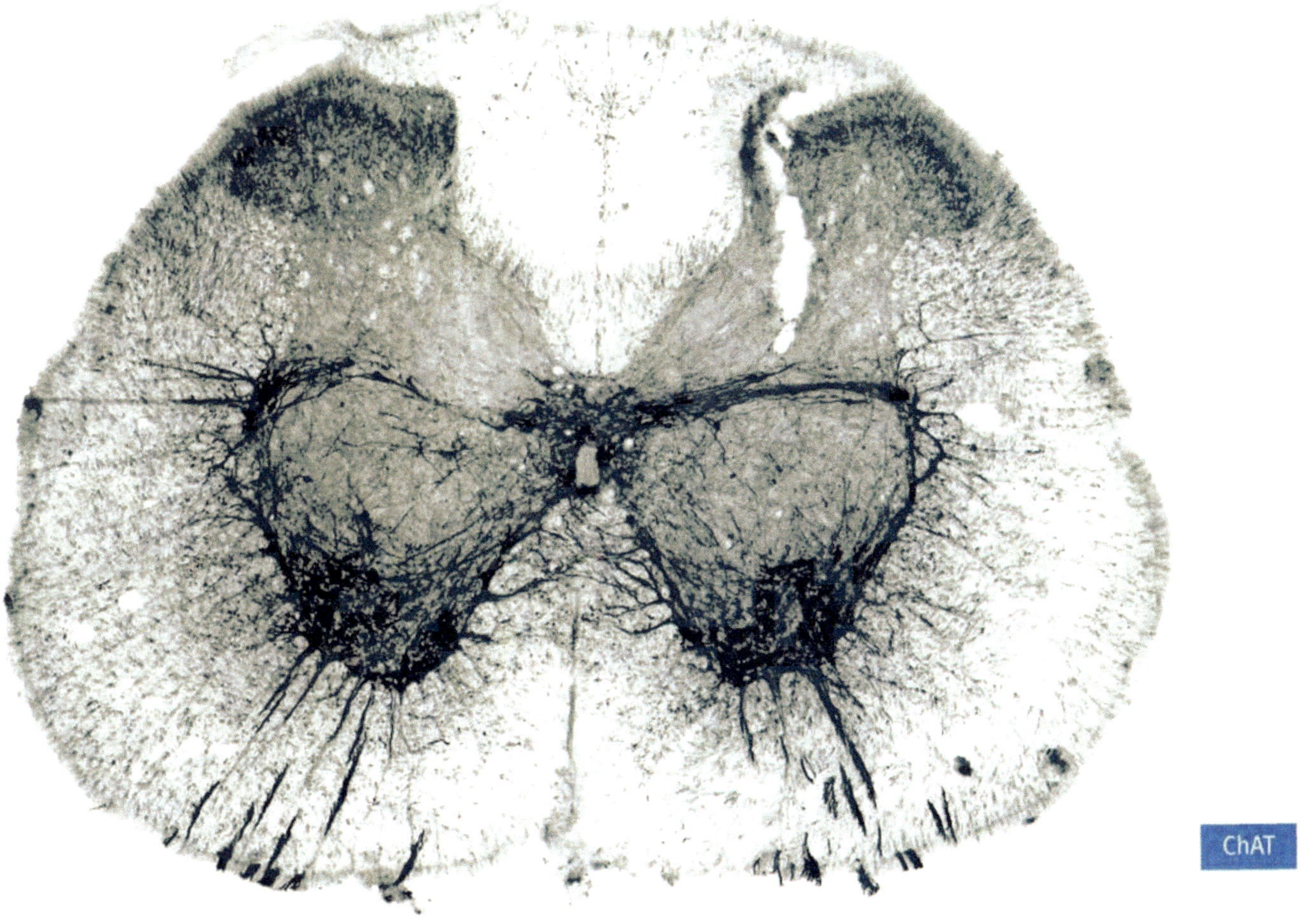

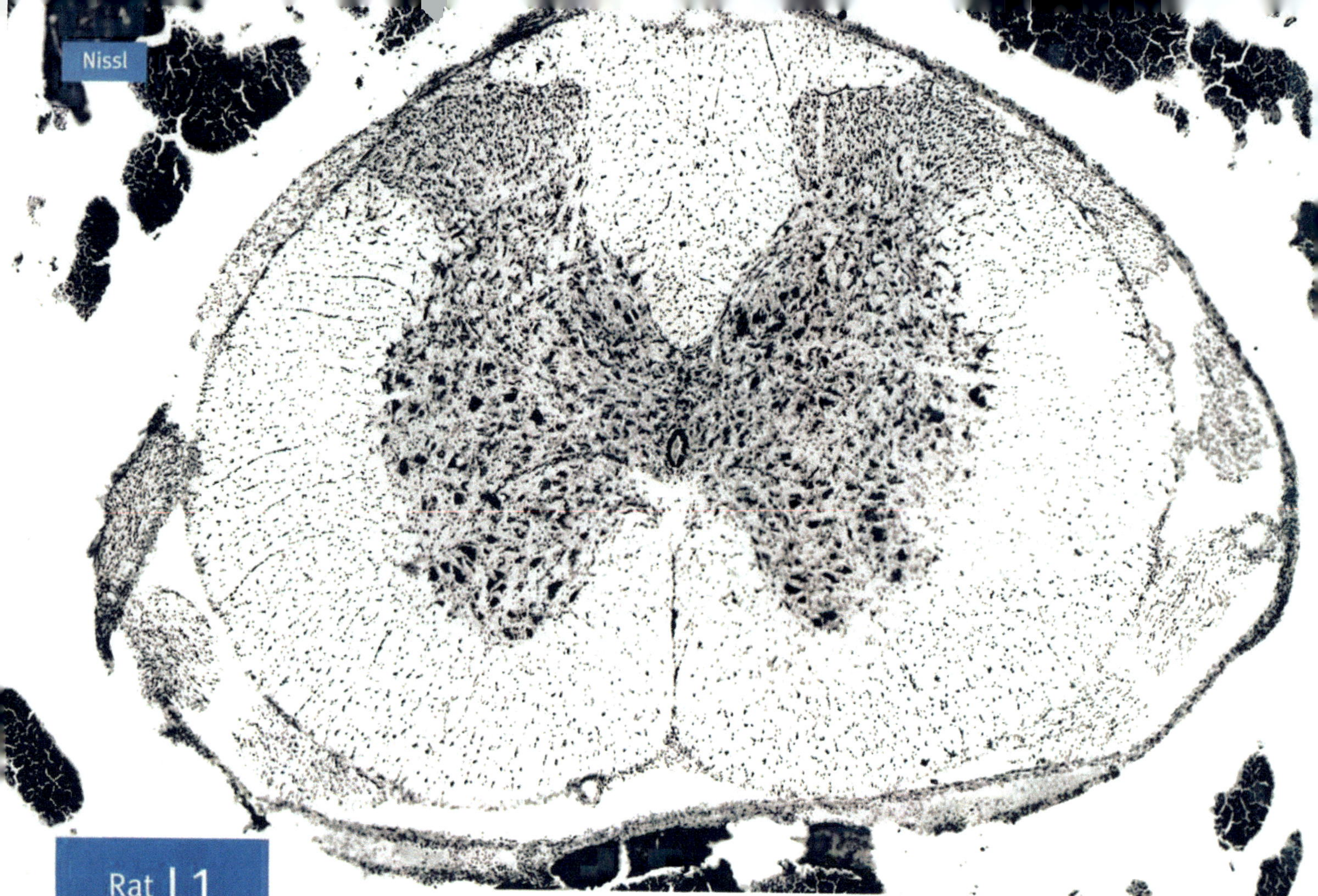
Nissl

Rat L1

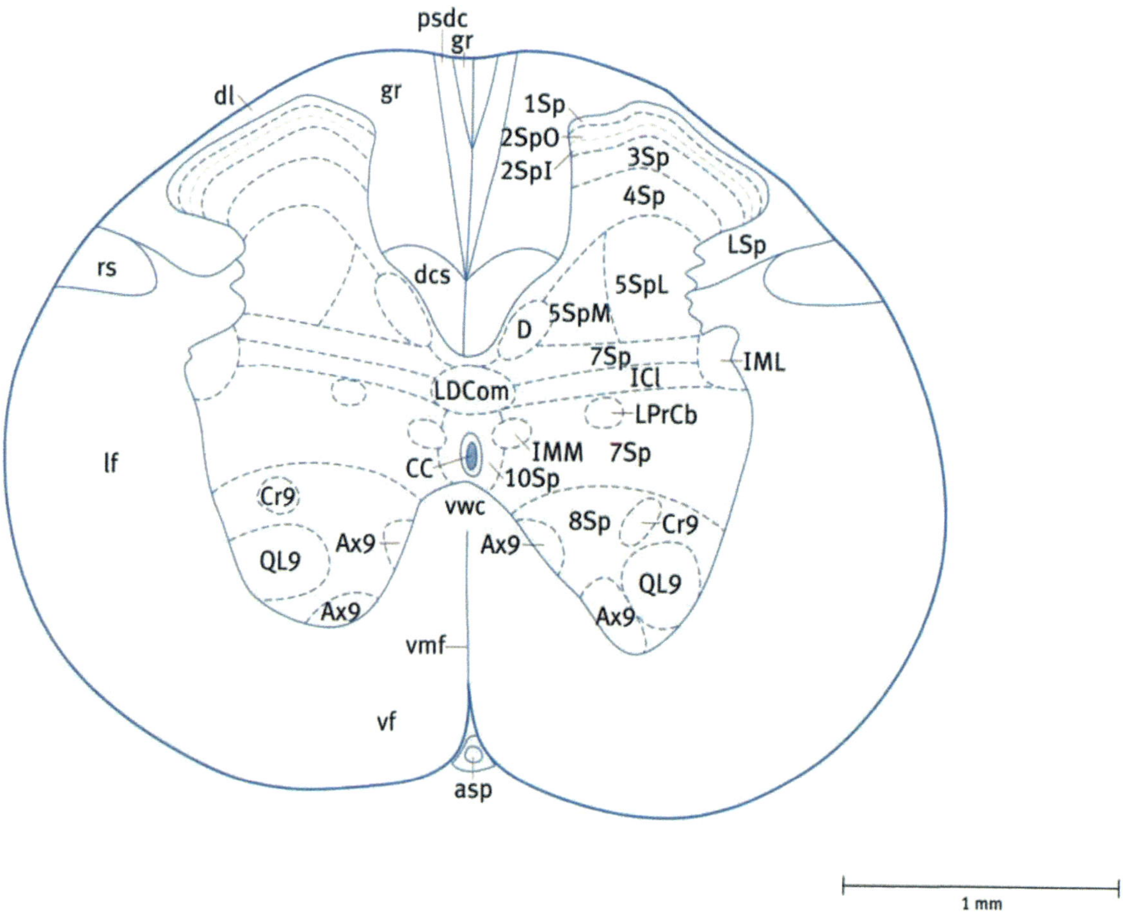
psdc
gr
dl
gr
1Sp
2SpO
2SpI
3Sp
4Sp
LSp
rs
dcs
5SpL
5SpM
D
7Sp
IML
LDCom
ICl
LPrCb
IMM
7Sp
lf
CC
10Sp
Cr9
vwc
8Sp
Cr9
Ax9
Ax9
QL9
QL9
Ax9
Ax9
vmf
vf
asp
1 mm

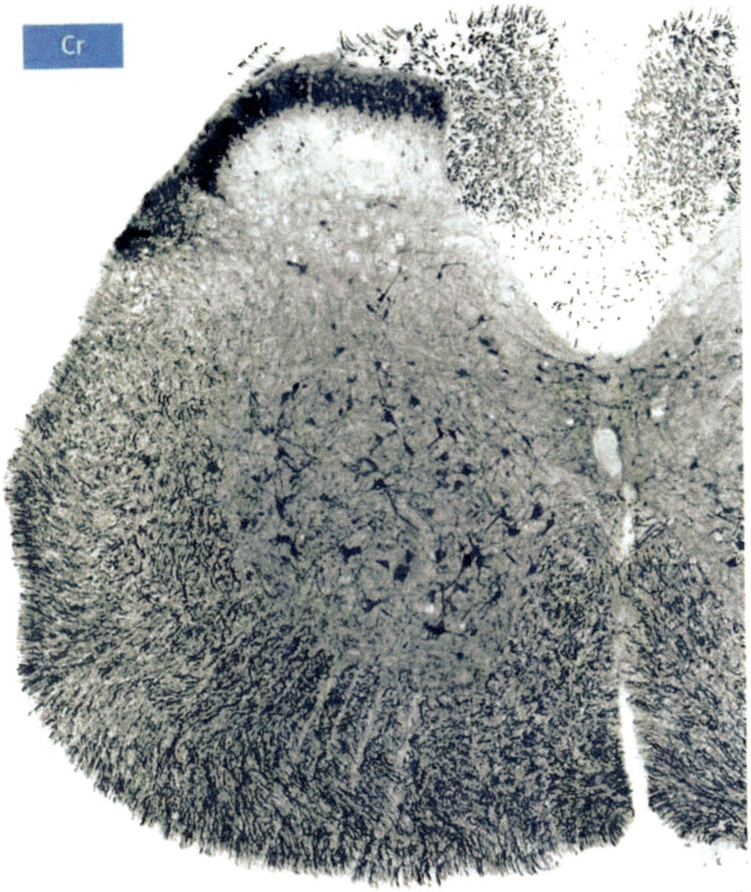

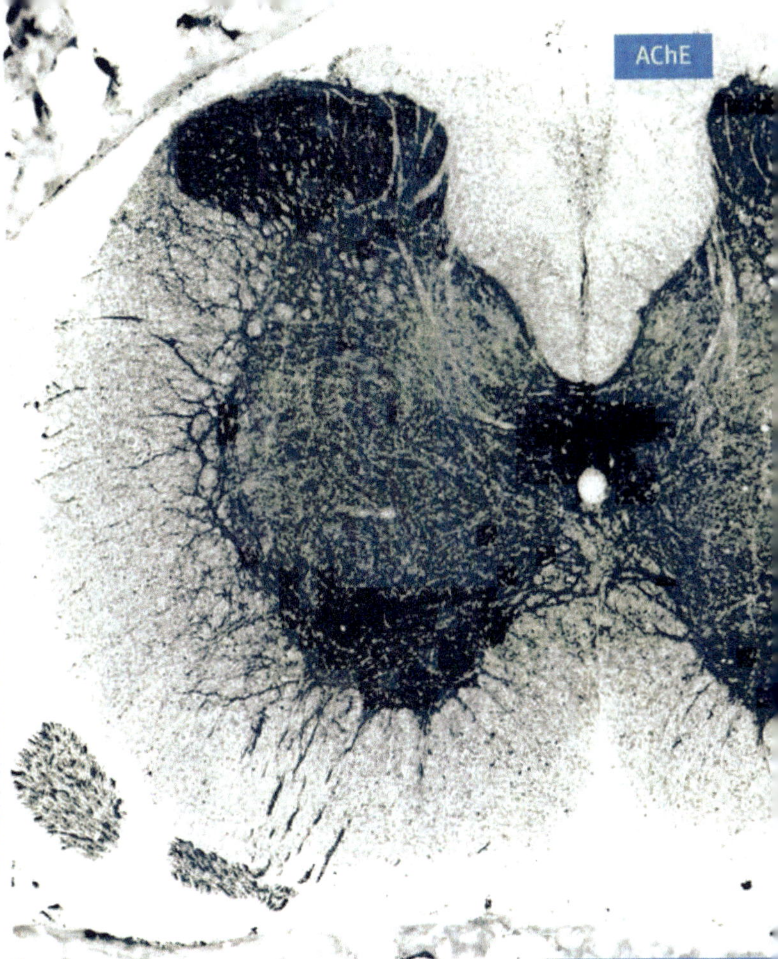

1Sp lamina 1
2SpI lamina 2, inner
2SpO lamina 2, outer
3Sp lamina 3
4Sp lamina 4
5SpL lamina 5, lat
7Sp lamina 7
8Sp lamina 8
10Sp lamina 10
asp ant spinal art
Ax9 axial muscles
CC central canal
Cr9 cremaster
D dorsal nucleus
dcs dors corticosp
dl dorsolat fasc
gr gracile fascic
IML intermediolat
IMM intermediomed
LDCom lum dors com
LPrCb lumbar preCb
LSp lat spinal nu
psdc postsyn dors col
QL9 quad lumborum
rs rubrospinal tr
vmf vent med fiss
vwc vent white com

Rat L1

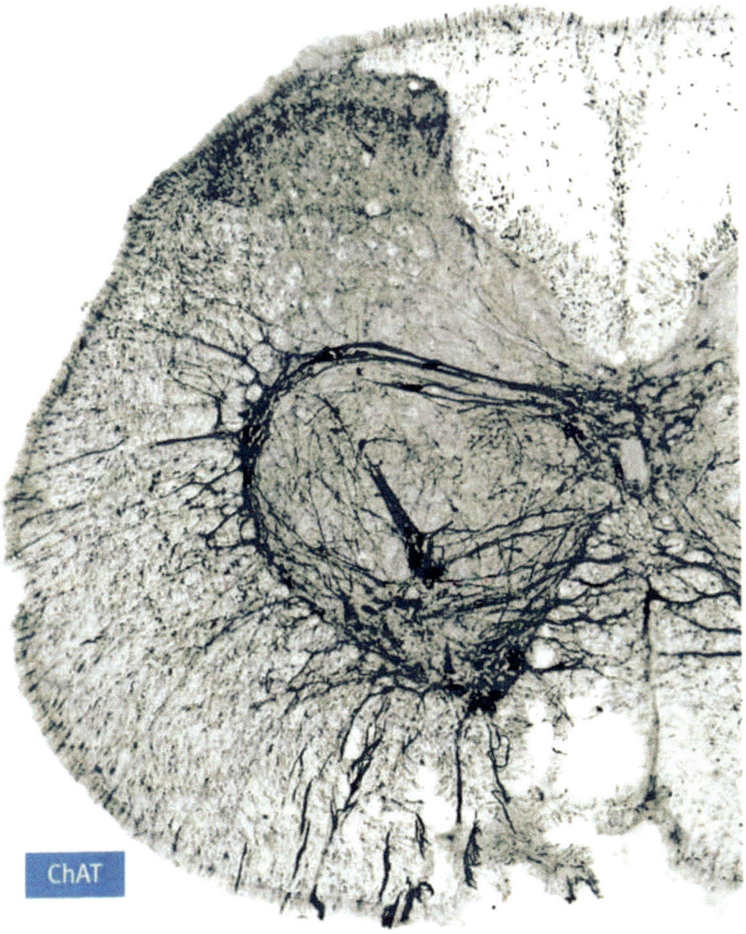

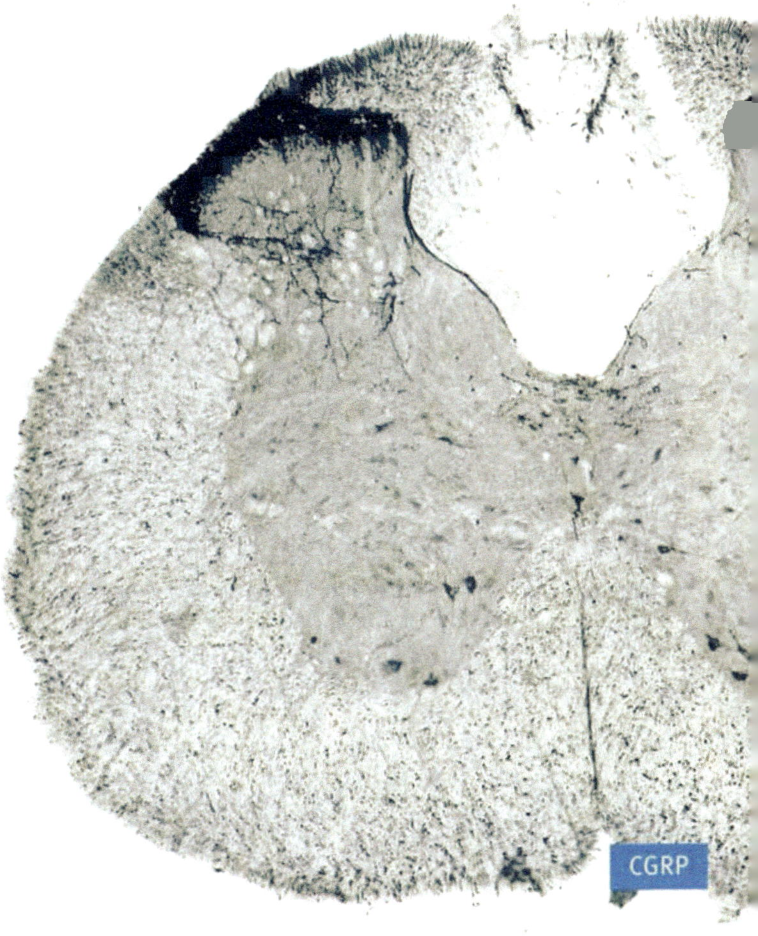

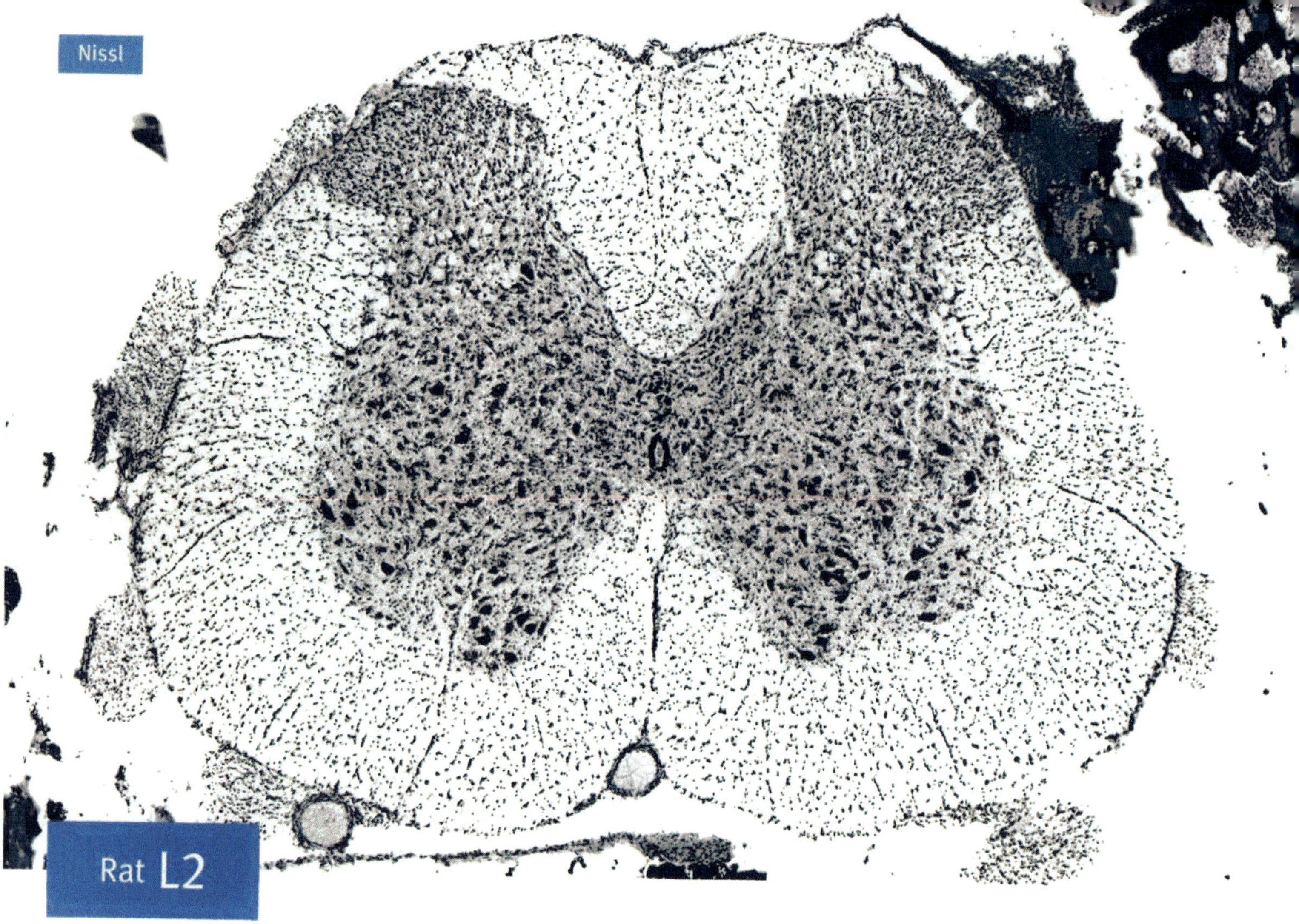
Nissl
Rat L2

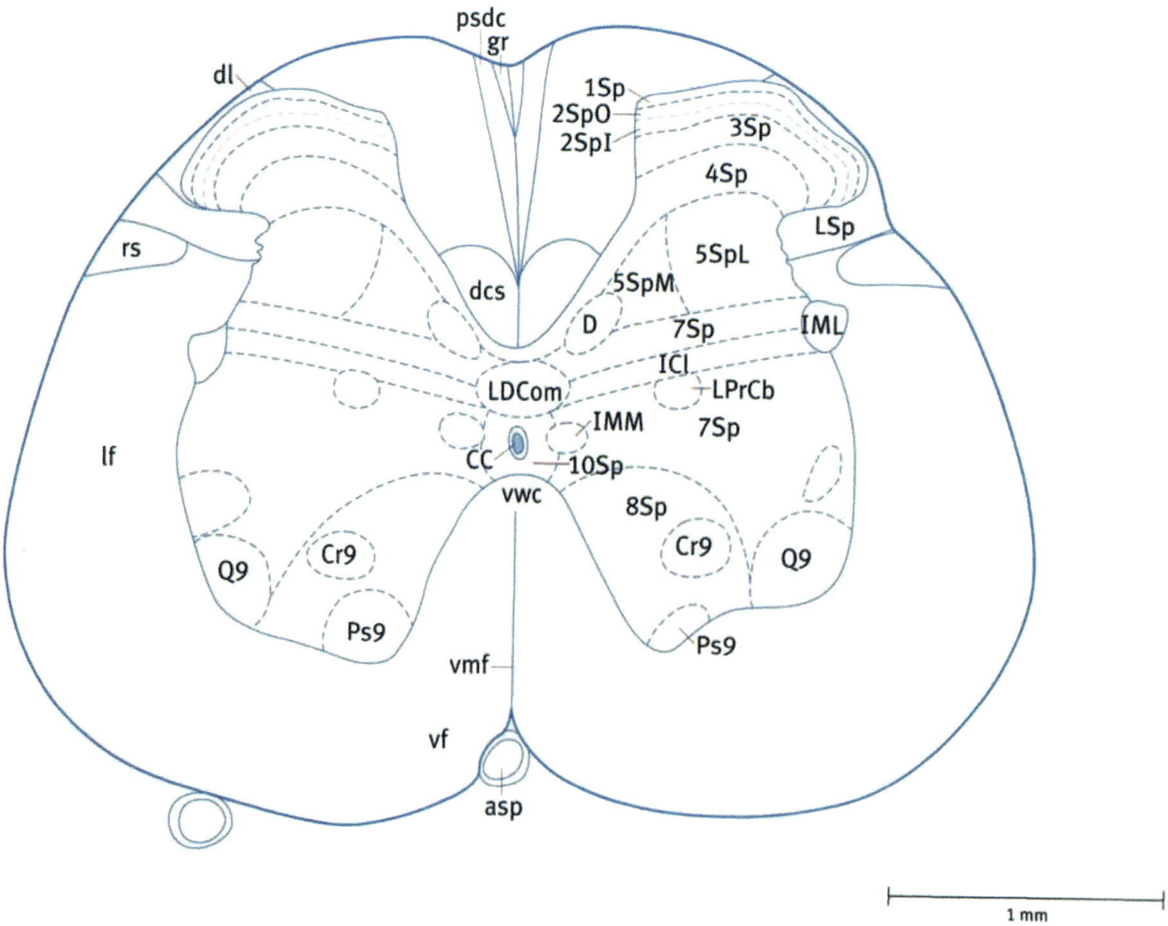
psdc
gr
dl
1Sp
2SpO
2SpI
3Sp
4Sp
rs
LSp
5SpL
dcs
5SpM
D
7Sp
IML
ICl
LDCom
LPrCb
IMM
7Sp
lf
CC
10Sp
vwc
8Sp
Q9
Cr9
Cr9
Q9
Ps9
Ps9
vmf
vf
asp
1 mm

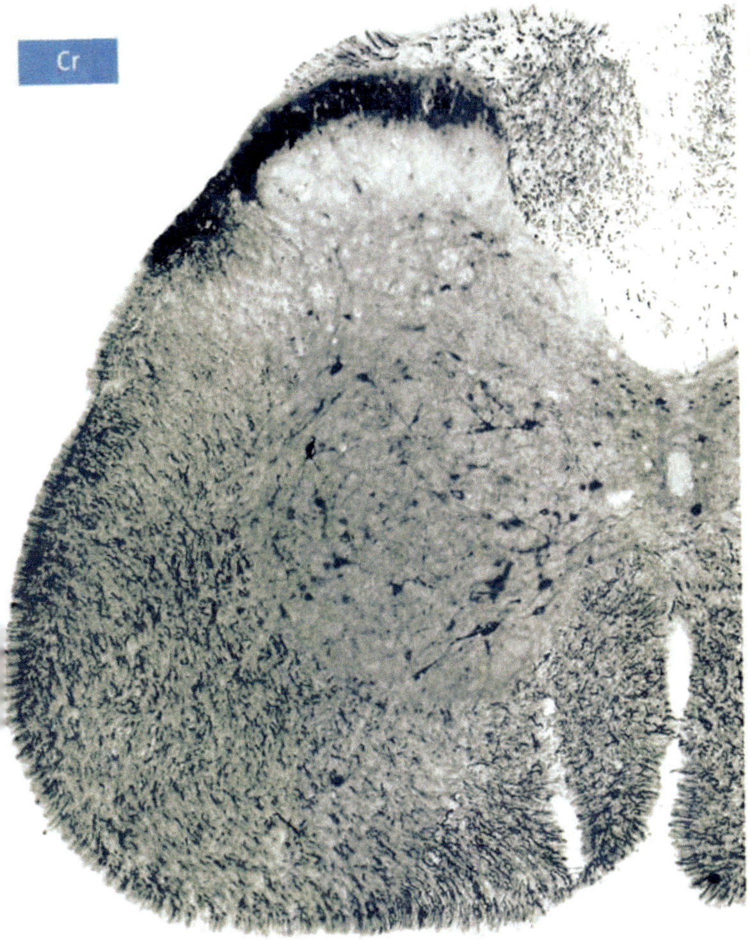

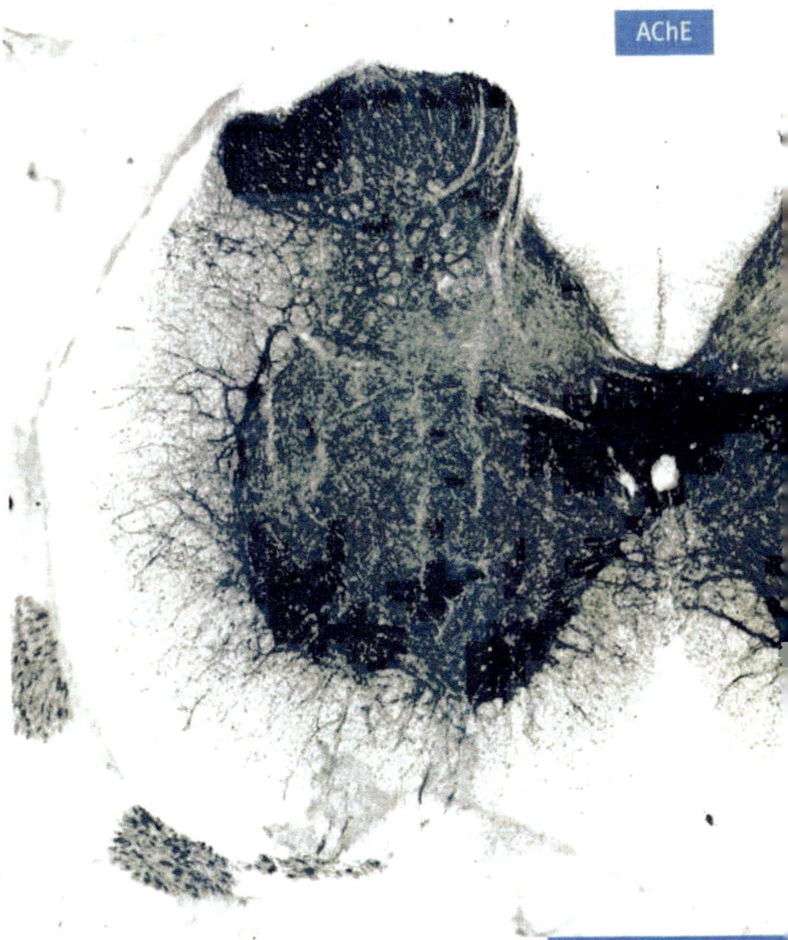

Rat L2

- 1Sp lamina 1
- 2SpI lamina 2, inner
- 2SpO lamina 2, outer
- 3Sp lamina 3
- 4Sp lamina 4
- 5SpL lamina 5, lat
- 5SpM lamina 5, med
- 7Sp lamina 7
- 8Sp lamina 8
- 10Sp lamina 10
- asp ant spinal art
- CC central canal
- Cr9 cremaster
- D dorsal nucleus
- dcs dors corticosp
- dl dorsolat fasc
- gr gracile fascic
- ICl intercalated nu
- IML intermediolat
- IMM intermediomed
- LDCom lum dors com
- LPrCb lumbar preCb
- LSp lat spinal nu
- Ps9 psoas
- psdc postsyn dors col
- Q9 quadriceps
- rs rubrospinal tr
- vmf vent med fiss
- vwc vent white com

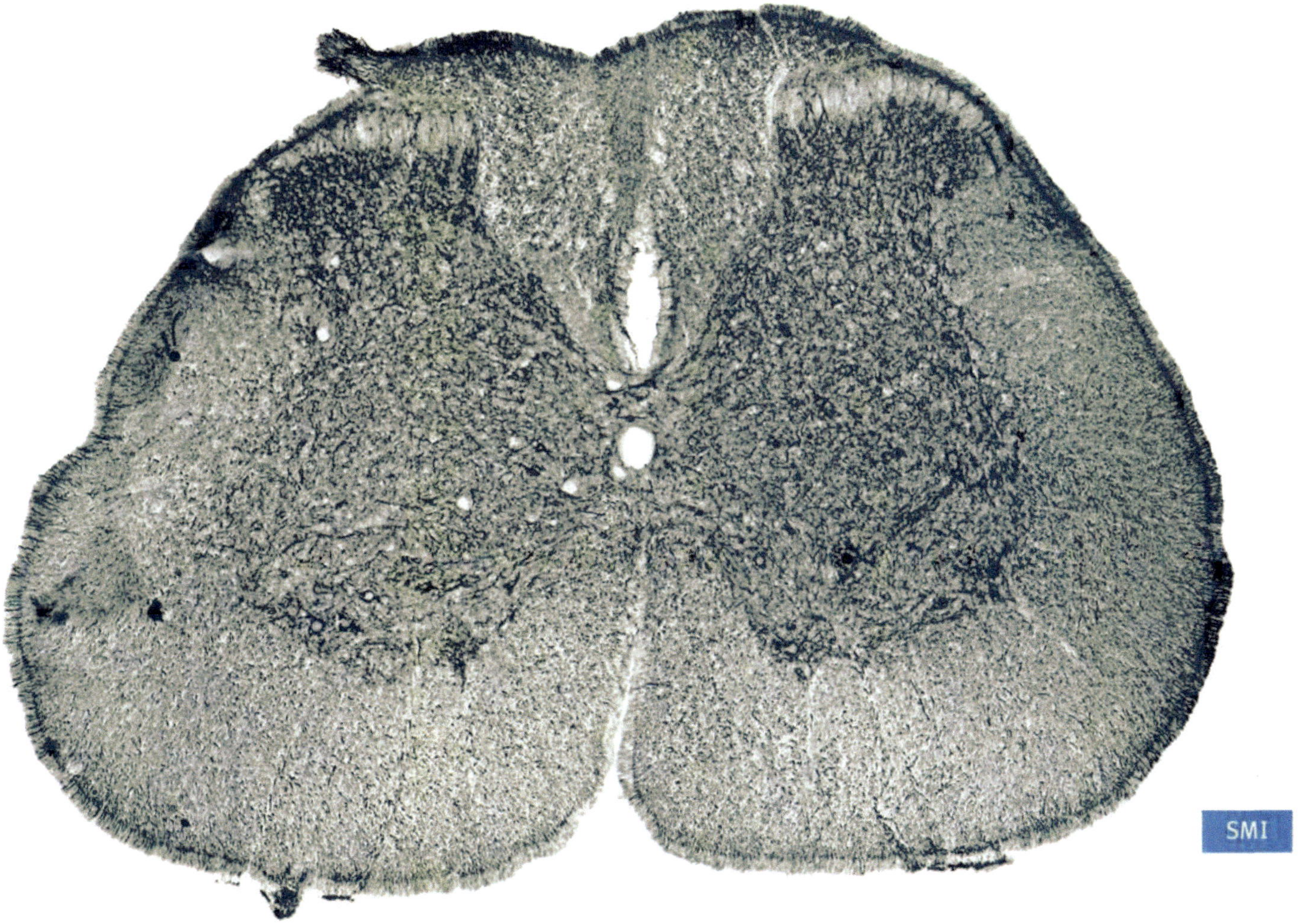

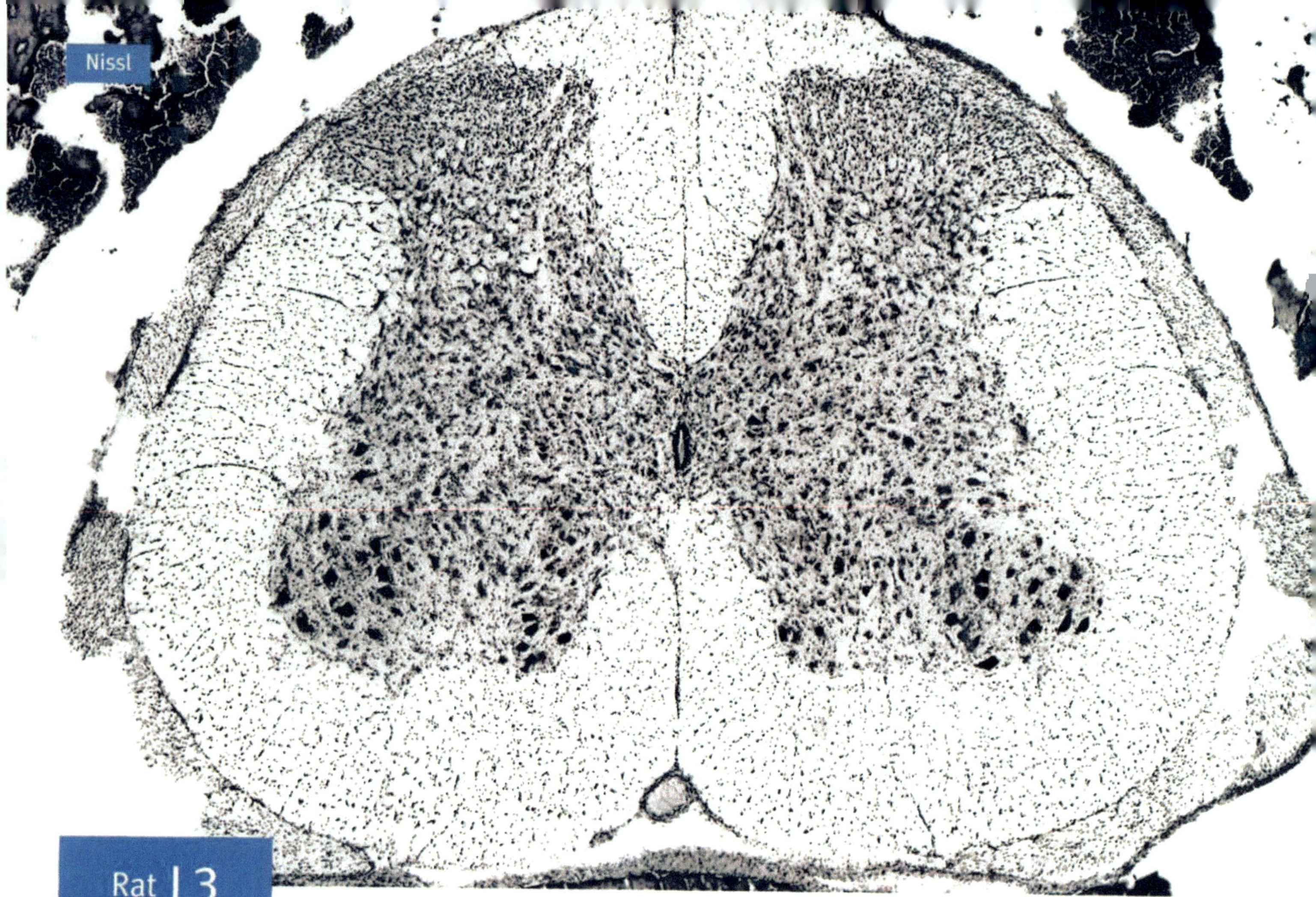
Nissl

Rat L3

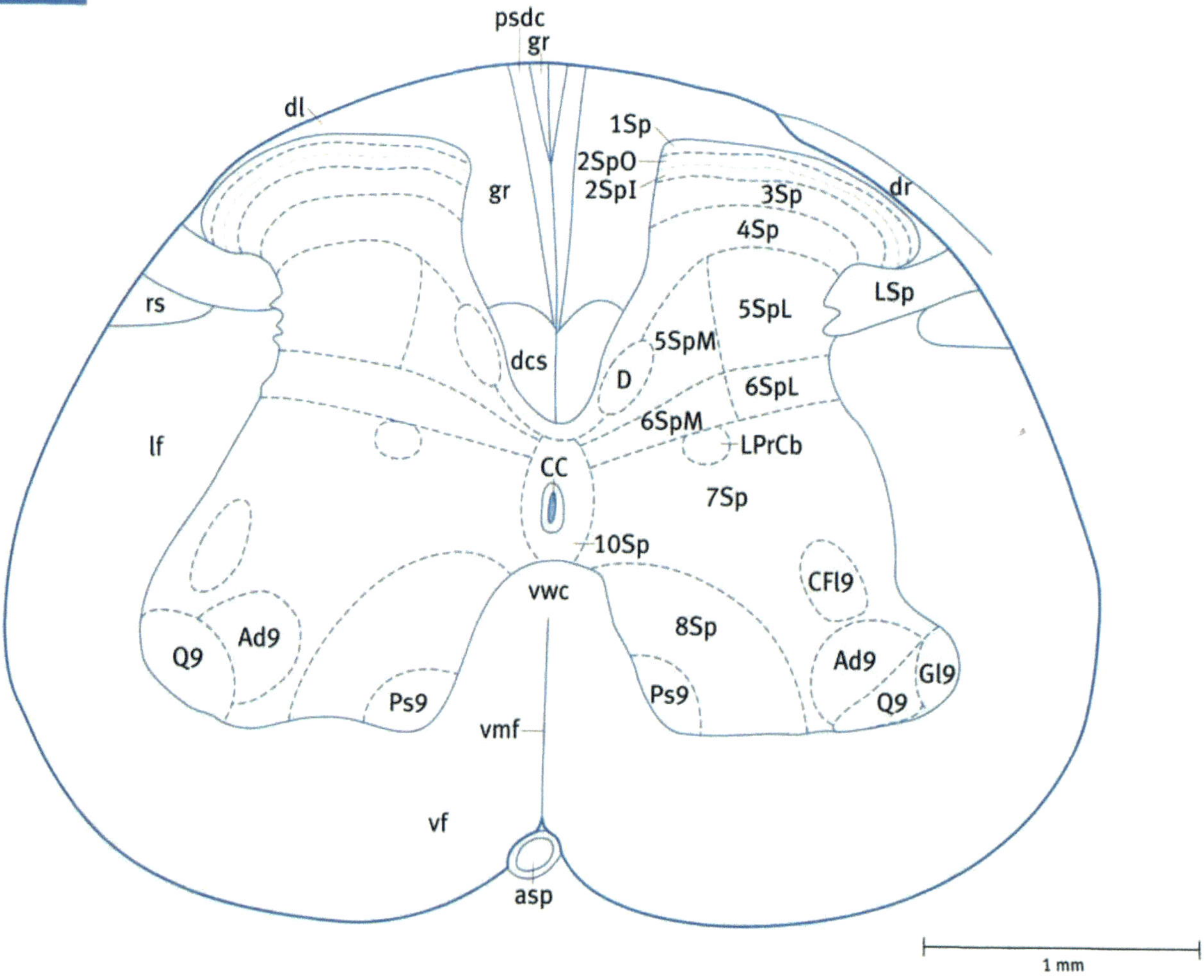
psdc
gr
dl
1Sp
2SpO
2SpI
gr
3Sp
dr
4Sp
rs
LSp
5SpL
5SpM
dcs
D
6SpL
6SpM
lf
LPrCb
CC
7Sp
10Sp
CFl9
vwc
8Sp
Ad9
Ad9
Gl9
Q9
Q9
Ps9
Ps9
vmf
vf
asp
1 mm

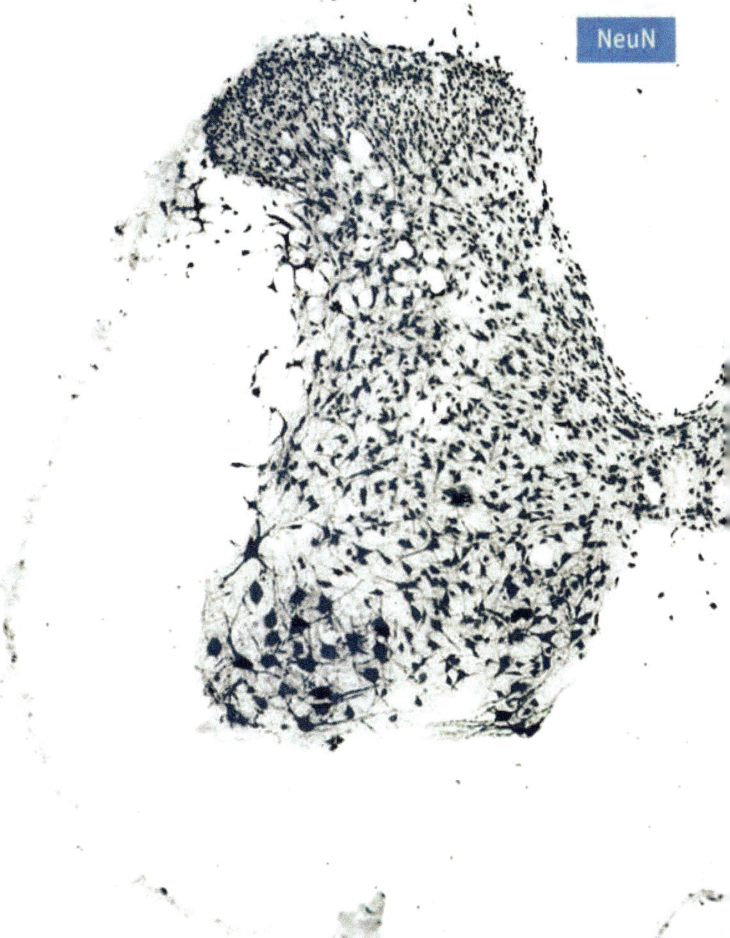

1Sp lamina 1
2SpI lamina 2, inner
2SpO lamina 2, outer
3Sp lamina 3
4Sp lamina 4
5SpL lamina 5, lat
5SpM lamina 5, med
6SpL lamina 6, lat
6SpM lamina 6, med
7Sp lamina 7
8Sp lamina 8
10Sp lamina 10
Ad9 adductors
asp ant spinal art
CC central canal
CFl9 crural flexors
D dorsal nucleus
dcs dors corticosp
dl dorsolat fasc
dr dorsal root
Gl9 gluteal muscles
gr gracile fascic
LSp lat spinal nu
Ps9 psoas
psdc postsyn dors col
Q9 quadriceps
rs rubrospinal tr
vmf vent med fiss
vwc vent white com

Rat L3

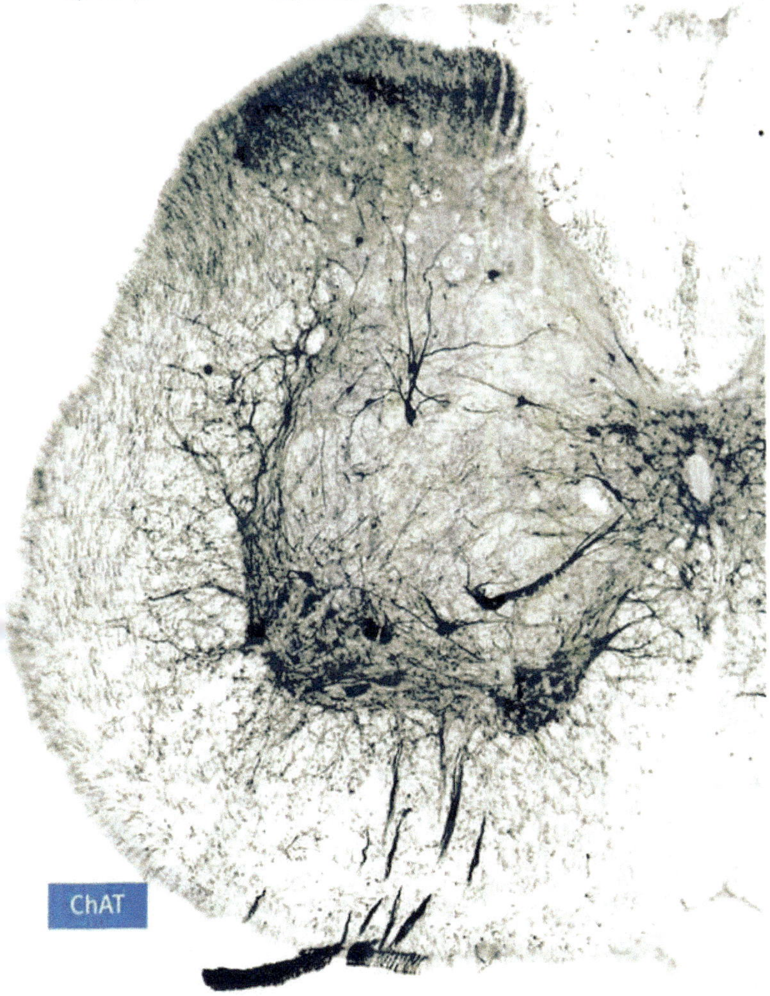

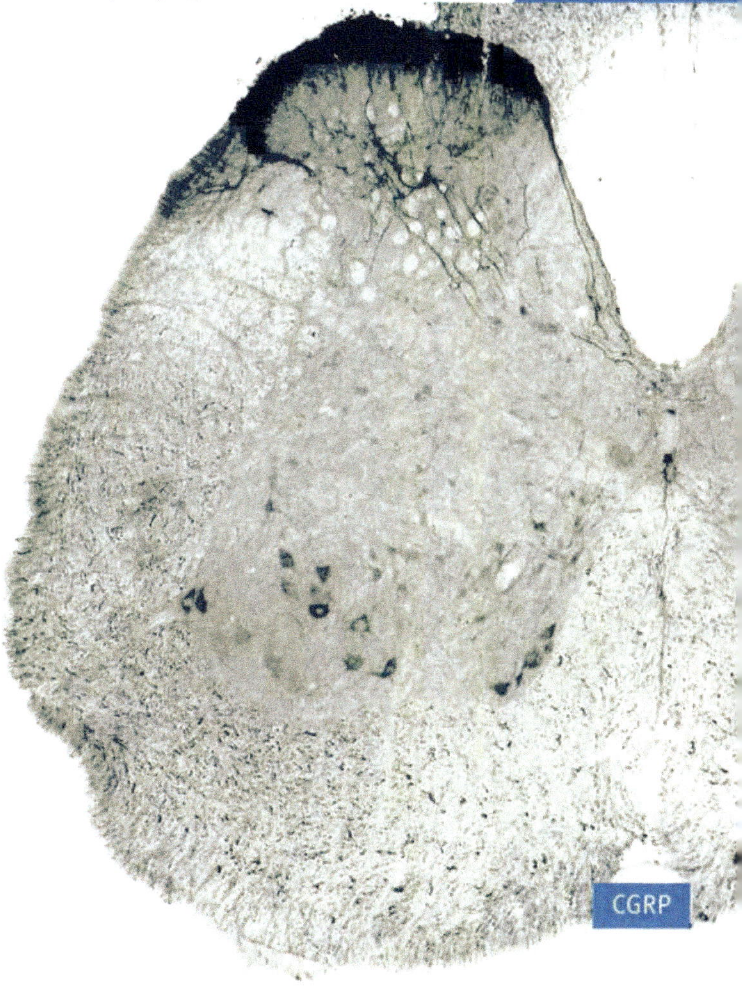

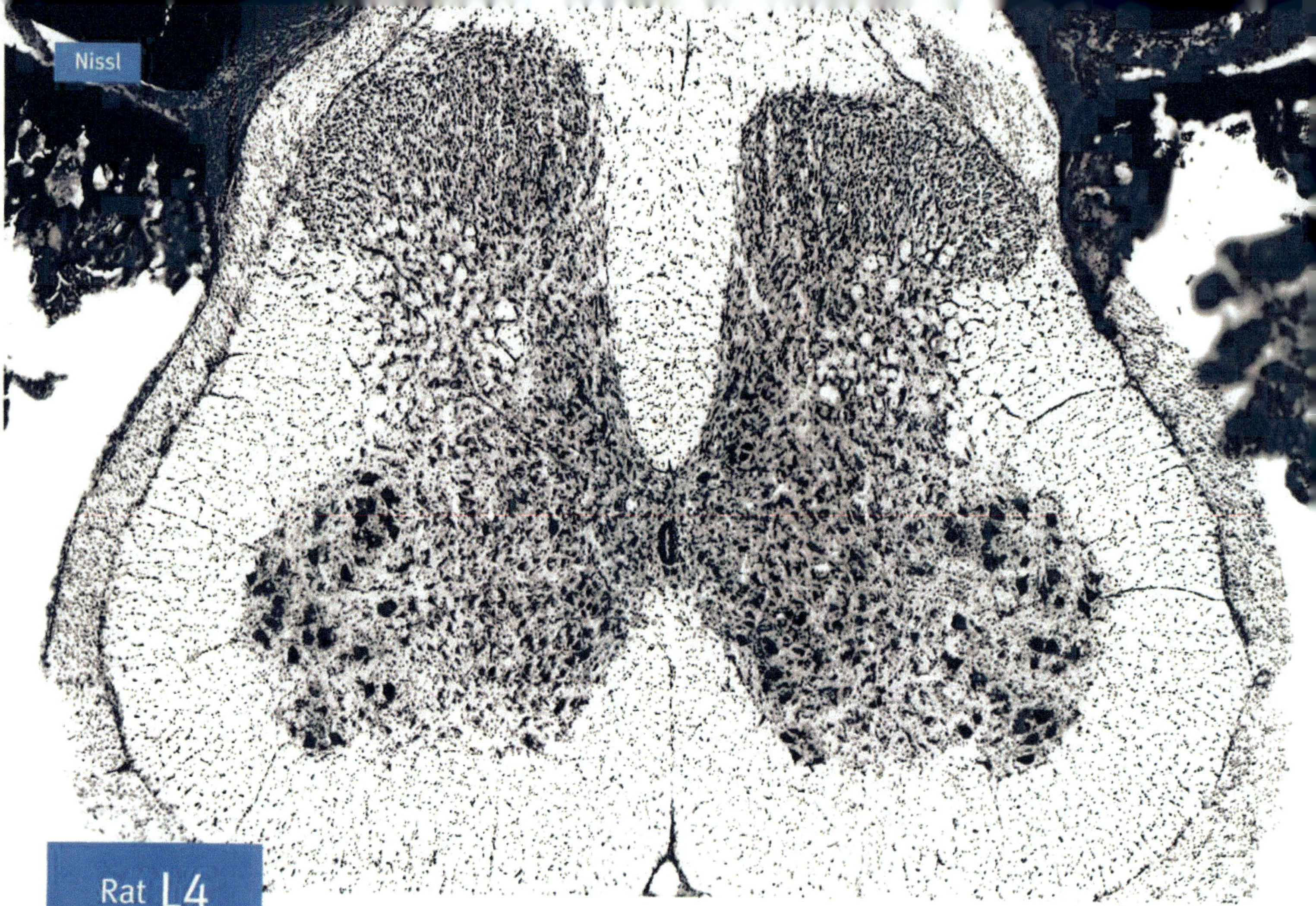
Nissl

Rat L4

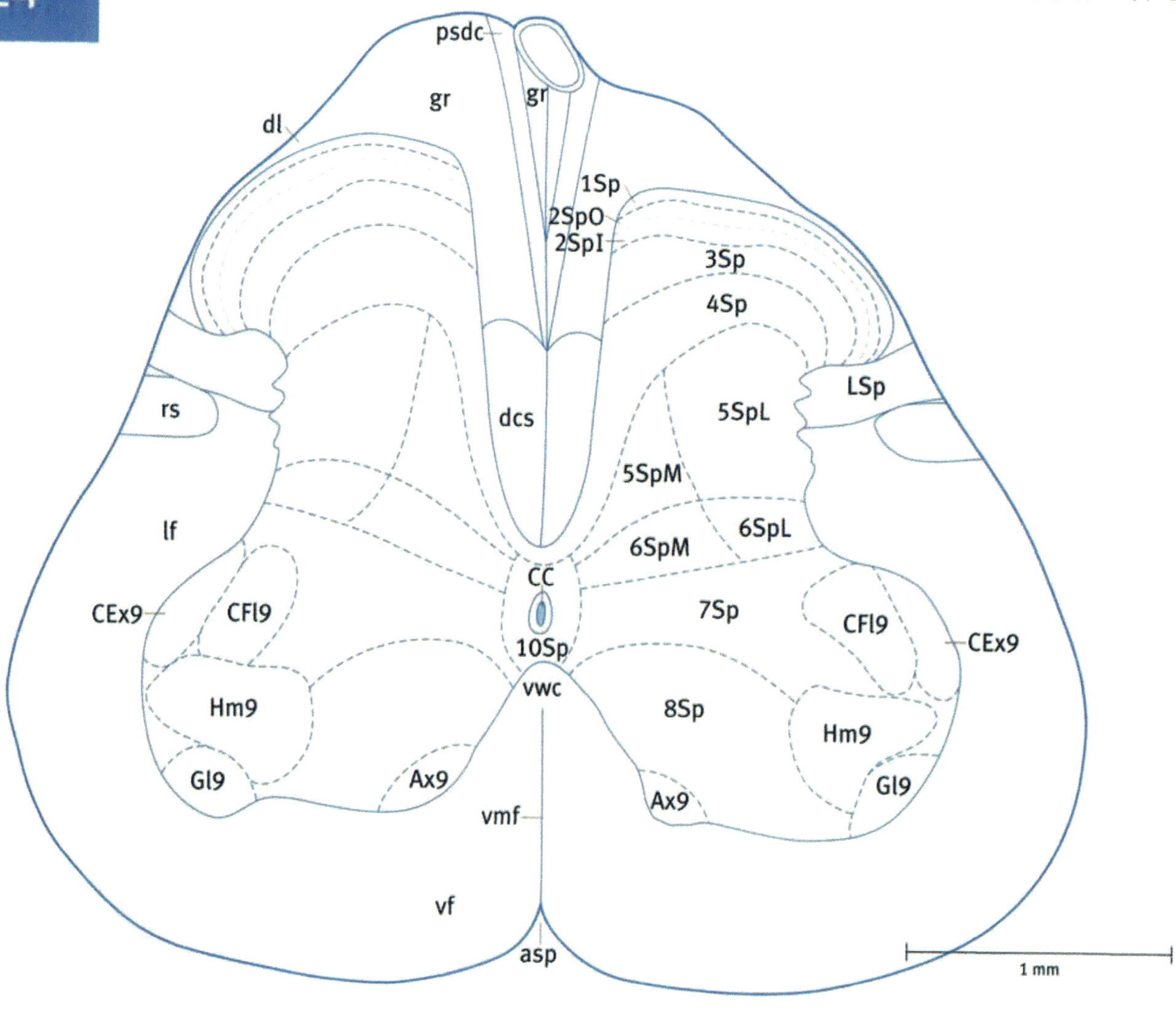
psdc
gr
gr
dl
1Sp
2SpO
2SpI
3Sp
4Sp
LSp
rs
dcs
5SpL
5SpM
lf
6SpL
6SpM
CC
CEx9
CFl9
7Sp
CFl9
CEx9
10Sp
vwc
Hm9
8Sp
Hm9
Gl9
Ax9
Ax9
Gl9
vmf
vf
asp
1 mm

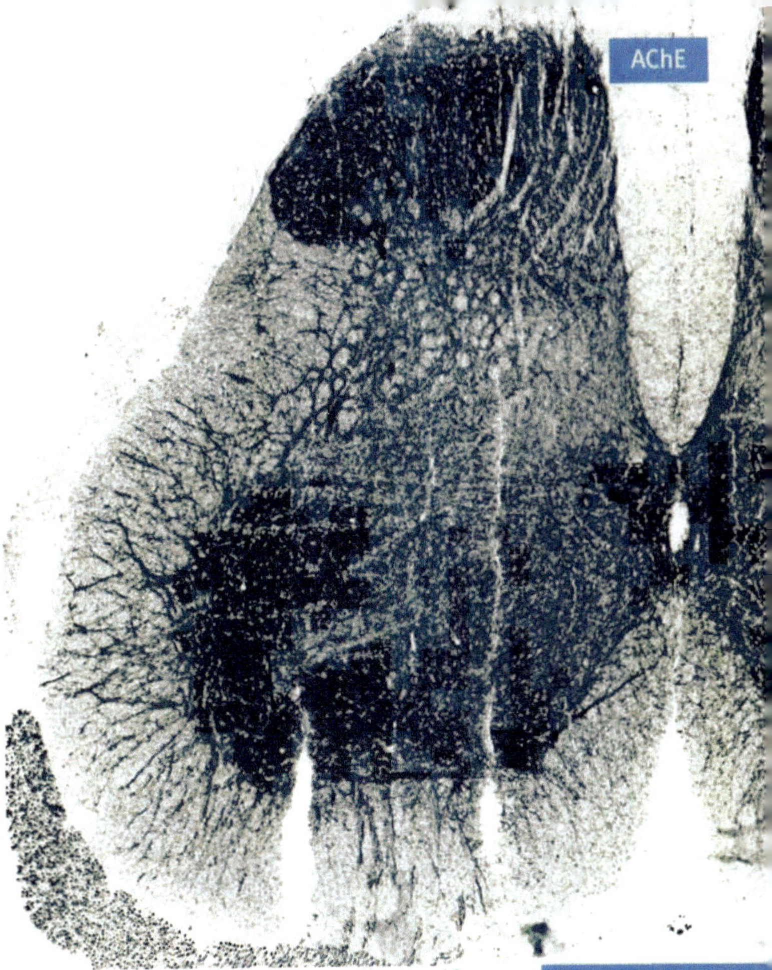

1Sp lamina 1	**5SpL** lamina 5, lat	**8Sp** lamina 8	**CEx9** crual extensors	**gr** gracile fascic	**vmf** vent med fiss
2SpI lamina 2, inner	**5SpM** lamina 5, med	**10Sp** lamina 10	**CFl9** crural flexors	**Hm9** hamstrings	**vwc** vent white com
2SpO lamina 2, outer	**6SpL** lamina 6, lat	**asp** ant spinal art	**dcs** dors corticosp	**LSp** lat spinal nu	
3Sp lamina 3	**6SpM** lamina 6, med	**Ax9** axial muscles	**dl** dorsolat fasc	**psdc** postsyn dors col	
4Sp lamina 4	**7Sp** lamina 7	**CC** central canal	**Gl9** gluteal muscles	**rs** rubrospinal tr	

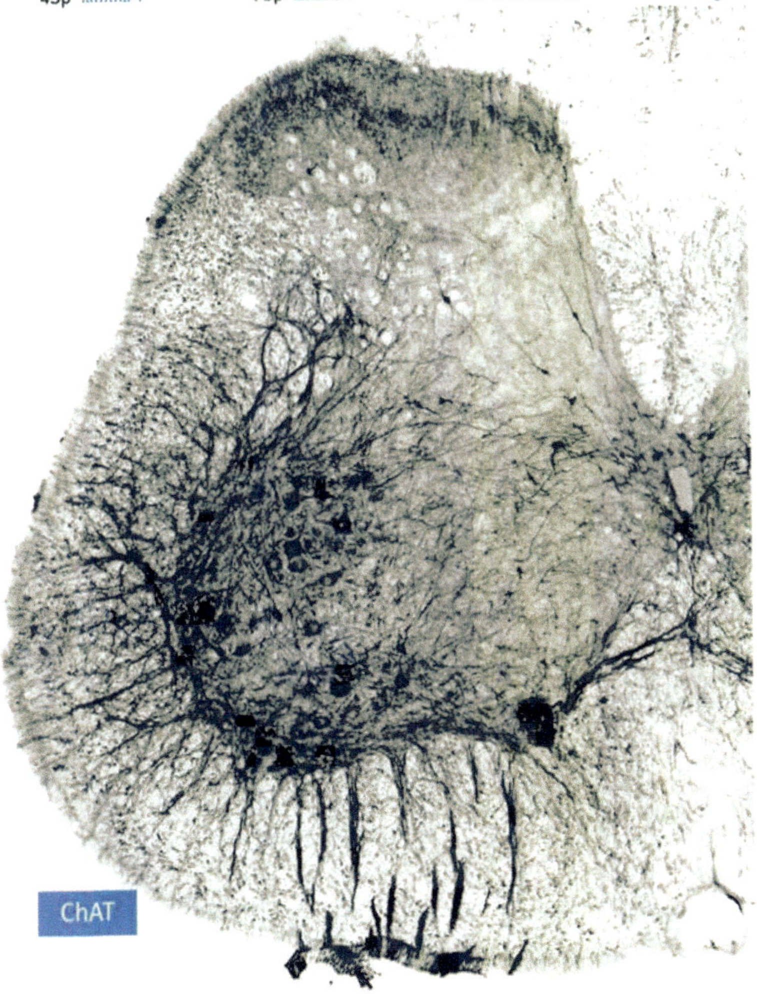

Nissl

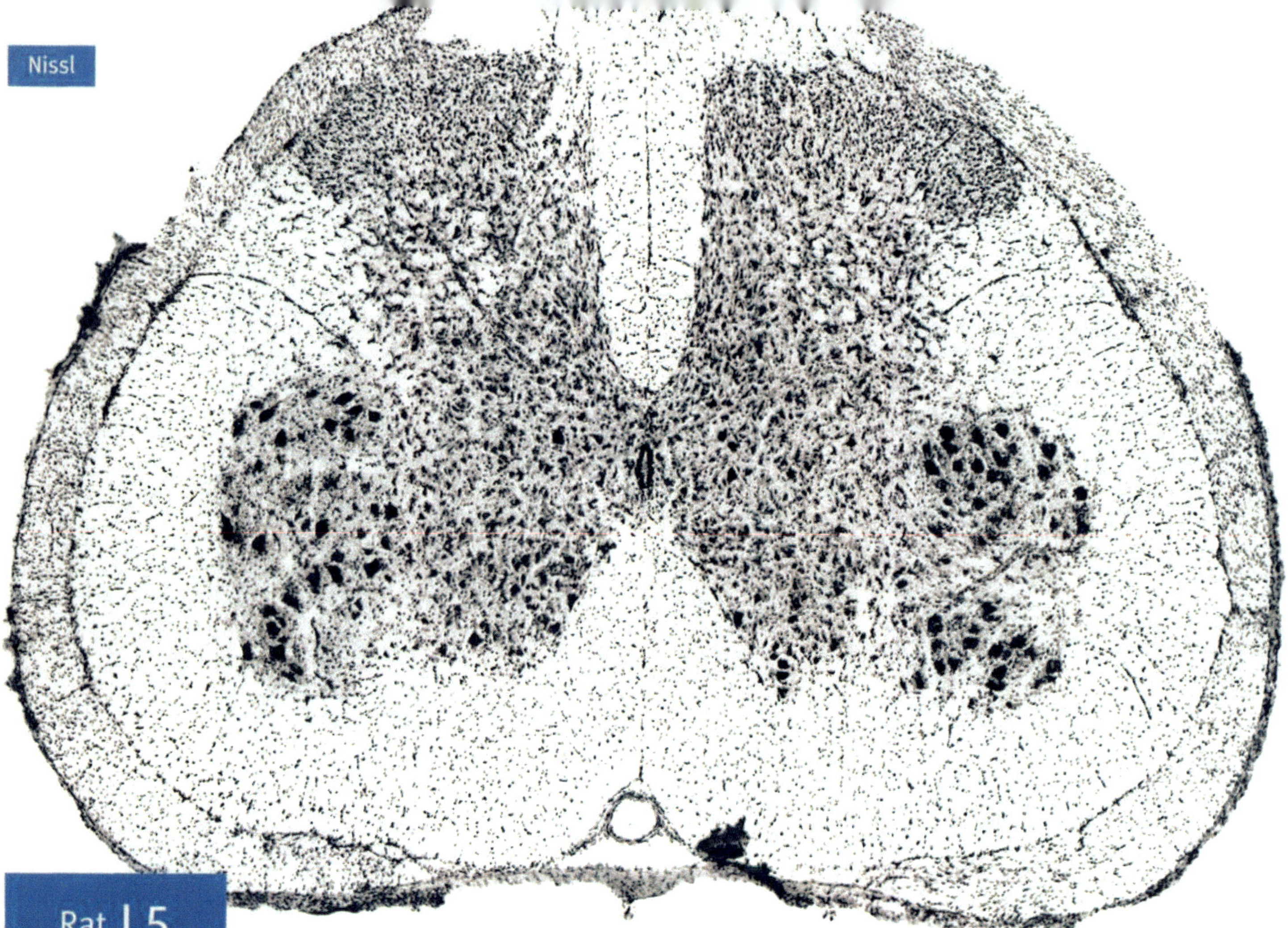

Rat L5

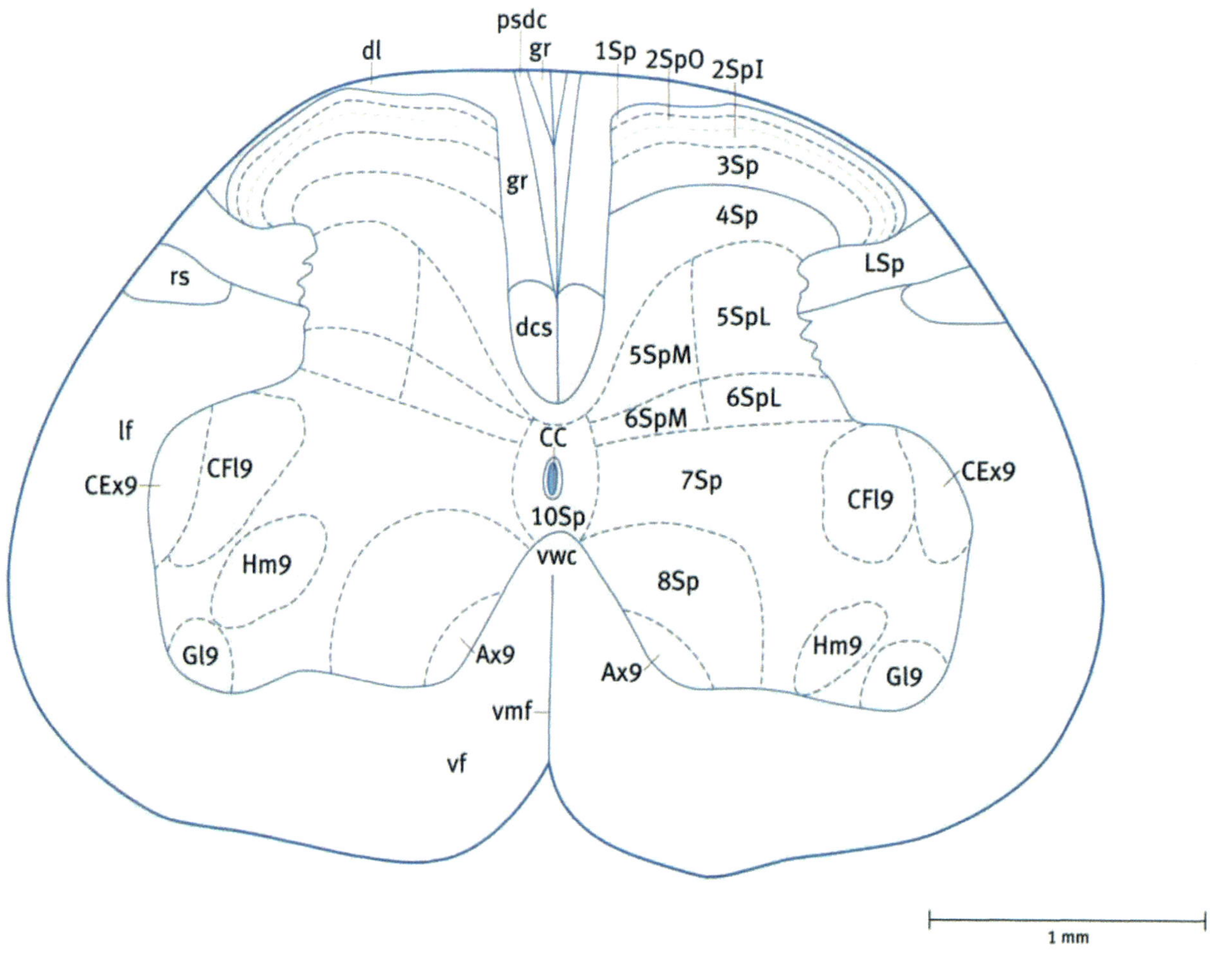
psdc
dl
gr
1Sp
2SpO
2SpI
3Sp
gr
4Sp
rs
LSp
5SpL
dcs
5SpM
6SpL
6SpM
lf
CC
CEx9
CFl9
7Sp
CEx9
CFl9
10Sp
vwc
Hm9
8Sp
Hm9
Gl9
Ax9
Ax9
Gl9
vmf
vf
1 mm

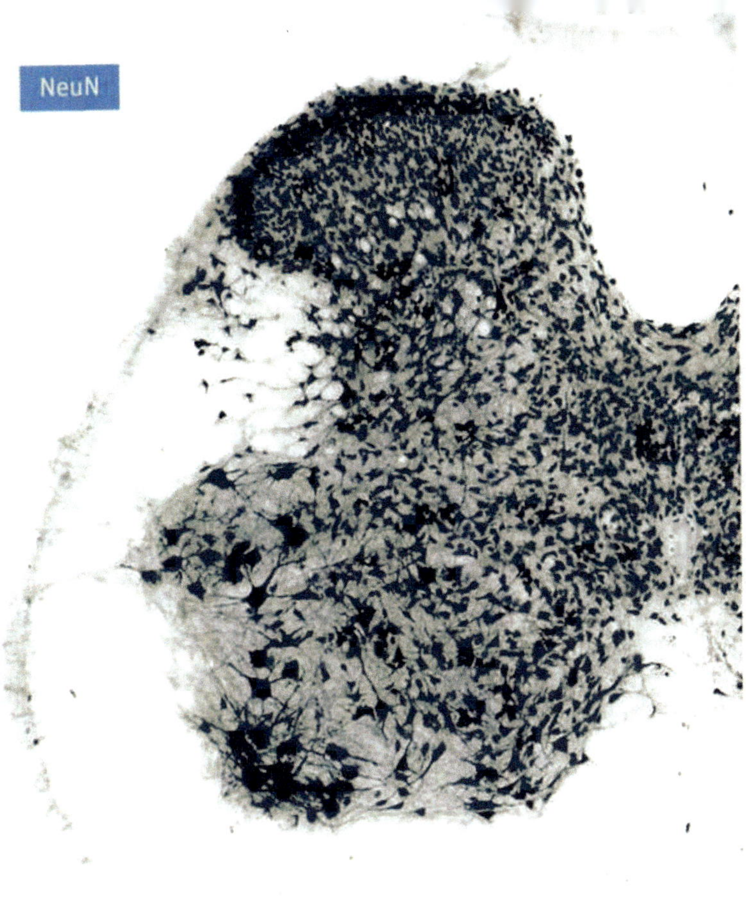

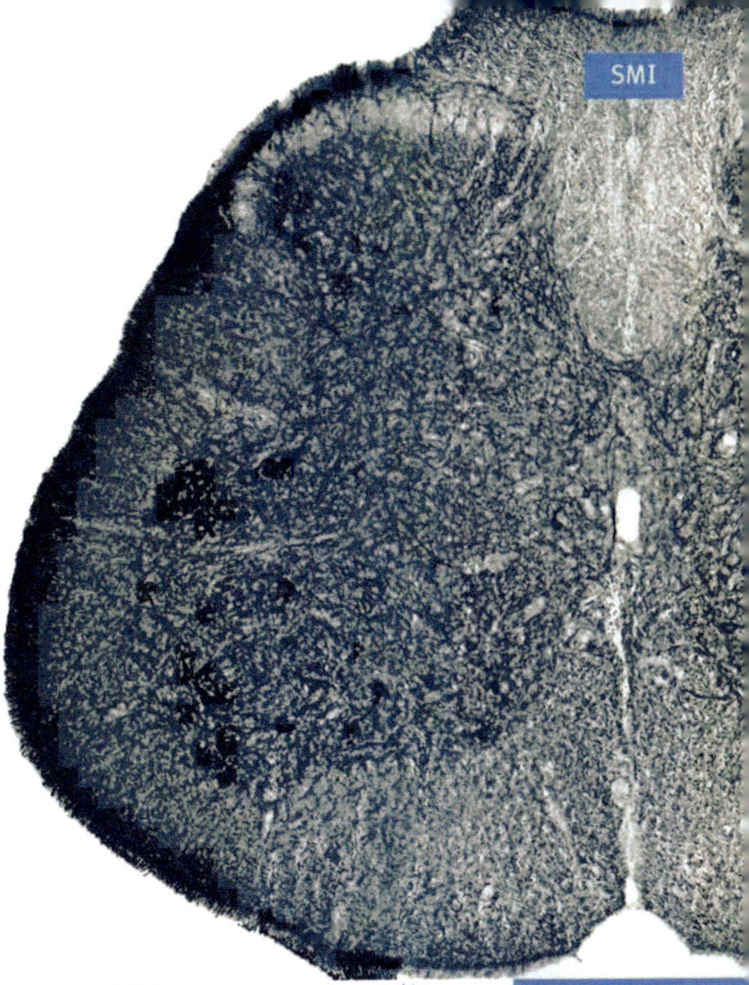

1Sp lamina 1
2SpI lamina 2, inner
2SpO lamina 2, outer
3Sp lamina 3
4Sp lamina 4
5SpL lamina 5, lat
5SpM lamina 5, med
6SpL lamina 6, lat
6SpM lamina 6, med
7Sp lamina 7
8Sp lamina 8
10Sp lamina 10
Ax9 axial muscles
CC central canal
CEx9 crual extensors
CFl9 crural flexors
dcs dors corticosp
dl dorsolat fasc
Gl9 gluteal muscles
gr gracile fascic
Hm9 hamstrings
LSp lat spinal nu
psdc postsyn dors col
rs rubrospinal tr
vmf vent med fiss
vwc vent white com

Rat L5

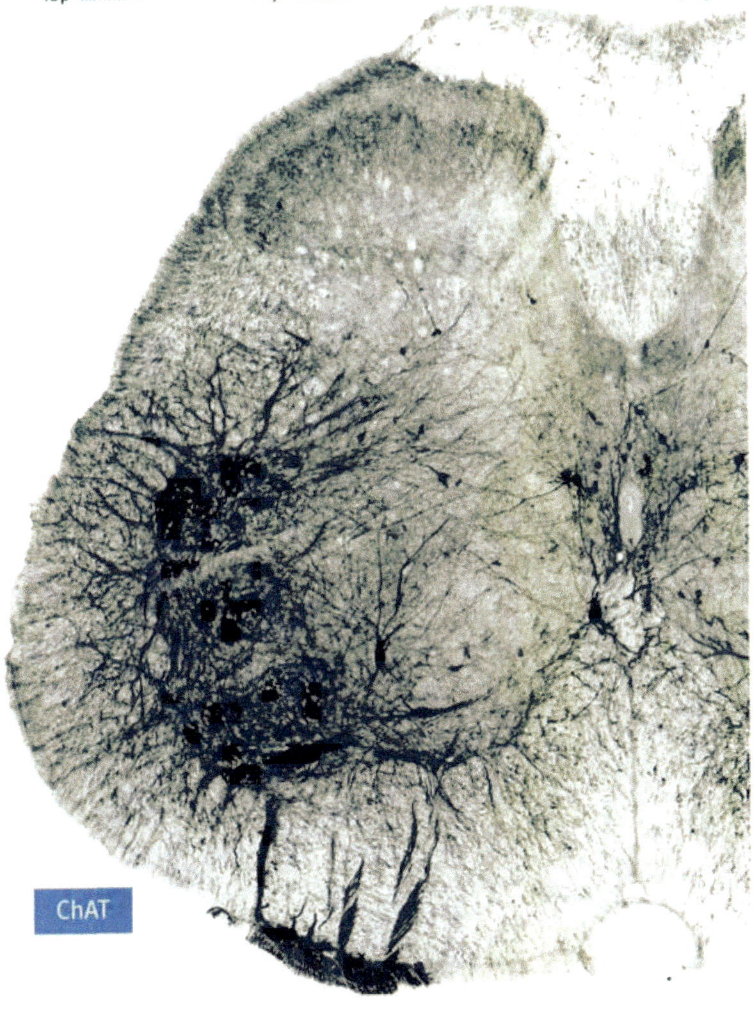

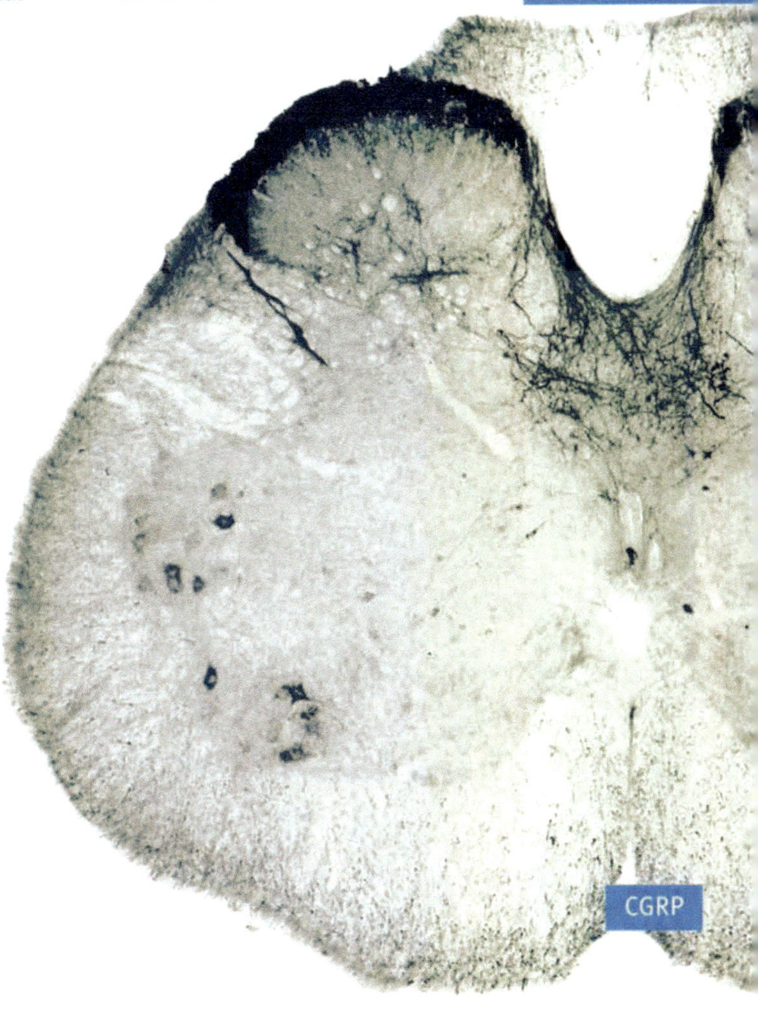

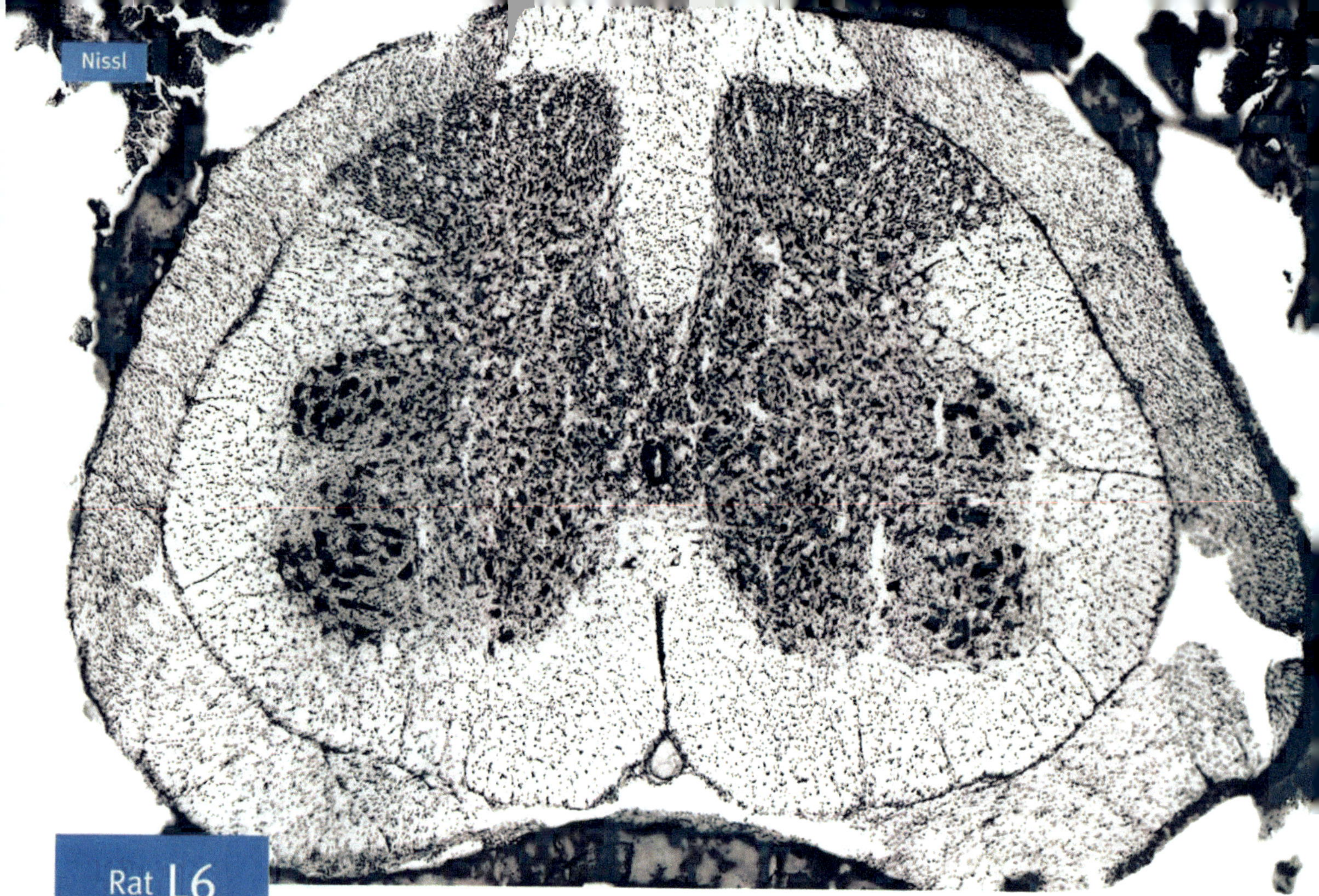
Nissl

Rat L6

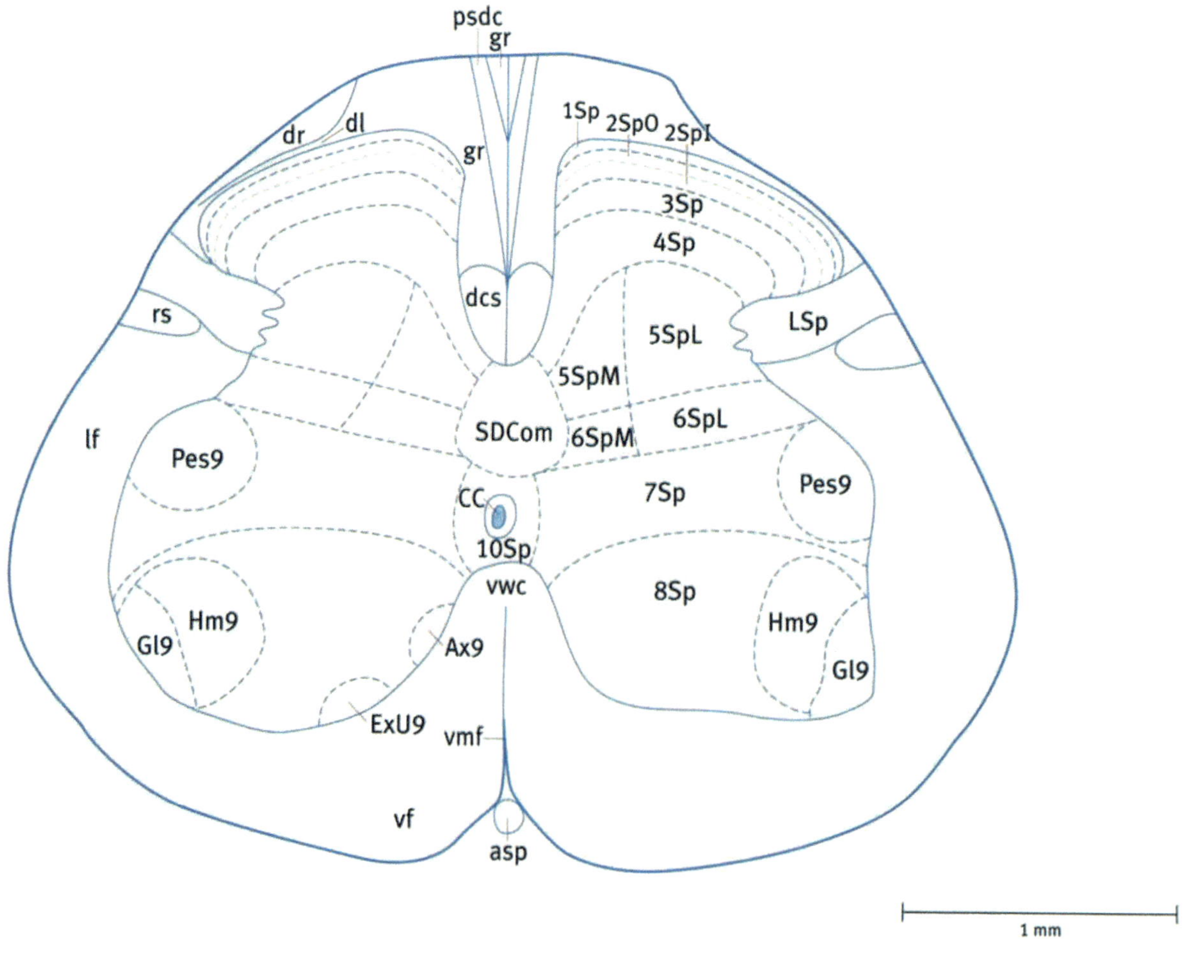
psdc
gr
dr
dl
gr
1Sp
2SpO
2SpI
3Sp
4Sp
dcs
rs
LSp
5SpL
5SpM
SDCom
6SpL
6SpM
lf
Pes9
Pes9
7Sp
CC
10Sp
vwc
8Sp
Hm9
Hm9
Gl9
Gl9
Ax9
ExU9
vmf
vf
asp
1 mm

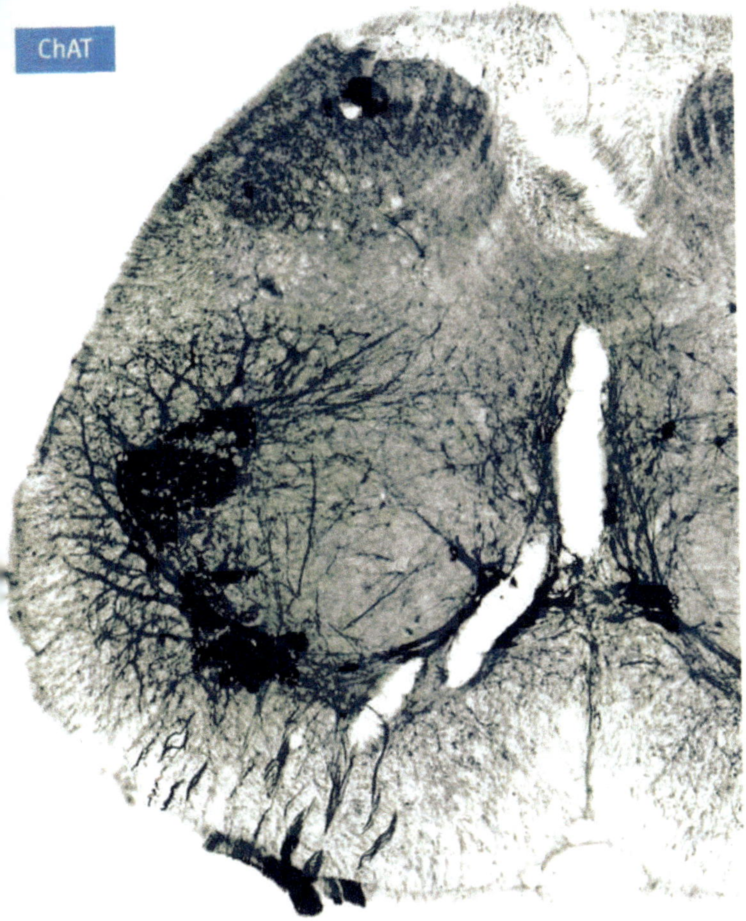

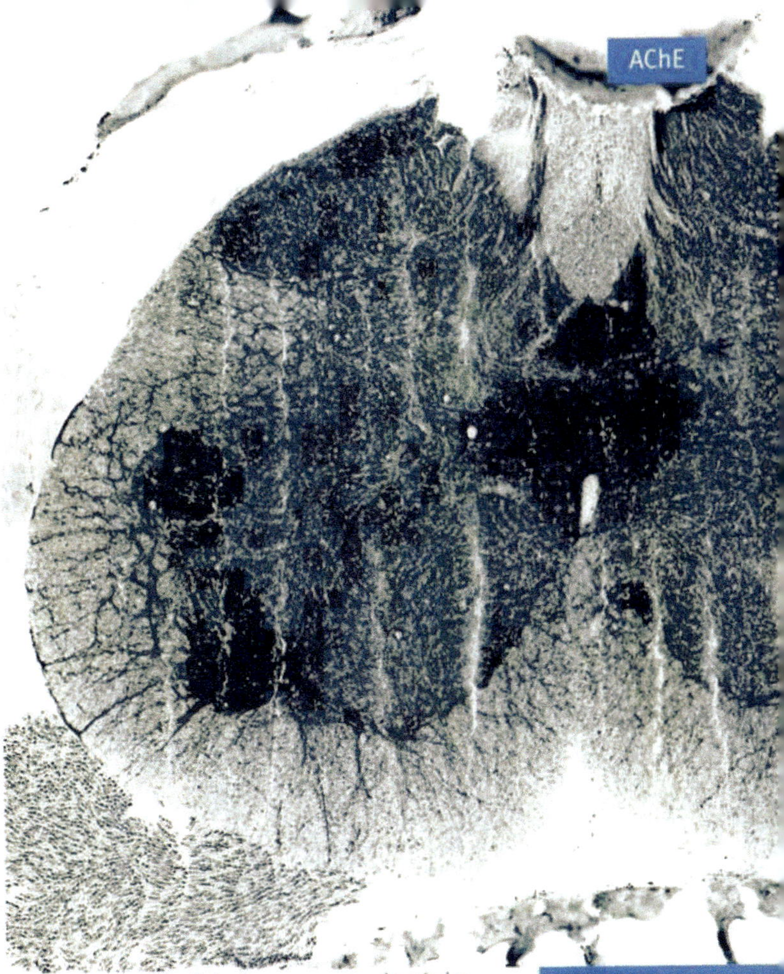

1Sp lamina 1
2SpI lamina 2, inner
2SpO lamina 2, outer
3Sp lamina 3
4Sp lamina 4
5SpL lamina 5, lat
5SpM lamina 5, med
6SpL lamina 6, lat
6SpM lamina 6, med
7Sp lamina 7
8Sp lamina 8
10Sp lamina 10
asp ant spinal art
Ax9 axial muscles
CC central canal
dcs dors corticosp
dl dorsolat fasc
dr dorsal root
ExU9 ext ureth sphin
Gl9 gluteal muscles
gr gracile fascic
Hm9 hamstrings
LSp lat spinal nu
Pes9 pes
psdc postsyn dors col
rs rubrospinal tr
SDCom sac dors comm
vmf vent med fiss
vwc vent white com

Rat L6

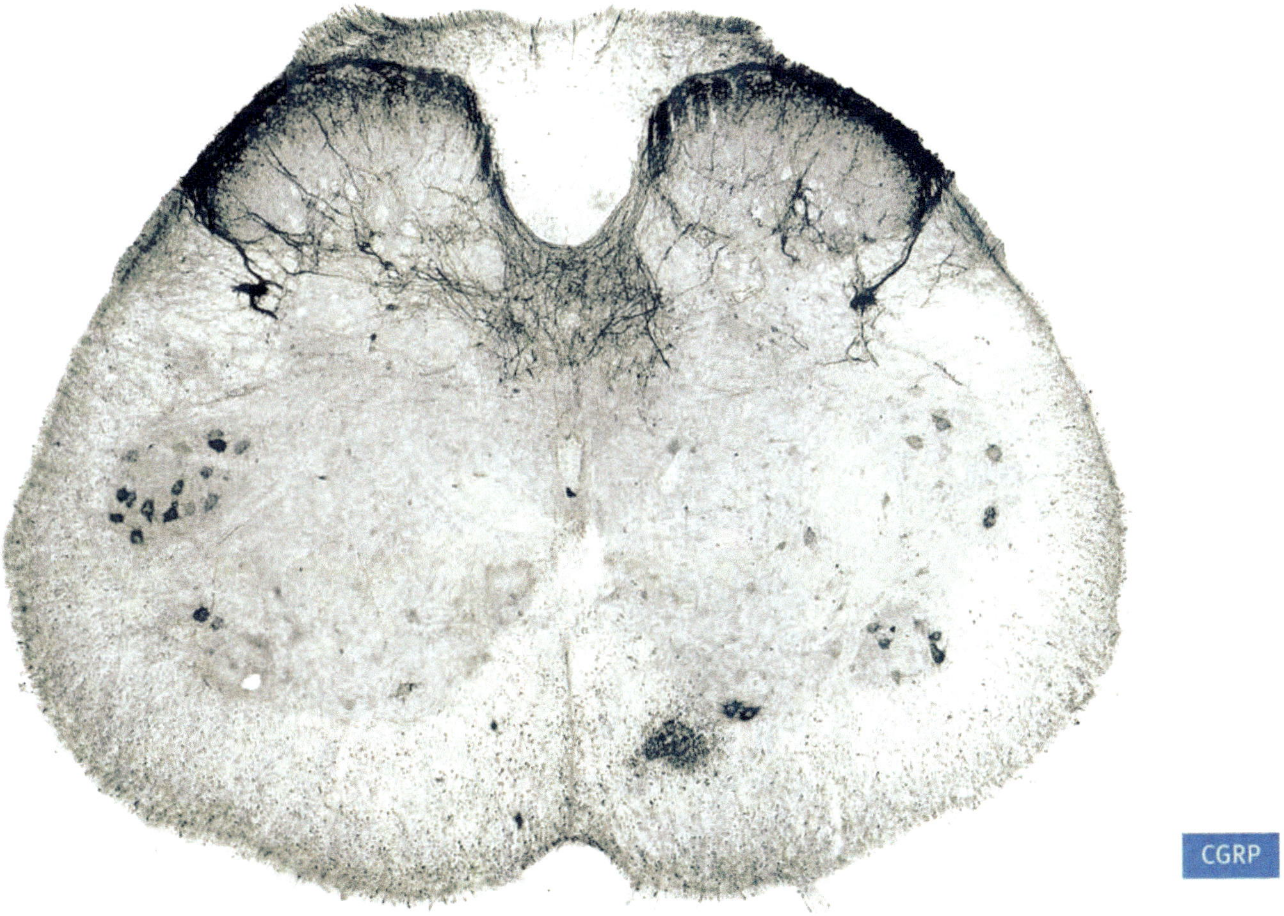

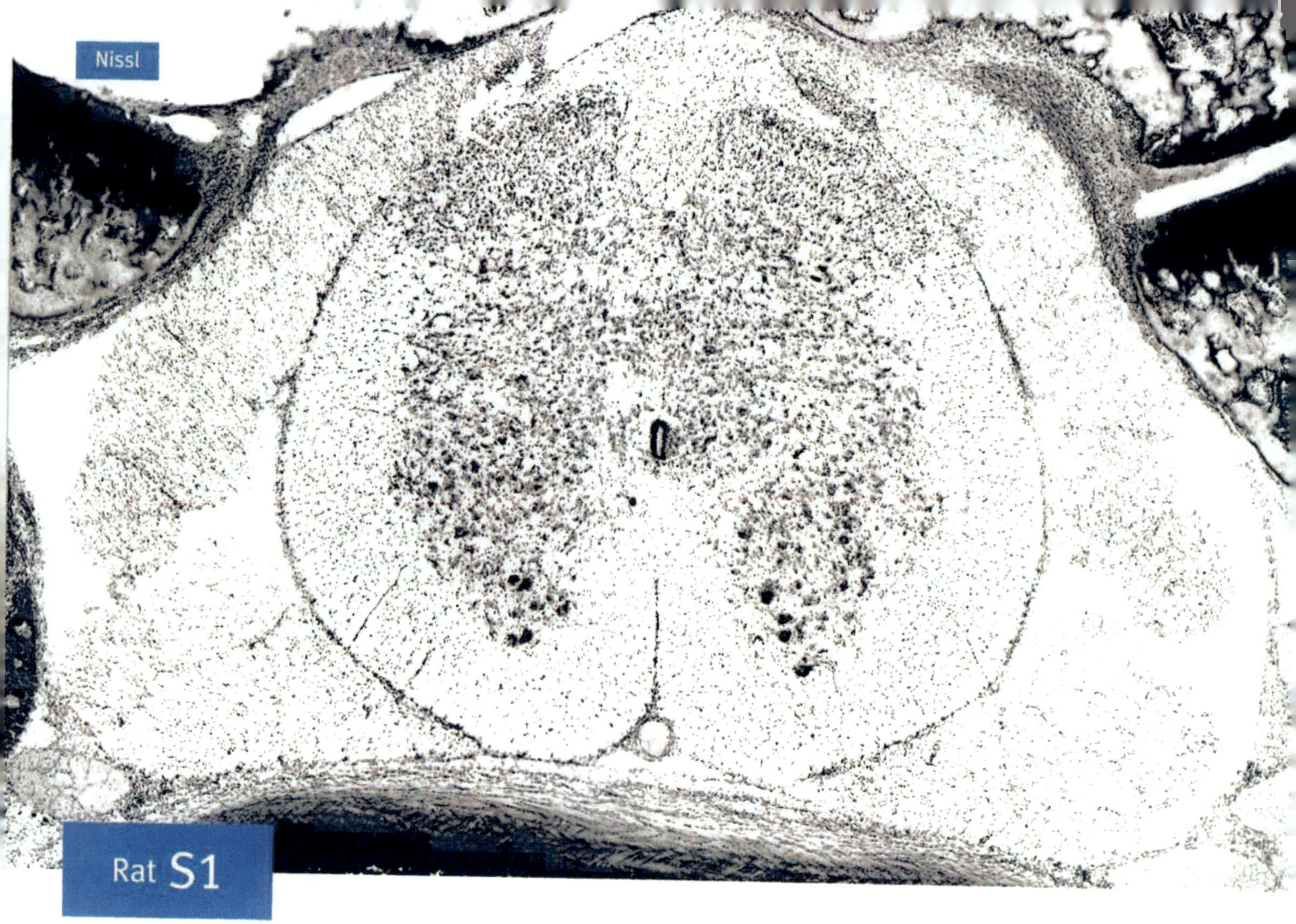
Nissl
Rat S1

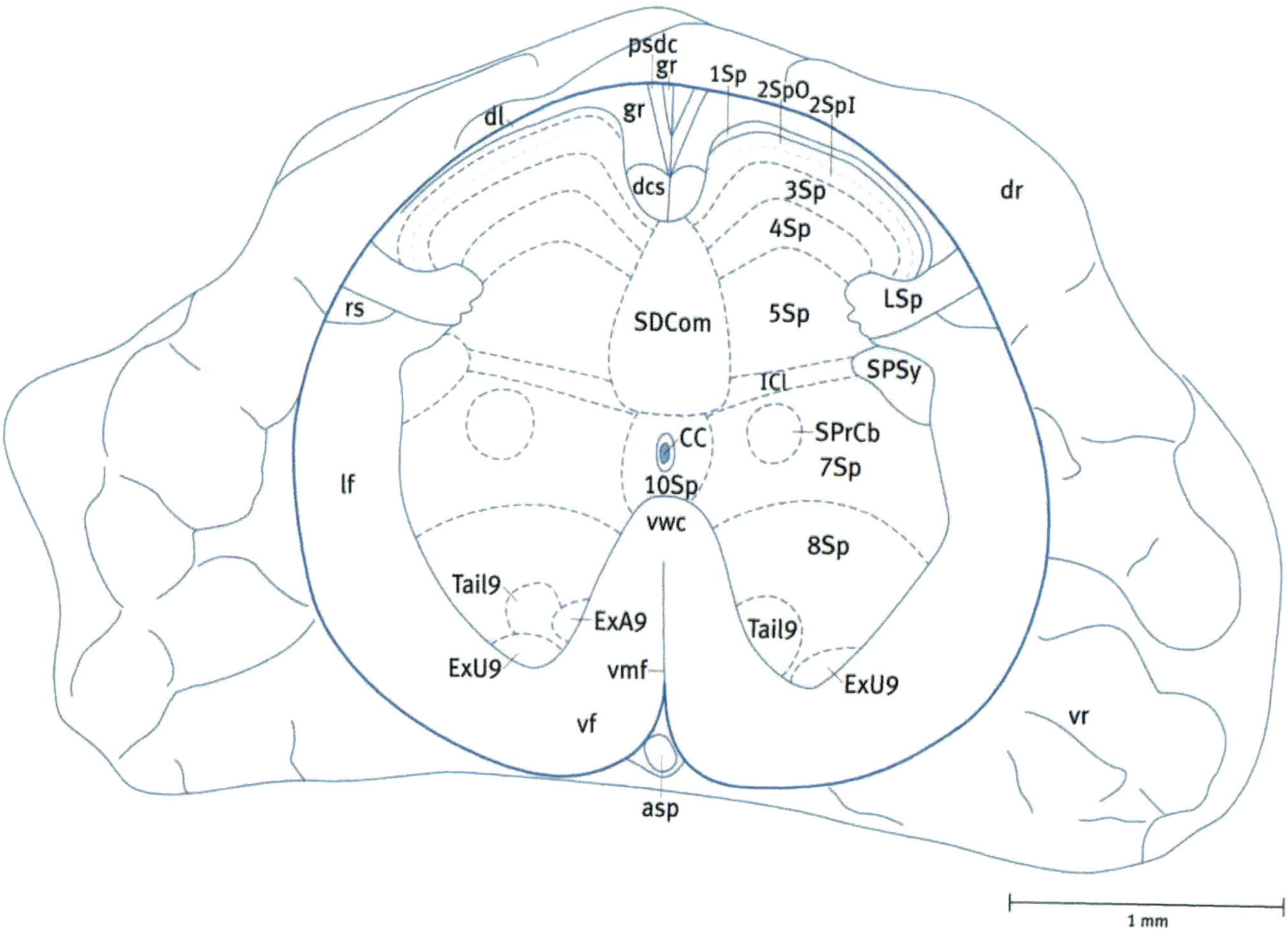
psdc
gr
1Sp
2SpO
2SpI
dl
gr
dcs
3Sp
dr
4Sp
rs
SDCom
5Sp
LSp
ICl
SPSy
CC
SPrCb
7Sp
lf
10Sp
vwc
8Sp
Tail9
ExA9
Tail9
ExU9
vmf
ExU9
vf
vr
asp
1 mm

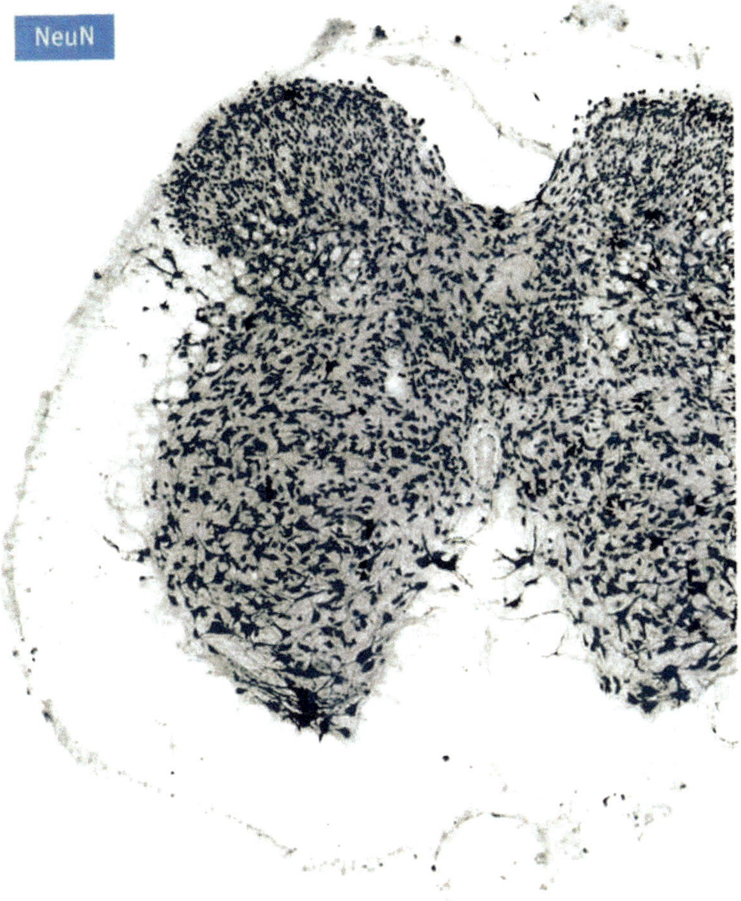

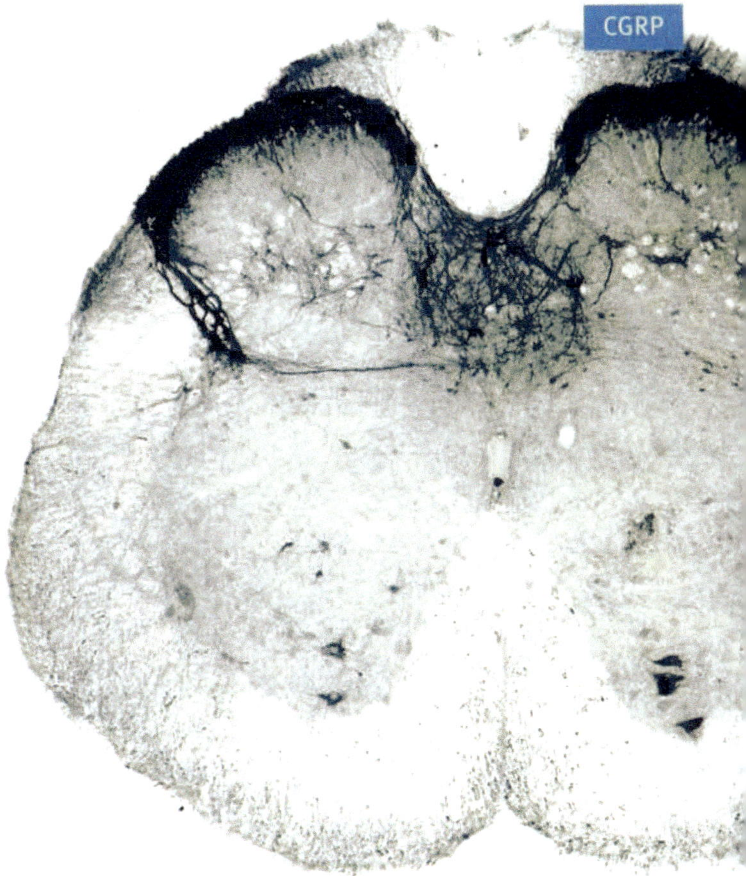

1Sp lamina 1
2SpI lamina 2, inner
2SpO lamina 2, outer
3Sp lamina 3
4Sp lamina 4
5Sp lamina 5
7Sp lamina 7
8Sp lamina 8
10Sp lamina 10
asp ant spinal art
CC central canal
dcs dors corticosp
dl dorsolat fasc
dr dorsal root
ExA9 ext anal sphin
ExU9 ext ureth sphin
gr gracile fascic
ICl intercalated nu
LSp lat spinal nu
psdc postsyn dors col
rs rubrospinal tr
SDCom sac dors comm
SPrCb sacral preCb
SPSy sacral parasym
Tail9 tail
vmf vent med fiss
vr ventral root
vwc vent white com

Rat S1

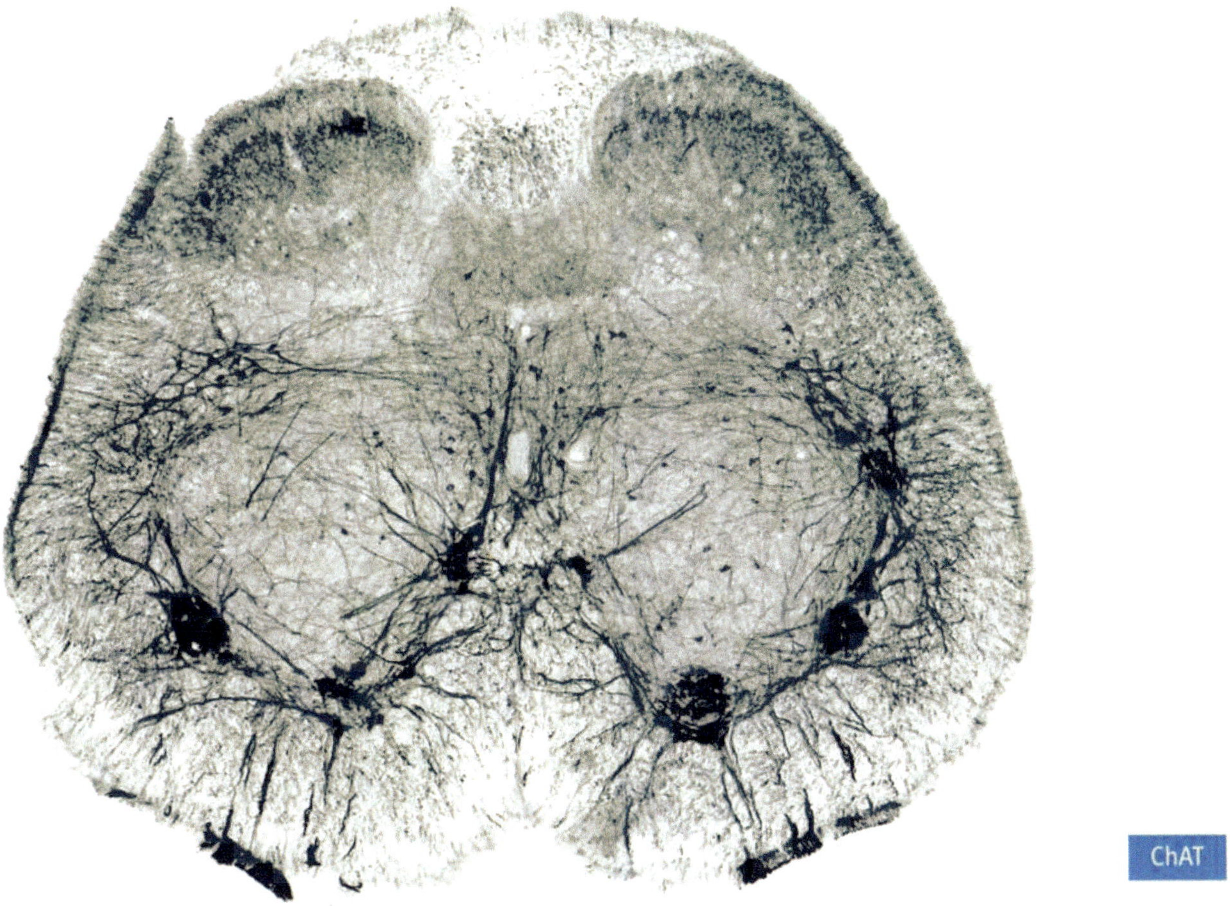

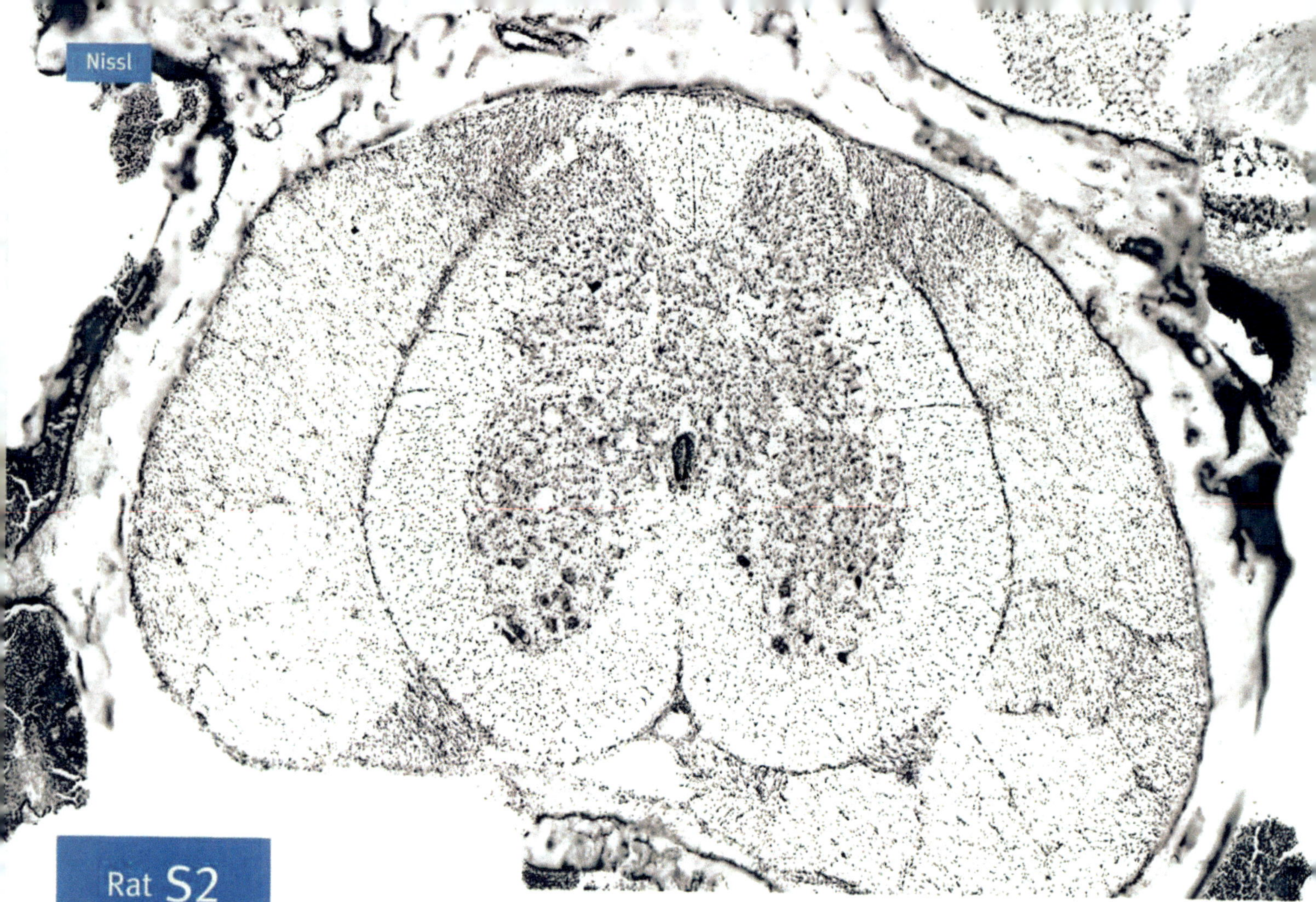
Nissl

Rat S2

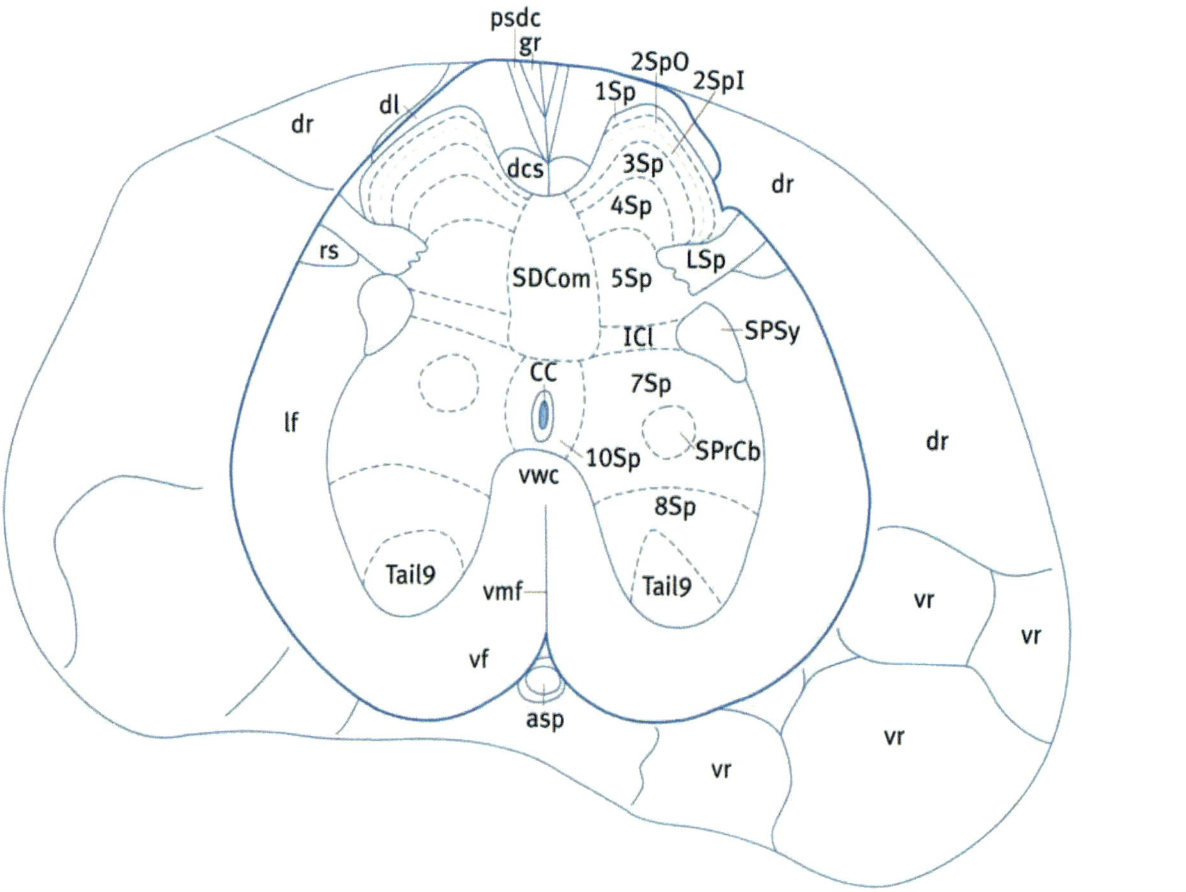
psdc
gr
2SpO
2SpI
1Sp
dl
dr
dcs
3Sp
dr
4Sp
rs
LSp
SDCom
5Sp
SPSy
ICl
CC
7Sp
lf
10Sp
SPrCb
dr
vwc
8Sp
Tail9
vmf
Tail9
vr
vr
vf
asp
vr
vr
1 mm

SMI

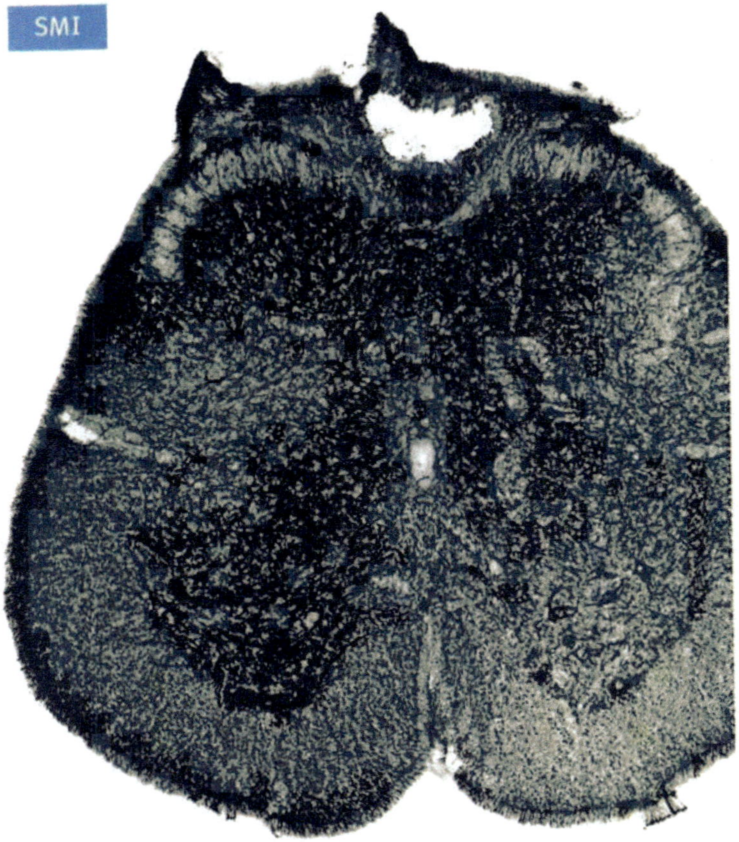

CGRP

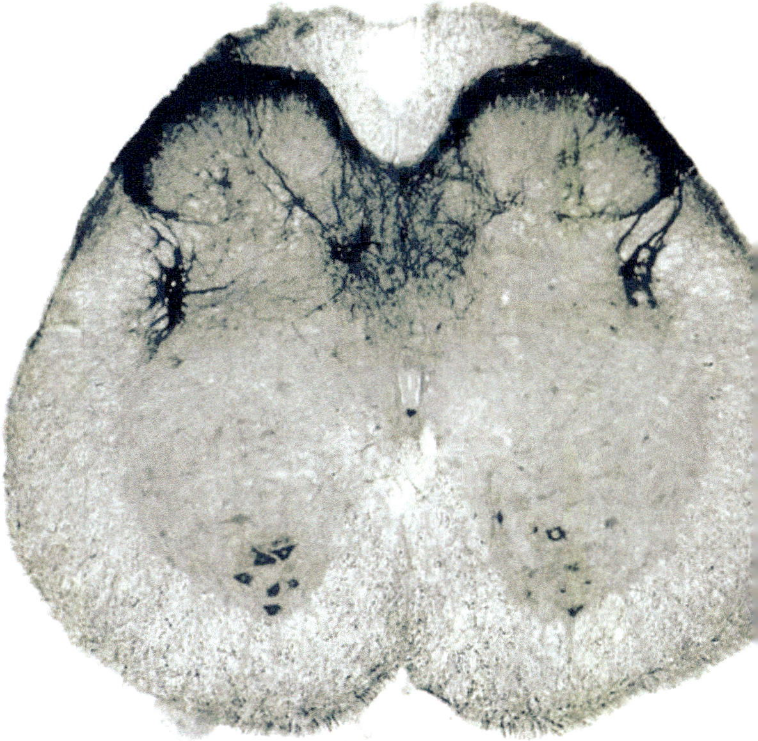

1Sp lamina 1
2SpI lamina 2, inner
2SpO lamina 2, outer
3Sp lamina 3
4Sp lamina 4
5Sp lamina 5
7Sp lamina 7
8Sp lamina 8
10Sp lamina 10
asp ant spinal art
CC central canal
dcs dors corticosp
dl dorsolat fasc
dr dorsal root
gr gracile fascic
ICl intercalated nu
LSp lat spinal nu
psdc postsyn dors col
rs rubrospinal tr
SDCom sac dors comm
SPrCb sacral preCb
SPSy sacral parasym
Tail9 tail
vmf vent med fiss
vr ventral root
vwc vent white com

Rat S2

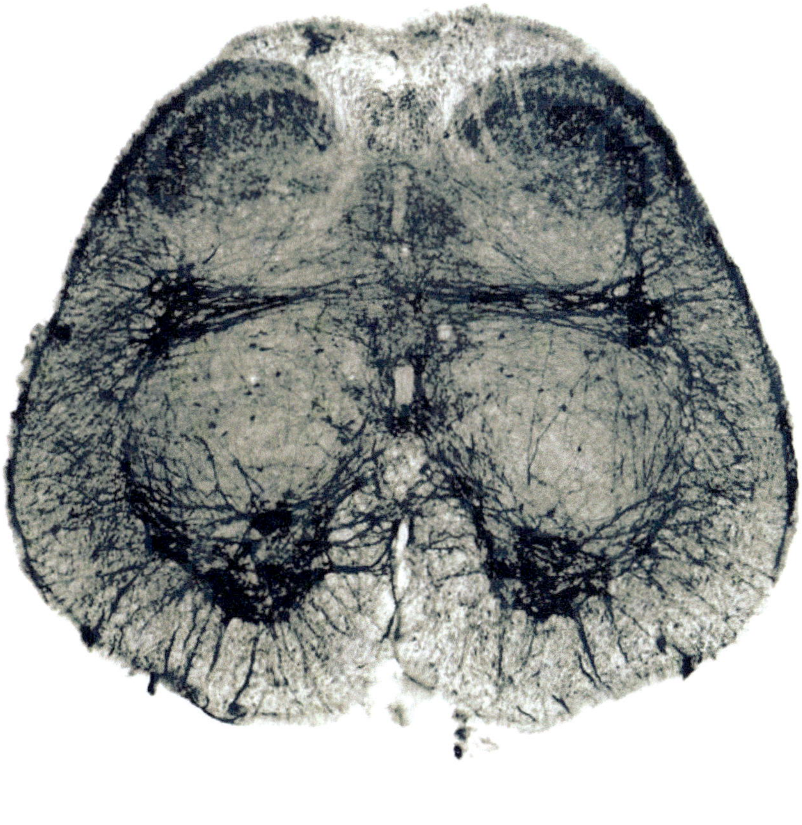

ChAT

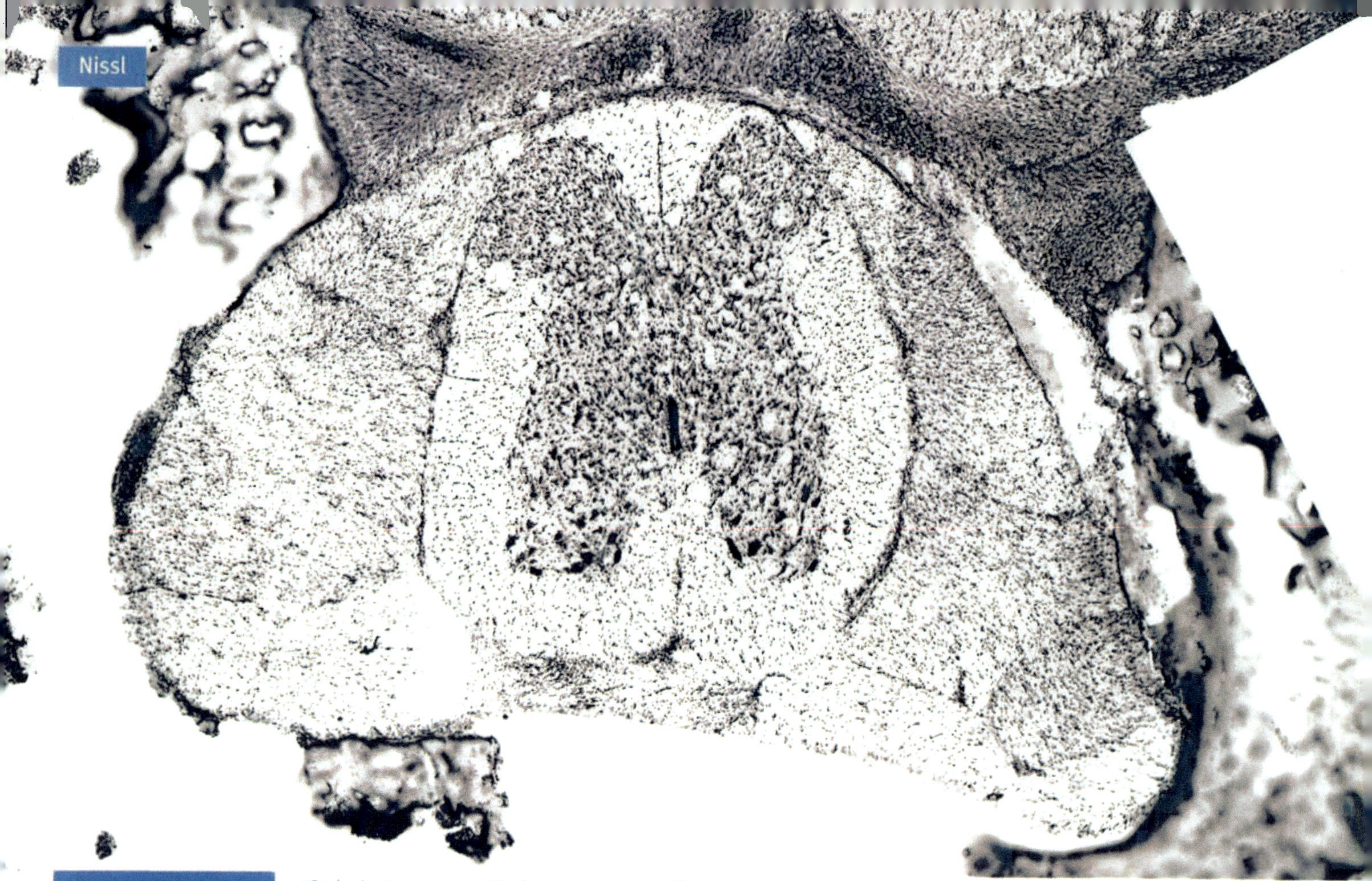

Rat S3

1Sp lamina 1	5Sp lamina 5	CC central canal	psdc postsyn dors col	vwc vent white com
2SpI lamina 2, inner	7Sp lamina 7	dcs dors corticosp	SDCom sac dors comm	
2SpO lamina 2, outer	8Sp lamina 8	dl dorsolat fasc	Tail9 tail	
3Sp lamina 3	10Sp lamina 10	gr gracile fascic	vmf vent med fiss	
4Sp lamina 4	asp ant spinal art	LSp lat spinal nu	vr ventral root	

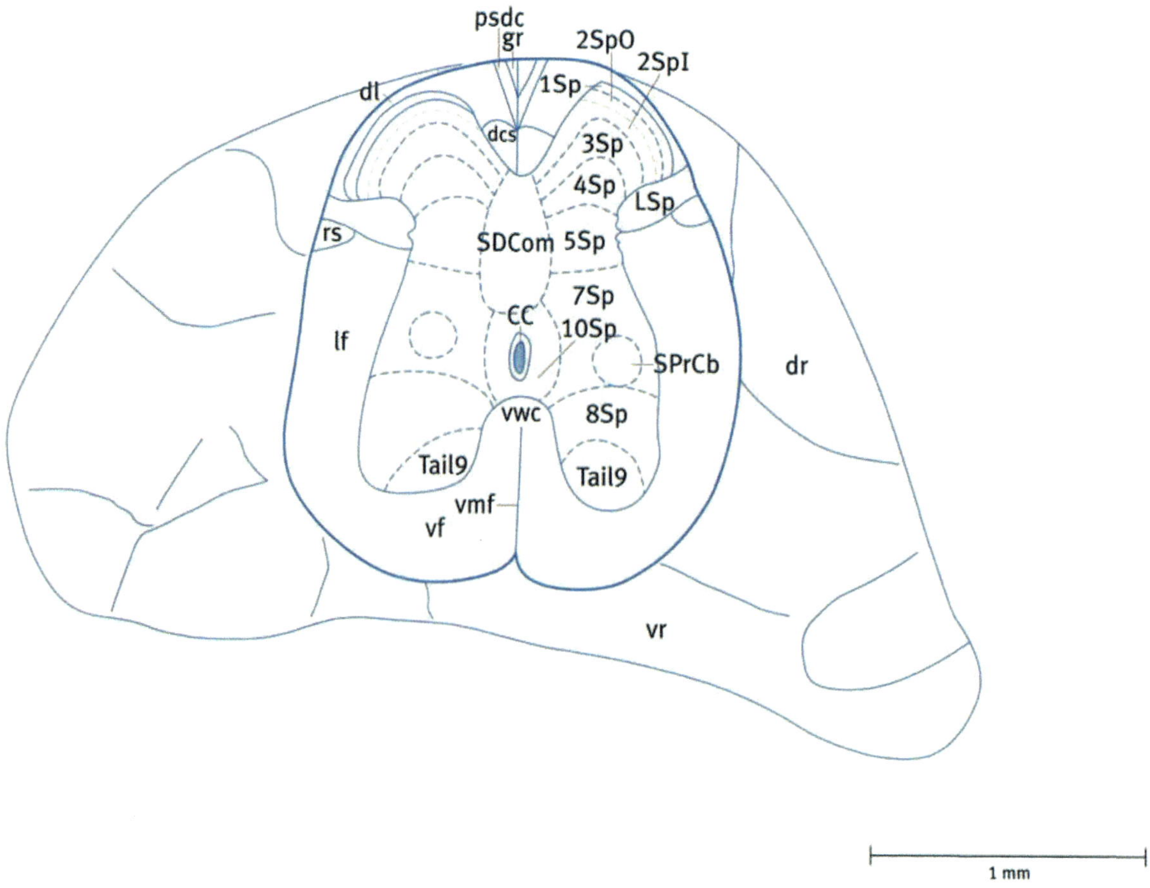

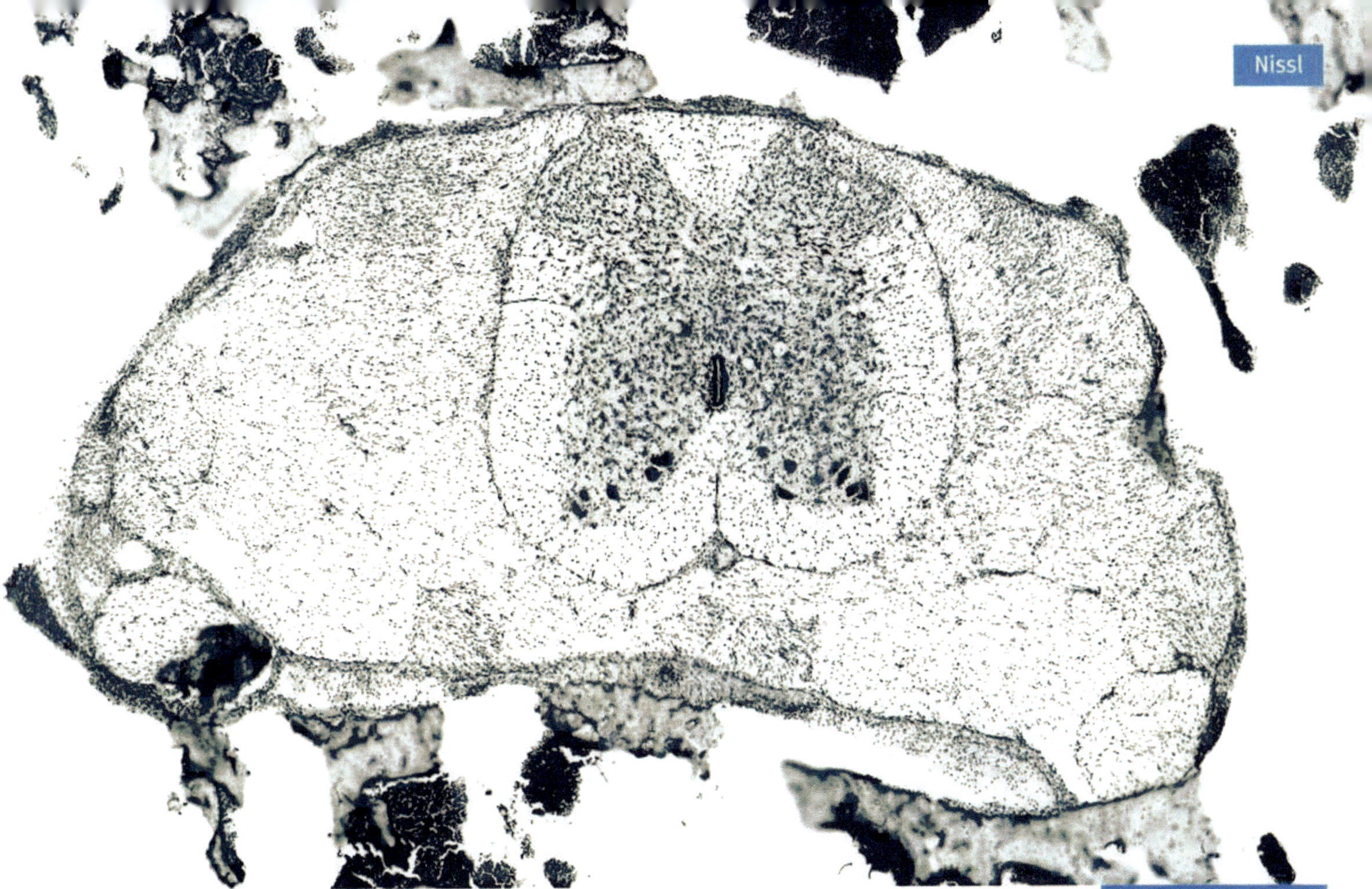

2SpI lamina 2, inner
3Sp lamina 3
4Sp lamina 4
5Sp lamina 5
7Sp lamina 7
8Sp lamina 8
10Sp lamina 10
asp ant spinal art
CC central canal
dcs dors corticosp
dl dorsolat fasc
dr dorsal root
gr gracile fascic
LSp lat spinal nu
psdc postsyn dors col
rs rubrospinal tr
SDCom sac dors comm
SPrCb sacral preCb
Tail9 tail
vmf vent med fiss
vr ventral root
vwc vent white com

Rat S4

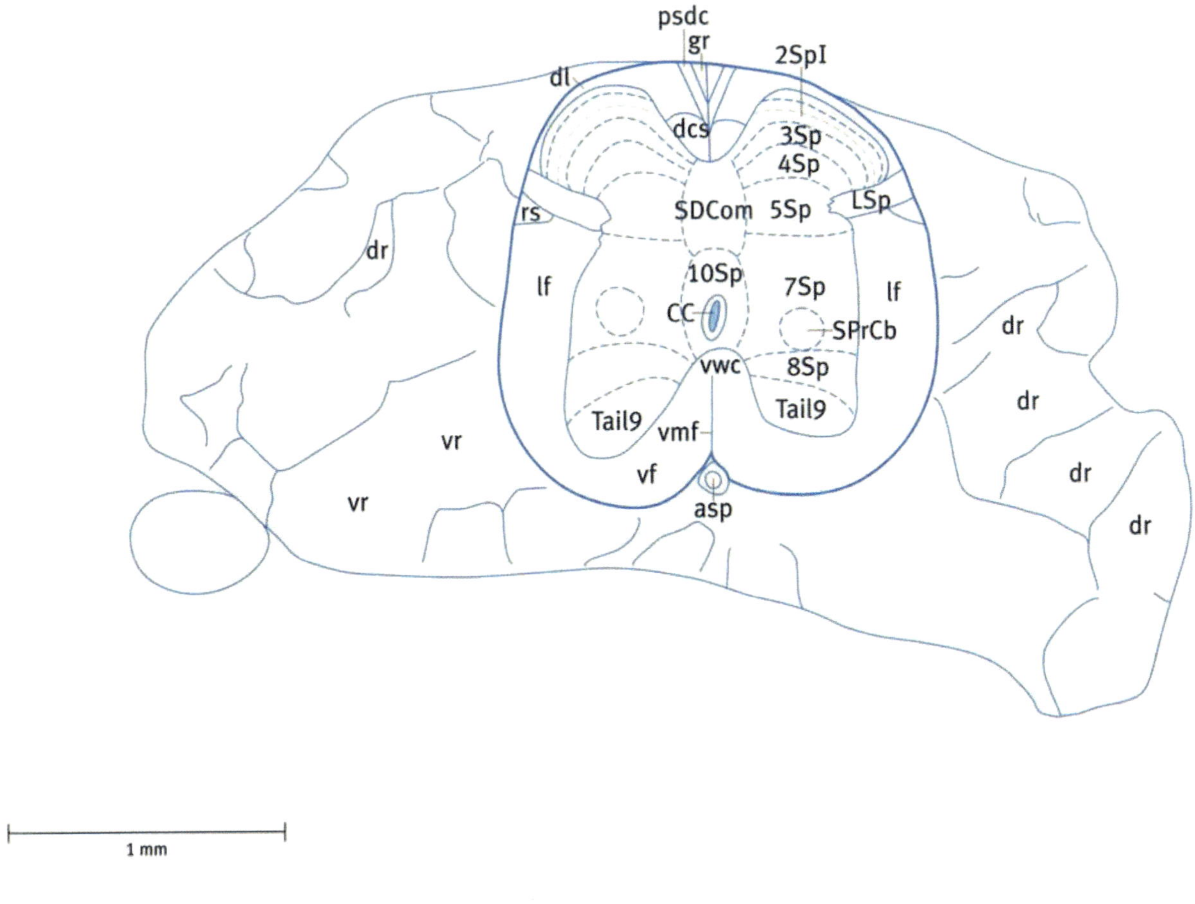

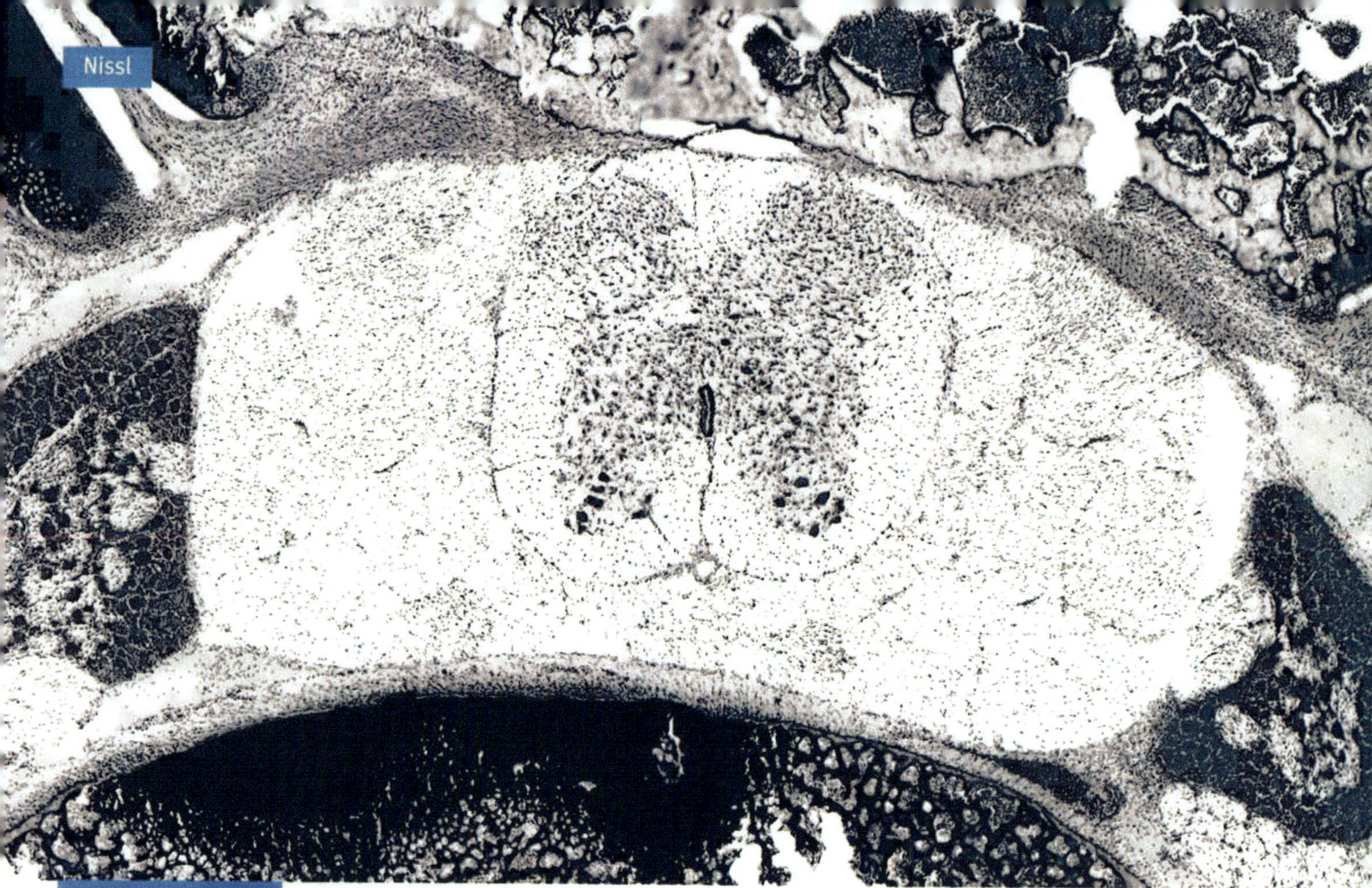

Rat Co1

1Sp lamina 1
2SpI lamina 2, inner
2SpO lamina 2, outer
3Sp lamina 3
4Sp lamina 4
5Sp lamina 5
7Sp lamina 7
8Sp lamina 8
10Sp lamina 10
asp ant spinal art
CC central canal
dcs dors corticosp
dl dorsolat fasc
gr gracile fascic
LSp lat spinal nu
psdc postsyn dors col
SDCom sac dors comm
Tail9 tail
vmf vent med fiss
vr ventral root
vwc vent white com

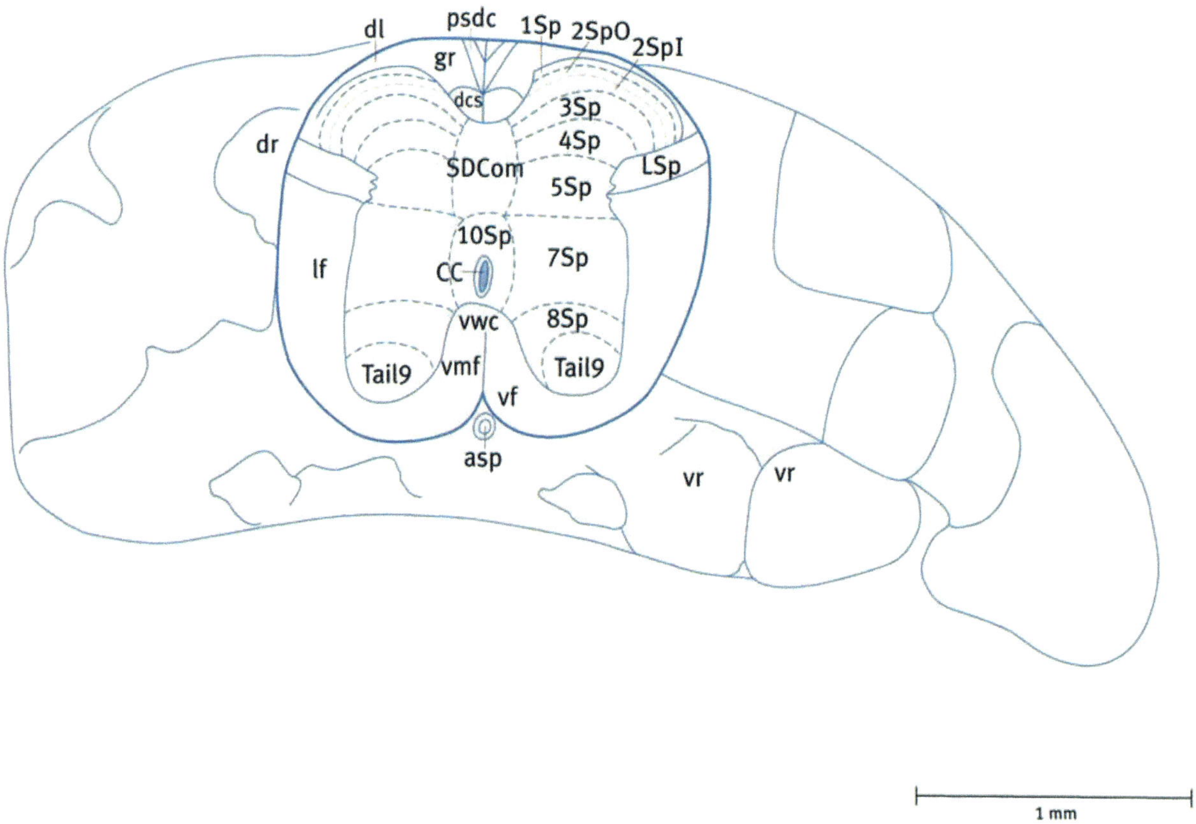

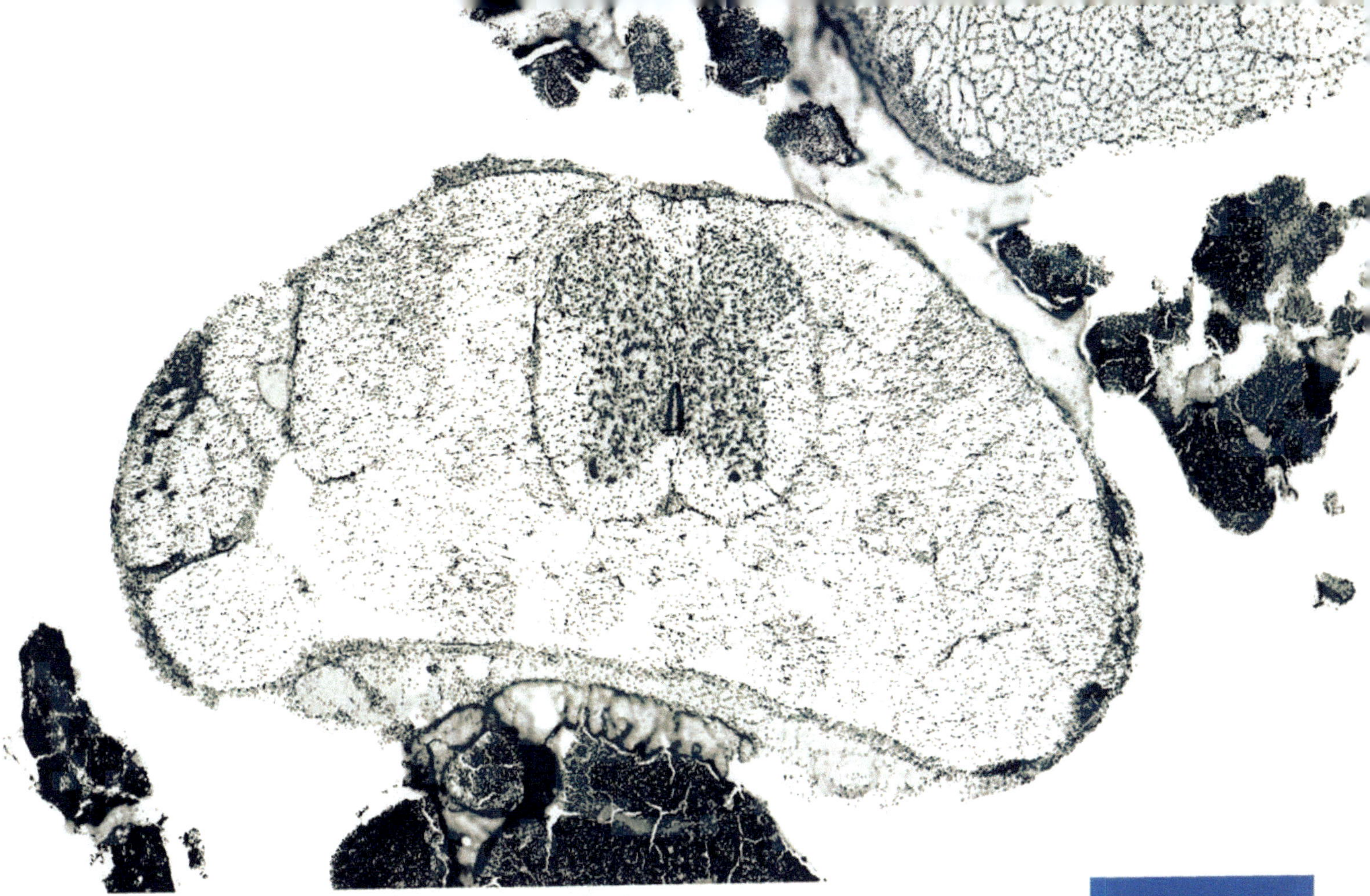

3Sp lamina 3
4Sp lamina 4
5Sp lamina 5
7Sp lamina 7
8Sp lamina 8
10Sp lamina 10
asp ant spinal art
CC central canal
dl dorsolat fasc
dr dorsal root
gr gracile fascic
SDCom sac dors comm
Tail9 tail
vmf vent med fiss
vwc vent white com

Rat Co2

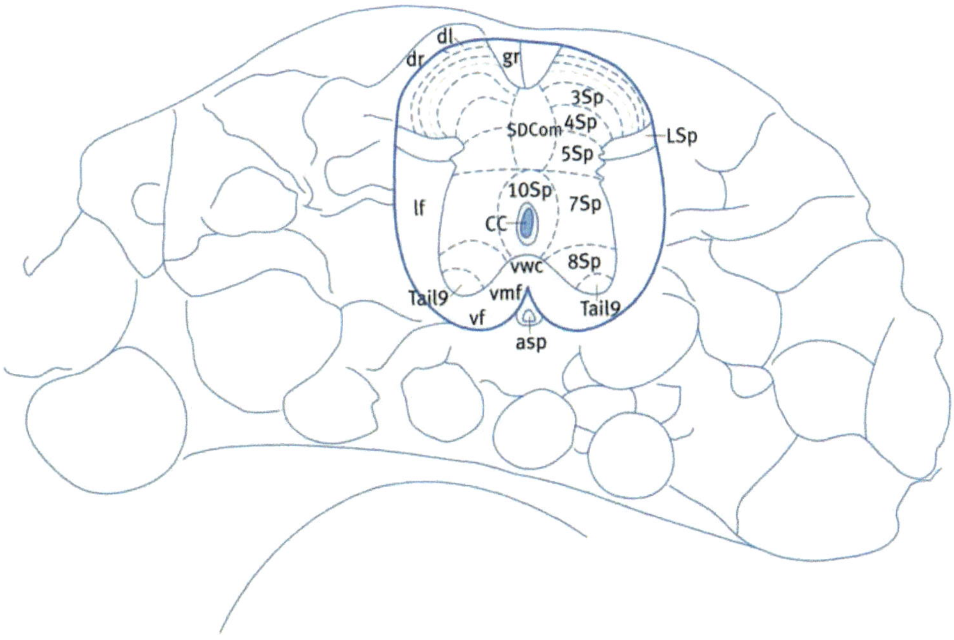

1 mm

Nissl

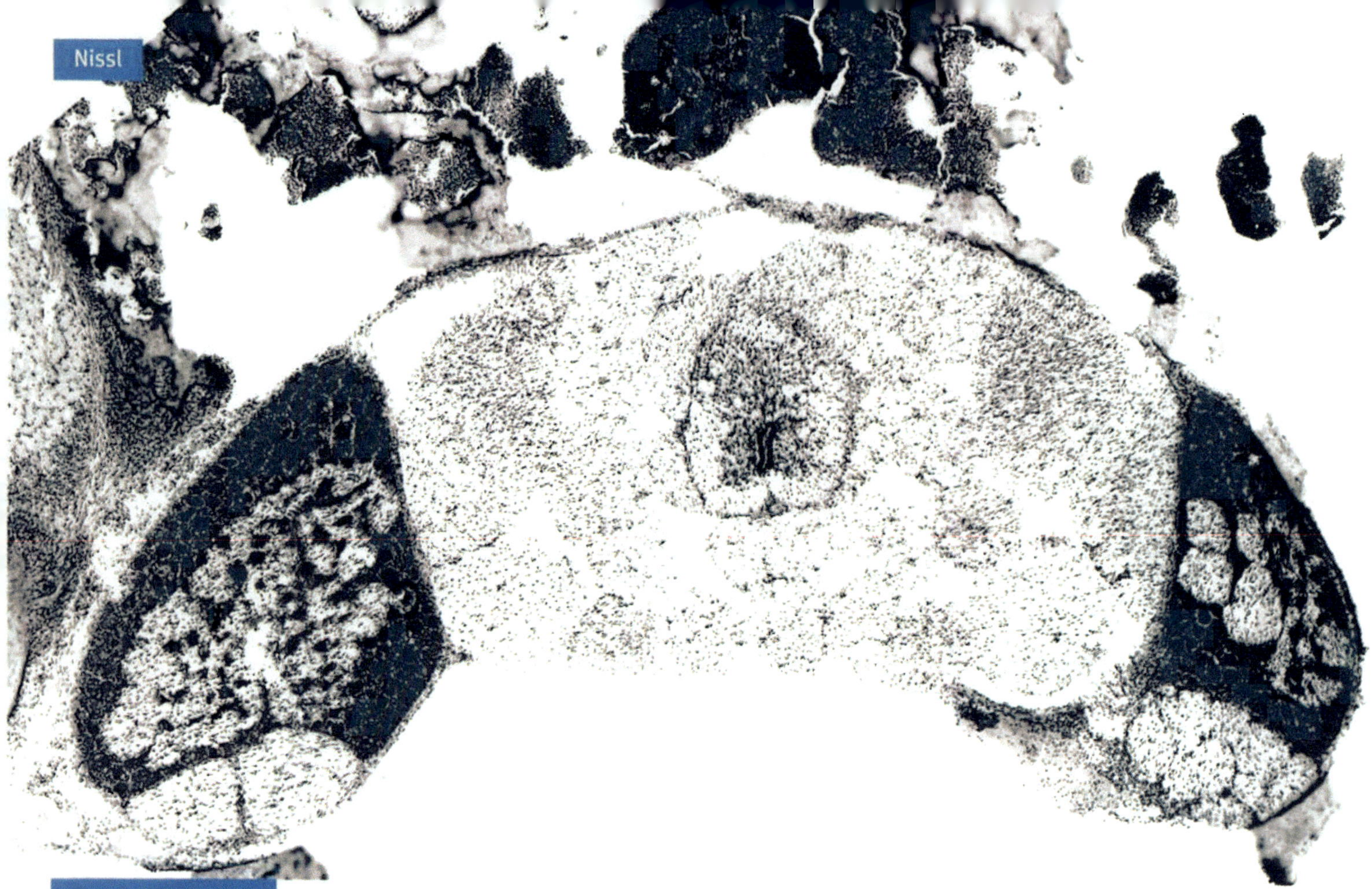

Rat Co3

2Sp lamina 2
3Sp lamina 3
4Sp lamina 4
5Sp lamina 5
7Sp lamina 7
8Sp lamina 8
10Sp lamina 10
asp ant spinal art
dr dorsal root
LSp lat spinal nu
SDCom sac dors comm
vr ventral root

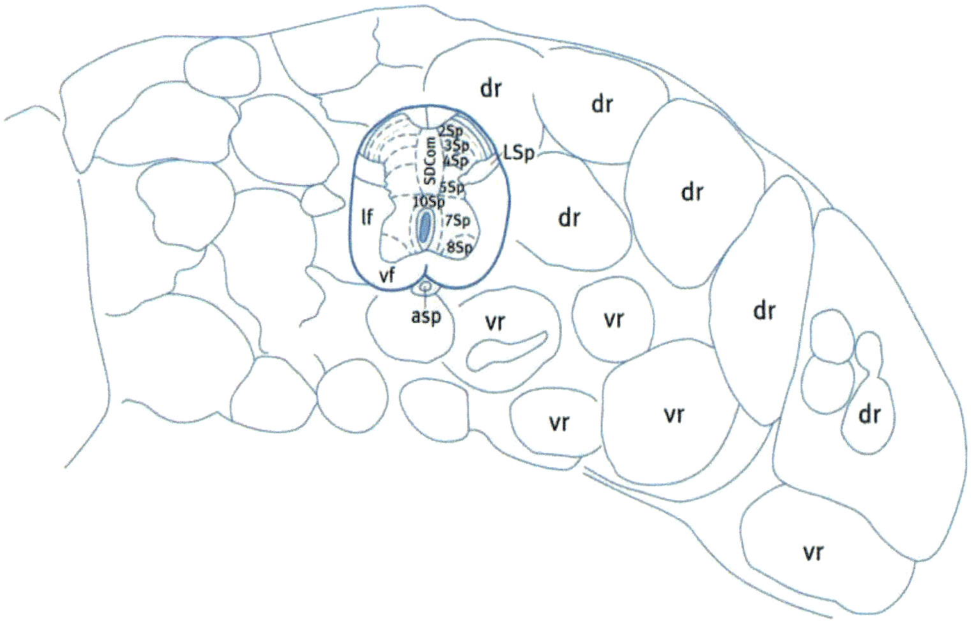

1 mm

16 Atlas of the Mouse Spinal Cord

Charles Watson, George Paxinos, Gulgun Kayalioglu and Claire Heise

Introduction

The reader is referred to the introduction to Chapter 15 (*Atlas of the Rat Spinal Cord*) for issues and procedures that are common to the preparation of rat and mouse spinal cord sections. Below we describe matters that are specific to the *Atlas of the Mouse Spinal Cord.*

Methods

Spinal cords from three C57BL/6J male mice were used for this atlas. As in the case of the rat, the left-hand side of the open book depicts a Nissl stained transverse section (the representative of the master set) and a diagrammatic depiction of this section. All the Nissl sections depicted on the left side of the open book were obtained from unfixed material from a single mouse weighing 28 g which was decapitated and its vertebral column dissected with the spinal cord in situ. The column and cord were sectioned on a cryotome at 40 μm thickness after the tissue was blocked in four parts. The sections were stained either for Nissl or acetylcholinesterase (AChE). The segmental level of each Nissl section in the master set was verified by comparison with sections taken from fixed spinal cord in which the spinal roots were exposed before the spinal cord was excised.

On the top of the right hand page we present a pair of Nissl and AChE sections obtained from one mouse, which was sectioned at the Allen Institute for Brain Science (see below). The bottom of the right-hand page features sections obtained from a fixed mouse for which the spinal cord was dissected free and sectioned and stained as per the rat (see Chapter 15).

Mouse spinal cord sections provided by the Allen Institute for Brain Science

We have been fortunate in being able to display a set of sections from a special series prepared by the Allen Institute for Brain Science. The Institute has prepared a large series of spinal cord sections to display gene patterns revealed by in situ hybridization (ISH), and the set presented here is a reference set for study of normal anatomy of the spinal cord segments. The importance of the display of this Allen set in the atlas is that it was prepared in the same way as the Allen ISH sets, and should, therefore, be of assistance to those who study the Allen ISH material. Moreover, the sections are taken from a set of serial sections from the whole spinal cord, so that intervening sections for study could be made available in the future by the Allen Institute for Brain Science.

We must also note that the histological quality of these sections is excellent, and the sections, enhance the value of our atlas. It is worth noting that the Allen Institute slides were also collected with the CryoJane Tape Transfer System.

The Allen Institute spinal cord sections were taken from a single spinal cord. All sections from the entire cord were collected and stained serially with Nissl, AChE, and haematoxylin and eosin (H&E). One Nissl, and one AChE section is displayed for each spinal cord segment on the top half of the right-hand side of an open-book page of this atlas.

The spinal cord was taken from a C57BL/6J male mouse 6 weeks of age and weighing 25 g (reference number 343-0929). The mouse was euthanized with 0.5% isoflurane. The fresh spinal cord was isolated by laminectomy of C1-T5 followed by extrusion. The fresh cord was cut into four pieces, frozen at -80°C, and embedded into a single OCT block. The blocks were sectioned with a cryostat at 20 μm, and the sections were placed on slides in rounds of three in order to create separate sets of Nissl, AChE, and H&E. A total of about 1900 sections were collected on 72 slides (24 Nissl slides, 24 AChE slides, and 24 H&E slides). After allowing the sections to air dry on the slides for a minimum of 30 min, the tissue was fixed in 4% neutral buffered paraformaldehyde (PFA) for 20 minutes and rinsed for 3 minutes in 1x PBS.

Nissl staining. Sections were defatted with xylene and hydrated through a graded series containing 100%, 95%, 70%, and 50% ethanol. After incubation in water, the sections were stained in 0.25% thionin. The sections were differentiated and dehydrated in water and a graded series containing 50%, 70%, 95%, and 100% ethanol before transfer to xylene. The slides were coverslipped with DPX.

Imaging and identification of spinal cord segments. All the Allen Institute slides were imaged with the Aperio system and viewed with Imagescope software, which is provided free by Aperio (www.aperio.com). The images are taken at a very high resolution and the total file size for the 72 slides (1900 sections) is about 80 GB. We extracted individual images with the 'extract region' tool in Imagescope. The extracted images were approximately 100 MB for each section, but when converted to grayscale, this was reduced to about 25 MB for each section.

Identification of levels of Allen Institute sections. When the spinal cord was removed from the spine, all spinal nerve roots were torn away, so there were no external landmarks to assist with segment identification. Segments were identified by comparison with the master set on the left-hand side of the open-book display for that segment. Identification of segments was greatly assisted by the availability of AChE sections adjacent to the Nissl sections. The most important landmarks in AChE sections were the phrenic nucleus, the upper limb motoneuron groups, the junction of T1 and T2 (the commencement of the intermediolateral column), the junction of L1 and L2 (the end of the intermediolateral column), the lower limb motoneuron groups, and the sacral parasympathetic nucleus in S1 and S2. Between T2 and T13, the segmental boundaries were estimated by dividing the length of this region by 12. The sections that were selected for display were taken from the center of the segment identified in each case. The Nissl and AChE sections displayed on the top half of the right-hand side of the open-book display for each segment level in the atlas are adjacent sections in all cases.

Photography and diagrams

Photographs were taken through an Olympus Provis microscope with a 16 megapixel digital camera.

Sections sustained different degrees of shrinkage through different stains. We were guided by the Nissl standard because the Nissl sections were applied directly to the microscope slides through the CyroJane apparatus. We resized the images of the other stains to approximately match the size of the Nissls.

The scale bar refers to the diagrams. The photographs are slightly larger than the diagrams to permit easier detection of structures.

Basis of delineation of structures

We identified laminae in the spinal cord gray matter according to the schema developed by Rexed (1952). A description of the cytoarchitecture of these laminae is presented in Chapter 6. We used the prefix Sp to indicate each lamina so as to allow computer searching systems to distinguish between this set of digits and those used in other areas (e.g. layers of cerebral cortex).

Certain nuclei have been identified within the laminae of Rexed. Some of these are well known (such as the dorsal nucleus), but some others are less well recognized. In the latter category are the nuclei of origin of the ventral spinocerebellar tract. We have identified these as the lumbar and sacral precerebellar nuclei (LPrCb and SPrCb). See further discussion of these cell groups in Chapter 6.

The preganglionic nuclei of the sympathetic and parasympathetic nervous systems are clearly identified, and a description is presented in Chapter 8. We have labeled the sympathetic preganglionic cells as the intermediolateral cell group (IML), but we chose the name 'sacral parasympathetic nucleus' for the parasympathetic preganglionic neuron groups. We have not separately identified the lateral outlying parts of either of these nuclei, but it should be noted that some authors name these lateral outlying cells in the thoracic region as the funicular part of the IML.

We have attempted to resolve an area of nomenclatural confusion involving the central part of the lumbar and sacral intermediate gray area. The name 'dorsal commissural nucleus' has been applied by different groups of authors to two completely different nuclei, one lumbar and one sacral. We have named the former the lumbar dorsal commissural nucleus (LDCom) and the latter the sacral dorsal commissural nucleus (SDCom). See Chapters 6 and 8 for further discussion of these nuclei.

The motoneuron clusters in lamina 9 have been named for the muscles or muscle groups they supply. Chapter 7 gives a full account of the evidence for these designations. We have not attempted to subdivide these motonuclei into groups that supply individual muscles because there is considerable overlap and we felt that the drawings should be kept as clear as possible. Those who wish to identify the motoneurons supplying individual muscles should consult the references cited in Chapter 7.

We were not able to identify the majority of long tracts in these spinal cord sections. However, their presumed position can be located with reference to Chapters 10, 11, and 12. The tracts we were able to identify were the dorsolateral fasciculus, the gracile fasciculus, the cuneate fasciculus, the postsynaptic dorsal column pathway, the rubrospinal tract, and the dorsal corticospinal tract. See Chapters 10 and 11 for more detail on these tracts.

References

Franklin KBJ, Paxinos G (2007) *The Mouse Brain in Stereotaxic Coordinates.* 3rd edn., Academic Press, San Diego.

Paxinos G, Watson C (2007) *The Rat Brain in Stereotaxic Coordinates.* 6th edn., Academic Press, San Diego.

List of structures

Long name	Abbreviation
accessory nerve nucleus	11N
spinal accessory nerve	11n
lamina 1 of the spinal gray	1Sp
lamina 2 of the spinal gray	2Sp
lamina 2 of the spinal gray, inner part	2SpI
lamina 2 of the spinal gray, outer part	2SpO
lamina 3 of the spinal gray	3Sp
lamina 4 of the spinal gray	4Sp
lamina 5 of the spinal gray	5Sp
lamina 5 of the spinal gray, lateral part	5SpL
lamina 5 of the spinal gray, medial part	5SpM
lamina 6 of the spinal gray	6Sp
lamina 6 of the spinal gray, lateral part	6SpL
lamina 6 of the spinal gray, medial part	6SpM
lamina 7 of the spinal gray	7Sp
lamina 8 of the spinal gray	8Sp
lamina 9 of the spinal gray	9Sp
lamina 10 of the spinal gray	10Sp
adductor motoneurons of lamina 9	Ad9
anterior spinal artery	asp
axial muscle motoneurons of lamina 9	Ax9
biceps motoneurons of lamina 9	Bi9
central canal	CC
cauda equina	CE
central cervical nucleus	CeCv
crural extensor motoneurons of lamina 9	CEx9
crural flexor motoneurons of lamina 9	CFl9
coccygeal vertebra	CoVert
cremaster motoneurons of lamina 9	Cr9
cuneate fasciculus	cu
dorsal nucleus (Clarke)	D
dorsal corticospinal tract	dcs
deltoid motoneurons of lamina 9	De9
dorsal funiculus	df
dorsolateral fasciculus (Lissauer)	dl
dorsal root	dr
dorsal root ganglion	DRG
external anal sphincter motoneurons of lamina 9	ExA9
external urethral sphincter motoneurons of lamina 9	ExU9
forearm extensor motoneurons of lamina 9	FEx9
forearm flexor motoneurons of lamina 9	FFl9
gluteal motoneurons of lamina 9	Gl9
gracile fasciculus	gr
hamstring motoneurons of lamina 9	Hm9
internal basilar nucleus	IB
intercalated nucleus	ICl
intercalated nucleus, paraependymal part	IClPe
intercostal muscle motoneurons of lamina 9	ICo9
infrahyoid muscle motoneurons of lamina 9	IH9
intermediolateral column	IML
intermediomedial column	IMM
iliopsoas motoneurons of lamina 9	IPs9
lateral cervical nucleus	LatC
latissimus dorsi motoneurons of lamina 9	LD9
lumbar dorsal commissural nucleus	LDCom
lateral funiculus	lf
lumbar precerebellar nucleus	LPrCb
levator scapulae motoneurons of lamina 9	LS9
lateral spinal nucleus	LSp
manus motoneurons of lamina 9	Man9
pectoral muscle motoneurons of lamina 9	Pec9
pes motoneurons of lamina 9	Pes9
phrenic motoneurons of lamina 9	Ph9
psoas motoneurons of lamina 9	Ps9
postsynaptic dorsal column pathway	psdc
quadriceps motoneurons of lamina 9	Q9
quadratus lumborum motoneurons of lamina 9	QL9
rhomboid muscle motoneurons of lamina 9	Rh9
rubrospinal tract	rs
scalene muscle motoneurons of lamina 9	Sc9
sacral dorsal commissural nucleus	SDCom
supraspinatus and infraspinatus motoneurons of lamina 9	SI9
sternomastoid motoneurons of lamina 9	SM9
spinous process of vertebra	SpPr
sacral precerebellar nucleus	SPrCb
sacral parasympathetic nucleus	SPSy
serratus anterior motoneurons in lamina 9	Sr9
tail muscle motoneurons of lamina 9	Tail9
thoracoabdominal wall muscle motoneurons of lamina 9	ThAb9
triceps motoneurons of lamina 9	Tr9
trapezius motoneurons of lamina 9	Tz9
trapezius and sternomastoid motoneurons of lamina 9	TzSM9
vertebral arch	VAr
vertebral body	VB
ventral funiculus	vf
ventral median fissure	vmf
ventral root	vr
ventral white commissure	vwc

Permission to reproduce figures

Acknowledgements

We are very grateful to the Allen Brain Institute for collaborating with us by providing the Nissl/AChE set of sections depicted on the top right of each atlas page. We owe particular thanks to Dr Ralph Puchalski, the manager of the Allen spinal cord project. We are greatly indebted to Charles Scouten (Myneurolab.com, Leica Biosystems St. Louis LLC, 5918 Evergreen Blvd, St. Louis MO 63134) their donation of a CryoJane Tape Transfer System. This system enabled us to section the spinal cord with significantly less distortion than otherwise, and we recommend it to those who section brain or spinal cord. Financial support for the construction of *Atlas of the Mouse Spinal Cord* was provided by the Christopher and Dana Reeve Foundation.

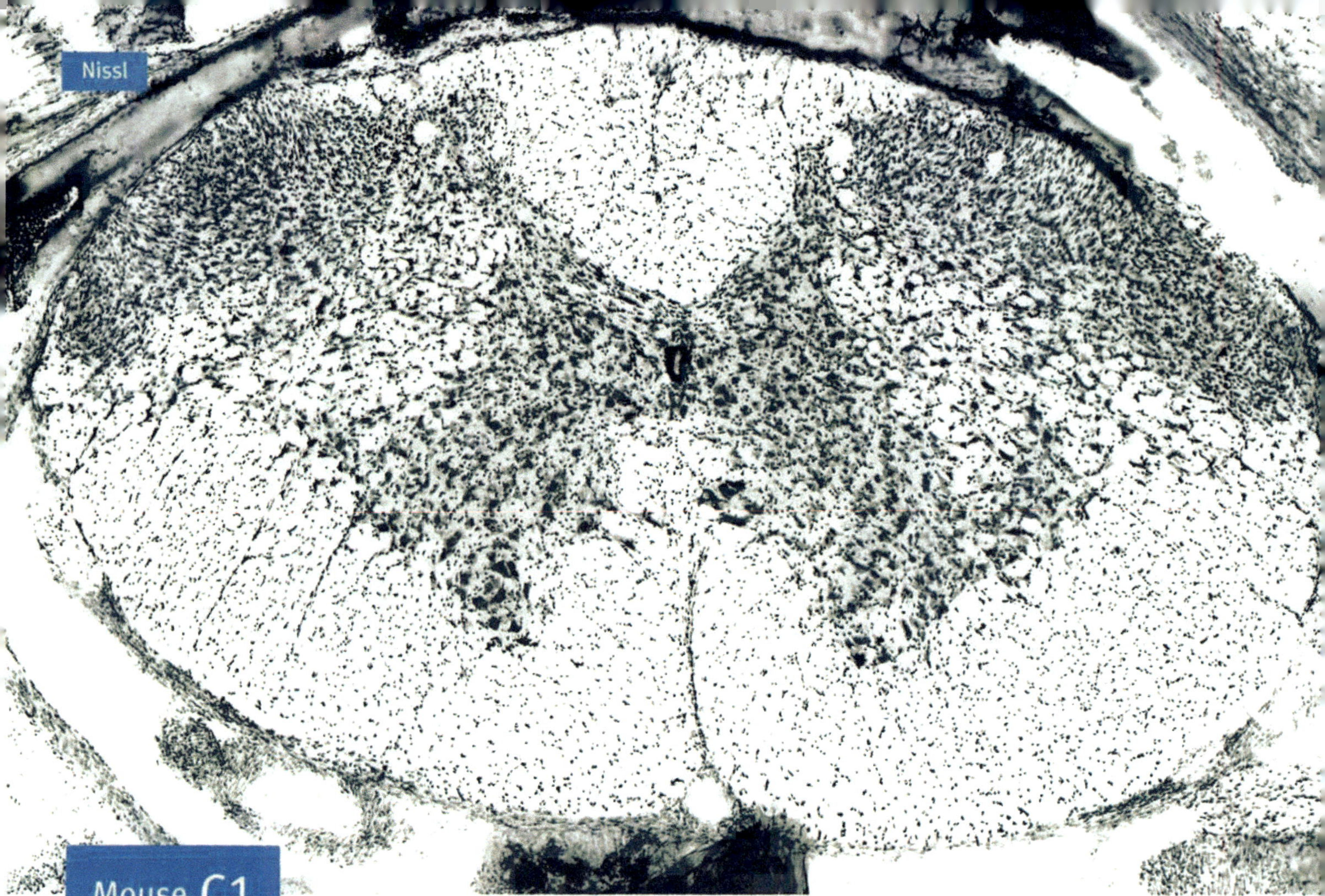
Nissl

Mouse C1

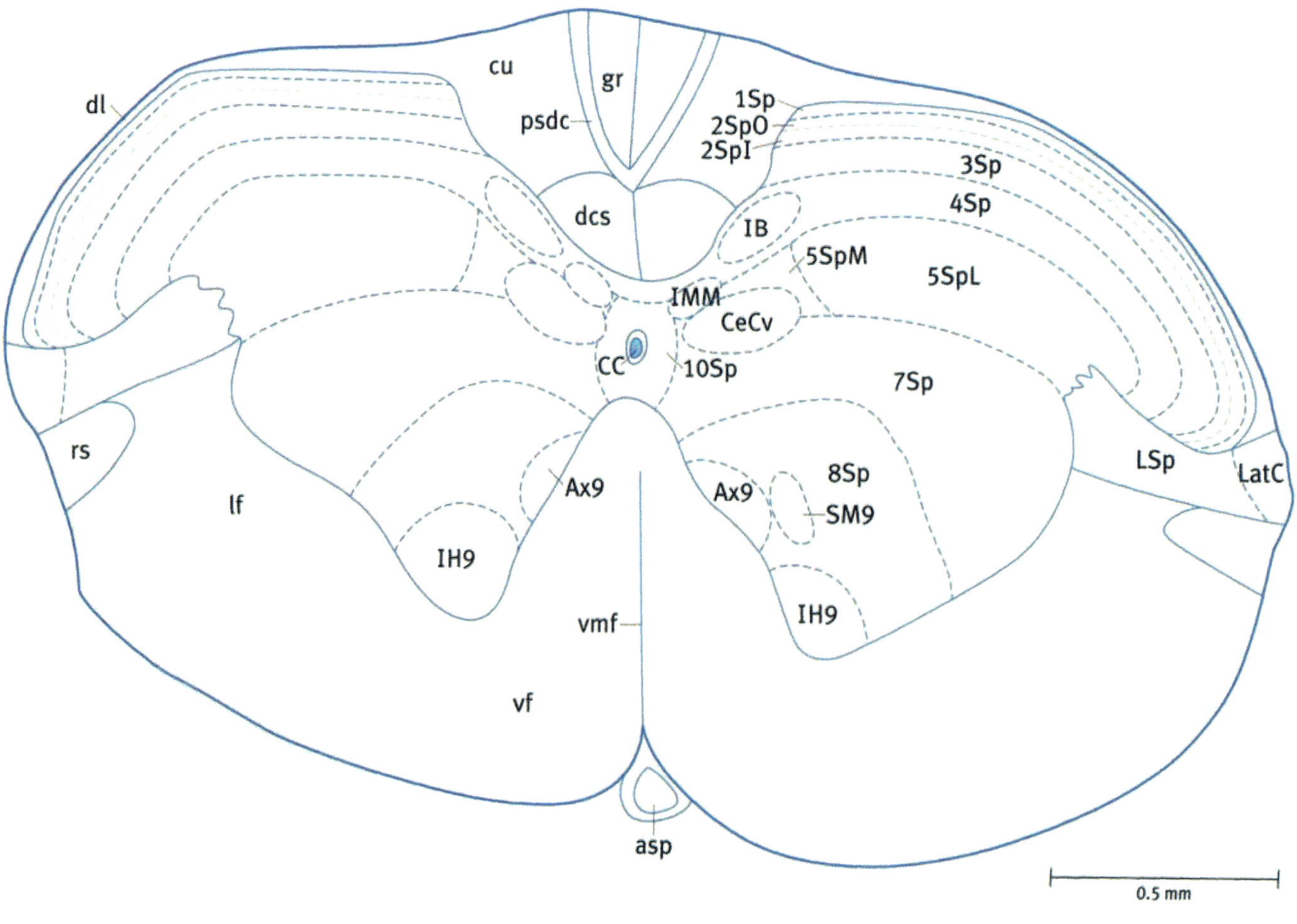
cu
gr
dl
psdc
1Sp
2SpO
2SpI
3Sp
4Sp
dcs
IB
5SpM
5SpL
IMM
CeCv
CC
10Sp
7Sp
rs
lf
Ax9
Ax9
8Sp
SM9
LSp
LatC
IH9
IH9
vmf
vf
asp
0.5 mm

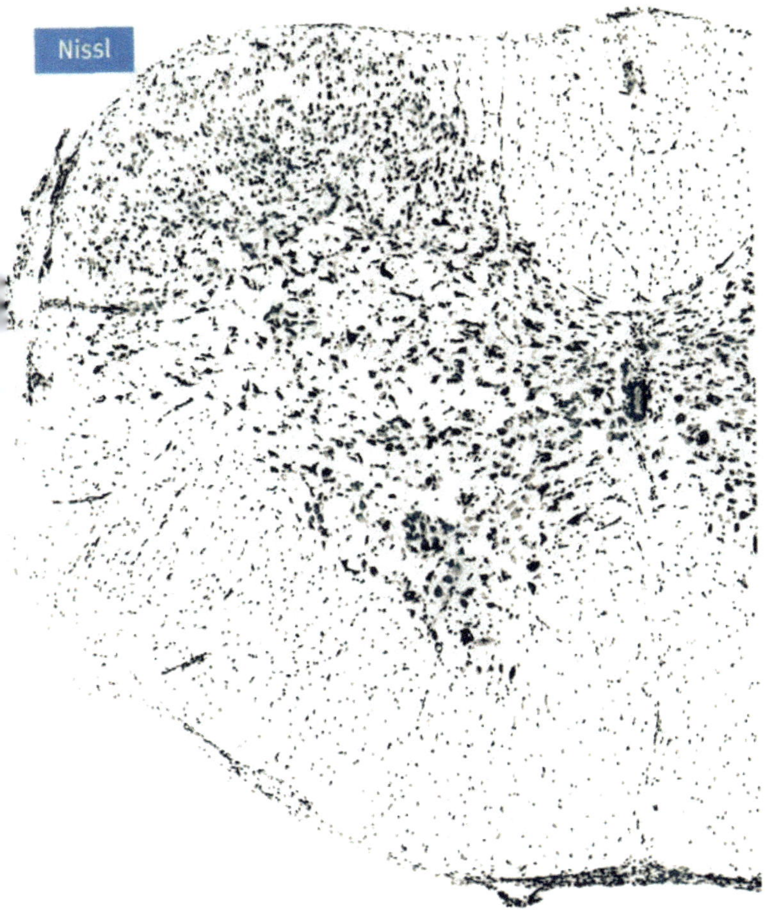

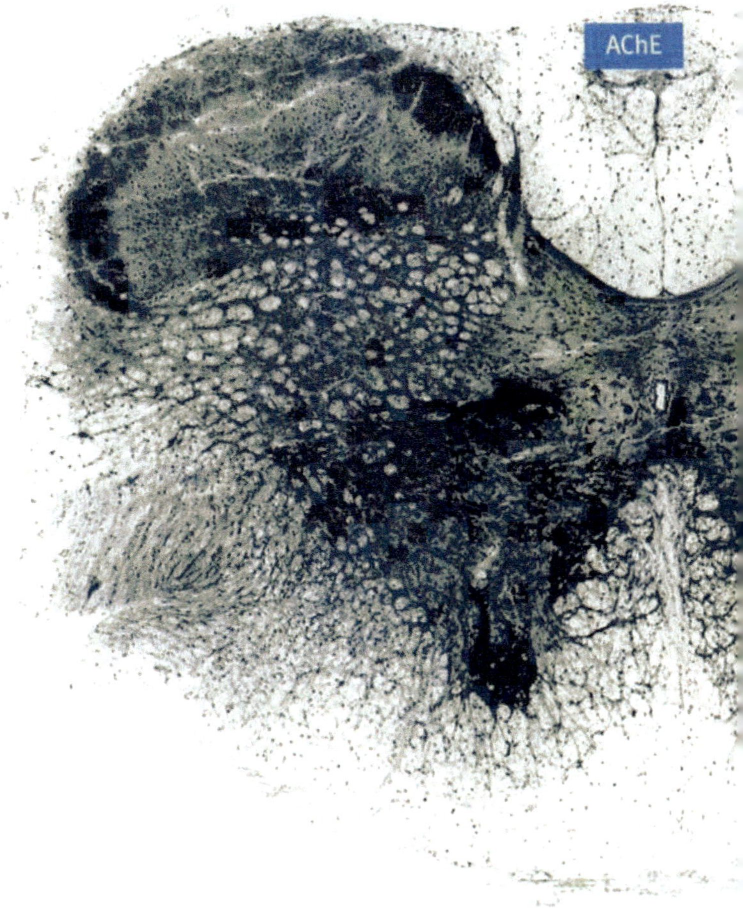

1Sp lamina 1
2SpI lamina 2, inner
2SpO lamina 2, outer
3Sp lamina 3
4Sp lamina 4
5SpL lamina 5, lat
5SpM lamina 5, med
7Sp lamina 7
8Sp lamina 8
10Sp lamina 10
asp ant spinal art
Ax9 axial muscles
CC central canal
CeCv cent cervical nu
cu cuneate fasc
dcs dors corticosp
dl dorsolat fasc
gr gracile fascic
IB int basal nu
IH9 infrahyoid musc
IMM intermediomed
LatC lat cervical nu
LSp lat spinal nu
psdc postsyn dors col
rs rubrospinal tr
SM9 sternomastoid
vmf vent med fiss

Mouse C1

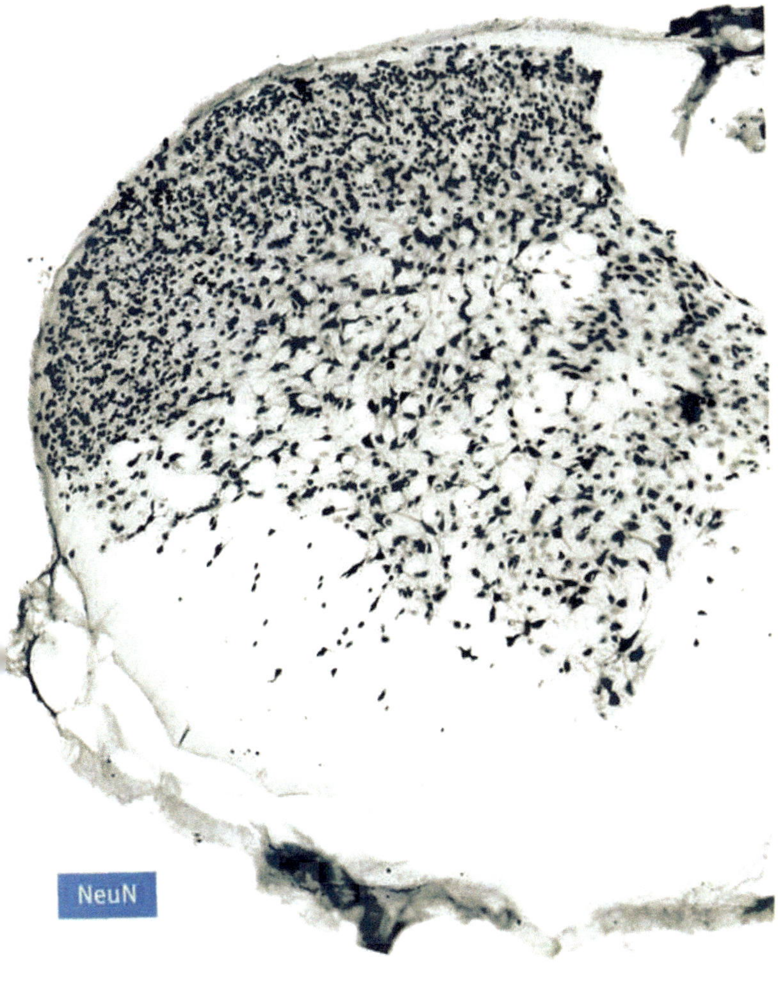

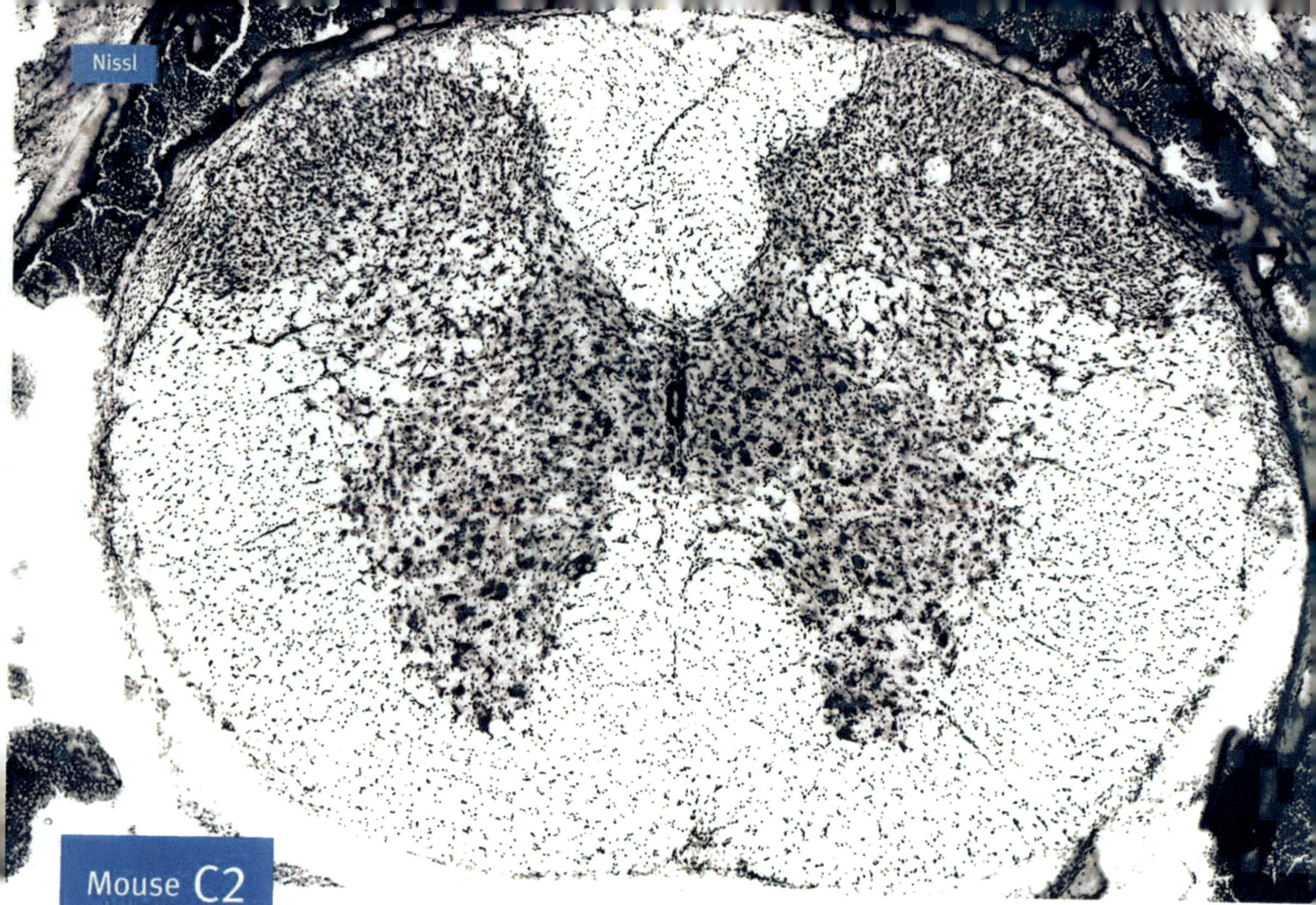
Nissl

Mouse C2

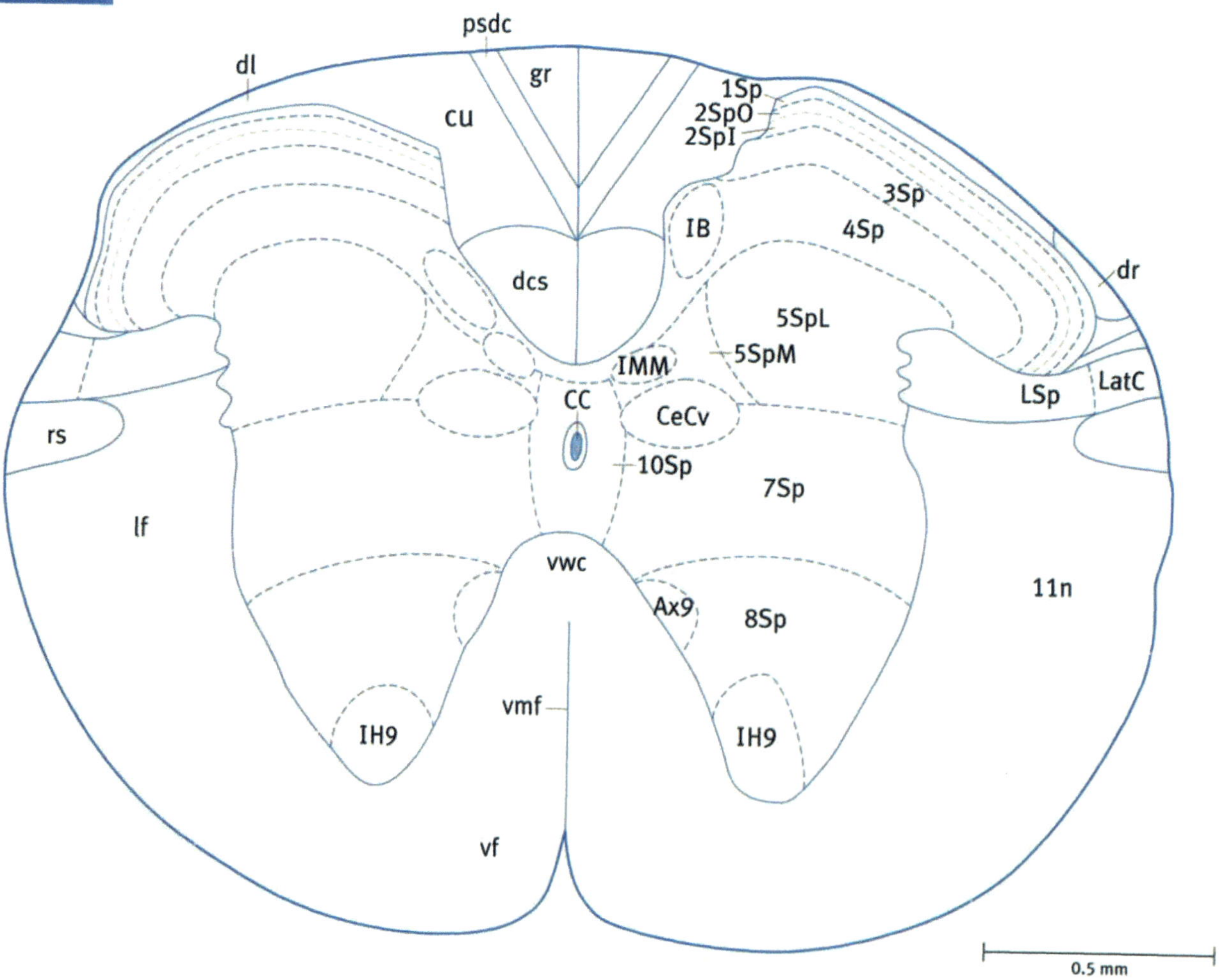
psdc
dl
gr
cu
1Sp
2SpO
2SpI
3Sp
IB
4Sp
dcs
dr
5SpL
5SpM
IMM
CC
LatC
LSp
CeCv
rs
10Sp
7Sp
lf
vwc
11n
Ax9
8Sp
vmf
IH9
IH9
vf
0.5 mm

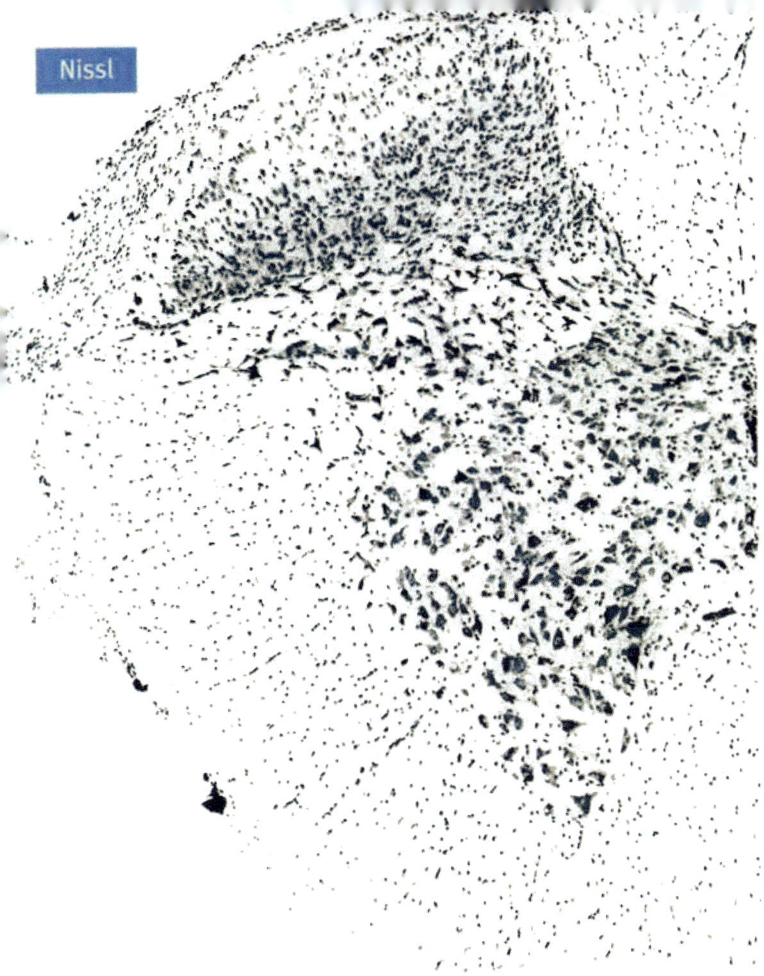

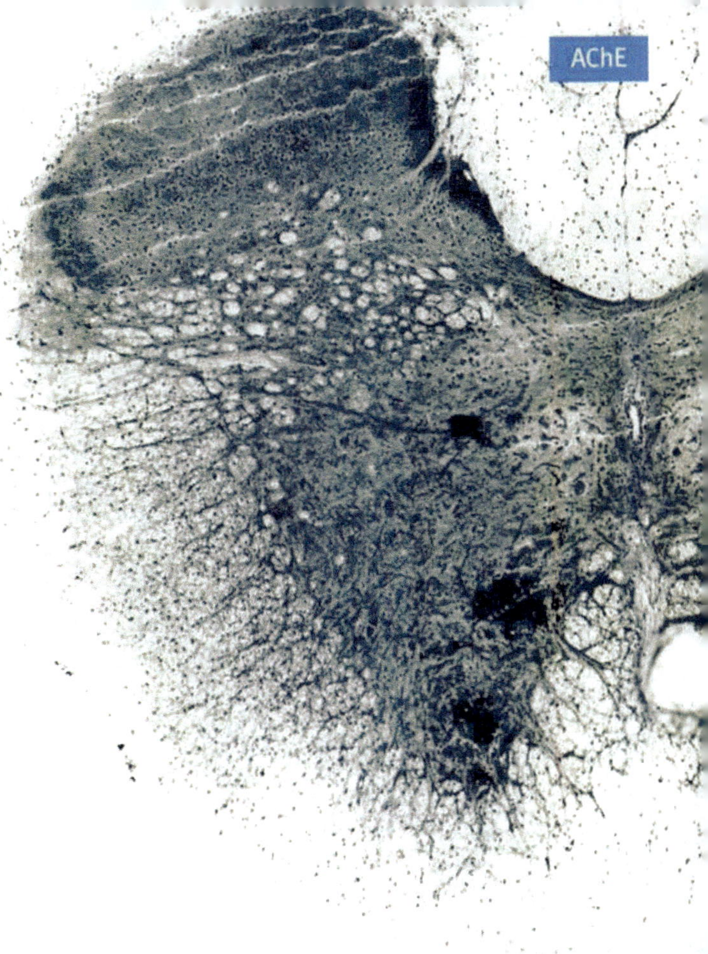

1Sp lamina 1
2SpI lamina 2, inner
2SpO lamina 2, outer
3Sp lamina 3
4Sp lamina 4
5SpL lamina 5, lat
5SpM lamina 5, med
7Sp lamina 7
8Sp lamina 8
10Sp lamina 10
11n spinal access n
Ax9 axial muscles
CC central canal
CeCv cent cervical nu
cu cuneate fasc
dcs dors corticosp
dl dorsolat fasc
dr dorsal root
gr gracile fascic
IB int basal nu
IH9 infrahyoid musc
IMM intermediomed
LatC lat cervical nu
LSp lat spinal nu
psdc postsyn dors col
rs rubrospinal tr
vmf vent med fiss
vwc vent white com

Mouse C2

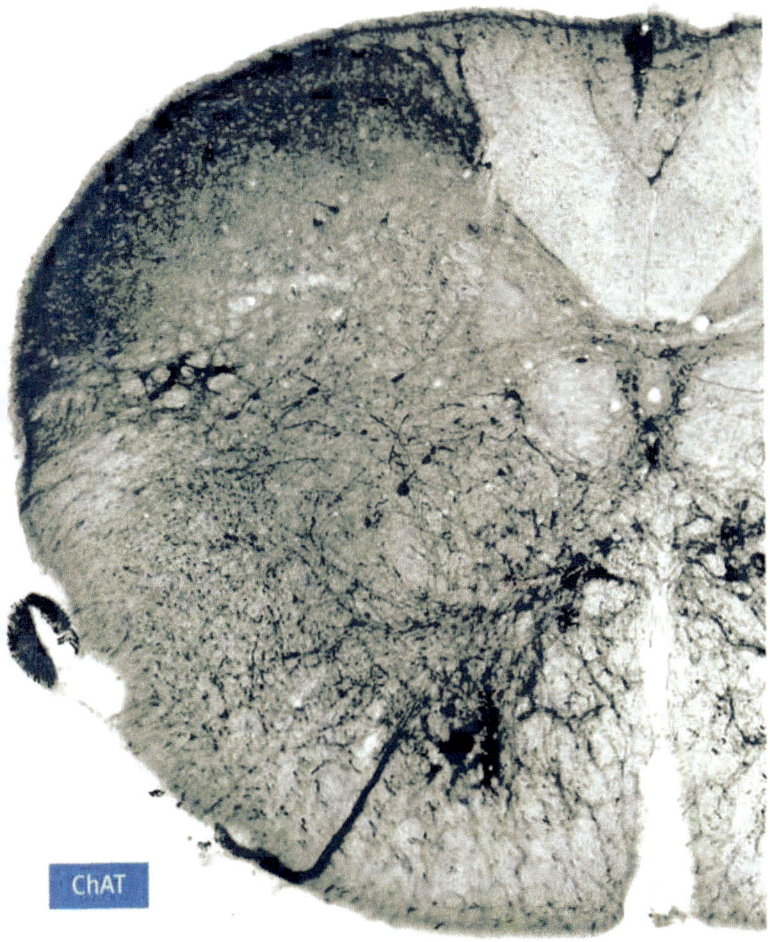

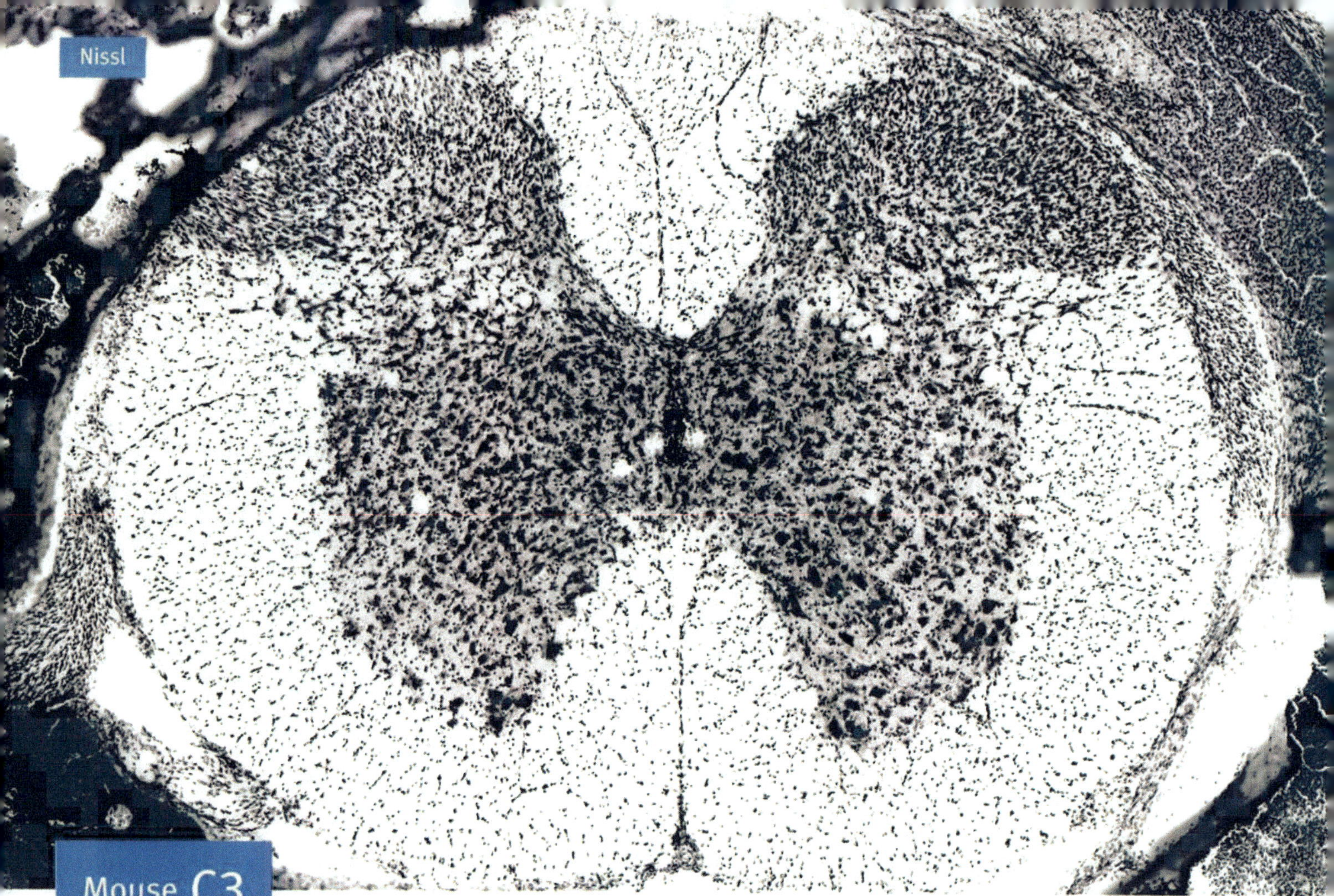
Nissl

Mouse C3

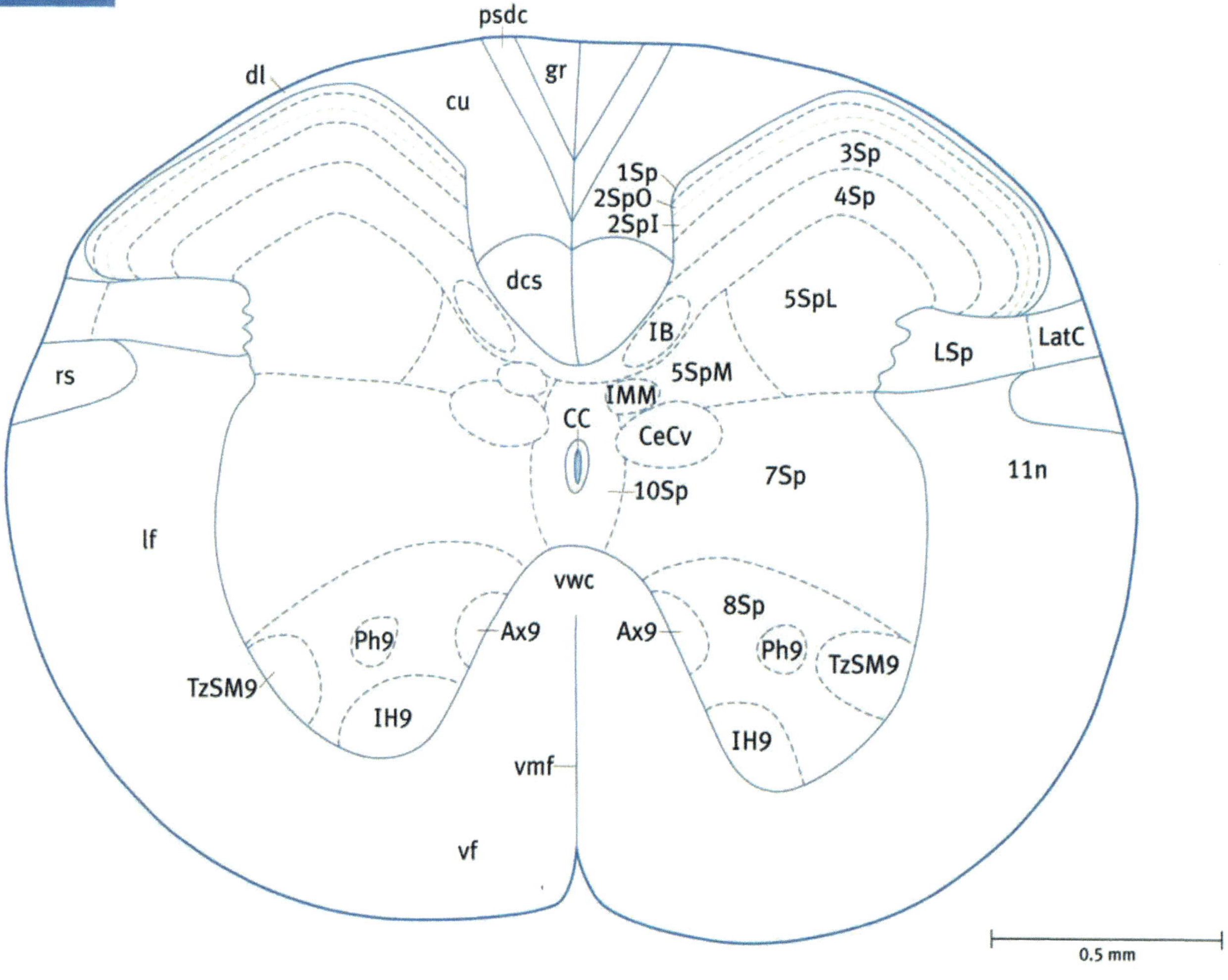
psdc
dl
cu
gr
1Sp
2SpO
2SpI
3Sp
4Sp
dcs
5SpL
IB
rs
LSp
LatC
5SpM
IMM
CC
CeCv
7Sp
11n
10Sp
lf
vwc
8Sp
Ph9
Ax9
Ax9
Ph9
TzSM9
TzSM9
IH9
IH9
vmf
vf
0.5 mm

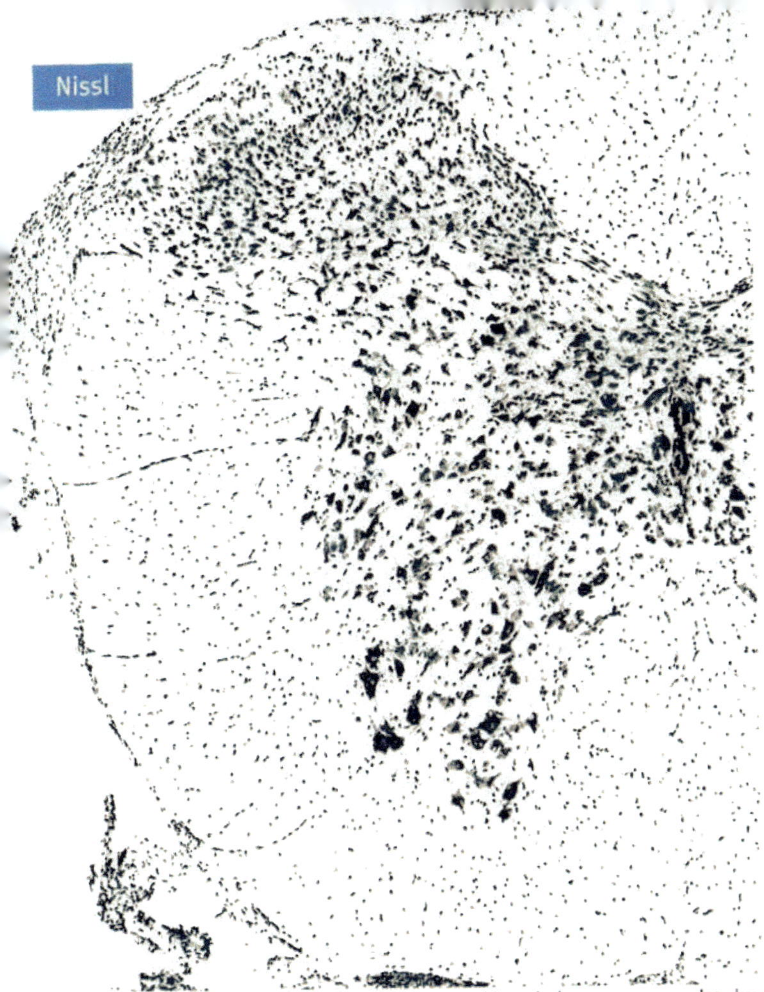

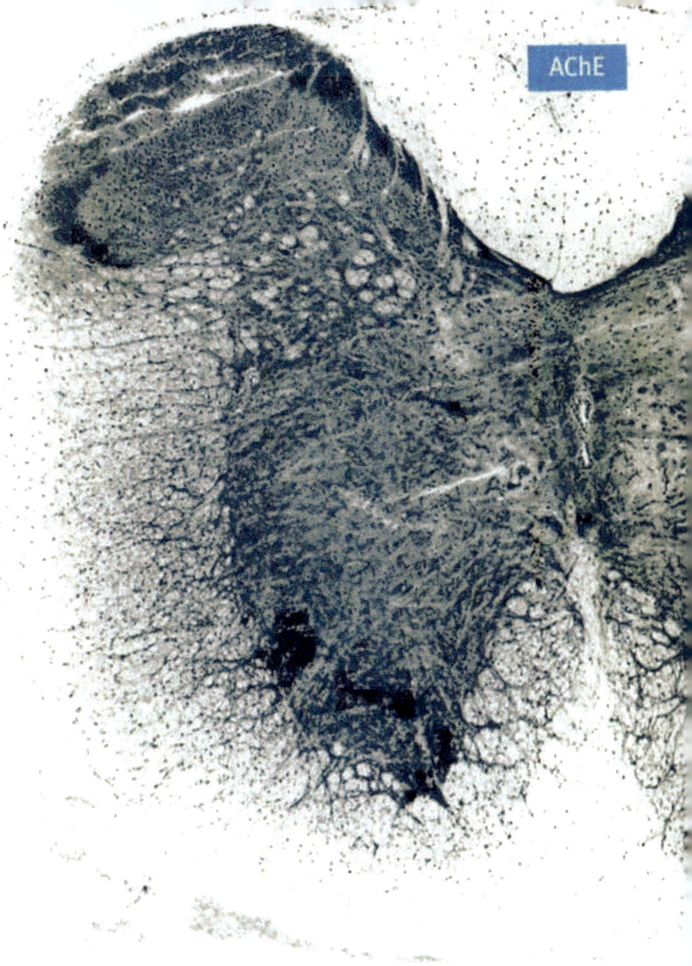

1Sp lamina 1
2SpI lamina 2, inner
2SpO lamina 2, outer
3Sp lamina 3
4Sp lamina 4
5SpL lamina 5, lat
5SpM lamina 5, med
7Sp lamina 7
8Sp lamina 8
10Sp lamina 10
11n spinal access n
Ax9 axial muscles
CC central canal
CeCv cent cervical nu
cu cuneate fasc
dcs dors corticosp
dl dorsolat fasc
gr gracile fascic
IB int basal nu
IH9 infrahyoid musc
IMM intermediomed
LatC lat cervical nu
LSp lat spinal nu
Ph9 phrenic
psdc postsyn dors col
rs rubrospinal tr
TzSM9 trap/sternom
vmf vent med fiss
vwc vent white com

Mouse C3

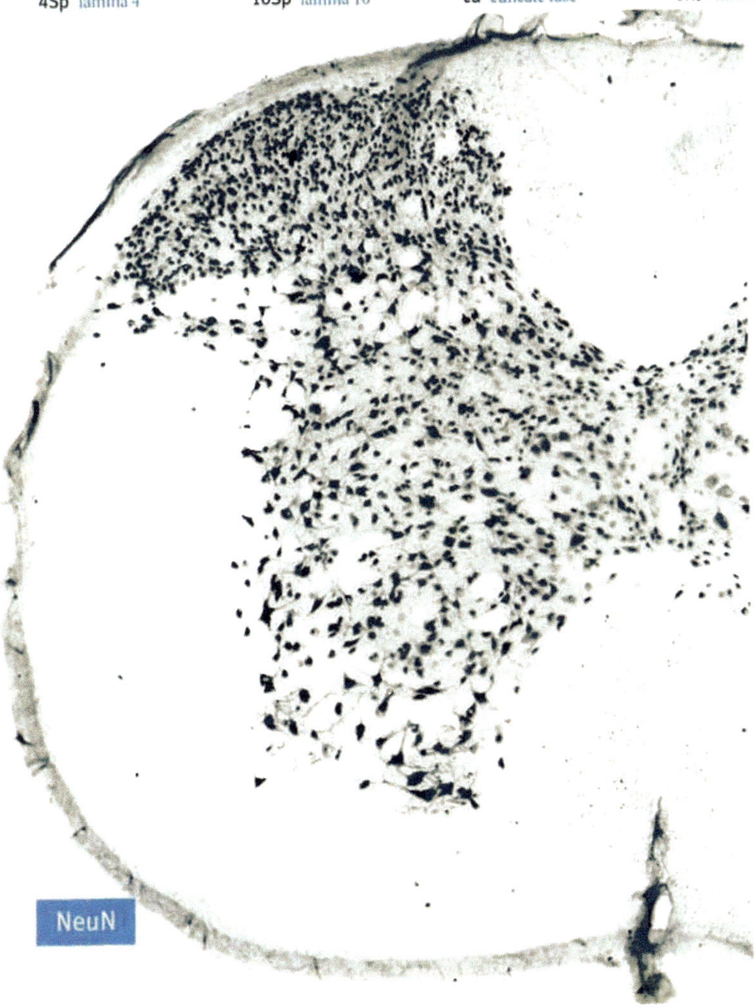

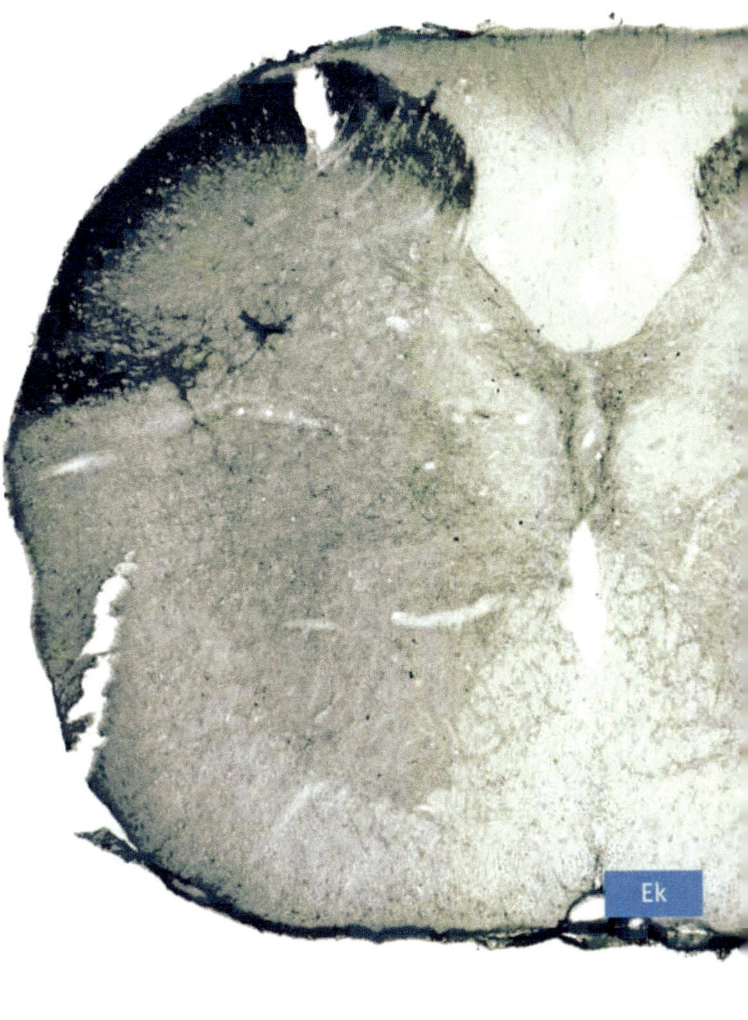

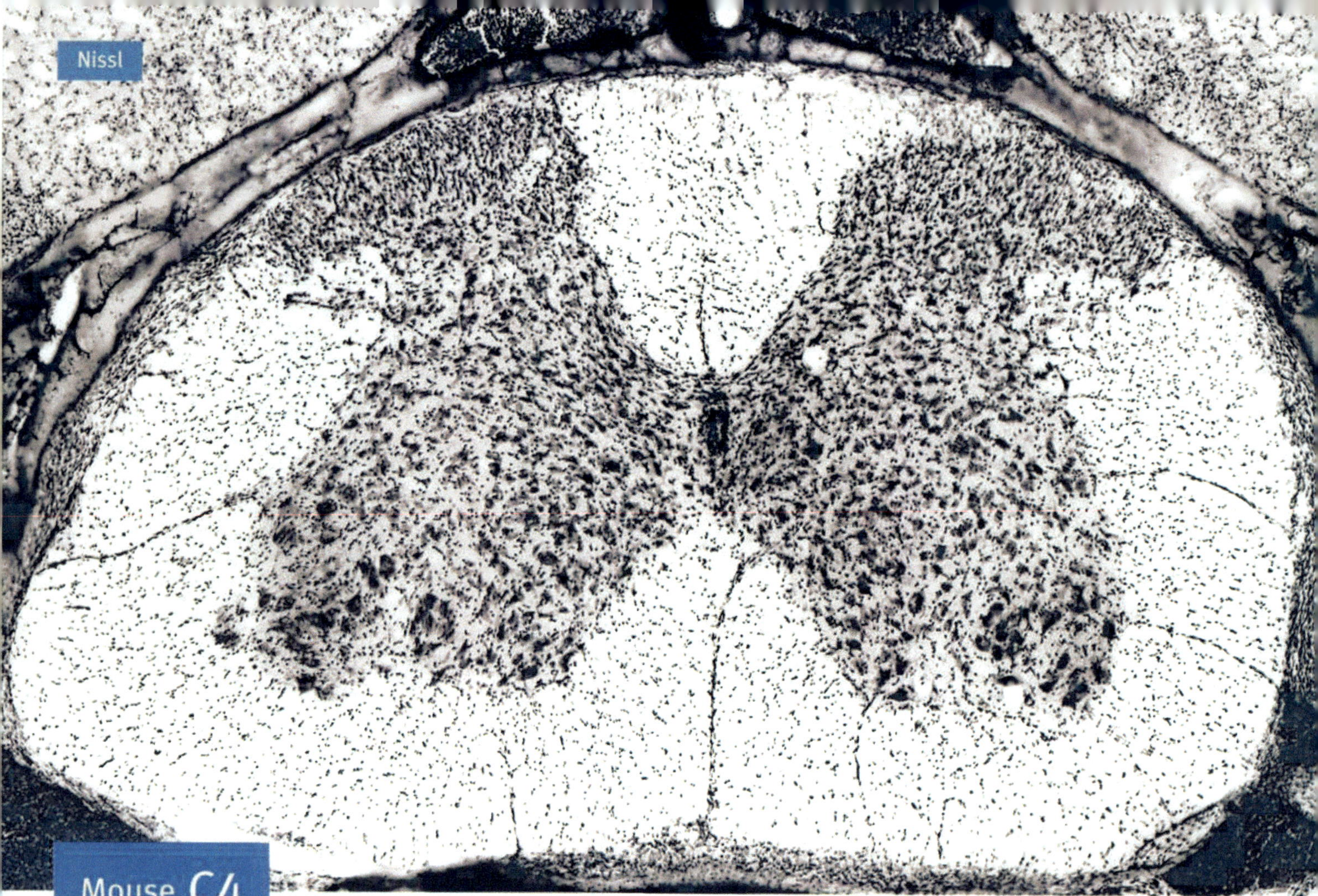
Nissl

Mouse C4

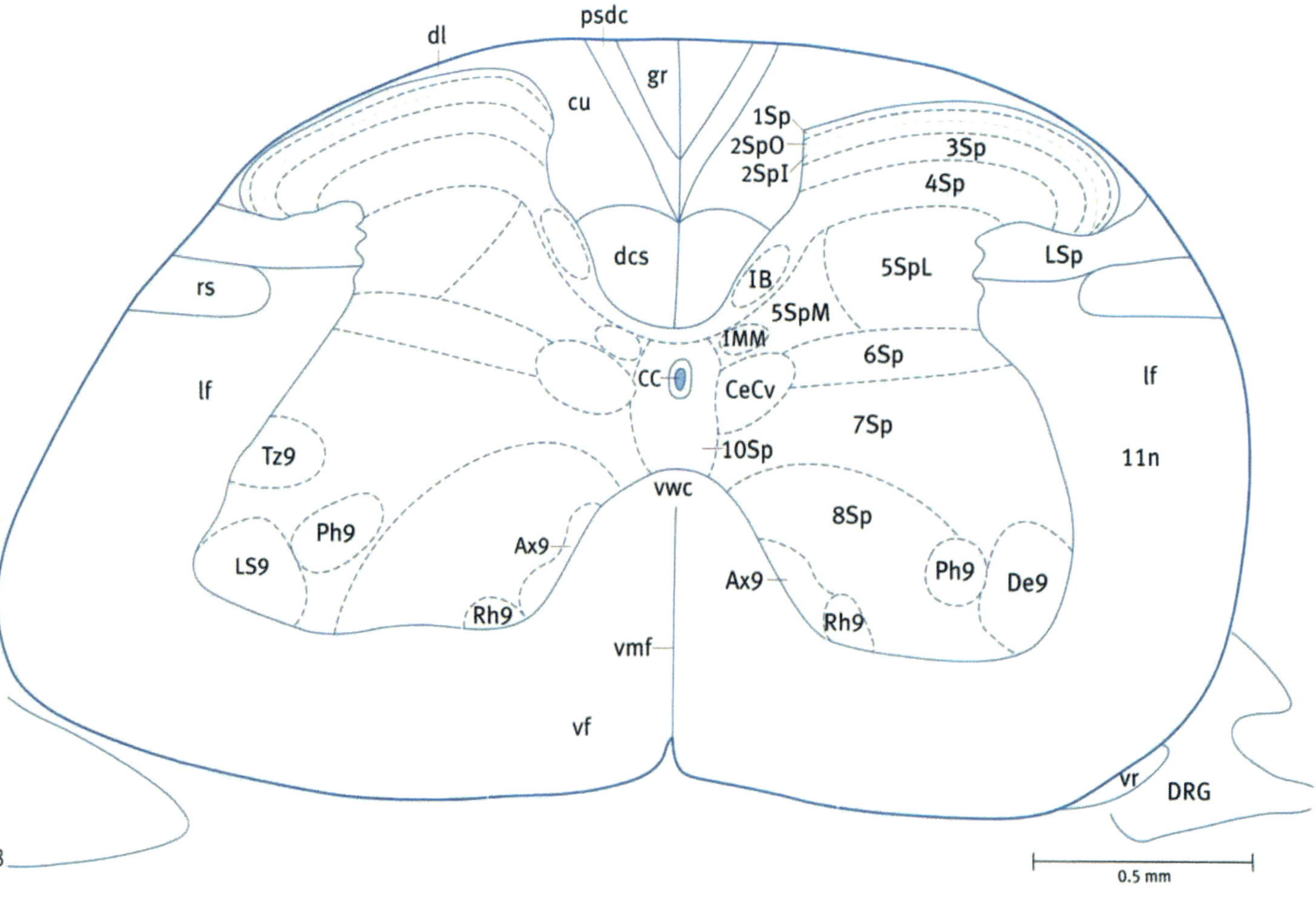
psdc
dl
gr
cu
1Sp
2SpO
2SpI
3Sp
4Sp
dcs
IB
5SpL
LSp
rs
5SpM
IMM
6Sp
CC
CeCv
lf
lf
7Sp
10Sp
Tz9
11n
vwc
8Sp
Ph9
Ax9
LS9
Ax9
Ph9
De9
Rh9
Rh9
vmf
vf
vr
DRG
0.5 mm

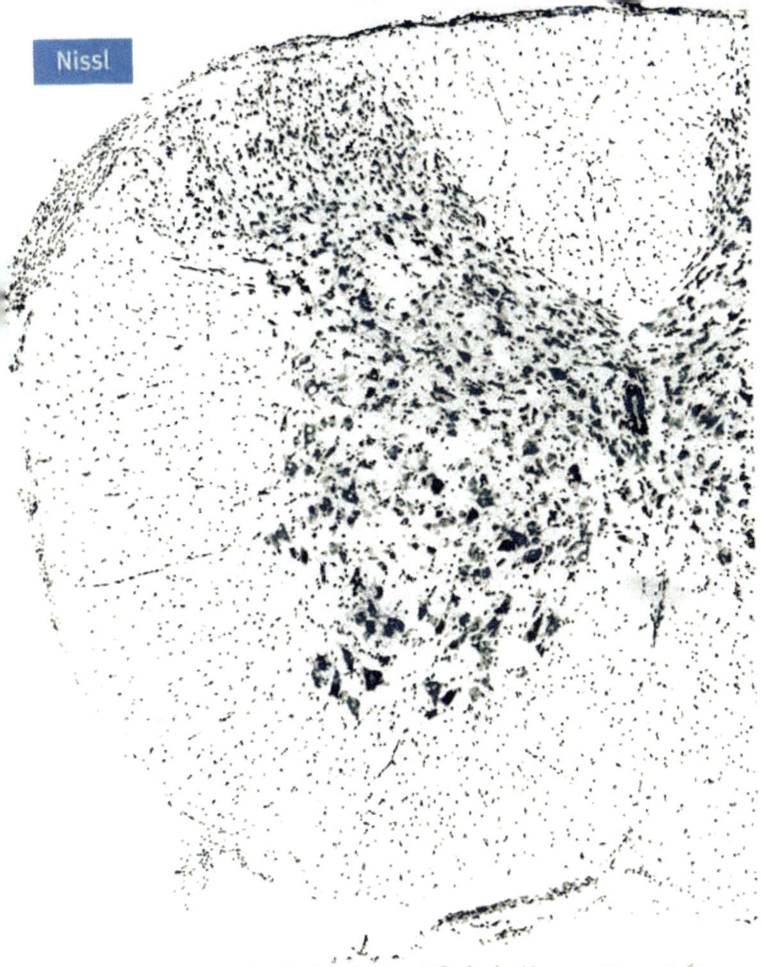

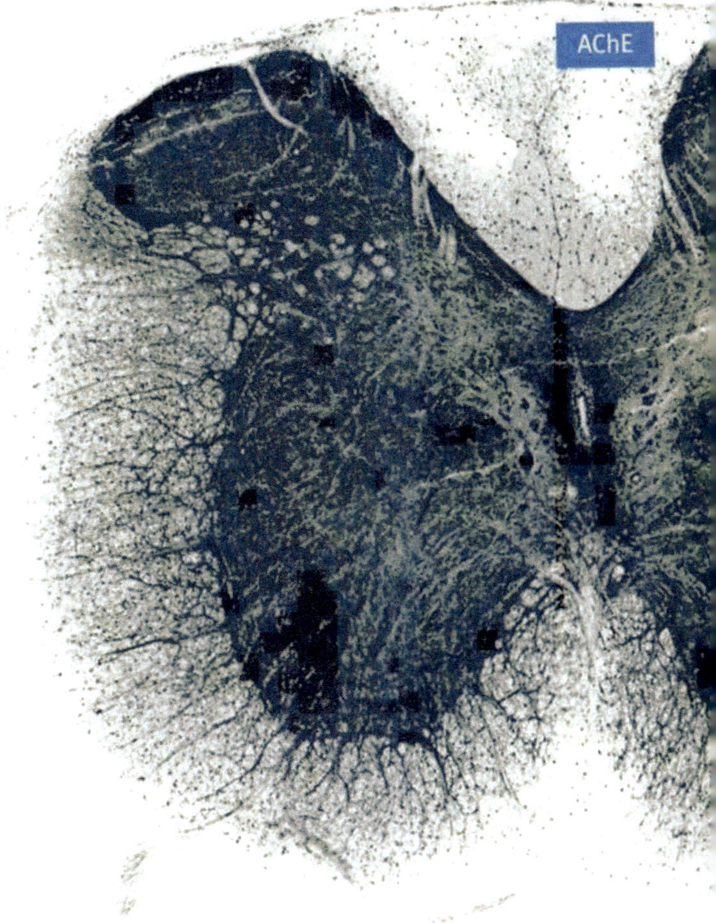

1Sp lamina 1
2SpI lamina 2, inner
2SpO lamina 2, outer
3Sp lamina 3
4Sp lamina 4
5SpL lamina 5, lat
5SpM lamina 5, med
6Sp lamina 6
7Sp lamina 7
8Sp lamina 8
10Sp lamina 10
11n spinal access n
Ax9 axial muscles
CC central canal
CeCv cent cervical nu
cu cuneate fasc
dcs dors corticosp
De9 deltoid
dl dorsolat fasc
gr gracile fascic
IB int basal nu
IMM intermediomed
LS9 levator scap
LSp lat spinal nu
Ph9 phrenic
psdc postsyn dors col
Rh9 rhomboid musc
rs rubrospinal tr
Tz9 trapezius
vmf vent med fiss
vr ventral root
vwc vent white com

Mouse C4

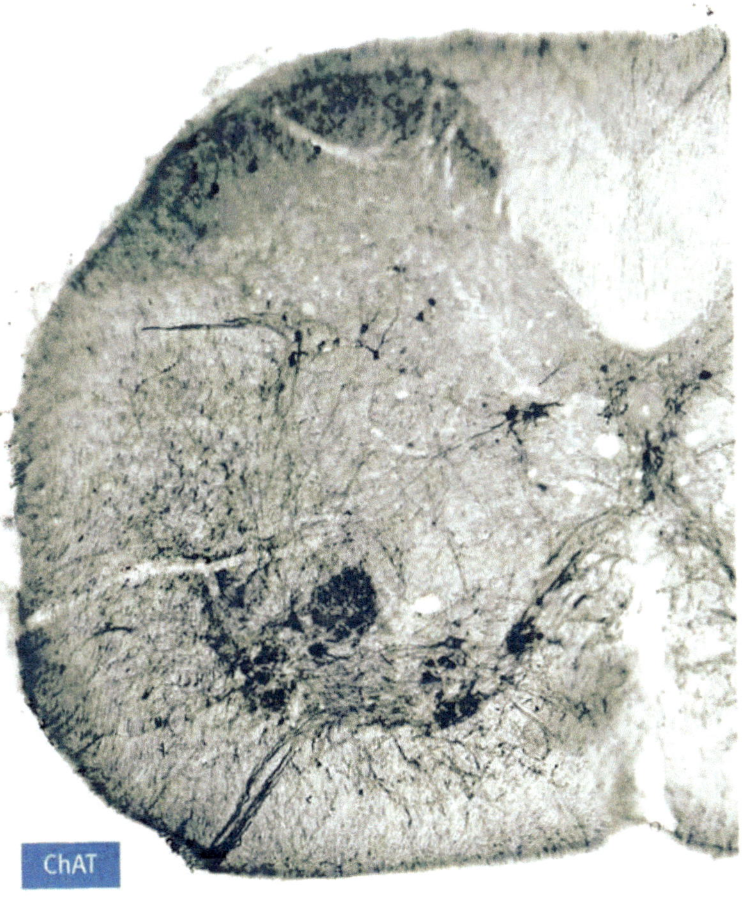

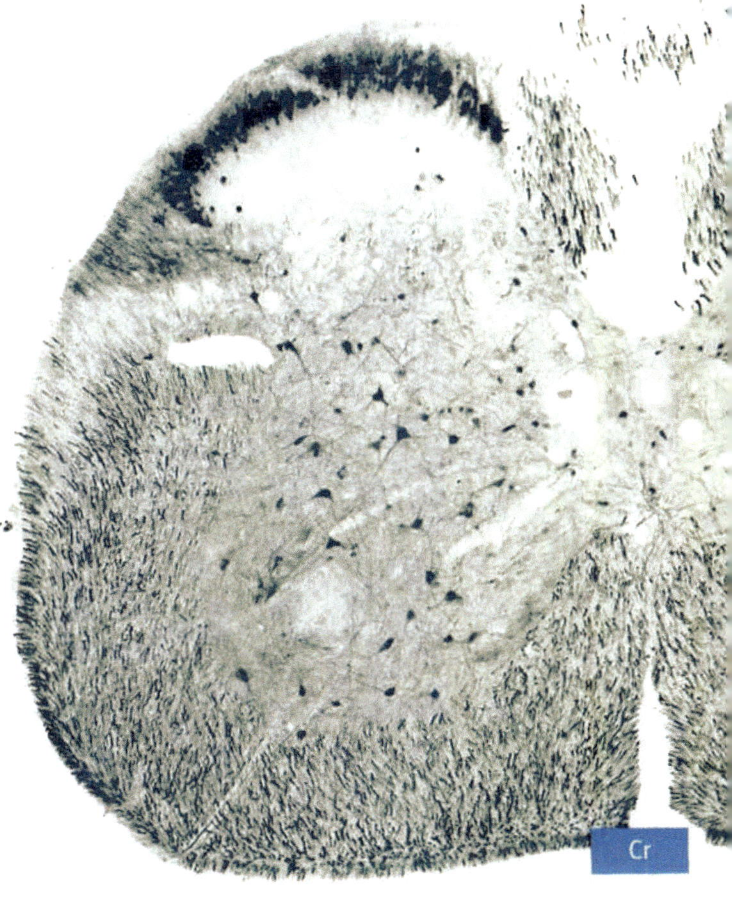

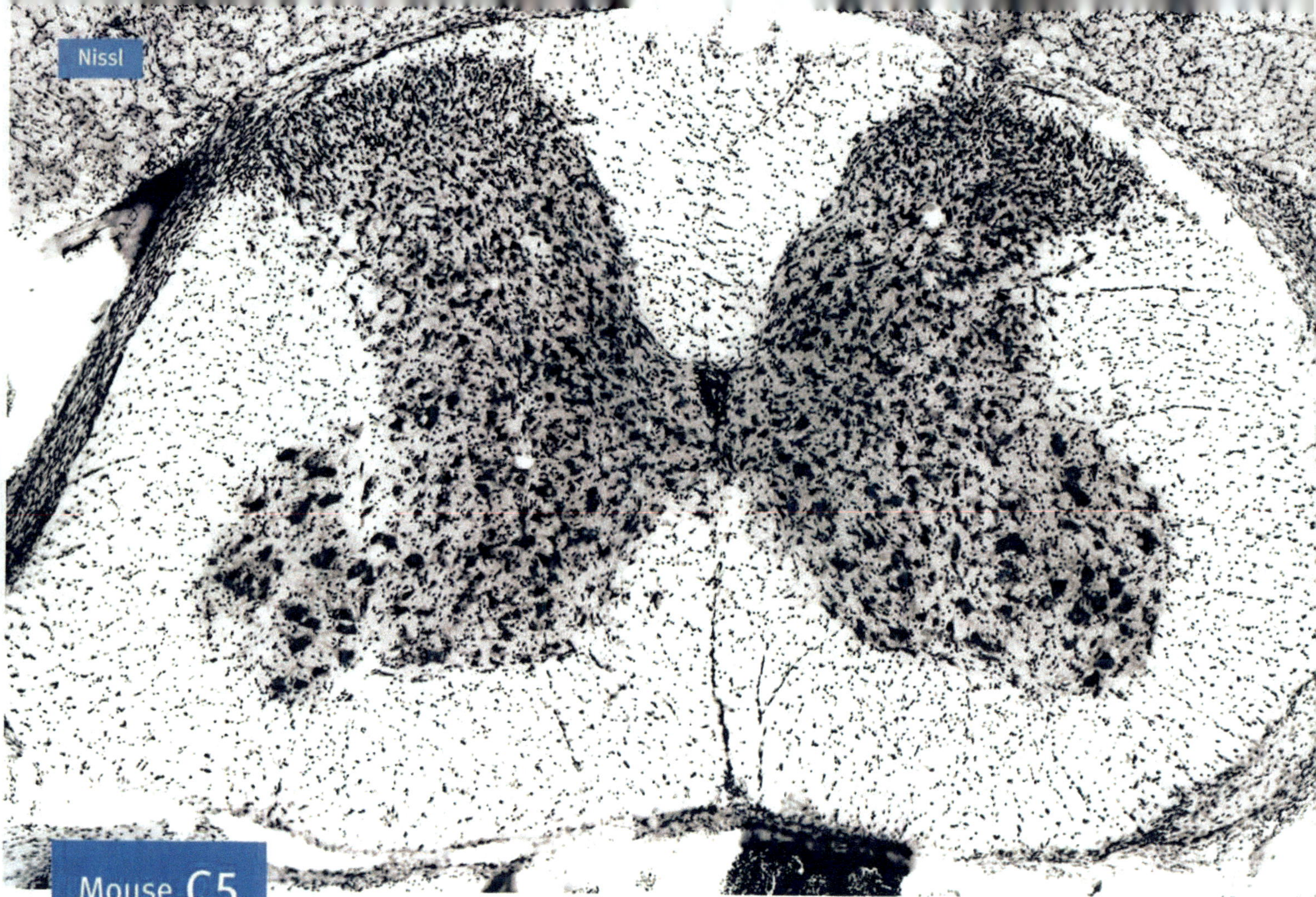
Nissl

Mouse C5

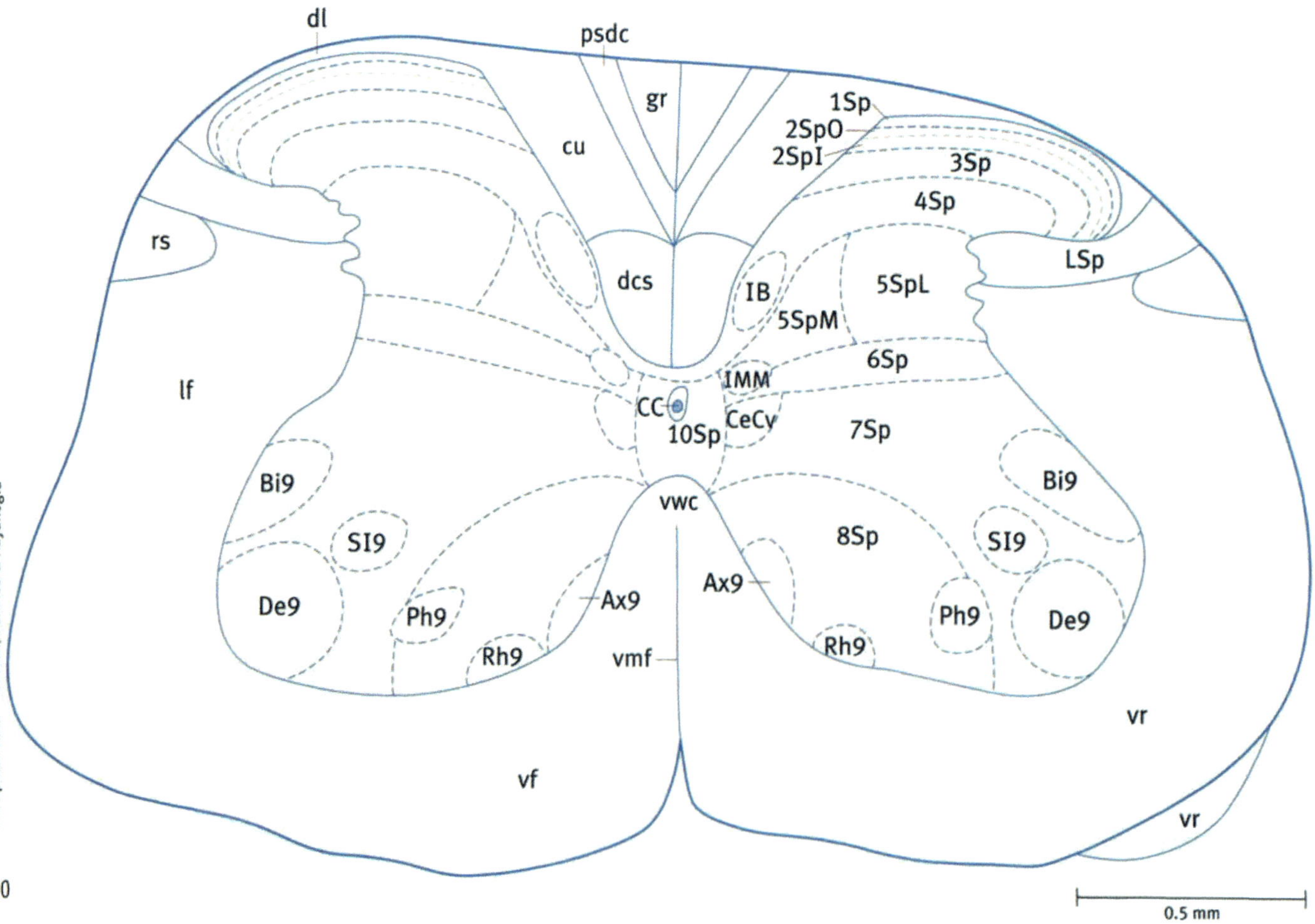
dl
psdc
gr
cu
1Sp
2SpO
2SpI
3Sp
4Sp
rs
LSp
dcs
IB
5SpL
5SpM
6Sp
IMM
lf
CC
CeCv
10Sp
7Sp
Bi9
Bi9
vwc
SI9
8Sp
SI9
Ax9
Ax9
De9
Ph9
Ph9
De9
Rh9
Rh9
vmf
vr
vf
vr
0.5 mm

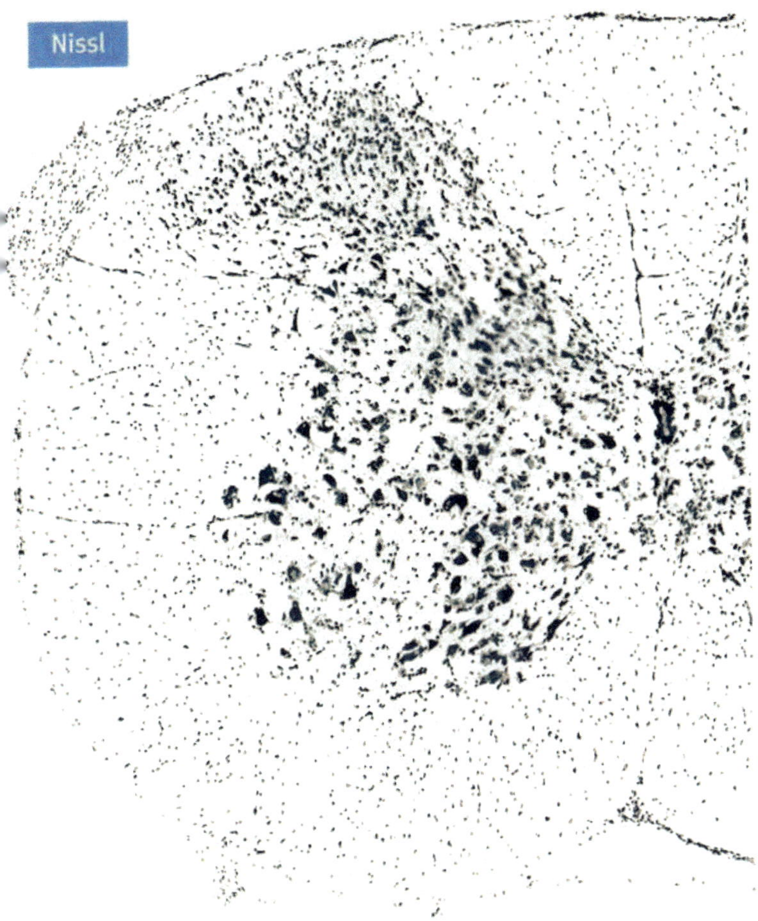

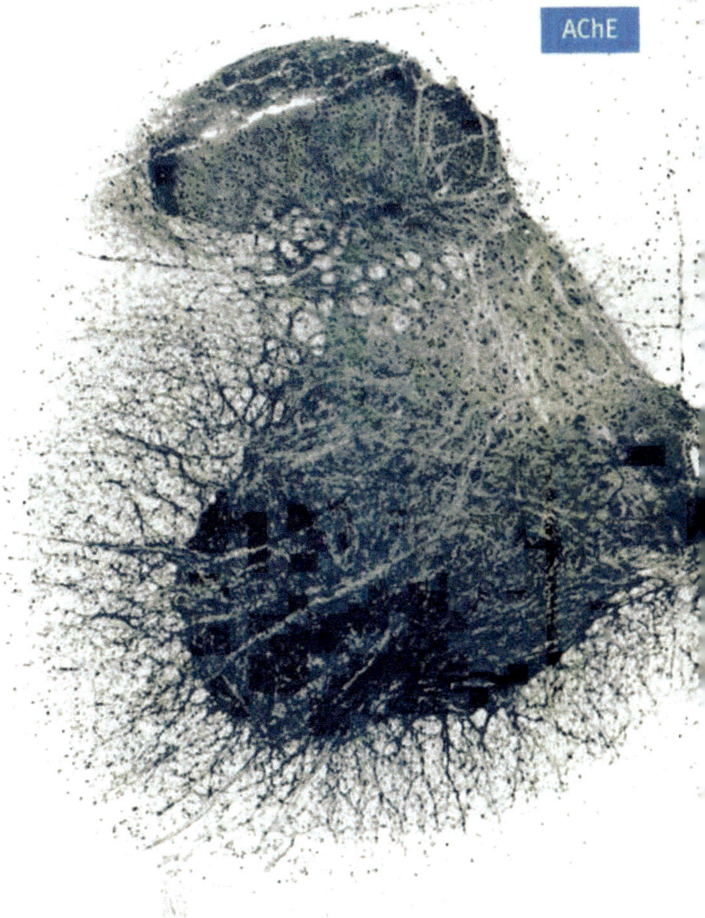

Mouse C5

1Sp lamina 1
2SpI lamina 2, inner
2SpO lamina 2, outer
3Sp lamina 3
4Sp lamina 4
5SpL lamina 5, lat
5SpM lamina 5, med
6Sp lamina 6
7Sp lamina 7
8Sp lamina 8
10Sp lamina 10
Ax9 axial muscles
Bi9 biceps
CC central canal
CeCv cent cervical nu
cu cuneate fasc
dcs dors corticosp
De9 deltoid
dl dorsolat fasc
gr gracile fascic
IB int basal nu
IMM intermediomed
LSp lat spinal nu
Ph9 phrenic
psdc postsyn dors col
Rh9 rhomboid musc
rs rubrospinal tr
SI9 supraspinatus
vmf vent med fiss
vr ventral root
vwc vent white com

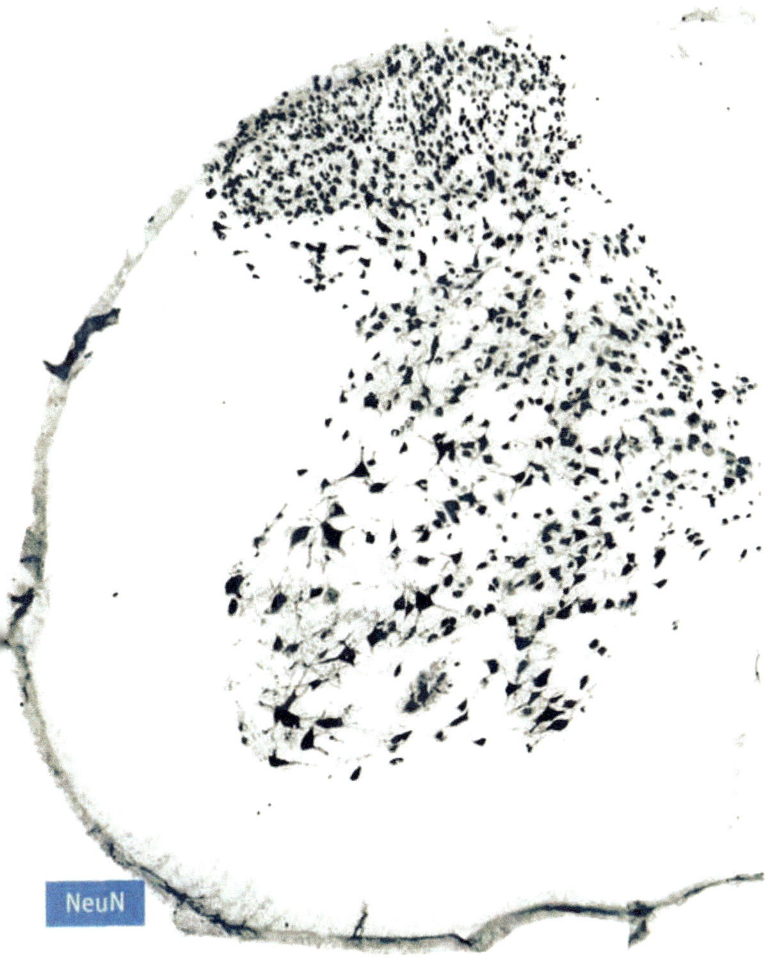

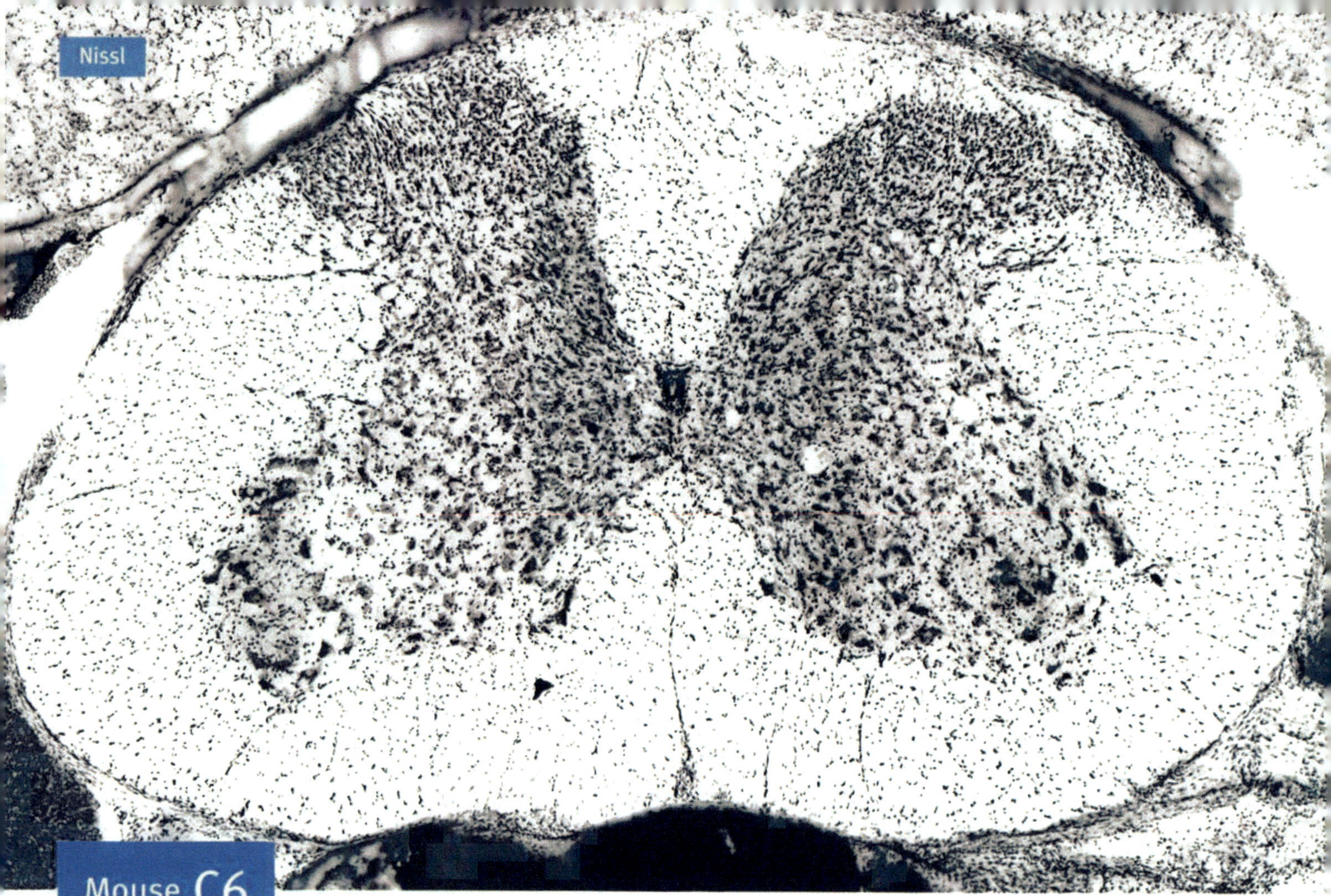
Nissl

Mouse C6

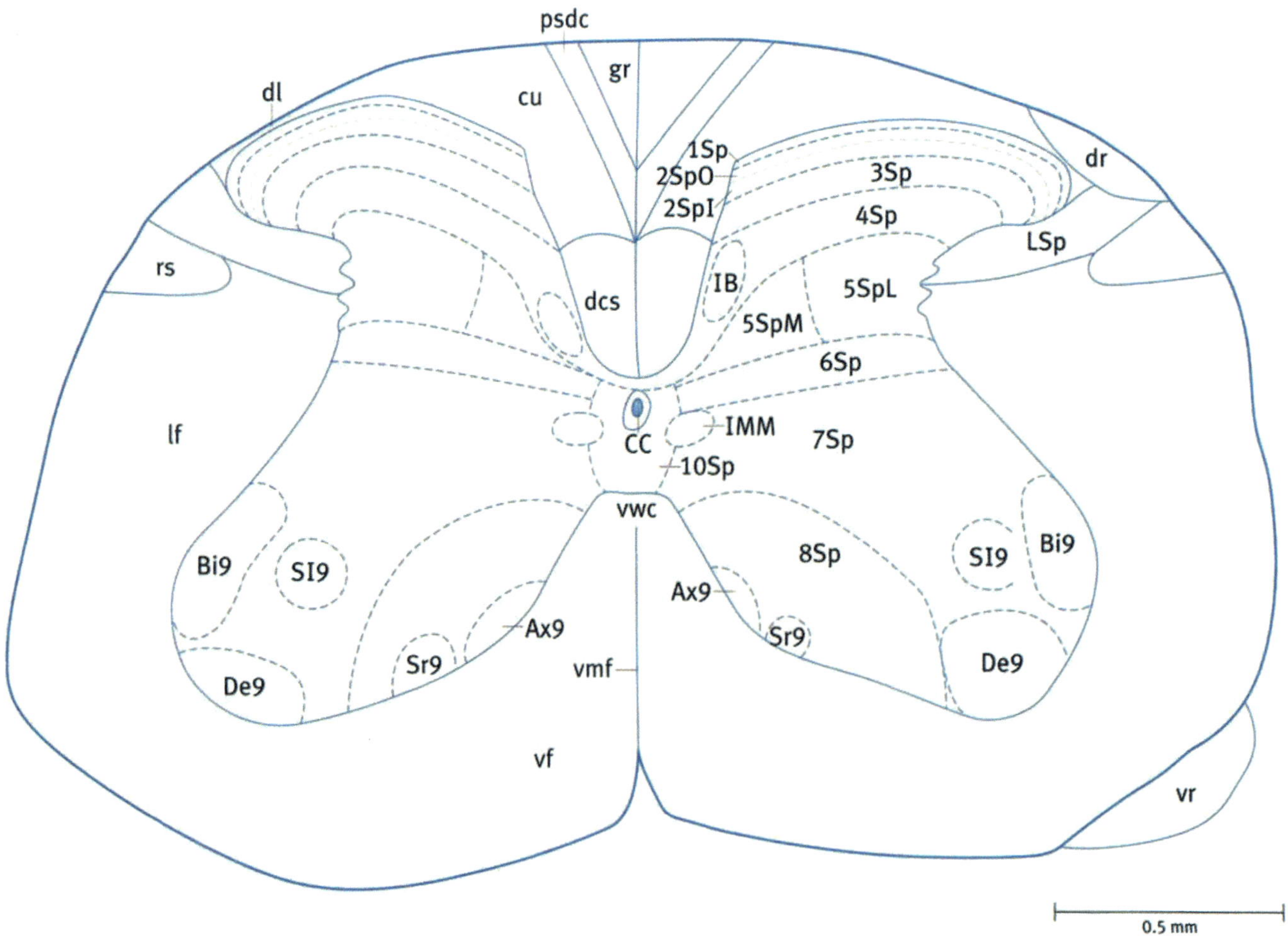
psdc
gr
cu
dl
1Sp
2SpO
2SpI
3Sp
dr
4Sp
LSp
rs
IB
dcs
5SpL
5SpM
6Sp
lf
IMM
CC
7Sp
10Sp
vwc
Bi9
SI9
8Sp
SI9
Bi9
Ax9
Ax9
Sr9
Sr9
vmf
De9
De9
vf
vr
0.5 mm

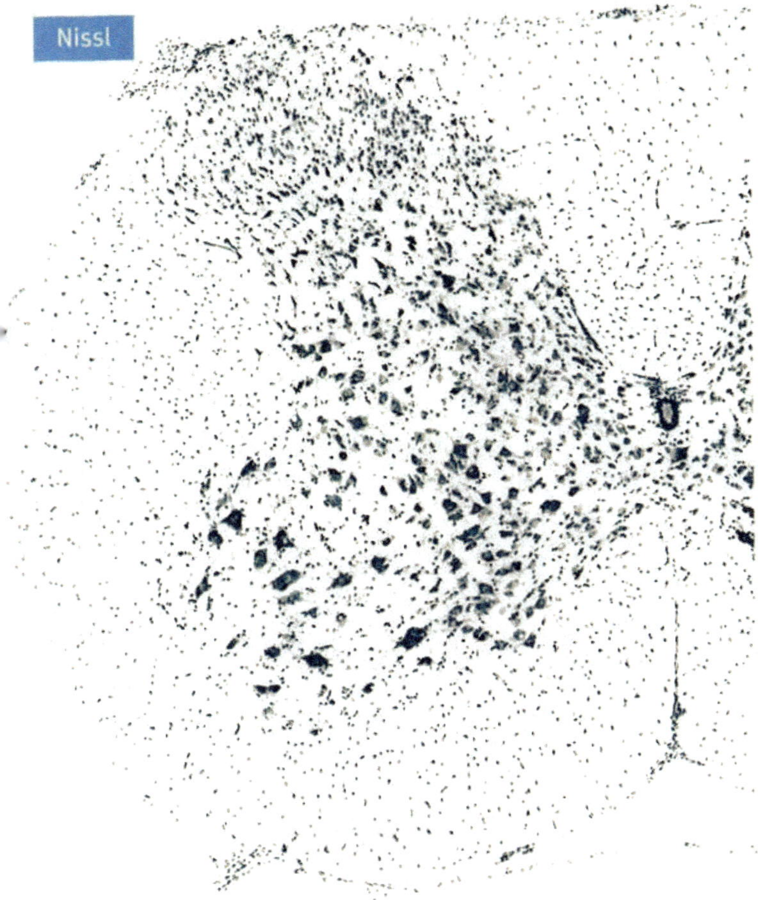

1Sp lamina 1
2SpI lamina 2, inner
2SpO lamina 2, outer
3Sp lamina 3
4Sp lamina 4
5SpL lamina 5, lat
5SpM lamina 5, med
6Sp lamina 6
7Sp lamina 7
8Sp lamina 8
10Sp lamina 10
Ax9 axial muscles
Bi9 biceps
CC central canal
cu cuneate fasc
dcs dors corticosp
De9 deltoid
dl dorsolat fasc
dr dorsal root
gr gracile fascic
IB int basal nu
IMM intermediomed
LSp lat spinal nu
psdc postsyn dors col
rs rubrospinal tr
SI9 supraspinatus
Sr9 serratus ant
vmf vent med fiss
vr ventral root
vwc vent white com

Mouse C6

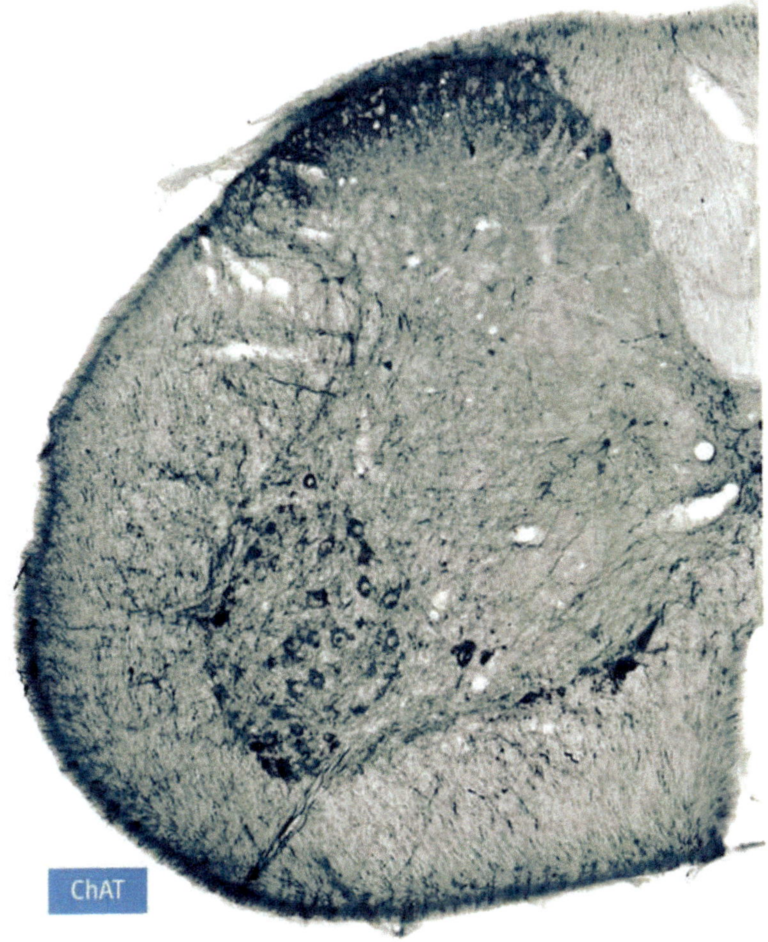

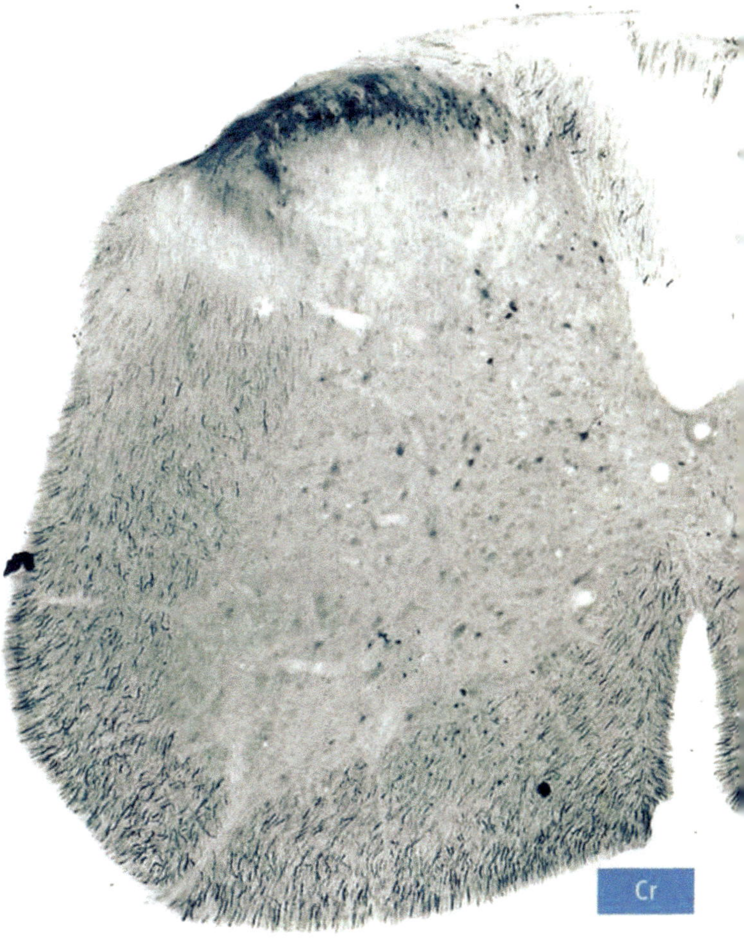

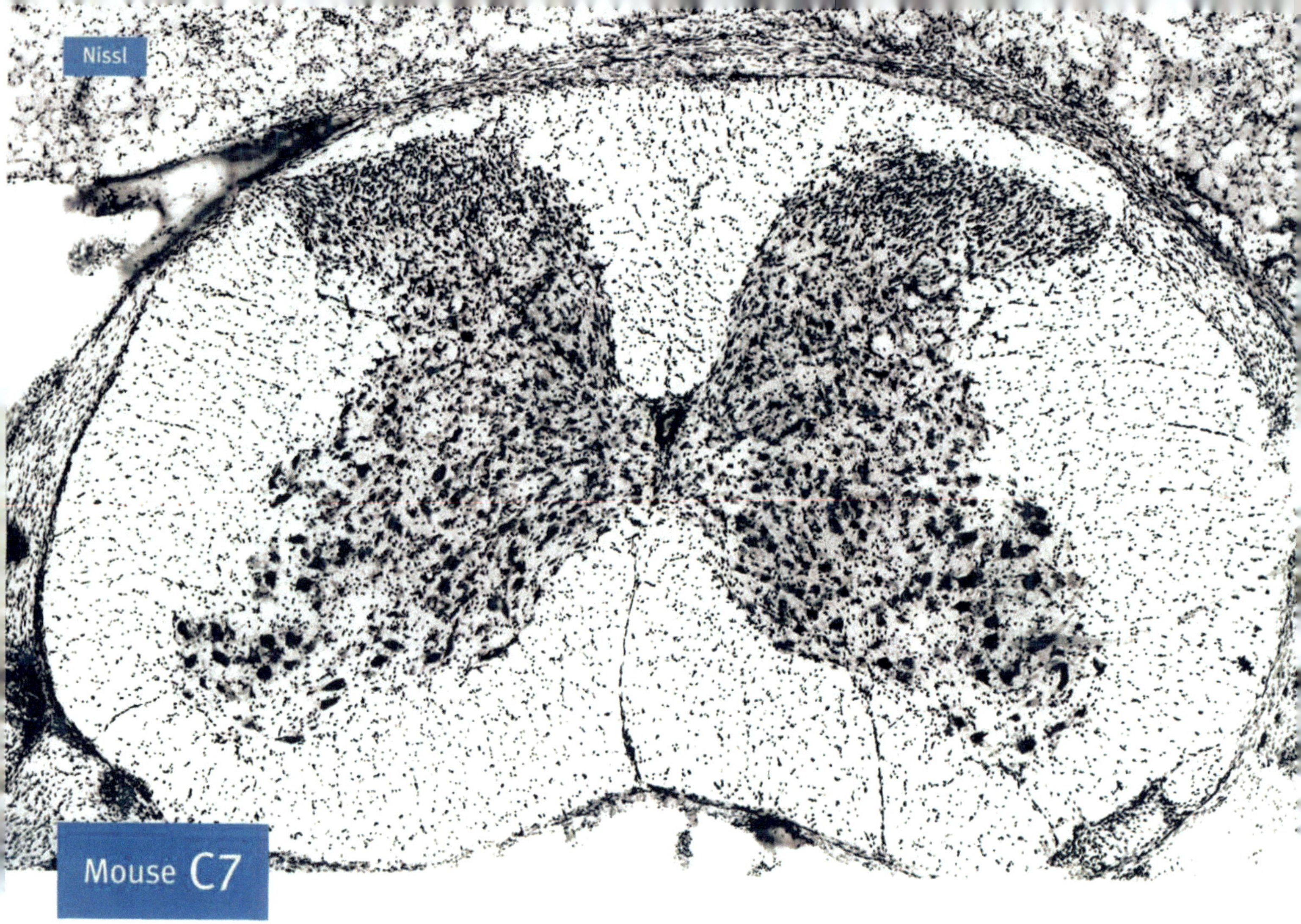
Nissl
Mouse C7

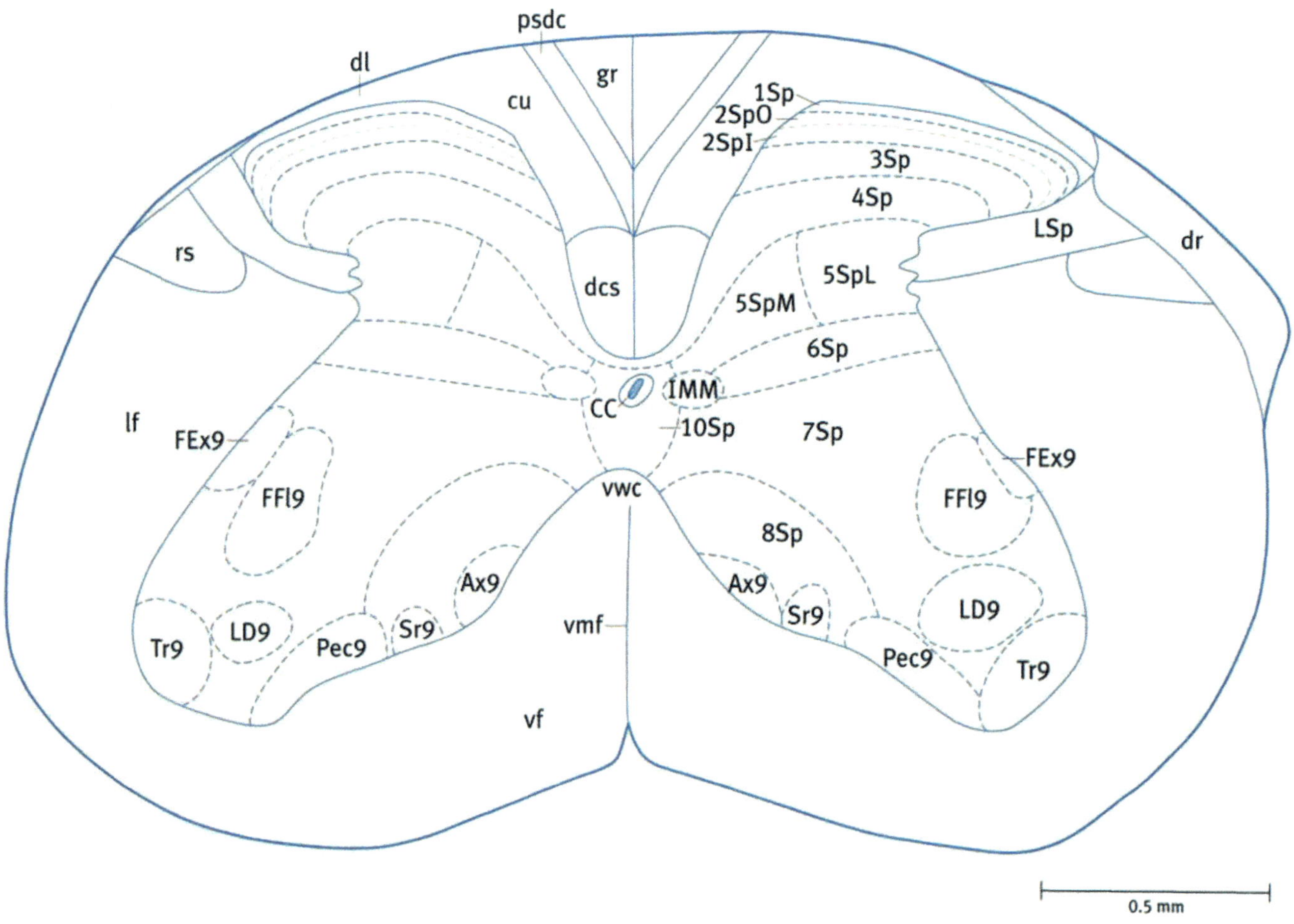
psdc
dl
gr
cu
1Sp
2SpO
2SpI
3Sp
4Sp
rs
LSp
dr
dcs
5SpL
5SpM
6Sp
IMM
CC
10Sp
7Sp
lf
FEx9
FFl9
FEx9
FFl9
vwc
8Sp
Ax9
Ax9
Sr9
Sr9
LD9
LD9
Tr9
Pec9
Pec9
Tr9
vmf
vf
0.5 mm

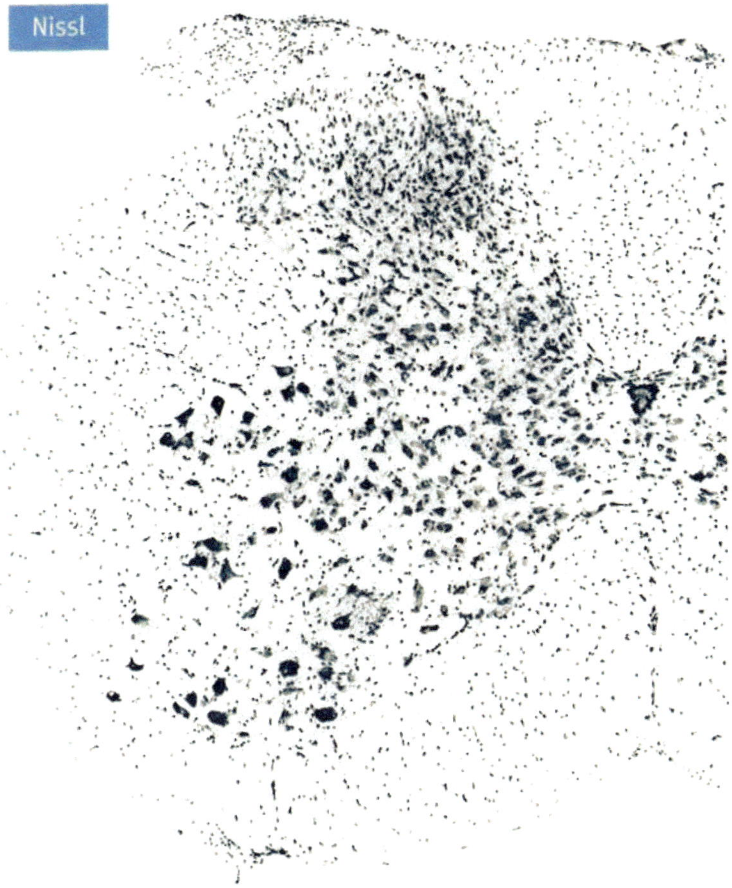

1Sp lamina 1
2SpI lamina 2, inner
2SpO lamina 2, outer
3Sp lamina 3
4Sp lamina 4
5SpL lamina 5, lat
5SpM lamina 5, med
6Sp lamina 6
7Sp lamina 7
8Sp lamina 8
10Sp lamina 10
Ax9 axial muscles
CC central canal
cu cuneate fasc
dcs dors corticosp
dl dorsolat fasc
dr dorsal root
FEx9 forearm ext
FFl9 forearm flex
gr gracile fascic
IMM intermediomed
LD9 latiss dorsi
LSp lat spinal nu
Pec9 pectoral musc
psdc postsyn dors col
rs rubrospinal tr
Sr9 serratus ant
Tr9 triceps
vmf vent med fiss
vwc vent white com

Mouse C7

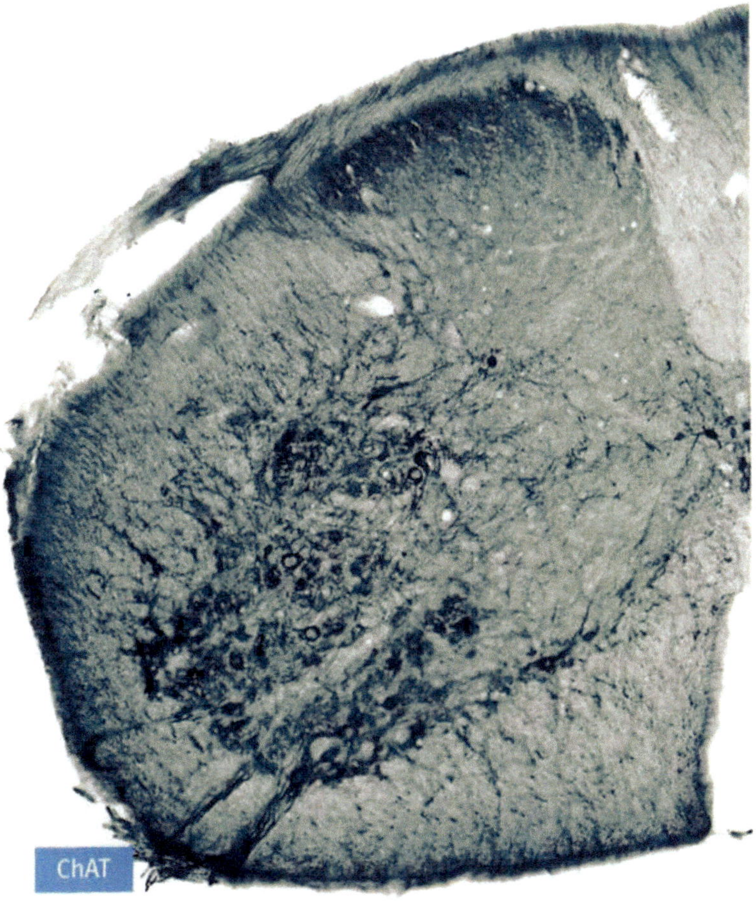

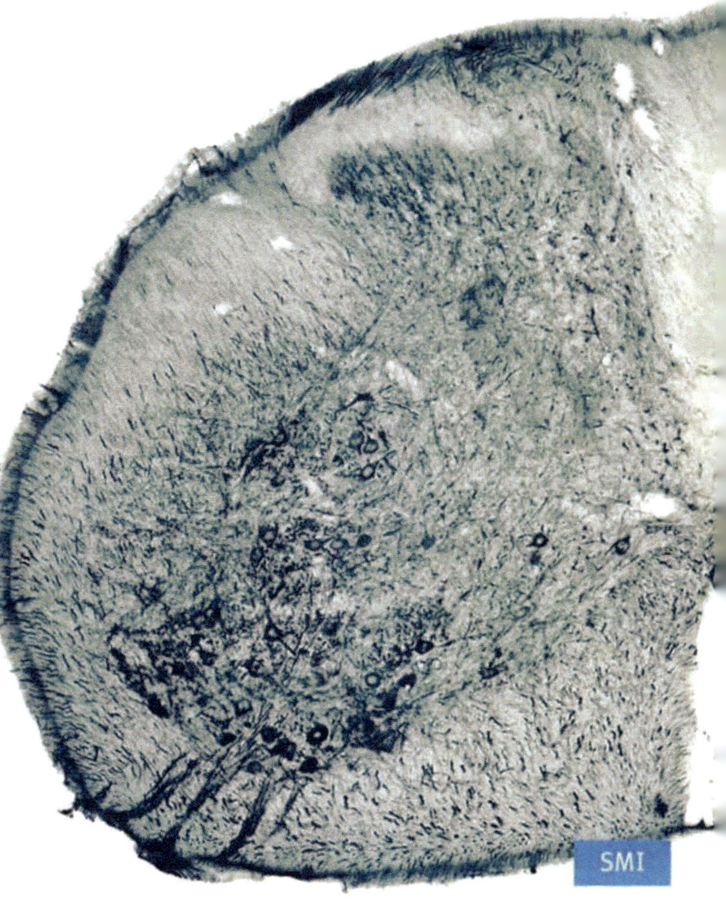

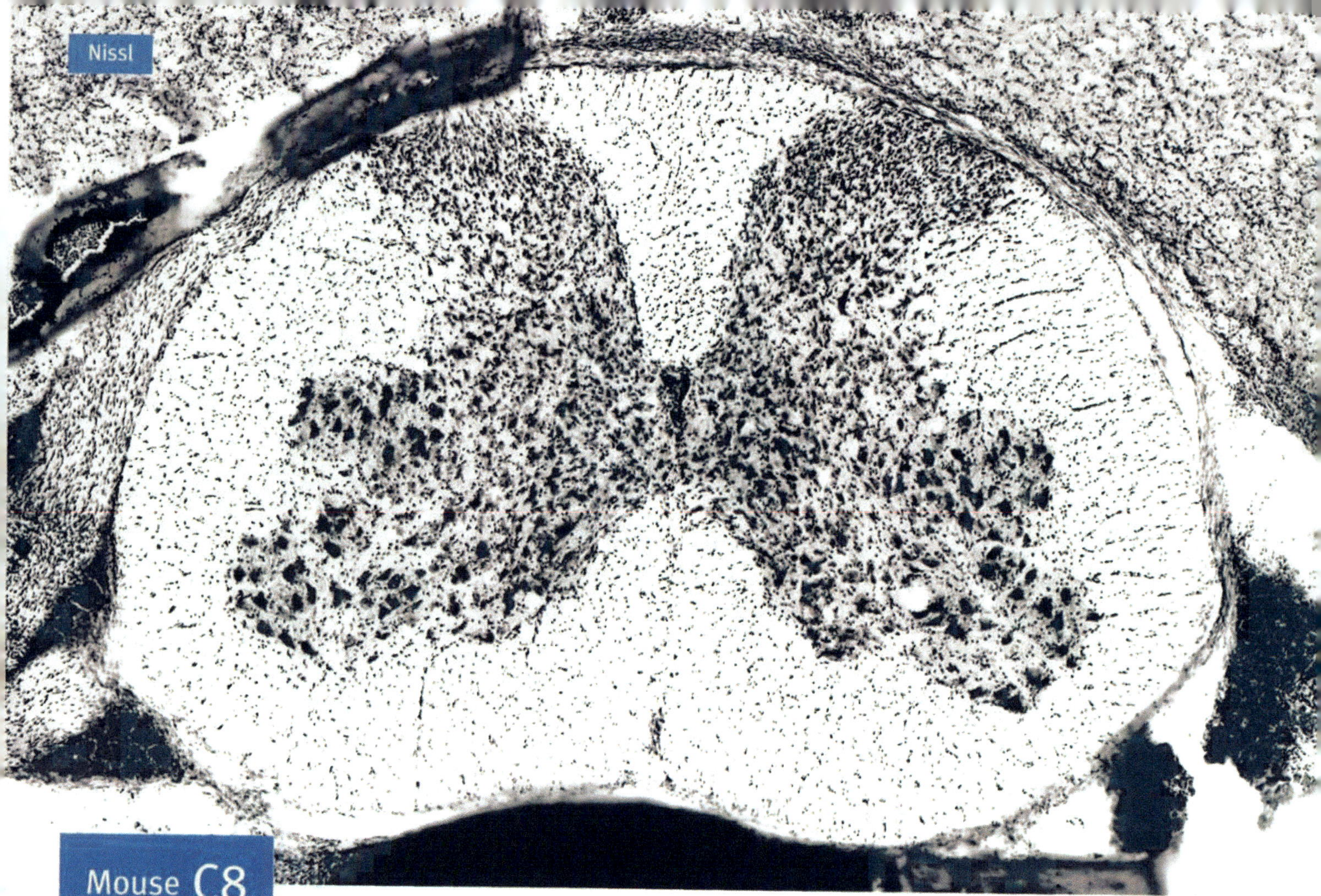
Nissl

Mouse C8

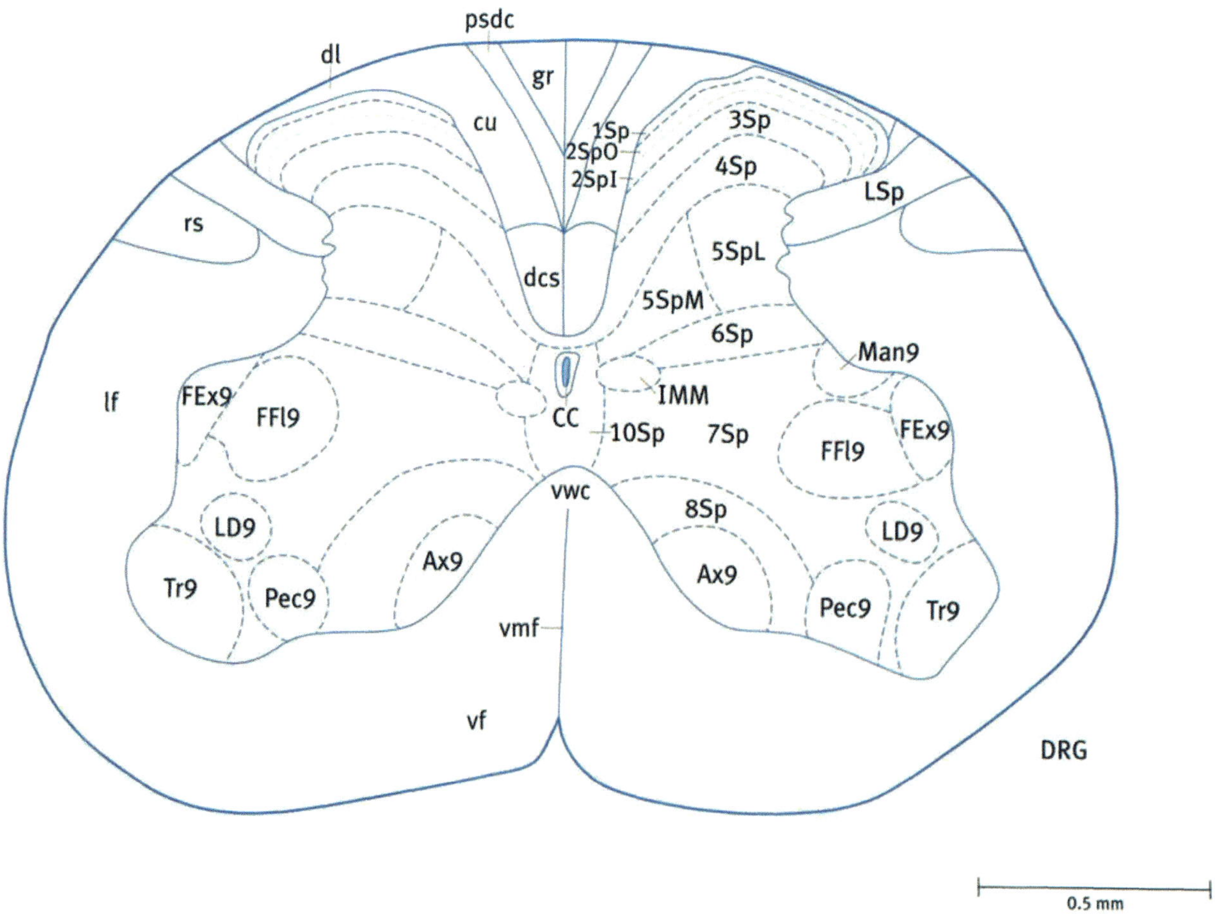
psdc
dl
gr
cu
1Sp
2SpO
2SpI
3Sp
4Sp
LSp
rs
5SpL
dcs
5SpM
6Sp
Man9
IMM
lf
FEx9
FFl9
CC
10Sp
7Sp
FFl9
FEx9
vwc
8Sp
LD9
LD9
Ax9
Ax9
Tr9
Pec9
Pec9
Tr9
vmf
vf
DRG
0.5 mm

Nissl

AChE

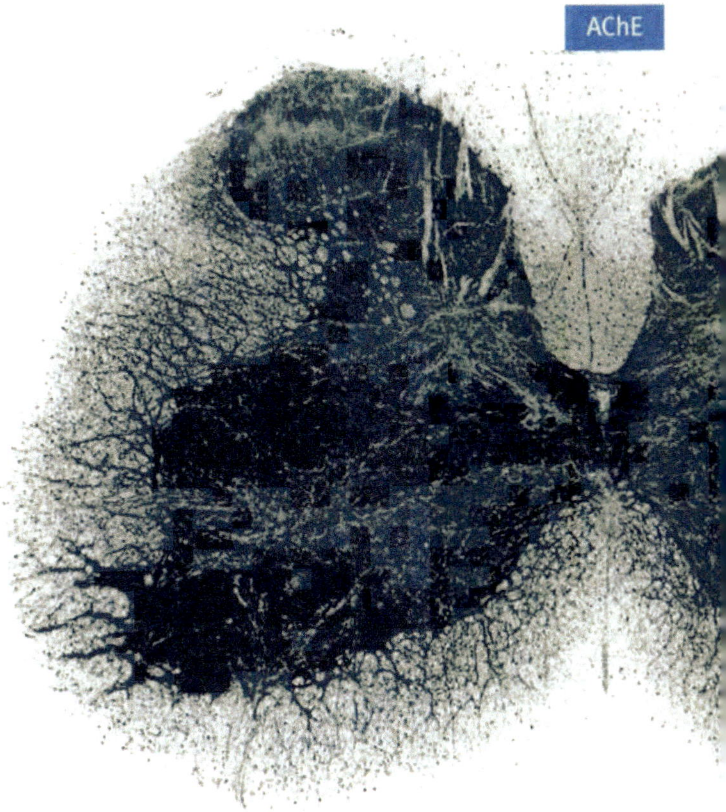

1Sp lamina 1
2SpI lamina 2, inner
2SpO lamina 2, outer
3Sp lamina 3
4Sp lamina 4
5SpL lamina 5, lat
5SpM lamina 5, med
6Sp lamina 6
8Sp lamina 8
10Sp lamina 10
Ax9 axial muscles
CC central canal
cu cuneate fasc
dcs dors corticosp
dl dorsolat fasc
FEx9 forearm ext
FFl9 forearm flex
gr gracile fascic
IMM intermediomed
LD9 latiss dorsi
LSp lat spinal nu
Man9 manus
Pec9 pectoral musc
psdc postsyn dors col
rs rubrospinal tr
Tr9 triceps
vmf vent med fiss
vwc vent white com

Mouse C8

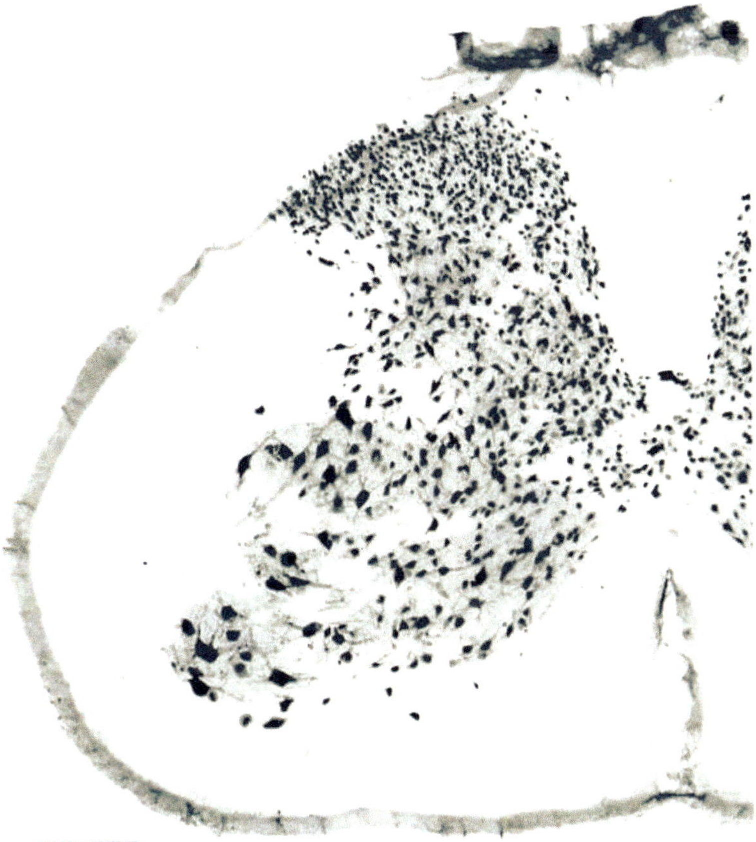

NeuN

Ek

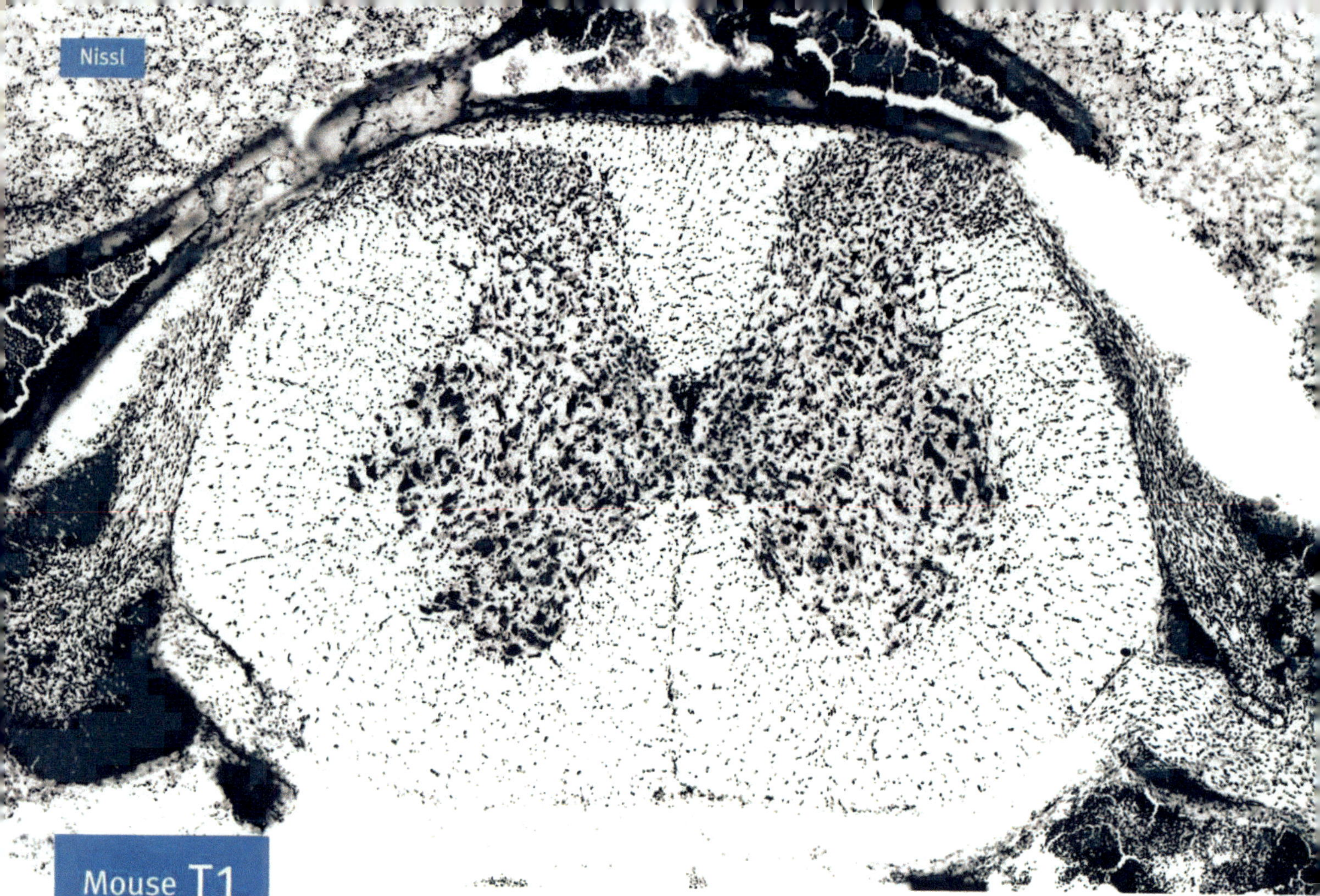
Nissl

Mouse T1

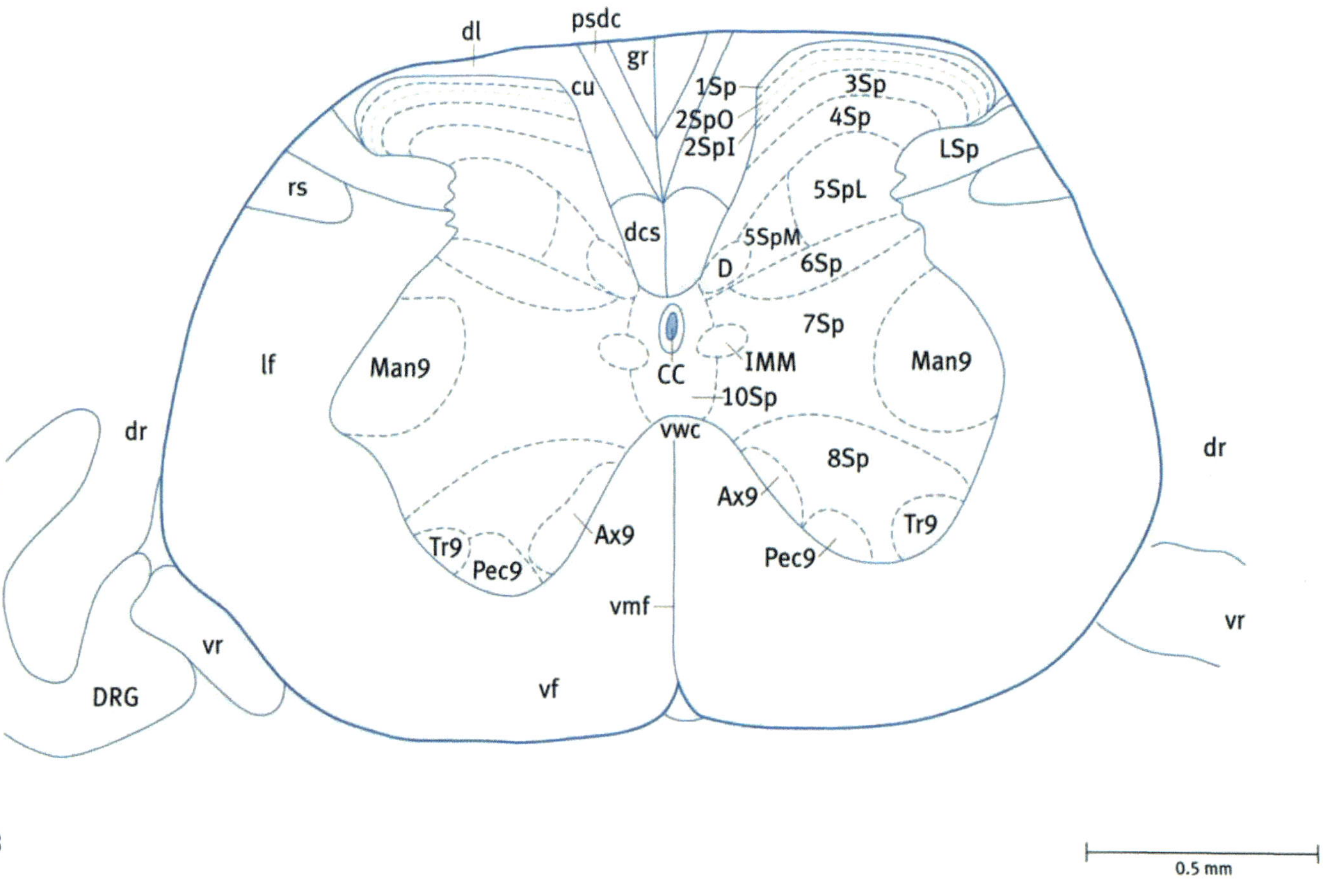
dl
psdc
gr
cu
1Sp
3Sp
2SpO
4Sp
2SpI
LSp
rs
5SpL
dcs
5SpM
D
6Sp
7Sp
lf
Man9
CC
IMM
Man9
10Sp
dr
vwc
dr
8Sp
Ax9
Ax9
Tr9
Tr9
Pec9
Pec9
vmf
vr
vr
DRG
vf
0.5 mm

Nissl

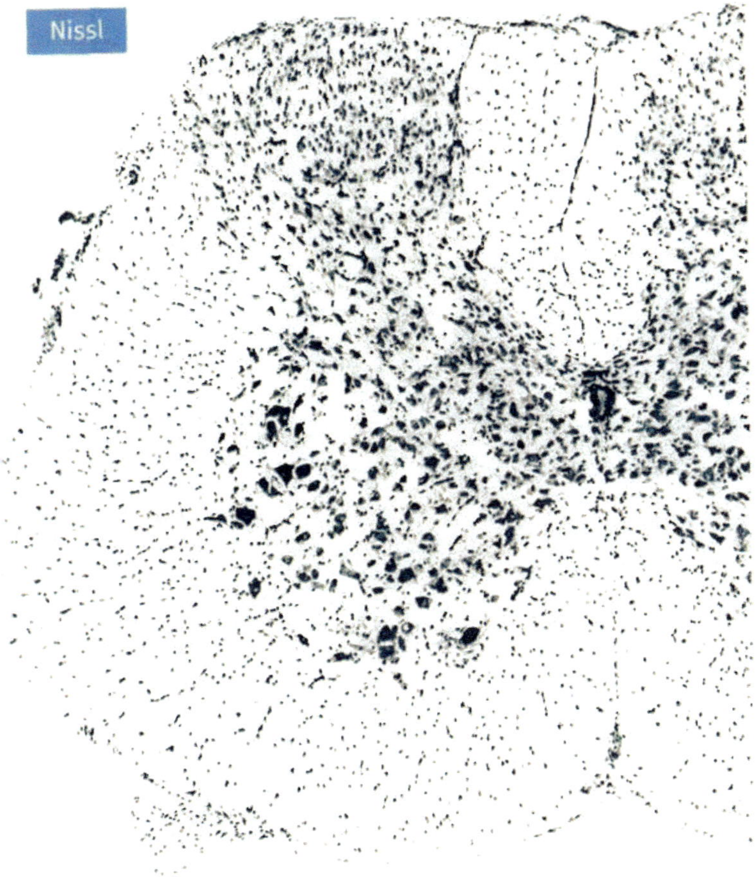

AChE

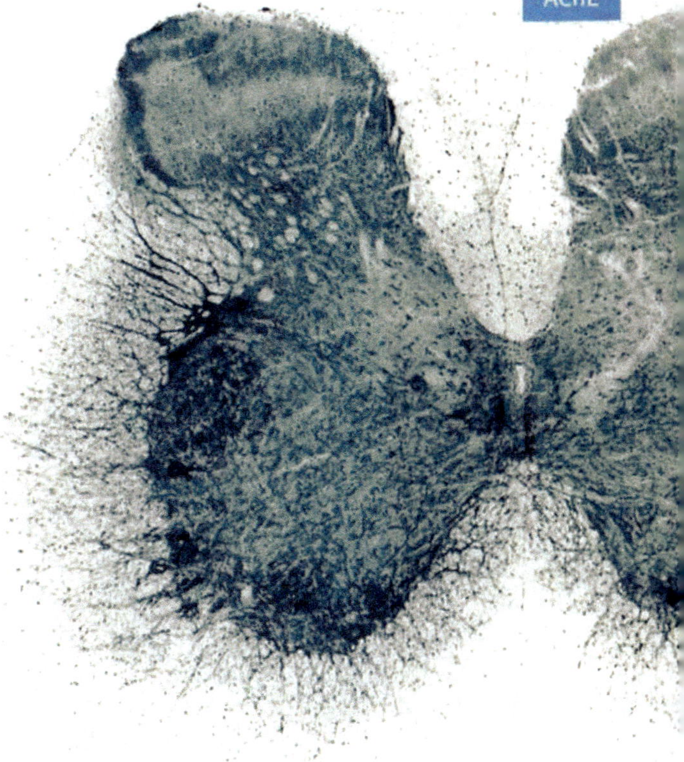

1Sp lamina 1
2SpI lamina 2, inner
2SpO lamina 2, outer
3Sp lamina 3
4Sp lamina 4
5SpL lamina 5, lat
5SpM lamina 5, med
6Sp lamina 6
7Sp lamina 7
8Sp lamina 8
10Sp lamina 10
Ax9 axial muscles
CC central canal
cu cuneate fasc
D dorsal nucleus
dcs dors corticosp
dl dorsolat fasc
dr dorsal root
gr gracile fascic
IMM intermediomed
LSp lat spinal nu
Man9 manus
Pec9 pectoral musc
psdc postsyn dors col
rs rubrospinal tr
Tr9 triceps
vmf vent med fiss
vr ventral root
vwc vent white com

Mouse T1

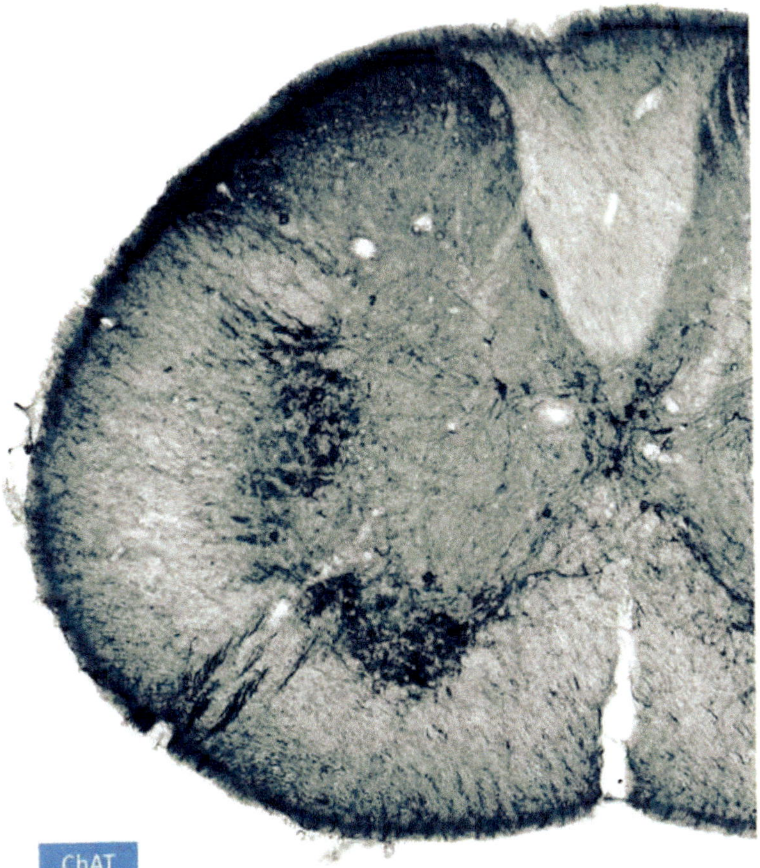

ChAT

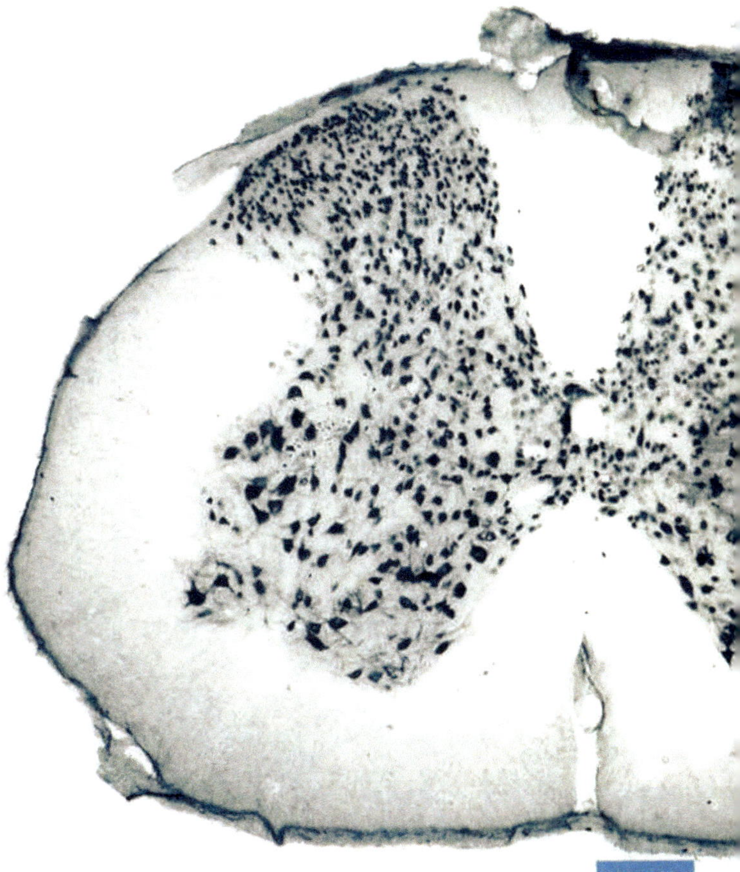

NeuN

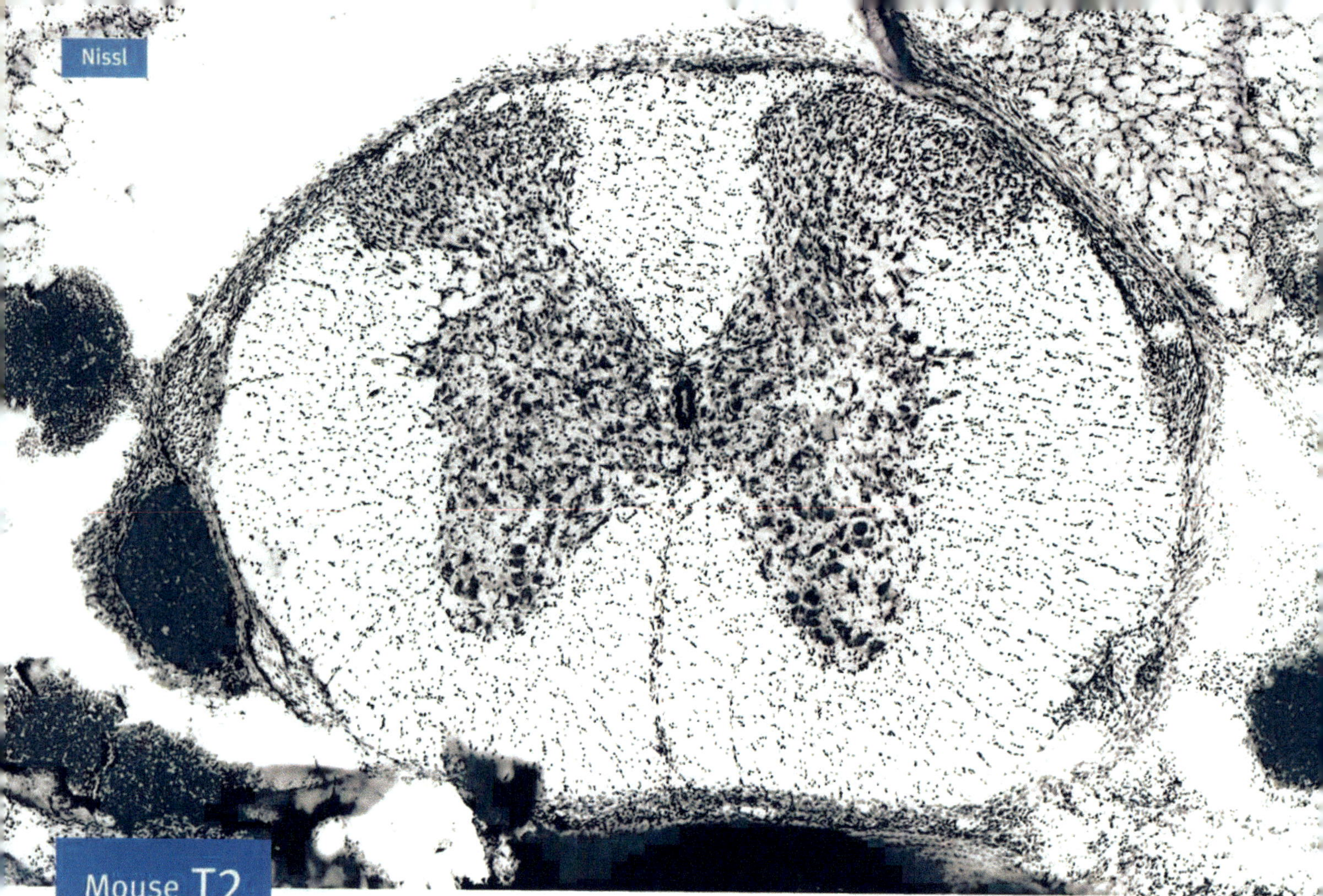
Nissl

Mouse T2

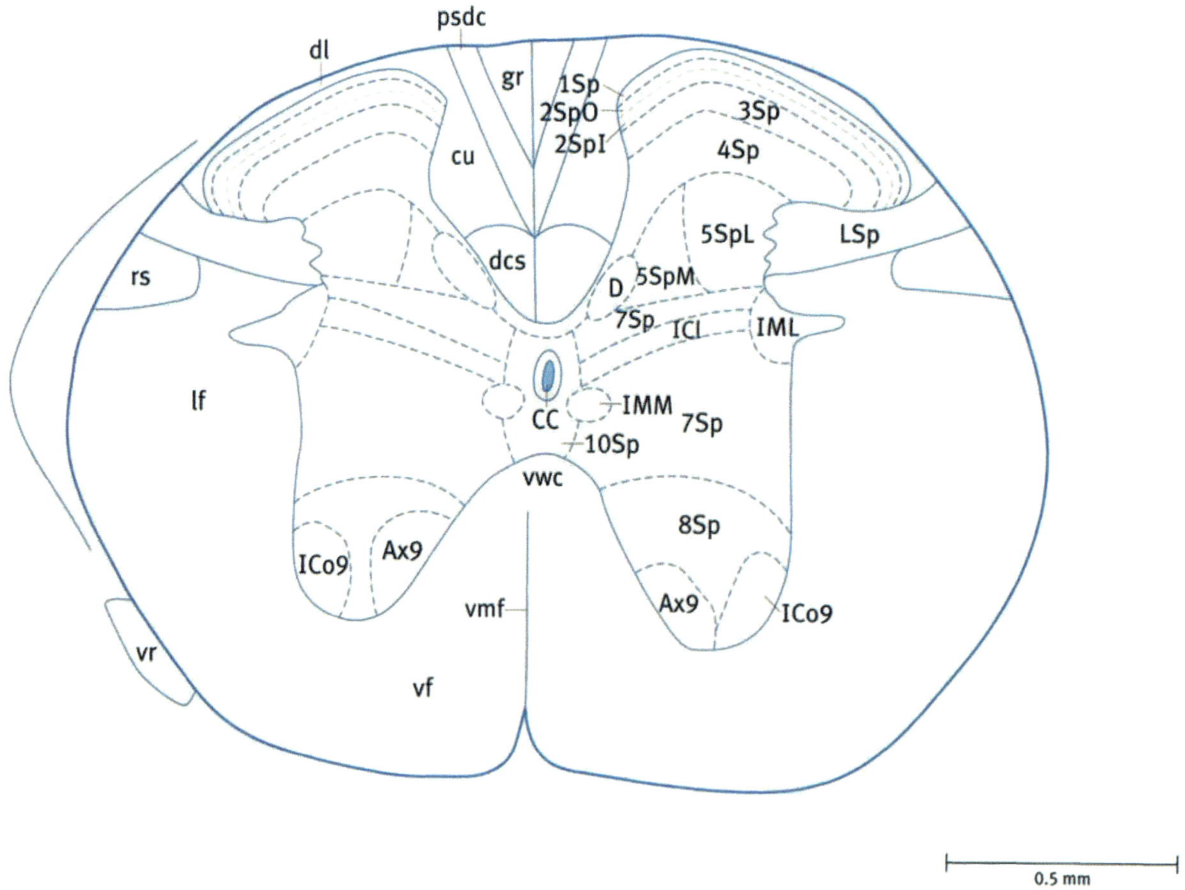
psdc
dl
gr
1Sp
2SpO
2SpI
3Sp
4Sp
cu
5SpL
LSp
dcs
rs
D
5SpM
7Sp
ICl
IML
lf
IMM
CC
7Sp
10Sp
vwc
8Sp
ICo9
Ax9
Ax9
ICo9
vmf
vr
vf
0.5 mm

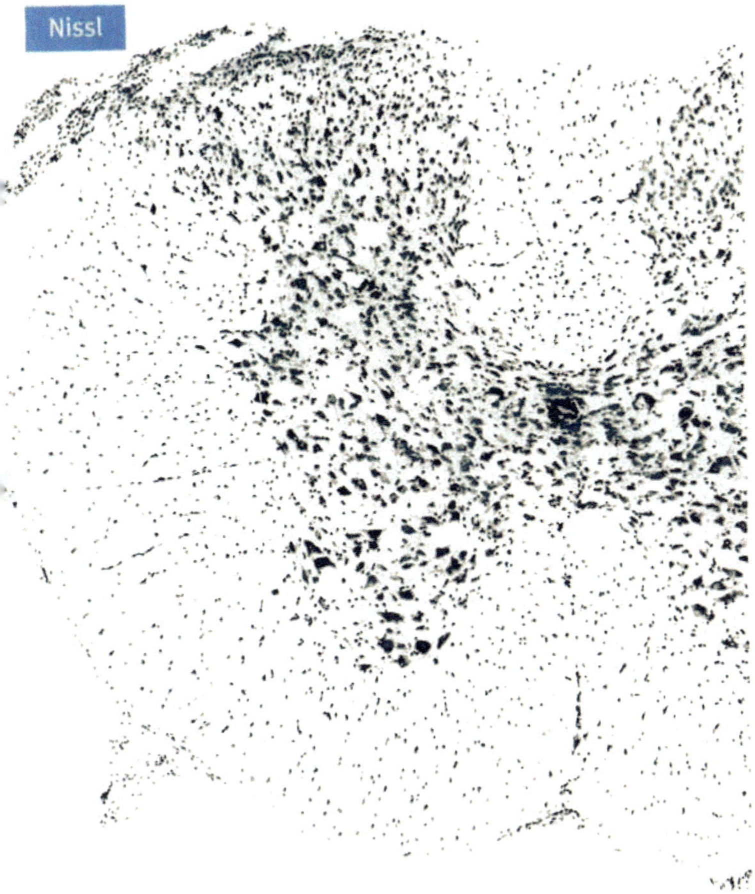

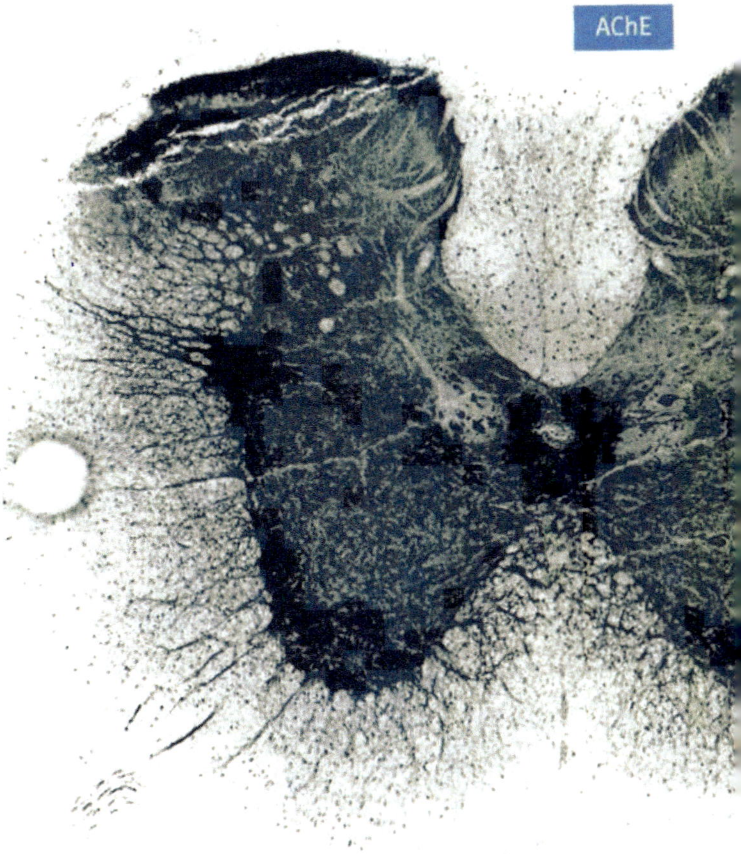

1Sp lamina 1
2SpI lamina 2, inner
2SpO lamina 2, outer
3Sp lamina 3
4Sp lamina 4
5SpL lamina 5, lat
5SpM lamina 5, med
7Sp lamina 7
8Sp lamina 8
10Sp lamina 10
Ax9 axial muscles
CC central canal
cu cuneate fasc
D dorsal nucleus
dcs dors corticosp
dl dorsolat fasc
gr gracile fascic
ICl intercalated nu
ICo9 intercostals
IML intermediolat
IMM intermediomed
LSp lat spinal nu
psdc postsyn dors col
rs rubrospinal tr
vmf vent med fiss
vr ventral root
vwc vent white com

Mouse T2

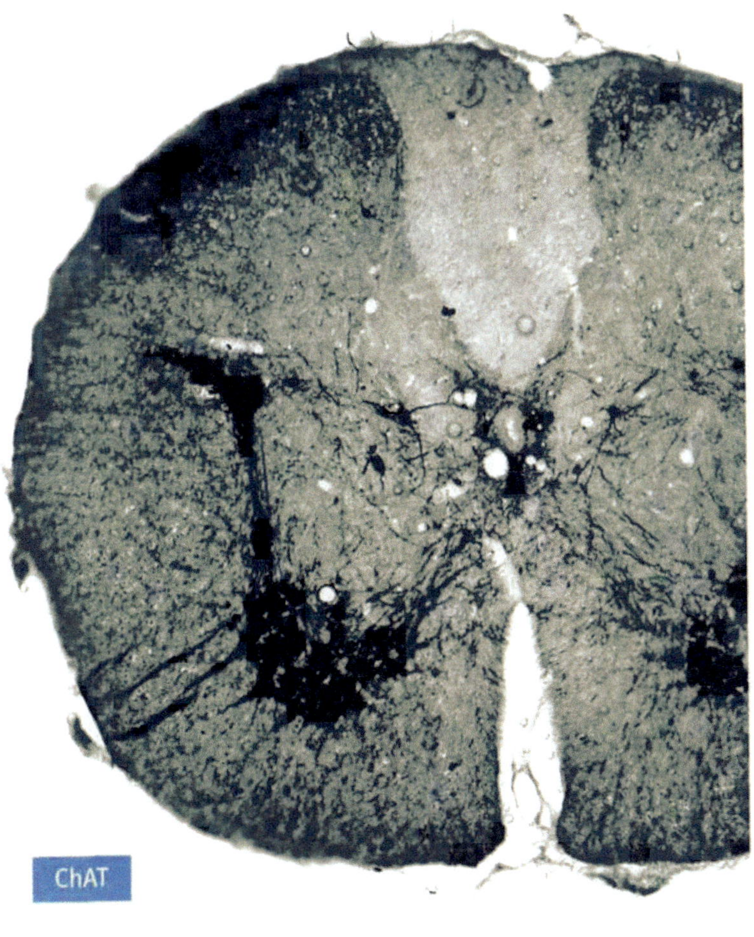

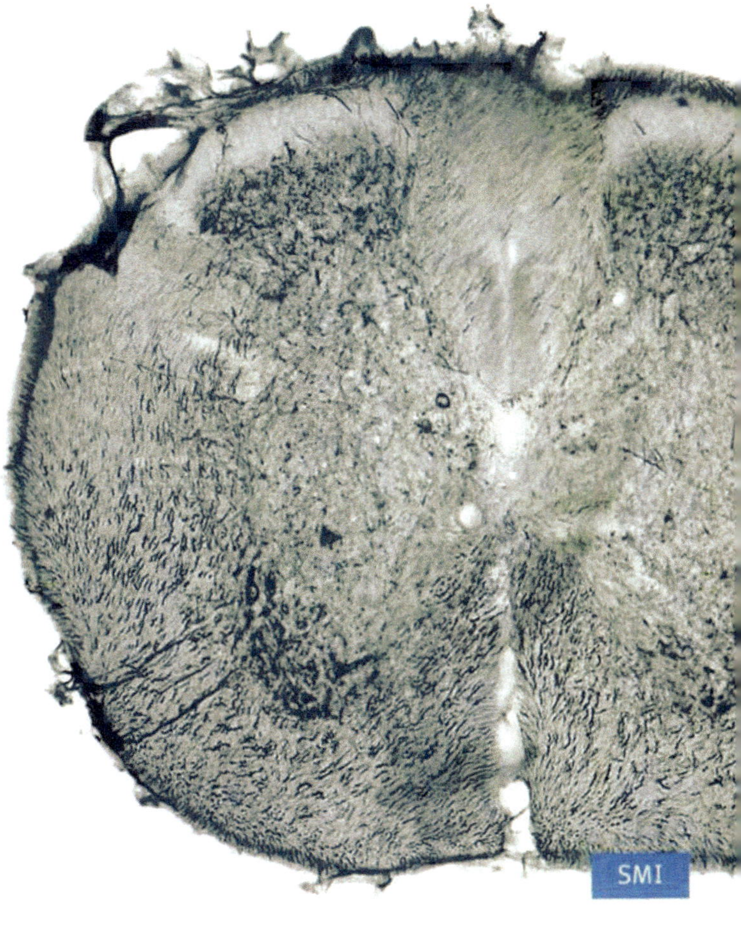

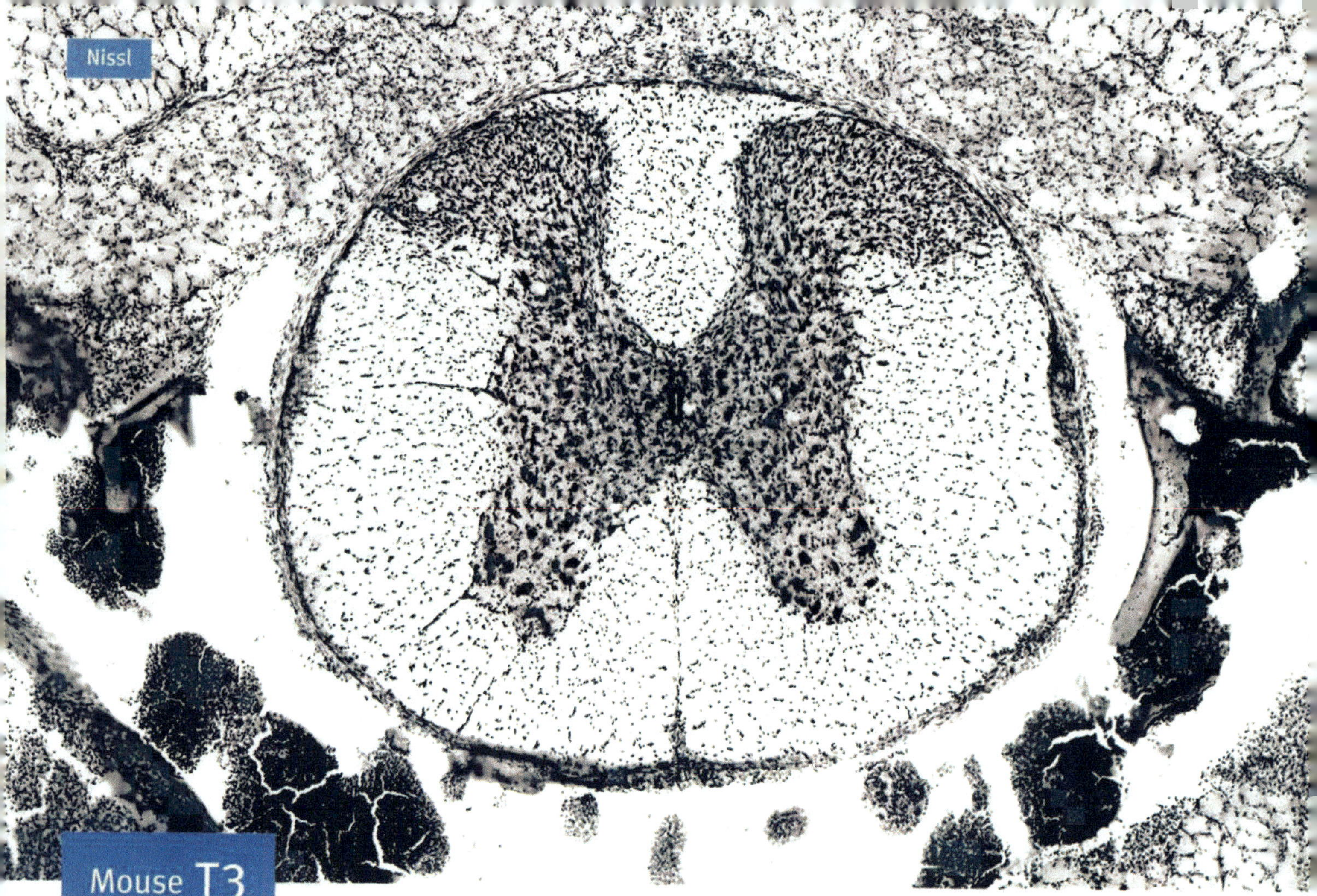
Nissl

Mouse T3

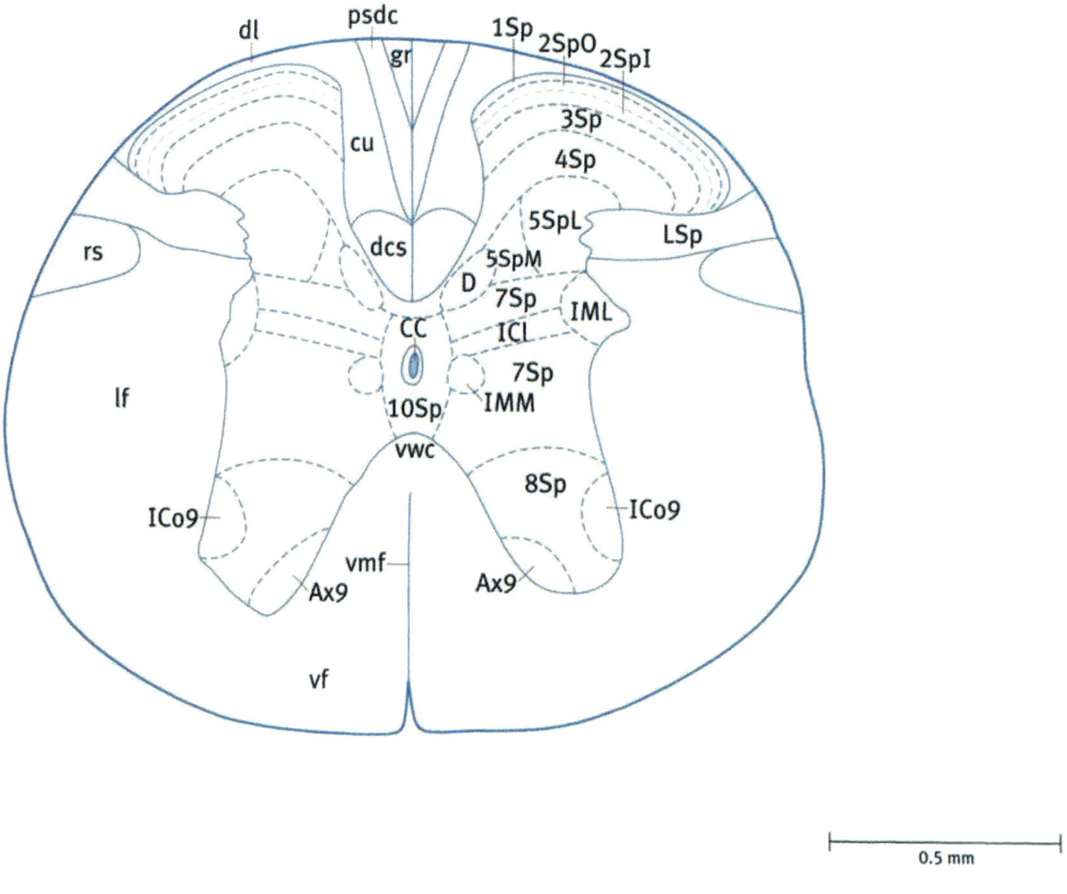
psdc
dl
gr
1Sp
2SpO
2SpI
3Sp
cu
4Sp
5SpL
LSp
rs
dcs
5SpM
D
7Sp
IML
CC
ICl
7Sp
lf
IMM
10Sp
vwc
8Sp
ICo9
ICo9
vmf
Ax9
Ax9
vf
0.5 mm

Nissl

AChE

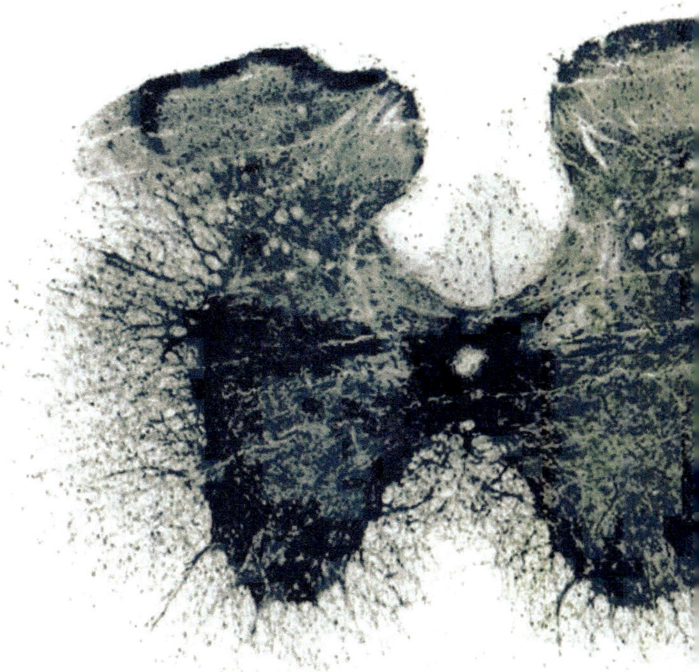

1Sp lamina 1
2SpI lamina 2, inner
2SpO lamina 2, outer
3Sp lamina 3
4Sp lamina 4
5SpL lamina 5, lat
5SpM lamina 5, med
7Sp lamina 7
8Sp lamina 8
10Sp lamina 10
Ax9 axial muscles
CC central canal
cu cuneate fasc
D dorsal nucleus
dcs dors corticosp
dl dorsolat fasc
gr gracile fascic
ICl intercalated nu
ICo9 intercostals
IML intermediolat
IMM intermediomed
LSp lat spinal nu
psdc postsyn dors col
rs rubrospinal tr
vmf vent med fiss
vwc vent white com

Mouse T3

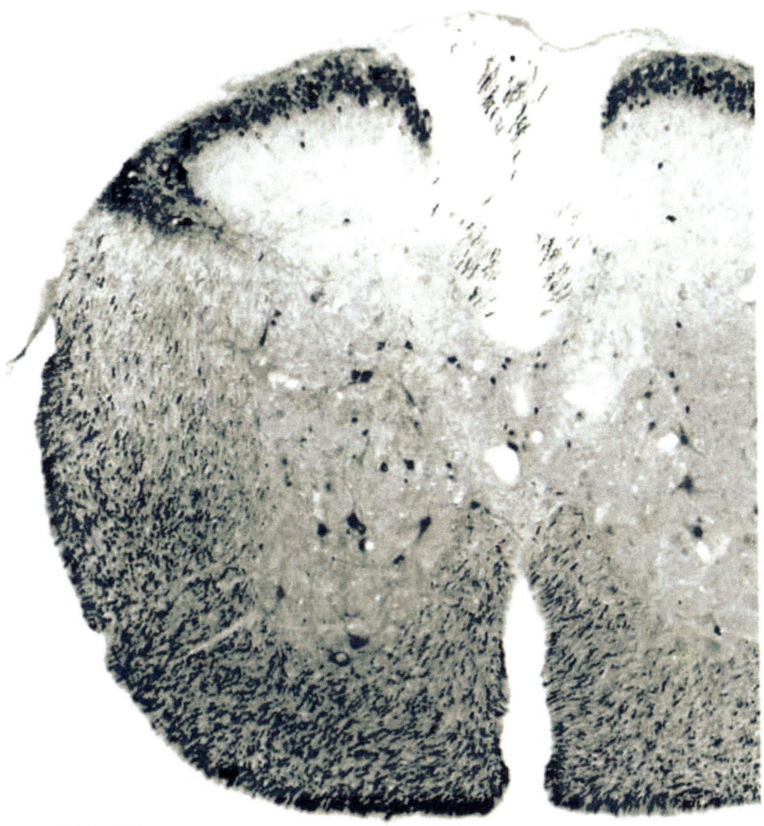

Cr

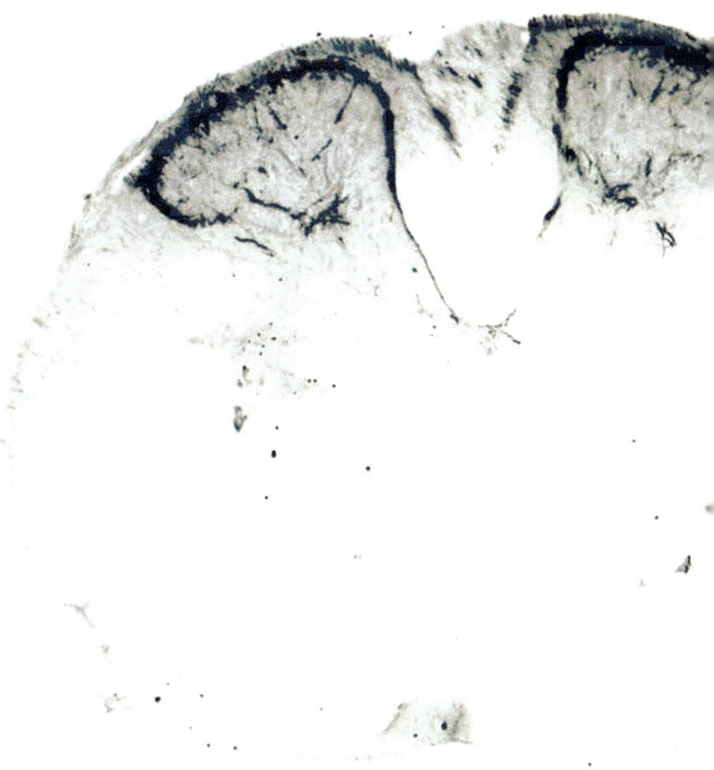

CGRP

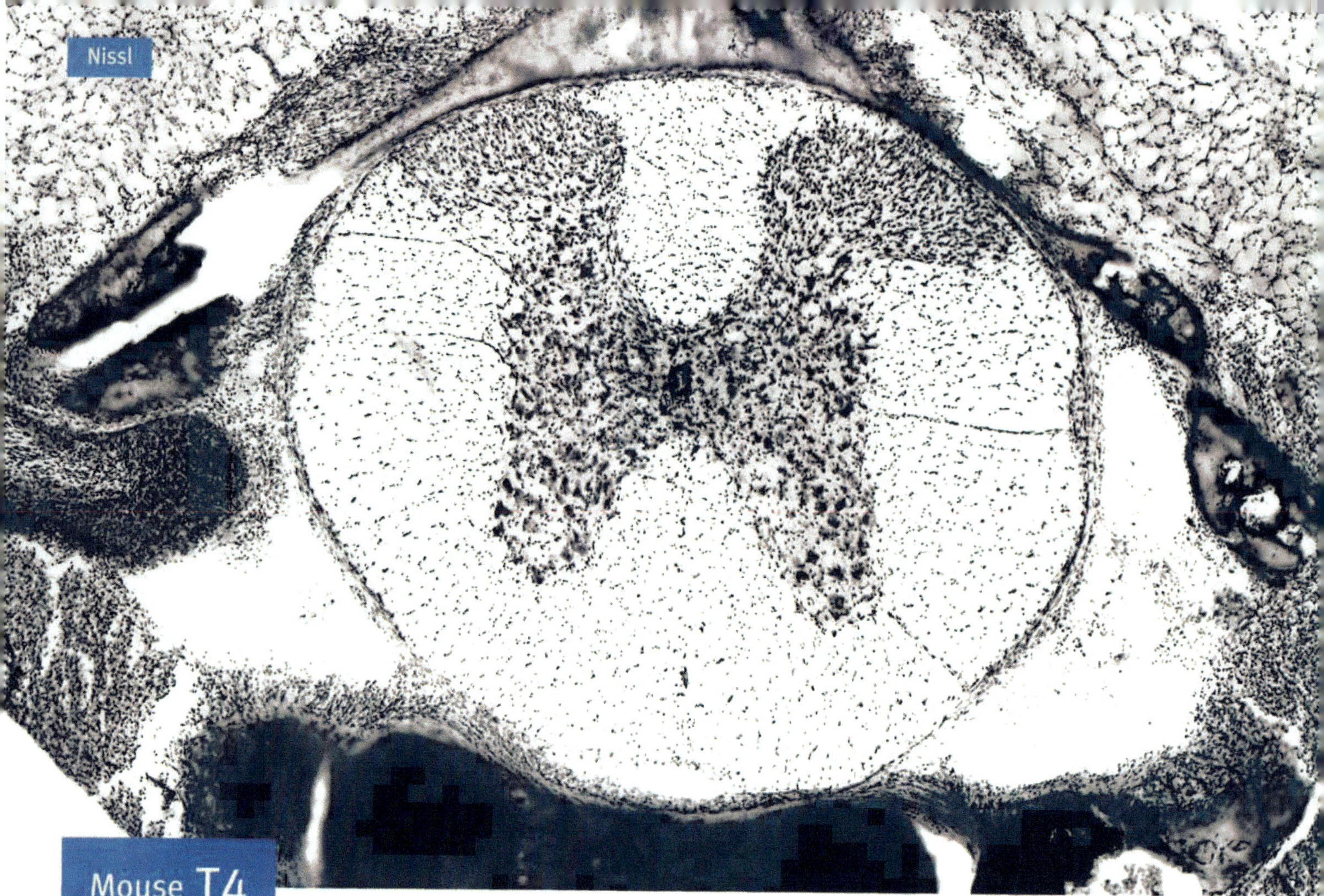

Mouse T4

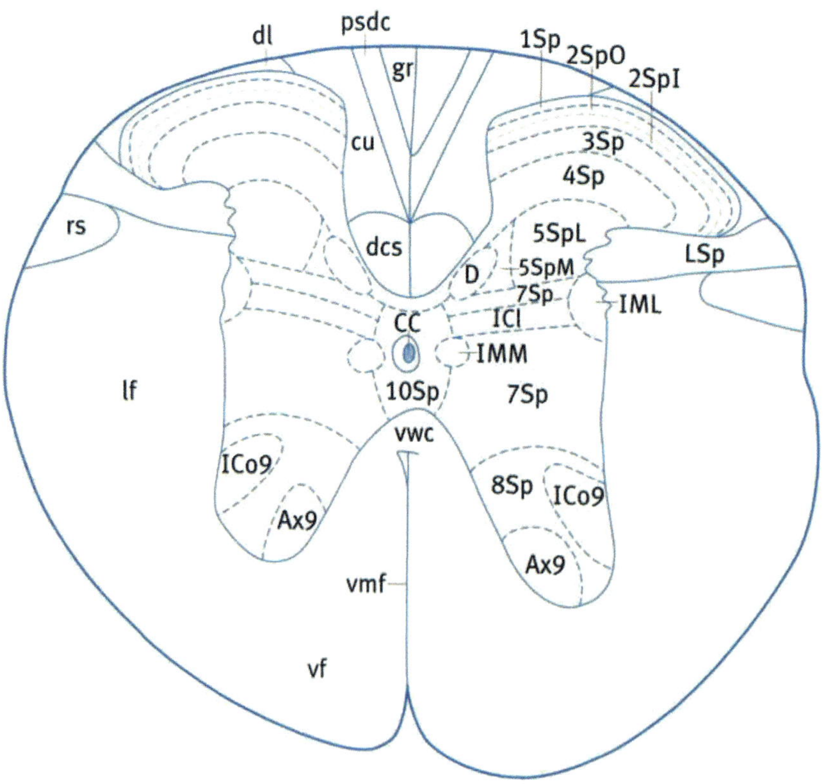

0.5 mm

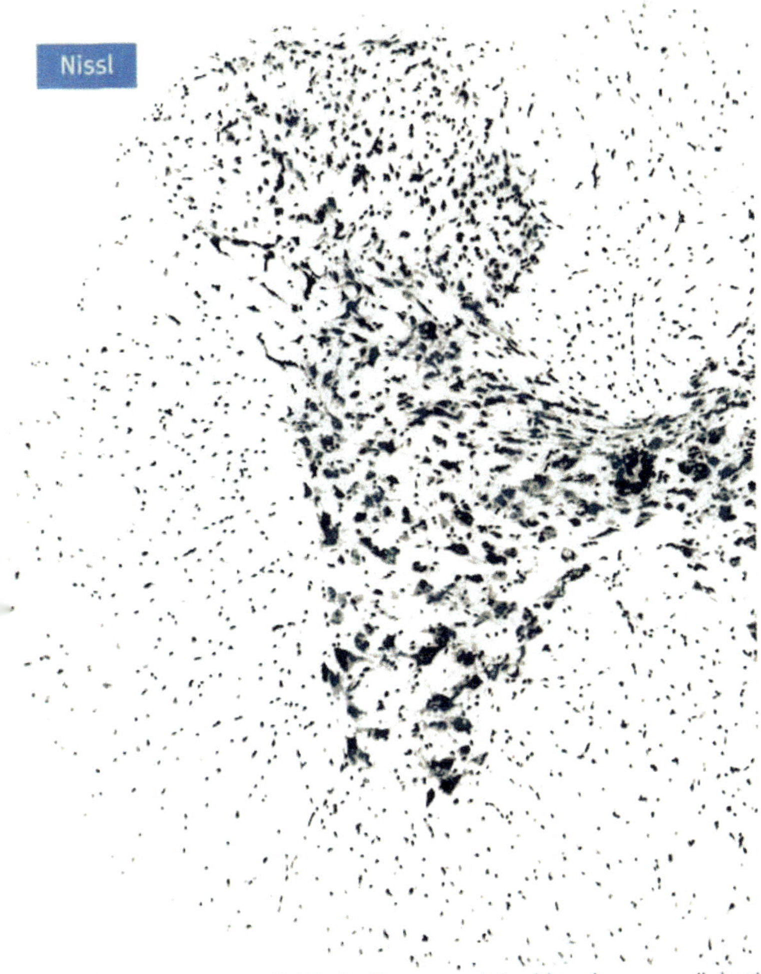

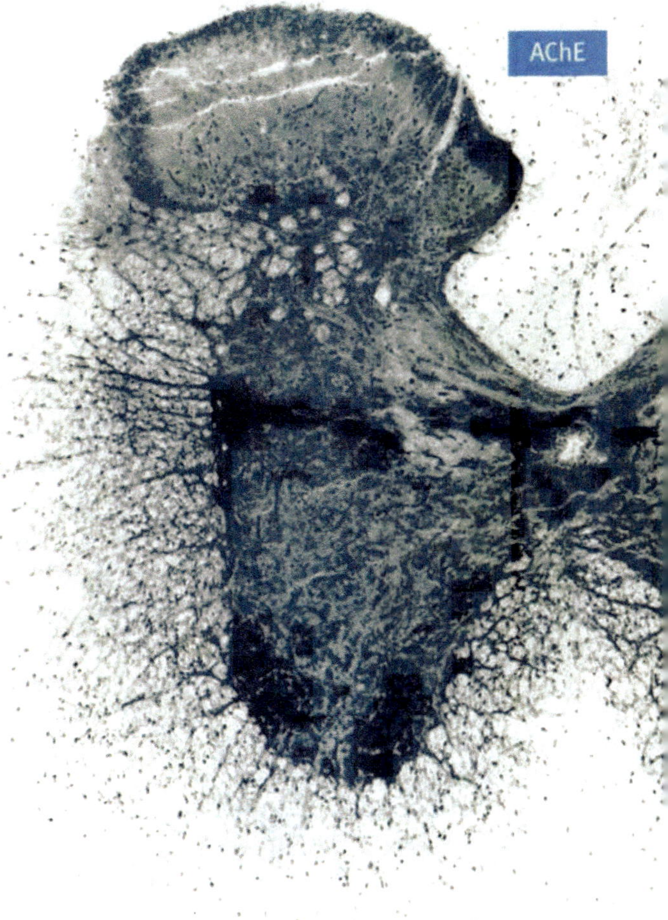

1Sp lamina 1
2SpI lamina 2, inner
2SpO lamina 2, outer
3Sp lamina 3
4Sp lamina 4
5SpL lamina 5, lat
5SpM lamina 5, med
7Sp lamina 7
8Sp lamina 8
10Sp lamina 10
Ax9 axial muscles
CC central canal
cu cuneate fasc
D dorsal nucleus
dcs dors corticosp
dl dorsolat fasc
gr gracile fascic
ICl intercalated nu
ICo9 intercostals
IML intermediolat
IMM intermediomed
LSp lat spinal nu
psdc postsyn dors col
rs rubrospinal tr
vmf vent med fiss
vwc vent white com

Mouse T4

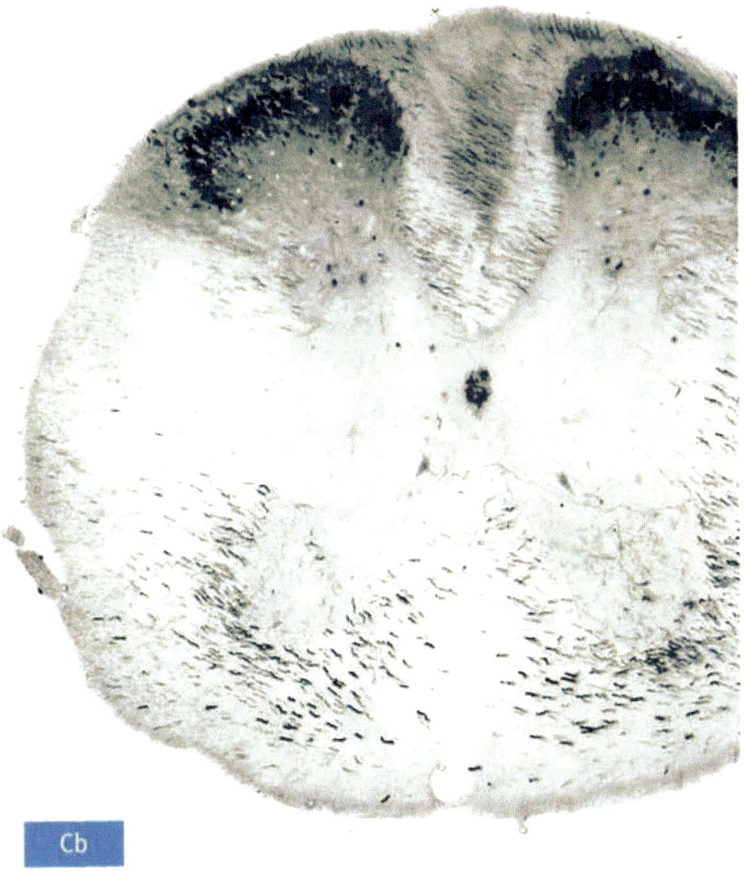

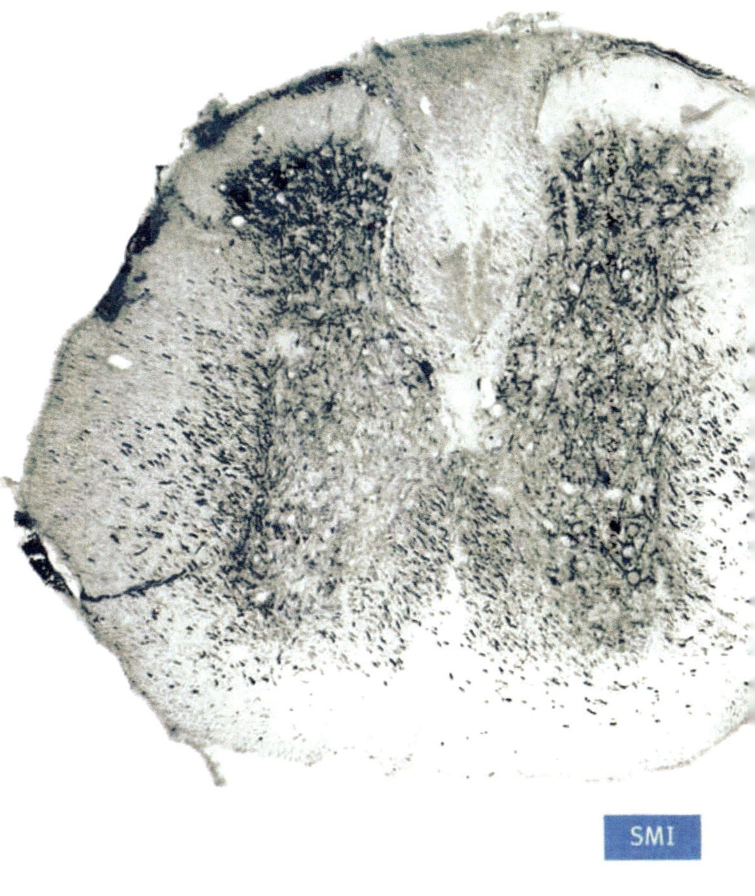

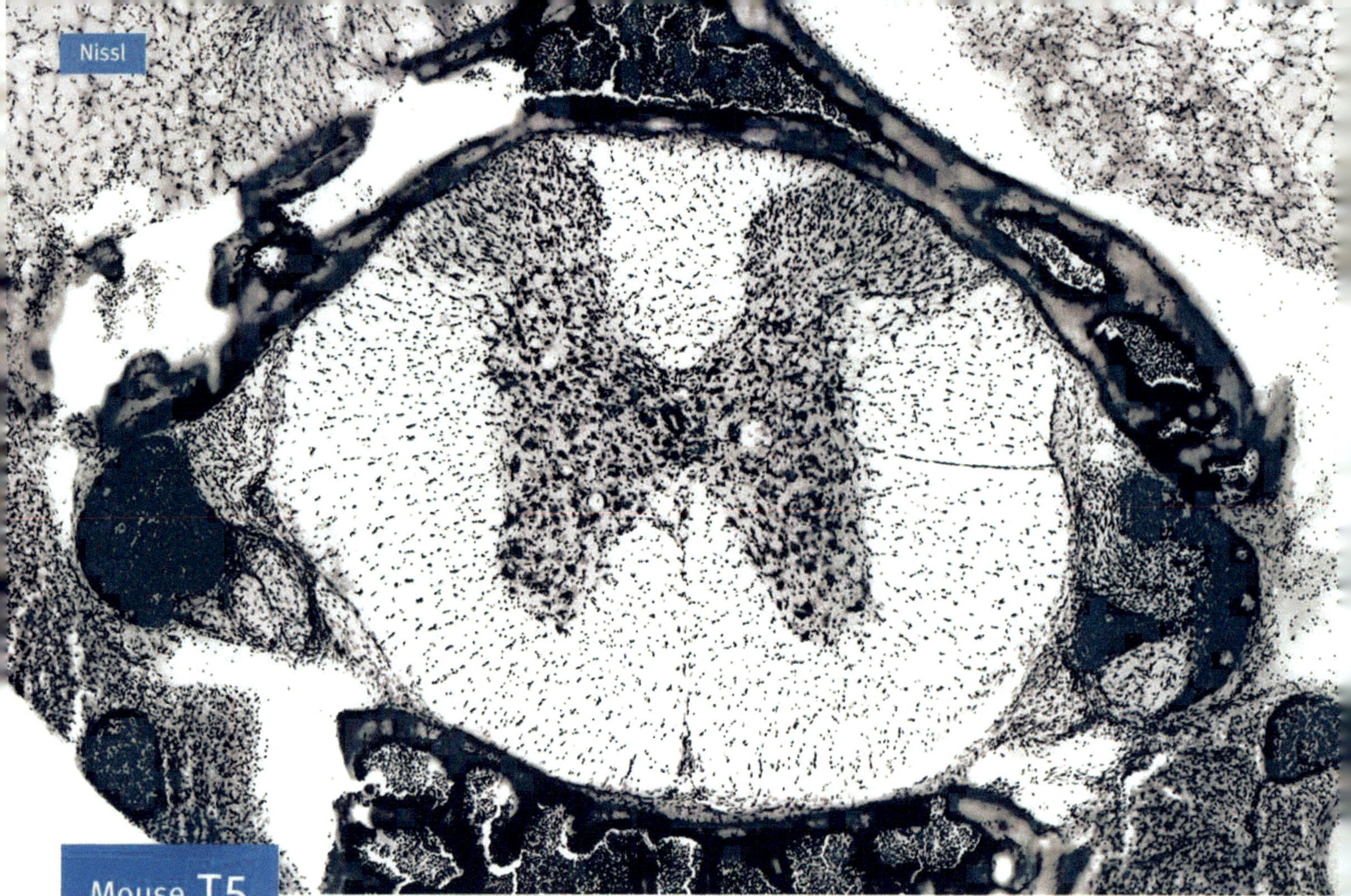
Nissl

Mouse T5

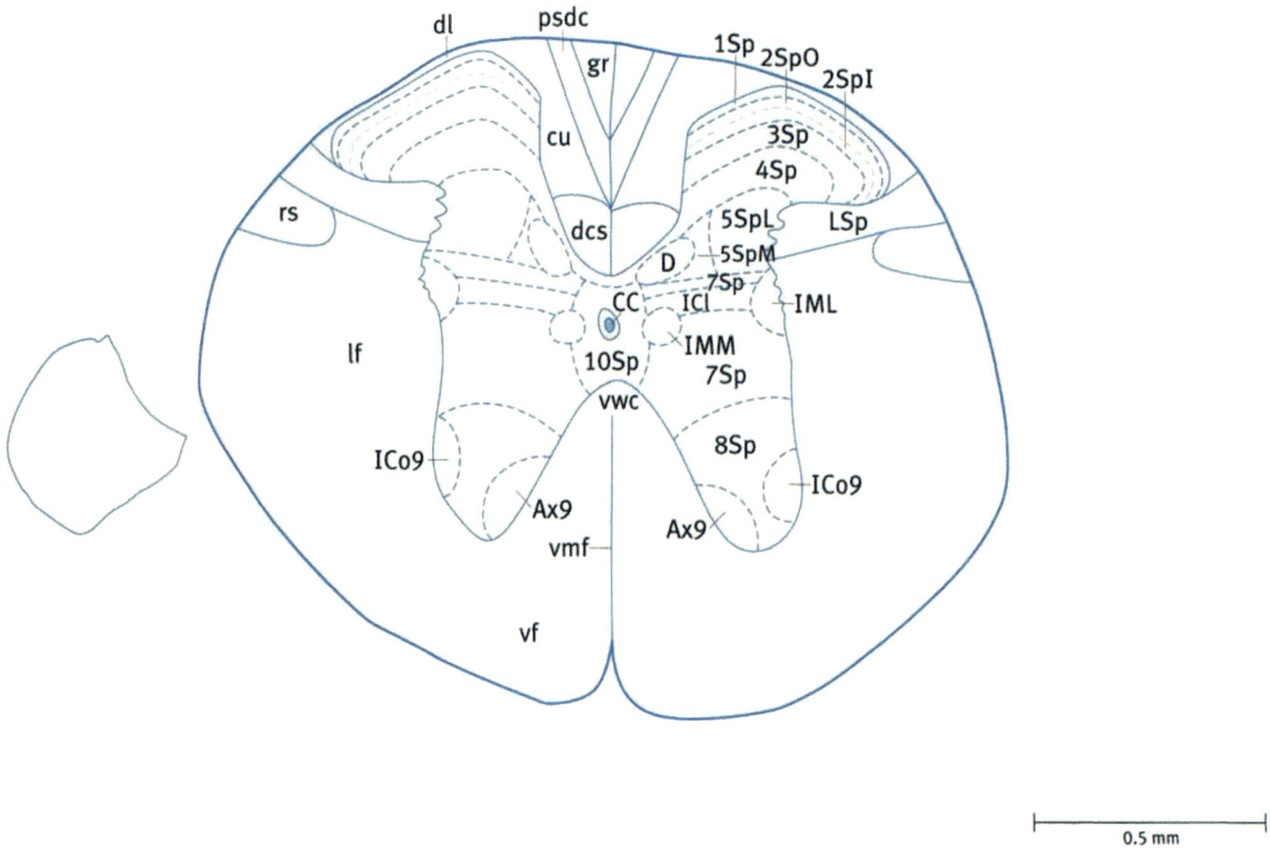
dl
psdc
gr
1Sp
2SpO
2SpI
cu
3Sp
4Sp
rs
5SpL
LSp
dcs
D
5SpM
7Sp
CC
ICl
IML
lf
10Sp
IMM
7Sp
vwc
8Sp
ICo9
ICo9
Ax9
Ax9
vmf
vf
0.5 mm

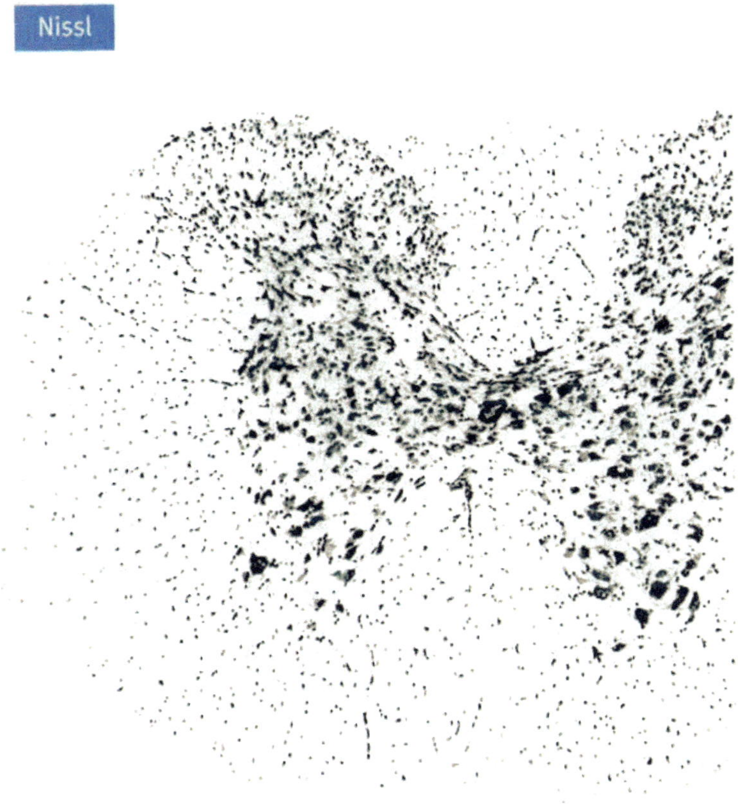

1Sp lamina 1
2SpI lamina 2, inner
2SpO lamina 2, outer
3Sp lamina 3
4Sp lamina 4
5SpL lamina 5, lat
5SpM lamina 5, med
7Sp lamina 7
8Sp lamina 8
10Sp lamina 10
Ax9 axial muscles
CC central canal
cu cuneate fasc
D dorsal nucleus
dcs dors corticosp
dl dorsolat fasc
gr gracile fascic
ICl intercalated nu
ICo9 intercostals
IML intermediolat
IMM intermediomed
LSp lat spinal nu
psdc postsyn dors col
rs rubrospinal tr
vmf vent med fiss
vwc vent white com

Mouse T5

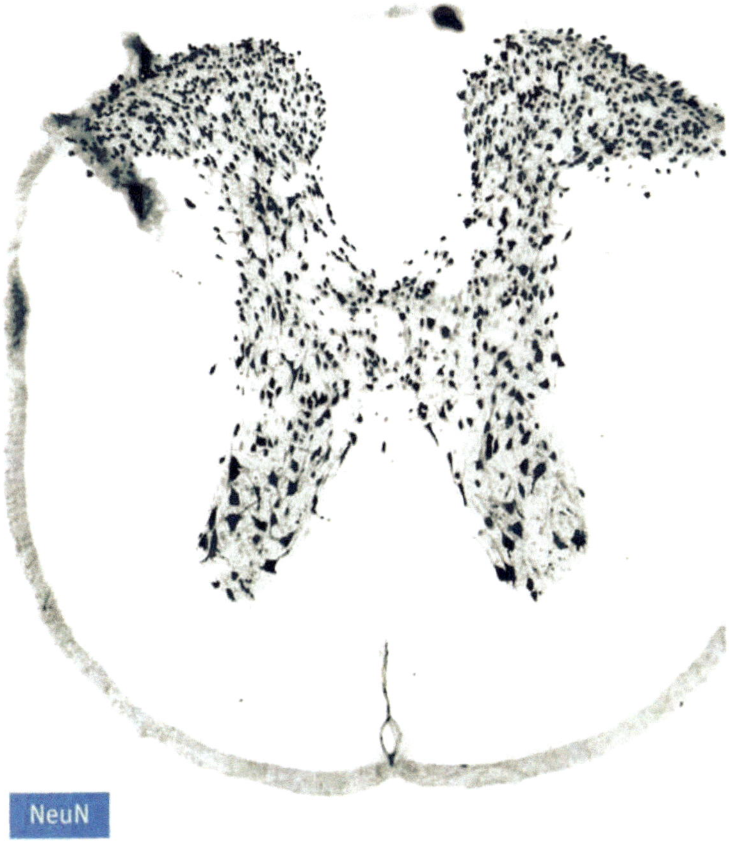

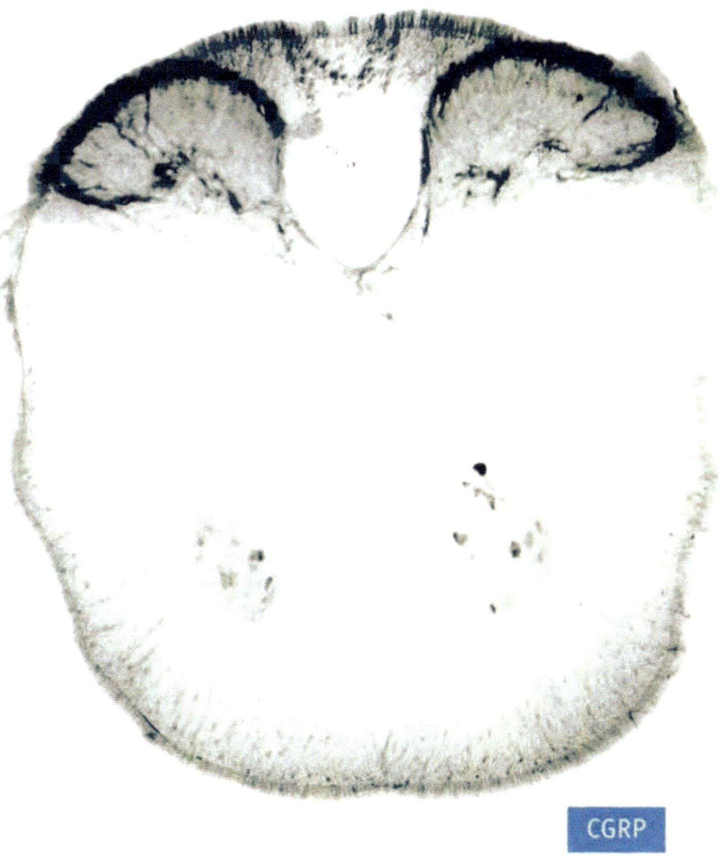

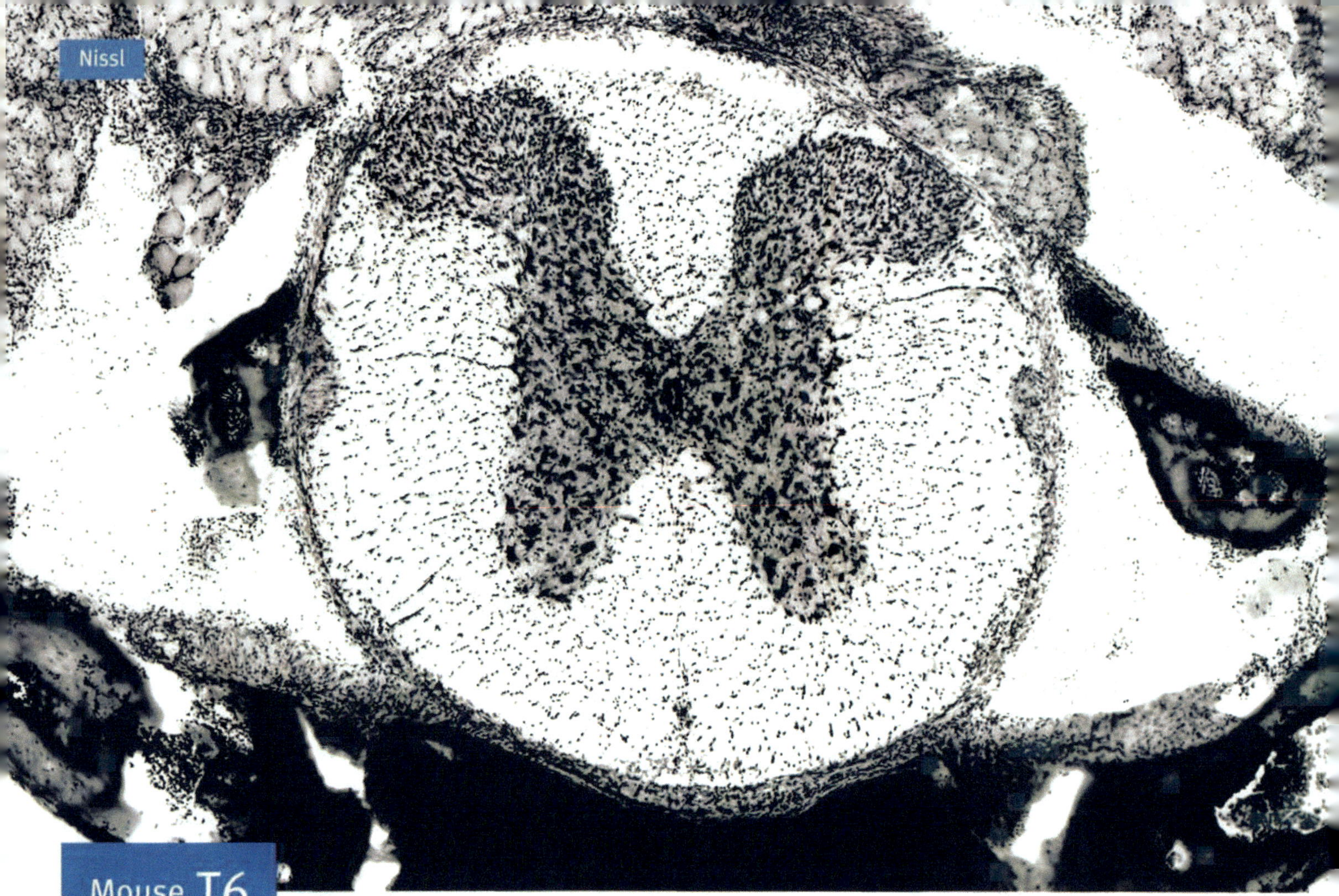
Nissl

Mouse T6

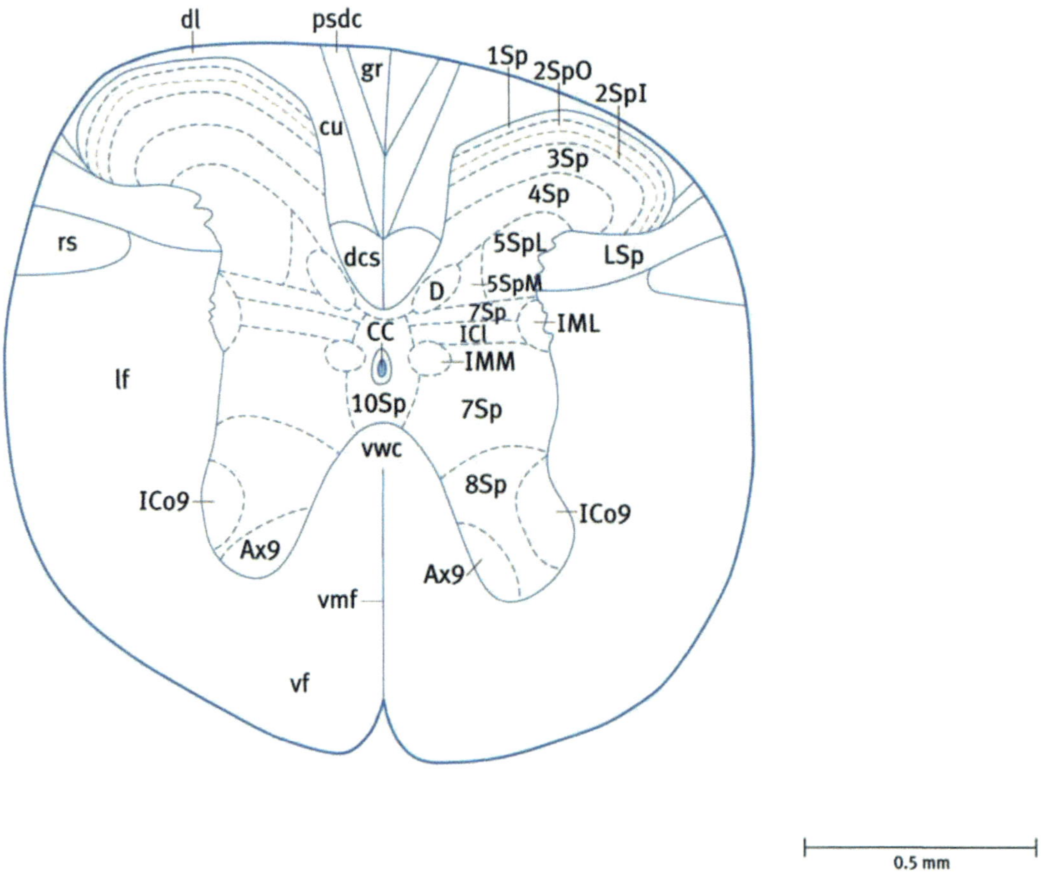
dl
psdc
gr
1Sp
2SpO
2SpI
cu
3Sp
4Sp
rs
dcs
5SpL
LSp
D
5SpM
7Sp
IML
CC
ICl
IMM
lf
10Sp
7Sp
vwc
8Sp
ICo9
ICo9
Ax9
Ax9
vmf
vf
0.5 mm

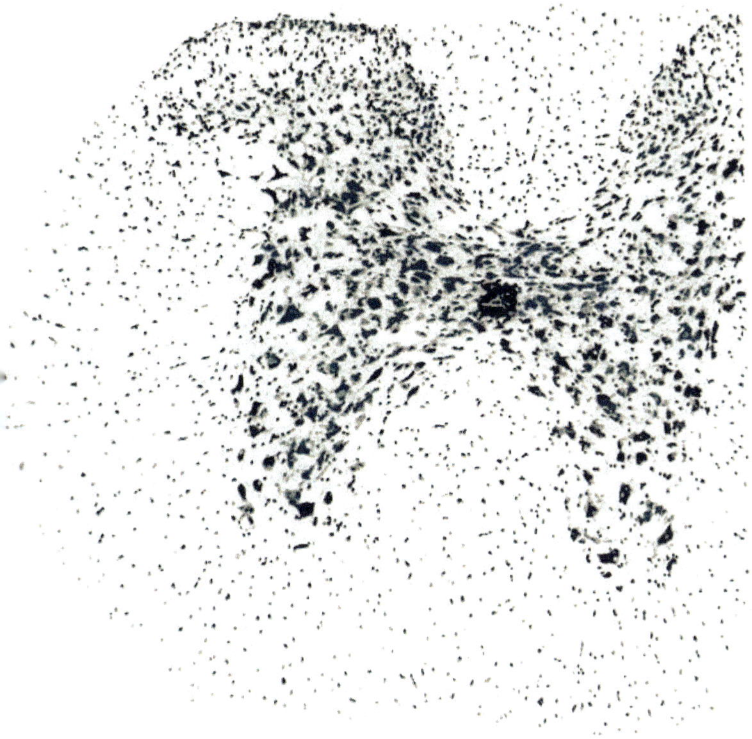

Nissl

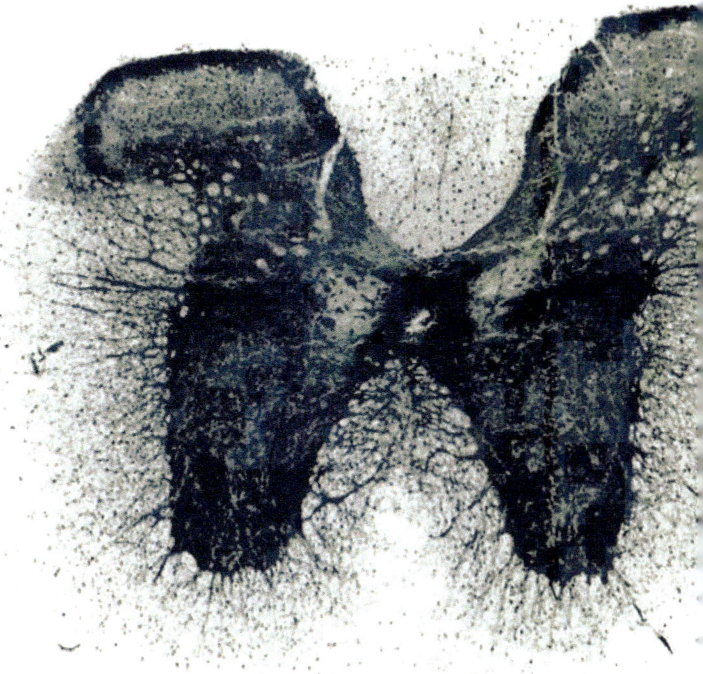

AChE

Mouse T6

1Sp lamina 1
2SpI lamina 2, inner
2SpO lamina 2, outer
3Sp lamina 3
4Sp lamina 4
5SpL lamina 5, lat
5SpM lamina 5, med
7Sp lamina 7
8Sp lamina 8
10Sp lamina 10
Ax9 axial muscles
CC central canal
cu cuneate fasc
D dorsal nucleus
dcs dors corticosp
dl dorsolat fasc
gr gracile fascic
ICl intercalated nu
ICo9 intercostals
IML intermediolat
IMM intermediomed
LSp lat spinal nu
psdc postsyn dors col
rs rubrospinal tr
vmf vent med fiss
vwc vent white com

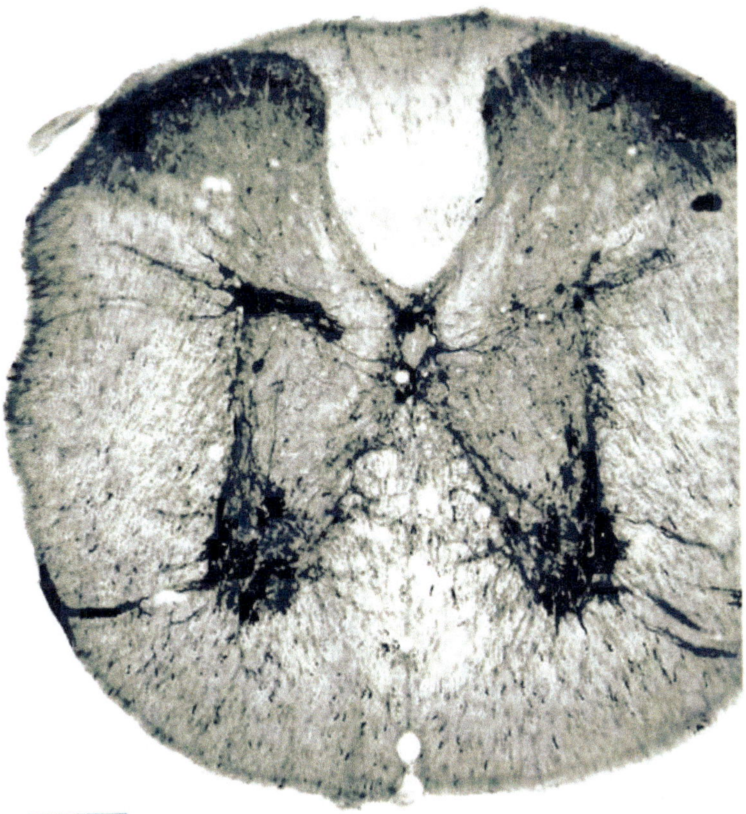

ChAT

Cb

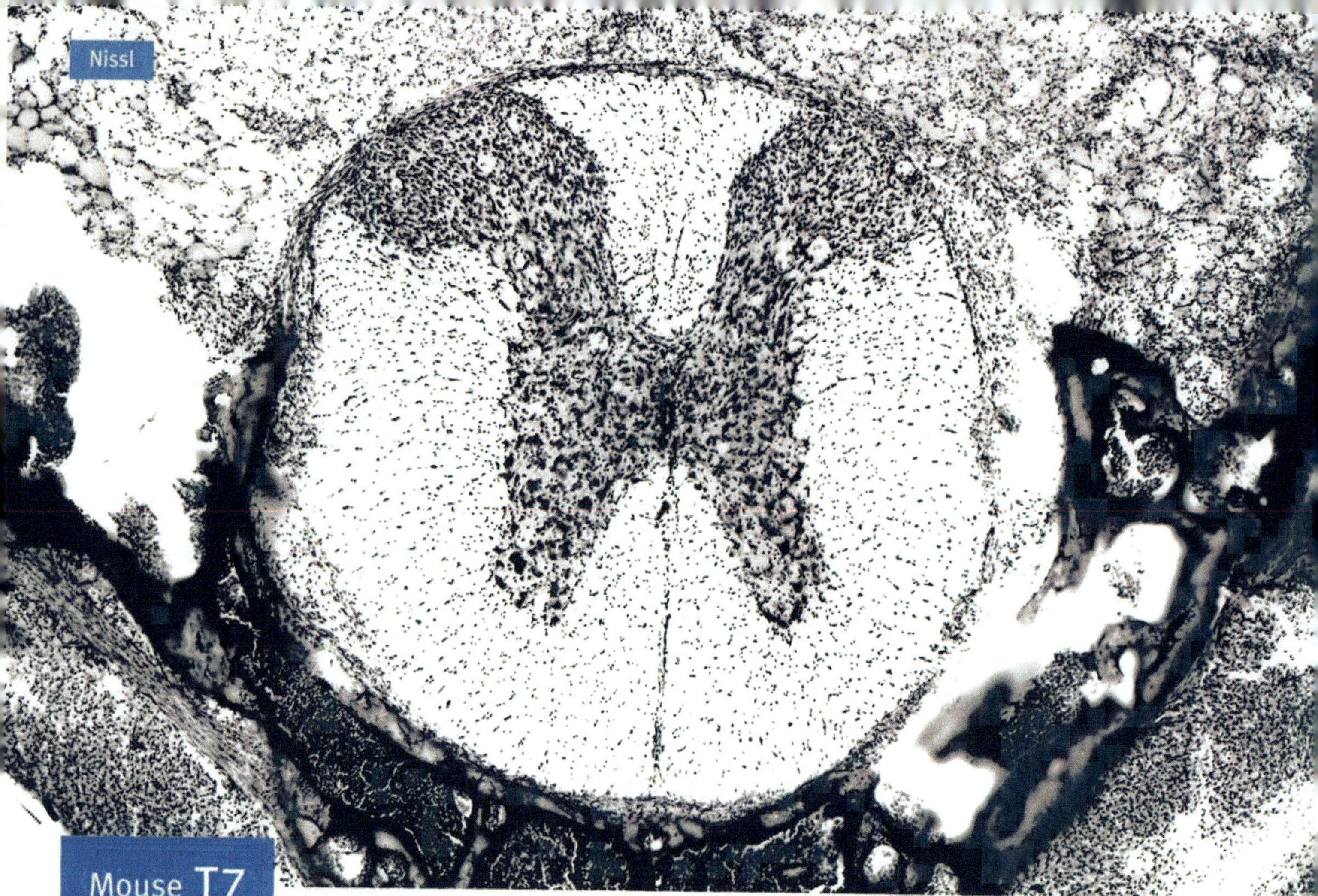
Nissl

Mouse T7

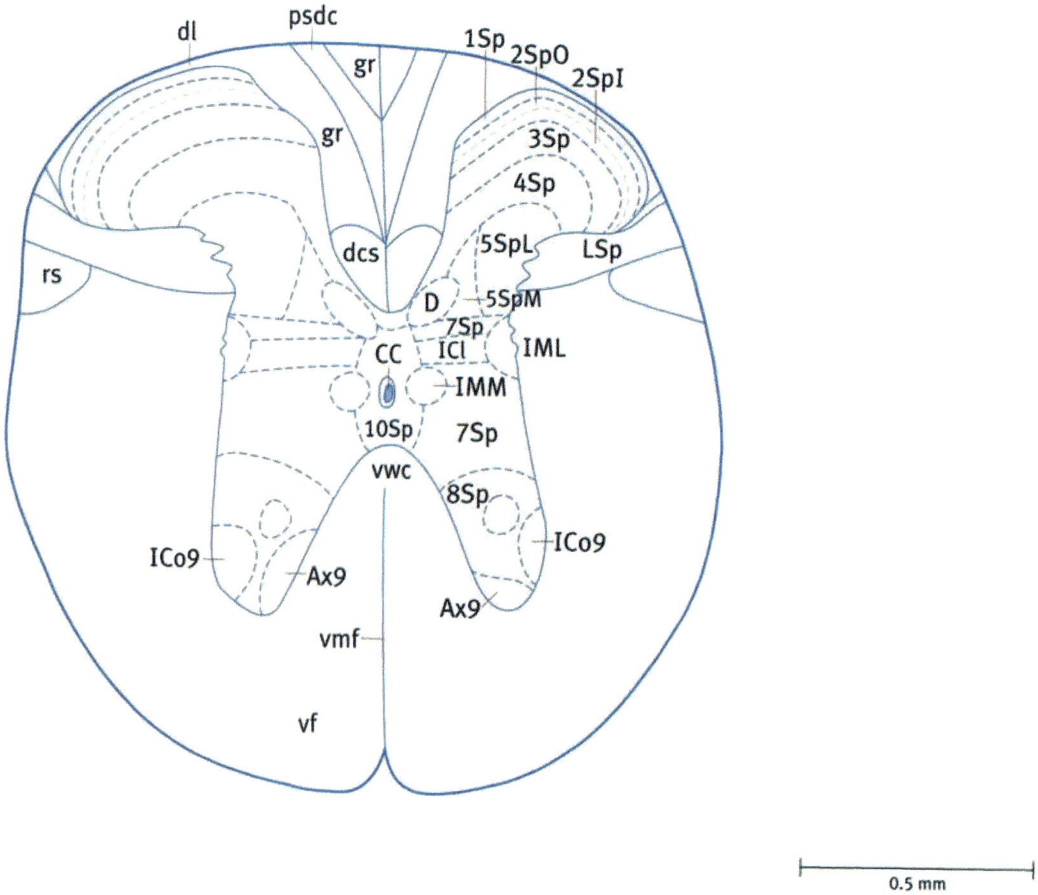
dl
psdc
gr
1Sp
2SpO
2SpI
gr
3Sp
4Sp
dcs
5SpL
LSp
rs
D
5SpM
7Sp
ICl
IML
CC
IMM
10Sp
7Sp
vwc
8Sp
ICo9
ICo9
Ax9
Ax9
vmf
vf
0.5 mm

Nissl

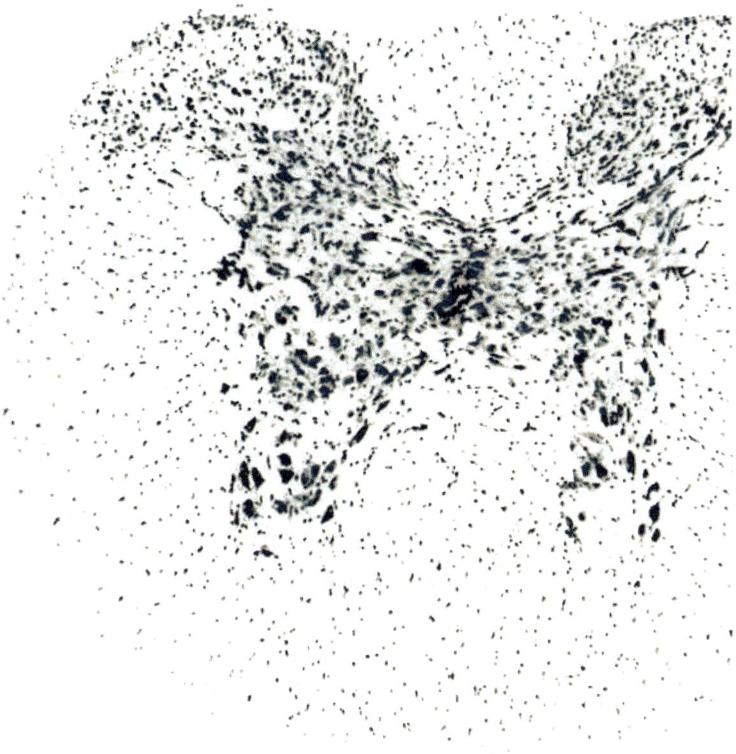

AChE

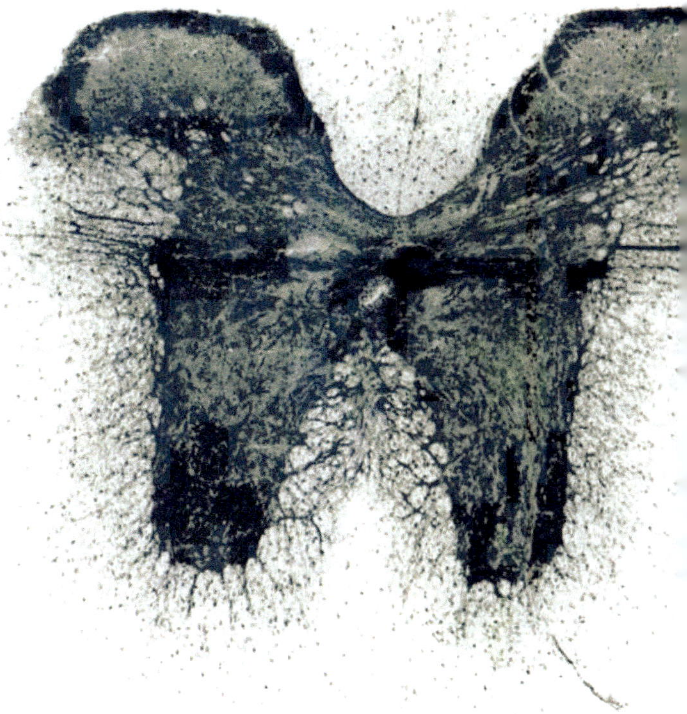

1Sp lamina 1	4Sp lamina 4	8Sp lamina 8	D dorsal nucleus	ICl intercalated nu	LSp lat spinal nu	vwc vent white com
2SpI lamina 2, inner	5SpL lamina 5, lat	10Sp lamina 10	dcs dors corticosp	ICo9 intercostals	psdc postsyn dors col	
2SpO lamina 2, outer	5SpM lamina 5, med	Ax9 axial muscles	dl dorsolat fasc	IML intermediolat	rs rubrospinal tr	
3Sp lamina 3	7Sp lamina 7	CC central canal	gr gracile fascic	IMM intermediomed	vmf vent med fiss	

Mouse T7

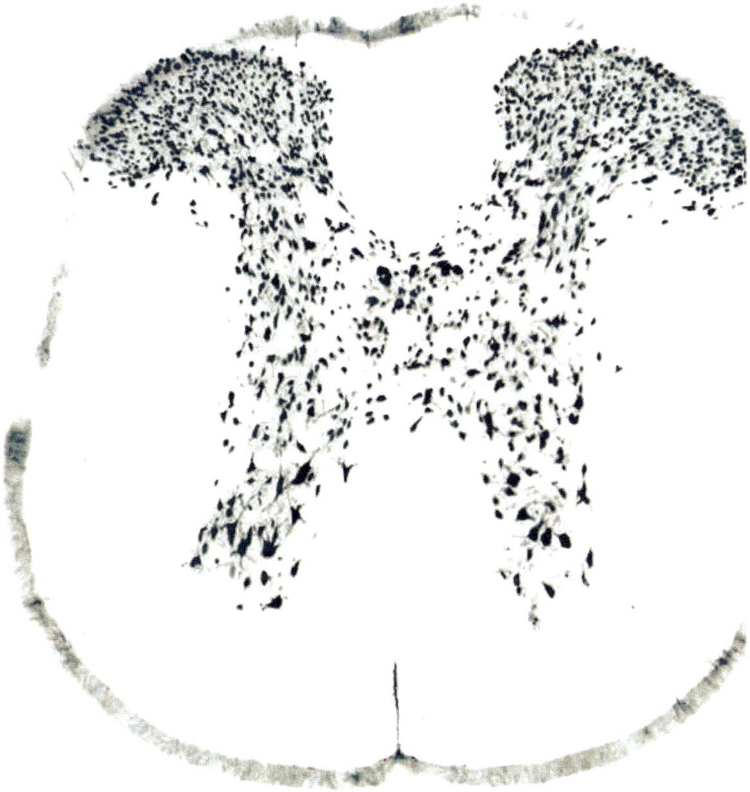

NeuN

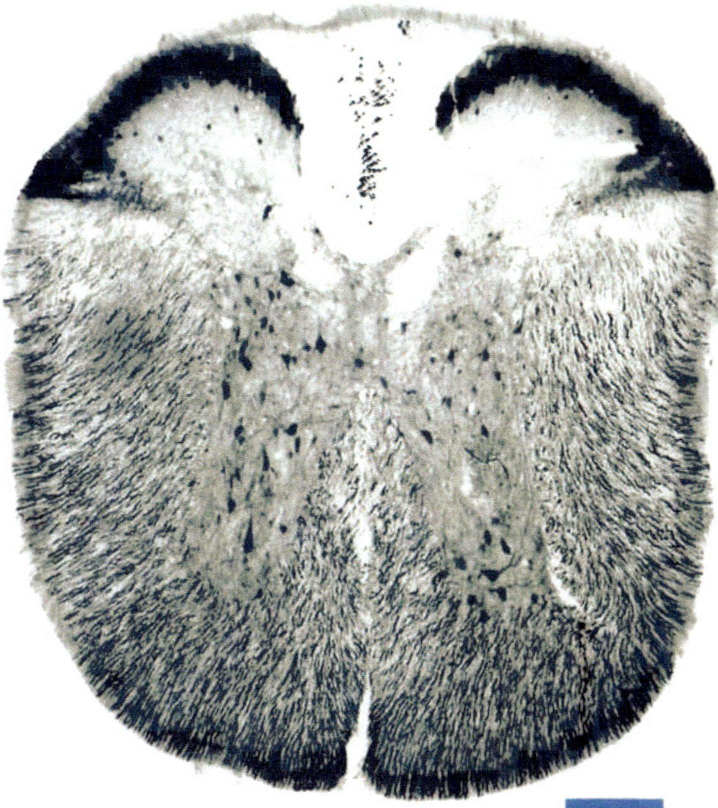

Cr

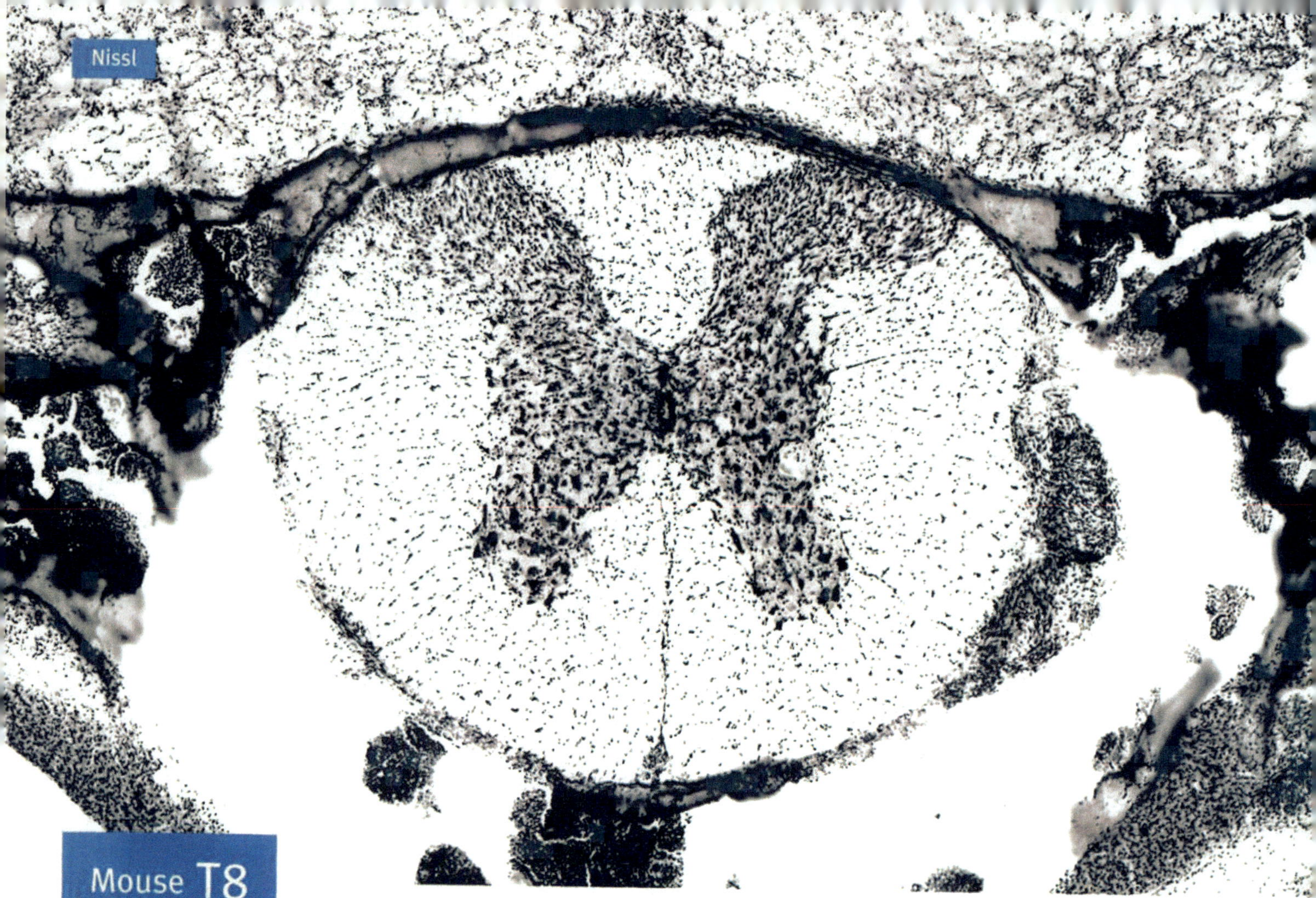
Nissl

Mouse T8

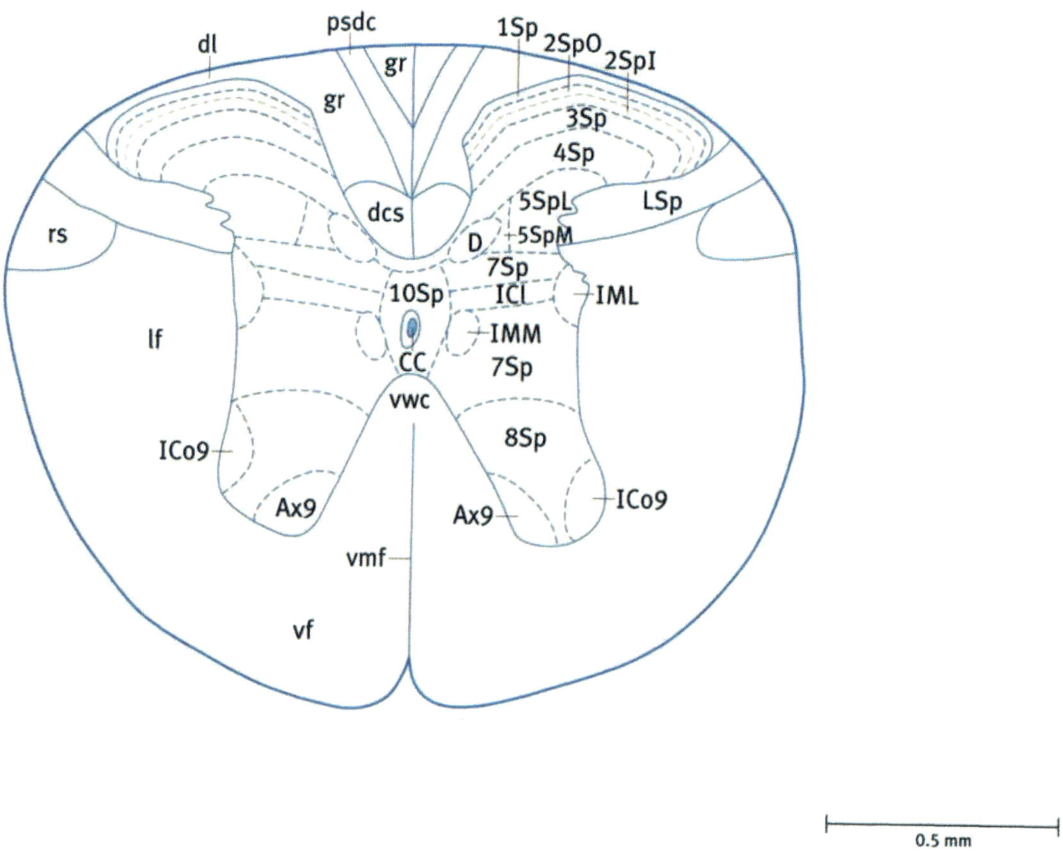
psdc
dl
1Sp
2SpO
2SpI
gr
gr
3Sp
4Sp
dcs
5SpL
LSp
rs
D
5SpM
7Sp
10Sp
ICl
IML
lf
IMM
CC
7Sp
vwc
8Sp
ICo9
Ax9
Ax9
ICo9
vmf
vf
0.5 mm

Nissl

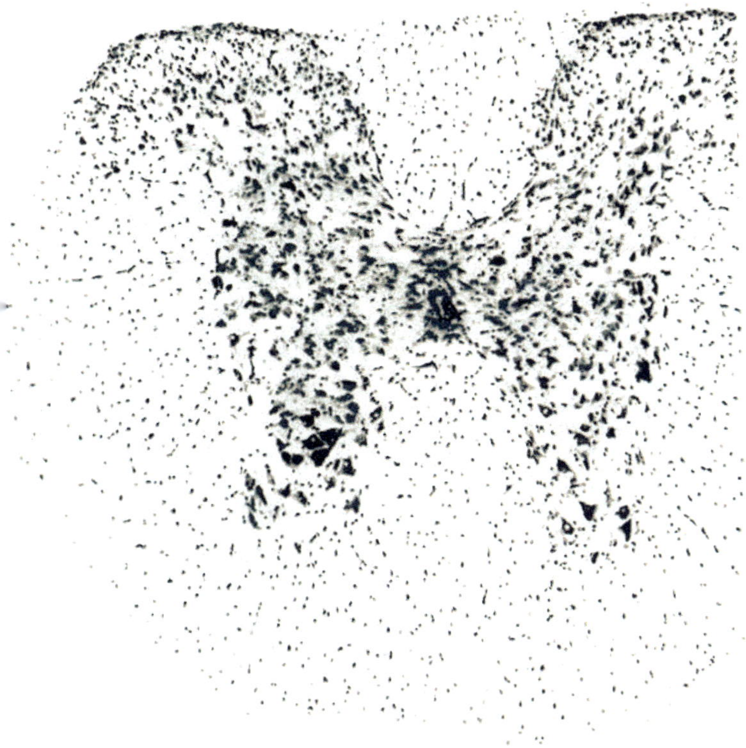

AChE

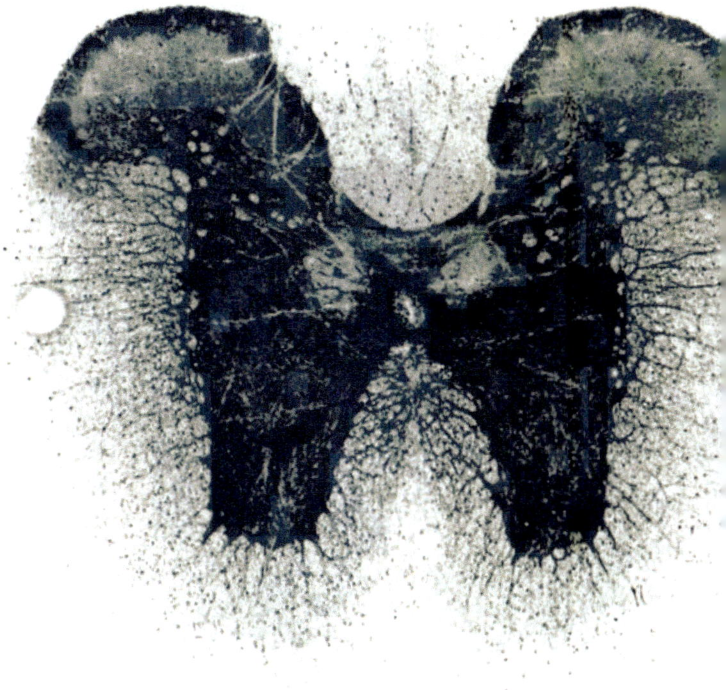

1Sp lamina 1
2SpI lamina 2, inner
2SpO lamina 2, outer
3Sp lamina 3
4Sp lamina 4
5SpL lamina 5, lat
5SpM lamina 5, med
7Sp lamina 7
8Sp lamina 8
10Sp lamina 10
Ax9 axial muscles
CC central canal
D dorsal nucleus
dcs dors corticosp
dl dorsolat fasc
gr gracile fascic
ICl intercalated nu
ICo9 intercostals
IML intermediolat
IMM intermediomed
LSp lat spinal nu
psdc postsyn dors col
rs rubrospinal tr
vmf vent med fiss
vwc vent white com

Mouse T8

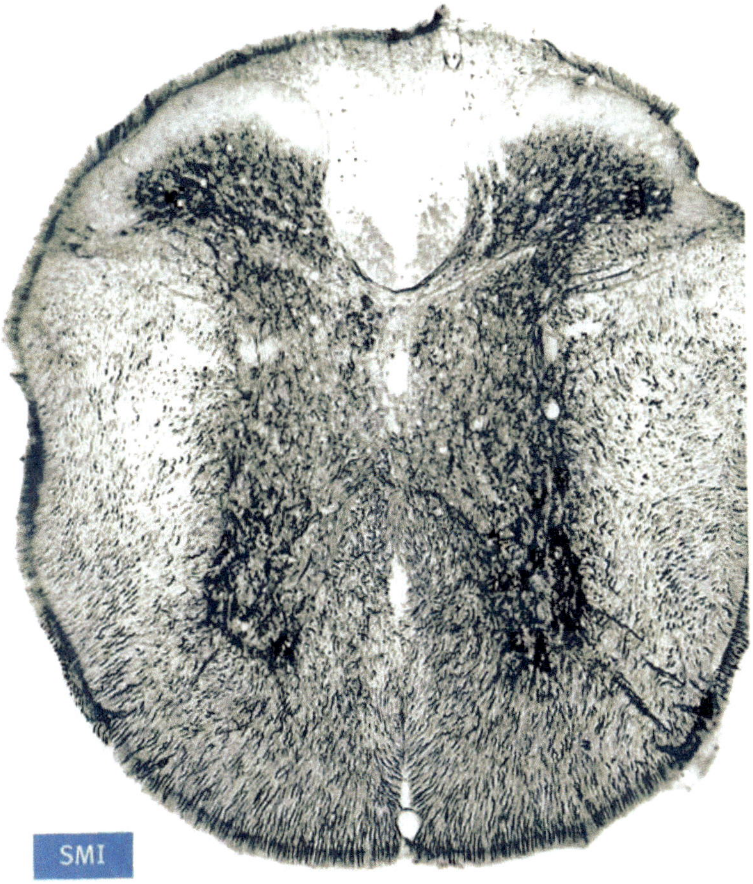

SMI

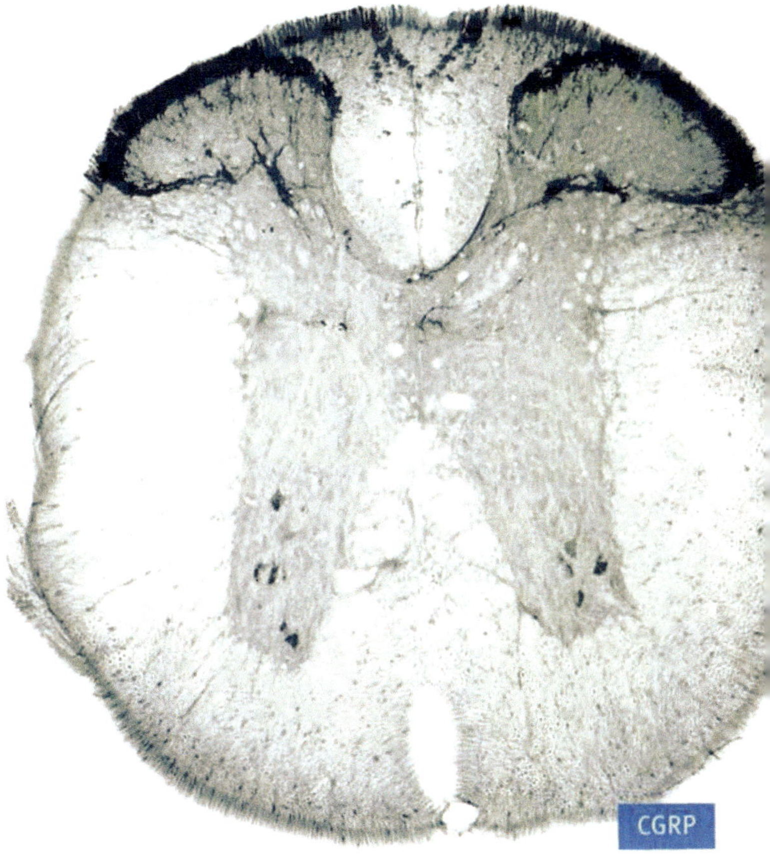

CGRP

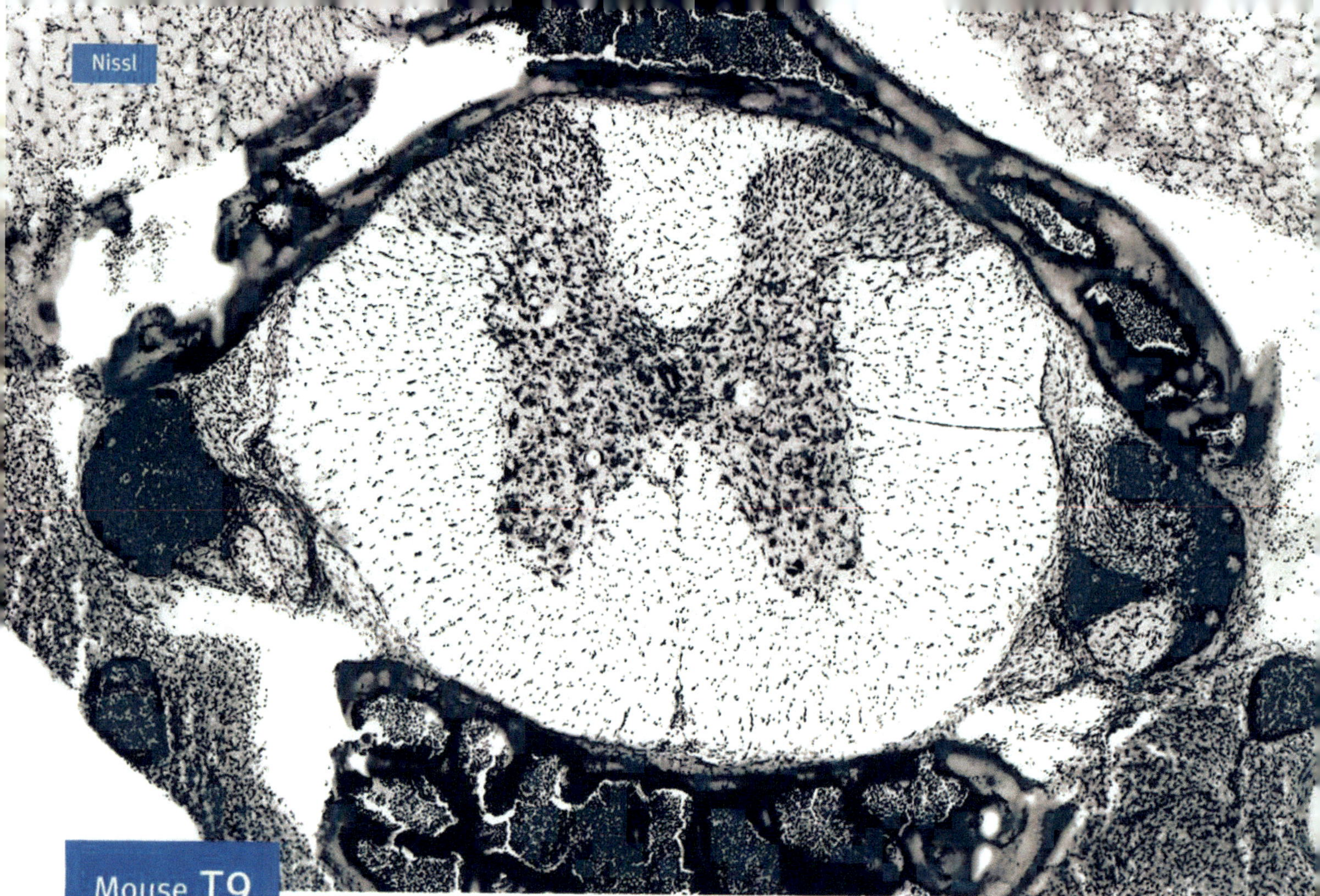
Nissl

Mouse T9

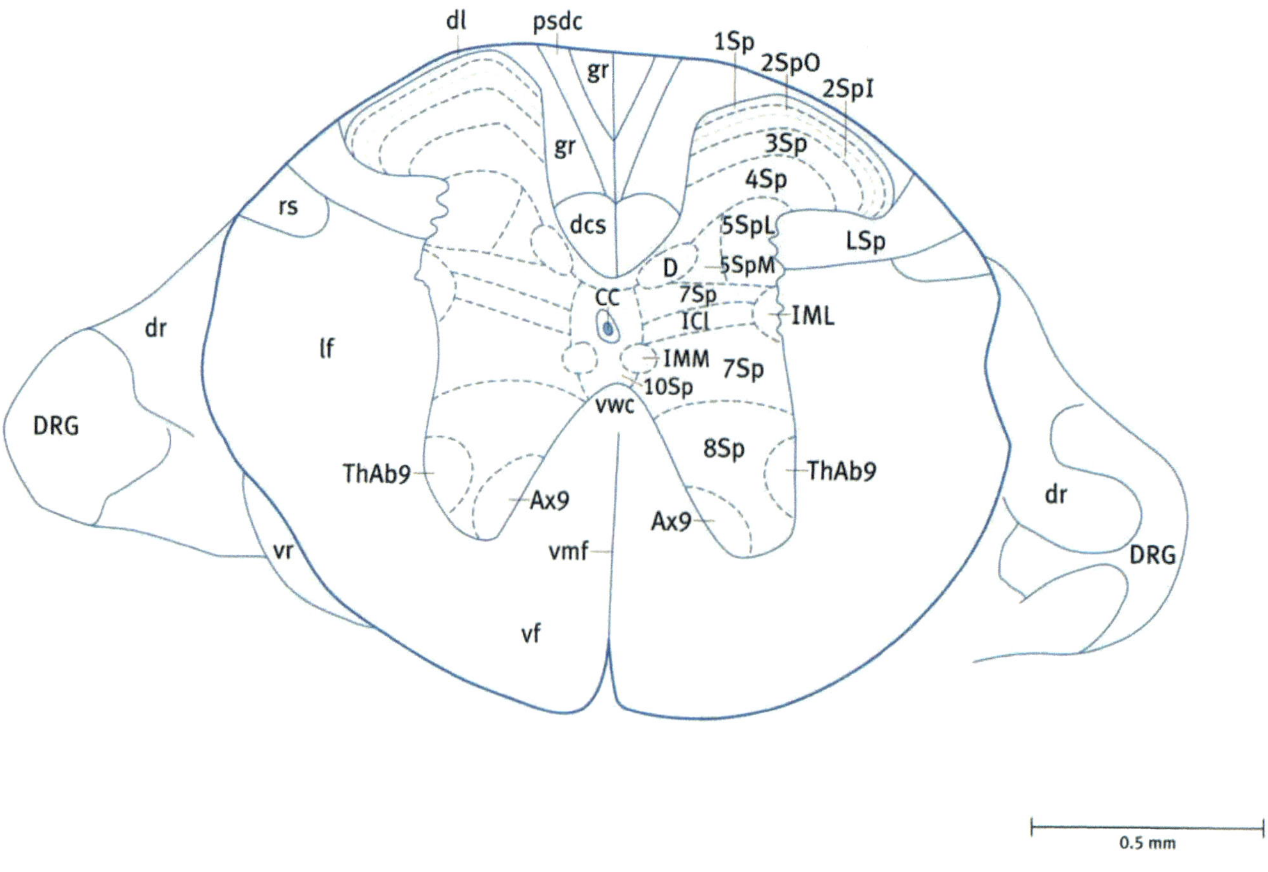
dl
psdc
gr
1Sp
2SpO
2SpI
gr
3Sp
4Sp
rs
dcs
5SpL
LSp
D
5SpM
CC
7Sp
dr
ICl
IML
lf
IMM
7Sp
10Sp
vwc
DRG
8Sp
ThAb9
ThAb9
Ax9
dr
Ax9
vr
vmf
DRG
vf
0.5 mm

Nissl

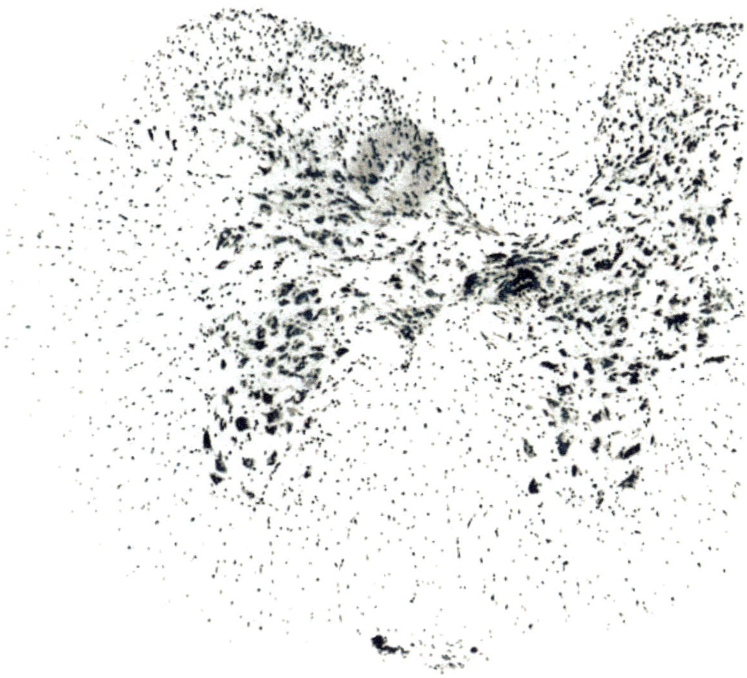

AChE

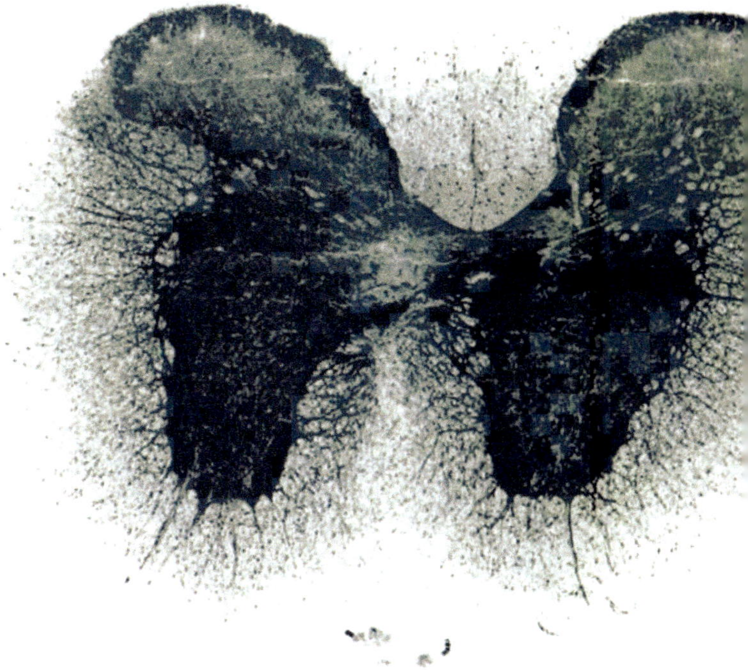

1Sp lamina 1
2SpI lamina 2, inner
2SpO lamina 2, outer
3Sp lamina 3
4Sp lamina 4
5SpL lamina 5, lat
5SpM lamina 5, med
7Sp lamina 7
8Sp lamina 8
10Sp lamina 10
Ax9 axial muscles
CC central canal
D dorsal nucleus
dcs dors corticosp
dl dorsolat fasc
dr dorsal root
DRG dors root gang
gr gracile fascic
ICl intercalated nu
IML intermediolat
IMM intermediomed
LSp lat spinal nu
psdc postsyn dors col
rs rubrospinal tr
ThAb9 thor-abd wall
vmf vent med fiss
vr ventral root
vwc vent white com

Mouse T9

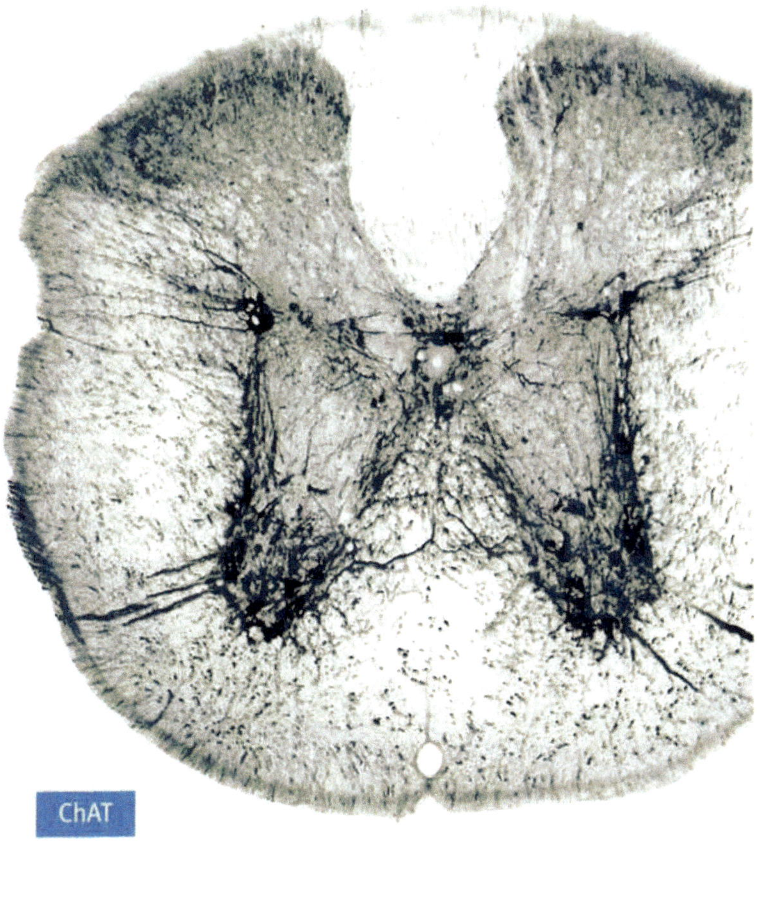

ChAT

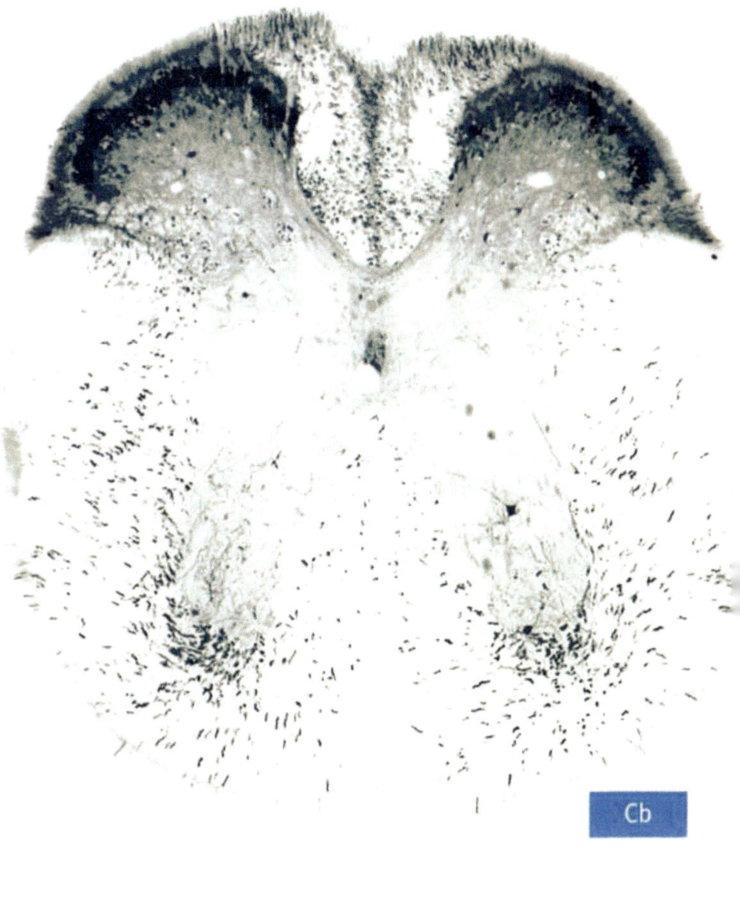

Cb

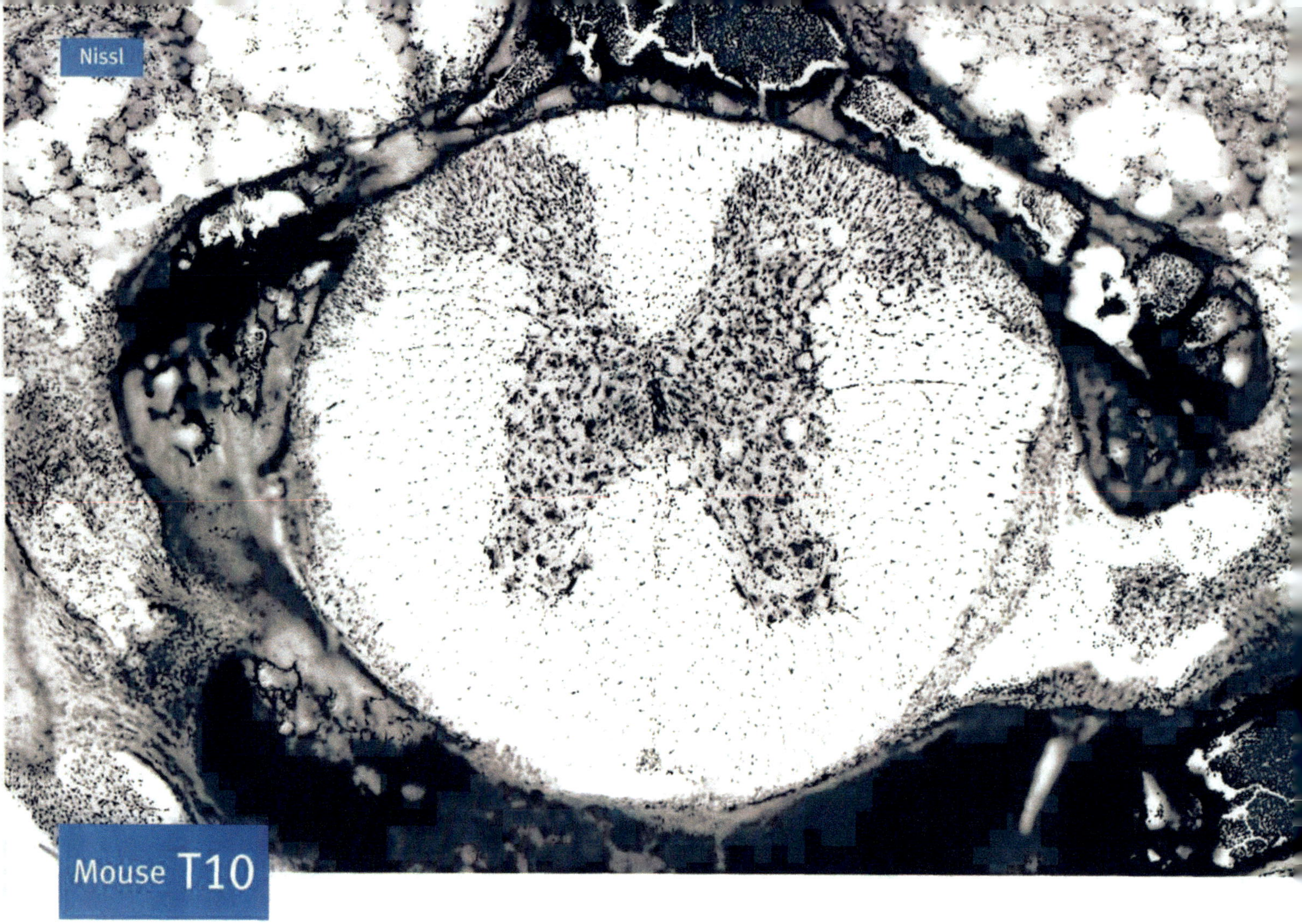
Nissl
Mouse T10

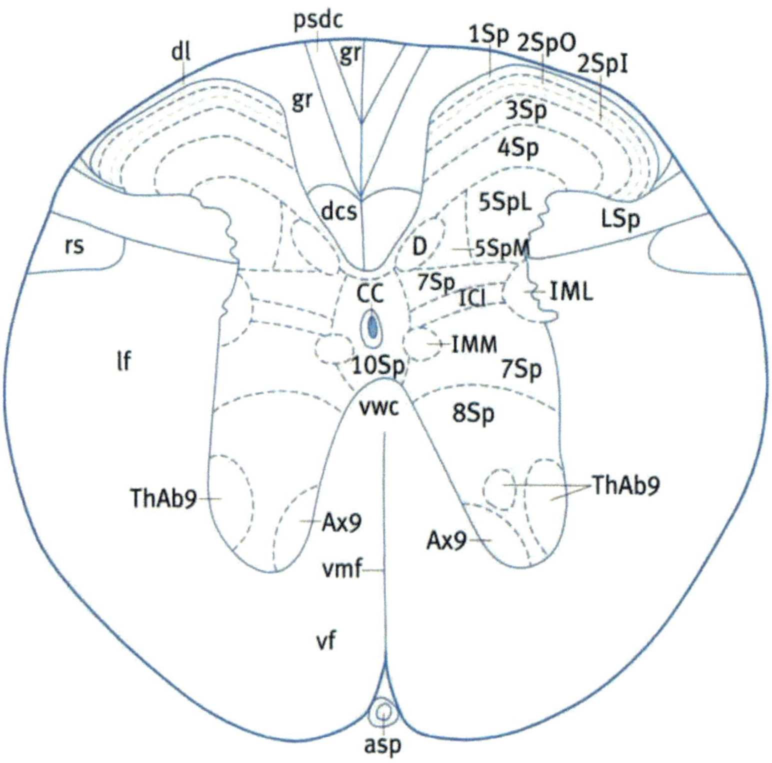
psdc
gr
dl
1Sp
2SpO
2SpI
gr
3Sp
4Sp
dcs
5SpL
LSp
rs
D
5SpM
7Sp
CC
ICl
IML
IMM
lf
10Sp
7Sp
vwc
8Sp
ThAb9
ThAb9
Ax9
Ax9
vmf
vf
asp

0.5 mm

Nissl

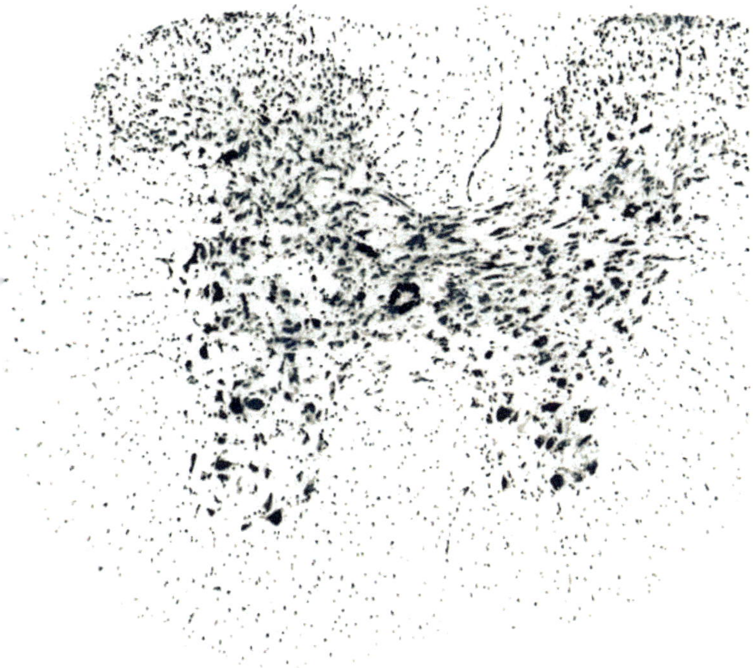

AChE

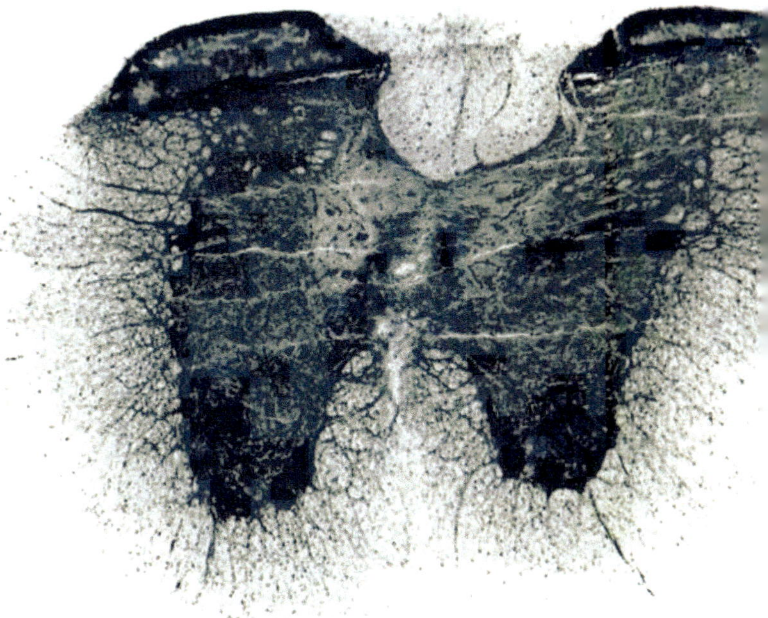

1Sp lamina 1
2SpI lamina 2, inner
2SpO lamina 2, outer
3Sp lamina 3
4Sp lamina 4
5SpL lamina 5, lat
5SpM lamina 5, med
7Sp lamina 7
8Sp lamina 8
10Sp lamina 10
asp ant spinal art
Ax9 axial muscles
CC central canal
D dorsal nucleus
dcs dors corticosp
dl dorsolat fasc
gr gracile fascic
ICl intercalated nu
IML intermediolat
IMM intermediomed
LSp lat spinal nu
psdc postsyn dors col
rs rubrospinal tr
ThAb9 thor-abd wall
vmf vent med fiss
vwc vent white com

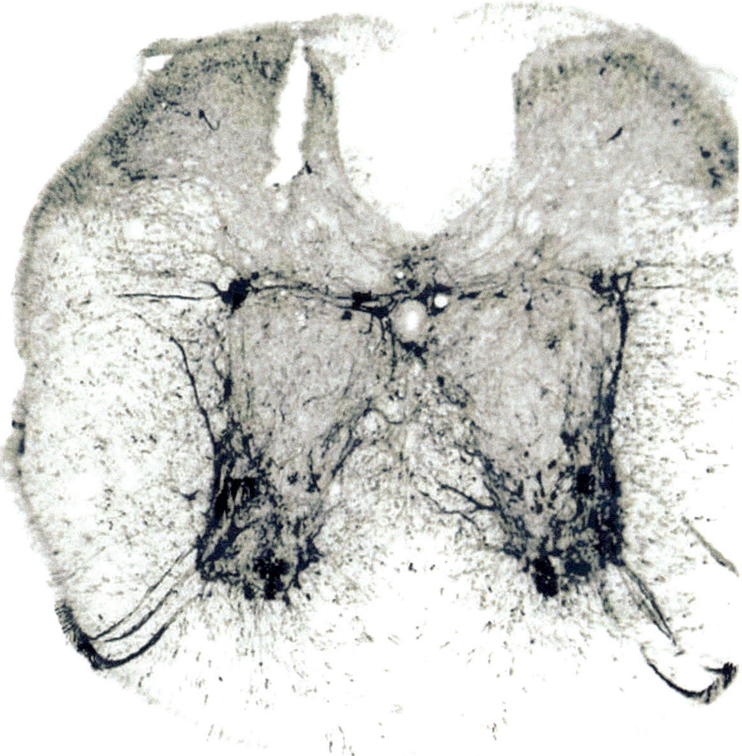

ChAT

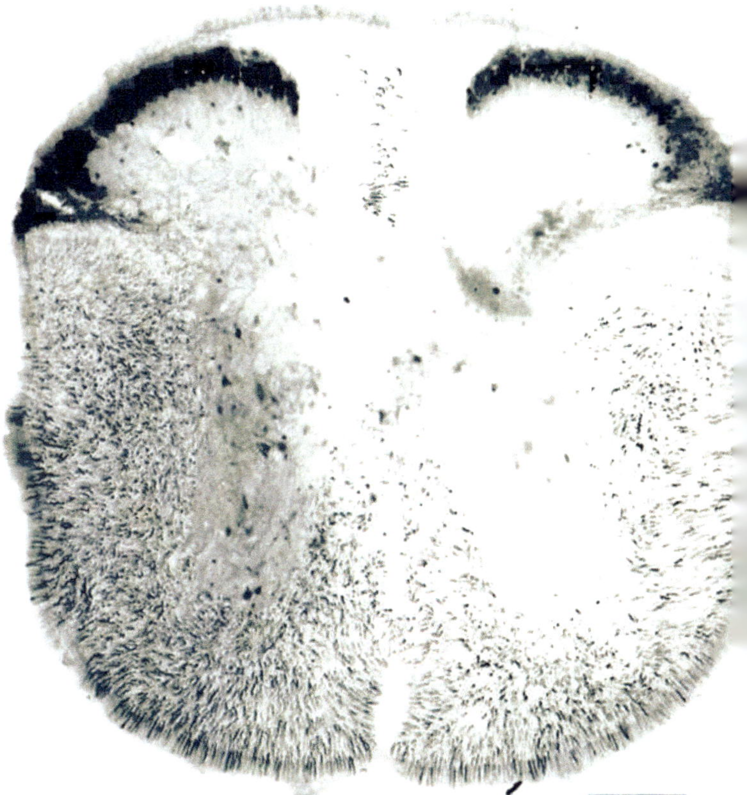

Cr

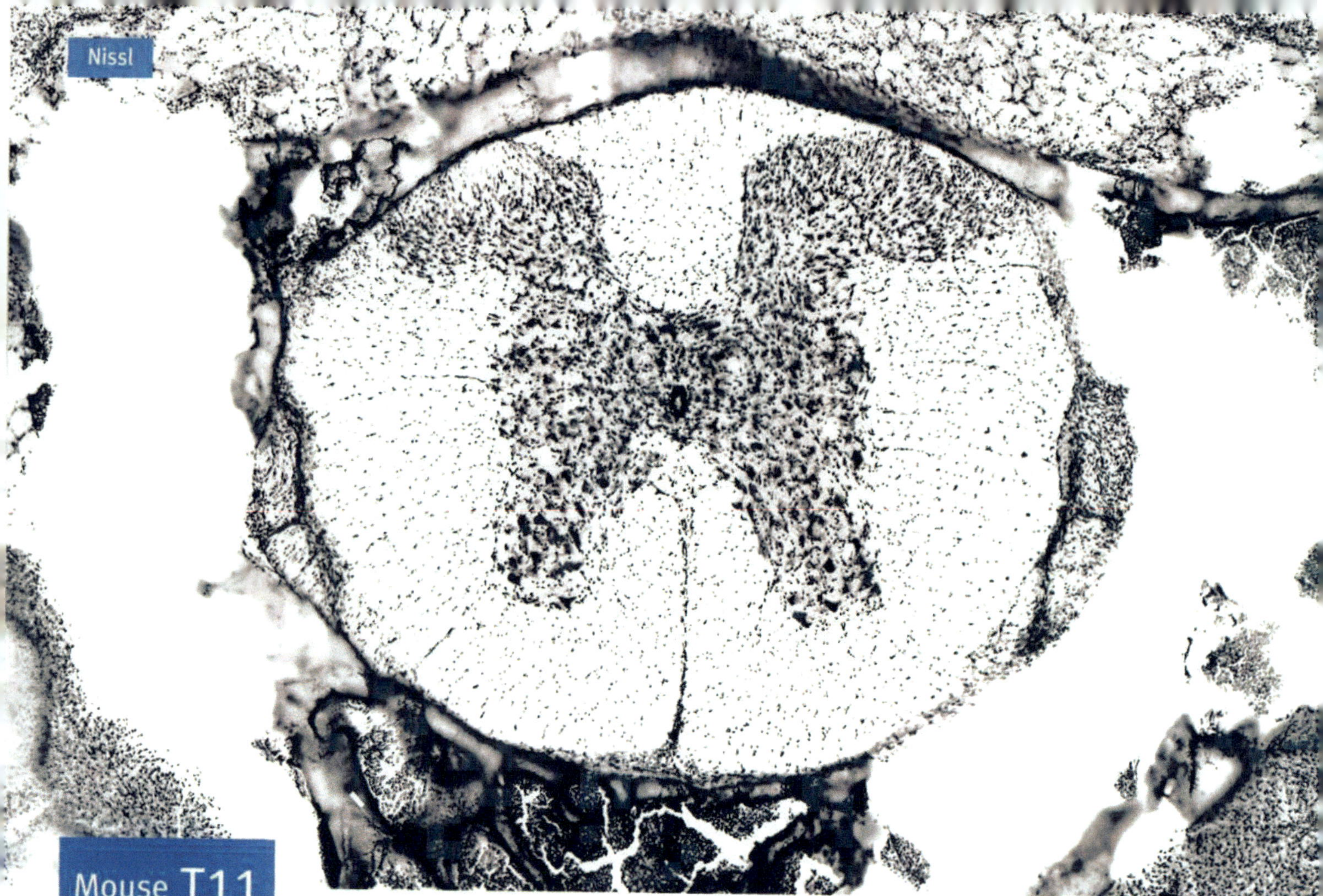
Nissl

Mouse T11

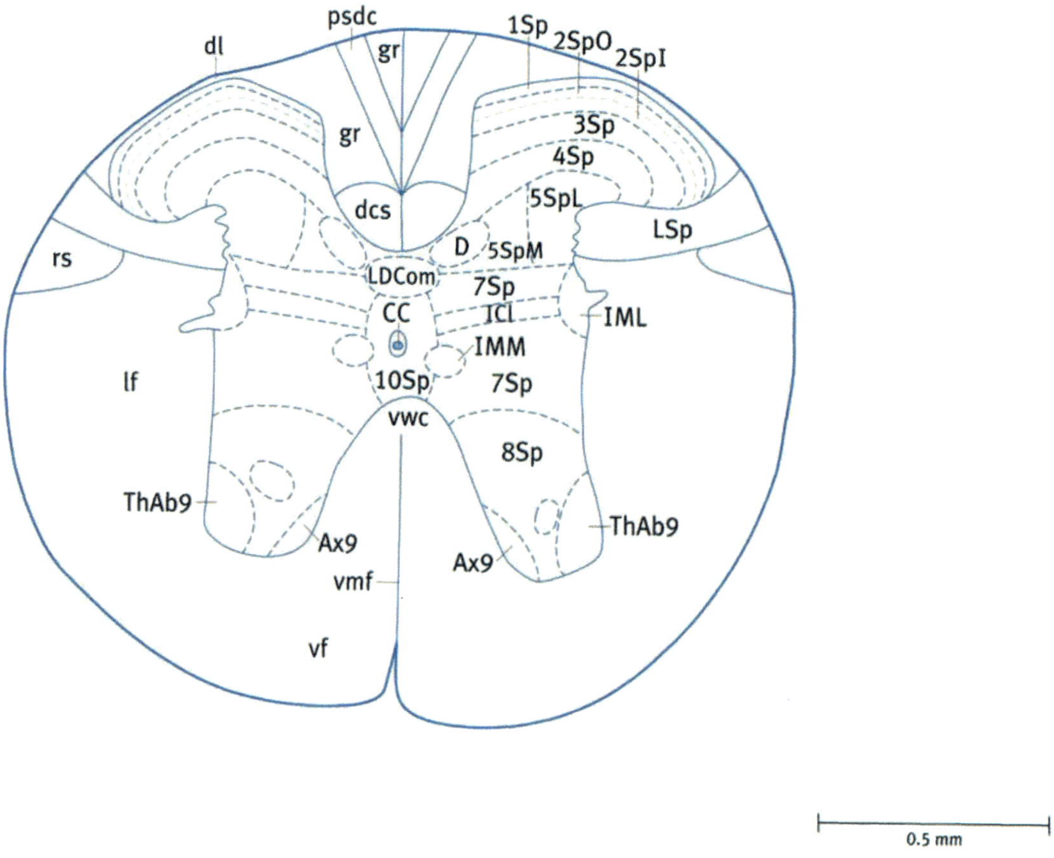
psdc
dl
gr
1Sp
2SpO
2SpI
gr
3Sp
4Sp
dcs
5SpL
LSp
rs
D
5SpM
LDCom
7Sp
CC
ICl
IML
IMM
lf
10Sp
7Sp
vwc
8Sp
ThAb9
ThAb9
Ax9
Ax9
vmf
vf
0.5 mm

Nissl

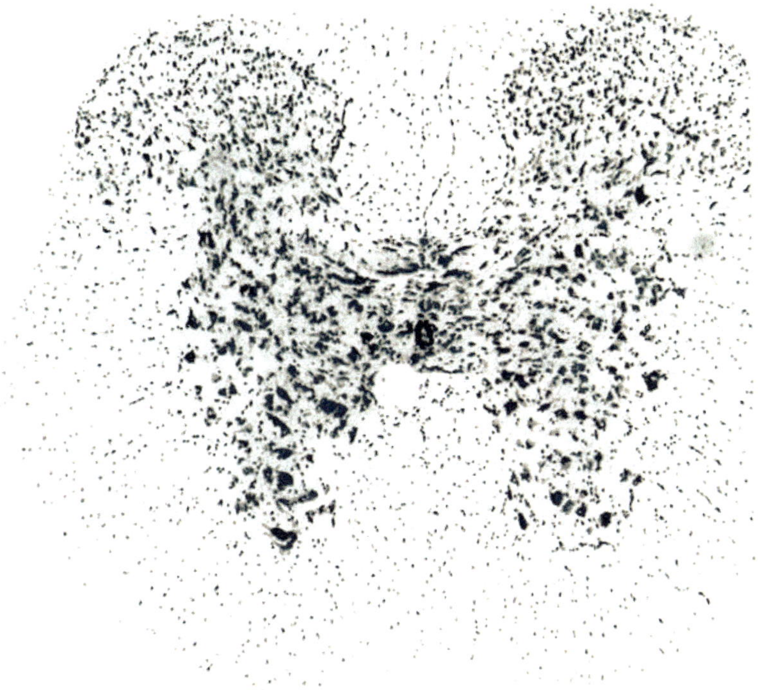

AChE

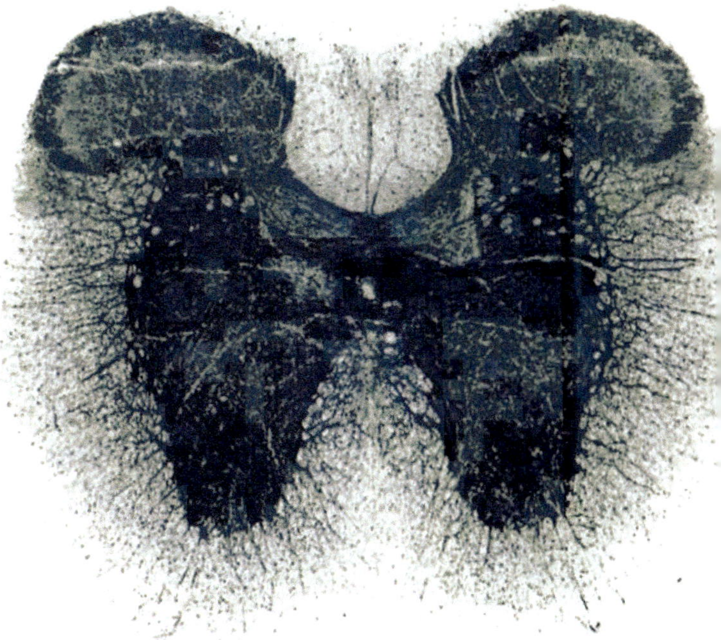

1Sp lamina 1
2SpI lamina 2, inner
2SpO lamina 2, outer
3Sp lamina 3
4Sp lamina 4
5SpL lamina 5, lat
5SpM lamina 5, med
7Sp lamina 7
8Sp lamina 8
10Sp lamina 10
Ax9 axial muscles
CC central canal
D dorsal nucleus
dcs dors corticosp
dl dorsolat fasc
gr gracile fascic
ICl intercalated nu
IML intermediolat
IMM intermediomed
LDCom lum dors com
LSp lat spinal nu
psdc postsyn dors col
rs rubrospinal tr
ThAb9 thor-abd wall
vmf vent med fiss
vwc vent white com

Mouse T11

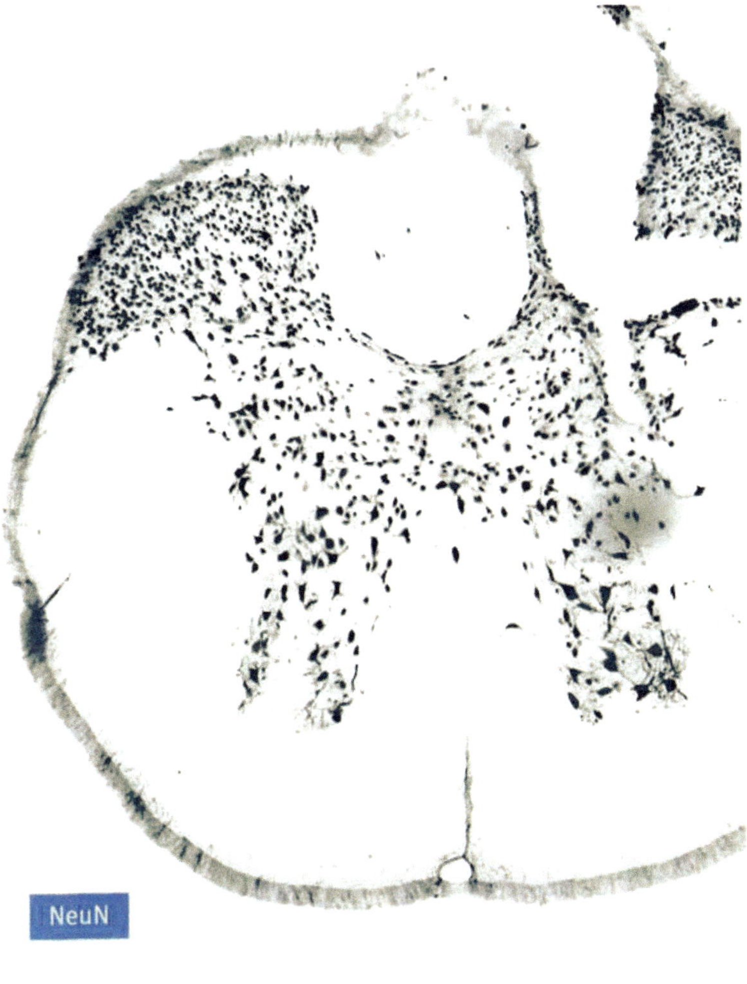

NeuN

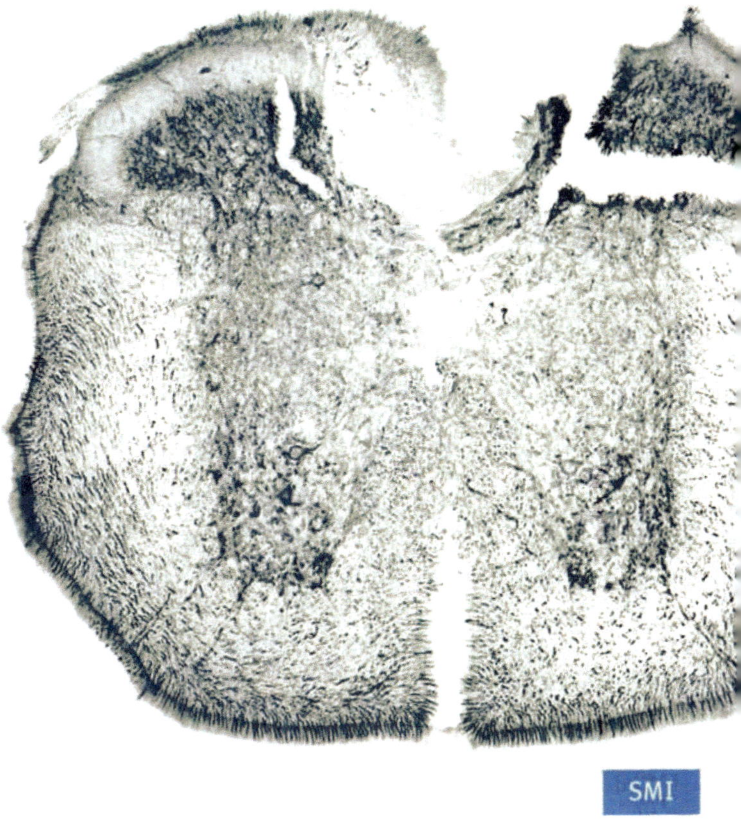

SMI

Nissl

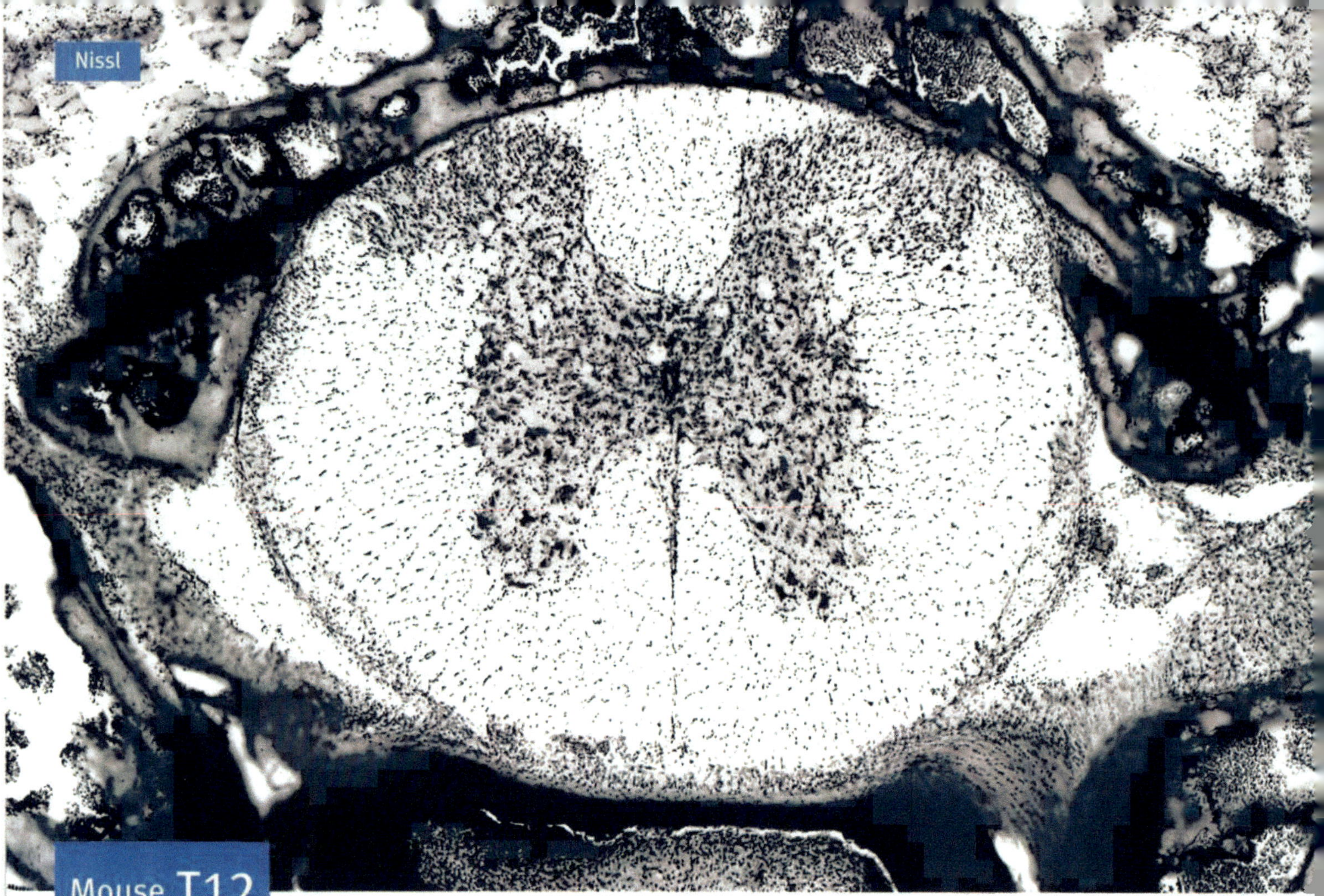

Mouse T12

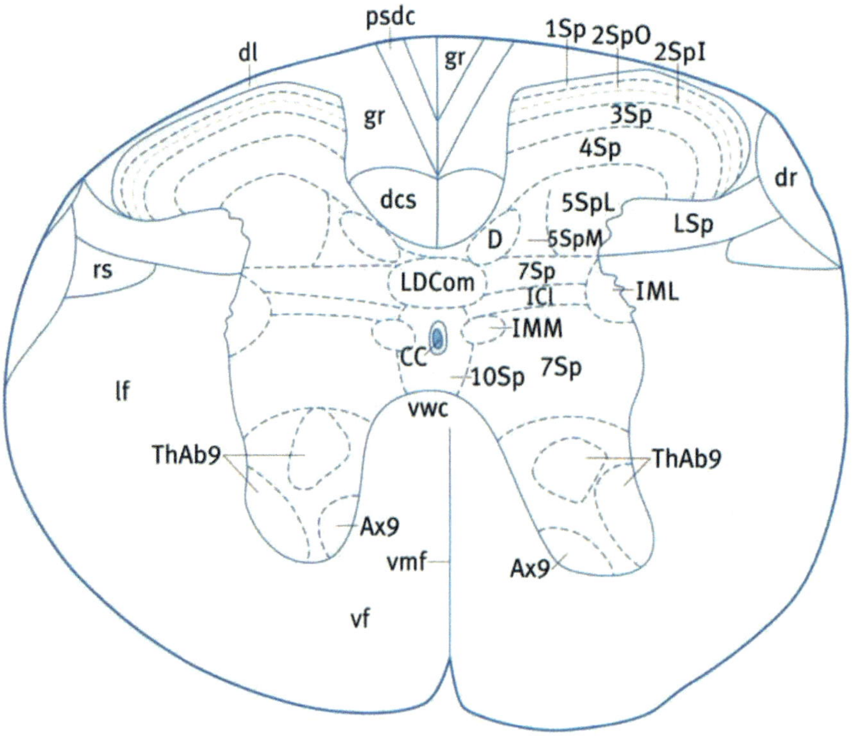

0.5 mm

Nissl

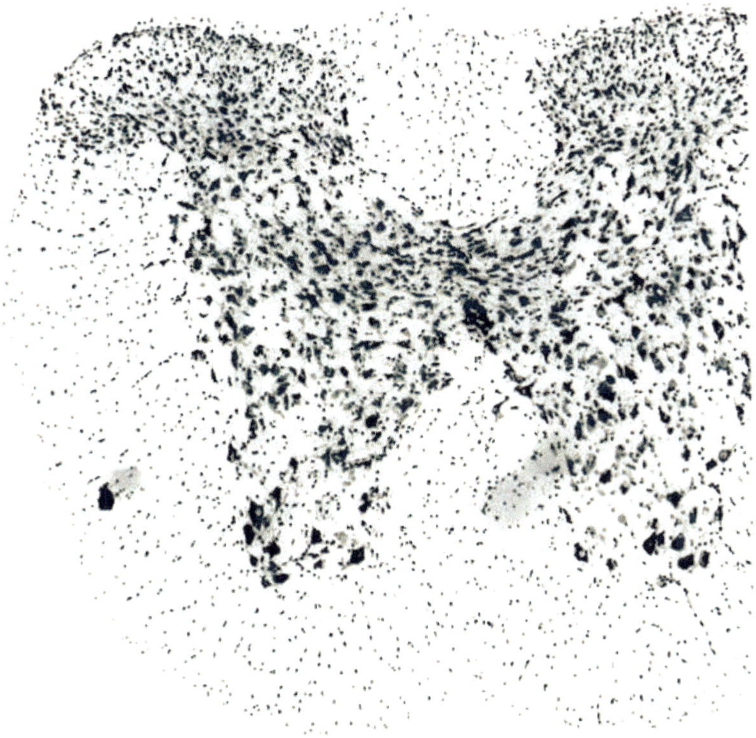

AChE

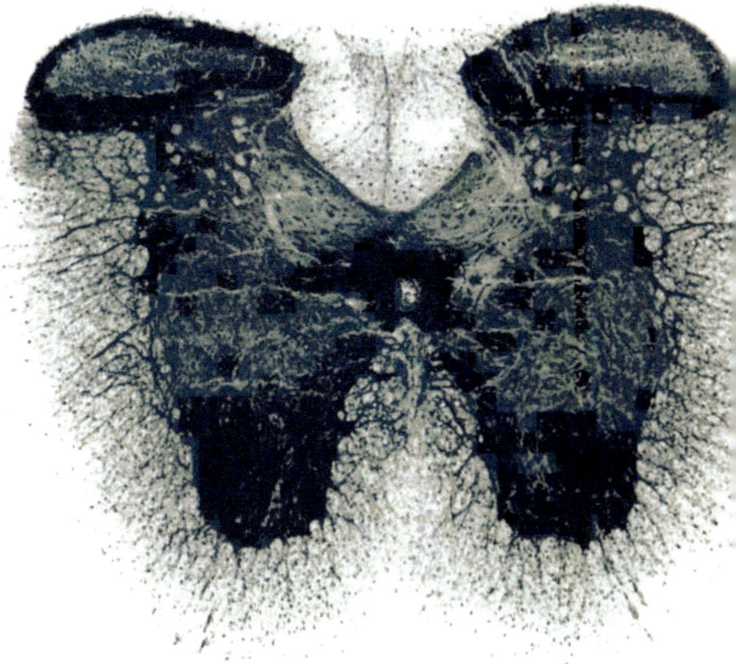

1Sp lamina 1
2SpI lamina 2, inner
2SpO lamina 2, outer
3Sp lamina 3
4Sp lamina 4
5SpL lamina 5, lat
5SpM lamina 5, med
7Sp lamina 7
10Sp lamina 10
Ax9 axial muscles
CC central canal
D dorsal nucleus
dcs dors corticosp
dl dorsolat fasc
dr dorsal root
gr gracile fascic
ICl intercalated nu
IML intermediolat
IMM intermediomed
LDCom lum dors com
LSp lat spinal nu
psdc postsyn dors col
rs rubrospinal tr
ThAb9 thor-abd wall
vmf vent med fiss
vwc vent white com

Mouse T12

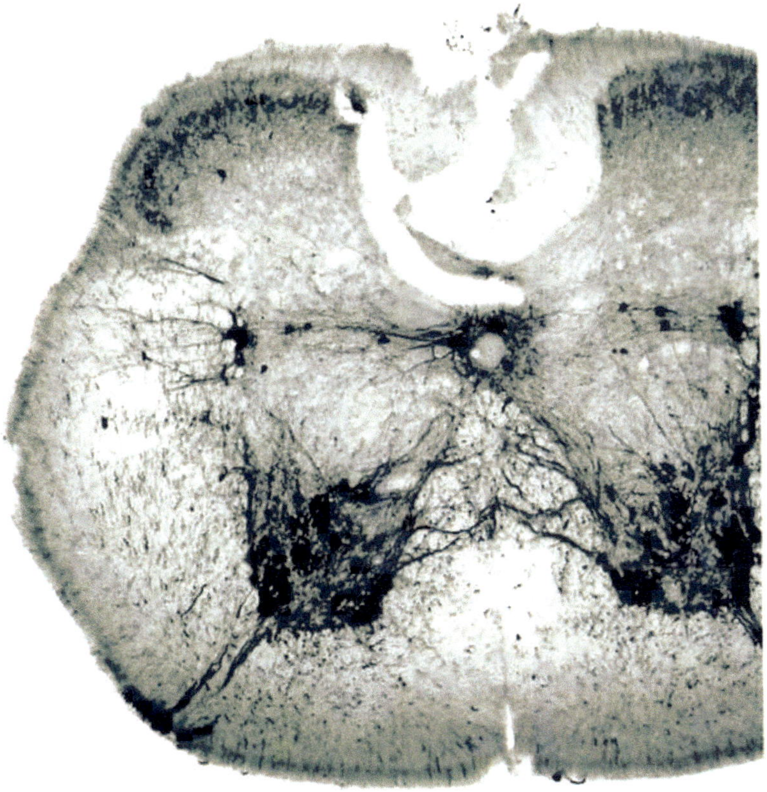

ChAT

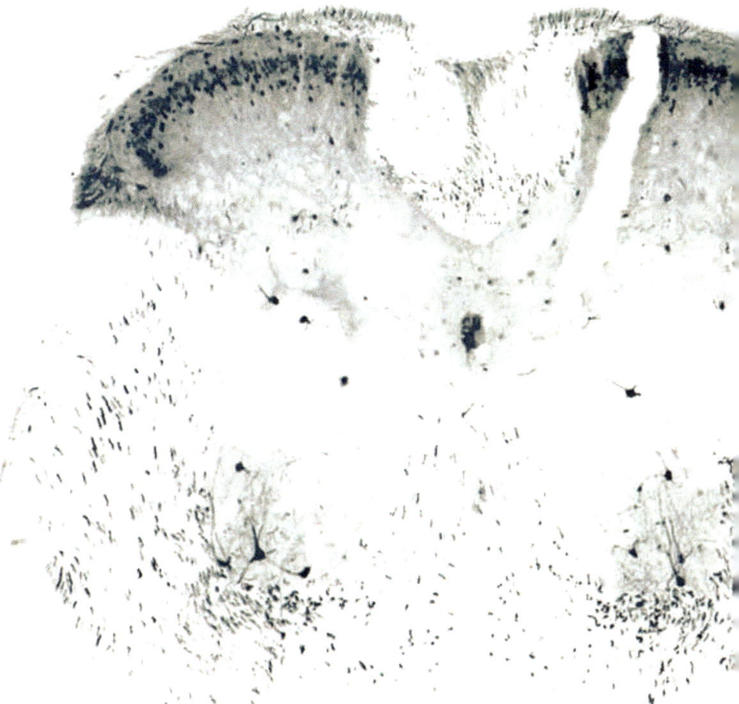

Cb

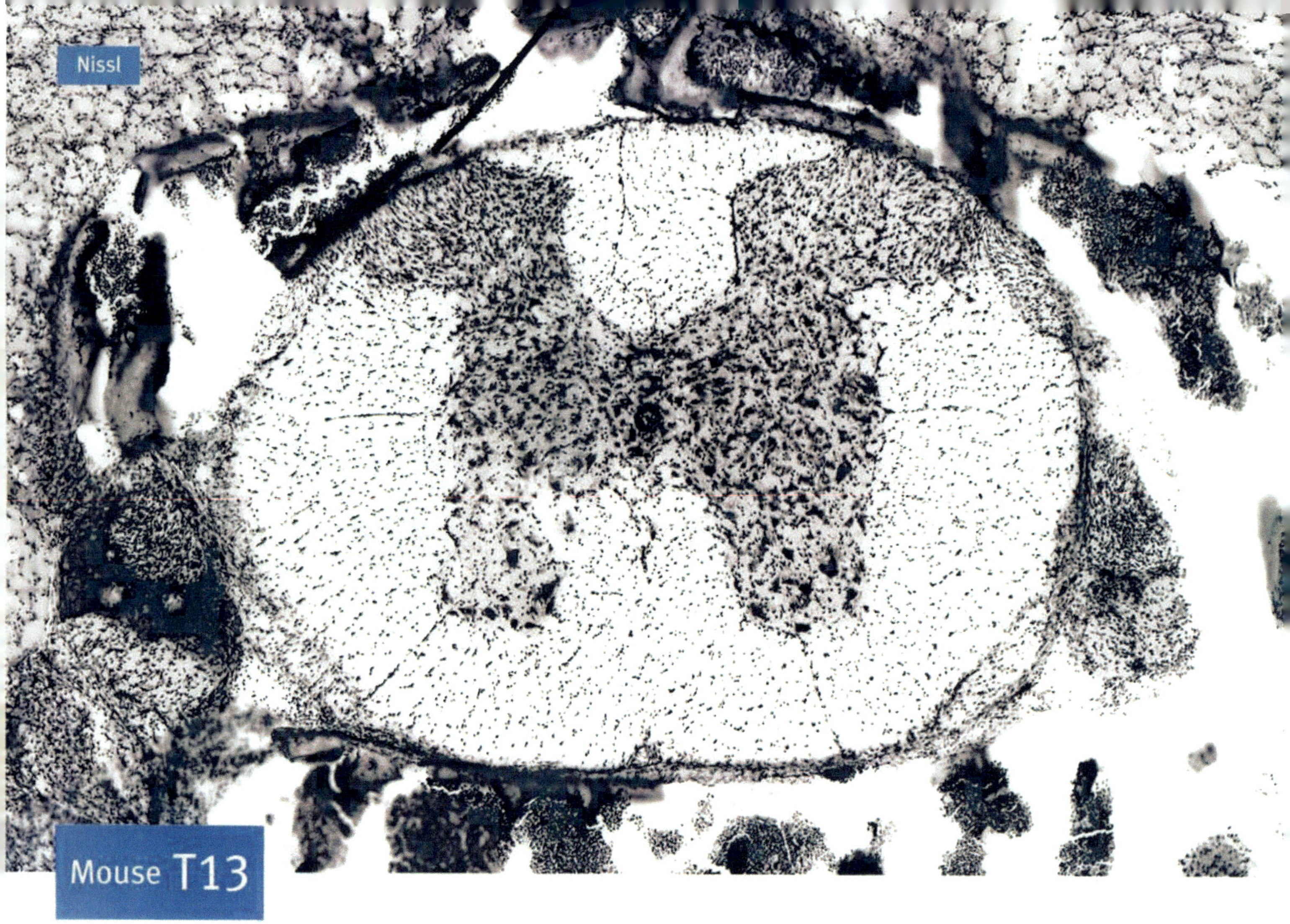
Nissl
Mouse T13

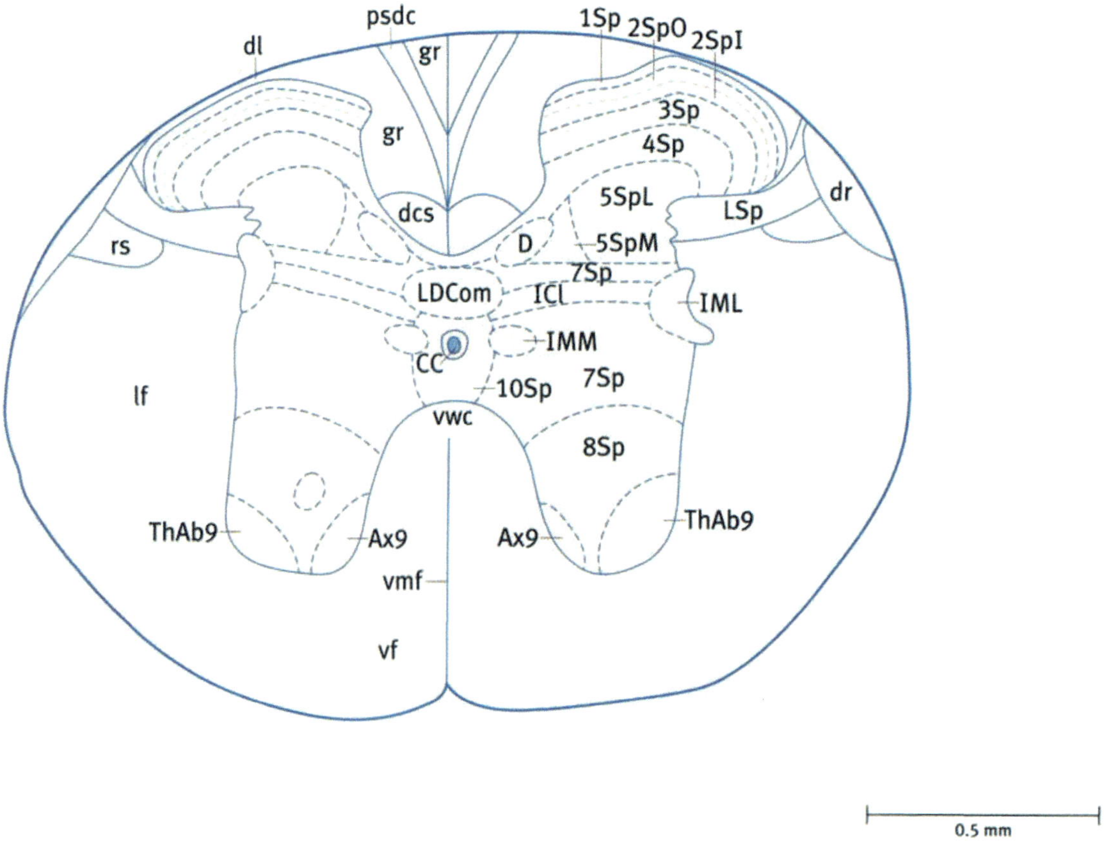
psdc
dl
gr
1Sp
2SpO
2SpI
3Sp
4Sp
gr
5SpL
dcs
LSp
dr
rs
D
5SpM
7Sp
LDCom
ICl
IML
IMM
CC
10Sp
7Sp
lf
vwc
8Sp
ThAb9
Ax9
Ax9
ThAb9
vmf
vf
0.5 mm

Nissl

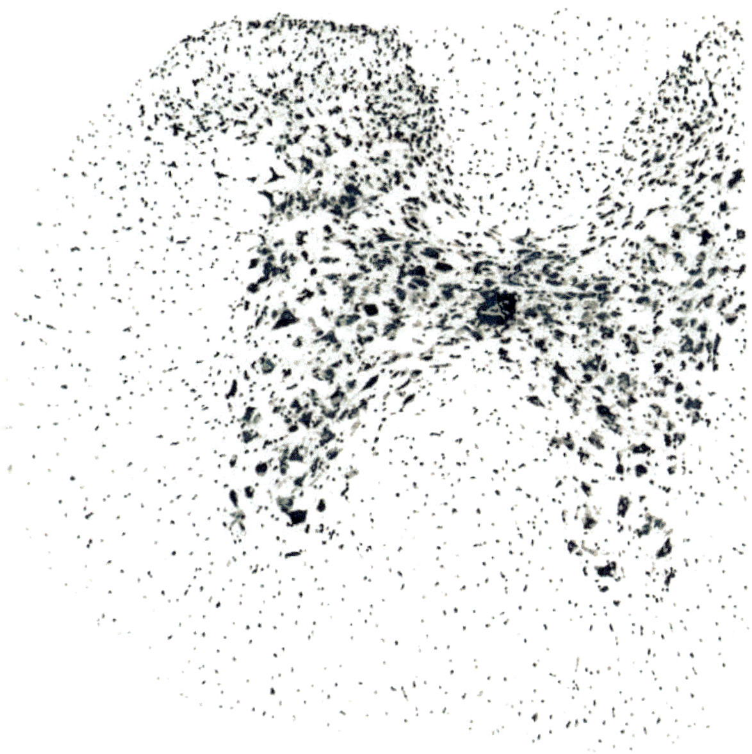

AChE

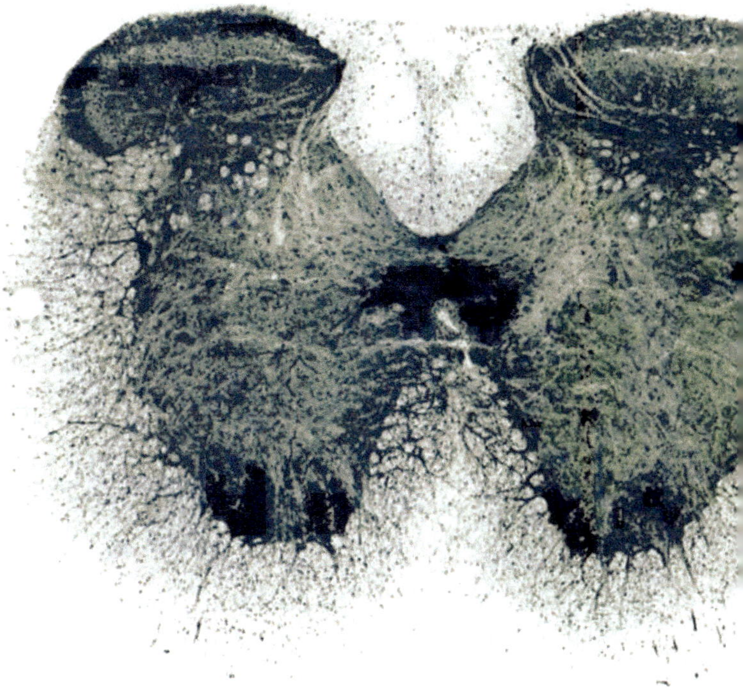

1Sp lamina 1
2SpI lamina 2, inner
2SpO lamina 2, outer
3Sp lamina 3
4Sp lamina 4
5SpL lamina 5, lat
5SpM lamina 5, med
7Sp lamina 7
8Sp lamina 8
10Sp lamina 10
Ax9 axial muscles
CC central canal
D dorsal nucleus
dcs dors corticosp
dl dorsolat fasc
dr dorsal root
gr gracile fascic
ICl intercalated nu
IML intermediolat
IMM intermediomed
LDCom lum dors com
LSp lat spinal nu
psdc postsyn dors col
rs rubrospinal tr
ThAb9 thor-abd wall
vmf vent med fiss
vwc vent white com

Mouse T13

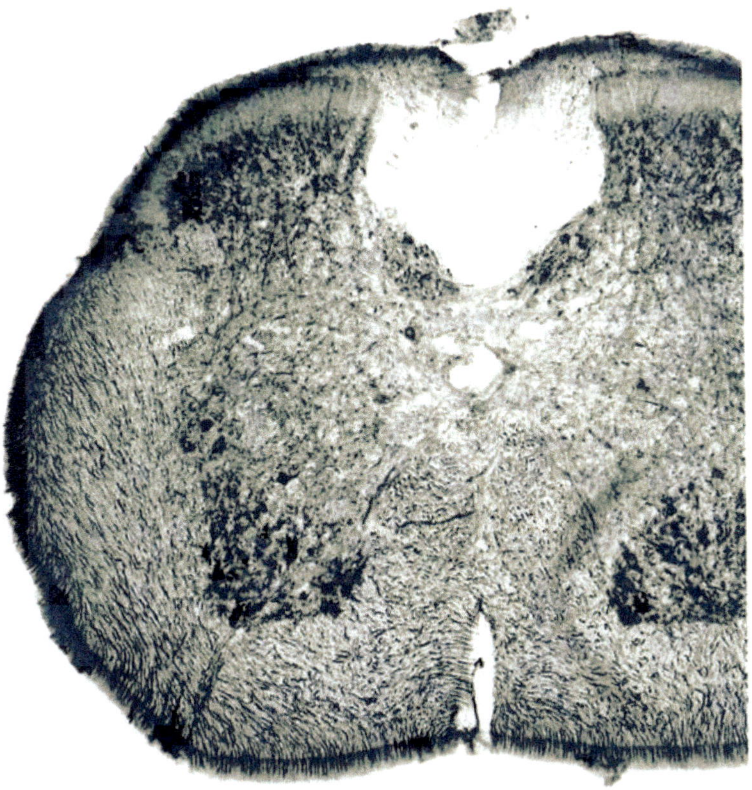

SMI

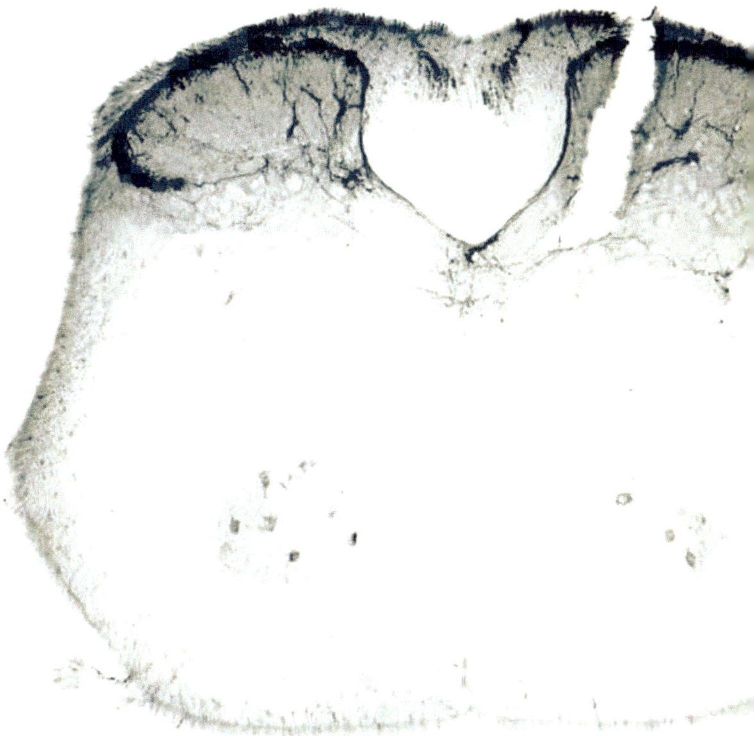

CGRP

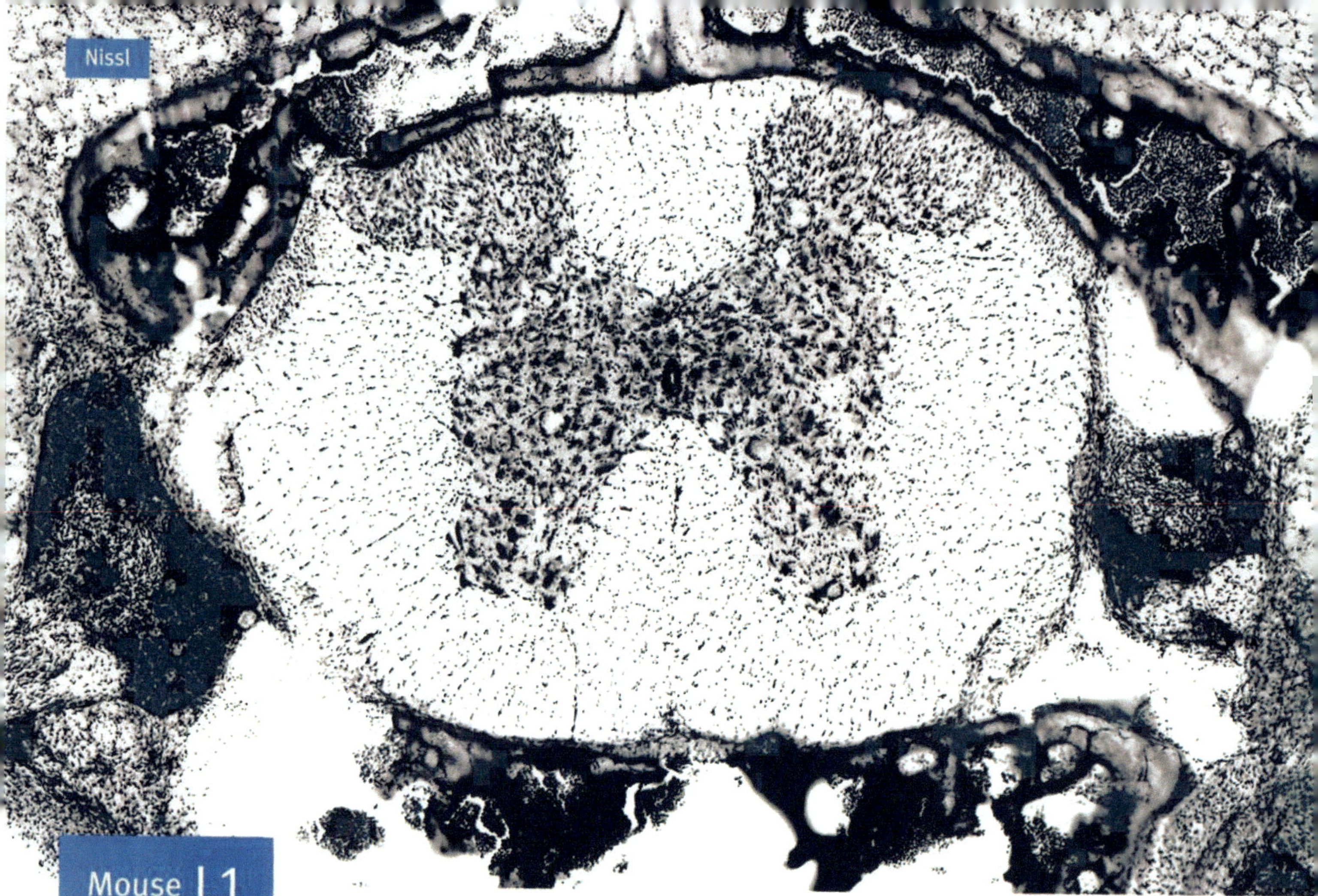
Nissl

Mouse L1

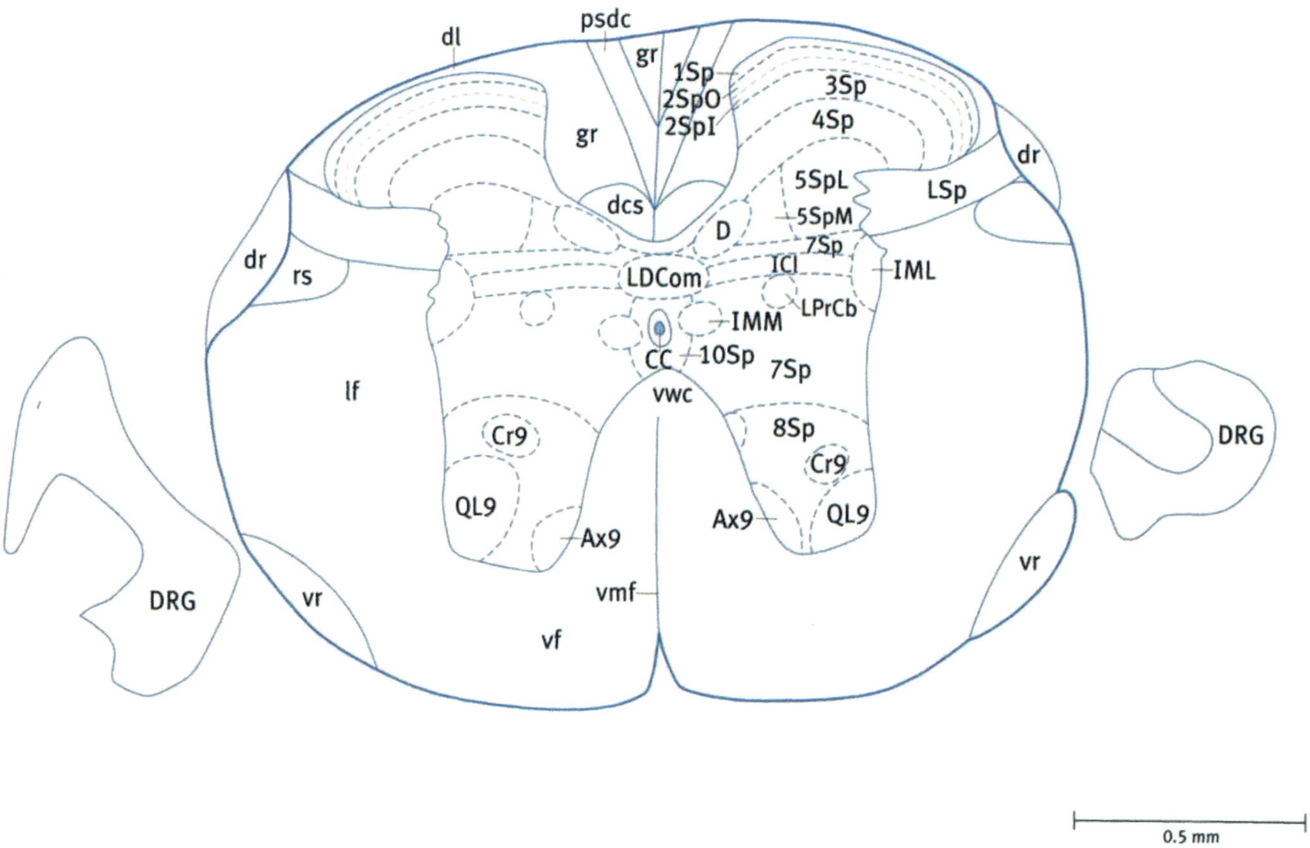
psdc
dl
gr
1Sp
2SpO
2SpI
3Sp
4Sp
gr
dr
5SpL
LSp
dcs
5SpM
D
7Sp
dr
rs
LDCom
ICl
IML
LPrCb
IMM
CC
10Sp
7Sp
lf
vwc
8Sp
Cr9
Cr9
DRG
QL9
Ax9
QL9
Ax9
vr
vmf
DRG
vr
vf
0.5 mm

Nissl

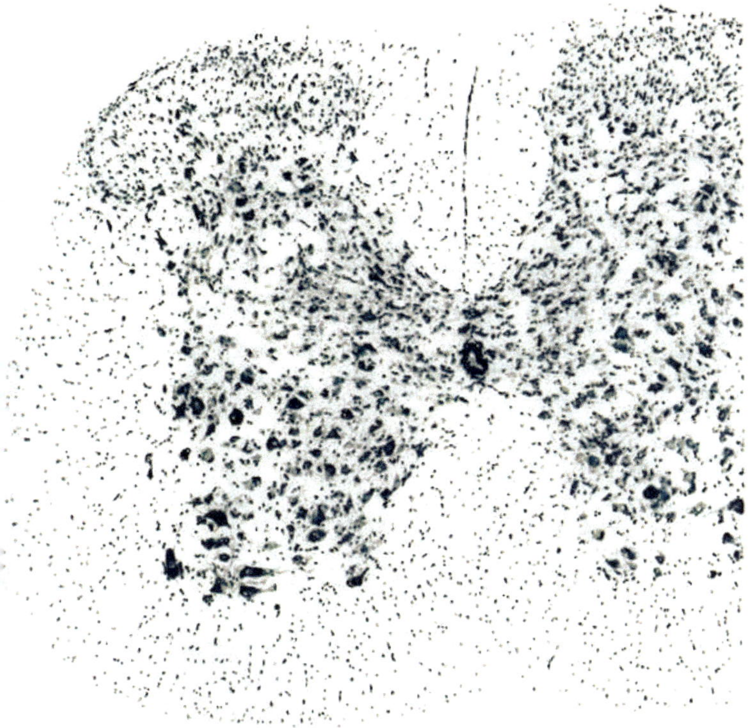

AChE

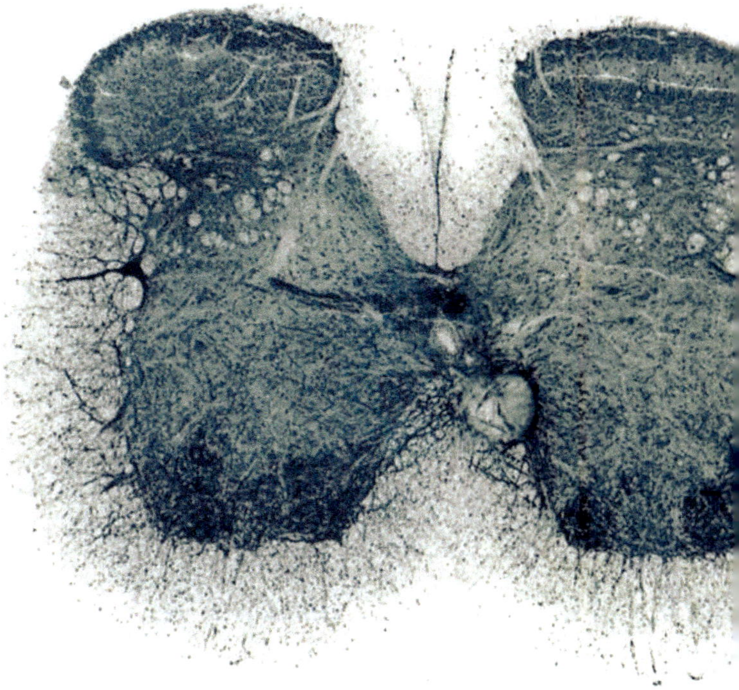

1Sp lamina 1
2SpI lamina 2, inner
2SpO lamina 2, outer
3Sp lamina 3
4Sp lamina 4
5SpL lamina 5, lat
5SpM lamina 5, med
7Sp lamina 7
8Sp lamina 8
10Sp lamina 10
Ax9 axial muscles
CC central canal
Cr9 cremaster
D dorsal nucleus
dcs dors corticosp
dl dorsolat fasc
dr dorsal root
DRG dors root gang
gr gracile fascic
ICl intercalated nu
IML intermediolat
IMM intermediomed
LDCom lum dors com
LPrCb lumbar preCb
LSp lat spinal nu
psdc postsyn dors col
QL9 quad lumborum
rs rubrospinal tr
vmf vent med fiss
vr ventral root
vwc vent white com

Mouse L1

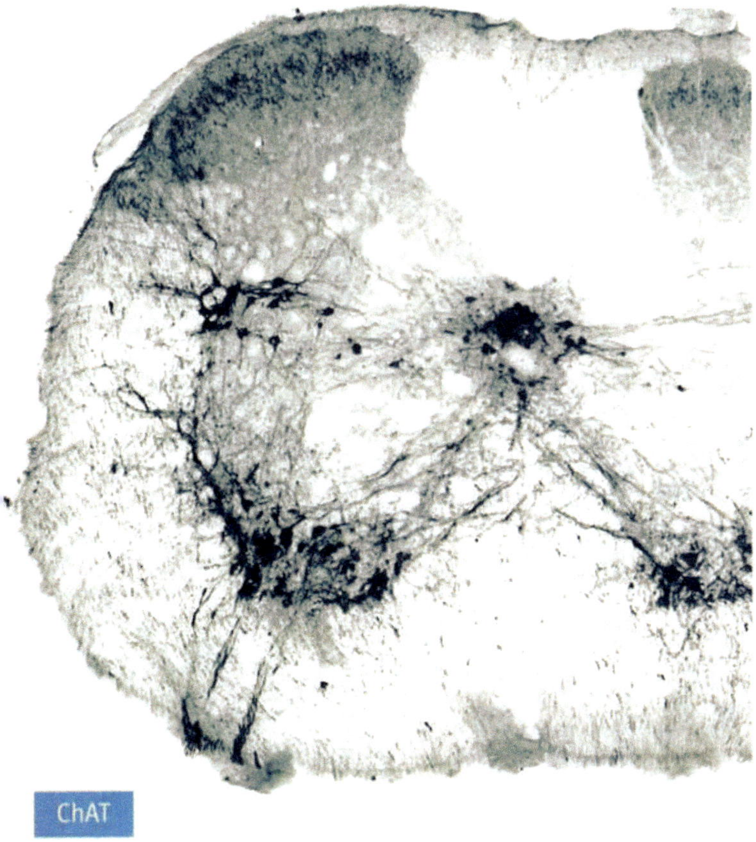

ChAT

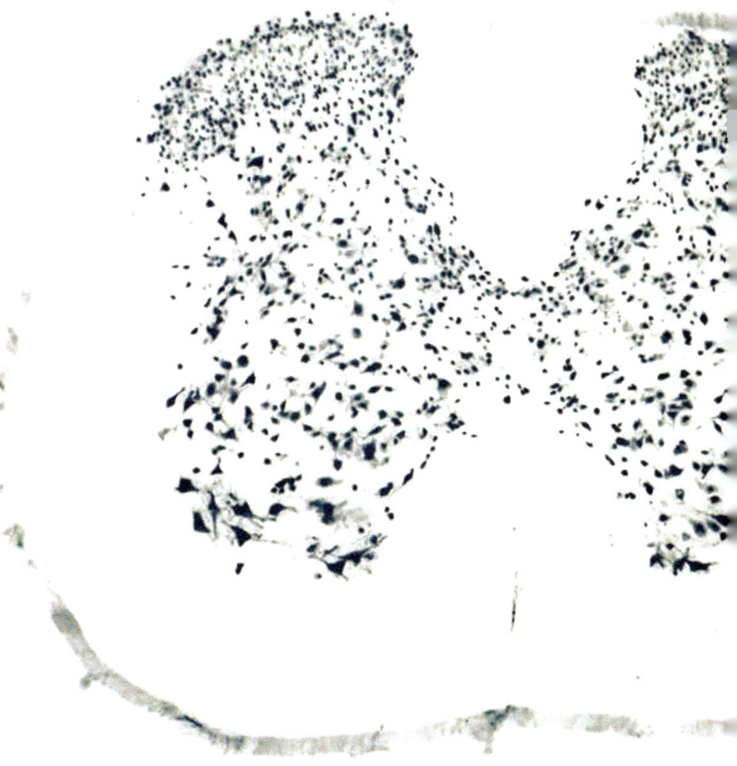

NeuN

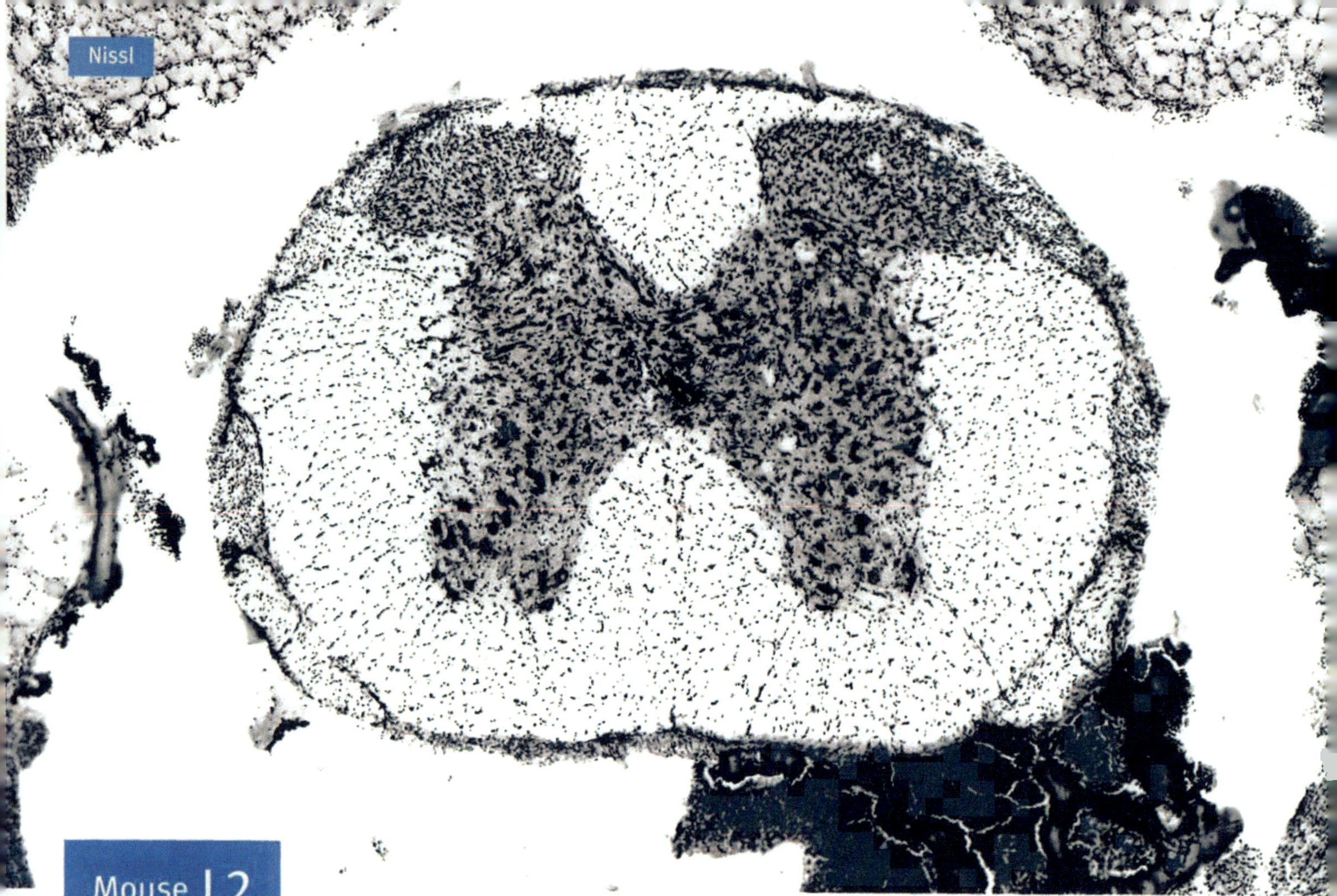
Nissl

Mouse L2

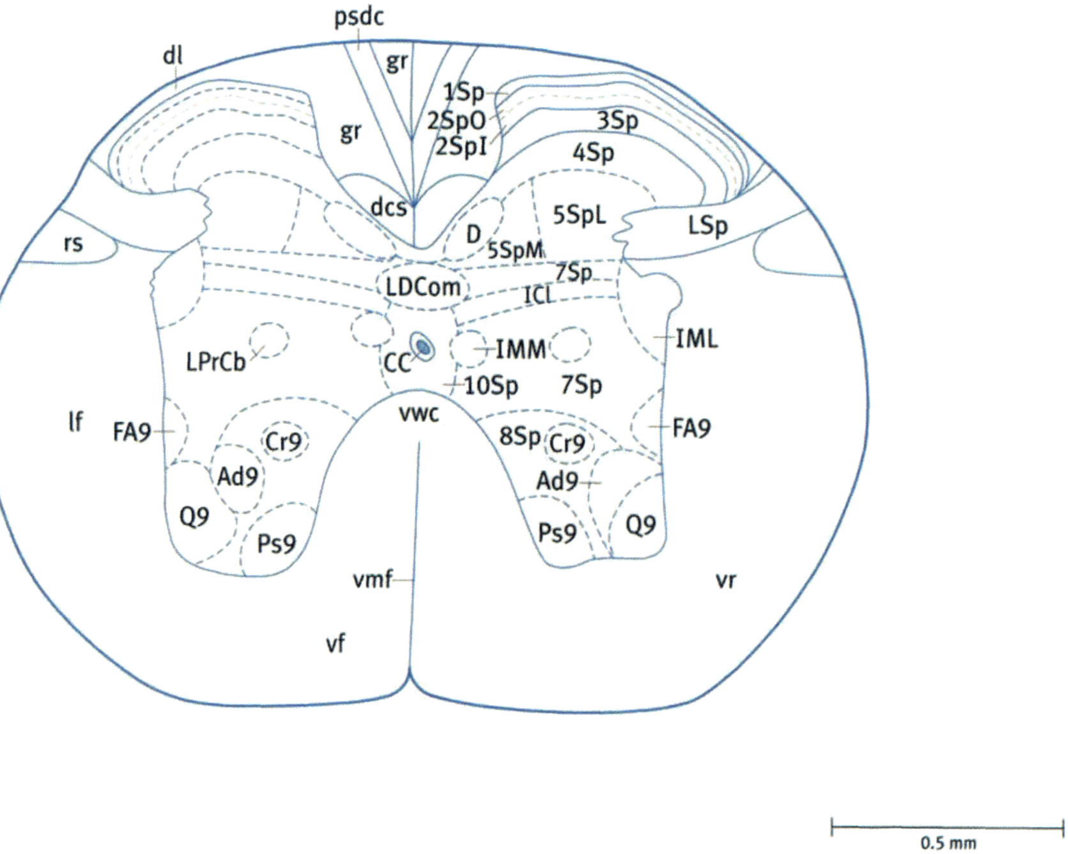
psdc
dl
gr
1Sp
2SpO
2SpI
3Sp
4Sp
gr
dcs
5SpL
rs
D
5SpM
LSp
7Sp
LDCom
ICl
IML
LPrCb
CC
IMM
10Sp
7Sp
lf
FA9
vwc
FA9
Cr9
8Sp
Cr9
Ad9
Ad9
Q9
Ps9
Ps9
Q9
vmf
vr
vf
0.5 mm

Nissl

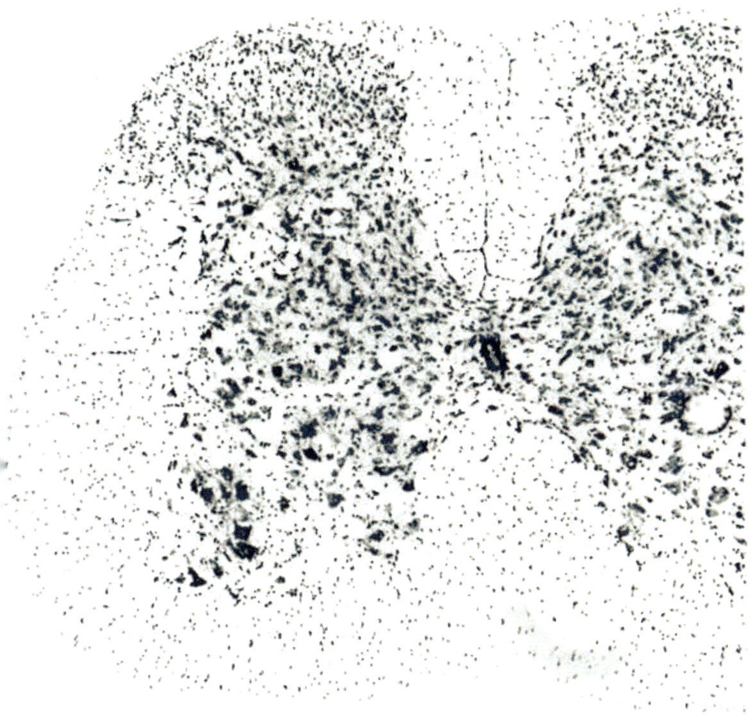

AChE

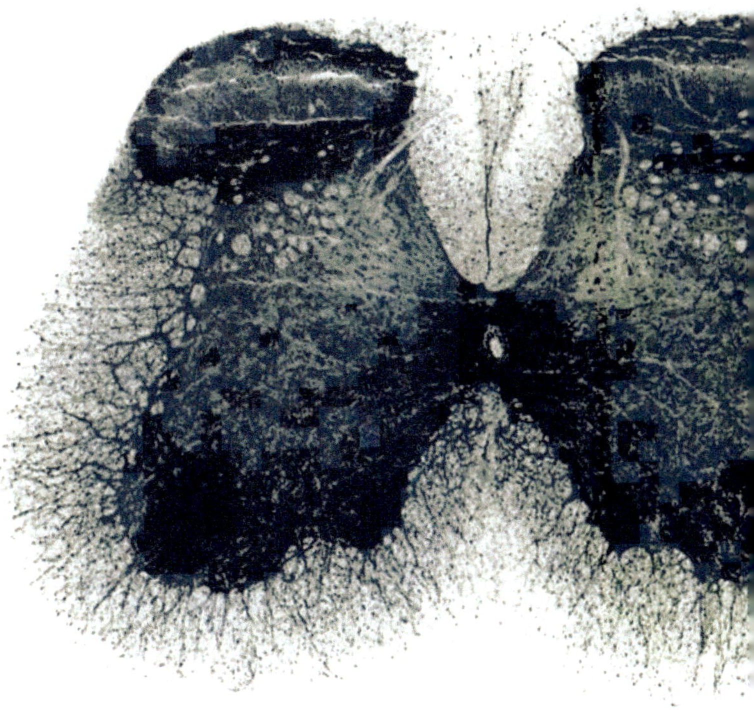

1Sp lamina 1
2SpI lamina 2, inner
2SpO lamina 2, outer
3Sp lamina 3
4Sp lamina 4
5SpL lamina 5, lat
5SpM lamina 5, med
7Sp lamina 7
8Sp lamina 8
10Sp lamina 10
Ad9 adductors
CC central canal
Cr9 cremaster
D dorsal nucleus
dcs dors corticosp
dl dorsolat fasc
gr gracile fascic
ICl intercalated nu
IML intermediolat
IMM intermediomed
LDCom lum dors com
LPrCb lumbar preCb
LSp lat spinal nu
Ps9 psoas
psdc postsyn dors col
Q9 quadriceps
rs rubrospinal tr
vmf vent med fiss
vr ventral root
vwc vent white com

Mouse L2

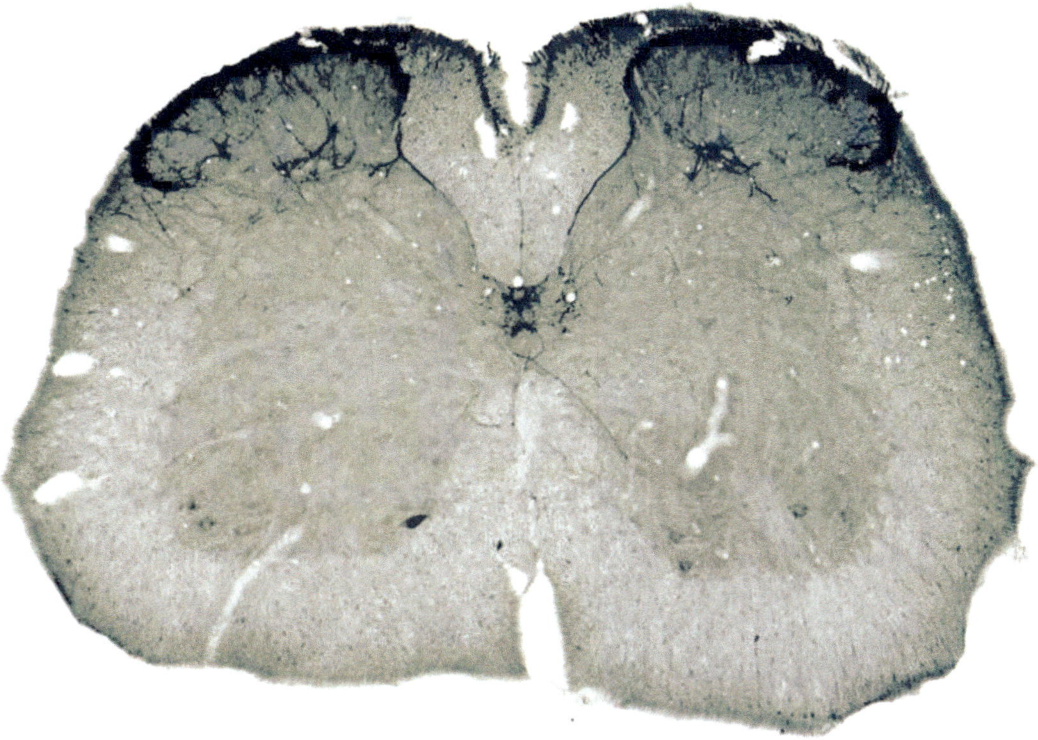

CGRP

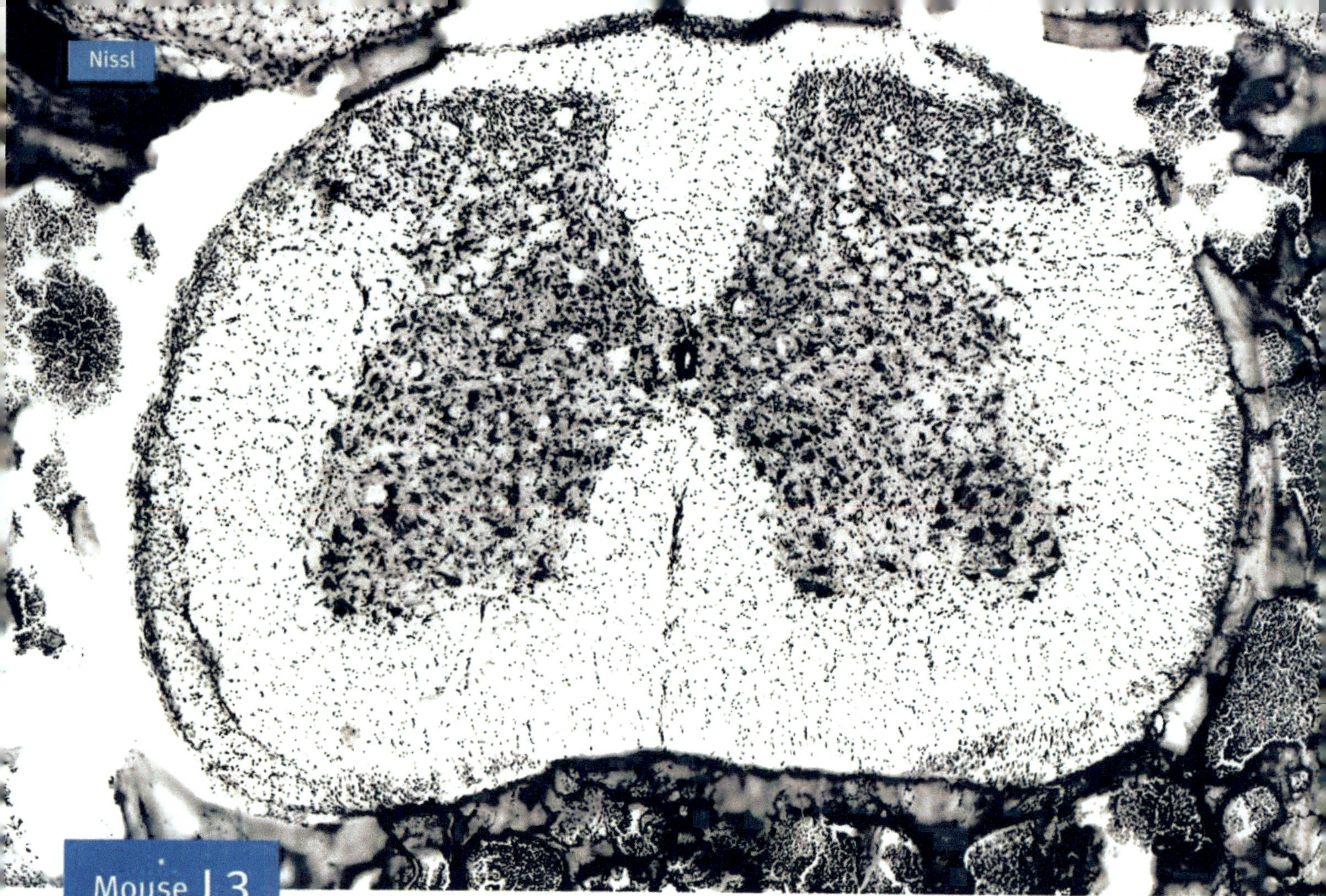
Nissl

Mouse L3

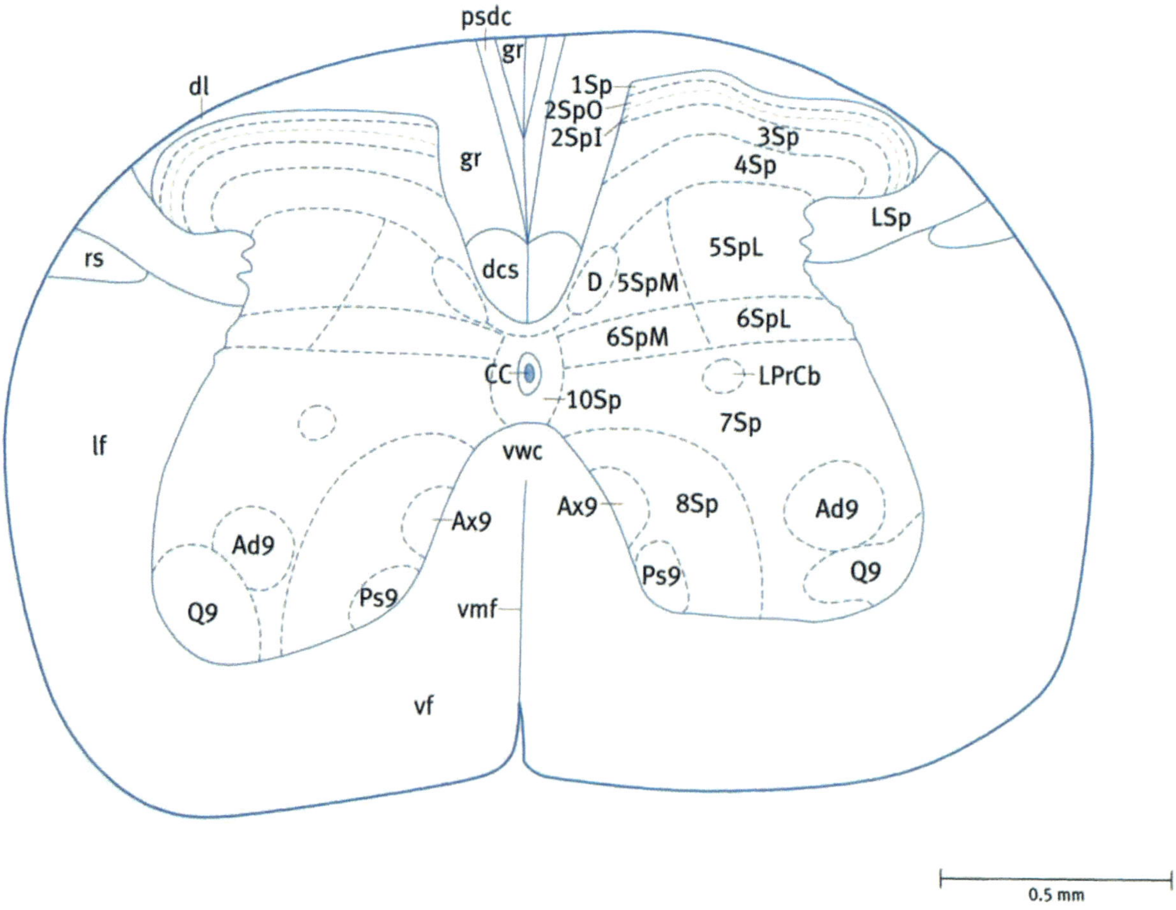
psdc
gr
dl
1Sp
2SpO
2SpI
3Sp
4Sp
gr
LSp
rs
dcs
5SpL
D
5SpM
6SpL
6SpM
CC
LPrCb
10Sp
lf
7Sp
vwc
Ax9
8Sp
Ad9
Ax9
Ad9
Ps9
Q9
Q9
Ps9
vmf
vf
0.5 mm

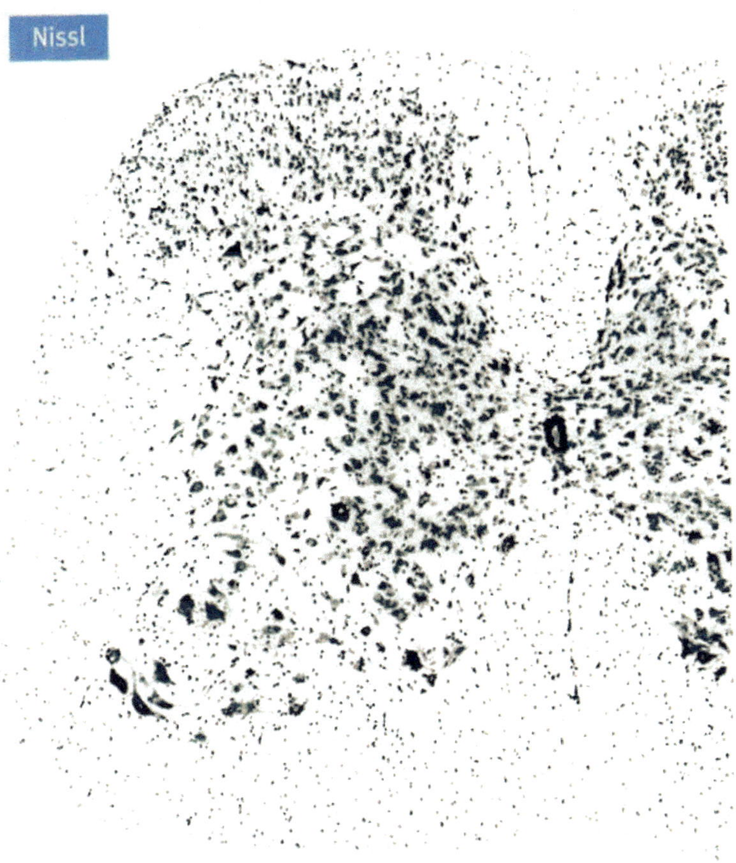

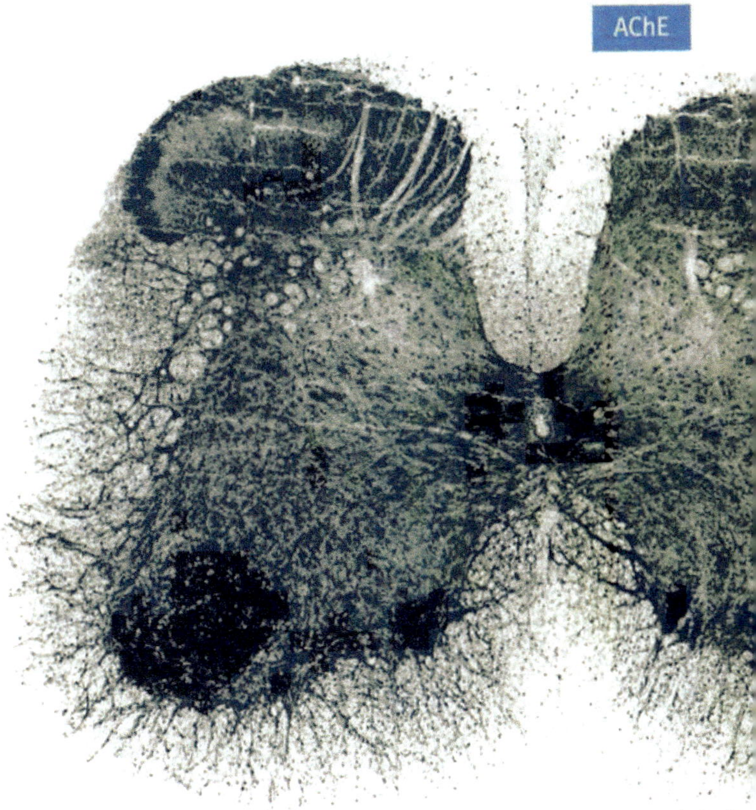

1Sp lamina 1
2SpI lamina 2, inner
2SpO lamina 2, outer
3Sp lamina 3
4Sp lamina 4
5SpL lamina 5, lat
5SpM lamina 5, med
6SpL lamina 6, lat
6SpM lamina 6, med
7Sp lamina 7
8Sp lamina 8
10Sp lamina 10
Ad9 adductors
Ax9 axial muscles
CC central canal
D dorsal nucleus
dcs dors corticosp
dl dorsolat fasc
gr gracile fascic
LPrCb lumbar preCb
LSp lat spinal nu
Ps9 psoas
psdc postsyn dors col
Q9 quadriceps
rs rubrospinal tr
vmf vent med fiss
vwc vent white com

Mouse L3

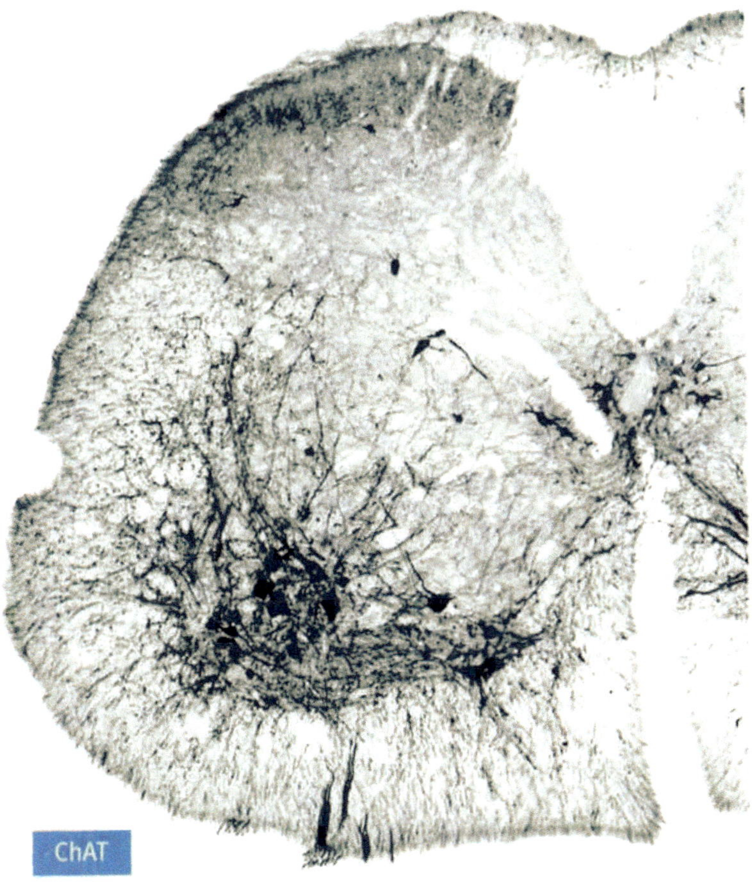

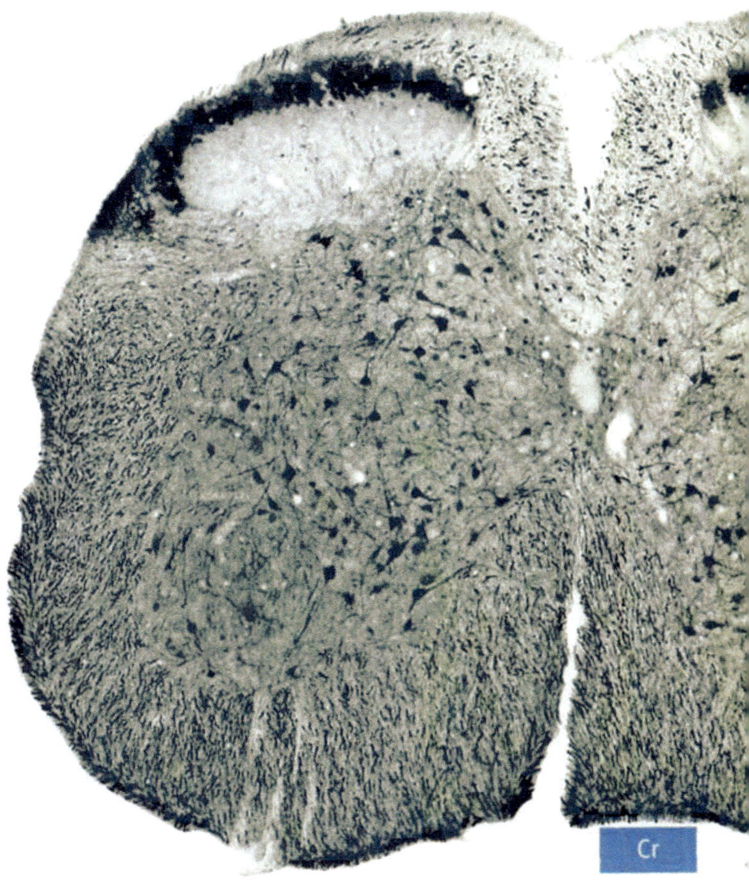

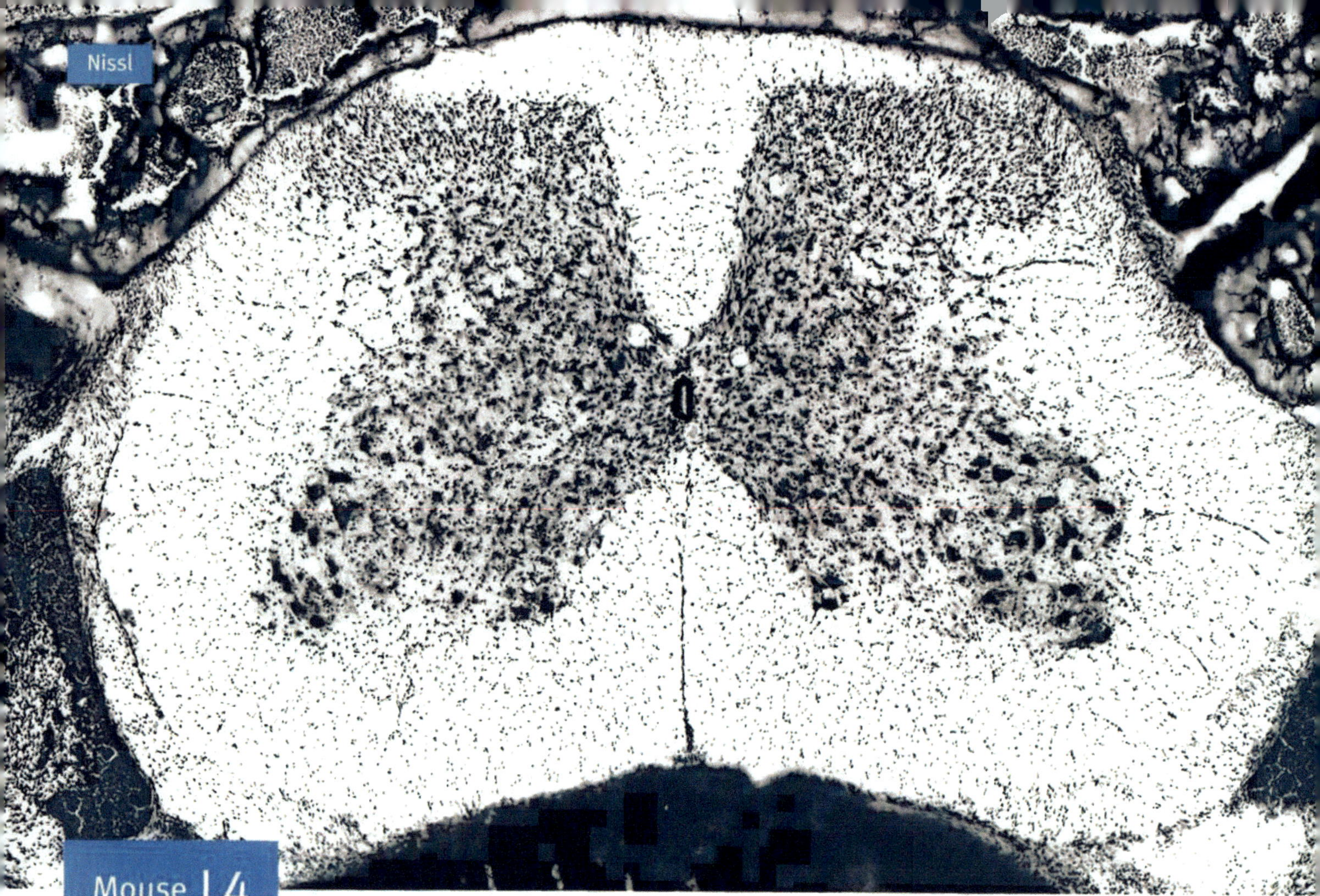
Nissl

Mouse L4

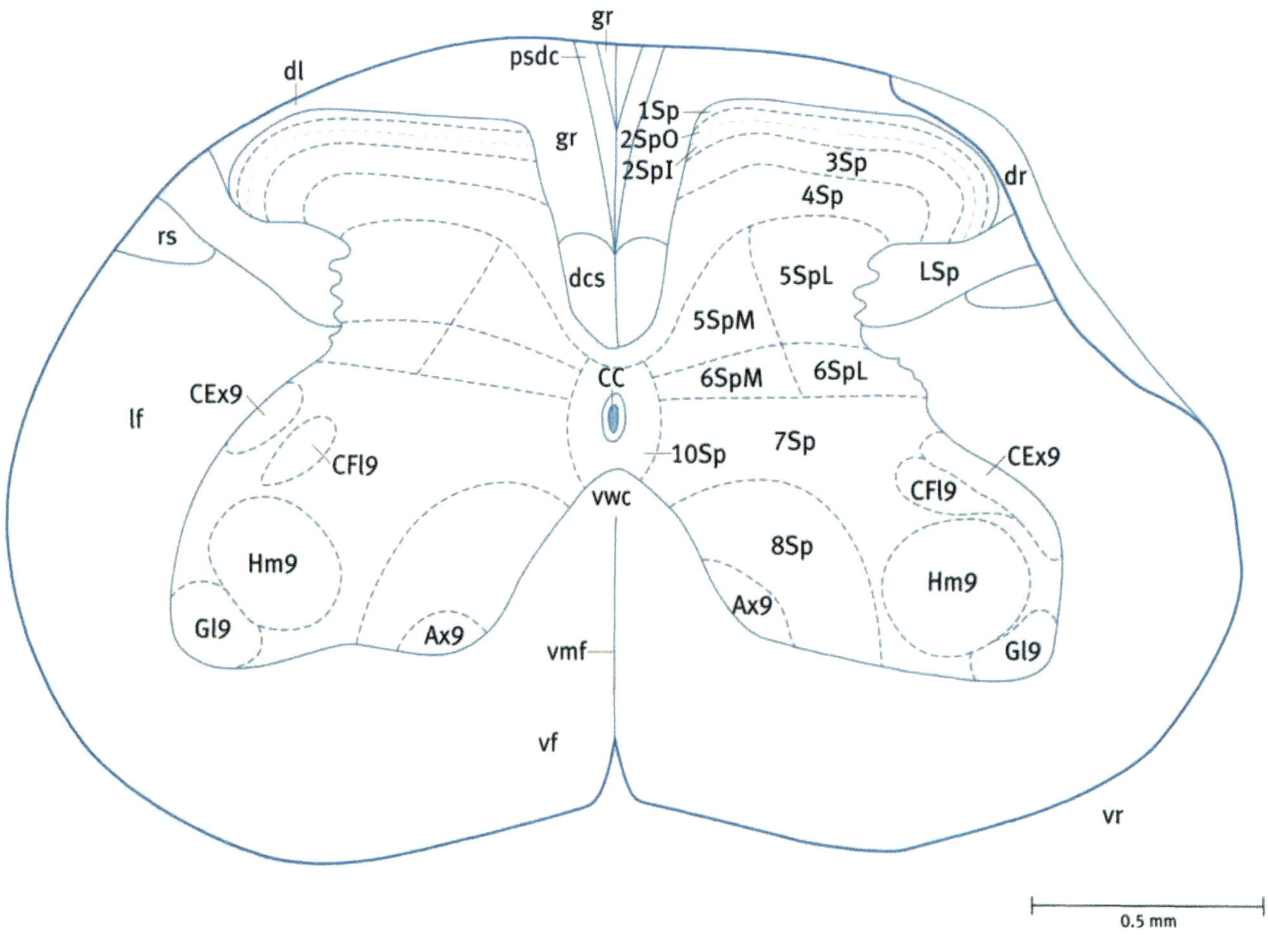
gr
psdc
dl
1Sp
gr
2SpO
2SpI
3Sp
dr
4Sp
rs
dcs
5SpL
LSp
5SpM
CC
6SpM
6SpL
CEx9
lf
10Sp
7Sp
CEx9
CFl9
CFl9
vwc
8Sp
Hm9
Hm9
Ax9
Gl9
Ax9
Gl9
vmf
vf
vr
0.5 mm

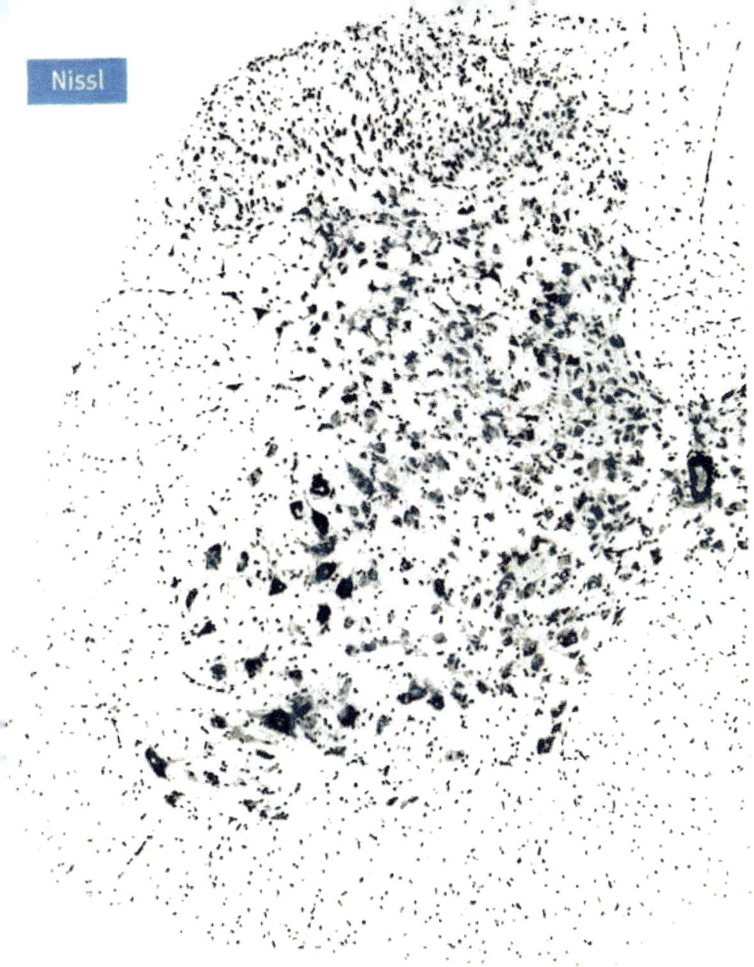

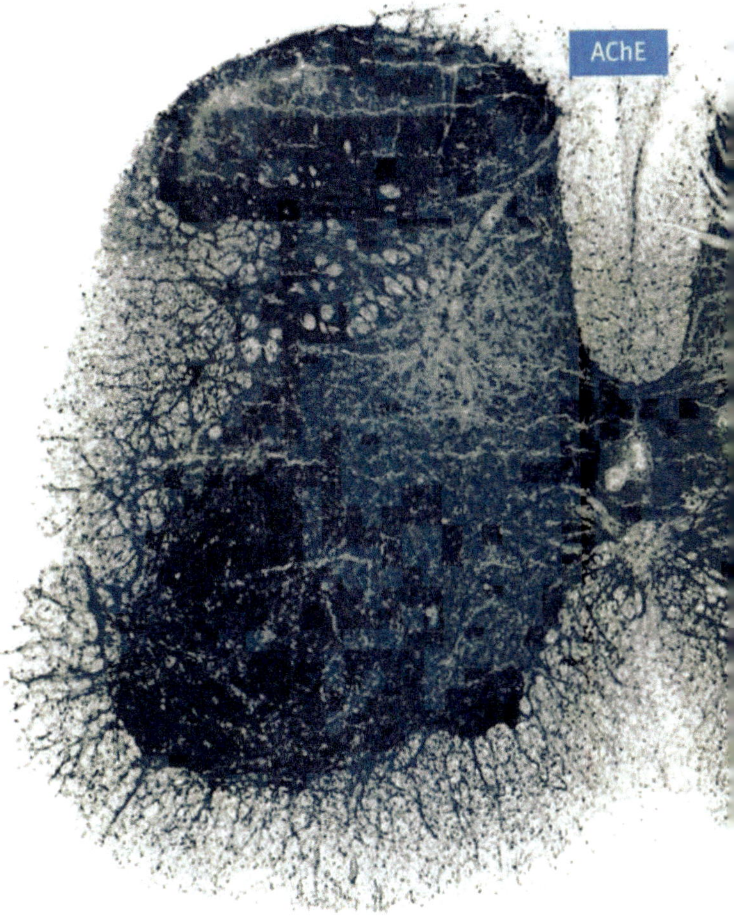

1Sp lamina 1
2SpI lamina 2, inner
2SpO lamina 2, outer
3Sp lamina 3
4Sp lamina 4
5SpL lamina 5, lat
5SpM lamina 5, med
6SpL lamina 6, lat
6SpM lamina 6, med
7Sp lamina 7
8Sp lamina 8
10Sp lamina 10
Ax9 axial muscles
CC central canal
CEx9 crual extensors
CFl9 crural flexors
dcs dors corticosp
dl dorsolat fasc
dr dorsal root
Gl9 gluteal muscles
gr gracile fascic
Hm9 hamstrings
LSp lat spinal nu
psdc postsyn dors col
rs rubrospinal tr
vmf vent med fiss
vr ventral root
vwc vent white com

Mouse L4

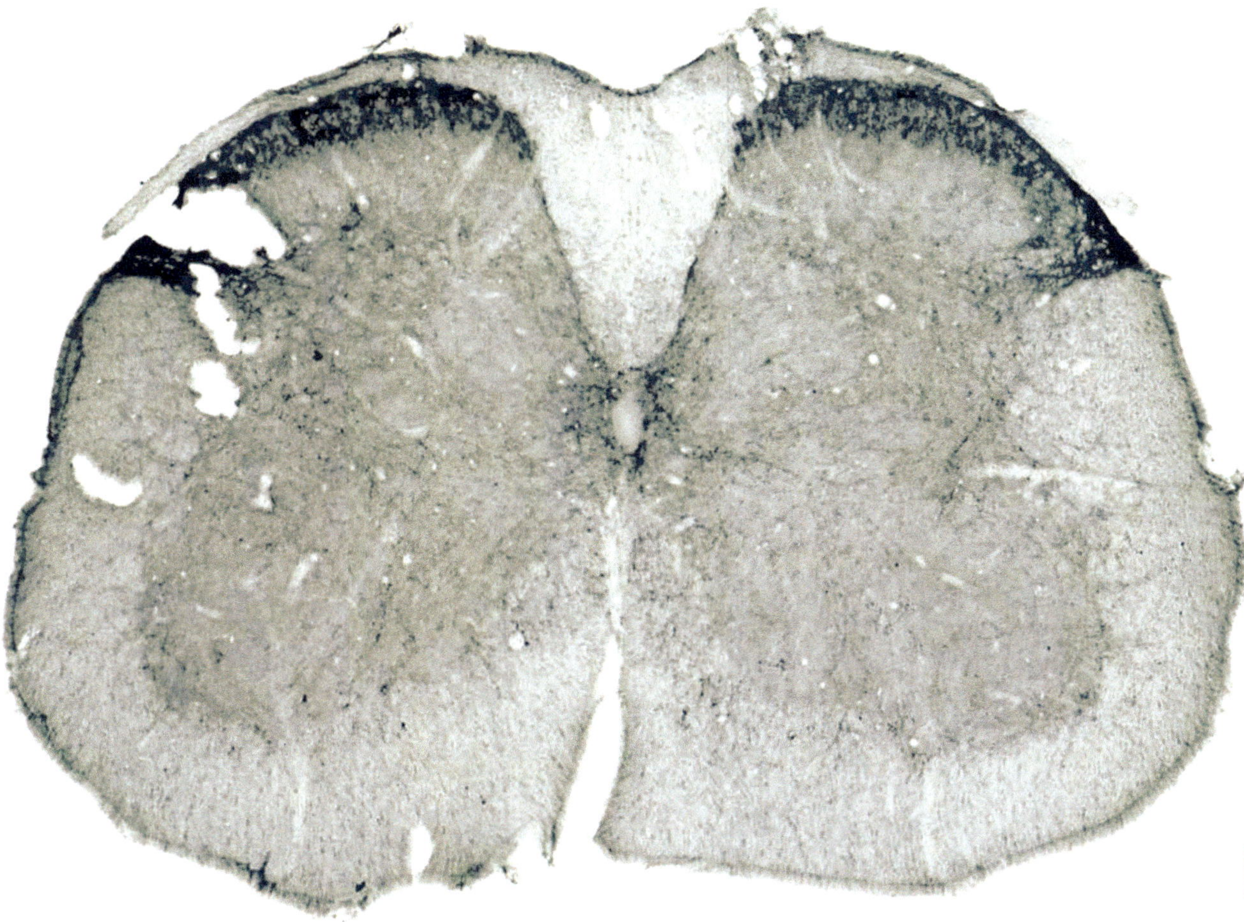

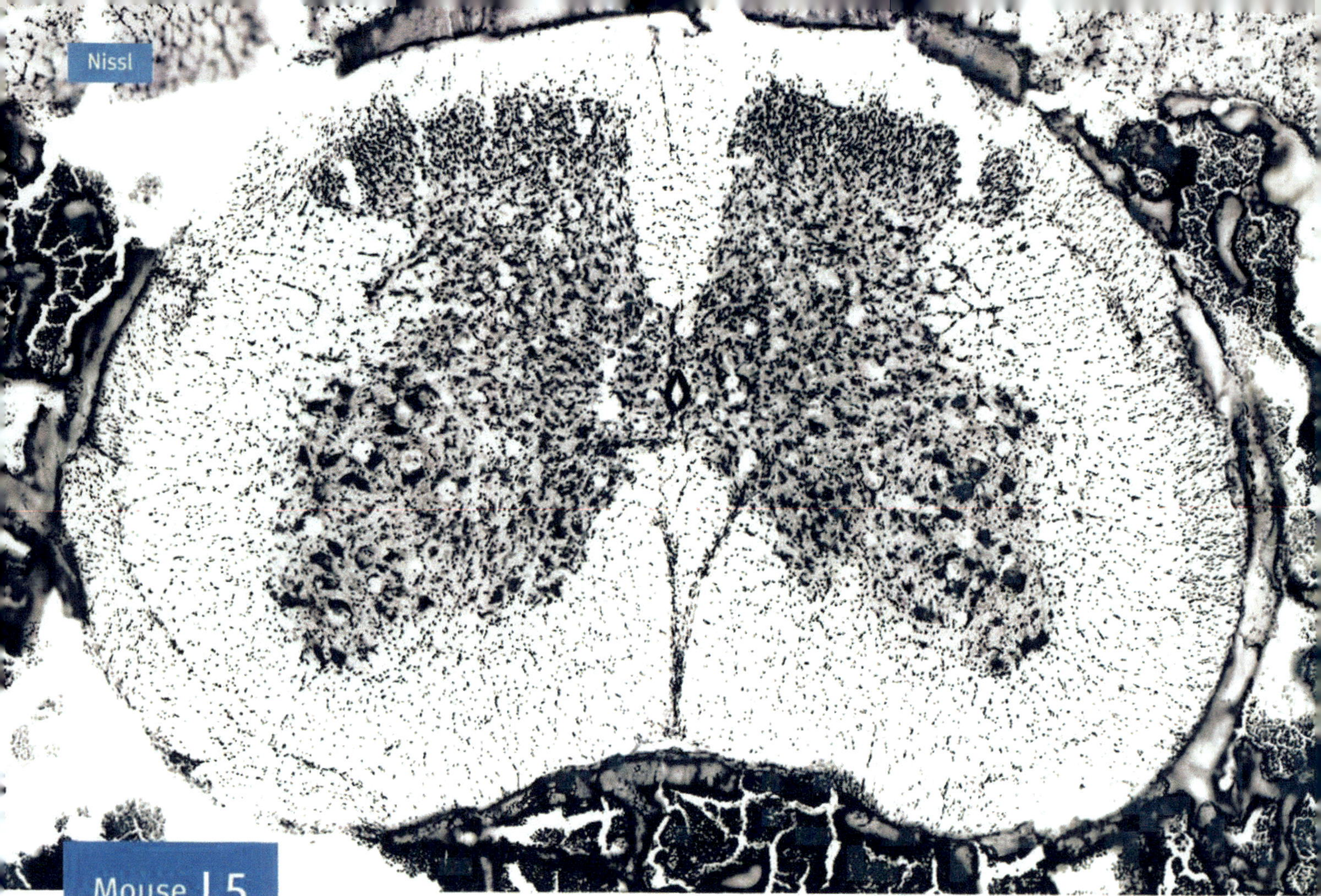

Mouse L5

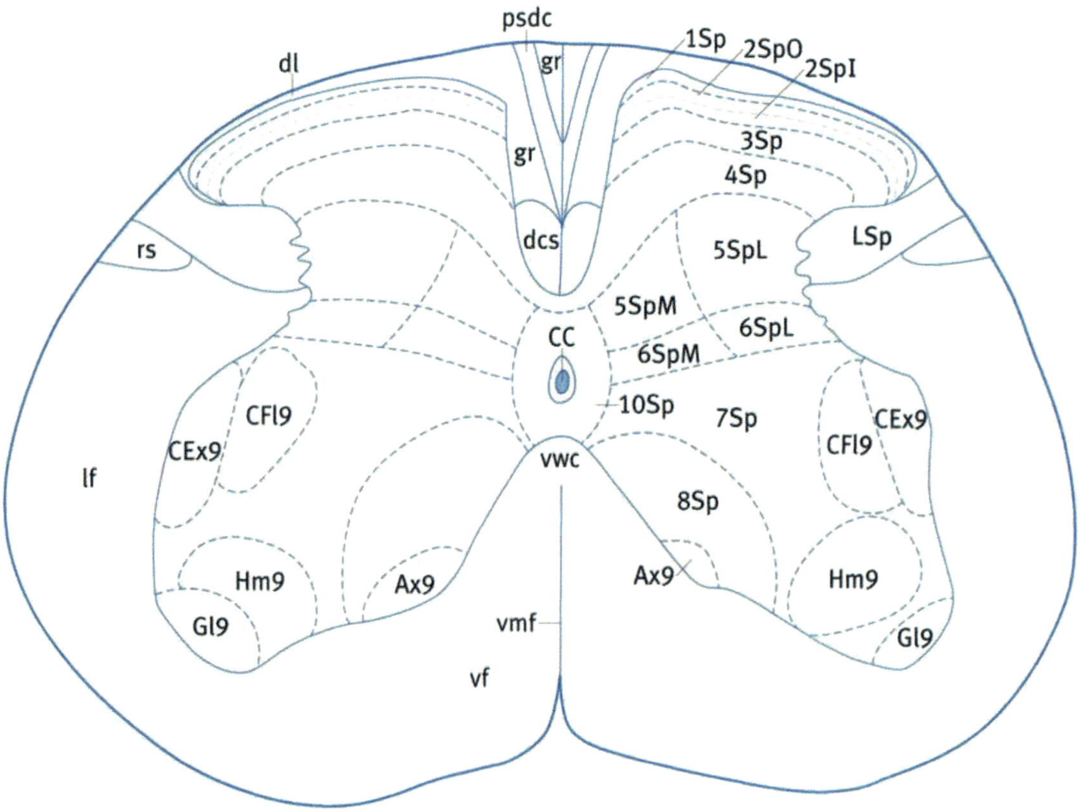

0.5 mm

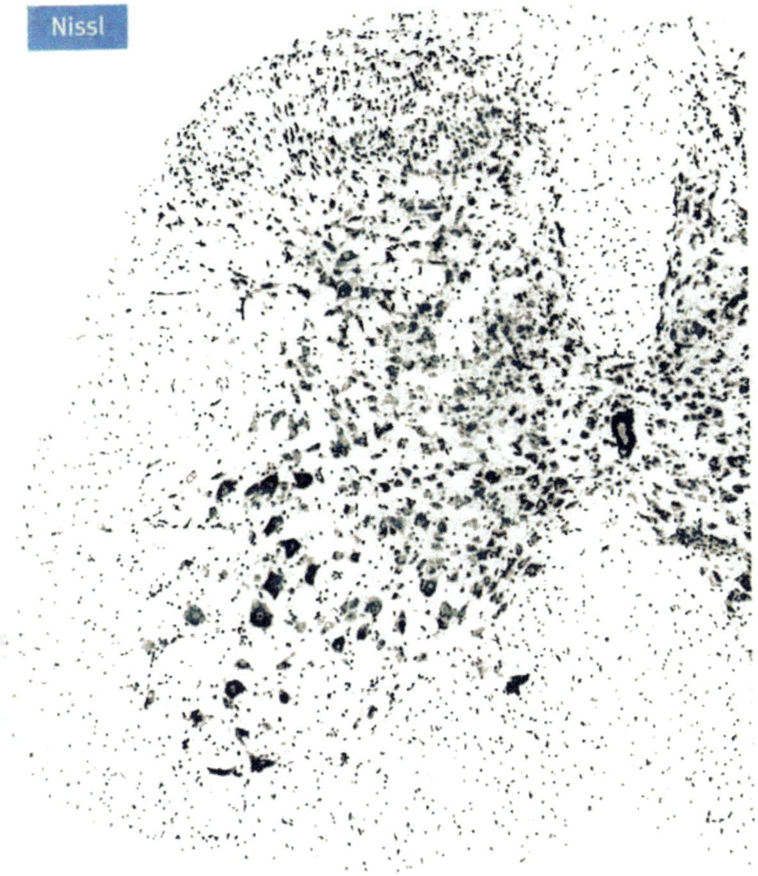

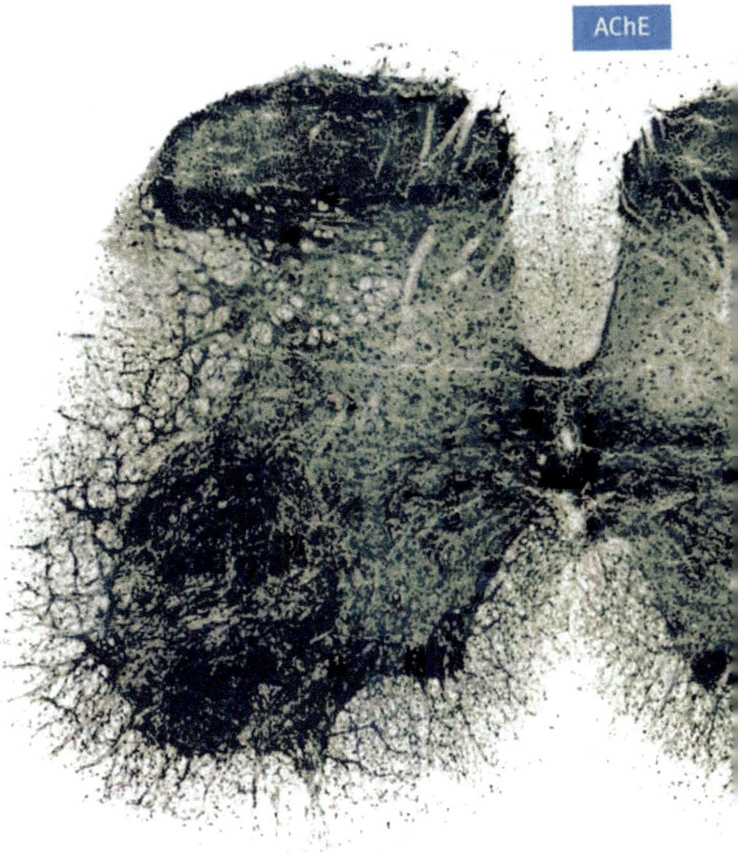

1Sp lamina 1
2SpI lamina 2, inner
2SpO lamina 2, outer
3Sp lamina 3
4Sp lamina 4
5SpL lamina 5, lat
5SpM lamina 5, med
6SpL lamina 6, lat
6SpM lamina 6, med
7Sp lamina 7
8Sp lamina 8
10Sp lamina 10
Ax9 axial muscles
CC central canal
CEx9 crual extensors
CFl9 crural flexors
dcs dors corticosp
dl dorsolat fasc
Gl9 gluteal muscles
gr gracile fascic
Hm9 hamstrings
LSp lat spinal nu
psdc postsyn dors col
rs rubrospinal tr
vmf vent med fiss
vwc vent white com

Mouse L5

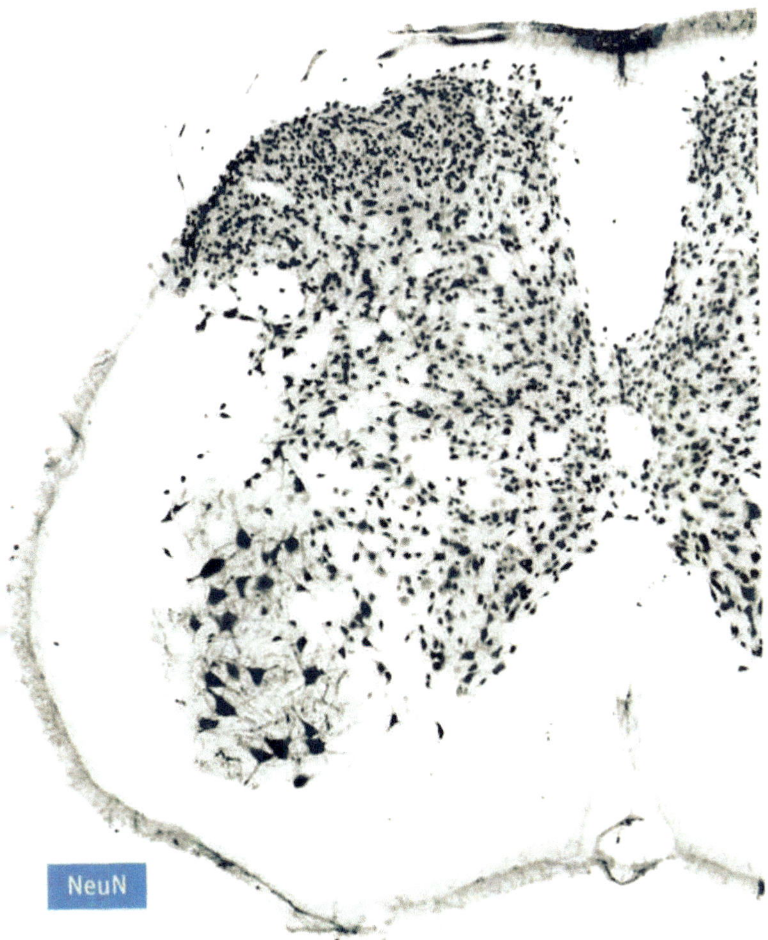

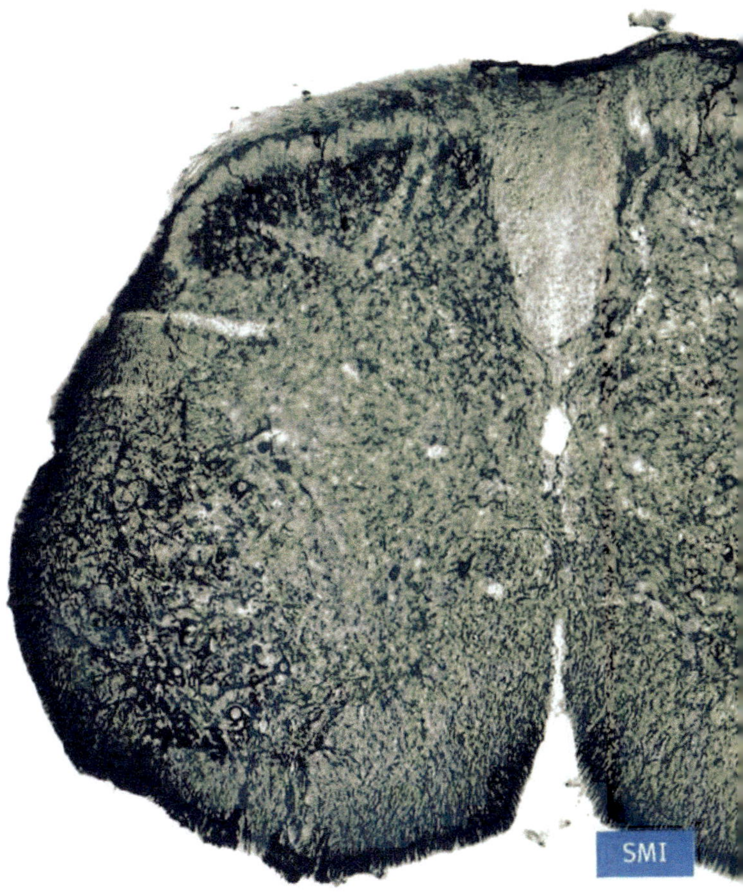

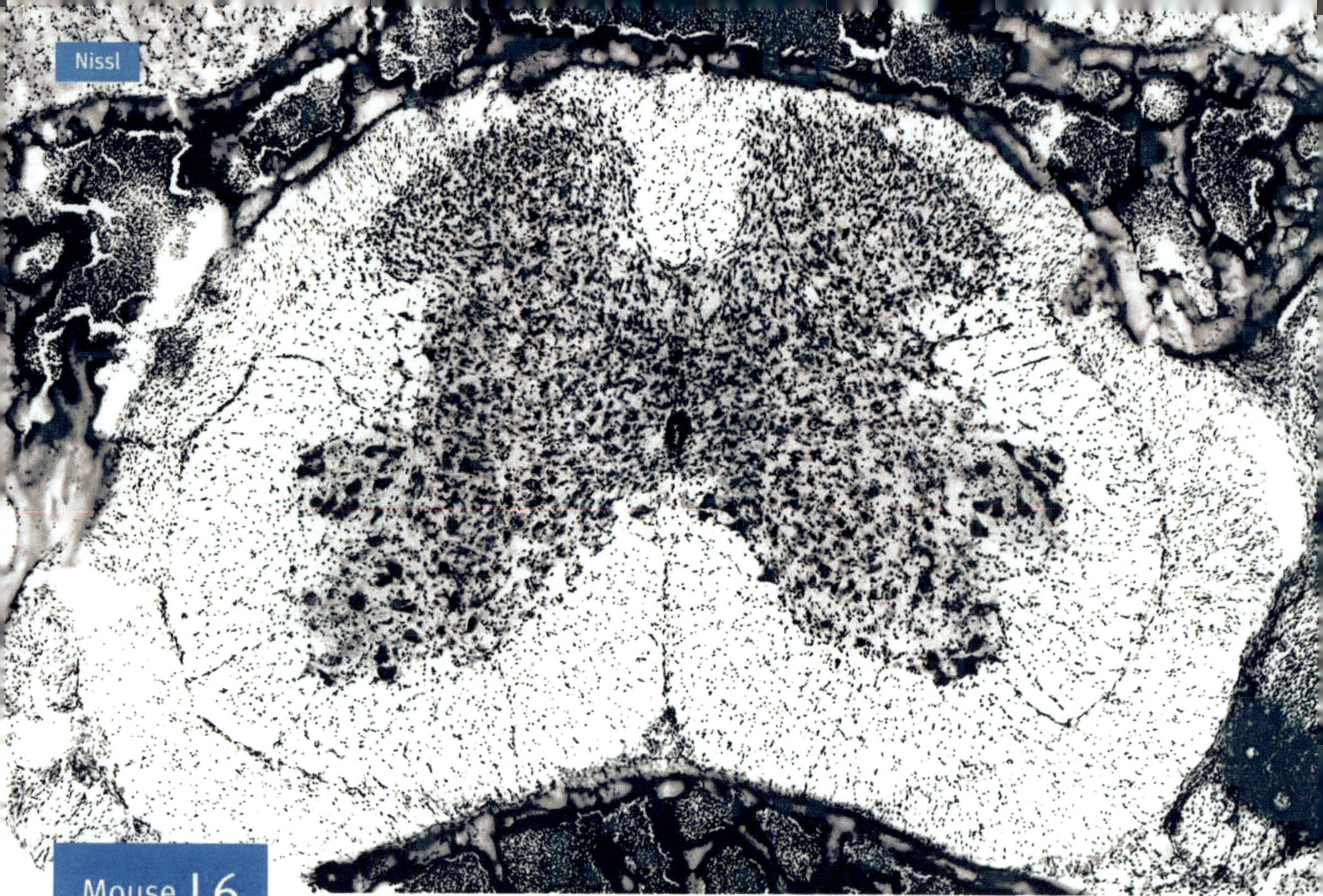
Nissl

Mouse L6

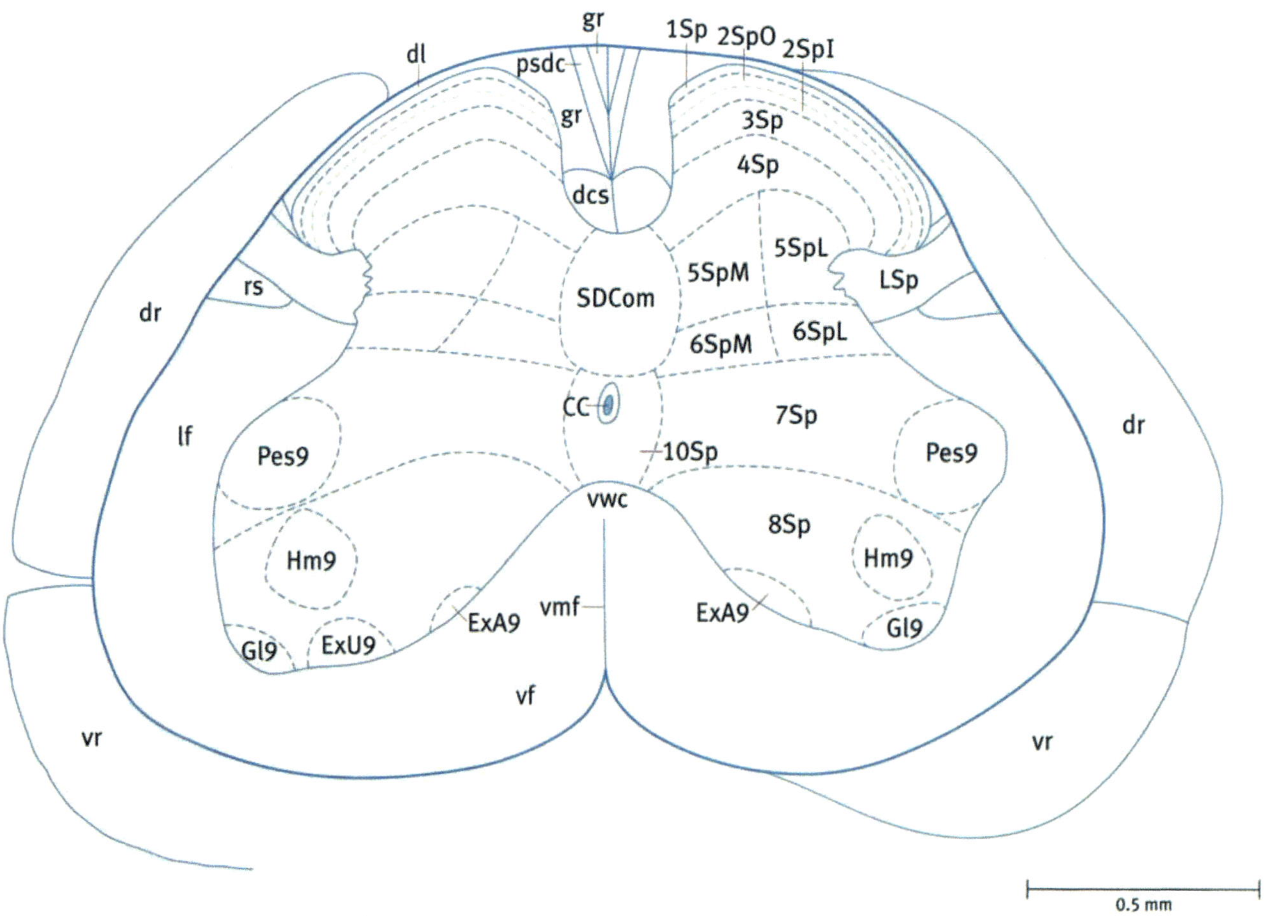
gr
1Sp
2SpO
2SpI
dl
psdc
gr
3Sp
4Sp
dcs
5SpL
rs
5SpM
LSp
dr
SDCom
6SpM
6SpL
CC
7Sp
lf
Pes9
10Sp
Pes9
dr
vwc
8Sp
Hm9
Hm9
ExA9
vmf
ExA9
Gl9
ExU9
Gl9
vf
vr
vr
0.5 mm

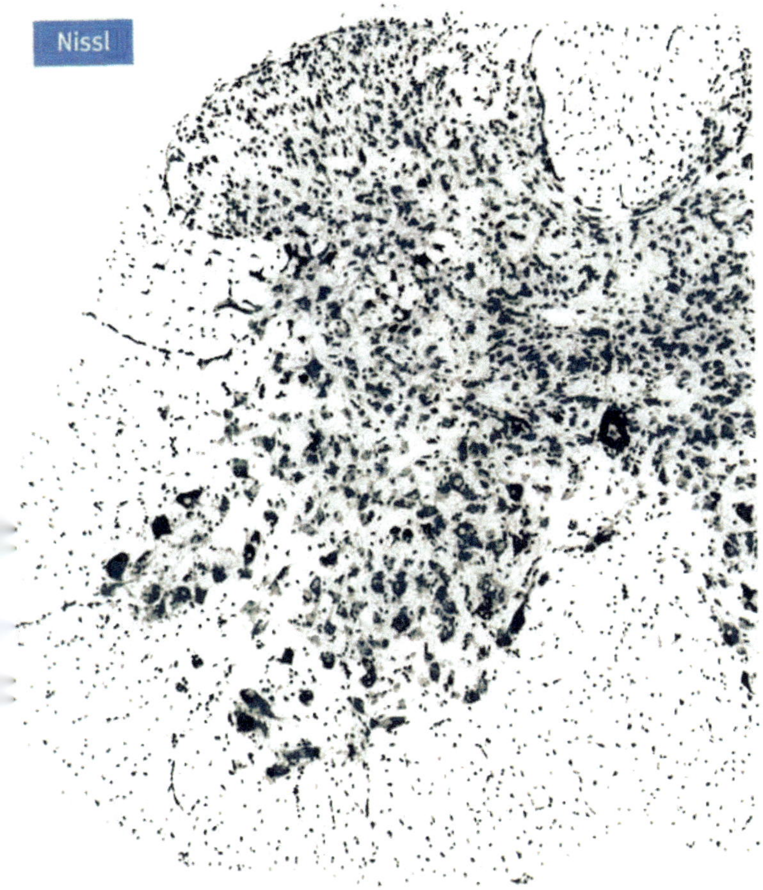

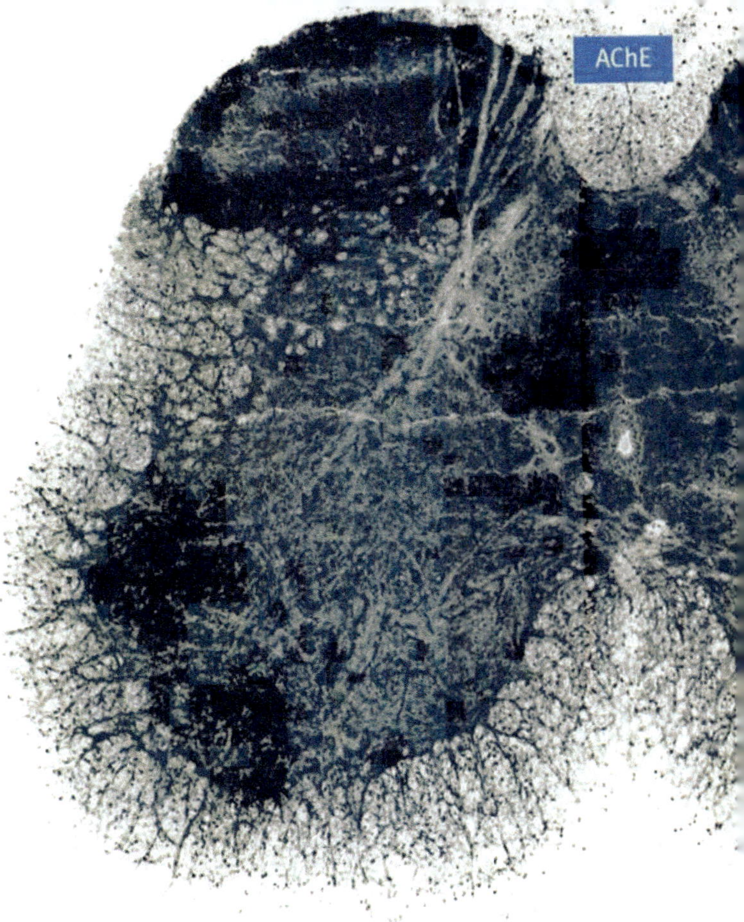

Mouse L6

1Sp lamina 1
2SpI lamina 2, inner
2SpO lamina 2, outer
3Sp lamina 3
4Sp lamina 4
5SpL lamina 5, lat
5SpM lamina 5, med
6SpL lamina 6, lat
6SpM lamina 6, med
7Sp lamina 7
8Sp lamina 8
10Sp lamina 10
CC central canal
dcs dors corticosp
dl dorsolat fasc
dr dorsal root
ExA9 ext anal sphin
ExU9 ext ureth sphin
Gl9 gluteal muscles
gr gracile fascic
Hm9 hamstrings
LSp lat spinal nu
Pes9 pes
psdc postsyn dors col
rs rubrospinal tr
SDCom sac dors comm
vmf vent med fiss
vr ventral root
vwc vent white com

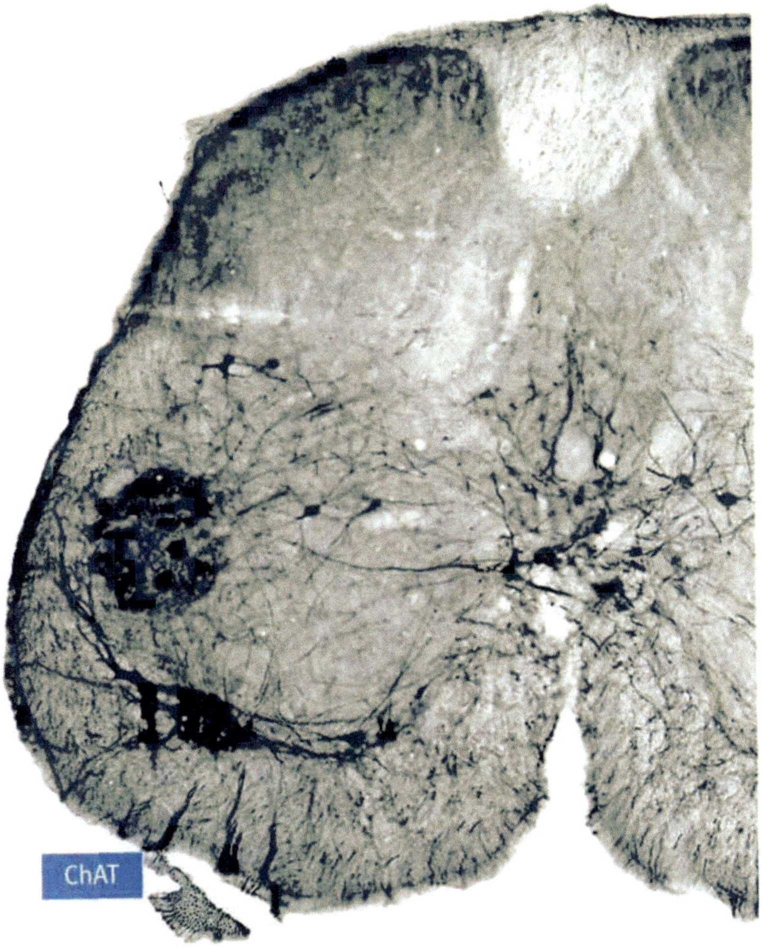

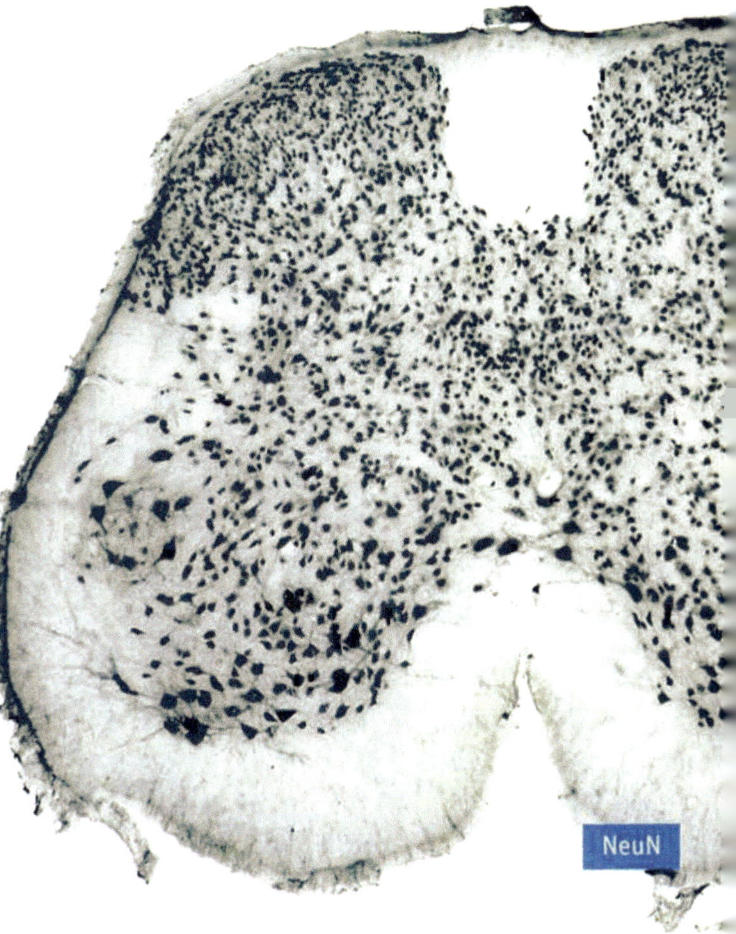

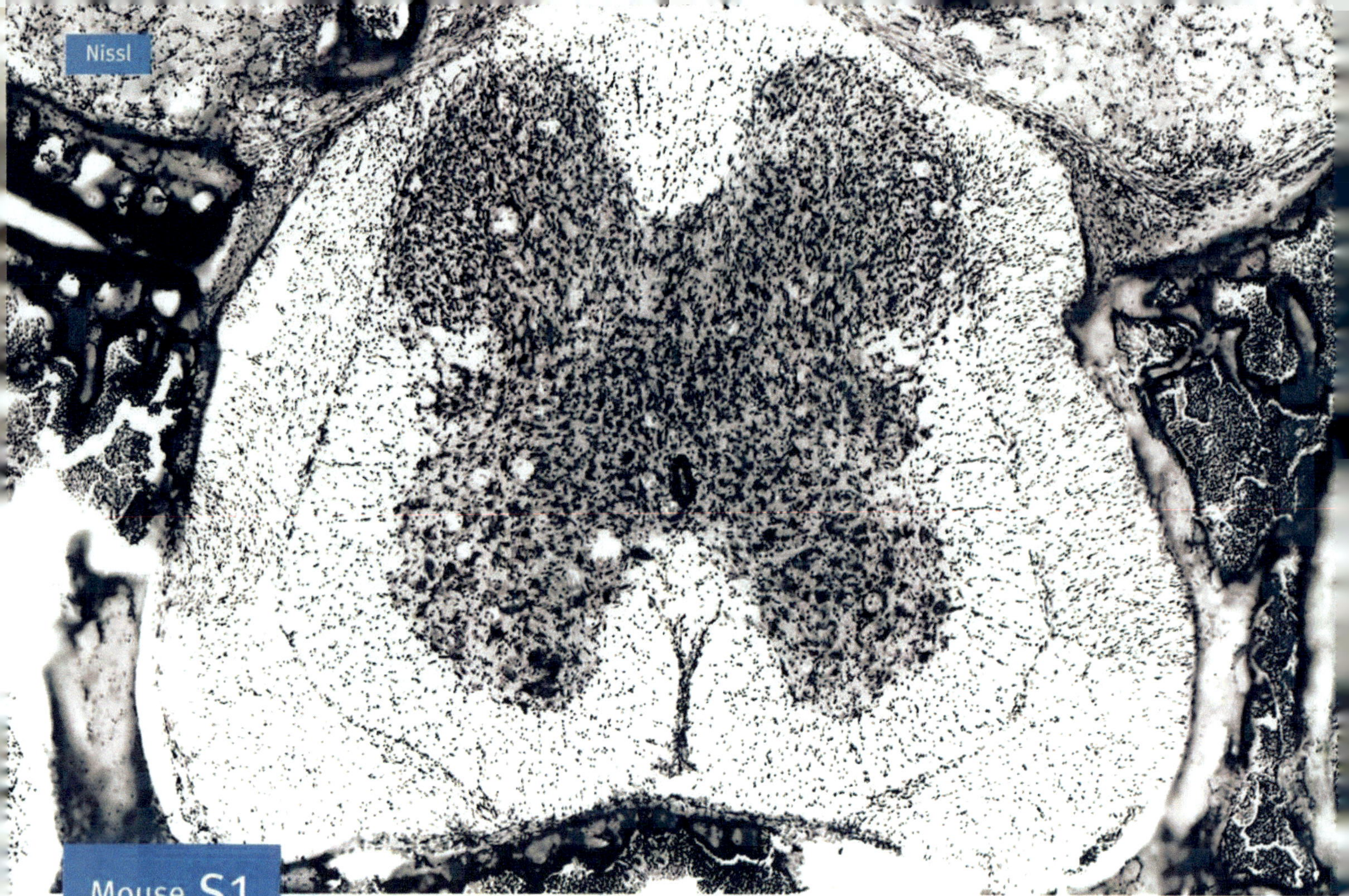
Nissl

Mouse S1

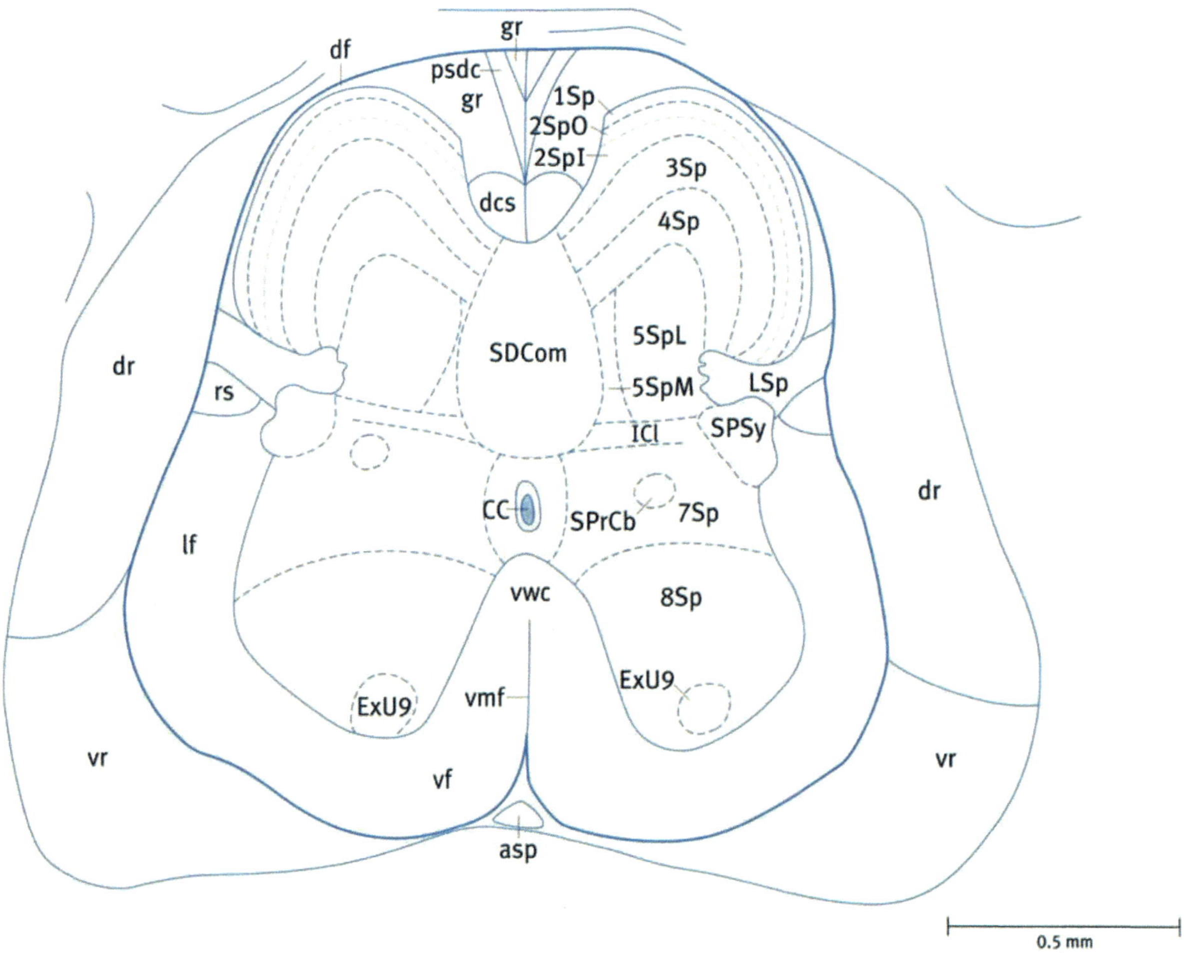
gr
df
psdc
gr
1Sp
2SpO
2SpI
3Sp
dcs
4Sp
SDCom
5SpL
dr
rs
5SpM
LSp
ICl
SPSy
CC
SPrCb
7Sp
dr
lf
vwc
8Sp
ExU9
ExU9
vmf
vr
vf
vr
asp
0.5 mm

Nissl

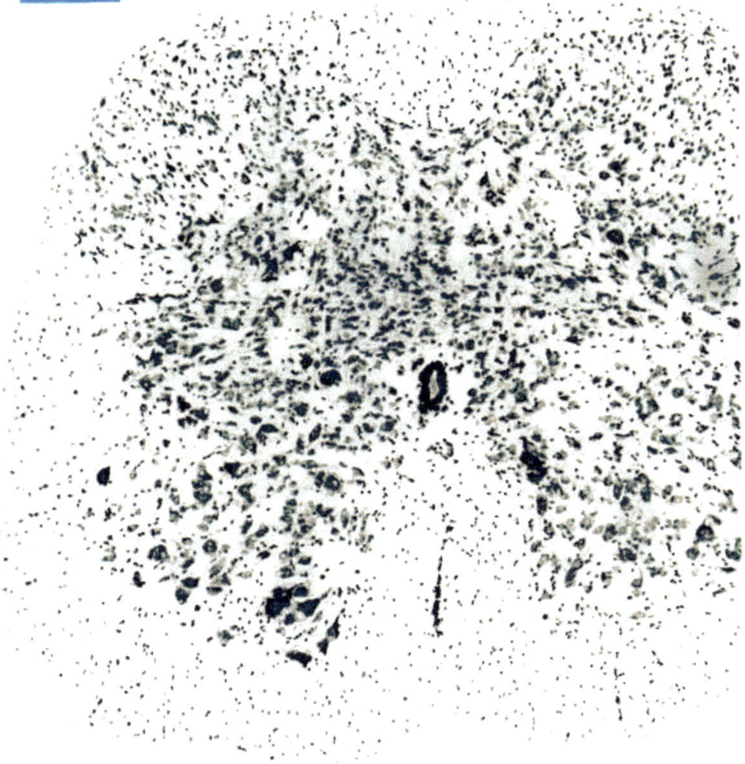

AChE

1Sp lamina 1
2SpI lamina 2, inner
2SpO lamina 2, outer
3Sp lamina 3
4Sp lamina 4
5SpL lamina 5, lat
5SpM lamina 5, med
7Sp lamina 7
8Sp lamina 8
asp ant spinal art
CC central canal
dcs dors corticosp
dr dorsal root
ExU9 ext ureth sphin
gr gracile fascic
LSp lat spinal nu
psdc postsyn dors col
rs rubrospinal tr
SDCom sac dors comm
SPrCb sacral preCb
SPSy sacral parasym
vmf vent med fiss
vr ventral root
vwc vent white com

Mouse S1

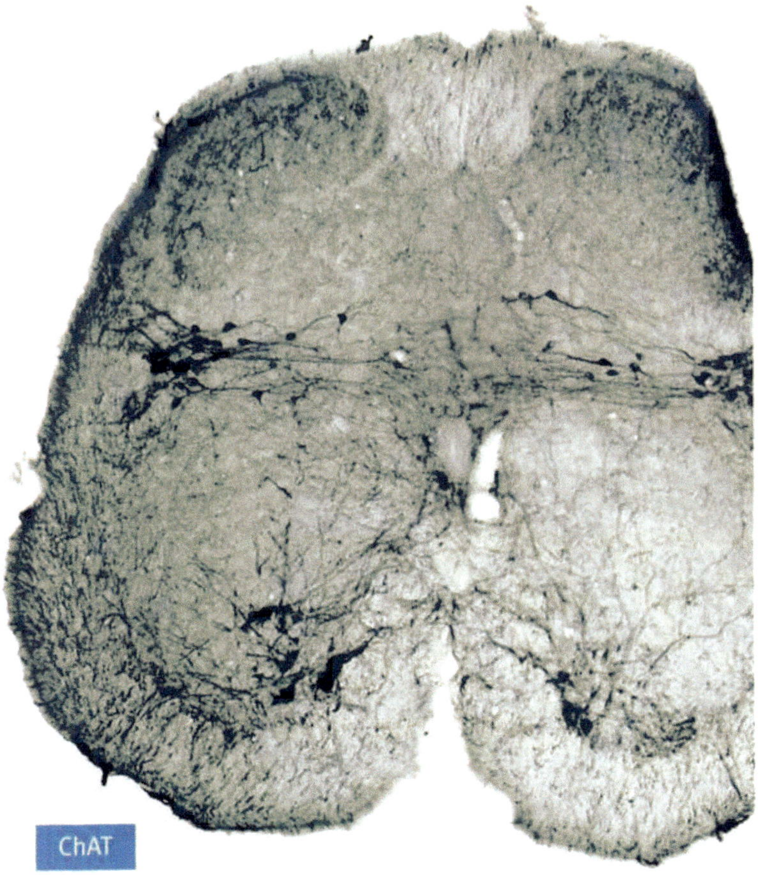

ChAT

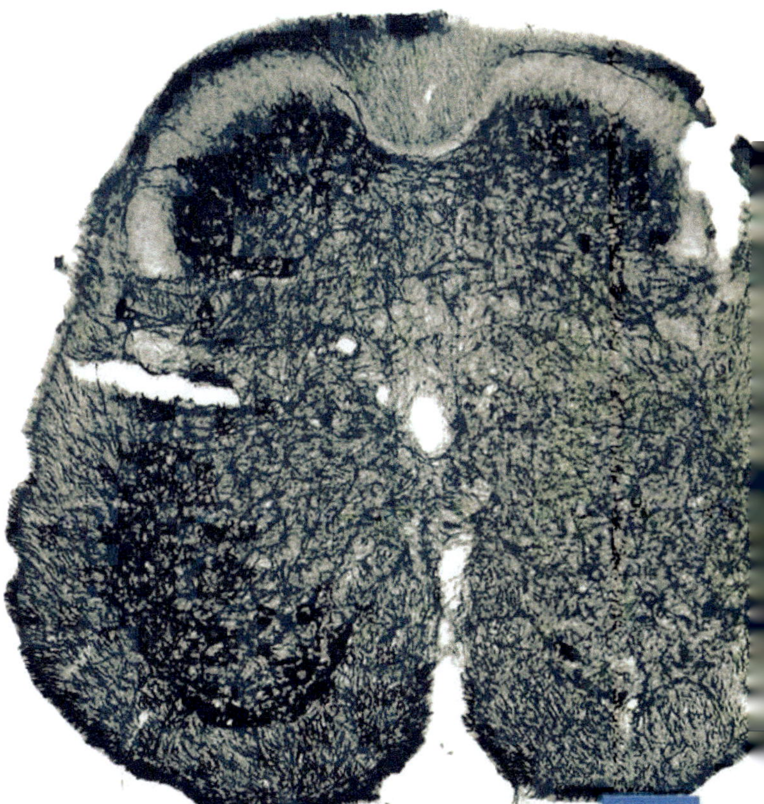

SMI

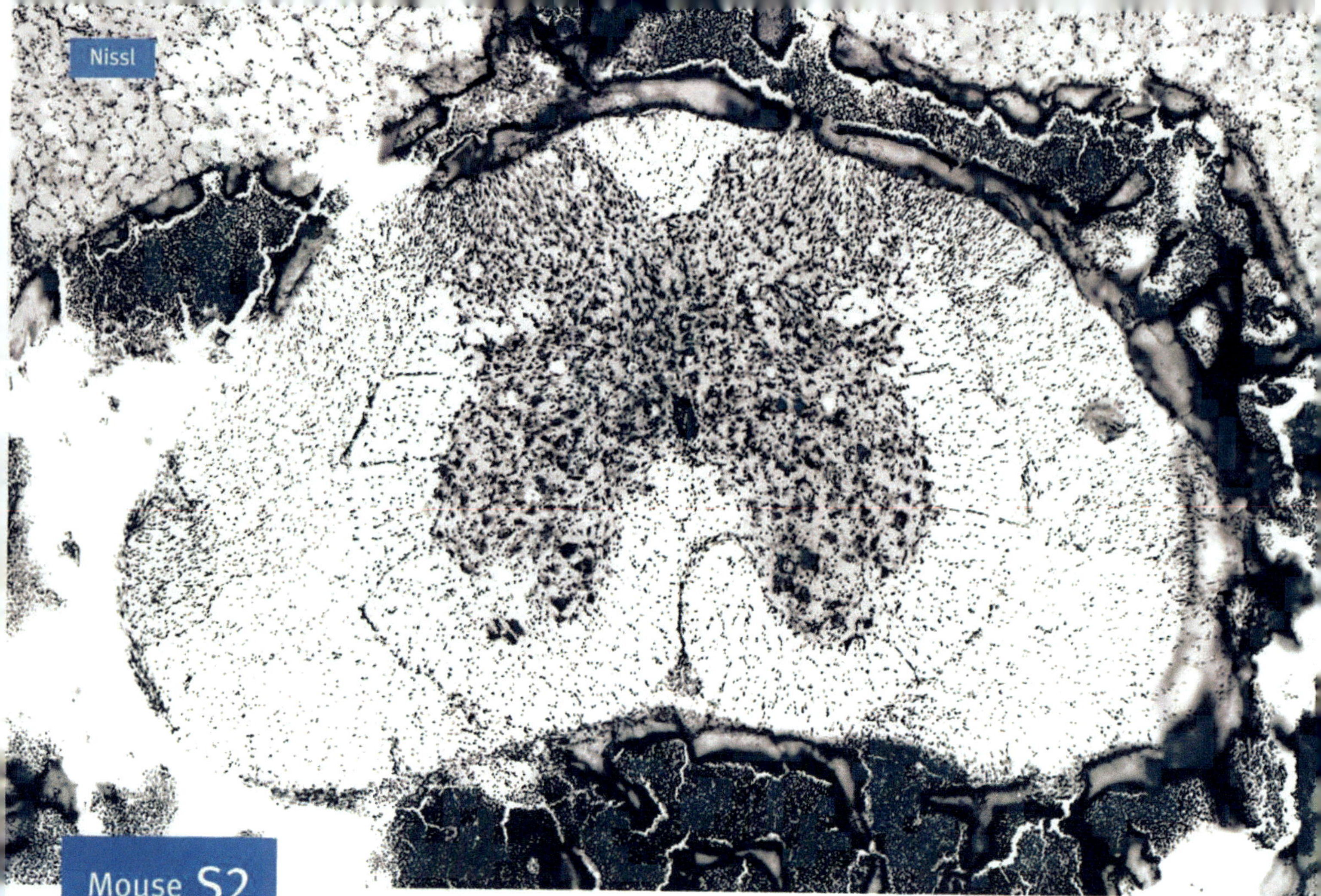
Nissl

Mouse S2

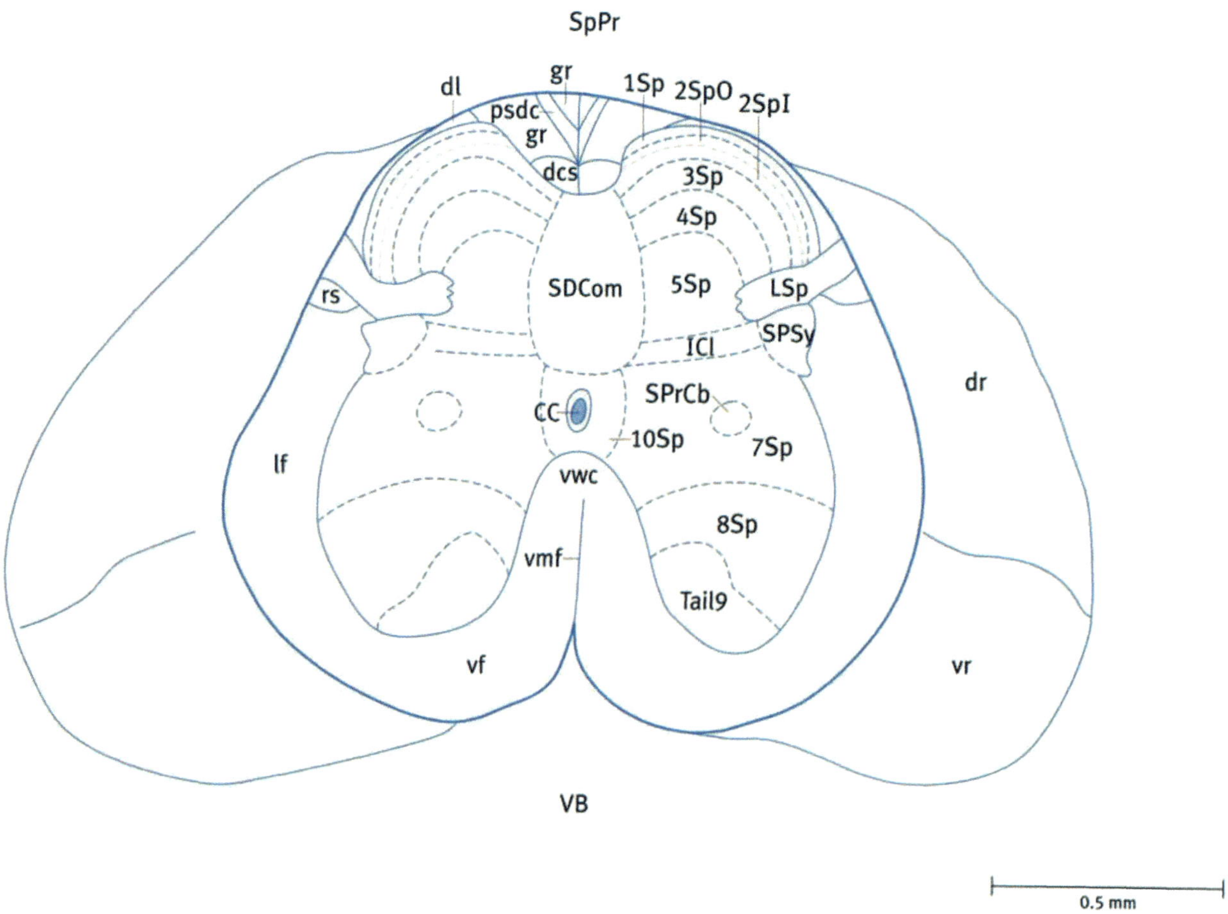
SpPr
dl
gr
psdc
gr
dcs
1Sp
2SpO
2SpI
3Sp
4Sp
SDCom
5Sp
rs
LSp
SPSy
ICl
SPrCb
CC
10Sp
7Sp
lf
dr
vwc
8Sp
vmf
Tail9
vf
vr
VB
0.5 mm

Nissl

AChE

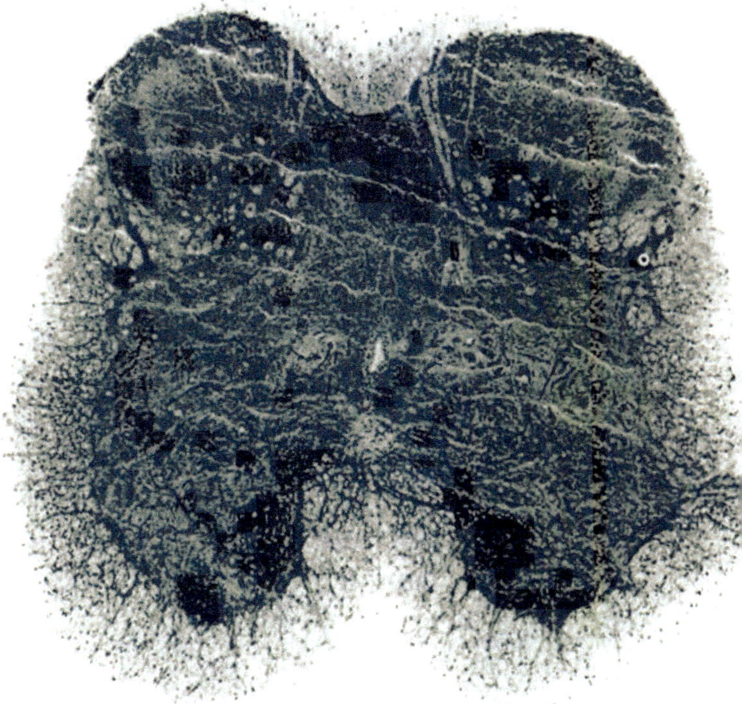

1Sp lamina 1
2SpI lamina 2, inner
2SpO lamina 2, outer
3Sp lamina 3
4Sp lamina 4
5Sp lamina 5
7Sp lamina 7
8Sp lamina 8
10Sp lamina 10
CC central canal
dcs dors corticosp
dl dorsolat fasc
dr dorsal root
gr gracile fascic
ICl intercalated nu
LSp lat spinal nu
psdc postsyn dors col
rs rubrospinal tr
SDCom sac dors comm
SpPr spin proc vert
SPrCb sacral preCb
SPSy sacral parasym
Tail9 tail
VB vertebral body
vmf vent med fiss
vr ventral root
vwc vent white com

Mouse S2

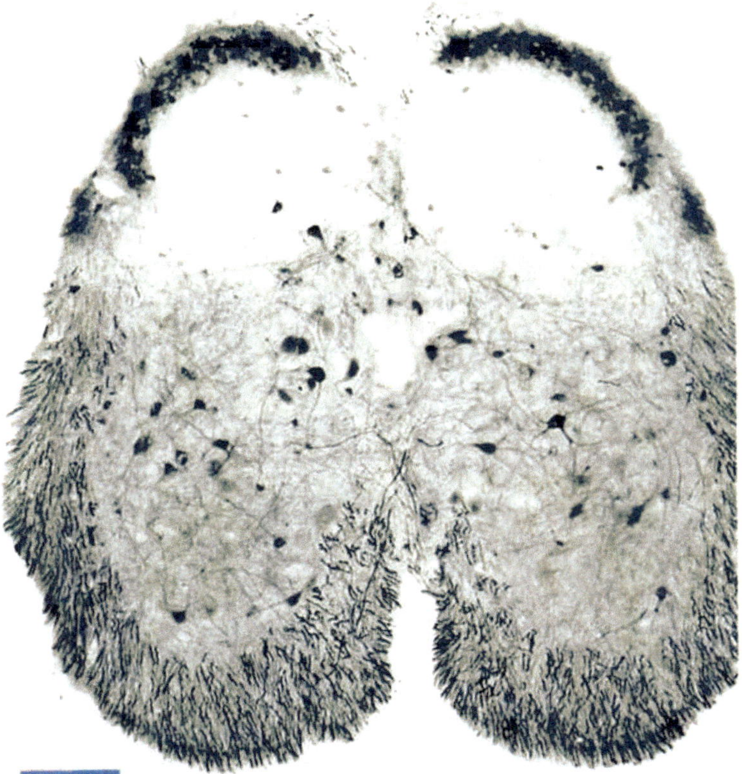

Cb

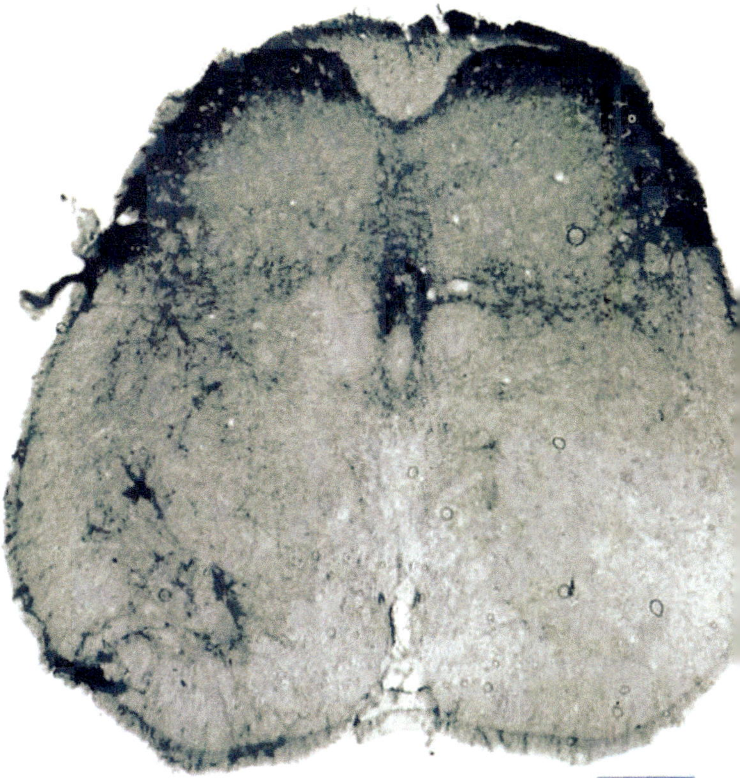

Ek

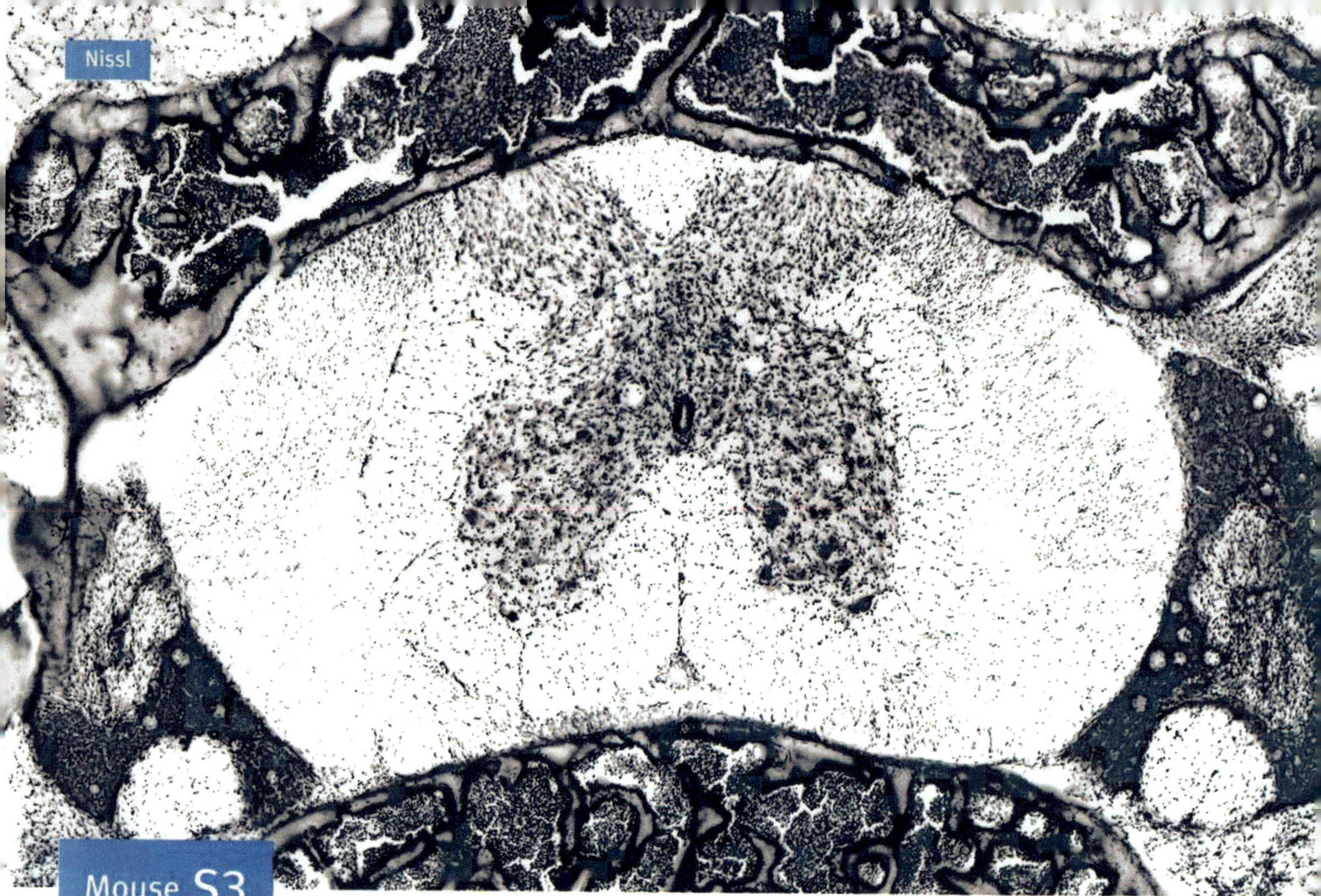
Nissl

Mouse S3

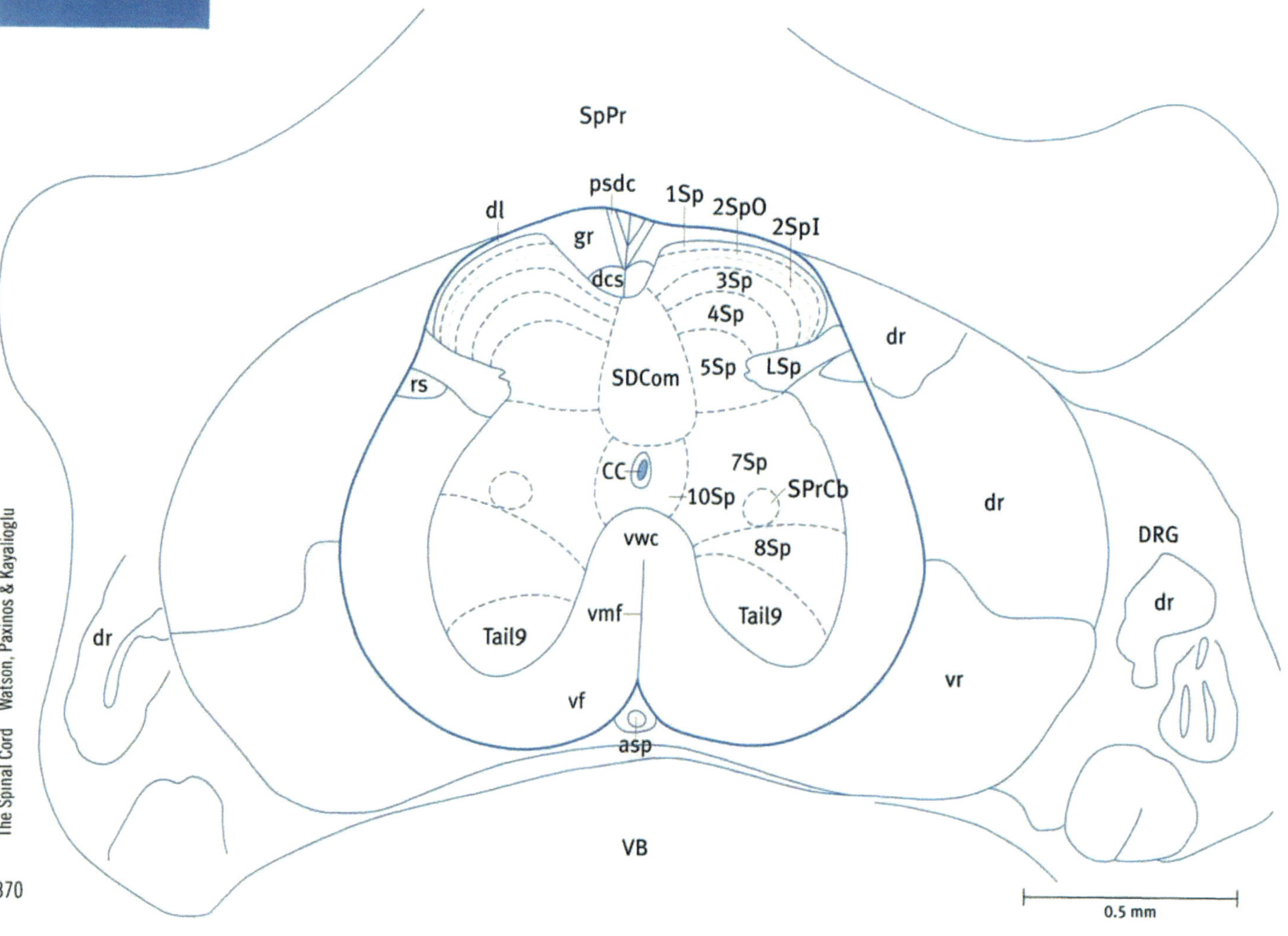
SpPr
psdc
1Sp
2SpO
2SpI
dl
gr
dcs
3Sp
4Sp
dr
SDCom
5Sp
LSp
rs
7Sp
CC
SPrCb
10Sp
dr
vwc
8Sp
DRG
dr
vmf
Tail9
Tail9
dr
vr
vf
asp
VB
0.5 mm

Nissl

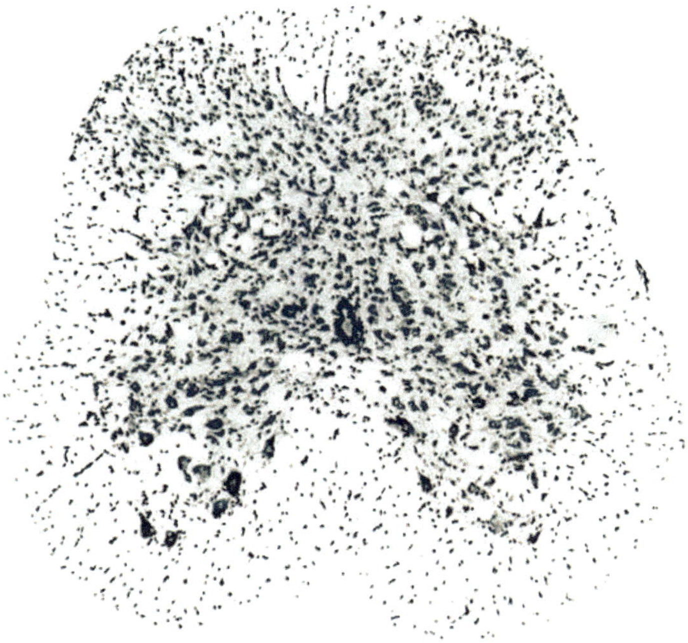

AChE

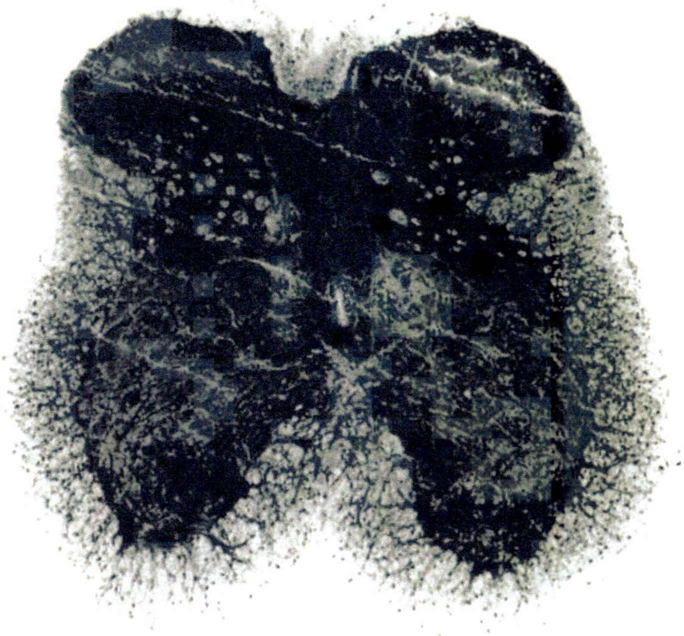

1Sp lamina 1
2SpI lamina 2, inner
2SpO lamina 2, outer
3Sp lamina 3
4Sp lamina 4
5Sp lamina 5
7Sp lamina 7
8Sp lamina 8
10Sp lamina 10
asp ant spinal art
CC central canal
dcs dors corticosp
dl dorsolat fasc
dr dorsal root
gr gracile fascic
LSp lat spinal nu
psdc postsyn dors col
rs rubrospinal tr
SDCom sac dors comm
SpPr spin proc vert
SPrCb sacral preCb
Tail9 tail
VB vertebral body
vmf vent med fiss
vr ventral root
vwc vent white com

Mouse S3

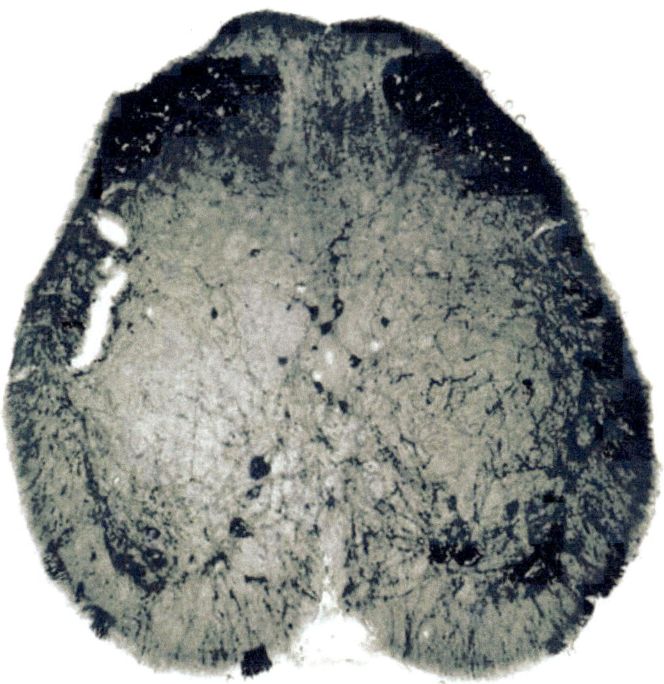

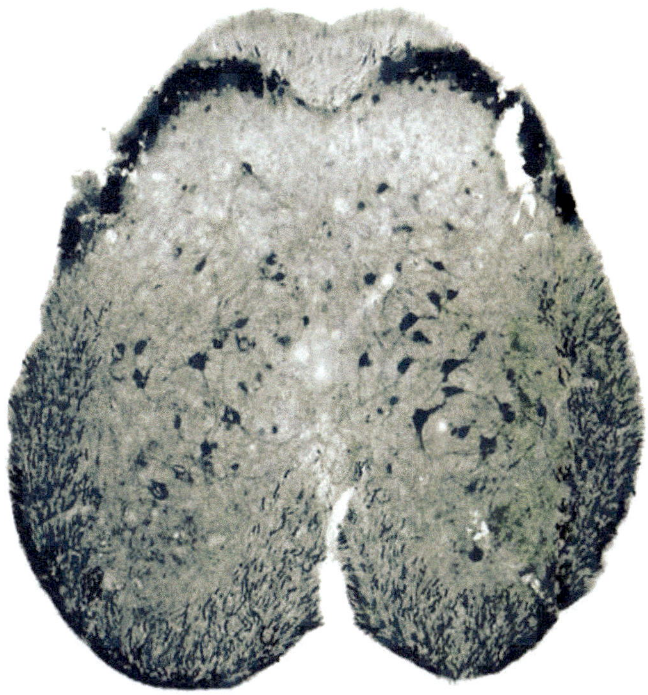

ChAT

Cr

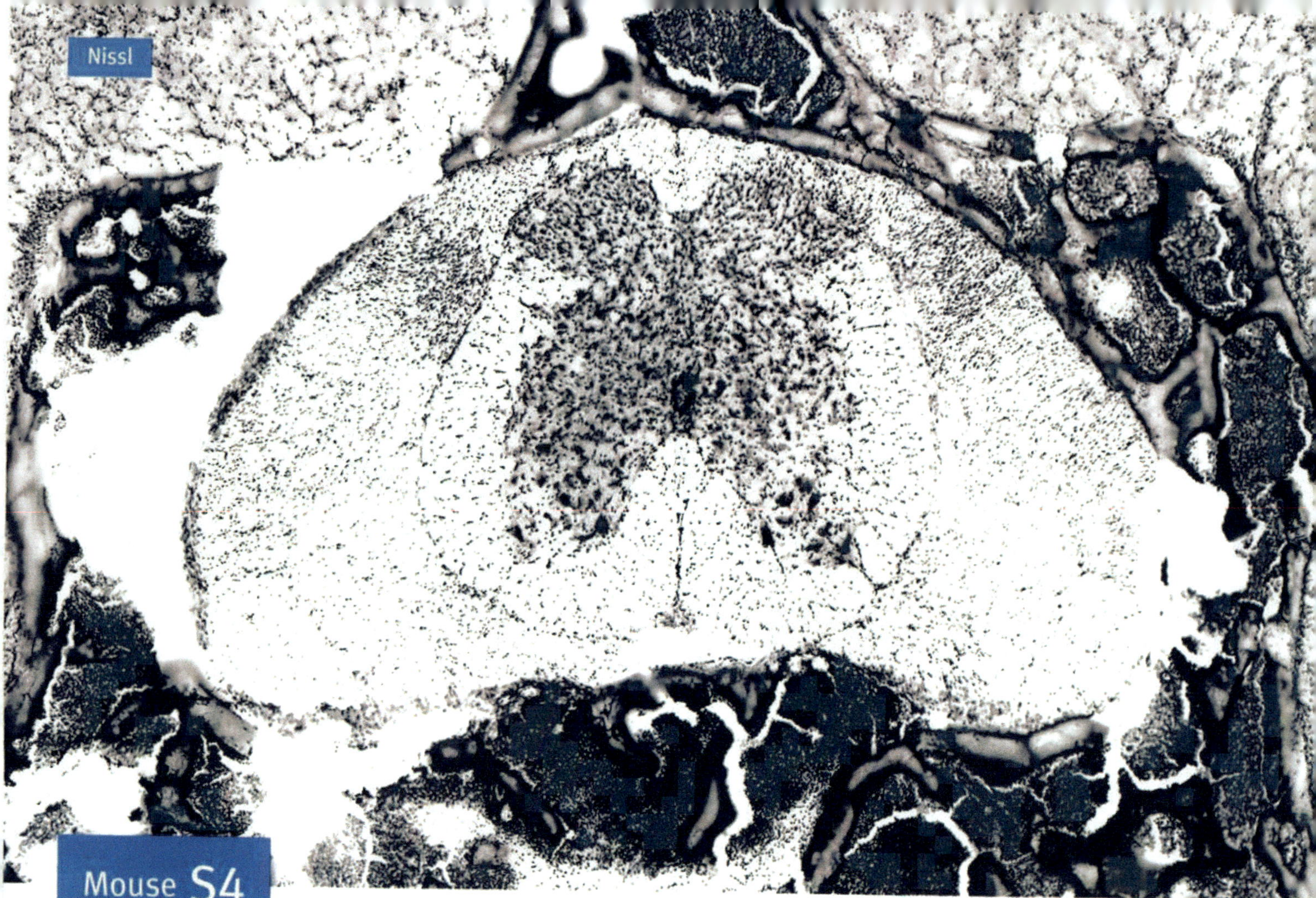
Nissl

Mouse S4

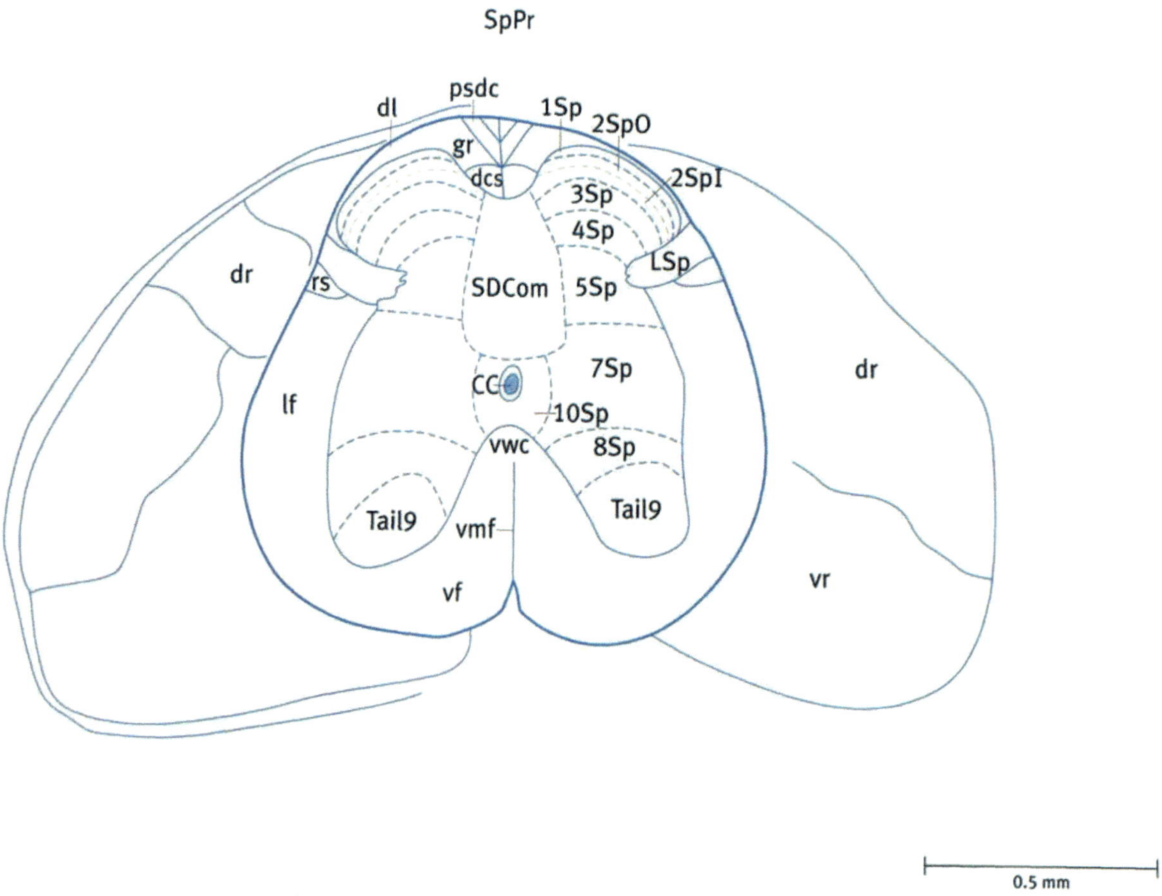
SpPr
psdc
dl
1Sp
2SpO
gr
dcs
2SpI
3Sp
4Sp
dr
rs
SDCom
5Sp
LSp
7Sp
dr
lf
CC
10Sp
vwc
8Sp
Tail9
vmf
Tail9
vf
vr
0.5 mm

Nissl

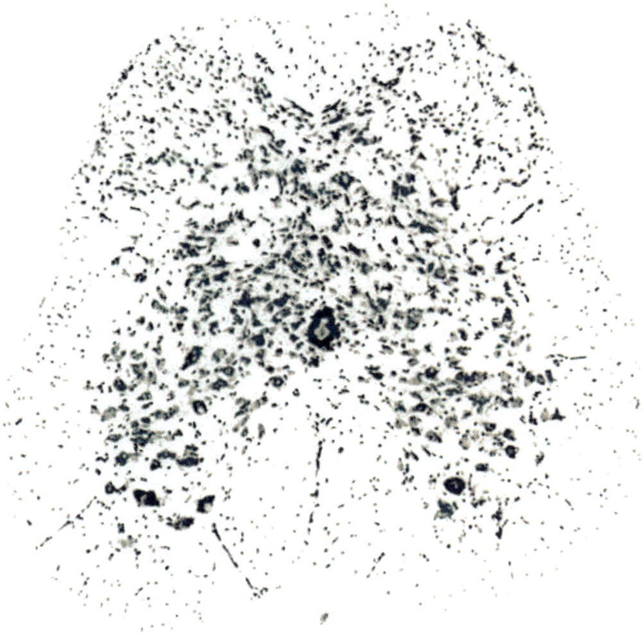

AChE

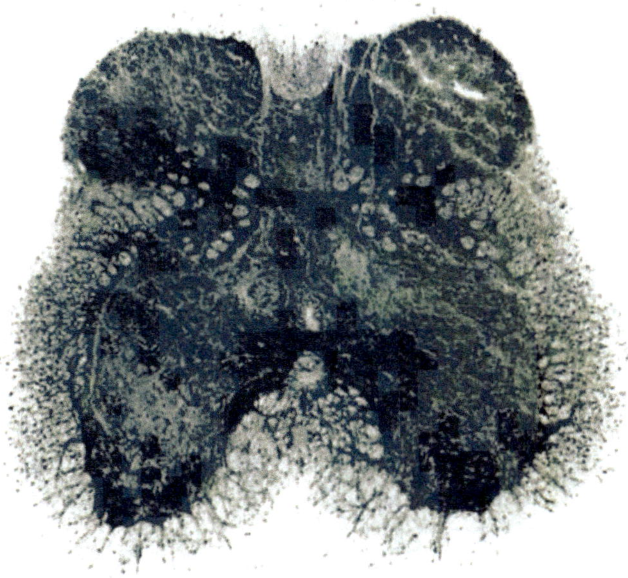

1Sp lamina 1
2SpI lamina 2, inner
2SpO lamina 2, outer
3Sp lamina 3
4Sp lamina 4
5Sp lamina 5
7Sp lamina 7
8Sp lamina 8
10Sp lamina 10
CC central canal
dcs dors corticosp
dl dorsolat fasc
dr dorsal root
gr gracile fascic
LSp lat spinal nu
psdc postsyn dors col
rs rubrospinal tr
SDCom sac dors comm
SpPr spin proc vert
Tail9 tail
vmf vent med fiss
vr ventral root
vwc vent white com

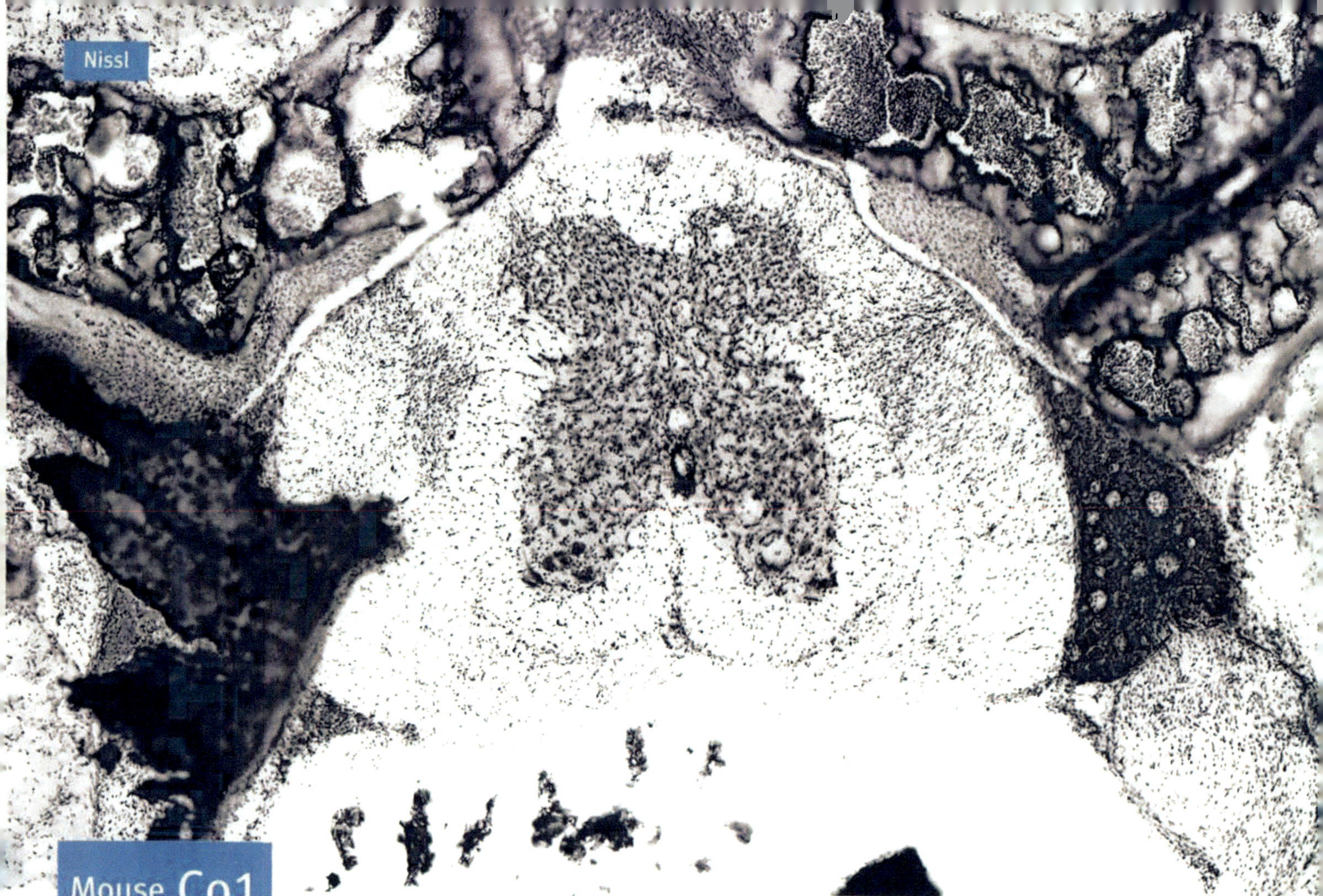
Nissl

Mouse Co1

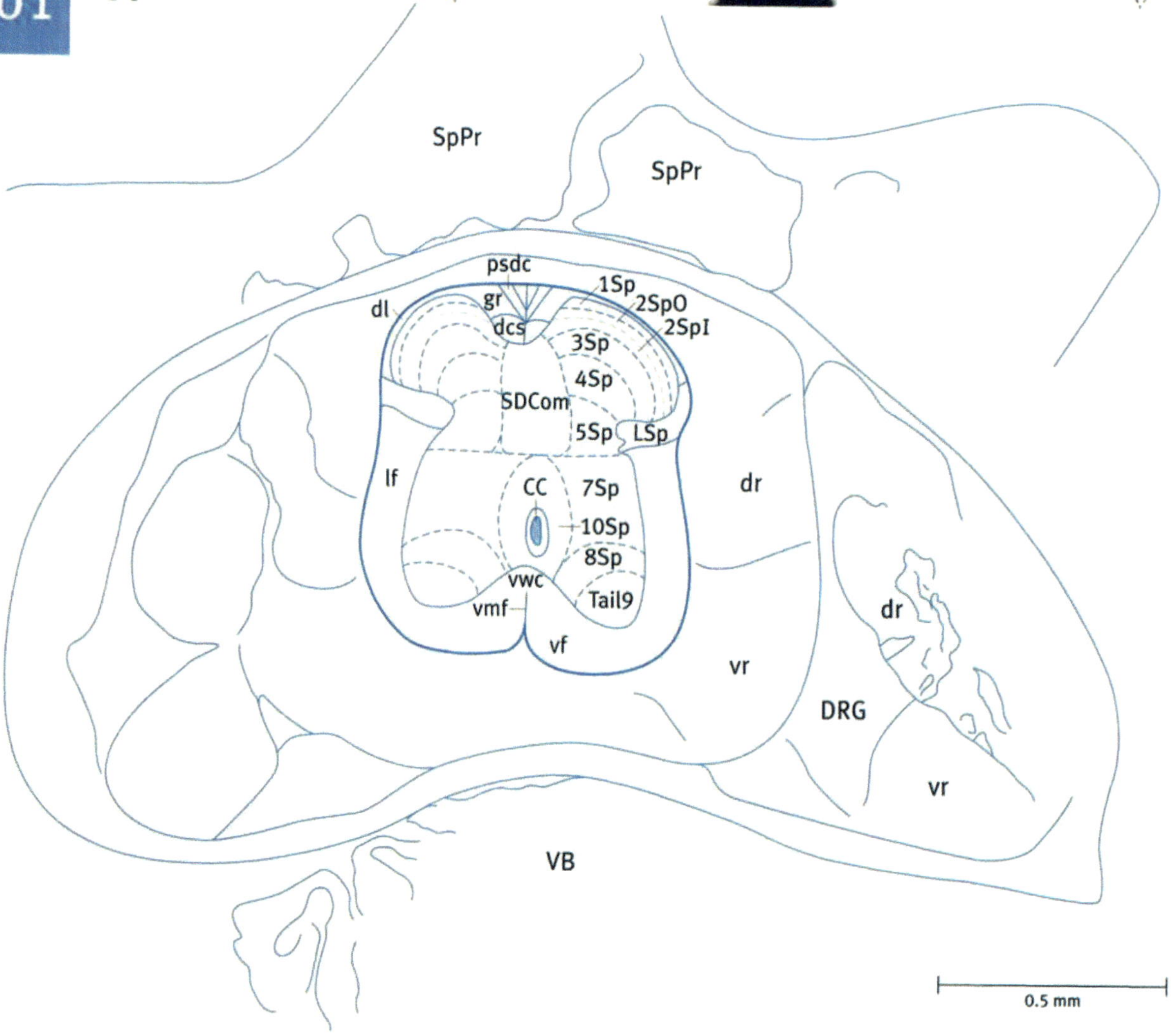
SpPr
SpPr
psdc
gr
1Sp
2SpO
2SpI
dl
dcs
3Sp
4Sp
SDCom
5Sp
LSp
lf
CC
7Sp
dr
10Sp
8Sp
vwc
vmf
Tail9
dr
vf
vr
DRG
vr
VB
0.5 mm

Nissl

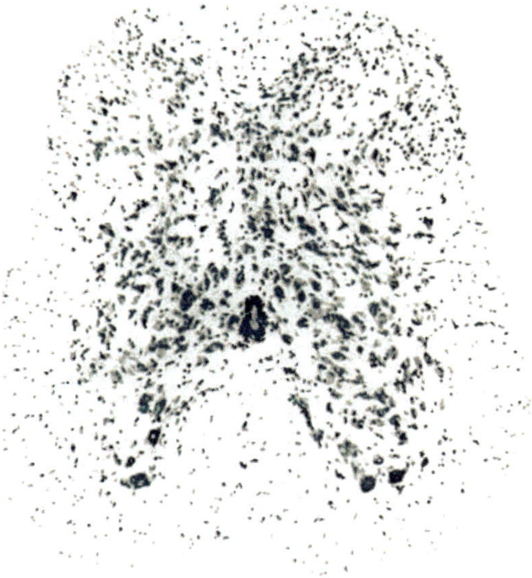

AChE

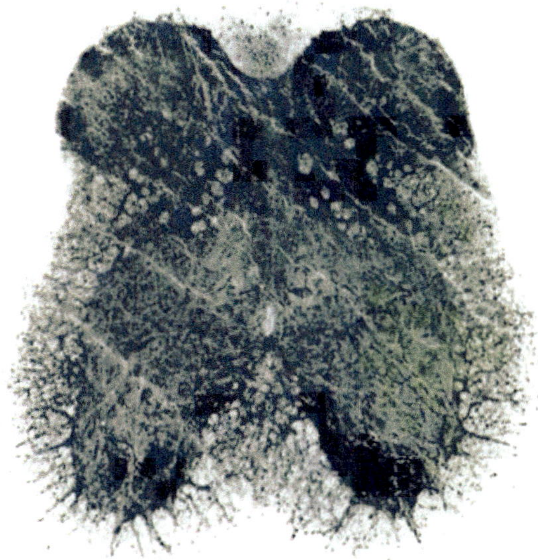

1Sp lamina 1
2SpI lamina 2, inner
2SpO lamina 2, outer
3Sp lamina 3
4Sp lamina 4
5Sp lamina 5
7Sp lamina 7
8Sp lamina 8
10Sp lamina 10
CC central canal
dcs dors corticosp
dl dorsolat fasc
dr dorsal root
DRG dors root gang
gr gracile fascic
LSp lat spinal nu
psdc postsyn dors col
SDCom sac dors comm
SpPr spin proc vert
Tail9 tail
VB vertebral body
vmf vent med fiss
vr ventral root
vwc vent white com

Mouse Co2

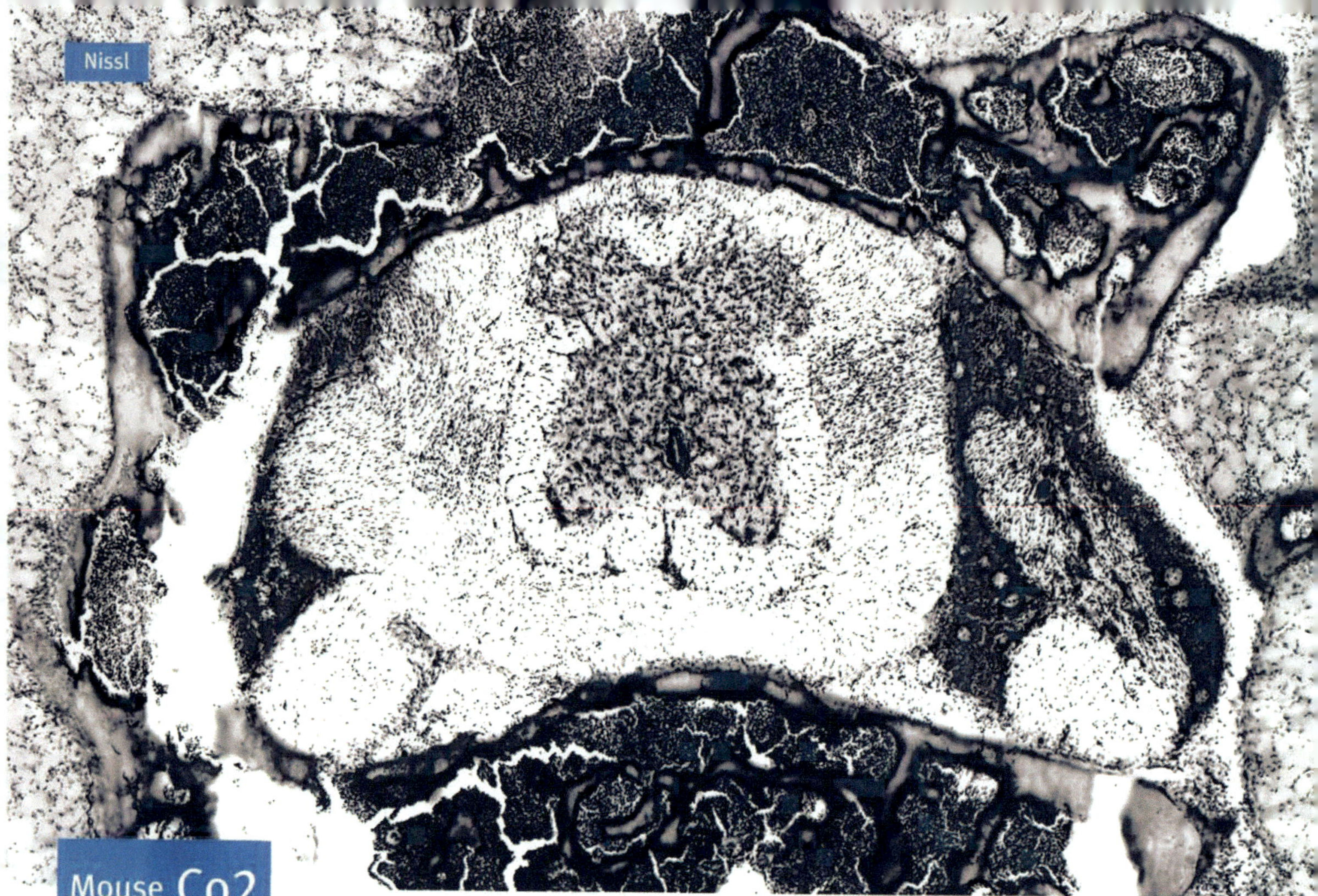

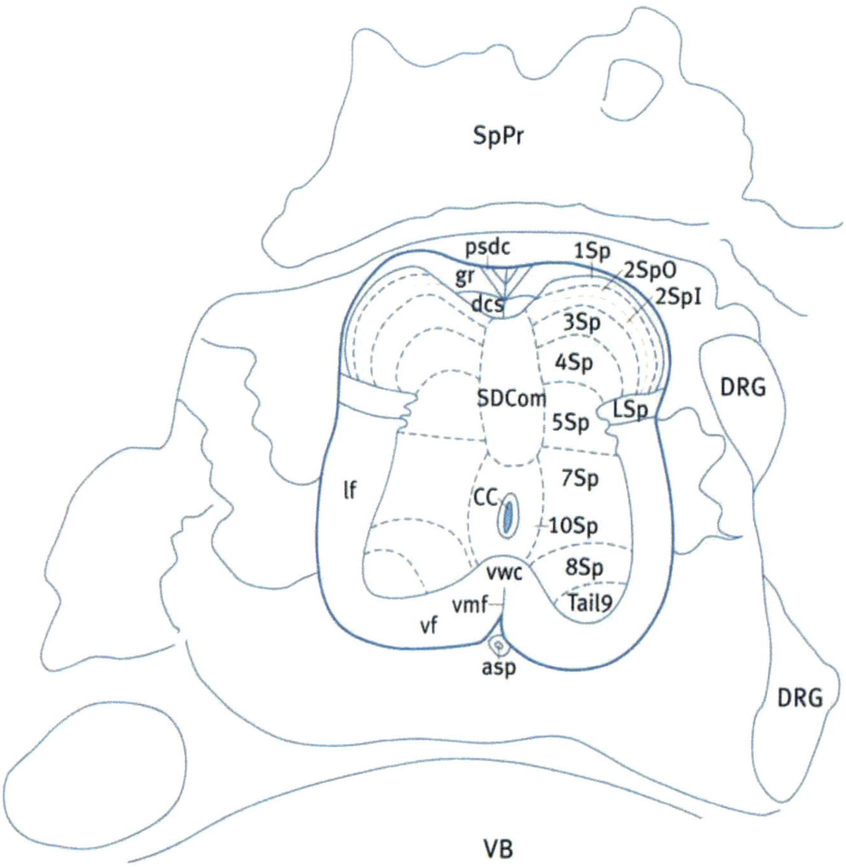

0.5 mm

Nissl

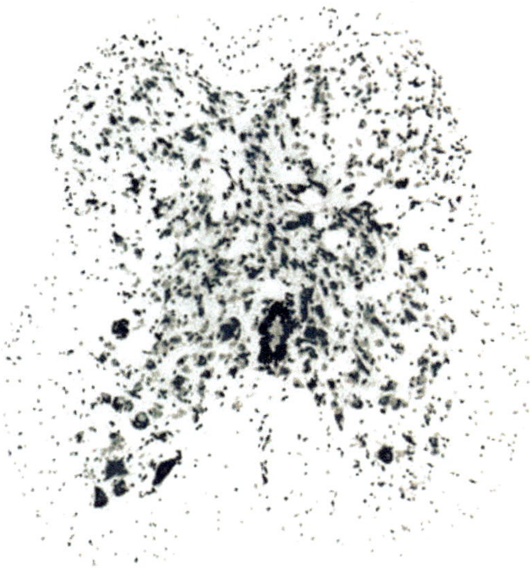

AChE

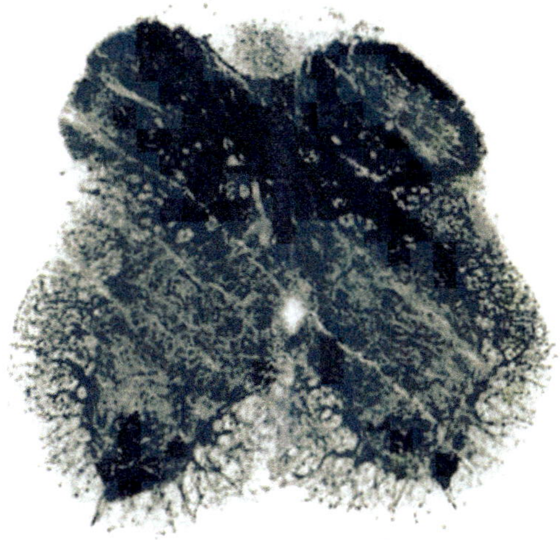

1Sp lamina 1
2SpI lamina 2, inner
2SpO lamina 2, outer
3Sp lamina 3
4Sp lamina 4
5Sp lamina 5
7Sp lamina 7
8Sp lamina 8
10Sp lamina 10
asp ant spinal art
CC central canal
dcs dors corticosp
DRG dors root gang
gr gracile fascic
LSp lat spinal nu
psdc postsyn dors col
SDCom sac dors comm
SpPr spin proc vert
Tail9 tail
VB vertebral body
vmf vent med fiss
vwc vent white com

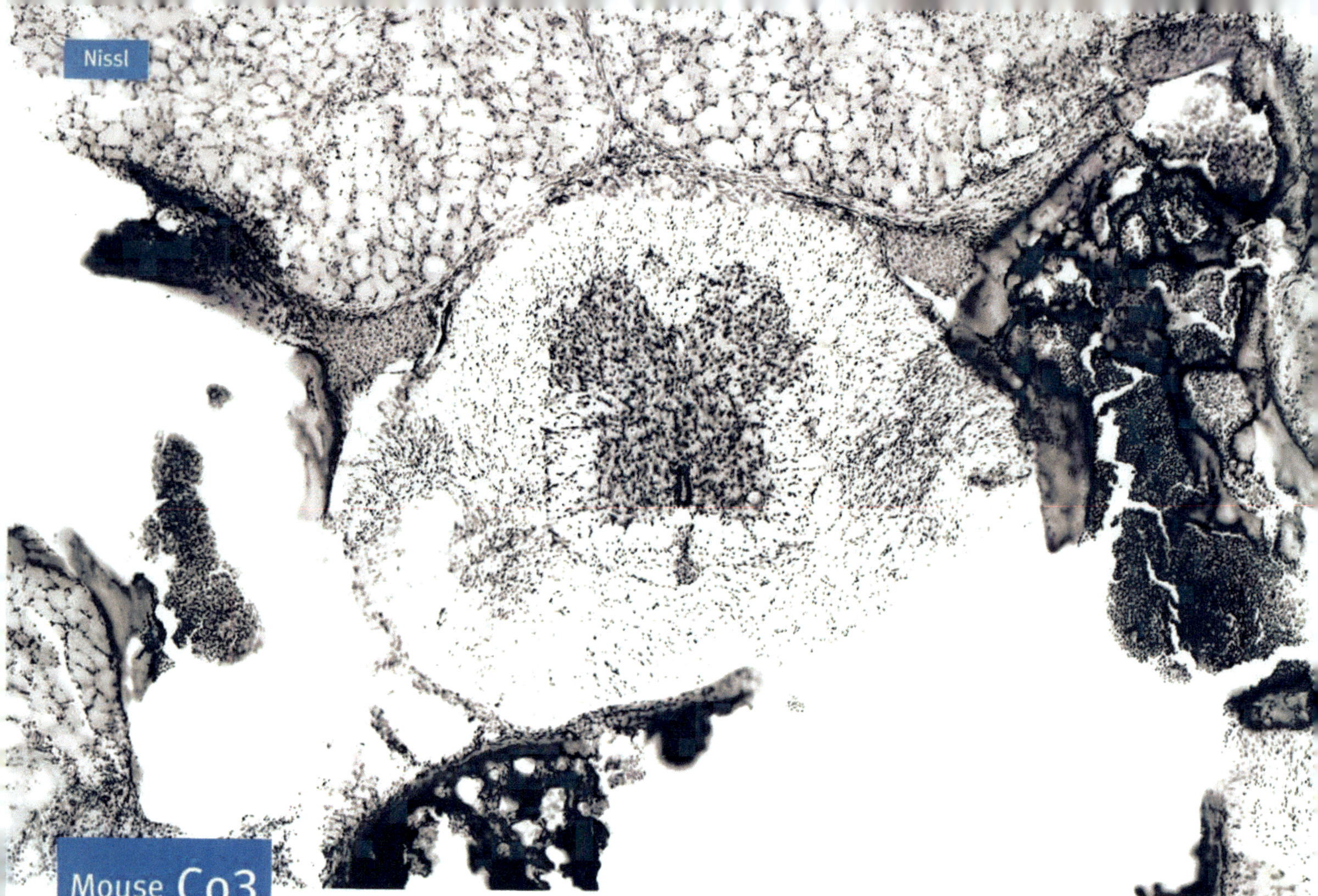

Mouse Co3

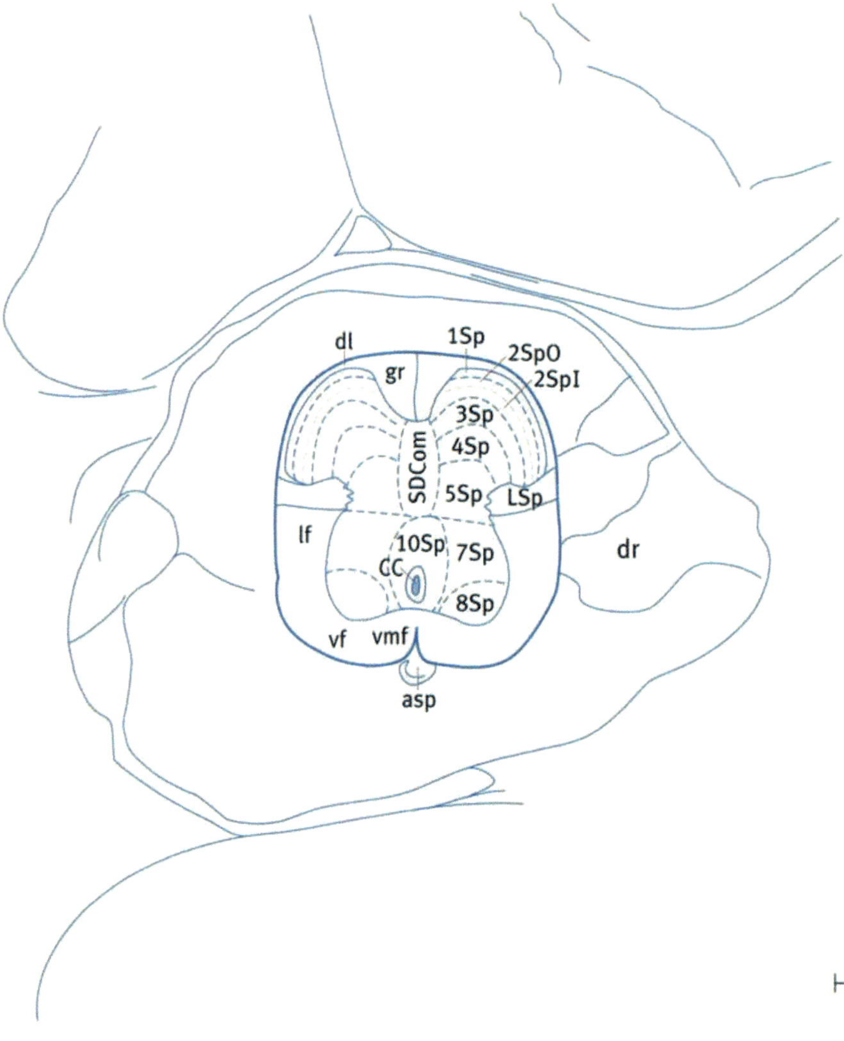

0.5 mm

Nissl

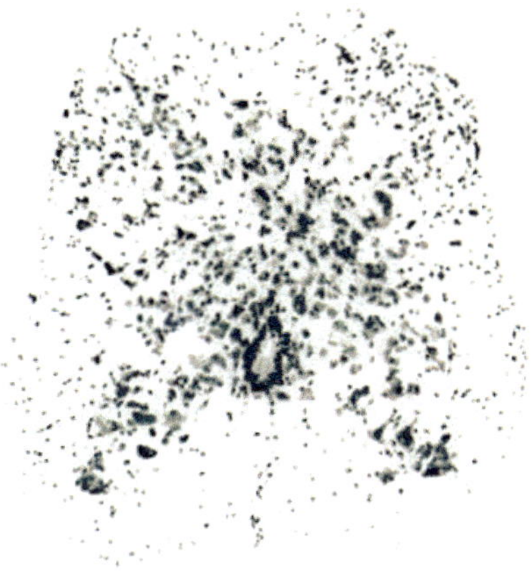

AChE

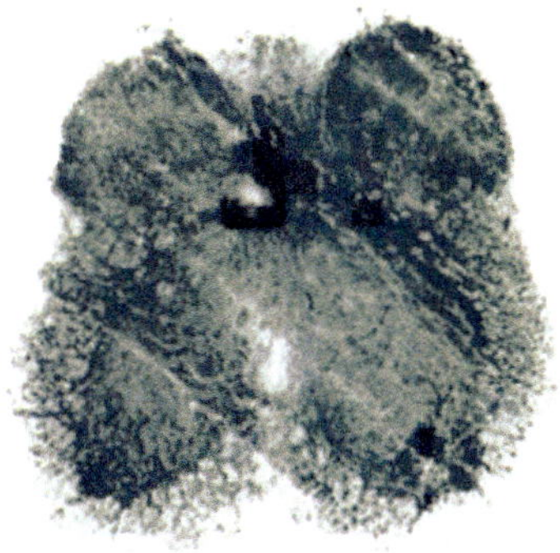

1Sp lamina 1
2SpI lamina 2, inner
2SpO lamina 2, outer
3Sp lamina 3
4Sp lamina 4
5Sp lamina 5
7Sp lamina 7
8Sp lamina 8
10Sp lamina 10
asp ant spinal art
CC central canal
dl dorsolat fasc
dr dorsal root
gr gracile fascic
LSp lat spinal nu
SDCom sac dors comm
vmf vent med fiss

17 Toward a Spinal Cord Ontology

Charles Watson and Amandeep Sidhu

This chapter is a speculative introduction to the creation of a new nomenclatural hierarchy for the spinal cord – an ontology. We have recently been engaged in the construction of mammalian brain ontology based on the avian schema presented by Puelles *et al.*, (2007). During this study, we were struck by the fact that the current spinal cord nomenclature is inconsistent with the developmental subdivisions of the spinal cord revealed by modern molecular genetics (Raff, 2000; Carpenter, 2002; Cohn and Tickle, 1999; Guthrie, 2004; Dasen *et al.*, 2005).

What is an ontology?

The concept of ontology was borrowed from the realm of philosophy by artificial-intelligence researchers and has since become a subject of interest to computer and information scientists in general. In computer science literature, the term takes on a new meaning, but one that is not entirely unrelated to its philosophical counterpart. There are many different ontology definitions in the computer and information science literature (Pretorius, 2004), but all researchers agree on the importance of ontology research in terms of the necessary mechanisms to represent, share, and reuse the existing domain knowledge (Gómez-Pérez *et al.*, 2003). A frequently cited definition of ontology is that of Thomas R. Gruber (Gruber, 1993). He states that:

"A body of formally represented knowledge is based on a conceptualization... A conceptualization is an abstract, simplified view of the world that we wish to represent for some purpose. Every knowledge base, knowledge-based system, or knowledge-level agent is committed to some conceptualization, explicitly or implicitly. An ontology is an explicit specification of a conceptualization."

The difference between ontology and a knowledge base can be described in terms of their different objectives. Ontology aims to capture the conceptual structures of a domain, while a knowledge base aims to specify a concrete state of the domain. The purpose of ontology is to describe facts assumed to be always true by a community of users. A generic knowledge base may also describe facts and assertions related to a particular state of affairs. For an agent, a shared ontology describes a vocabulary for communicating about a domain. In contrast, a knowledge base contains the knowledge needed to solve problems or answer queries about such a domain by committing to an ontology. Ontology is, in the first approximation, a table of categories, in which the collection of nodes within a hierarchical tree captures every type of entity within that domain.

Regional subdivisions in the spinal cord

The traditional regional subdivision of the spinal cord is based on the levels of cervical, thoracic, lumbar, sacral, and coccygeal vertebrae. The boundaries of these vertebral regions correlate roughly with the functional regions of the cord, but the discrepancies are significant. For example, the expanded ventral horn which houses the forelimb motoneurons begins at the fourth or fifth cervical spinal cord segment in most mammals and the root of the lowest segment emerges not below the last cervical vertebra, but below the T1 vertebra. Likewise, the region containing sympathetic preganglionic cells in rodents usually begins at the segment whose nerves emerge belowT2 (not T1) and ends at the segment whose roots emerge below L1 or L2, not below T13. In both cases, the functional transition level disregards the 'vertebral' transition level. On the other hand, the functional regions correlate well with regions of *Hox* gene expression in the limb enlargements and in the sympathetic preganglionic region (Raff, 2000; Carpenter, 2002; Guthrie, 2004; Dasen *et al.*, 2005). The *Hox* gene expression territories can be conveniently observed in acetyl cholinesterase (AChE) or choline acetlyltransferase (ChAT) preparations, which highlight the anatomy of the somatic and preganglionic motoneurons (see Chapters 15 and 16).

A new regional classification based on development

We propose a different regional classification based on the distinct areas of gene expression and AChE/ChAT staining that define the limb enlargements and the sympathetic and parasympathetic preganglionic groups as an alternative to the 'vertebral' classification. This new scheme clearly clashes with the traditional cervical, thoracic, lumbar, sacral, coccygeal designations, but is true to the developmental processes. Unlike the traditional classification, it can be applied to a range of species that have different numbers of cervical, thoracic or lumbar vertebrae, and takes into account intraspecific variation, such as 'prefixing' and 'postfixing' of limb enlargements.

The system is founded on the assumption that the basic plan in mammals and birds is to have a group of five major spinal cord segments supplying the forelimb, followed by a long group of segments with sympathetic preganglionic neurons, which is in turn followed by a group of five major spinal cord segments supplying the hindlimb. The hindlimb group is immediately followed by two or three segments containing the parasympathetic neurons. In most mammals, the forelimb group is preceded by four segments, which contain the branchial motoneurons of the accessory nerve and the phrenic nucleus. The parasympathetic group is succeeded by the part of the spinal cord, which supplies the tail muscles.

We suggest that the new subdivisions be called prebrachial, brachial, interramal, crural, postcrural, and caudal.

- 'Prebrachial' refers to the region between the caudal end of the brainstem and the brachial enlargement. In mammals this is represented by spinal cord segments C1 to C4. This region contains the branchial motoneurons of the accessory nerve that supply trapezius and sternocleidomastoid. It also contains most of the phrenic nucleus motoneurons.
- 'Brachial' refers to the forelimb (brachial) enlargement, typically five segments long in birds and mammals. It is characterized by the presence of a lateral motor column (LMC). In almost all mammals it extends from segment C5 to T1.
- 'Interramal' refers to the region between the limb enlargements. This region is characterized by absence of an LMC the presence of preganglionic sympathetic neurons. In rodents it extends from segment T2 to L1.
- 'Crural' refers to the hindlimb (lumbar) enlargement. It is typically 5 segments long. It is characterized by the presence of an LMC. In rodents it typically extends from segment L2 to L6, and in humans from segment L2 to S1 (humans normally have only five lumbar segments).
- 'Postcrural' refers to the two or three sacral segments that contain preganglionic parasympathetic neurons. In rodents this region extends from segment S1 to S2; in humans it extends from segment S2 to S4.
- 'Caudal' refers to the part of the spinal cord caudal to the sacral preganglionic parasympathetic neurons. This region supplies the tail muscles. In rodents it extends from segment S3 to the last coccygeal segment.

An ontological outline of spinal cord nomenclature

The following is an outline of the way our proposed spinal cord ontology is arranged. The ontology has 6 levels, with the sixth level representing individual neuron groups. Level 1 is the level of the brain vesicles and the spinal cord. At the suggestion of Luis Puelles, there is at Level 4 a division of the developing neural tube into roof plate, alar plate, liminal region, basal plate, and floor plate (Puelles *et al.*, 2007). These groupings work well in the subdivision of the prosomeres, midbrain, isthmus and rhombomeres, but we are as yet unsure of their complete application to the spinal cord ontology.

The abbreviations for cell laminae or cell clusters (the components of level 6 of the ontology) are those that are represented in the plates and drawings of the rat and mouse spinal cord atlases (see Chapters 15 and 16).

Six levels in the spinal cord ontology

In this section we present the level 1 and 2 subdivisions of the whole spinal cord, and the level 3, 4, 5, and 6 components of only a single segment (C8) to show the way the tree extends down to local cell groups at level 6 (see Chapters 15 and 16 for explanation of abbreviations used here). See Figure 17.1.

Subdividing the limb enlargements into rostral and caudal groups

A further step in this classification could be to divide each of the limb enlargements into rostal and caudal parts, according to anatomical differences (see Chapters 15 and 16) and gene expression (e.g. Dasen et al. 2005). Five segments supply most of the input to the upper and lower limb enlargements in most mammals (segments C5 to T1 for the upper limb and segments L2 to L6 for the lower limb in rodents). In each case, the rostral two segments are notably different from the caudal three segments. Note that for the purposes of this argument, we have chosen to ignore the small contribution of segment C4 to the upper limb enlargement, and the contribution of segment L1 to the lower limb enlargement.

The similarities between the gray matter in the upper and lower limb enlargements are striking. If one examines the ventral horn in isolation, it is easy to confuse segment C6 with L3, C8 with L5, and so on.

Detailed similarities between the arrangement of motoneuron groups in the brachial and lumbar enlargements

While each limb enlargement in mammals consists primarily of five segments that contain limb motoneurons (C5 to T1 for the upper limb and L2 to L6 for the lower limb in rodents), there is a distinct difference between the rostral two segments and the lower three segments. In the case of the upper limb, there are only two main clusters of motoneurons in the upper two segments – the dorsally placed biceps group and the

Figure 17.1 Six levels in the spinal cord ontology

Level 1
Spinal cord

- **Level 2**
  Prebrachial (segments C1 to C4) This is the region supplying the neck
  Brachial (C5 to T1) The forelimb enlargement
  - **Level 3**
    C8
    - **Level 4**
      Roof plate
      Alar plate
      - **Level 5**
        lamina 1
        - 1Sp

        lamina 2
        - 2Sp
        - 2SpO
        - 2SpI

        lamina 3
        - 3Sp

        lamina 4
        - 4Sp

        lamina 5
        - 5Sp
        - 5SpM
        - 5SpL
    - **Level 4**
      Basal plate
      - **Level 5**
        lamina 6
        - 6Sp
        - IMM

        lamina 7
        - 7Sp

        lamina 8
        - 8Sp

        lamina 9
        - 99Tr (triceps motoneuron group)
        - 9Pec (pectoralis motoneuron group)
        - 9LD (latissimus dorsi motoneuron group)
        - 9FFl (forearm flexor motoneuron group)
        - 9FEx (forearm extensor motoneuron group)
        - 9Ax (axial muscle motoneuron group)

        lamina 10
        - 10Sp
    - **Level 4**
      Floor plate

  Interramal (segments T2 to L1) This is the region between the two limb enlargements, which contains the sympathetic preganglionic neurons.

  Crural (segments L2 to L6) – this is the hindlimb enlargement

  Postcrural (segments S1 to S2) – this is the sacral parasympathetic zone

  Caudal (segments S3 to Co3) – the region that innervates the tail

ventrally placed deltoid group. In the case of the lower limb, the picture is very similar; segments L2 and L3 have only two main groups of motoneurons – the quadriceps/iliopsoas group ventrally and the adductor/gracilis group dorsally.

In the lower three segments of the brachial region (C7 to T1), four major motoneuron groups supplying the limb can be identified – a ventromedial group supplying the pectoralis major, a ventrolateral group supplying the triceps, a dorsolateral group supplying the forelimb extensors, a dorsomedial group supplying the forelimb flexors. Four groups of in a similar pattern can be identified in the lumbar segments L4 to L6 of the lower limb enlargement – a ventromedial group supplying the hamstrings, a ventrolateral group supplying the gluteal muscles, a dorsolateral group supplying the crural extensors, and a dorsomedial group supplying the crural flexors. In the brachial and lumbar enlargements, the dorsolateral and dorsomedial groups coalesce in the most caudal segment (T1 in the upper limb; L6 in the lower limb) to form a distinct round group of motoneurons that supply the manus and pes respectively.

The study of gene expression in the chick brachial enlargement by Dasen *et al.*, (2005), supports a subdivision into rostral and caudal regions on the basis of *Hox* gene and *Hox* transcription factor expression patterns. Ryan *et al.*, (1998) have demonstrated that the pattern of motoneuron clusters and their relationship to forelimb muscle groups is highly conserved in vertebrate evolution. They demonstrated that the forelimb motorneuron patterns in a reptile, a bird, and a mammal were very similar.

Similarities between the segments that immediately precede the upper and lower limb enlargements

Another feature shared by the two limb enlargements is the presence of a specialized motor nucleus in the segment immediately rostral to the enlargement. In the case of the brachial enlargement, this is the phrenic nucleus, with a very compact group of cells lying in the centre of the dorsal horn at segment C4, with extensions into segments C3 and C5. In the case of the lumbar enlargement, there is a nucleus of strikingly similar appearance to the phrenic in the same relative position. This is the cremaster nucleus, mainly located in L1 but with extensions into L2 and T13. The similarity between the phrenic and cremaster nuclei is most clearly seen in horizontal ChAT sections, which emphasize the tight packing of the motoneurons and the bundling of their dendrites.

Does this spinal cord ontology have any practical application?

In a perfect world, new logical concepts would rapidly replace the old. But in reality, old uses that have served us well are hard to displace. The failure of the Dvorak keyboard to replace the QWERTY keyboard is an excellent example of such resistance (Gould, 1991). Because of this, we do not think that there will be a rapid movement toward the use of 'interramal' in preference to 'thoracic'. However, it is possible that developmental biologists will appreciate the utility of an ontology that is independent of species variation in vertebral morphology.

We also acknowledge that clinical neurologists and other clinicians are no more likely to adopt this scheme (even if they heard about it) than they are to recognize the rhombomeric subdivisions of the hindbrain and the existence of the isthmus (see Puelles *et al.*, 2007 for discussion of these issues). But in the end, serious researchers should be drawn to nomenclatural usage that has heuristic value and serves to clarify the issues they are investigating. We are optimistic.

References

Carpenter EM (2002) Hox genes and spinal cord development. Dev Neurosci, 24, 24-34

Cohn MJ and Tickle C (1999) Developmental basis of limblessness and axial patterning in snakes. Nature 399, 474-479.

Dasen JS, Tice BC, Brenner-Morton S, Jessell TM (2005) A Hox regulatory network establishes motor neuron pool identity and target-muscle connectivity. Cell 123, 477-491.

Gómez-Pérez, A, Fernández-López, M & Corcho, O (2003) *Ontological Engineering* Springer-Verlag, London.

Gould SJ (1991) The Panda's Thumb of Technology. In: *Bully for Brontosaurus*. Norton, New York.

Gruber TR (1993) A translation approach to portable ontology specifications. Knowledge Acquisition 5, 199-220.

Guthrie S (2004) Neuronal development: putting motor neurons in their place. Curr Biol 14, R166-R168.

Puelles L, Martinez-de-la-Torre M, Paxinos G, Watson C, and Martinez S (2007) *The chick brain in stereotaxic coordinates. An atlas featuring neuromeric subdivisions and mammalian homologies.* Elsevier Academic Press, San Diego.

Pretorius AJ (2004) Lexon visualization: visualizing binary fact types in ontology bases. In: *Banissi E, et al. (eds.), Information Visualisation* (Proceedings Eighth International Conference, IV'04, London.

Raff RA (2000) Evo-devo: the evolution of a new discipline. Nat Rev Genet 1, 74-79.

Ryan JM, Cushman J, Jordan B, Samuels A, Frazer H, Baier C (1998) Topographic position of forelimb motoneuron pools is conserved in vertebrate evolution. Brain Behav Evol 51, 90-99.

Index

A

B

C

D

E

F

G

H

I

L

M

N

O

P

R

S

T

U

V

W

Z

9780123742476